Shock Waves
in Condensed Matter—1981
(Menlo Park)

AIP Conference Proceedings
Series Editor: Hugh C. Wolfe
Number 78

Shock Waves in Condensed Matter—1981
(Menlo Park)

Editors
W. J. Nellis
Lawrence Livermore National Laboratory
L. Seaman
SRI International
R. A. Graham
Sandia National Laboratories

American Institute of Physics
New York   1982

Copying fees: The code at the bottom of the first page of each article in this volume gives the fee for each copy of the article made beyond the free copying permitted under the 1978 US Copyright Law. (See also the statement following "Copyright" below). This fee can be paid to the American Institute of Physics through the Copyright Clearance Center, Inc., Box 765, Schenectady, N.Y. 12301.

Copyright © 1982 American Institute of Physics

Individual readers of this volume and non-profit libraries, acting for them, are permitted to make fair use of the material in it, such as copying an article for use in teaching or research. Permission is granted to quote from this volume in scientific work with the customary acknowledgment of the source. To reprint a figure, table or other excerpt requires the consent of one of the original authors and notification to AIP. Republication or systematic or multiple reproduction of any material in this volume is permitted only under license from AIP. Address inquiries to Series Editor, AIP Conference Proceedings, AIP, 335 E. 45th St., New York, N. Y. 10017.

L.C. Catalog Card No. 82-70014
ISBN 0-88318-177-0
DOE CONF- 810684

PREFACE

The 1981 Topical Conference on Shock Waves in Condensed Matter was held in Menlo Park, California, 23-25 June, 1981.

The objective of this conference was to provide a format for scientific interactions on questions of physico-chemical properties and processes of condensed matter under shock compression. Typical conditions include pressures ranging from a few gigapascal (1 GPa = 10 kbar) to a few hundred gigapascals with accompanying temperature increases from a few degrees to a few tens of thousands of degrees. Shock pressure pulses last from a nanosecond to a few microseconds. The shock-wave technique produces unique and extreme states of condensed matter for scientific investigation.

This topical conference was the second on this subject sponsored by the American Physical Society. The first was held in 1979 at Washington State University. For about fifteen years prior to these two meetings, the community met in conjunction with general meetings of the American Physical Society. These proceedings are the first to be published by this community. Abstracts of papers were published in the Bulletin of the American Physical Society, Volume 26, 1981, page 647 and following.

These topical conferences represent a maturation of the field of shock waves in condensed matter in terms of the breadth of material properties under investigation, the diagnostic techniques used to measure these properties, and the size of the community of workers in the field. The conference was attended by 200 scientists with international participants from Canada, France, Great Britain, Israel, Japan, Peoples Republic of China, Soviet Union, and West Germany.

Special features included a poster session on shock-wave facilities and a panel discussion on Perspectives on the Chemistry of Shock-Compressed Matter. Invited papers were presented on constitutive modeling, electrical properties, metallurgical effects, microstructural modeling, optical properties, and ultrahigh pressure.

The Conference is especially grateful to SRI International for making their facilities available for the conference.

W. J. Nellis, L. Seaman, and R. A. Graham
Editors

CONFERENCE ORGANIZATION

General Chairman: W. J. Nellis

Program Committee:
- W. J. Nellis, Chairman
- T. J. Ahrens
- J. R. Asay
- R. J. Clifton
- G. R. Fowles
- R. A. Graham
- J. W. Hopson
- D. E. Mikkola
- L. Seaman
- G. K. Straub
- H. C. Vantine
- E. Zimet

Local Chairman: L. Seaman

Publication Committee:
- W. J. Nellis
- L. Seaman
- R. A. Graham

TABLE OF CONTENTS

PREFACE

CHAPTER I: Chemistry of Shock Compression--Material Synthesis

Perspectives on Inorganic Chemistry
S. S. Batsanov 1

Shock-Induced Inorganic Chemistry
B. Morosin and R. A. Graham....................................4

Inorganic Synthesis under Shock-Wave Compression
S. S. Batsanov...14

The Shock-Wave Chemistry of Organic Substances
A. N. Dremin and L. V. Babare 27

Shock-Induced Organic Chemistry
B. W. Dodson and R. A. Graham..................................42

The Electrical-to-Chemical Connection
R. A. Graham 52

Low Pressure Hugoniot Cusp in Polymeric Materials
S. A. Sheffield and D. D. Bloomquist 57

An Exploratory Study of Reactivity in Organic Compounds
Subjected to Shock Loading
B. W. Dodson 62

Analysis of Capsules for Recovery of Shock-Compressed Matter
Lee Davison, D. M. Webb, and R. A. Graham.....................67

Catalytic Activity of Shock-Loaded TiO_2 Powder
John Golden, Frank Williams, B. Morosin,
E. L. Venturini, and R. A. Graham 72

Paramagnetic Defects in Shock-Loaded TiO_2
E. L. Venturini, B. Morosin, and R. A. Graham..............77

Shock-Induced Chemical and Structural Modification of
Zirconia, Lead Oxide and their Mixed Powders
J. D. Keck, D. L. Hankey, R. A. Graham, and B. Morosin.......82

Shock Synthesis Experiments of Nb-Si System
Y. Syono, T. Goto, W. K. Wang, H. Iwasaki, S. Ohshima,
and T. Wakiyama 87

Theoretical Studies of Shock Dynamics in Two-Dimensional
Structures V. Microscopic Constraints on Shock-Induced
Signals
  A. M. Karo, F. E. Walker, W. G. Cunningham, and
  J. R. Hardy 92

Experimental and Theoretical Studies on Shock Compression
of Liquid Carbon Monoxide
  F. H. Ree, W. J. Nellis, M. van Thiel, and
  A. C. Mitchell 97

CHAPTER II: Microstructural Properties

Metallurgical Effects of Shock Loading
  D. E. Mikkola and R. N. Wright 98

Development of Computational Models for Microstructural
Features
  L. Seaman 118

Twinning in Iron by Laser Generated Shock Waves
  F. Cottet and J. P. Romain......................................130

Stress and Temperature-Driven Nucleation of Microscopic
Voids in Metals
  D. R. Curran 135

Structural Deformation of Experimentally Shock-Loaded
Periclase (MgO)
  H. Schneider and I. Jung..140

CHAPTER III: Ultra High Pressure

Ultra High Pressure Laser-Driven Shock Wave Experiments
  R. J. Trainor, N. C. Holmes, and R. A. Anderson 145

Time and Space Resolved Measurements of the Dynamics of
Laser Fusion Targets
  Robert H. Price, Mordecai D. Rosen, David L. Banner,
  Neil C. Holmes, Marian Kobierecki, James R. Zickuhr,
  and Harlow G. Ahlstrom 155

Impedance-Match Experiments Using High Intensity Lasers
  N. C. Holmes, R. J. Trainor, R. A. Anderson,
  L. R. Veeser, and G. A. Reeves 160

Contrasts in One- and Two-Dimensional Hydrocode
Calculations of Laser-Generated Shock Waves in Disk Targets
  R. J. Harrach, Y. T. Lee, R. J. Trainor, N. C. Holmes,
  M. D. Rosen, D. L. Banner, and R. J. Olness 164

Shock Compression Measurements at Pressure > 1 TPa
    C. E. Ragan, B. C. Diven, M. Rich, E. E. Robinson, and
    W. A. Teasdale 169

Shock Hugoniot Experiments Using an Electric Gun
    K. E. Froeschner, H. Chau, G. Dittbenner, R. S. Lee,
    K. Mikkelson, D. Steinberg, and R. C. Weingart 174

Railguns for Equation-of-State Research
    R. S. Hawke, A. L. Brooks, A. C. Mitchell,
    C. M. Fowler, D. R. Peterson, and J. W. Shaner 179

Enhanced Performance of a Two-Stage Light-Gas Gun
    A. C. Mitchell, W. J. Nellis, and B. Monahan 184

Shock Effects in Particle Beam Fusion Targets
    M. A. Sweeney, F. C. Perry, J. R. Asay, and
    M. M. Widner 188

CHAPTER IV: Equation of State

Some Techniques and Results from High-Pressure Shock-Wave Experiments Utilizing the Radiation from Shocked Transparent Materials
    R. G. McQueen and J. N. Fritz 193

Theoretical Equations of State for Metals
    G. I. Kerley 208

A New Full-Range Equation of State for Copper
    K. S. Long, D. Young, and F. H. Ree 213

Accurate Self-Consistent-Field Isotherms for NaCl to 30 GPa
    M. S. T. Bukowinski 218

The Shock Compression of Liquid H_2 to 10 GPa (100 kbar)
    W. J. Nellis, M. Ross, M. van Thiel, A. C. Mitchell,
    G. J. Devine, and N. Brown 223

Equation of State Experiments and Theory Relevant to Modeling the Major Planets
    Marvin Ross and William J. Nellis 226

One-Dimensional Isentropic Compression
    Gregory A. Lyzenga and Thomas J. Ahrens 231

A Method of Determining Points on the Principal Isentropes of Molecular Liquids
    M. B. Boslough and T. J. Ahrens 236

Molecular Dynamics Study of Sodium Using a Model
Pseudopotential
    Richard E. Swanson, Galen K. Straub, and
    Brad Lee Holian 241

Characterization of Thermomechanical Response of Porous
Vanadium
    R. E. Tokheim and A. B. Lutze 246

CHAPTER V: Electrical and Optical Properties

Behavior of Ferroelectric Ceramics and PVF_2 Polymers
under Shock Loading
    F. Bauer 251

Optical Pyrometry at High Shock Pressures and Its
Interpretation
    G. A. Lyzenga 268

Microwave Dielectric Constant of Shock-Loaded Lithium
Niobate
    J. K. Hartman, J. L. Wise, R. A. Graham,
    R. O. Johnson, G. E. Clark, and T. J. Burns 277

Effect of Shock Waves on the Absorption Spectrum of Ruby
    R. S. Hixson, P. M. Bellamy, G. E. Duvall, and
    C. R. Wilson 282

Temperature Measurements of Shocked Translucent Materials
by Time-Resolved Infrared Radiometry
    William G. Von Holle 287

Time Resolved Spectroscopy of Shock Compressed Liquids
    K. Ogilvie and G. E. Duvall 292

The Resistivity of Liquid Carbon Disulfide During Shock
Compression
    C. R. Wilson, G. E. Duvall and K. Ogilvie 296

Electrical and Optical Measurements on Fused Quartz under
Shock Compression
    K. Kondo, T. J. Ahrens, and A. Sawaoka 299

Shock-Compression Temperature Rise Determined from
Resistivity of Embedded Metal Foils
    D. D. Bloomquist and S. A. Sheffield 304

Reflection of a Laser-Generated Optical Signal from a Shock
Front in Water
    P. Harris and H. N. Presles 309

CHAPTER VI: Phase Transitions

High Temperature Phase Transformation in the Titanium Alloy
Ti-6Al-4V
    J. E. Shrader and M. D. Bjorkman 310

Glass Transition in Shock Loaded Ceramics
    Y. Horie 315

Shock-Induced Phase Transition in GaP
    Tsuneaki Goto and Yasuhiko Syono 320

Shock Compression and Phase Transformation of AlN and BP
    K. Kondo, A. Sawaoka, K. Sato, and M. Ando 325

Crystallographic Properties of Lithium Niobate Shock-Loaded
from 3.7 to 20 GPa and Preserved for Post-Shock Study
    B. Morosin and R. A. Graham 330

Shock-Induced Defects in Cadmium Sulfide
    B. W. Dodson and E. L. Venturini 335

The α-β Transition in T=0 High Pressure Beryllium
    A. K. McMahan 340

Basic Features of Dynamic Phase Transition in Finite Samples
    N. Salansky, H. Mar, D. Hawken, and Y. Kleiman 345

CHAPTER VII: Constitutive Modeling

On Constitutive Modelling for the Shock Physicist
    W. Herrmann 346

Dynamic Stress-Strain Curves at Plastic Shear Strain Rates
of 10^5 s^{-1}
    C. H. Li and R. J. Clifton 360

Interpretation of Shock-Wave Data for Beryllium and Uranium
with an Elastic-Viscoplastic Constitutive Model
    Daniel J. Steinberg and Richard W. Sharp, Jr. 367

On a Criterion for Thermo-Plastic Shear Instability
    T. J. Burns, D. E. Grady, and L. S. Costin 372

Dynamic Compaction of Elastic-Visco-Plastic Porous
Materials under Shock
    K. Kim and S. I. Oh 376

Numerical Analysis of Hopkinson Bar Experiments on
Dislocation Dynamics
    Minao Kamegai 381

CHAPTER VIII: Strength of Compressed Solids

Quasi-Elastic High-Pressure Waves in 2024 Al and Copper
C. E. Morris, J. N. Fritz, and Brad Lee Holian............382

Shock Release of 2024-T351 Aluminum in the 10 GPa Range
Y. Partom, D. Yaziv, and Z. Rosenberg 387

Shock-Wave Compression of a Borosilicate Glass up to 170 kbar
J. Cagnoux 392

Determination of Mean and Deviatoric Stresses in Shock Loaded Solids
P. F. Chartagnac 397

Symmetric Rod Impact Technique for Dynamic Yield Determination
D. C. Erlich, D. A. Shockey, and L. Seaman................402

Re-examination of the Precursor Decay Anomaly
R. J. Clifton 407

A Thermal-Viscous Model for Heterogeneous Yielding in Aluminum
D. B. Hayes and D. E. Grady 412

Shock Compression of Beryllium
J. L. Wise, L. C. Chhabildas, and J. R. Asay 417

Reshock and Release Behavior of Beryllium
L. C. Chhabildas, J. L. Wise, and J. R. Asay 422

Strain Rate Effects of Beryllium under Shock Compression
J. R. Asay, L. C. Chhabildas, and J. L. Wise 427

Shock Wave Propagation in Beryllium at Small Impact Stresses and Elevated Temperatures
M. D. Bjorkman and J. E. Shrader 432

Large Amplitude Compression and Shear Wave Propagation in an Elastomer
Y. M. Gupta, W. J. Murri, and D. Henley 437

CHAPTER IX: Fracture

Spallation by Ductile Void Growth
J. N. Johnson 438

Analyses of Ductile Flow and Fracture in Two Dimensions
M. E. Kipp and Lee Davison 442

Nucleation Threshold Stresses for the Dynamic Fracture of a
Low-Alloy Ni-Cr Steel
    G. L. Moss, P. H. Netherwood, Jr., and L. Seaman 446

An Investigation of Incipient Fracture in Shock-Loaded
Lamellar Cobalt-Aluminum Eutectic
    William E. Thompson and William W. Predebon 451

Fragment Size Prediction in Dynamic Fragmentation
    D. E. Grady 456

Fracture Model for High Energy Propellant
    W. J. Murri, D. R. Curran, and L. Seaman 460

Calculations of Cratering Experiments with the Bedded Crack
Model
    L. G. Margolin 465

Fracture Initiation Using Tailored-Pulse Loading
    A. S. Kusubov and R. P. Swift 470

Correlation of Impulsive Strain Effects in Rocks with
Detonation Parameters of the Strain-Generating Explosive
Charges
    J. Roth 475

Dynamic and Stress Intensity in Elastic Strips
    W. G. Hoover and B. Moran 480

CHAPTER X: Shock Phenomena and Experimental Technique

Ramp-Wave Generator Studies
    M. Germain-Lacour and M. de Gliniasty 481

Experiments on the Attenuation of Shock Waves in Condensed
Matter
    Ch. Klee, M. Kroh, and D. Ludwig 486

Calibration of Piezoresistive Gauges
    G. L. Nutt and J. O. Hallquist 491

Electromagnetic Gauge for Measuring the Radial Particle
Velocity in 2-D Flow
    G. Rosenberg, D. Yaziv, and M. Mayseless 495

Effects of Thin Glue Bonds on Shock Waves in LiF
    P. Majewski 500

Ejection of Material from Shocked Surfaces of Tin,
Tantalum, and Lead-Alloys
    P. Andriot, P. Chapron, and F. Olive 505

Shock Attenuation in an Inertial Confinement Fusion Reactor
L. A. Glenn 510

Cavitation in Water Induced by the Reflection of Shock Waves
P. L. Marston and G. L. Pullen 515

Shock Wave Stability
G. R. Fowles 520

Piezoelectric Shear Stress Gage for Dynamic Loading
Y. M. Gupta and W. J. Murri 525

Analysis and Modeling of Piezoresistance Response
Y. M. Gupta 526

A Theory for the Shock-Loading Response of an
Alumina-Filled Epoxy Mixture
D. S. Drumheller 527

CHAPTER XI: High Velocity Penetration

Numerical Modeling of Oblique Hypervelocity Impact Using
Two-Dimensional Plane Strain Models
William T. Brown 529

Measurement Problems in High Velocity Impact Experiments
William Lawrence 534

Experimental and Numerical Investigations Concerning the
Dynamics of Penetration Processes
H. Senf, U. Hornemann, H. Rothenhäusler, F. Scharpf,
A. Poth, and W. Pfrang 539

TOODY-WONDY Calculations of Penetration Events
D. L. Hicks, F. R. Norwood, and T. G. Trucano 544

Hypervelocity Impact Response of Ti and Be
S. J. Bless 548

CHAPTER XII: High Explosives

Free-Surface Velocity Measurements of Plates Driven by
Reacting and Detonating RX-03-BB and PBX-9404
L. M. Erickson, H. G. Palmer, N. L. Parker, and
H. C. Vantine 553

Thin Pulse Initiation of PBX-9404
H. C. Vantine, J. Chan, and L. M. Erickson 558

High Explosive Detonations in Varying Oxygen Atmospheres
F. R. Kovar, K. R. Trigger, L. G. Guymon, and
J. R. Harvey 563

Calculated Shock Pressures in the Aquarium Test
  J. N. Johnson 568

Effect of Charge Diameter on Detonation Pressure Measured
by Aquarium Technique
  Kang Xu, De-yang Yu, Yun-xiang Xu, and Xiung-fei Zeng 573

Shock Waves in Fresh Water Generated by Detonation of
Pentolite Spheres
  T. P. Liddiard and J. W. Forbes 578

Comparison of Axial Longitudinal Velocity Measurements
Determined Ultrasonically and by a Weak Shock Velocity
Technique on an Aluminized Melt Cast Explosive
  J. W. Forbes and W. L. Elban 583

Grüneisen Parameter Measurements for High Explosives
  George H. Bloom 588

Pressure Dependent Vibronic Relaxation in Shocked Explosives
  M. J. Frankel 593

Shock-Induced Instability Due to Crack-Like Defects in a
Solid Propellant
  B. M. Belgaumkar 598

The Divergent Quasistationary Detonation Wave
  G. Damamme 603

Radiographic Study of Impact in Polymer-Bonded Explosives
  Erik Fugelso, J. D. Jacobson, Robert R. Karpp, and
  Russ Jensen 607

Plane Shock Initiation of Gamma-Irradiated PETN Single
Crystals
  J. J. Dick 612

CHAPTER XIII: Experimental Facilities for Shock Compression Research

The Lawrence Livermore National Laboratory Two-Stage
Light-Gas Gun
  A. C. Mitchell, W. J. Nellis, and R. J. Trainor 613

Shock Wave Physics Group M-6
  Charles E. Morris 616

The Sandia Shock Thermodynamics Applied Research Facility
  Lalit C. Chhabildas 621

Shock-Wave Experiments Using Explosives and Light-Gas Gun
Facilities
    R. Cheret, P. Andriot, P. Chapron, C. Le Drean,
    J. M. Lezaud, R. Loichot, J. Martineau, and F. Olive 626

Shock Wave Apparatus for Studying Minerals at High Pressure
and Impact Phenomena on Planetary Surfaces
    Thomas J. Ahrens, Mark B. Boslough, Warren G. Ginn,
    Mario S. Vassiliou, Manfred A. Lange, J. Peter Watt,
    Ken-ichi Kondo, Robert F. Svendsen, Sally M. Rigden,
    and Edward M. Stolper 631

Boeing Shock Physics Laboratory
    R. M. Schmidt..634

Investigations of Hydrodynamic Stability Using Electron and
Ion Beams
    F. C. Perry 639

Nuclear-Explosive-Driven Experiments
    C. E. Ragan..644

High Energy Laser Facilities at Lawrence Livermore National
Laboratory
    N. C. Holmes 648

Shock Wave Facilities at Poulter Laboratory of SRI
International
    W. J. Murri 652

Sandia 25-Meter Compressed Helium/Air Gun
    R. E. Setchell...657

Plate Impact Facility at Brown University
    R. J. Clifton..661

NSWC/WO Light Gas Gun and Explosive Facility
    E. R. Lemar, J. W. Forbes, J. O. Erkman, and
    J. W. Watt 663

Impact Physics Facilities at the University of Dayton
Reseach Institute
    S. J. Bless 668

ISL Shock Wave Facilities
    F. Bauer...674

Facilities for the Study of Shock Induced Decomposition of
High Explosives
    J. E. Vorthman 680

Lawrence Livermore National Laboratory Single-Stage 101 mm Gun
    Leroy Erickson 685

Rail Gun Development for EOS Research
    C. M. Fowler, D. R. Peterson, R. S. Hawke, and
    A. L. Brooks 686

Performance of a 100 KV, 78 KJ Electric Gun System
    H. Chau, G. Dittbenner, K. Mikkelsen, R. Weingart,
    K. Froeschner, and R. Lee 691

Shock-Wave Facility at Tokyo Institute of Technology
    A. Sawaoka and K. Kondo 696

Shock Wave Facilities for High-Pressure Experiments at Tohoku University
    Yasuhiko Syono and Tsuneaki Goto 701

Rafael Terminal Ballistics Laboratory-Facility and Capability
    Gideon Rosenberg 706

Impact Facilities at the Ernst-Mach-Institute
    A. J. Stilp, V. Hohler, E. Schneider, R. Tham,
    M. Hülsewig, G. Kuscher, and W. Junckermann 711

Author Index 713

CHAPTER I: Chemistry of Shock Compression--Material Synthesis

PERSPECTIVES ON INORGANIC CHEMISTRY

S. S. Batsanov
Institute of Physical-Technical Measurements, Moscow

ABSTRACT

As the text of a presentation presented during a panel discussion on Perspectives on the Chemistry of Shock-Compressed Matter, this paper describes the motivations and some of the principal observations obtained from over twenty years of persistent study of chemical effects in shock-compressed inorganic substances.

PERSPECTIVES

The principal problem of chemistry--creation of substances with useful properties--has both analytical and synthetic solutions. In the first case one finds the relations between the structure and properties of substance; in the second case, one changes the composition and then investigates the consequences of the change.

Accordingly, most of the physical methods used in chemistry have analytical character while synthesis-oriented chemists still use the methods of Bacon and Paracelsus. The technique of high pressure is one of the few (if not the only) physical method which makes it possible to intentionally change the chemical composition. However, high static pressure generation is difficult and the useful cell volume diminishes gradually as the upper pressure limit grows.

Twenty to thirty years ago new synthesis opportunities became available with the advent of high-pressure shock-compression techniques. This method has a number of advantages, namely: the experiments are inexpensive and do not take much time to perform, higher values of pressures and temperatures can be obtained, and there are fewer restrictions on the size of the samples to be treated. Psychologically, the explosive method is attractive for young men because of its romantic nature.

These factors led to widespread use of chemical studies by the dynamic high-pressure method in advanced countries. After the progress of so many years it is now revealing to provide a perspective of this new branch of science--shock-induced inorganic chemistry.

What are the major achievements made by shock chemistry in inorganic syntheses? It is clear that with the shock compression technique we can realize all the principal types of chemical reactions: elemental synthesis, double exchange, disproportionation, complex formation, and oxidation-reduction.

What advantages has the shock-compression technique as compared with traditional methods of chemistry? It is known that a chemical reaction initiates when the atoms and molecules are mixed and drawn closer to one another. As a rule it is necessary to melt or dissolve substances. Some substances, however, cannot be melted under normal pressure; for example, most of complex compounds, halogenides of platinum metals, ammonium salts, crystal solvates, and so on. Some substances such as silicates, carbides, borides, many elemental covalent crystals, diamond, and so on cannot be dissolved or are difficult to dissolve. Therefore, the chemical synthesis of such substances is difficult or even impossible. The synthesis of carbides of metals occurs at high temperatures so that carbon under such conditions is in a graphite form. Possible this circumstance is the reason for the absence of interaction between Pb, Sn and C. The shock compression technique can realize the reaction with participation of high pressure phases, i.e., achieve the chemical individuality of different polymorphic modifications.

The second principal peculiarity of shock chemistry consists in the possibility of the different thermodynamic action on components of one reaction mixture. While in the traditional chemistry all reagents totally dissolve or melt, i.e., are in the same thermodynamic conditions, the propagation of a chock wave compresses the individual components of a mixture differently (according to their compressibility) and therefore heats them differently as well. Consequently, the unloaded system is a mosaic of hot and cold pieces with temperature differences of hundreds of degrees. Such pairs of elements as B and Sb, C and I, C and S and so on, having equal electronegativities and therefore zero heats of interaction, cannot ordinarily combine. For example, the compounds BSb is unknown. If the elements of each pair are under different thermodynamic conditions, then their electronegativities are no longer equal and combination becomes possible.

In our laboratory we have devised an experimental ampule for such selective compression, in which one component of reaction is expelled onto the target of another one. In this case, the differential temperature and pressure can reach extremely high magnitudes. With this method we have obtained CrO from Cr and Cr_2O_3, a reaction which cannot be induced by common methods.

The third peculiarity I wish to emphasize concerns the method of heat transmission through the sample. In fact, it is an intuitive hypothesis which I cannot deduce or definitely prove, but nevertheless, I attach importance to the following difference. With conventional heaters, heat from a source must penetrate into the sample step by step with a limiting speed defined by the thermal conductivity of the substance. The thermal pulse induces chemical and physical transformations and reactions as it moves through the system. On the other hand, shock compression heats all the particles of the sample simultaneously to a certain temperature without heat transfer from one particle to another. It is evident that the behavior of a mixture of substances must be quite different in two cases.

I shall illustrate the latter statement with the reaction between Cu and S. By heating the mixture in a furnace one can obtain Cu_2S, CuS or CuS_2 according to the ratio between the amounts of the starting components. But if copper is dispersed in the sulfur by so-called electric explosion (i.e., an electric discharge passing through the mixture of copper and sulfur powders) then only one product is formed irrespective of the ratio of the reactants. It has the stoichiometry $Cu_{1.96}S$; just the same as the formula of mineral digenite which is the most stable of the copper sulfides and thus the most widespread in nature.

This identification of a composition of material prepared by the shock-wave method with a naturally occurring one is more than accidental coincidence. In my opinion shock-wave inorganic chemistry is a good method for modeling the natural processes taking place in the depths of the earth or on other planets. Extremely interesting is the behavior of a solid body exposed simultaneously to shock compression and radiation.

All reactions performed by dynamic methods up to now involve only solid substances. But quite recently it became possible to generalize the method and compress such systems as gas/solid and liquid/solid. The first such experiments carried in our laboratory are rather encouraging. For example, we have studied the reactions of water with metals situated in the electrochemical row on either side of hydrogen. It is well known that under high pressure water is highly ionized. Hence it would dissolve all metals standing before hydrogen in the electrochemical row and leave unchanged the metals standing after hydrogen if the reaction is under high dynamic pressure. This scheme was confirmed by experiments in which Zn, Ti, Cr, Mn and Fe converted into oxides, while Cu and V remained metallic.

In the text above I have treated explosion effects on the substance only from experimental point of view. Nothing has been said of the essential theoretical problem, namely: on what stage of shock compression of solid bodies do the chemical reactions occur? The three possible answers are: either in the high pressure region, in the shock-wave itself, or in the zone of post-shock effects.

The residual temperatures measured by Ahrens and coworkers appear to be so high that any chemical reaction would progress rapidly after the shock compression.

Another alternative, that is, reaction in the shock wave, meets with the following difficulty: the high pressure of explosion is maintained for only an extremely short period of time. If we assume that the reaction is completed during this time, then the rate of diffusion in the reaction area must be incredibly high.

All the information available up to now leads to a conclusion that a reaction can proceed under high shock pressure if, and only if, at least one of the reagents undergoes a phase transition. When the crystal structure of a substance is deeply rearranged, the mobility of atoms may increase to values otherwise impossible for solid body. It is well known that in the moment of phase transition atoms can move 1 Å in the time about 10^{-12} seconds, or 0.1mm in 1 microsecond. Such speed is enough to provide the chemical interaction in powdered materials.

The applied aspects of inorganic shock chemistry were described by others in this panel discussion. I want only to add that, to my opinion (as far as I can judge) the application of explosion to catalyst activation is of no less importance than explosive welding, forming and compaction. The shock wave causes grinding of grains, generates defects (especially vacancies) and electronic defects and thus sharply increases the catalytic activity of the sample.

In this paper I have described those problems of shock synthesis that seem most interesting to me. Surely, in inorganic chemistry there are many important and rewarding problems, say, diamond or boron nitride synthesis. Nothing, however, can give a chemist a more heightened sensation of joy than holding in his hands a chip of matter that nobody ever has held before.

SHOCK-INDUCED INORGANIC CHEMISTRY*

B. Morosin and R. A. Graham
Sandia National Laboratories,† Albuquerque, New Mexico 87185

ABSTRACT

The fundamentals of chemical processes occurring in shock-compressed inorganic materials are reviewed. The materials survey presented complements previous reviews on this topic and examines specific topics from a different perspective. Future directions for research in this area are suggested.

INTRODUCTION

This paper briefly reviews the fundamentals of chemical processes occurring in shock-compressed inorganic materials and provides the reader with a source of current references in the field.

The embryonic conditions controlling the chemistry of shock-compressed nonmetallic elements and compounds result from conditions which force matter into states quite different from those which are normally encountered. The resulting restructured atomic arrangements may drastically alter chemical bonding, as is the case even for simple elements, e.g. the transformation of graphite into diamond. More importantly, defects are introduced in large concentration and alter the standard thermodynamic state of the material, generally enhancing reactivity and strongly influencing chemical processes.

The chemistry which occurs under shock compression is of fundamental importance in realistic descriptions of shock processes and in descriptions of solid state reactivity. Such fundamental understanding should prove beneficial to specific synthesis problems or preparation of materials with enhanced reactivity properties for subsequent processing.

Current diagnostic tools are beginning to probe chemical changes in substances even though no material is preserved for post-shock examination. However, most work in this field is based upon post-shock analysis. Ryabinin reported the first studies of substances which were preserved after shock-compression in 1956. This effort reported the decomposition of shock-loaded $CuSO_4$, $MgCO_3$, $Pb(NO_3)_2$, as well as the polymorphic transformation of sulfur. Of more technological significance was the subsequent demonstration by DeCarli and Jamieson on the phase transformation of graphite into diamond, which was the first observation of diamond synthesis without the addition of a catalyst.[2] In 1963, Kimura reported the first synthesis of a compound, $ZnFe_2O_4$, from the corresponding zinc and and iron oxides.[3] In 1965, DeCarli and Milton[4] reported synthesis of stishoite from quartz with shock-compression and shortly thereafter, Bergmann and Barrington[5] demonstrated greatly enhanced solid state reactivity in shock-loaded ceramic powders. About this same time, Horiguchi and

*This work sponsored by the U.S. DOE under Cont. DE-AC04-76-DP00789.
†A U.S. DOE facility.

Nomura[6] demonstrated enhanced catalytic activity in shocked acetylene black. Even though this early work has led to some specific commercial applications of shock synthesis, e.g. diamond synthesis, the fundamental questions raised by these observations were difficult, and the major effort required to systematically develop this field was not carried out outside the Soviet Union. Soviet scientists,[7] however, have carried out a large, consistent effort to understand and utilize shock-induced chemical effects, and that work now serves as a firm foundation for further progress.

Batsanov and coworkers have carried out much of the shock chemistry work, and he has reviewed this work thoroughly.[8-10] Other principal groups include Dremin and Breusov and coworkers,[12,13] and Adadurov and Gol'danskii and coworkers.[14,15] Aspects related to mineralogy which involve chemical and physical changes have been reviewed by Stoffer.[16] A recent review of structural transitions is given by Duvall and Graham,[17] and a recent comprehensive review of shock-compression in materials is given by Davison and Graham.[18]

In the remainder of this paper some unique features of shock deformation will be considered, and then specific results on different inorganic systems will be reviewed.

SHOCK COMPRESSION PROCESSES

Perhaps the most unique consequence of shock compression of solids is that of introduction of large numbers of lattice defects.[8,17] These defects are dislocations, crystallographic shear planes, vacancies, interstitials and substitutional defects formed by the enormous stress, velocity, and temperature gradients within and in the vicinity of the shock front. In extreme cases, such high dislocation and defect concentrations render the materials in an amorphous or glassy state. As the shock pulse is removed, some of these defects may be annealed if the temperature is excessive. Substantial numbers may remain, however, contributing to the reactivity of the solid.

In single crystals which have been carefully studied, shock-induced defects appear over the background of previous defect motiff or pattern as long as no phase transition occurs.[9,10] Usually such phase transitions tend to remove or sweep out the original defect pattern. Studies of defects in single crystals yield valuable detailed information which, of course, applies directly to powders since they are really randomly oriented crystallites and typically of large dimension relative to the shock wave front.

Comminution resulting from shock loading also plays an important role, and it is unlike that of conventional processes.[8,9,12] From a practical point, when one normally grinds a solid, the fracturing process involves fractures along defects present in the crystallites. Thus, as comminution continues, the material decreases crystallite size at the expense of defect concentration. The strength of crystalline materials and the lack of cleavage planes and of ion defect concentration may prevent size reduction to levels aiding chemical processing and reactivity. On the other hand with shock loading, the original defect concentration is not important for comminution - only the pressure gradients, together with the cleavage or other crystal structure properties. Thus, not only is the surface area increased, but a large defect concentration is produced in the bulk.

Shock loading may reduce the material to tiny crystallites which often resemble mosaic blocks - hundreds of Å in size. Such fracturing is usually not self-healed by subsequent temperature rises because of small rotations which many of these mosaic blocks undergo. Further shocking of such comminuted crystallites does not appear to yield further size reduction - suggesting that there is a limiting crystallite size which can be attained through shock loading.[8,9]

Line broadening of X-ray diffraction patterns serves as a qualitative measure of crystal lattice defect concentration as well as for comminution resulting from the shock wave. Defects result in elastic strained lattices and broaden the high two-theta lines of the diffraction patterns. Comminution reduces the size of the diffracting media and hence broadens the low two-theta lines when crystallite sizes are smaller than about 2000 Å.

Our own studies[19] suggest that there is a threshold with respect to pressure in order to achieve comminution below thousands of Å. Also, the elastic properties of materials under study yield a large variation in lattice strain broadening. For example, on our capsule shocked to about 20 GPa, powdered CdS and single crystal $LiNbO_3$ were recovered with sharp high two theta lines[19] while a host of other oxides as well as FeS_2[20] showed broadened lines. For ZrO_2 such broadening is sufficiently severe that even qualitative information is difficult to extract.[21]

The most important consequence of introduction of large numbers of defects and comminution to small crystallite sizes is the enhanced solid state reactivity of the materials. This is dramatically illustrated by data reported by Adadurov and coworkers[22] on Nd_2O_3 exposed to atmospheric water vapor. Ordinary calcined material reacts extremely slowly and only after 400 hrs. can discernible quantities be measured. Material carefully ground in a mortar has significant water uptake in 40-60 hrs, while material subjected to shock-loading begins to hydrate immediately. Furthermore, it is well known that many inorganics spontaneously react with oxygen when particle sizes are extremely small. Shock comminution provides such small particles, thereby increasing contact surfaces between reactants; the high defect concentration provides ionic mobility.

Extremely high pressures are accessible to the shock-compression experimenter. Of course, the temperature attained in such dynamic experiments will vary over a wide range, depending on the effective density, initial temperature, and the details of the shock experiment itself. As a result, one is able to achieve compression states of matter inaccessible by other means. Such states are characterized by strong compression, drastic reduction in interatomic separations and possibly large deformation of the electron shells of atoms and particularly bonding and antibonding orbitals. Typically, cations are reported to show reduced valence states following such shock compression.

Even though such large compressions are important thermodynamically, the material should not necessarily be viewed as merely highly compressed matter, a benign description involving a pressure, volume, and temperature relationship. As the shock waves interact with materials of different impedances, wave reflections are caused which amplify as well as diminish shock peak pressures in adjacent submicron regions of the material. Such reverberations are responsible for the rotation of comminuted crystallites so that annealing might be impossible and for displacing highly defective, and hence reactive, grains against each other, enhancing reactivity. Within each reverberating wave, shock-induced mass motion may provide greatly enhanced mixing and blending with coalescence of materials.

The experimenter is able to control to some degree the pulse function, shape, and magnitude of the pressure pulse depending on the design and geometry of the experimental capsules and the properties of the material under examination. Such control may require rather detailed studies of the mechanical properties and shock compression properties of the material. With the passage of the pressure pulse, high pressure or high temperature phases as well as metastable reaction products may be quenched and stabilized by the defects.

Shock compression is a convenient method of producing various glasses. The quenching achieved by the relief wave cools far below the liquidus line with sharply increased viscosity so that crystallization cannot be initiated. Glasses of densities far exceeding those normally attained can be produced. For example, quartz glass with a density exceeding that of conventional fused quartz by 20% can be formed by shock compression.[18]

The remaining unique property is shock enhanced mass transfer, which previously has been mentioned as possibly occurring at contact points between grains. Several measurements show atomic migration over distances of tens of microns - four orders of magnitude greater than in normal circumstances.[23-27] Such mass transfer greatly enhances reactivity. It also may pose as a potential problem as, for example, the typical contamination from the containment capsules observed in some of the very high pressure synthesis studies.

In summary, the action of shock compression is that of a super mixmaster - chopping to small size, activating by forming defects, injecting neighboring material, all in a high pressure and temperature environment which can be quickly turned on and quickly quenched. These conditions provide an environment in which chemical and structural transformations can proceed with extraordinary rapidity.

SHOCK-INDUCED MODIFICATION OF INORGANIC MATERIALS

A review of previous studies can conveniently catalog materials studied either as homogeneous or heterogeneous systems, i.e., the pure material or mixtures subjected to shock compression. The distinquishing feature of chemical reactions in heterogeneous systems is that mass motion must occur over the many micron dimensions of the particles in the shocked state.

The earliest sample preservation studies demonstrated that materials may decompose under shock compression.[1] The importance of thermal contributions to shock-induced chemistry for materials is illustrated by the study[15] of sixty-one oxides which have been tabulated as to their stability when placed in low percentages in fused quartz powder and subjected to shock compression. A reasonable correlation with respect to the energy required for oxygen removal is found. Adadurov et al[15] review the area of decomposition very well; recently the dissociation of Al_2SiO_5 has been reported.[28]

Various oxides, such as V_2O_5, Nb_2O_5, Ta_2O_5, and Nb_xO_2 with x = 0.1 to 0.8, transform to structure types which demand large cation vacancies.[29-33] For example, Ta_2O_5 transforms to the tetragonal rutile structure type. In this structure, only 4/5 of the cation sites are occupied since TiO_2 requires a 1:2 cation:anion ratio. Other structure types result in different ratios, but are, in principle, very similar to our example. These solids can properly be termed structurally defective solids. The solid state chemistry and physics of such materials have not, to our knowledge, been examined in detail; however, they should prove to be scientifically very interesting. From a crystal structure viewpoint, they parallel the class of materials known as solid electrolytes or superionic conductors in that in the high conductivity phase, the crystal structures requirement is that there are sufficiently more lattice sites than cations available. This is the case for these shock compressed materials. For example the rutile structure type mentioned above would be deficient 1/5 of the required cations.

Little has appeared on shock activation of catalysts since the early work of Horiguchi and Nomura[6] and Boreskov and coworkers.[34,35] This topic has been under investigation in our laboratories with special emphasis on detailed understanding of defects. Our early studies were on pyrite,[20] specifically "Robena pyrite," which is a U.S. Steel byproduct when sulfur is removed from metallurgical coal for steel production. This material contains various impurities and has been found to increase liquefaction yields when employed as a catalyst for coal liquefaction.[36] Our studies demonstrated various material alteration of the Robena pyrite from shock compression, such as the formation of pyrrhotite, smythite and Fe_2C, the introduction of lattice defects as well as magnetic changes. Liquid yields in coal liquifaction were not improved by using shock activated pyrite additives. However, recent understanding of this complex reaction problem suggest the catalyst is not the pyrite phase.

Recently, our attention has shifted to TiO_2, in which defects are being studied in quantitative detail. Our results[37] suggest defects other than the Ti^{3+} suggested by Batsanov, et al[34] may be responsible for catalytic activity.

Heterogenous systems span simple combinations of elements, such as Horiguchi and Nomura's[38,39] pioneering synthesis of the carbides of Ti, W & Al as well as those for Mo recently reported,[40] to complex ternary systems such as CaO, Cr_2O_3 and SiO_2.[12] Shock compression has made possible combinations previously unattainable, such as alloying W and Fe with unlimited miscibility.[43] Further alloys with vastly different melting points are possible such as combining W and Mn. Batsanov[8-11] has reported solid solution of various alkali halides, some combinations being reported for the first time.

Interestingly, other combinations which were attempted without success suggested the importance of the ratio of cation to anion radii, particularly when the combinations may have low melting points and may be compressible. Our studies on solid solution resulting from shock compression suggest CaO and ZnO are mutually soluble at low concentration in each other.

Recent oxidation reduction reaction studies include the reaction of SiC with oxygen to form a silica containing carbon substitutionally for part of the silicon in the lattice.[41] In another study, the usual equilibrium constants appear reversed as AgCl reacts with iodine to form AgI and Cl_2 under shock compression.[41]

Since Kimura's pioneering work on the reaction of oxides,[3] many systems have been studied and previously documented in reviews mentioned above. Recent studies include Al_2SiO_5[41] formed from Al_2O_3 and SiO_2, as well as the examination of the $PbO \cdot ZrO_2$ system in our laboratories.[22] $PbZrO_3$ has been formed directly under shock.

Numerous rare earth (hereafter RE) chalcogenides have been examined under shock compression and are particularly interesting since there is a nearly a linear relationship as to size of radii with respect to atomic number even though the chemistry is usually rather similar. Differences are present, for example, the decomposition temperatures of $RE(OH)_3$ systematically decrease from the value for La. Batsanov[8] found that the corresponding oxides would combine with water under shock compression depending on the radius of the RE ion, the larger ions forming $RE(OH)_3$, the intermediate $REO(OH)$ while the smaller Dy through Lu as well as Y not reacting. The combination of the RE-oxides with the corresponding fluorides yielded the oxyfluorides.[42] These proved to be very sensitive to small deviations in stoichiometry as to the resulting structure type. For example, a 1:1 $SmF_3:Sm_2O_2$ mixture yields the cubic fluorite structure; a small excess of SmF_3 yields a tetragonal structure form while a small deficiency yields a rhombohedral form. Repeated shock loading of the cubic material which is a disordered structure does not result in any symmetry change unless the stoichiometry is altered by addition of one of the components. The corresponding chalcogenides result in some different structure types and repeated shock loading in these cases alter their structure type.

The reaction of titania with the RE_2O_3 has been extensively studied with shock pressure being very important as to which structure type is obtained.[43,44] Typically, shock loaded compounds are are reported to have 2 - 3% smaller lattice parameters than those obtained by normal synthesis. The size of the RE appears to influence the crystal structure type obtained. Sm serves as a convenient example since its hexagonal phase field covers some 7 mole % in a linear fashion, the lower concentration being fluorite cubic while the richer combination is orthorhombic. For smaller RE ions, one obtains the monoclinic structure type rather than orthorhombic form. Further, the region or width of the hexagonal phase decreases as the ionic radius of the RE is decreased. Under particular shock conditions, unique solid solutions can occur. For example, Er and Yb form the $RE_2Ti_2O_7$ pyrochlore structure-type. The cell is twice that for fluorite and over this compositional range a linear relation is found for the lattice constants; further the super-lattice reflections gradually diminish and vanish.

Recent work on the corresponding RE oxides reacting with HfO_2 and ZrO_2 has appeared also with complex interesting relationships.[45]

The greatly enhanced reactivity demonstrated in numerous shock-activated sintering studies[47-50] and synthesis of superconductors[51-59] are areas of significant technological promise. The role of inert additives which may act as heat sinks or cause localized shock reverberations in powder mixtures is critical. The careful studies on many different substances added to the mixtures used in preparing cubic BN show a rather striking dependence of yield on shock impedance of the additives.[46] Inert additives, such as H_2O, may prevent decomposition of various oxides and carbonates, significantly altering the thermal properties,[60,61]

Electrochemical behavior of shock-loaded substances [62,63,64] as well as decomposition products behind the shock point[65] suggest dissociation into ionic species. The mobility of H^+ and OH^- ions appear responsible for conductivity increases in aqueous solutions.[66]

The stability of defects formed and the environments they experience after the shock-loading event is important. The shocked material must be removed from recovery capsules for further study. Their removal may drastically alter the defects which the shock wave introduces. Explosively synthesized diamonds,[61] which were extracted by acid treatment after shock compression to 70 GPa, are reported to have highly developed specific surface area, covered with hydroxyl, carbonyl and carboxyl groups, to the extent of 20% of the surface area or expressed as total weight, almost 10% of the bulk. Thus, the shock compressed material was drastically altered in the recovery procedure. Not as drastic effects have been suggested by our results[37] on TiO_2 shock-activated catalyst powders.

FUTURE DIRECTIONS

The chemistry of shock-compressed inorganic solids is of fundamental interest for its understanding of shock processes. Future thrusts need to define shock-induced defects on an atomic level and their influence on enhanced reactivity; these defects undoubtedly hold the key to understanding mechanisms for chemical conversion.

In the large majority of reactions, the final product is observed; this has been termed the integral effect. We do not know precisely when a particular transformation occurred, whether the shock loading, the relief wave or longer time processes are responsible. We know shock reverberations are important in forming the high density phase of BN and in promoting other reactions. The addition of inert additives to materials can modify chemical conversion, though the reasons are not clear.

Alterations of loading with isotropic compression techniques appear to provide an important new dimension to material modification. The role of temperature and its control is extremely important and capsule design and temperature control, both initial, during the course as well as following the shock experiment, are steps which much be achieved. Several of these questions will require using real time techniques rather than relying on post-shock examination. Picosecond spectroscopy and clever thermometry[62] techniques should have an excellent future. Finally, more realistic modeling of material response is necessary along with supportive experimental studies which define the role defects, comminution, shock-induced diffusion, and shock-enhanced diffusion play in shock-induced solid state reactivity.

REFERENCES

1. Yu. M. Ryabinin, Sov. Phys. Tech. Phys. 1, 2575 (1956).
2. P. S. DeCarli and J. C. Jamieson, Science 133, 1821 (1961).
3. Y. Kimura, Japan J. Appl. Phys. 2, 312 (1963).
4. P. S. DeCarli and D. J. Milton, Science 147, 144 (1965).
5. O. R. Bergmann and J. Barrington, J. Amer. Ceramic Soc. 49, 502 (1966).
6. Y. Horiguchi and Y. Nomura, Carbon 2, 436 (1965).
7. Various symposia proceedings published in the Soviet Union to be referenced include (translations in parenthesis): a) Proceedings, First-All Union Symposium on Shock Pressures, Vol. 2, October 24-26, 1973, Moscow, Edited by S. S. Batsanov, Moscow 1974. (SNL Report SAND80-6009, April 1980); b) Physics of Shock Pressure, Proceedings, Second All-Union Symposium on Shock Pressures, October 19-22, 1976, Moscow, edited by S. S. Batsanov, Moscow (1979). (SNL Report SAND81-6006, February 1981); c) Proceedings, Third All-Union Symposium on Impulsive Pressures. October 16-18, 1979, Moscow, edited by S. S. Batsanov, Moscow (1979). (SNL Report SAND80-6004, March 1980); d) Detonation, Critical Phenomena, Physicochemical Transformations in Shock Waves, edited by F. I. Dubovitskii, Chernogolovka (1978). (UCRL Report 11444, December 1978); e) High Pressures and Properties of Materials: Materials of the Third Republican Scientific Seminar (1980), Selected Papers. Ukranian Academy of Sciences SSR. Institute of Problems of Materials, Kiev. (SNL Report RS3140/81/31, February, 1981).
8. S. S. Batsanov, J. Engineering Physics 12, 59 (1967).
9. S. S. Batsanov, Inorganic Materials 6, 615 (1970).
10. S. S. Batsanov in 7a, p.4.
11. S. S. Batsanov, in 7d, p.197.
12. A. N. Dremin and O. N. Breusov, Russian Chemical Reviews 37, 392 (1968).
13. A. N. Dremin and O. N. Breusov, Prioda, No. 12, 10 (1971). Translation in SNL Report SAND80-6003, March 1980.
14. G. A. Adadurov, V. I. Gol'danskii and P. A. Yampolskii, Mendeleev Chemistry Journal 18, 92 (1973).
15. G. A. Adadurov, T. V. Bavina, O. N. Breusov, A. N. Dremin, E. N. Klopova and V. F. Tatsii in First International Symposium on Explosive Cladding, 1970, Marianske Lazne, Czechoslovakia (1971), p. 223. (SNL Report SAND80-6008, April 1980)
16. D. Stoffer, Fortschr. Mineral 49, 50 (1972).
17. G. E. Duvall and R. A. Graham, Rev. Mod. Phys., 49, 523 (1977).

18. L. Davison and R. A. Graham, Phys. Rep. 55, 255 (1979).
19. See papers by the present authors in this Proceedings.
20. R. A. Graham, B. Morosin, P. M. Richards, F. V. Stohl and B. Granoff, abstract in Chemistry and Physics of Coal Utilization - 1980. (APS, Morgantown) edited by B. R. Cooper and L. Petrakis, American Institute of Physics, New York (1981), p.464.
21. J. Keck, D. L. Hankey, R. A. Graham and B. Morosin, this Proceedings.
22. G. A. Adadurov, O. N. Breusov, A. N. Dremin and V. N. Drobyshev, Russian J. Inorganic Chemistry 16, 1073 (1971).
23. S. V. Zemsky, Ye. A. Ryabchikov and G. N. Epshteyn, Phys. Met. Metall. 46, 171 (1979).
24. V. V. Gal', V. S. Ivanov, Yu. L. Krasulin and E. E. Spirirdonov in 7c, p. 224.
25. S. S. Batsanov, I. A. Ovsyannikova and N. A. Shestakova, Fizikaii Khimiya Obrabotki Materialov, No.1, 1116 (1974). Translation in SNL Report RS 3140/80/72, June 1980.
26. G. N. Epshtein, in 7e, p.7.
27. S. V. Zemskii, E. A. Ryabchikov, T. G. Ryabchenko and G. N. Epshtein, in 7e, p.13.
28. H. Schneider and U. Hornemann, J. Mater. Sci. 16, 45 (1981).
29. G. A. Adadurov, O. N. Breusov, A. N. Dremin, V. N. Drobyshev and S. V. Pershin, Combustion, Explosion and Shock Waves, 4, 503 (1971).
30. G. A. Adadurov, O. N. Breusov, V. N. Drobyshev and S. V. Pershin Translation in Third All-Union Symposium on Combustion and Explosion (1972) Report AD/A - 002962 p. 540.
31. O. N. Breusov, A. N. Dremin, V. N. Drobyshev and S. V. Pershin, Russian J. of Inorganic Chemistry 18, 157 (1973).
32. V. N. Drobyshev, in 7a, p. 25.
33. A. V. Anan'in, O. N. Breusov, A. N. Dremin, V. N. Drobyshev and S. V. Pershin, Russian J. of Inorganic Chemistry 20, 319 (1975).
34. S. S. Batsanov, G. K. Boreskov, G. U. Grisasova, N. P. Keier, L. M. Kefeli, V. M. Kudinov, V. I. Mali and I. S. Sazonova, Kinetics and Catalysts 8, 1140 (1967).
35. G. Boreskov, I. Sazonova, N. Keyer, V. Kudinov, G. Gridasova, V. Mali and L. Kefely in Behavior of Dense Media Under High Dynamic Pressures, edited by J. Berger, Gordon and Breach, N.Y. (1968), p. 389.
36. B. Granoff and P. A. Montano, "Mineral Matter Effects in Coal Compression" in Chemistry & Physics of Coal Utilization, 1980 (APS, Morgantown) Edited by B. R. Cooper and L. Petrakis Am. Inst. of Phys. N.Y. (1981) p. 291.
37. J. Golden, F. Williams, B. Morosin, E. L. Venturini and R. A. Graham, this Proceedings.
38. Y. Hoiguchi and Y. Nomura, Bull. Chem. Soc. Japan 36, 486 (1963).
39. Y. Horiguchi and Y. Nomura, J. Less-Common Metals 11, 378 (1966).
40. V. P. Alekseevskii, L. P. Isacur, V. I. Kovtun, M. D. Smolin, I. I. Timofeeva and V. V. Yarosh in 7e, p. 40.
41. S. S. Batsanov, this Proceedings.

42. E. D. Ruchkin, L. N. Travkina and S. S. Batsanov in 7a, p. 62.
43. L. G. Shcherbakova, A. V. Kolesnikov and O. N. Breusov, Inorganic Materials 15, 1724 (1979).
44. A. V. Kolesnikov, L. G. Scherbakova and O. N. Breusov, Doklady, Academy Nauk, SSSR, Physical Chemistry 251, 172 (1980).
45. A. V. Kolesnikov, L. G. Shcherbakova and O. N. Breusov, Doklady, Academy Nauk, SSSR, Physical Chemistry, 256, 113 (1980) (Russian).
46. G. A. Adadurov, T. V. Bavina, O. N. Breusov, A. N. Dremin and S. V. Pershin, in 7a, p.34.
47. For references on shock-activation and dynamic consolidation on ferroelectrics see reference 21.
48. G. A. Adadurov, O. N. Breusov, A. N. Dremin and V. N. Drobyshev, Soviet Powder Metallurgy and Metal Ceramics, No. 3 (99), 227 (1971).
49. G. A. Adadurov, O. N. Breusov, A. N. Dremin and V. F. Tatsii, ibid. No. 11 (1907), 859 (1971).
50. A. V. Anan'in, O. N. Breusov, A. N. Dremin, V. B. Ivanova, S. V. Pershin, V. F. Tatsii and F. A. Fekhretdinov, ibid No. 8 (140), 662 (1974); ibid No 10 (142), 100 (1974); ibid No. 9 (165), 714 (1976); ref. 7a, p. 1.
51. G. Otto, O. Y. Reece and U. Roy, Appl. Phys. Lett. 18, 418 (1971).
52. I. M. Barskii, V. Ya. Dikovskii, and A. I. Matytsin, Combustion, Explosion and Shock Waves, 8, 578 (1972).
53. G. H. Otto, U. Roy and O. Y. Reece, J. Less-Common Metals 32, 355 (1973).
54. V. M. Pan, V. P. Alekseevskii, A. G. Popov, Yu. I. Beletskii, L. M. Yupko and V. V. Yarosh, JETP Lett. 21, 228 (1975).
55. V. N. Pan, A. G. Popov, V. P. Alekseevskii, O. G. Kulik and V. V. Yarosh, Fiz. Nizk. Temp. 3, 801 (1977).
56. D. Dew-Hughes and V. D. Linse, J. Appl. Phys. 50, 3500 (1979).
57. L. N. Fedotov, I. I. Kirshenina, Yu. K. Konov, and L. I. Belova, in 7b, p. 185.
58. B. Olinger and L. R. Newkirk, Sol. State Comm. 37, 613 (1981).
59. Y. Syono, T. Goto, S. Ohshima, N. Sone, M. Saito and T. Wakiyama, Bull. Am. Phys. Soc. 26, 661 (1981).
60. A. N. Dremin, V. P. Ivanov and A. N. Mikhailov, Combustion, Explosion and Shock Waves, 9, 784 (1973).
61. D. D. Bloomquist and S. A. Sheffield, J. Appl. Phys. 51, 5260 (1980).
62. A. C. Mitchell, N. I. Kovel, W. J. Nellis and R. N. Keeler in High Pressure Science and Technology. Vol. 2, edited by B. Vodar and Ph. Marteau, Pergamon, NY (1980), p. 1048; Ref. therein.
63. V. V. Yakushev and A. N. Dremin; Russian J. of Phys. Chem 45 50 (1971).
64. V. M. Shunin, A. N. Dremin and V. V. Yakushev, Doklady Acad. Nauk, SSSR, Phys Chem 251, 246 (1980).
65. A. N. Dremin and V. V. Yakushev, Acta Astronautica 1 885 (1974).
66. O. N. Breusov, A. N. Dremin, V. N. Kochnev, O. K. Rosanov and V. V. Yakushev, Soviet Electrochem 7 395 (1970); ibid 5 719 (1979)

INORGANIC SYNTHESIS UNDER SHOCK-WAVE COMPRESSION

S. S. Batsanov
Institute of Physical-Technical Measurements,
State Committee of Standards, Moscow

ABSTRACT

Studies of the synthesis of inorganic substances with high pressure shock loading are summarized.

INTRODUCTION

The present paper reviews the studies of chemical interactions between inorganic substances subjected to shock-wave compression. The review includes results published to date with special emphasis on the work performed in this laboratory. Befor discussing the chemical data I shall outline the experimental technique used for the synthesis studies.

RECOVERY FIXTURES

In most cases we have used cylindrical recovery fixtures and Figure 1 shows the simpliest design. It is a steel tube with walls 1 to 3 mm thick and 8 to 12 mm in diameter. The steel tube is surrounded by an explosive whose detonation velocity ranges from 4 to 8 km/s. As indicated in Figure 2, the detonation

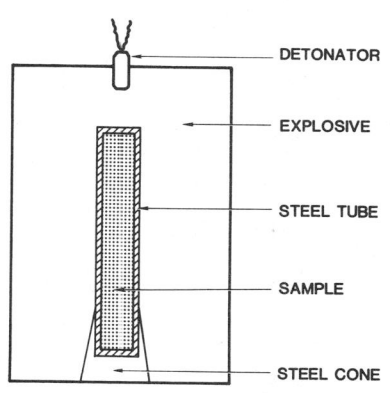

Figure 1. Cylindrical recovery fixture

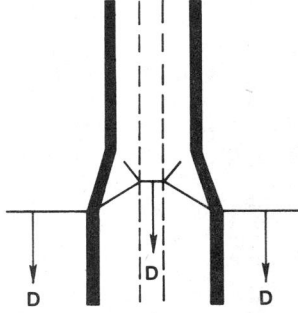

Figure 2. Mach wave configuration formed in cylindrical fixture

of the high explosive leads to a three-wave configuration, called a Mach wave, which propagates from the top to the bottom of the ampule with a speed equal to the detonation velocity, D, of the high explosive and independent of the contents of the ampule. The diameter of the Mach cylinder is about 10 to 50% of the outside diameter of the ampule. Thus, if the ratio of the outside diameter of the ampule to its inside diameter ranges from 2 to 10, the Mach cylinder can occupy the entire inside space of the ampule and the cylindrical compression subjects the sample to planar, one-dimensional compression. The same one-dimensional mode of compression can be obtained by reducing the diameter of the explosive charge with an accompanying increase in its height. In this case, as shown in Figure 3, the explosive surrounding the ampule acts as a steel ring and the cylindrical ampule produces planar conditions.

To increase the shock pressure and temperature the cylindrical configuration is modified to include an inert material (such as formed cast plastic as shown in Figure 4) which changes the direction in which the detonation wave propagates and results in a colliding shock wave configuration. The surrounding layer of steel powder increases the pressure pulse duration.

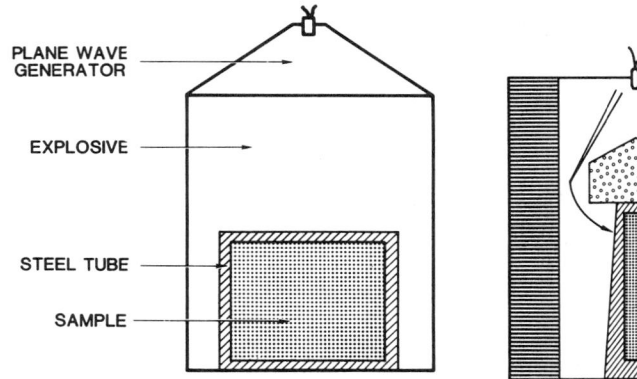

Figure 3. Cylindrical fixture with reduced explosive diameter

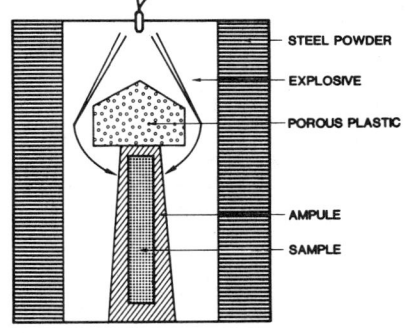

Figure 4. Cylindrical fixture with inert wave intensifier

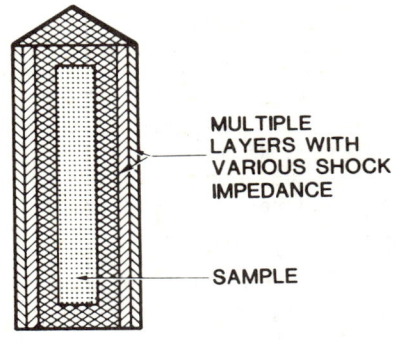
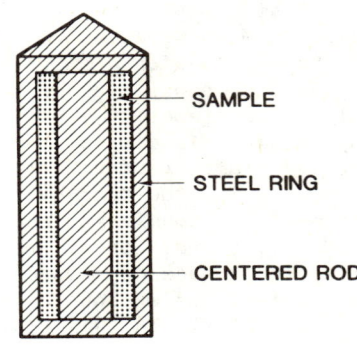

Figure 5. Cylindrical ampule with layered walls.

Figure 6. Cylindrical ampule with inert center wall.

The increase in the temperature caused by the explosive compression can be reduced in several ways. The layered walls of the ampule in Figure 5 cause the shock wave to be split into a series of smaller amplitude waves whose compression approaches isentropic conditions. Systems have been utilized with more than 100 steps in the wave risetime. Another method is to install a coaxial metallic rod within the ampule as shown in Figure 6. The diameter of the rod is chosen to be greater than the Mach cylinder.

A fixture to mix hot, shocked samples with cold, unshocked samples is shown in Figure 7. The shocked sample is accelerated by the Mach wave and thrown into the cold unshocked material. The later can be cooled with liquid nitrogen. Upon impact between the stream and the target a rapid cooling occurs proportional to the mass ratio of the target to stream. Other techniques will be described below.

CHEMICAL REACTION MECHANISMS

Shock compression causes a comminution of crystallites and generation of lattice defects. The size of coherent diffracting domain of the shocked crystallites is about 100Å while the microstrains indicated by x-ray diffraction line broadening is about 10^{-3}. The combination of fine grinding and numerous defects causes a lowering of the crystal destruction energy (endo-effects) and an increase in the energy of exo-effects upon combination.

To illustrate this effect, heating curves of the Cr+Se system are presented in Figure 8. Shock compression which is insufficient to cause chemical reaction decreases the endo-effect

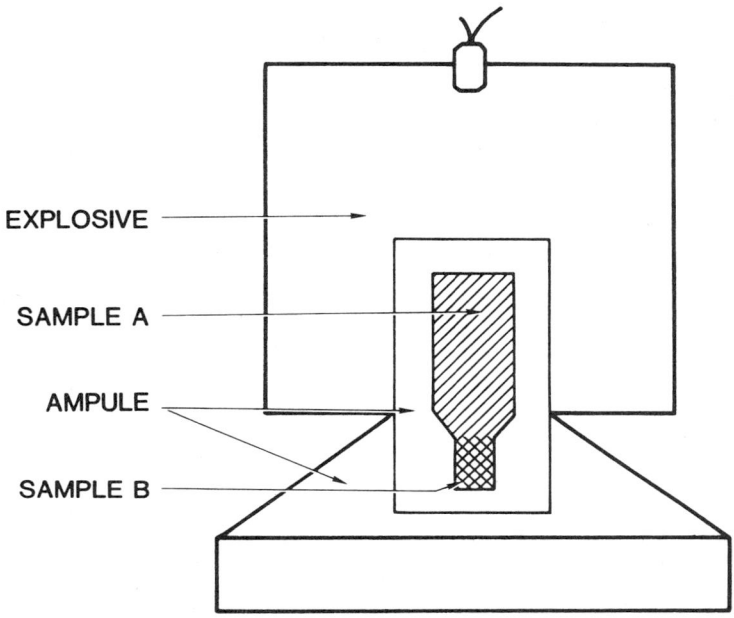

Figure 7. Ampule for mixing hot shocked sample (A) with cold unshocked sample (B).

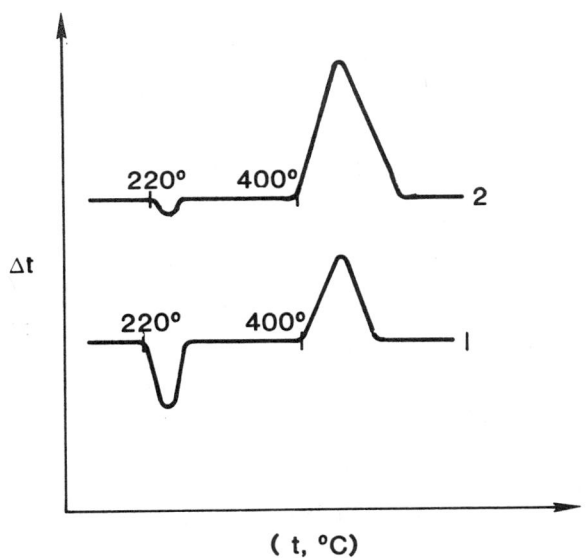

Figure 8. Heating curves for Cr + Se powders.
1 is the initial mixture and 2 is the shocked material.

while increasing the exo-effect.

The causes of chemical interactions induced by shock loading are the decrease of activation energy of the reaction (due to partial destruction of the lattice), the increase in contact surface between grains in the compressed samples and formation of numerous active centers. Finally, the simultaneous action of high temperature and pressure is a significant factor which can produce interesting chemical effects.

CHEMICAL REACTIONS

The simplest chemical reaction is the formation of solid solutions and a summary of those obtained with shock compression is given in Table I.

The most interesting features of shock-wave formation of solid solutions are the increase in miscibility of W-Fe, the formation of a high pressure phase of $NH_4Br-CsBr$ and the overall simplicity of the method which permitted us to obtain a series of new solid solutions. The most essential condition for the solid solution formation is probably the high temperature since less dense packing of the starting powders always results in a greater number of mixed crystals.

Examples of chemical reactions of oxides are shown in Table II. While early work in development of shock-induced chemical synthesis was only concerned with observing the existence of chemical reactions, the work has now become of practical significance.

If one of the reagents is a volatile substance, e.g., water, the stoichiometric coefficients of the reaction can be easily determined. The recovery capsule is preserved when all the water is absorbed by the product of the reaction; otherwise, the remainder of the water evaporates and blows up the ampule. In this manner we have made a preliminary determination of the composition of hydroxides yielded by reactions between rare earth elements and water. The lanthanides of the Ce group form trihydrates while those of the Y group form monohydrates.

In addition to molecular synthesis, the shock compression method has been used to carry out studies of the reactions of numerous elements. A summary of such work is shown in Table III.

The reaction between tin and chalcogenides may be described in a more detailed manner in temperature measurements in ampules. Experiments in which S, Sn, SnS and Sn+S were shock compressed under identical conditions and temperature measurements 0.1 second after compression gave values of 110, 120, 130 and 1100°C, respectively. The excess increase in temperature in the last case is due to the exothermic reaction of Sn+S. The theoretical calculation from thermochemical data gives 1400°C.

We have also studied the combination of tin chalcogenides in explosive welding conditions. Powdered S, Se or Te was placed on a Sn plate and a shock wave was applied perpendicular to the layer. The magnitude of the impulse was varied and the depth

of diffusion was measured by the x-ray microspectral method
perpendicular to the joint. The measurements show the impulse
H (in kgs) vs depth of diffusion d (in μm) data as H = 3.2, d =
2; H = 5.4, d = 4; H = 7.8, d = 6.

The formation of SnS in the middle of the diffusion layer
was indicated by x-ray diffraction while SnS_2 was found on the
boundary and near it.

When the samples were cooled in liquid nitrogen before the
explosion, the diffusion distance was reduced by 1.5 times. In
this case there is no region indicating the formation of SnS.

Finally, the thermal diffusion on Sn and S shocked plates
was studied at atmospheric pressure. Shocked and unshocked
plates were heated in close contact for 50 hours and the x-ray
microanalysis was carried out. In the shocked plate (H = 3.2)
the diffusion depth was 9 μm where it was only 4 μm in the
control plate.

The last synthesis group to be reported here is the interaction of the type $M + MX_n = MX_m$, where m<n, M is a metal and
X a nonmetal. Examples of such reaction studied in this
laboratory are shown in Table IV.

The decrease in activation energy results in lowering the
temperature of reaction and rigid experimental ampules make decomposition difficult. Another factor promoting chemical reactions under shock compression results from inhomogeneous compression of a mixture of components with different compressibi-
lities. Upon release of pressure these components have different
residual temperatures; this difference may amount to hundreds
of degrees under pressures of tens of GPa. This provides an
additional thermodynamic stimulus for the reaction. To check
this hypothesis we performed an electrical explosion of the
$Ln + LnF_3$ mixture during which the metal evaporated while
dielectric LnF_3 remained cold. This experiment also yielded
LnF_2.

CONCLUSION

In conclusion let us consider the question: at what stage
of the shock compression does chemical reaction take place? Most
of the chemical transformations can be ascribed to high temperatures generated by shock wave. But in a few cases one can decide
unequivocally that the reaction is performed under high pressure,
i.e., during the time of microseconds. Thus the next task is
to explain the extremely high mobility of atoms needed for the
enormous rate of diffusion needed for such reactions.

If we compare the structures of NaCl, ZnS and SiO_2 with those
of Na, Zn and Si, respectively, we see that in each pair of the
structures the metal-to-metal distances and coordination numbers
are practically the same. So the chemical reaction may be
presented as the intrusion of non-metal atoms into octahedral or
tetrahedral interstices between metal atoms. To make the intrusion
possible, halogen or chalcogen molecules must previously dissociate

(under high pressure) into atoms. This process corresponds to the phase transition from dielectric into metal[40,41].
Taking into consideration that LnF_3 under high pressure converts into a CaF_2-type structure[42], Ln may interact with LnF_3 to form LnF_2 at the very moment of the phase transition as well

We may generalize this fact and state that chemical reaction under shock compression can proceed <u>only</u> if one of the components undergoes phase transition under such conditions. High speed of atomic movements and bond breaking during phase transition apparently explain the enormous rates of atomic diffusion needed for chemical interaction in shock waves.

REFERENCES

1. S. S. Batsanov, Izvestia, Siberian Academy of Sciences, Chemical Section, No. 6, 676 (1967) {Russian original Izvestia Sibirskogo Otdeleniya Akademii Nauk SSSR, Seriya Khimicheskikh Nauk, No 6 pp 22-35, November-December, 1967}
2. E. M. Moroz, S. V. Ketchik and S. S. Batsanov, Russian Journal of Inorganic Chemistry 17, 921 (1972). {Russian original Zhurnal Neorganicheskoi Khimii Vol. 17, No. 6. pp 1775-1776. 1972}
3. S. S. Batsanov, G. S. Doronin, E. M. Moroz, I. A. Ovsyannikova and O. I. Ryabinina, Combustion, Explosion and Shock Waves 5, 193 (1969). {Russian original Fizika Goreniya i Vzryva, Vol. 5, No 2 pp 283-285, 1969}
4. S. S. Batsanov, E. M. Moroz, E. D. Ruchkin and E. V. Lazareva, Academy of Sciences of the USSR, Bulletin, Division of Chemical Science 22, 2263 (1973). {Russian original Izvestiya Akademii Nauk SSSR, Seriya Khimicheskaya, No. 10 pp 2323-2324, October, 1973.}
5. S. S. Batsanov, L. I. Kopaneva and E. V. Lazareva, in Proceedings First All-Union Symposium on Shock Pressures, Vol 2, October 24-26, 1973, Moscow. Edited by S. S. Batsanov, Moscow (1974). Translation in Sandia National Laboratories SAND80-6009, April 1980, p 25.
6. V. N. Drobyshev, in Proceedings, First All-Union Symposium on Shock Pressures, Vol. 2, loc cit
7. K. I. Kozorezov, L. I. Mirkin and N. F. Skugorova, Sov. Phys. Dok 18, 426 (1973). {Russian original Dokl. Akad. Nauk SSSR 210 No. 5, 1067-1070, June 1973.}
8. Y. Kimura, Japan. J. Appl. Phys. 2, 312 (1963)
9. G. A. Adadurov, G. V. Novikov, N. S. Ovanesyan, V. A. Trukhtanov and V. M. Shekhtman, Soviet Physics, Solid State 11, 1601 (1970). {Russian original Fizika Tverdogo Tela, Vol. II, No. 7, pp 1988-1990, July 1969}
10. Y. Horiguchi, J. Amer. Ceramic Soc. 49, 519 (1966)
11. A. N. Dremin and O. N. Breusov, Russian Chemical Reviews, 37, 392 (1968). {Russian original Uspekhi Khimii 37, 898, 1968.}

12. S. S. Batsanov, V. P. Bokarev, L. I. Kopaneva, J. A. Nechaeva and I. N. Temnitskii in Third All-Union Symposium Impulsive Pressures, 1979. Edited by S. S. Batsanov. Translation, Sandia National Laboratories Report SAND80-600A, March 1980 pp 161-163.
13. A. I. Rogacheva, A. V. Anan'in O. N. Breusov, V. N. Drobyshev and S. V. Pershin, in Proceedings, First All Union Symposium on Shock Pressures, Vol. 2, loc cit, pp 53-61.
14. A. A. Deribas and A. M. Staver, Compustion, Explosion and Shock Waves 6, 116 (1970). {Russian original Fizika Gorniya i Veryra, Vol 6. No 1, pp 122-123, 1970.}
15. A. A. Martynova (sic, should be A. I. Martynov), I. N. Temnitskii, A. A. Artemova, L. I. Kopaneva and S. S. Batsanov, Inorganic Materials, 11 626 (1975). {Russian original Izvestiya Akademii Nauk SSSR, Neorganicheskie Materialy, Vol 11. No. 4, pp 730-732, April 1975.}
16. S. S. Batsanov, A. A. Deribas and G. N. Kustova, Russian J. of Inorganic Chemistry, 12 1206 (1967). {Russian original Zhurnal Neorganicheskoi Khimii, Vol 12 p 2283, 1967.}
17. E.D. Ruchkin, L. N. Travkina and S. S. Batsanov in Proceedings, First All-Union Symposium on Shock Pressures, Vol 2, loc cit p 62-65.
18. S. S. Batsanov, V. P. Bokarev, L. I. Kopaneva, I. A. Nechaeva and I. N. Temnitskii in Proceedings, Third All-Union Symposium on Impulsive Pressures, 1979, loc cit pp 161-163.
19. O. N. Breusov, A. V. Kolesnikov and L. G. Shcherbakova in Proceedings, Third All-Union Symposium on Impulsive Pressures, 1979, loc cit pp 163-165.
20. L. G. Shcherbakova, A. V. Kolesnikov and O. N. Breusov Inorganic Materials 15, 1724 (1979). {Russian original Izvestiya Akademii Nauk SSSR, Neorganicheskie Marterialy, Vol 15, No. 12 pp 2195-2201, December 1979.}
21. A. V. Kolesnikov, L. G. Shcherbakova and O. N. Breusov Academy of Sciences of the USSR, Proceedings Physical Chemistry Section, 251, 172 (1980). {Russian original Doklady Akademii Nauk SSSR, Vol 251, No 1, pp 142-144, March 1980).}
22. A. V. Koleshikov, L. G. Shcherbakova, and O. N. Breusov, Doklady, Akademii Nauk SSSR, Vol 256 No. 1, pp 113-117, 1981. In Russian.
23. Y. Horiguchi and Y. Nomura, Bull. Chemical Soc. of Japan 36, 486 (1963).
24. Y. Horiguchi and Y. Nomura, J. Less Common Metals 11, 378 (1966)
25. Y. Horiguchi and Y. Nomura, Chemistry and Industry October 23, 1965, p 1791 (1965)
26. S. S. Batsanov and E. S. Zolotova, Academy of Sciences of USSR, Proceedings Chemistry Section 180, 383 (1968). {Russian original Doklady Akademii Nauk SSSR, Vol 180, No. 1, pp 93-94, May 1968.}

27. S. S. Batsanov, N. A. Shestakova, V. P. Stupnikov, G. S. Litvak and V. M. Nigmatullina, Academy of Sciences of USSR, Proceedings Chemistry Section 185, 174 (1969). {Russian original Doklady Akademii Nauk SSSR, Vol 185 No. 2, pp 330-331, March 1969.}
28. H. Suzuki, H. Yoshida and Y. Kimura, J. Ceram. Assoc. Japan 77, 278 (1969).
29. G. Otto, O. Y. Reece and U. Roy, Appl. Phys. Lett. 18, 418 (1971).
30. G. H. Otto, U. Roy, and O. V. Reece, J. Less Common Metals 32, 355 (1973).
31. L. N. Fedotov, I. I. Kirshenina, Yu. K. Konov and L. I. Belova in Proceedings, Second All-Union Symposium on Impulsive Pressures, October 19-22, 1976, edited by S. S. Batsanov, Moscow (1979). Translation Sandia National Laboratories Report SAND 81-6006, February 1981, pp. 185-189.
32. S. S. Batsanov, et al, to be published, Zhurnal Neorganicheskoi Khimii.
33. S. S. Batsanov, E. V. Lazareva and L. I. Kopaneva, Russian J. of Inorganic Chemistry, 24, 1253 (1979) {Russian original Zhurnal Neorganicheskoi Khimii 24, 2258-2261, 1969.}
34. S. S. Batsanov and E. M. Moroz, Fiz. Khim. Obr. Mater. No. 6, 127 (1972), in Russian.
35. S. S. Batsanov, V. P. Bokarev, E. V. Lazareva, L. I. Kopaneva, K. A. Tleulieva and A. I. Martynov, Russian Journal of Inorganic Chemistry 22, 492 (1977). {Russian original Zhurnal Neorganicheskoi Khimii 22, 888-892, 1977.}
36. L. I. Kopaneva and E. V. Lazareva in Proceedings, Second All-Union Symposium on Shock Pressures, loc cit, p. 276-280.
37. S. S. Batsanov, E. V. Lazareva and L. I. Kopaneva, Russian Journal of Inorganic Chemistry 23, 148 (1978). {Russian original Zhurnal Neorganicheskoi Khimii, 23, 262-263, 1978.}
38. S. S. Batsanov, L. N. Travkina and E. G. Ippolitov, Russian Journal of Inorganic Chemistry 20, 1870 (1975). {Russian original Zhumal Neorganicheskoi Khimii, 20, 3381-3383, 1975.}
39. S. S. Batsanov, V. A. Egorov and Yu. B. Khvostov, Academy of Sciences of USSR, Proceedings Chemistry Section 227, 251 (1976). {Russian original Doklady Akademii Nauk SSSR, Vol 227, pp. 860-862, April 1976.}
40. S. S. Batsanov, in First All-Union Symposium on Impulsive Pressures, loc cit, pp. 1-10.
41. K. Dunn and F. Bundy, J. Chem Phys 72, 2936 (1980).
42. E. Ya. Atabaeva and N. A. Bendeliani, Geokhimiya No. 1., 136 (1980). (In Russian).

Table I. Formation of Solid Solutions

Compounds	Structure	Remarks	Reference
KCl-KBr	NaCl	Increased solid solution density	(1,2)
KCl-CsCl	CsCl	Exists only at high temperature	Present work
RbCl-CsCl	NaCl	Composition independent of starting stoichiometry	(3)
NH_4-CsCl	NaCl	—	(4)
NH_4-CsBr	CsCl	Obtained first by shock	(2)
NH_4I-CsI	NaCl	—	(4)
NH_4X-KX	NaCl	Miscibility increases in the series X = I, Br, Cl	(5)
Nb_xO_2	TiO_2	Vacancies in cation sublattice ✓ ✓ ✓ ✓	(6)
Ta_xO_2	TiO_2		(6)
LnF_2-CaF_2(a)	CaF_2	Obtained first by shock	Present work
ZnS-ZnSe	ZnS	Sphalerite structure, 0 to 50% S, Wurtite structure, 50 to 100% S.	Present work
W-Fe	Fe	W concentration seven times previous limit	(7)

a.) Ln = rare earth element

Table II. Reaction of Oxides

Mixture	Synthesized Substance	Authors	Reference
$ZnO + Fe_2O_3$	$ZnFe_2O_4$	Kimura, Adadurov, et al	(8,9)
$ZnO + SiO_2$	Zn_2SiO_4	Horiguchi	(10)
$Al_2O_3 + SiO_2$	$Al_2(SiO_3)_2$	Dremin and Breusov	(11)
$Al_2O_3 + SiO_2$	Al_2SiO_5	Batsanov, et al	(12)
$MgO + SiO_2$	$MgSiO_3$	Rogacheva, et al	(13)
$CaO + SiO_2$	$CaSiO_3$	Rogacheva, et al	(13)
$BaCO_3 + TiO_2$	$BaTiO_2$	Deribus and Staver	(14)
$PbO + TiO_2$	$PbTiO_3$	Martynov, et al	(15)
$PbO + ZrO_2$	$PbZrO_3$	Martynov, et al	(15)
$PbO + ZrO_2 + TiO_2$	$Pb(Ti, Zr)O_3$	Martynov, et al	(15)
$Si + BaO_2$	Ba_2SiO_4	Batsanov, et al	(12)
$Ln_2O_3 + H_2O(a)$	$LnOOH$ or $Ln(OH)_3$	Batsanov, et al	(16)
$Ln_2O_3 + LnF_3(b)$	$LnOF$	Ruchkin, et al	(17)
$SmO + TiO_2$	$SmTiO_3$	Batsanov, et al	(18)
$Ln_2O_3 + TiO_2(c)$	Ln_2TiO_5, Ln_2TiO_7	Breusov, et al	(19)
$Ln_2O_3 + TiO_2(c)$	Ln_2TiO_5, $Ln_2Ti_2O_7$	Shcherbakova	(20)
$Ln_2O_3 + TiO_2(d)$	Ln_2iO_5	Kolesnikov, et al	(21)
$Ln_2O_3 + Zr(Hf)O_2(d)$	$Ln_2Zr(Hf)_2O_7$	Kolesnikov, et al	(22)

a.) Ln = rare earth elements La, Nd, Sm, Eu, Gd
b.) Ln = rare earth elements, Gd, Dy, Ho, Er. Tm, Yb, Lu and Y
c.) Ln = rare earth elements, Sm through Lu series and Y
d.) Ln = rare earth elements, La through Lu series and Y

Table III. Reactions of Elements

Mixture	Synthesized Substance	Reference
Ti+C	TiC	Horiguchi and Nomura (23)
W+C	WC, W$_2$C	Horiguchi and Nomura (24, 25)
Al+C	Al$_4$C$_3$	Horiguchi and Nomura (24)
Cr+S, Se, Te	CrX	Batsanov and Zolotova (26)
Sn+S, Se, Te	SnX	Batsanov, et al (27)
Ti, Zr, Si+C	MC	Suzuki, et al (28)
Sn + Nb	Nb$_3$Sn	Otto, et al (29)
Al + Nb	Al$_3$Nb	Otto, et al (30)
Ho+S	Ho$_2$S$_3$	Ruchkin, et al (17)
Sm+Se(Te)+SmF$_3$	SmXF	Ruchkin, et al (17)
Pb + Mo + S	PbMo$_5$S$_6$	Fedotov, et al (31)
Zn + S, Se, Te	ZnX	Batsanov, et al (32)
AgCl + I	AgI + Cl	Batsanov, et al (33)

Table IV. Reactions with Decreasing Valence

Mixture	Synthesized Substance	Remarks	Reference
$Cu + CuBr_2$	$CuBr$	Diminished cell parameter	(34)
$Cu + CuO$	Cu_2O	Common modification	(35)
$Ti + TiO_2$	TiO	NaCl structure	(36)
$Ti + SiO_2$	$TiO_2, TiSi_2$	—	(36)
$Ti + GeO_2$	$TiO_2, FeGe_2(a)$	—	(36)
$Ti + SnO_2$	Sn, SnO, TiO_2, Ti_6Sn_5	Common modification	(36)
$Mn + Cr_2O_3$	$MnCr_2O_4$	Diminished cell parameter	(37)
$Cr + Cr_2O_3$	Cr_3O_4	Obtained in pure form	Present work

THE SHOCK WAVE CHEMISTRY OF ORGANIC SUBSTANCES

A. N. Dremin and L. V. Babare
Institute of Chemical Physics (Branch) USSR Academy
of Sciences, Moscow Region
Chernogolovka, 142432 USSR

ABSTRACT

A review and assessment of principal results is presented for the field of the chemistry of shock-compressed organic substances.

INTRODUCTION

In the middle 1950s papers were published dealing with the effect of explosive shock waves on various condensed substances. In the first stage of the work, mainly inorganic substances were investigated. The research has been possible only due to the development of special sample recovery ampules.[1,2]

In 1964 the polymerization of organic monomers under the action of shock waves was discovered in our Institute.[3,5] It was this discovery that was the onset of shock-wave organic chemistry. As a consequence, the following advancements have been achieved:

1. Elaboration of recovery techniques as well as the study of shock-compressed states of liquid and crystalline organic substances.[6-13]

2. Investigation of various chemical reaction mechanisms.[4,5,13-39]

3. Study of physico-chemical states of polymeric materials.[10, 40-44]

4. Study of the feasibility of an organic product shock industry; namely, the production of polymers with new properties, and the modification of physico-chemical and mechanical properties in various polymer materials.[28,40,45]

5. Investigation of the role of shock in creation of primary biopolymers in the earth's prebiological evolution.[10,27,46,47]

SHOCK-INDUCED REACTIONS

As shown in Table I, the first experiments have shown that various reactions can be brought about in organic substances by shock waves. They include: chemical bond rupture (destruction), substitution, addition, and isomerization.[1] Not all of the above mentioned reactions have been studied to the same extent.

TABLE I

Typical Shock-Induced Reactions

1. Destruction reaction:

   benzene $\xrightarrow{11 \text{ to } 15 \text{ kbar}}$ $[-HC\!:\!CH\!-\!HC\!:\!CH\!-\!HC\!:\!CH-]$ ⟶ products

   $CCl_4 \xrightarrow{P \sim 50 \text{ kbar}} \cdot CCl_3 + \cdot Cl + :CCl_2 \longrightarrow C_2Cl_6 + Cl_2C\!:\!CCl_2 + [-Cl_2C-CCl_2-]_n$

2. Substitution reaction:

   benzene + $CCl_4 \xrightarrow{P \sim 50 \text{ kbar}}$ chlorobenzene + $HCCl_3$

3. Addition reaction:

   $H_2C\!:\!CH\;|\;C_6H_5 \xrightarrow{P \sim 30 \text{ kbar}} [-H_2C-CH(C_6H_5)-]_n$

   $H_3C-(CH_2)_3-HC\!:\!CH_2 + CCl_4 \xrightarrow{P \sim 50 \text{ kbar}} H_3C-(CH_2)_3-HC(Cl)-CH_2(Cl)$ (with CCl_2 group)

4. Isomerization reaction:

   cis-stilbene $\xrightarrow{P \sim 30 \text{ to } 60 \text{ kbar}}$ trans-stilbene

As can be seen from Table I, the reactions are unusual in nature. The rupture of molecular bonds and formation of active fragments are the common features of these reactions. Therefore, it is not surprising that the destruction reactions have been studied in much more detail. Two large groups of organic compounds, namely aromatic and aliphatic, have been investigated. It has been shown that the shock destruction process for aromatic compounds differs from that observed in thermal equilibrium.

As shown in Table II, it is known that in thermal equilibrium, destruction of aliphatic compounds occurs at lower temperatures than the aromatic compounds. This phenomenon is in agreement with bond energies of the two organic compound groups. It should be noted that the thermal destruction of aromatic compounds proceeds without the benzene ring breaking. During the process, ring condensation and hydrogen splitting off take place. Complete graphitization occurs at the temperatures 1800-2000°C.[2,3]

Under shock-wave conditions aromatic and aliphatic compounds behave differently. Under the effect of rather weak shock waves (11 to 15 kbar, incident shock) destruction of aromatic rings (benzene, naphtalene) occurs, with the formation of specific products of destruction. (Table II). Under the same conditions, the aliphatic compounds absolutely do not decompose. A considerable shock pressure increase up to 30-50 kbar gives rise to the common thermal destruction products. The same results are observed with preheated samples.[4]

According to Block and Weir's data,[50] benzene is unchanged chemically at static pressures of 40 kbar and temperatures of 600°C. Therefore, the shock wave data mentioned above can not be interpreted only by the effects of pressure and temperature.

From the modern point of view, the regularities of organic compound destruction are determined by the shock loading process. The point is that, as it is well known now,[6,7] shock-loading of the substance takes place in a very narrow shock front discontinuity zone (10^{-12} to 10^{-9} sec.), the times being comparable to or even less than vibrational relaxation times of the majority of organic compounds. Obviously, this is the cause of nonequilibrium destruction.

The vibrational relaxation times of aliphatic compounds as measured by the molecular acoustics technique[8] are of the order of 10^{-13} sec. Such a short relaxation time implies that the aliphatic compounds, even under the shock conditions, decompose in an equilibrium manner. The compounds do not decompose under weak shock action since the equilibrium shock temperature is low. Under the same conditions the temperature of the aromatic compounds is also low; yet, as it has been shown experimentally, destruction of the compounds takes place. As the vibrational relaxation times of the compounds are of the order of 10^{-10} sec., and clearly more than the shock rise time, destruction of the compounds proceeds in a nonequilibrium manner.

It is interesting to note that the destruction of solid organic substances under the shock effect has certain similarities with these same substances subjected to high static pressure with

TABLE II
Comparison of Equilibrium and Shock Reactions

Thermal Equilibrium Destruction

a) **Aliphatic compounds**

Bond energy: $C_{al} - C_{al}$ = 79 kcal/mole; $C_{al} - H_{al}$ = 98 kcal/mole

(C_{al} is aliphatic carbon)

$$\text{(aliphatic chain)} \xrightarrow{T \sim 400 \text{ to } 600°C} CH_4 + C_2H_6 + C_2H_2 + C_3H_8$$
(gaseous products)

b) **Aromatic compounds**

Bond energy: $C_{ar} - C_{ar}$ = 116 kcal/mole; $C_{ar} - H_{ar}$ = 100 kcal/mole

(C_{ar} is aromatic carbon)

$$\text{(benzene)} \xrightarrow{T \sim 900 \text{ to } 1400°C} \text{(biphenyl)} + H_2 \longrightarrow \text{(terphenyl)} + H_2$$

$$\xrightarrow{T \sim 2000°C} \text{complete graphitization}$$

Shock Destruction

a) **Aliphatic compounds**   do not decompose at 11 to 40 kbar

b) **Aromatic compounds**

The total yield of destruction products ~ 0.4 to 1%

(gaseous products)

$$\text{(benzene)} \xrightarrow{P = 11 \text{ to } 15 \text{kbar}} [-HC:CH-]_{3n} + CH_4 + H_2 + \text{(other gases)} + \text{soot}$$

n=5-9)   80-85% 5-6% 8-15%

shear. It has been discovered by Academician N. S. Enikolopyan and his co-workers[10,11] that application of static pressure alone up to 100 kbar does not cause destruction of aromatic compounds. However, the destruction has been observed under the action of similar pressures with shear; the products being of the same kind as in the case of the weak shock effects.[11-15]

SHOCK-INDUCED POLYMERIZATION

The second question studied is polymerization reaction. It has been found that a number of compounds are polymerized under the shock wave effect[12-18]. Table III shows common monomers, forming common polymers (e.g., acrylamide, styrene, trioxane), and common monomers producing nontrivial polymer products (e.g., tetrahydrofuran). Due to strong de-hydrogenization, a dark brown product with a melting temperature of 500°C has been obtained from tetrahydrofuran under shock-wave loading instead of the white polymer catalytically produced which has a melting temperature 40°C.

Table IV demonstrates that polymers, which are hard to polymerize due to steric effects (e.g., tetrachlorethylene, stilbene, tolane, maleic anhydride, oil acid nitrile, and benzaldehyde) are polymerized under shock loading. This table also demonstrates a polymer forming from the specific reaction of benzene and naphtalene aromatic rings.

It appears that shock polymerization can proceed with various types of chemical bonds (Table III): double C=C bond (acrylamide, styrene, acrylonitrile); triple C≡C bond (tolane); triple C≡N bond (acrylonitrile); C=O bond (benzaldehyde); heterocycles (trioxane, tetrahydrofuran)...

The usual polymerization process under static conditions utilizes various catalysts which easily form the active centers and stimulate chain growth. However, under shock wave loading the polymerization of various monomers proceeds without any catalysts. While studying the polymerization process, the principal question was whether hundreds of thousands of molecules join together during shock loading times of 10^{-6} sec or whether the shock wave only initiates the process. In the latter case, the process proceeds after shock compression during times of the order of seconds, minutes and hours which is 8 or 9 orders of magnitude greater than the shock compression times. To solve the problem, the most convenient monomers to be investigated, trioxane and acrylamide, were selected. They produce a large yield and strongly changed polymer product characteristics, such as chain length and chain number, depending on shock wave parameters and loading conditions.

Trioxane and acrylamide differ strongly from each other by the nature of their solid phase polymerization process. The crystal structure of trioxane promotes the polymerization process. In large perfect crystals trioxane polymerizes better than in small ones, but it does not polymerize at all in the liquid state. The chain growth process takes place along the crystallographic axes c and a. A similar effect of the crystal structure on the polymerization process is maintained even under the shock-wave effect.

Table III
Shock-Induced Polymerization:
Readily Polymerized Substances

Monomers	Formula	Polymer structure	Pressure (kbar)	Yield %	$M_v, 10^3$ The average viscous mole. weight
Acrylamide	$H_2C : CH - C\begin{smallmatrix}O\\NH_2\end{smallmatrix}$	$(-CH_2-CH-)_n$ $\quad\quad\quad O:C$ $\quad\quad\quad\quad NH_2$	20 to 80	6 to 57	30 to 500
Styrene	$H_2C : CH$ $\quad\quad C_6H_5$	$[-H_2C-CH-]_n$ $\quad\quad\quad C_6H_5$	88 to 167 (in copper)	3 to 17	88 to 90
Acrylonitrile	$H_2C : CH - C\equiv N$	(ring structure)$_{n/2}$	30 to 70	1.5 to 12	15 to 70
Trioxane	(trioxane ring)	$(-CH_2-O-)_{3n}$	30 to 110	2 to 18	200 to 1300
*Tetrahydrofuran	(THF ring)	$(-CH:CH-CH:CH-O-)_n$ $T_{(melt)} \sim 500°C$	108 to 220	0.3 to 4	2.8 to 17

*The catalytic polymerization:

(THF ring) $\xrightarrow{SbCl_5}$ $(-CH_2-CH_2-CH_2-CH_2-O-)_n$

$M_v > 1.0 \times 10^6 \quad\quad T_{(melt)} = 42 \text{ to } 45°C$

Table IV
Shock-Induced Polymerization:
Difficult to Polymerize Substances

Monomers	Formula	Polymer structure	Pressure (kbar)	Yield %	$M_v, 10^3$ The average viscous mole. weight
Tetrachlor-ethylene	$Cl_2C = CCl_2$	$(-Cl_2C - CCl_2-)_n$	108 to 238 (in copper)	1.1 to 6	−
Stilbene	C_6H_5H $C=C$ HC_6H_5	$(-HC-CH-)_n$ $C_6H_5C_6H_5$	200 to 300 (in copper)	0.5 to 2	−
Tolane	$C_6H_5 - C \equiv C - C_6H_5$	$(-C=C-)_n$ $H_5C_6C_6H_5$	30 to 40	7 to 10	−
Maleic anhydride	$HC = CH$ $\vert\vert$ $O=CC=O$ $\backslash O/$	$[-HC=CH-]$ $\vert\vert$ $O=CC=O$ $\backslash O/]_n$	80 to 100	2.7 to 4	1.62
Oil acid nitride	$C_3H_7 - C \equiv N$	$(-C=N-)_n$ $\vert$ C_3H_7	42 to 225 (in steel)	1.7 to 4.2	−
Benzaldehyde	$HC=O$ $\vert$ C_6H_5	$[-HC-O-]_n$ $\vert$ C_6H_5	50 to 170 (in steel)	2.7 to 10	−
Benzene	(benzene ring)	$[-HC=CH-HC=CH-HC=CH-]$ $n = 5$ to 10	180 to 225 (in steel)	0.1 to 4	0.27 to 0.4
Naphtalene	(naphthalene structure)	$[-HC-CHHC-CH-]$ (with benzene ring)	108 to 225 (in steel)	0.5 to 6	0.28 to 0.6

Therefore, the polymerization investigation has been carried out with monocrystal specimens. The energy emitted while joining like monomers to the growing chain accounts for 1.5 kcal/mol.[19,20]

Unlike trioxane, successful polymerization of acrylamide requires that the crystal structure be destroyed. Therefore, acrylamide polymerizes better in small defective crystals than in large ones and the process proceeds best in the liquid state, when the molecules have the greatest mobility. This mobility contributes to bond orientation along which the process occurs. The energy emitted upon the monomer joining like segments accounts for 15 kcal/mol; an order of magnitude more than in the trioxane case. Considerable energy release means that to a certain extent the transformation can continue by the thermal explosion mechanism.[21,22]

The passage of a single shock-wave with an amplitude of 50 kbar in an acrylamide specimen does not cause polymerization. If the conserved specimen is heated to melting temperature after the shock, a certain quantity of polymer is formed. This means that the shock wave initiates the polymerization active centers by a destruction mechanism, but because of insufficient mobility of monomer molecules, the chain growth process is suppressed.[13]

It is interesting to note that similar heating of the samples and preliminary irradiation by penetrating radiation also gives rise to additional polymerization.[23]

One succeeds in providing the necessary acrylamide molecular mobility when the specimen is subjected to the multiple action of reflected shock-waves. It has been observed that at the same resulting pressure as in the case of a single shock-wave passage, under conditions when the polymer was not detected, that multiple loading by a number of successively reflected shock-waves, causes the polymer to form even though the temperature in the latter case is clearly less. This observation means that the molecular mobility has not been provided by thermal effect and is a mechanical action.[13]

Trioxane exhibits a significant difference in its polymerization under the action of a shock. The energy of joining of the elementary link to the polymer chain for trioxane is rather insignificant. Therefore, the post-effect of the process by a thermal explosion mechanism, which is characteristic of acrylamide, is out of the question here.

We were able to prove that the trioxane polymerization process proceeds at the time of shock-wave passage and is not a post-effect. Thus, As shown in Fig. 1, at the same shock wave pressure but a different pulse duration the same number of polymer chains form, but the chain length becomes much longer with increasing pulse duration. This polymer products characteristics, such as the chain length and chain numbers, directly depend on the amplitude and time action of the shock-wave.

Increasing the chain growth rate with higher pressure at the same pulse duration as shown in Figure 2, also indicates that the process is proceeding during the time of shock-wave action.

One more fact supportive of the conclusion on the trioxane polymerization process proceeding in time of shock action is the dependence of conformation and characteristics of the polymer chains upon

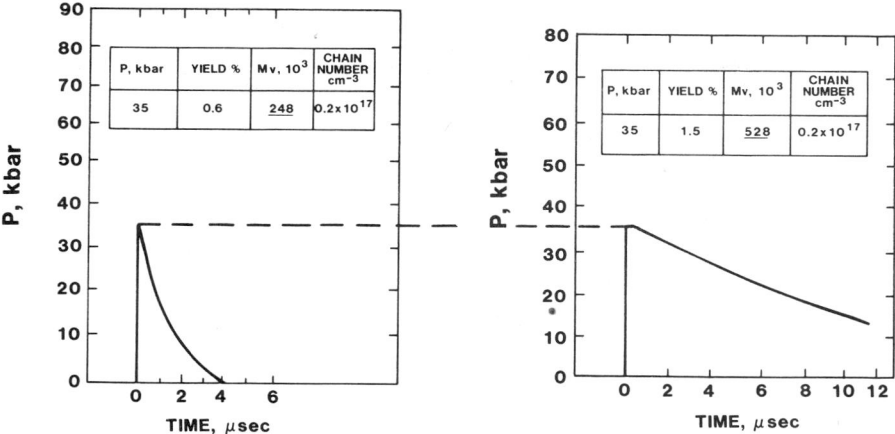

Fig. 1 Shock-induced polymerization of trioxane is found to be strongly influenced by pulse duration. Longer pulse durations increase the yield and molecular weight even though the number of polymer chains remains the same.

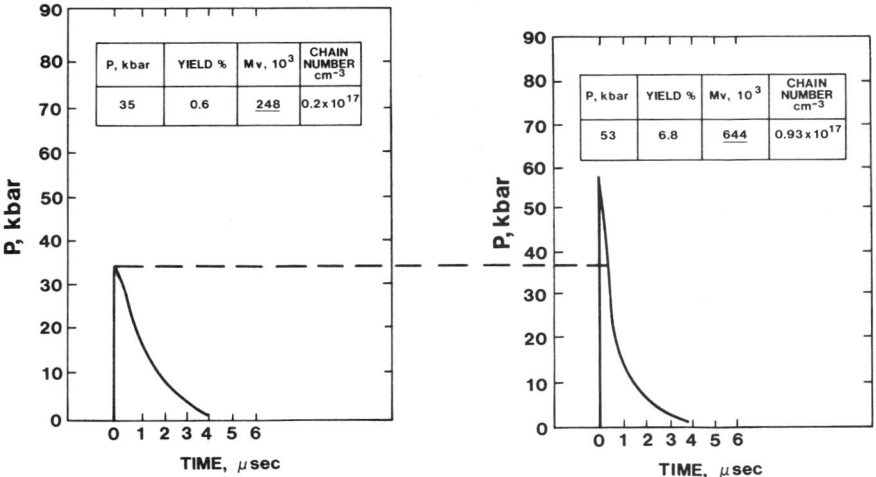

Fig. 2 Increase in shock pressure from 35 to 53 kbar is found to increase the polymerization yield of trioxane and increase the molecular weight.

orientation of the trioxane single crystal axes relative to the direction of shock-wave movement. Thus, as shown in Fig. 3 and Fig. 4, when the c-axis is normal to the wave propagation direction, the chain conformation resembles closely the classical spiral. This indicates that the stress is practically absent at the chain growth. On the other hand, when the c-axis is parallel to the wave propagation direction, the polymer chains appear to be strongly deformed.

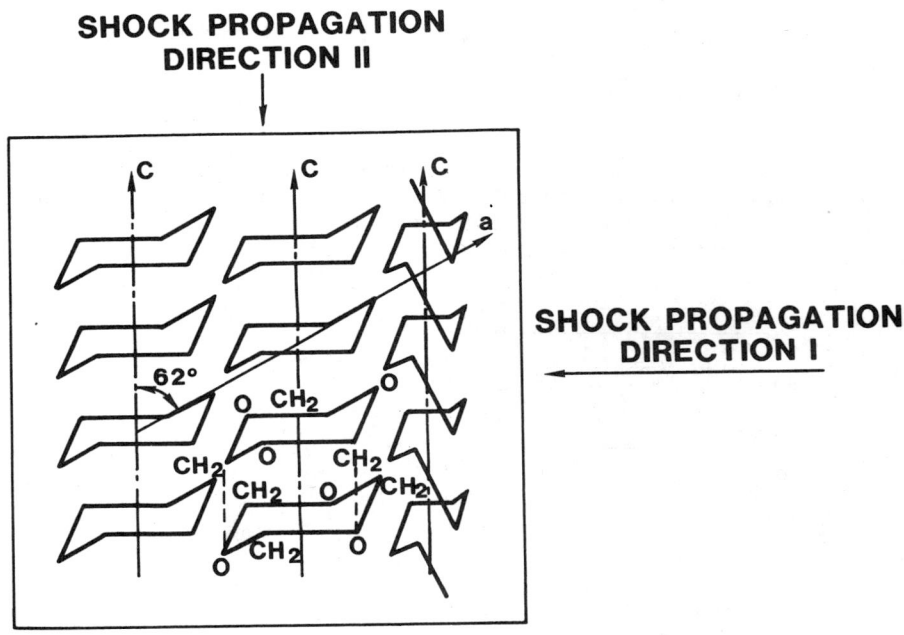

Fig. 3 Shock loading of trioxane crystals along the c-axis and normal to the c-axis is carried out to investigate residual efforts in the shock-polymerized samples.

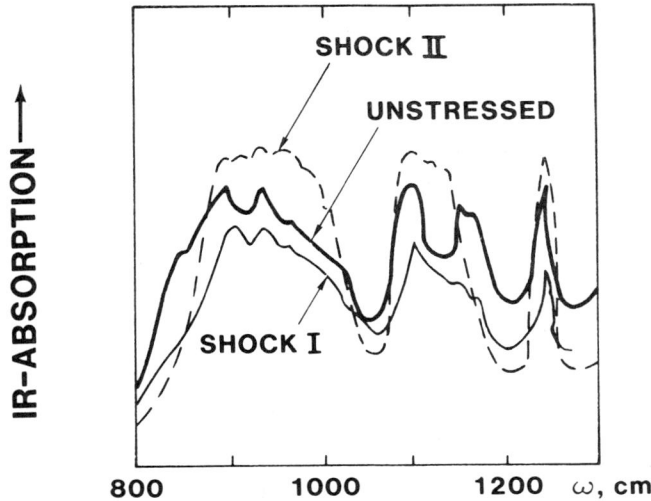

Fig. 4 IR absorption measurements on unstressed and shock-polymerized polyoxymethylene show significant differences.

It is necessary also to note, that deformation of the trioxane molecule which is observed under static compression, and detected by the IRS technique directly during compression, disappears when the pressure is relieved.

Consequently, forming of polymer chains of various conformation means that at the time of chain formation the monomer was deformed differently due to polymer chain growth in the crystal matrix at the time of shock-wave action. If the process proceeded in the post-effect, these conformation differences would not be possible. It appears that during the time of shock-wave compression thousands and tens of thousands of molecules have a chance to join together. This observation implies that the conditions of shock-wave loading strongly accelerate the reactions.

REFERENCES

1. Yu.N.Riabinin, Soviet Physics, Doklady, 1, 424 (1956). {Russian original, Dokl.Acad.Nauk SSSR, 109, 289 (1956)}
2. Yu.N.Riabinin, Soviet Physics, Technical Physics 1, 2575 (1956). {Russian original, Zh.Techn.Fiz., 26, 2661 (1956).}
3. G.A. Adadurov, I.M. Barkalov, V.I. Gol'danskii, A.N. Dremin, T.N.Ignatovich, A.N. Mikhailov, V.L. Tal'roze and P.A.

Yampol'skii, Discovery no. 125 of 13.03.1973, priority of 23.06.1964.
4. G.A. Adadurov, I.M. Barkalov, V.L. Tal'roze, V.I. Gol'danskii, A.N.Dremin, T.N. Ignatovich, A.N. Mikhailov and P.A. Yampol'skii, Polymer Science, USSR, $\underline{7}$, 196 (1965). {Russian original, Vysokomol. Soedineniya, $\underline{7}$, 180 (1965)}
5. G.A. Adadurov, I.M. Barkalov, V.I. Gol'danskii, A.N. Dremin, A.N.Mikhailov, V.L. Tal'roze and P.A. Yampol'skii, Academy of Sciences of USSR, Proceedings, Physical Chemistry Section, $\underline{165}$, 835 (1965). {Russian original, Dokl. Akad. Nauk SSSR, $\underline{165}$, 851 (1965)}
6. G.A. Adadurov, A.N. Dremin, G.I. Kanel and S.V. Pershin, Combustion Explosion and Shock Waves, $\underline{3}$, 175 (1967). {Russian original, Fizika Goreniya i Vzryva, $\underline{3}$, 281 (1967)}
7. S.V. Pershin, G.I. Kanel, "The method of recovery of shock compressed samples and analysis of the stress pattern," Moscow, Dep. VINITI no. 1446-70, (1970).
8. S.S. Batsanov, Journal of Engineering Physics, $\underline{12}$, 59 (1967). {Russian original, Inzhenerno-Fizich.Zh., $\underline{12}$, 104 (1967)}
9. G.A. Adadurov, V.V. Gustov and P.A. Yampol'skii, Combustion, Explosion and Shock Waves $\underline{7}$, 243 (1971). {Russian original, Fizika Goreniya i Vzryva, $\underline{7}$, 284 (1971)}
10. G.A. Adadurov, V.I. Gol'danskii and P.A. Yampol'skii, Mendeleev Chemistry Journal $\underline{18}$, 912 (1973). {Russian original, Zh. Vses. Khim. Obshch. D.I. Mendeleev, $\underline{18}$, 80 (1973)}
11. G.A. Adadurov, V.V. Gustov, V.S. Zhuchenko, M.Yu. Kosygin and P.A.Yampol'skii, Combustion, Explosion and Shock Waves, $\underline{9}$, 499 (1973). {Russian original, Fizika Goreniya i Vzryva, $\underline{9}$, 576 (1973)}
12. G.A. Adadurov, V.A. Zhuchenko, M.Yu. Messinev, and P.A. Yampol'skii, Authors license no. 442945, Bull.Isobreteni no. 7 (1976).
13. L.V. Babare, A.N. Dremin, S.V. Pershin, V.V. Yakovlev, in Report on International Conference on the Use of Explosion Energy in Chemistry and Chemical Industry," ChSSR, Pardubize, 1970, p. 248.
14. V. A. Veretennikov, A.N. Dremin and A.N. Mikhailov, Combustion, Explosion and Shock Waves, $\underline{2}$, 58 (1966). {Russian original, Fizika Goreniya i Vzryva, $\underline{2}$, 95 (1966)}
15. I.M. Barkalov, V.I. Gol'danskii, V.L. Tal'roze and P.A. Yampol'skii, Soviet Physics JETP Letters, $\underline{3}$, 200 (1966). {Russian original, JETPh (SSSR), $\underline{3}$, 309 (1966)}
16. G.A. Adadurov, A.N. Dremin and A.N. Mikhailov, Combustion, Explosion and Shock Waves, $\underline{3}$, 254 (1967). {Russian original, Fizika Goreniya i Vzryva $\underline{3}$, 412 (1967)}
17. I.M. Barkalov, G.A. Adadurov, A.N. Dremin, V.I. Gol'danskii, T.N.Ignatovich, A.N. Mikhailov, V.L. Tal'roze and P.A. Yampol'skii, J.Polym.Sci. Part C, no. 16, 2597 (1967).
18. L.V. Al'tshuler, I.M. Barkalov, I.N. Dulin, V.N. Zubarev, T.N. Ignatovich and P.A. Yampol'skii, High Energy Chemistry $\underline{3}$, 73 (1968). {Russian original, Khimiya Vys. Ener. $\underline{3}$, 88 (1967)}

19. L.V. Babare, A.N. Dremin, A.N. Mikhailov and V.V. Yakovlev, Vysokomolekuly Soedineniya Ser. B, 9, 642 (1967). Translation in Sandia Natl. Laboratories Report RS3140/79/176. September, 1979
20. L.V. Babare, A.N. Dremin, S.V. Pershin and V.V. Yakovlev, Proceedings, Academy of Sciences of USSR, Physical Chemistry, 184, 105 (1969). {Russian original, Dokl.Acad.Nauk. SSSR, 184, 120 (1969)}
21. L.V. Babare, A.N. Dremin, S.V. Pershin and V.V. Yakovlev, Combustion, Explosion and Shock Waves, 5, 364 (1969). {Russian original, Fizika Goreniya i Vzryva, 5, 528 (1969)}
22. I.N. Dulin, V.N. Zubarev, A.G. Kazakevich and P.A. Yampol'skii, High Energy Chemistry, 3, 340 (1969). {Russian original, Khimiya Vys. Ener. 3, 372 (1969)}
23. L.V. Babare, A.N. Dremin, S.V. Pershin and V.V. Yakovlev, Combustion, Explosion and Shock Waves, 5, 364 (1969). {Russian original, Fiz. Goreniya i Vzryva, 5, 534 (1969)}
24. L.V. Babare, A.N. Dremin, S.V. Pershin and V.V. Yakovlev, Abstracts of Reports "The Second All- Union Symposium on Combustion and Flame," Erevan 1969, p. 3050.
25. P.A. Yampol'skii and T.N. Ignatovich, High Energy Chemistry, 4, 62 (1970). {Russian original, Khimiya Vys. Ener. 4, 74 (1970)}
26. T.N. Ignatovich, I.M. Barkalov, I.N. Dulin, V.N. Zubarev and P.A.Yampol'skii, High Energy Chemistry, 4, 394 (1970). {Russian original, Khimiya Vys., Ener. 4, 443 (1970)}
27. T.N. Ignatovich, P.A. Yampol'skii and L.M. Bragintseva, Vysokomol.Soedineniya, 12B, 506 (1970). Translation in Sandia Natl. Laboratories Report SAND80-6010. June 1980
28. A.N. Dremin and O.N. Breusov, Priroda, no. 1, 10 (1971). Translation in Sandia Natl. Laboratories Report SAND80-6003, March 1980.
29. L.V. Babare, Abstracts of Reports "The Third All-Union Symposium on Combustion and Flame. Leningrad, 1971," p. 228.
30. I.N. Dulin, V.N. Zubarev, Yu.N. Novikov and M.E. Volpin, Russian Journal of Physical Chemistry, 45, 1642 (1971). {Russian original, Zh. Fiz. Khimiya, 45, 2904 (1971)}
31. G.A. Adadurov, V.V. Gustov, M.Yu. Kosygin and P.A. Yampol'skii, in "Gorenie i Vzriv," Moscow, Nauka, 1972, p. 529.
32. G.A. Adadurov, V.V. Gustov, A.N. Kaplan, M.Yu Kosygin, and P.A. Yampol'skii, Combustion, Explosion and Shock Waves, 8, 465 (1972).{Russian original, Fiz. Goreniya i Vzryva, 8, 566 (1972)}
33. G.A. Adadurov, V.I. Gol'danskii, V.V. Gustov, M.Yu. Kosygin, and P.A. Yampol'skii, High Energy Chemistry, 7, 492 (1973). {Russian original, Khimiya Vys. Ener., 7, 554 (1973)}
34. L.V. Babare, T.K. Goncharov, A.N. Dremin and V.P. Roshchupkin, Polymer Science, USSR 16, 118 (1979). {Russian original, Vysokomol.Soedineniya, A16, 969 (1974)}
35. L.V. Babare, A.N. Dremin and V.Yu. Klimenko, Abstracts of Reports "The Fifth All-Union Symposium on Mechanoemission and

Mechanochemistry of Solids" Tallin, 1975, p. 74. Translation, Sandia National Laboratories.
36. V.Yu. Klimenko, A.N. Dremin and L.V. Babare, Abstracts of Reports "The Fifth All-Union Symposium on Mechanoemission and Mechanochemistry of Solids" Tallin, 1975, p. 75. Translation, Sandia National Laboratories.
37. V.V. Yakushev, S.S. Nabatov and O.B. Yakusheva, Academy of Sciences of USSR, Proceedings, Physical Chemistry, $\underline{214}$, 136 (1973). {Russian original, Dokl.Acad. Nauk SSSR, $\underline{214}$, 879, (1974)}
38. M.Yu. Messinev, G.A. Adadurov, V.V. Gustov and A.G. Kazakevich, High Energy Chemistry, $\underline{11}$, 406 (1977). {Russian original, Khimiya Vys. Ener. $\underline{11}$, $\overline{376}$ (1977)}
39. A.G. Kazakevich, M.Yu. Messinev and G.A. Adadurov, Khimiya Vys. Ener. $\underline{14}$, 92 (1980), in Russian.
40. I.M. Barkalov, V.I. Gol'danskii, V.V. Gustov, A.N. Dremin, A.M. Mikhailov, V.L. Tal'roze and T. (sic. P) A. Yampol'skii, Academy of Sciences of USSR, Proceedings, Physical Chemistry, $\underline{167}$, 217 (1966). {Russian original, Dokl.Akad.Nauk SSSR; $\underline{167}$ $\overline{1077}$ (1966)}
41. P.A. Yampol'skii, I.M. Barkalov, V.I. Gol'danskii, $\overline{V}$.V. Gustov, I.N. Dulin, V.N. Zubarev and A.G. Kazakevich, Polymer Science, USSR, $\underline{A10}$, 929 (1968). {Russian original, Vysokomol. Soedineniya, $\underline{A10}$, 799 (1968)}
42. V.A. Kargin, $\overline{I.Yu}$. Tzarevskaja, V.N. Zubarev, V.I. Gol'danskii and P.A. Yampol'skii, Polymer Sciences USSR $\underline{10}$, 3019 (1968). {Russian original, Vysokomol. Soedineniya $\underline{A10}$, 2600 (1968)}
43. V.A. Kargin, G.A. Andrianova, I.Yu. Tsarevskaja, V.I. Gol'danskii, and P.A. Yampol'skii, J.Polym.Sci., $\underline{A29}$, 1061, (1971).
44. G.A. Adadurov, V.V. Gustov, L.Yu. Zlatkevitch, V.G. Nikolskii, I.Yu. Tzarevskaja and P.A. Yampol'skii, Moscow, Dep. VINITI no. 168-74.
45. L.V. Babare, F.I. Dubovitskii and A.N. Dremin, Author's License no. 434761 (SSSR) and the statement no. 16931155 (1971). USA Patent No. 3839290 (1973).
46. L.A. Baratova, V.I. Gol'danskii, M.Yu. Kosygin and P.A. Yampol'skii, Doklady, Biochemistry $\underline{35}$, 105 (1970). {Russian original, Biokhimiya $\underline{35}$, 1216 (1970)}
47. V.I. Gol'danskii, T.N. Ignatovitch and P.A. Yampol'skii, Dokl.Acad. Nauk SSSR, $\underline{207}$, 218 (1972) in Russian.
48. K.P. Kinni, in "Petroleum Hydrogen Chemistry" $\underline{II}$, 93 (1958).
49. R.Z. Magaril, "The Mechanism and Kinetics of Hydrocarbon Homogeneous Thermal Transformations" Moscow, Khimiya (1970).
50. S. Block, C.E. Weir, and P.J. Piermarini. Science, $\underline{169}$, 586 (1970).
51. S.B. Kormer, Soviet Physics Uspekhi, $\underline{11}$, 224 (1968). {Russian original, Uspekhi Fiz. Nauk, $\underline{94}$, 641 $\overline{(1968)}$}
52. V.Yu. Klimenko and A.N. Dremin, Soviet Physics, Doklady, $\underline{24}$, 984 (1979). {Russian original, Dokl.Akad.Nauk SSSR, $\underline{249}$, 840, (1979)}
53. R. Holmes, G.R. Jones and R. Lawrence, Trans. Faraday Soc., $\underline{62}$, 46 (1966).

54. P.K. Khabibulaev, M.G. Khaliulin, M.I. Shachparonov et al., in "Physics and Physico-Chemistry of Liquids" Moscow, Moscow University, part I, 1972, pp. 37-93.
55. N.P. Chistotina, A.A. Zharov, Yu.V. Kissin and N.S. Enikolopyan, Academy of Sciences of USSR, Proceedings Physical Chemistry $\underline{191}$, 265 (1969). {Russian original, Dokl.Acad. Nauk SSSR, $\underline{191}$, $\underline{632}$, (1970)}
56. V.G. Dzamukashvili, A.A. Zharov, N.P. Chistotina, Yu.V. Kissin and N.S. Enikolopyan, Academy of Sciences of USSR, Proceedings Physical Chemistry, $\underline{215}$, 233 (1974). {Russian original, Dokl. Acad. Nauk SSSR, $\underline{215}$, 127 (1974)}
57. N.S. Enikolopyan and S.A. Volfson, "Chemistry and Technology of Formaldehyde" Moscow, Khimiya, (1968).
58. K.J. Hayashi, S. Okamura. Khimiya i Tekhnol.Polimerov, no. 4, 89 (1964).
59. M.A. Bruk. in "Advances of Polymer Chemistry" Moscow, Kimiya, (1968) p. 78, 91.
60. V.A. Kargin, V.A. Kabanov, Zh.Vses.Khim.Obshch. D.I. Mendeleev, $\underline{9}$, 620, (1964).
61. B.Baysal, D.Adler, D.Ballantine and R.Colombo, J.Polym.Sci. $\underline{44}$, 117 (1960).
62. M.Bradbury, S.Hamman, M.Hinton, Austral.J.Chem. $\underline{23}$, 511 (1970).

SHOCK-INDUCED ORGANIC CHEMISTRY*

B. W. Dodson and R. A. Graham
Sandia National Laboratories,† Albuquerque, New Mexico 87185

ABSTRACT

A wide variety of chemical rections are known to occur in organic compounds subjected to shock loading. An overview is presented of the evidence that a) shock loading can cause non-thermal reactions, i.e., reactions which do not take place under equivalent combinations of static temperature and pressure; b) unique and/or unusual reactions can occur under shock; and c) reactions can occur in the shock front itself. Rules obtained for reactivity of organic compounds under shock loading are used to address the problem of initiation sensitivity of aromatic explosives.

INTRODUCTION

The traditional view of shock waves in condensed media is based largely on continuum concepts or microscopic interactions in perfect lattices. Such a view has been termed a "benign shock concept."[1] A large body of experimental observations which appear "anomalous" within this framework has accumulated.[2,3] A review of many such observations is given by Davison and Graham.[4] Graham[1] has proposed that the "anomalous" effects may be natural results of a "catastrophic shock concept," in which the shock wave creates large densities of defects, breaks chemical bonds mechanically, and produces relative mass motions of different species within the shock front. Unique shock-induced chemical effects would be an obvious outgrowth of such processes. To the extent that catastrophic shock concepts are needed to explain observations, more attention must be directed to microscopic processes occurring within shock fronts than has been the case. A good case can be made for the existence of conditions in the shock front which are both non-equilibrium and severe enough to break chemical bonds. In fact, extrapolations from existing mechanochemical studies on polymers,[5] which are much less severe both in rate and intensity, indicate that such conditions are to be expected.

It is the object of this paper to provide an expository review of prior work on shock-induced organic chemistry and to provide a framework for a unified view of chemical reactivity under shock loading. Work is in progress to provide a comprehensive critical review of the literature. Previous limited reviews have been published by Adadurov, et al.,[6] Dremin and Breusov,[7] and Babare, et al.[8] A listing of many pertinent references appears in Graham and Dodson.[9]

---

*This work sponsored by the U.S. DOE under Cont. DE-AC04-76-DP00789.
†A U.S. Department of Energy facility.

ORGANIC REACTIONS OCCURRING UNDER SHOCK LOADING

The various reactions produced by shock loading of condensed organic compounds include synthetic reactions, isomerizations, and polymerizations of many organics. Such reactions exhibit typical shock thresholds in the range of 1-10 GPa. These reactions are deduced from examinations of samples recovered for post-shock study. We describe examples of the shock induced reactions below.

a) Carbene formation:
A carbene is a carbon diradical group such as $Cl_2C:$, which is an extremely reactive entity and has only a transitory existence. A synthetic reaction which occurs under shock and clearly involves a carbene intermediate is described.[10] A solution of tetrachlorethylene and 1-hexene was shock loaded and the formation of a compound with the three-membered ring typical of carbene reactivity was observed.

$$CH_3-(CH_2)_3-CH=CH_2 + Cl_2C=CCl_2 \xrightarrow{SHOCK} CH_3-(CH_2)_3-\underset{\underset{CH_2}{\diagup}}{\overset{\overset{CCl_2}{\diagdown}}{CH}}$$

Obtaining such a product from the experiment implies that the initial reaction step is $CCl_2=CCl_2 \rightarrow 2Cl_2C:$, with subsequent reactions occurring in a more-or-less normal manner.

b) Isomerization:
Isomers are compounds that have identical chemical formulas but which differ in the spatial configuration of their atoms. An isomerization reaction transforms one isomer into another. Owing to the similarity between reactant and product, this is one of the simplest classes of organic reactions.

An isomerization reaction in stilbene has been reported.[11] A sample of trans-stilbene underwent multiple compression between explosively loaded steel plates to a pressure of 12 GPa.

Small amounts (5-10%) of cis-stilbene, as well as ~50% of a polystilbene compound were found in the recovered product. The transformation of trans- to cis-stilbene cannot be due to the temperature rise in the shocked sample (~100°K) as increased temperature favors the reverse process.

There are at least two possible mechanisms for this isomerization. In one, the C=C bond is changed into a single bond, forming a stilbene diradical. Rotation may then take place around the C-C bond, and reformation of the double bond will produce both trans- and cis-stilbene. The other mechanism involves complete scission of the C=C bond, producing two carbene-like diradicals, whose subsequent condensation can form both stilbene isomers.

c) Polymerization:

Polymerization is a chemical reaction under which two or more small molecules combine to form a larger molecule which contains repeating structural units of the monomers. Polymerization of many organic compounds is observed under shock loading.[12] Polymerization is found in compounds which conventionally polymerize in free-radical reactions (acrylamide), ionic reactions (trioxane), or not at all in solid state (maleic anhydride). Because of the wide variety of compounds that polymerize under shock, study of such reactions provides a powerful probe of shock-induced chemical reactions.

Free-radical polymerization proceeds through three separate processes: I) Initation, or the production of free radicals; II) Propogation, or radical chain growth; and III) Chain termination, the step which produces an inert polymer chain. These processes can be stated in terms of model reactions as below, where M represents a monomer unit, $R_n^{\cdot}$ is a free radical containing n monomer units, and P_n is an inactive polymer chain containing n monomer units.

I. Initiation: $M \xrightarrow{\text{SHOCK}} R_1^{\cdot}$

II. Propagation: $R_n^{\cdot} + M \longrightarrow R_{n+1}^{\cdot}$

III. Termination: $R_m^{\cdot} + R_n^{\cdot} \longrightarrow$ Inactive polymer

Clearly, the relative rates and time histories of these three processes will be reflected in the nature of the end polymer. Thus, detailed study of the polymerized product will provide useful information on the microscopic kinetics of reactions occurring under shock.

REACTIVITY OF ACRYLAMIDE UNDER SHOCK LOADING

a) Polymerization:

Acrylamide has been polymerized under multiple compression shocking.[13,14] These experiments involve the explosive loading of a polycrystalline acrylamide sample with dimensions of about 15 mm ⌀ and 1 mm thickness in a steel[13] or copper[14] capsule. The sample thus reaches the input pressure in the capsule through multiple reverberations of the shock wave.

When shocked as described above, polymerization begins at a threshold pressure of about 2 GPa, following the reaction mechanism

$$n \left[H_2C=CH-C\overset{\nearrow O}{\underset{\searrow NH_2}{}} \right] \xrightarrow{SHOCK} \left[-H_2C-CH- \atop \underset{NH_2 \diagdown O}{\overset{C}{|}} \right]_n$$

This polymer, a linear polyacrylamide, is water soluble.

At about 6 GPa, cross-linking of the polymer begins, probably through an imidization reaction, resulting in an insoluble product and evolution of ammonia (NH_3). A density transition corresponding to a 30% reductiuon in volume is seen between 9 and 11 GPa[15,16] (see Figure 1.) Above 18 GPa, the shocked acrylamide begins to degrade and carbonize.

Acrylamide polymerizes readily in conventional processes. It is therefore important to establish whether the polymerization is actually due to the catastrophic shock conditions, or to some more benign effect of the shock loading, such as residual temperature. Such questions have been considered in some detail. The presence of a radical inhibitor, hydroquinone, is observed to have no effect on the shock-induced polymerization reaction.[17] If polymerization were caused by melting of the acrylamide, the inhibitor would prevent any reaction. When cooled to 77°K and shocked, acrylamide still shows significant polymerization at 3 GPa, although the residual temperature is still well below room temperature.[18] By contrast, a slowly-pulsed pressure of 5-7 GPa applied for milliseconds at room temperature produces no polymer.[19] Enikolopyan[20] has studied polymerization under static pressure in over 100 monomers, including acrylamide, and found that a pressure of 10 GPa is not sufficient to induce polymerization. Hence the polymerization is not due simply to the benign application of temperature and pressure.

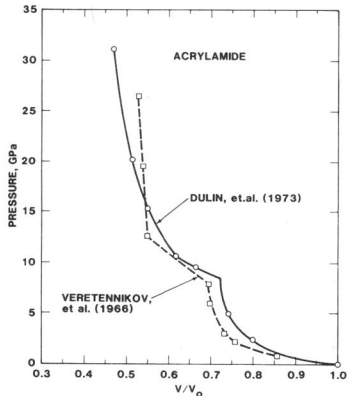

Figure 1. Pressure-volume relation for shocked acrylamide (Veretennikov, et al,[15] and Dulin, et al.[16]) Acrylamide has a 30% volume reduction between 9 and 11 GPa, which is correlated with a change in chemical reactivity.

b) Reactivity in the Shock Front:

The relevant data on acrylamide reactivity is summarized in Figures 2 and 3. Figure 2 shows polymer yield vs. molecular weight for single and multiple shock compression. Figures 3a and 3b show linear polyacrylamide molecular weight and polyacrylamide chain density, respectively, vs. shock pressure for multiple compression shock loading of acrylamide. The polymer chain density, M, is defined as the number of polymer chains per initial monomer unit. It is important to note that M is about 3×10^{-6} for all single compression experiments and for a 4 GPa multiple compression experiment, even though molecular weights and polymer yields are strongly dependent on the loading details.

If we assume a radical mechanism with simple termination processes (either combination, $R_m^• + R_n^• \rightarrow P_{m+n}$, or disproportionation, $R_m^• + R_n^• \rightarrow P_m + P_n$), the polymeric chain density corresponds to a radical density of about $3 \times 10^{16}/cm^3$. Similar radical densities have been found in shock loaded polymers.[21] The slow increase in M with pressure below 9 GPa for the multiple compression work (Fig. 3b) is probably due to reflected waves becoming strong enough to produce more free radicals.

The density transition near 9 GPa is strongly correlated with a large decrease in mean molecular weight of the polyacrylamide and a large increase in the polymer chain density. Both these changes are nearly two orders of magnitude. This bespeaks of a major change in the polymerization mechanism associated with the density transition.

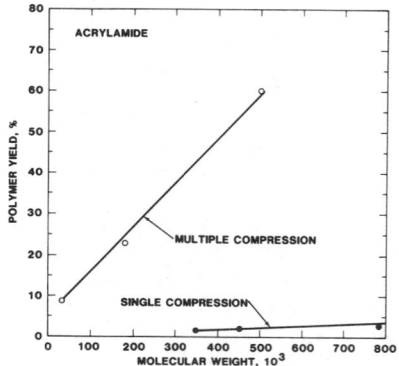

Figure 2. Total polyacrylamide yield vs. linear polyacrylamide molecular weight for single and multiple shock loadings. Based on data from Ref. 13

There are at least two possible mechanisms for the observed effects. One mechanism is that an additional species of radical is formed at a threshold pressure of about 9 GPa, thus increasing M and making simple termination reactions more likely, with a resulting decrease in molecular weight. However, the chemical makeup of the products does not change at this pressure, which is inconsistent with formation of a new radical species.

The other mechanism involves the activation of a new form of termination reaction: a transfer reaction. In this case, inactive polymer is formed by transferring the radical to a monomer unit, i.e., $R_m^• + M \rightarrow P_m + R_1^•$. If such a termination reaction dominates, the much higher density of monomer compared to radicals will produce a higher density of shorter polymer chains; just the observed behavior. We therefore suggest that a chemical process is taking place in the shock front which activates the transfer termination mechanism. The observations above are not so much intended to identify a specific mechanism as to show that by combining chemical analysis with conventional shock compression data one can consider such questions in quantitative detail.

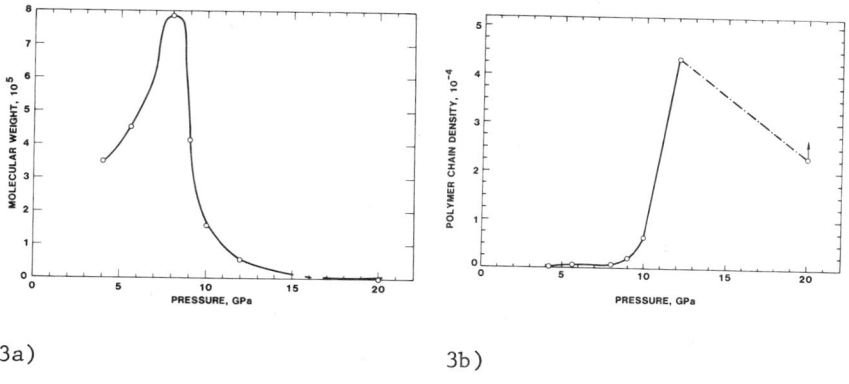

Figure 3. Figures 3a and 3b show linear polyacrylamide molecular weight and polyacrylamide chain density, respectively, vs. shock pressure for multiple compression shock loading of acrylamide, based on data from Ref. 13. (The 20 GPa point on Fig. 3b is a lower limit only.) The behavior of both quantities in the 9-11 GPa region imply a major change in the polymerization mechanism.

REACTIVITY OF BENZENES UNDER SHOCK

a) Aromatic ring-opening reactions; violation of Haber's rule:

Haber's rule of pyrolysis states that under thermal excitation, an aliphatic C-C bond is less stable than an aromatic C-H bond, and further that the aromatic carbon ring will survive successive dehydrogenations to the point of graphitization. Thus, aromatic hydrocarbons are thermally more stable than aliphatic hydrocarbons. In thermal pyrolysis of benzene, hydrogen is split off from the aromatic ring, beginning at around 500°C, and polyconjugate ring compounds are formed. At the temperatures of 2500 - 3000°C, the material is completely converted to graphite.

The behavior of benzene under shock loading is very different, providing strong evidence for non-thermal shock processes. Benzene shocked in a steel recovery capsule to pressures above 10 GPa in the steel reacts to form ~1% of a polyacetylene product.[22]

$$m\left[\bigcirc\right] \xrightarrow{SHOCK} \left[-CH=CH-CH=CH-CH=CH-\right]_m$$

This product results from opening of the aromatic ring and subsequent condensation of the resulting radicals. The molecular weight of the product is low (~400), and thus only 4 or 5 benzene rings combine to form the polymer. The density of polymer chains, M, is about 2×10^{-3}, which is very high compared to observed shock compression radical densities of a few $\times 10^{-6}$. Either benzene forms orders of magnitude more radicals under shock than acrylamide or a transfer termination reaction occurs.

It is interesting to compare the shock-induced reactivity of cyclohexane to that of benzene. Under identical shock conditions no evidence for reactivity was seen in cyclohexane under about 22.5 GPa in the steel, and under those conditions the yield was two orders of magnitude less than for benzene. (The products were traces of hexene and amorphous carbon.) This difference in reactivity may be due to either different ring-opening behaviors under shock or to a difference in lifetime of the free radicals, with the greater stiffness of a conjugated chain inhibiting the reformation of the benzene.

The observed reactivity of benzene under shock loading clearly violates Haber's rule and indicates that non-thermal processes have occurred.

b) Structural stability of substituted benzenes:

Shock-induced chemical reactivity of at least 18 benzene derivatives has been studied.[23] Barbare et al.[24] have presented rules of reactivity for substituted benzenes based on this data. Substituents significantly affect the shock stability of the benzene ring. Electron-donor groups increase the ring strength while electron-acceptors weaken the ring. Thus, the integrated density of the delocalized aromatic binding electrons is qualitatively proportional to the stability of the benzene ring against opening under shock compression.

A convenient measure of strength of the electronic effect of a substituent group on benzene is provided by the sign and magnitude of the proton NMR shift Δ in parts per million due to the substituent. An electron-donor group will produce a negative value of Δ. A relative indication of the electronic strength of a substituent may be obtained by assigning a value of 1/2 the NMR shift Δ for para-disubstituted benzene. All remaining aromatic protons are equivalent, and an unambiguous value may be assigned to the group.

c) Initiation sensitivity of aromatic explosives:

It is difficult to unambiguously order explosives in terms of their sensitivity to shock initiation. However, if one considers explosives with sufficiently different properties, such an ordering can be made. For the present discussion, consider the following series.[25]

TNB   TNT   DATB   TATB

LESS SENSITIVE →

Is there some connection between the rules for shock reactivity obtained by studying non-energetic materials and the problem of shock initiation of explosives? The first point to be made is that the nitro group, with an electronic strength of +0.57 ppm, is a strongly destabilizing substituent. In fact, dinitrobenzene, which has an oxygen balance similar to that of TNT, is an extremely insensitive explosive. The effect of destabilization of the benzene ring by the nitro substituents seems connected with the explosive properties. In addition, when electron-donor groups such as $-CH_3$ (-0.16 ppm) or $-NH_2$ (-0.48 ppm) are added to trinitrobenzene, the resulting explosive becomes much more difficult to initiate. This is consistent with the strengthening of the aromatic ring associated with electron-donors. These observations provide evidence of a connection between the sensitivity of an aromatic explosive to unique shock-induced chemical processes and the initiation sensitivity of that explosive. It is important to realize that these rules of shock-induced reactivity were discovered by post-shock chemical analysis, a procedure much easier to carry out on non-explosive compounds.

CONCLUSION

The literature examined in the present paper shows that a wide variety of chemical reactions occur in shock loaded organic compounds. The specific reactivity of a compound under shock compression depends upon the molecular structure. In addition, the shock pressure, pulse duration, sample geometry, and whether single or multiple compression is applied can strongly affect the reaction. Evidence has been presented that shock-induced organic reactions cannot be understood in terms of the benign shock concept, but rather must be caused by a microscopic environment unique to shock loading.

A good case can be made for the existence of processes within the shock front which are nonequilibrium and severe enough to initiate and control chemical reactions. Intramolecular vibrational relaxation times for organic molecules are picoseconds and longer.[26] If the stress gradient in the shock front is sufficiently large, vibrational excitation of the molecule may take place before internal equilibrium is attained. Such a process has been described, based on molecular dynamics calculations.[27] The initial distribution of energy within the molecule will depend on the coupling of the structure of the molecule to deformation within the shock front. Thus, given a suitable molecular structure, the vibrational energy may be localized in a specific part of the molecule. If the energy deposited in a particular bond is large enough, the bond will break before the molecule can redistribute the vibrational energy; a non-thermal effect.

The size of the stress gradient in the leading edge of a shock front is open to some question, but it is not difficult to accept that a substantial portion of the pressure rise occurs on a subnanosecond time scale.[28-31] Such rise times can produce stress gradients sufficiently large to excite the nonequilibrium vibrational states described earlier.

We therefore present the following picture of shock-induced chemical reactions in organic compounds. A shock wave excites a molecule on a short time scale. The resulting state is nonequilibrium and exhibits intramolecular localization of vibrational energy; the energy being deposited in specific bonds, whose location depends on the molecular structure. If the energy is sufficiently large, the bond will break before the molecule can redistribute the vibrational energy. This process forms organic free radicals, which undergo relative mass motion within the shock front to produce chemical reactions. These reactions will be intrinsically non-thermal in nature and may be the origin of the unique reactions obsrved under shock loading.

It appears that shock-induced chemistry is on the threshold of providing a major advance in the understanding of shock compression processes. By forcing matter into nonequilibrium conditions not encountered in other environments, shock compression is opening a new field of physical chemistry.

REFERENCES

1. R. A. Graham, J. Phys. Chem. 83, 3048 (1979).
2. R. A. Graham, Bull. Am. Phys. Soc. 25, 495 (1980).
3. R. A. Graham, in Shock Waves and High-Strain-Rate Phenomena in Metals, ed. M. A. Meyers and L. E. Murr, Plenum, New York, (1981), p. 375.
4. L. Davison and R. A. Graham, Phys. Reports 55, 255 (1979).
5. P. Ya. Butyagin, Russ. Chem. Rev. 40 901 (1971).
6. G. A. Adadurov, V. I. Gol'danskii, and P. A. Yampol'skii, Mendeleev Chem. Jour. 18, 92 (1973).
7. A. N. Dremin and O. N. Breusov, Priroda, 12, 10 (1971). Translation in Sandia National Laboratories Report SAND80-6003, March 1980.

8. L. V. Babare, A. N. Dremin, S. V. Pershin, and V. V. Yakovlev, 1st Int. Symp. Expl. Cladding, 1971, p. 239. Translation in Sandia National Laboratories Report SAND80-6006, April 1980.
9. R. A. Graham and B. W. Dodson, Bibliography on Shock Induced Chemistry, Sandia National Laboratories Report SAND80-1642, August 1980.
10. L. V. Babare, S. V. Pershin, and V. V. Yakovlev, 2nd All-Union Symposium on Combustion and Explosion, Chernogolovka, 1971. Translation in Sandia National Laboratories Report RS3140/79/43, May 1979.
11. I. N. Dulin, V. N. Zubarev, Yu. N. Novikov, and M. E. Vol'pin, Russ. J. Phys. Chem. $\underline{45}$, 1642 (1971).
12. L. V. Babare. A. N. Dremin, S. V. Pershin, and V. V. Yakovlev, 1st Int. Symp. Expl. Cladding, 1971. Translation in Sandia National Laboratories Report SAND80-6006, April 1980.
13. T. N. Ignatovich, I. M. Barkalov, I. N. Dulin, V. N. Zubarev, and P. A. Yampolskii, High Energy Chem. $\underline{4}$, 394 (1970).
14. B. W. Dodson, present proceedings.
15. V. A. Veretennikov, A. N. Dremin, and A. N. Mikhailov, Comb., Exp., and Shock Waves $\underline{2}$, 58 (1966).
16. I. N. Dulin, V. N. Zubarev, A. N. Shuikin, and P. A. Yampol'skii Russ. J. Phys. Chem. $\underline{47}$, 475 (1973).
17. A. V. Babare, A. N. Dremin, and A. N. Mikhailov, Comb. Exp. and Shock Waves $\underline{5}$, 401 (1969).
18. G. A. Adadurov, V. V. Gustov, and P. A. Yampol'skii, Comb., Exp., and Shock Waves $\underline{7}$, 243 (1971).
19. P. A. Yampol'skii and T. N. Ignatovich, High Energy Chem. $\underline{4}$, 62 (1970).
20. N. S. Enikolopyan, in Problems in Kinetic Physics, Conference Proceedings, 1979. Translation in Sandia National Laboratories Report RS3140/81/36, February 1981.
21. R. A. Graham, P. M. Richards, and R. D. Shrouf, J. Chem. Phys. $\underline{72}$, 3421 (1980).
22. L. V. Babare, A. N. Dremin, S. V. Pershin, and V. V. Yakovlev, Comb., Exp., and Shock Waves $\underline{5}$, 364 (1969).
23. See Ref. 10, Ref. 22, and V. Yu. Klimenko, A. N. Dremin, and L. V. Babare, in Abstracts of the Papers of the 5th All-Union Symposium on Mechanochemistry and Mechanoemission of Solid Bodies, Tallin, 1975. Translated by Sandia National Laboratories, 1981.
24. See Ref. 10.
25. F. E. Walker and R. J. Wasley, Prop. and Exp. $\underline{1}$, 73 (1976).
26. D. von der Linde, in Ultrashort Light Pulses (Springer-Verlag, Berlin, 1977), p. 204.
27. V. Yu. Klimenko and A. N. Dremin, in Sixth All-Union Symp. on Comb. and Exp., Alma Ata, September 23-26, 1980. Translation in Sandia National Laboratories Report RS3140/81/38, February 1981.
28. L. C. Chhabildas and J. R. Asay, J. Appl. Phys. $\underline{50}$, 2749 (1979).
29. D. B. Hayes, J. Appl. Phys. $\underline{45}$, 1208 (1974).
30. S. B. Kormer, Sov. Phys. Uspekii $\underline{11}$, 229 (1968).
31. P. Harris and H. N. Presles, J. Chem. Phys. $\underline{74}$, 6864 (1981).

THE ELECTRICAL-TO-CHEMICAL CONNECTION*

R. A. Graham
Sandia National Laboratories, Albuquerque, New Mexico 87185†

ABSTRACT

New data on shock-induced electrical polarization in polymers is presented for polysulfone, polystyrene, and polypyromellitimide. The polarization generated in polysulfone is the largest reported for any polymer. A strong correlation is observed between the number of benzene rings in the monomer repeat unit and the magnitude of the polarization.

INTRODUCTION

Self-generated electrical polarization signals from shock-loaded polymers have been known to exist for many years. These signals are anomalous in terms of processes in thermodynamic equilibrium and models which use either "benign shock concepts" involving no molecular damage, or "catastrophic shock concepts" involving molecular damage have been used to describe the phenomena. It is the object of this paper to present new experimental data on shock-induced polarization in polymers and explore the connection between these electrical signals and shock-induced chemical processes.

In the earliest published work on such signals from polymers by Eichelberger and Hauver[1] and Hauver[2], it was proposed that the shock-induced polarization signals resulted from alignment of electric dipoles toward the shock direction due to the acceleration in the shock front. This elastic dipole rotation model involves no molecular damage and is therefore benign in concept. The model provided a satisfactory description of the early experimental data but as more data accumulated on a variety of polymers, this model appeared to lack features necessary to explain the observations. Furthermore, it failed to account for the observed microsecond duration relaxation times. Considering the enormous acceleration forces within the shock front with large gradients of velocity and stress at the molecular level, mechanochemical studies on polymers[3,4] indicate that bonds will be broken with molecular structure strongly influencing the bond scission.

Based on strong evidence for mechanically-induced bond scission in polymers, I proposed that shock-induced polarization was a result of mechanically-induced bond scission within the shock front with subsequent separation of electrically charged species by relative mass motion within the front.[5] Since this model is based upon molecular damage, it is a catastrophic shock model.

---

*This work sponsored by U.S. DOE under Contract DE-AC04-76-DP00789.
†A U.S. DOE facility.

Since publication of the bond scission model, a large body of literature on shock-induced chemistry in organic substances has been uncovered which provides considerable detail on shock-induced organic reactions.[6,7] Most of these organic reaction studies are based on examination of samples after shock loading, and, although careful attention has been paid to separating post-shock processes from processes while the sample is under shock compression, more information on the relation between damage processes and molecular structure requires "real time" data on the material while under shock. If we can continue to develop confidence in the bond-scission model for electrical polarization, it can serve as a very effective probe of these damage processes. Even though this model is well supported and there is a scattering of data on many polymers, there is little comprehensive information on the various polymers over a full range of compression. Such information is necessary since the observed polarization shows characteristically different behavior with compression including: 1) a subthreshold region, 2) a strong generation region and 3) a saturation region.[5]

In the present paper we report new data on three polymers, polysulfone (Udel P-1700), polypyromellitimide (Vespel SP-1) and polystyrene. The polysulfone (PSF) data are the first available on that material. The polypyromellitimide (PPMI) study extends our previous work[5] into the high compression region where saturation of polarization occurs. In addition, acceleration pulse loading is used to demonstrate a well-defined threshold and to look for pulse rise-time influences. The polystyrene (PSR) study investigates the influence of molecular weight.

EXPERIMENTAL

Shock compression experiments were of two basic types: high pressure plane-wave explosive loading and precisely controlled planar impacts. The planar impact technique is described in our prior work.[5] In the explosive-loading experiments, P-40 plane-wave generators with 25.4 mm thick pads of the high explosives, Baratol, TNT, or Comp B were used to drive shock waves into 2024 aluminum plates 12.5 mm thick. The polymer samples were mounted on the aluminum plates on the flat surface opposite the explosive drivers and nominal pressures and particle velocity values in the polymer samples were computed from nominal system conditions with impedance mismatch solutions for conditions in the samples.

Two acceleration pulse loading experiments were carried out with impact loading into a pyroceramic acceleration pulse generator[8] 12.5 mm thick. This loading produced a final particle velocity of 0.6 km/s rising continuously from rest in about 150 nanoseconds.

Electrical response measurements were carried out as in our prior work.[5] Various sample thicknesses were employed to identify loss mechanisms such as dielectric relaxation or shock-induced conduction. All samples were constructed with a guard ring configuration.

The PPMI and PSF polymers were commercial products while the PSR was either common grade with a spread of molecular weight or research grade with a narrow molecular weight distribution.

RESULTS AND DISCUSSION

Because of limited publication space, only the principal features of the observations will be discussed. Detailed experimental results are shown in Table I.

The PPMI data to large compression shows that the polarization saturates to a significantly lower effective polarization value than PMMA even though the two polymers have the same polarization in the strong generation region. The observed current pulse wave shapes indicates little if any shock-induced conduction for experiment with Baratol drivers but experiments with TNT and Comp B drivers indicate major distortions in the current pulse with spikes of current at wave entry and wave exit. The highest pressure PPMI experiments subject the samples to pressures in excess of the phase transition.

The PSF was chosen for study so that the influence of complexity of the monomer repeat unit could be investigated. This polymer has the most complex monomer repeat unit yet studied under shock compression, containing four benzene rings. Because of our previously noted correlation between monomer repeat unit complexity and polarization it was expected that this polymer would exhibit the largest polarization of any studied to date. As shown in Figure 1, which compares the polarization with the number of benzene rings in the monomer repeat unit, the polarization of PSF is substantially larger than PPMI (and also PMMA which is not shown). Furthermore, there is a good correlation for all the aromatics investigated between the number of benzene rings in the monomer repeat unit and the polarization.

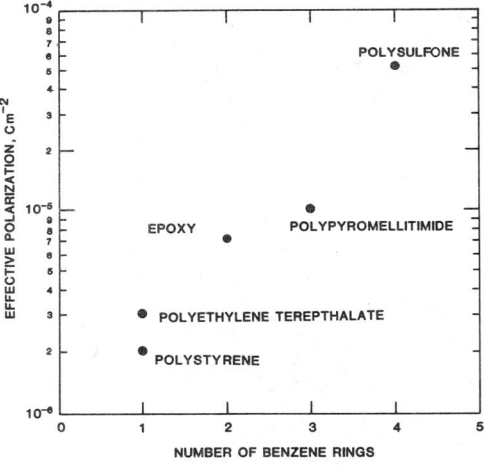

FIGURE 1

The shock-induced polarization at a compression of about 25% for all aromatic polymers studied to date shows a strong correlation to the number of benzene rings in the monomer repeat unit.

TABLE I. SUMMARY OF SHOCK-INDUCED POLARIZATION EXPERIMENTS

Experiment Number (a)	u_0 or P(b) km/s,(GPa)	ℓ(c) mm	$i_i \ell / A$(d) 10^{-3} amp/m	U(e) km/s	u(f) km/s	P_i(g) $\mu C/m^2$	Remarks(h)
Vespel SP-1							
1599	0.813(3.6)	1.23	19	3.6	0.71	2.7	i_R = 1.83
30G800	(9.0)	1.48	38	4.67	1.37	12	i_R = 2.5
15G800	(9.0)	2.96	41	4.67	1.37	12	i_R= 2.0
20G800	(9.0)	5.94	38	4.67	1.37	12	i_R = 2.4
25G800	(13.7)	1.48	—	5.4	1.83	—	pulse distortion
21G800	(13.7)	5.94	—	5.4	1.83	—	pulse distortion
26G800	(21.5)	1.49	—	—	2.53	—	pulse distortion
22G800	(21.5)	5.93	—	—	2.53	—	pulse distortion
Polysulfone							
1610	0.774(3.1)	0.203	76	3.2	0.69	30	—
1611	1.003(4.2)	0.241	250	3.7	0.89	90	—
31G800	(8.0)	3.01	620	4.62	1.42	196	i_R = 2.0
29G800	(8.0)	6.04	433	4.62	1.42	134	i_R = 2.4
34G800	(13.0)	3.01	1070	5.28	1.88	315	i_R = 2.8
Polystyrene							
1615	0.992(3.5)	1.74	~ 11*	3.68	0.92	4	Common grade, $i_R \approx 1.8$
1616	0.994(3.5)	2.37	12	3.68	0.92	4.3	Mw = 390,000 i_R = 1.8
1617	0.995(3.5)	2.31	13	3.68	0.92	4.7	Mw = 900,000 i_R = 1.6

a) Experiments with G indication are explosive loading; those with full numerals are compressed gas gun experiments. b) u_0 is the impact velocity. The impact configuration is copper on polymer. P is the pressure and is shown in ( ). Pressures in the vicinity of 8 to 9 GPa are Baratol drivers, in the vicinity of 13 GPa are TNT drivers and in the vicinity of 21 GPa are Comp B drivers. c) ℓ is the sample thickness. d) i_i is the initial current jump. A is the charge collecting area. e) U is the shock velocity in the polymers which is taken to be that reported in Marsh.9 f) u is the specimen particle velocity; also obtained from data in Marsh.9 g) P_i is the effective polarization which is equal to $i_i \ell / AU(1-u/U)$. h) i_R is the ratio of final current to first current jump. *Early portion of current pulse not recorded due to partial experimental failure.

The polarization of PSR is found to be independent of the molecular weight for the samples investigated. This observation agrees with prior work of Novitskii, et al.[10]

The response of PPMI to acceleration pulse loading shows evidence for a well characterized threshold in that no signal was observed upon arrival of the early part of the pulse; there was a definite delay in the electrical output until the compression reached a critical value. The peak output current agreed with that expected from shock-loading experiments. Hence, the lower acceleration value did not significantly alter the process leading to the polarization. Presumably the gradient-induced relative mass motion can be completed and cause charge separation for accelerations less than those existing in the shock front.

Shock-induced electrical effects which are the result of shock-induced chemical processes appear to have considerable potential for probing those chemical effects in, and in the vicinity of, shock fronts.

I wish to acknowledge the excellent technical assistance of G. T. Holman and J. R. Browning.

REFERENCES

1. R. J. Eichelberger and G. E. Hauver in "Les Ondes De De'tonation", Editions du Centre National de la Recherche Scientifique, 15 Quai Anatole France, Paris VII, 1962 p. 363.
2. G. E. Hauver, J. Appl. Phys. $\underline{36}$, 2113 (1965).
3. A. Casale and R. S. Porter, "Polymer Stress Reactions", Vol. 1, Academic Press, New York, 1978.
4. A. Casale and R. S. Porter, "Polymer Stress Reactions", Vol. 2, Academic Press, New York, 1979.
5. R. A. Graham, J. Phys. Chem. $\underline{83}$, 3048 (1979).
6. R. A. Graham and B. W. Dodson, Bibliography on Shock Induced Chemistry, Sandia National Laboratories Report SAND80-1642, August, 1980.
7. B. W. Dodson and R. A. Graham, this proceedings.
8. W. B. Benedick and J. R. Asay, Bull. Am. Phys. Soc. $\underline{21}$, 298 (1976).
9. S. P. Marsh, Editor, LASL Shock Hugoniot Data, University of California Press, Berkeley (1980)
10. E. Z. Novitskii, V. N. Mineev, A. G. Ivanov, E. S. Tyuun'kin, Z. I. Peshkova, N. P. Khokhlov and F. I. Tsyplenkov in Proceedings, Second All-Union Symposium on Combustion and Explosion. Erevan, October 25-30, 1969, edited by L. N. Stesik, Chernogolovka (1971) p. 269. Translation in Sandia National Laboratories Report RS-3140/79/132, May 1979.

LOW PRESSURE HUGONIOT CUSP IN POLYMERIC MATERIALS*

S. A. Sheffield and D. D. Bloomquist
Sandia National Laboratories‡
Albuquerque, New Mexico 87185

ABSTRACT

It has previously been shown that polymethyl methacrylate (PMMA) exhibits a cusp in the shock Hugoniot at about 2.0 GPa which corresponds with the beginning of shock-induced polarization and the beginning of an exothermic reaction measured in thermocouple and resistivity gauge temperature studies. We now report results we have recently obtained from an ongoing study which indicate that other polymers have similar behavior at about the same pressure. Quartz gauge impact experiments have been performed using polypyroellitimide (Vespel) and polysulfone impactors to obtain Hugoniot information and the stress history at the impact plane. In the case of Vespel a slight Hugoniot cusp was observed at about 1.8 GPa which coincides with the start of shock-induced polarization. Polysulfone does not appear to have a cusp but does show stress relaxation at the impact plane beginning at about 1.8 GPa, again coinciding with the start of shock-induced polarization. It has been suggested earlier that the abnormal behavior in PMMA is the result of a shock-induced chemical reaction. This new information suggests that a stress of about 2 GPa is a threshold for shock-induced chemical reaction in several polymers.

INTRODUCTION

At the Fourth Symposium on Detonation in 1965, Liddiard[1] pointed out that PMMA had a cusp in the shock Hugoniot at about 2.0 GPa. Although other data by Barker, Schuler, et al.[2,3] also showed this behavior, no explanation for it was given. Recently Sheffield and Bloomquist[4] pointed out that the PMMA Hugoniot cusp corresponds to the start of shock-induced polarization and to the beginning of an exotherm measured in thermocouple temperature measurement experiments.[5] Recent resistivity gauge temperature measurements support the earlier thermocouple information.[6] The reason for the anomalous behavior was suggested to be shock-induced bond scission and associated chemical reactions.

Shock-induced polarization is observed in many polymers

---

*This work sponsored by the U. S. Department of Energy under Contract DE-AC04-76-DP00789
†Shock-induced polarization is observed in many polymers

depending on their chemical makeup[7] and it is natural to suppose that behavior similar to that in PMMA could be expected from these polymer materials. A literature search indicated that very few good Hugoniot measurements have been made on polymers below 3 GPa although Carter and Marsh[8] have made a rather extensive study of many polymers at higher pressures.

A study was initiated to make low pressure Hugoniot measurements on several polymers including polypyromellitimide and polysulfone to determine if they had cusps similar to PMMA. Although this is an ongoing study some important results have been obtained which will be discussed in the following.

EXPERIMENTAL PROCEDURE

Impact experiments were performed using a light gas gun in which polymer specimens were impacted directly onto quartz gauge target assemblies as shown in Fig. 1. Experimental techniques are described in Ref. 9.

Output from the quartz gauge was recorded on an oscilloscope providing a stress history at the impact plane. By measuring the impact stress, the Hugoniot state in the polymer at impact could be determined using impedance matching techniques.

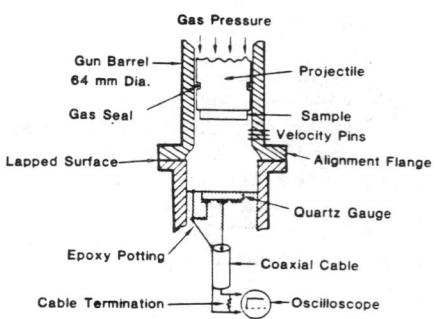

Fig. 1 Experimental Setup for quartz gauge impact experiments.

Polypyromellitimide used in the experiments was Vespel SP-1 which had an initial density of 1.425 Mg/m^3. Pulysulfone was Udel P-1700 with an initial density of 1.234 Mg/m^3.

RESULTS AND DISCUSSION

Vespel impactors were used in six experiments. The impact Hugoniot data are plotted in shock velocity-compression space in Figure 2. Also plotted are shock-induced polarization data for Vespel measured by Graham.[7,10] A Hugoniot cusp occurs at about 1.8 GPa and corresponds nicely to the beginning of shock-induced polarization. To demonstrate the problems associated with measuring the Hugoniot cusp accurately, all the available low pressure data on polypyromellitimide have been plotted in Figure 3 along with the data of this study.

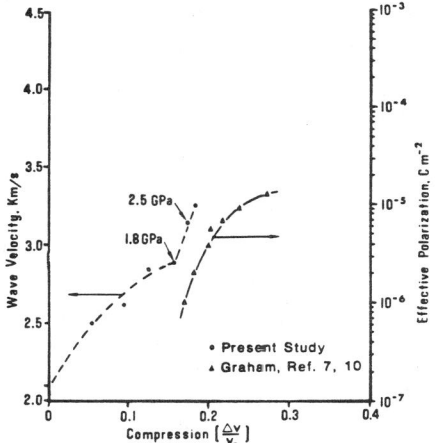
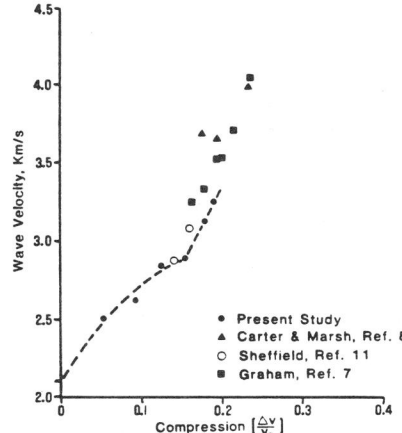

Fig. 2. Comparison of Hugoniot cusp and shock-induced polarization for Vespel.

Fig. 3. Low pressure Hugoniot data for polypyromellitimide (Vespel, Melden).

Information reported by Graham[7] and Carter and Marsh[8] were obtained by measuring the average shock velocity through the sample and using impedance matching techniques along with the jump conditions while those from this study were based on impact stress measurements. If reaction were occurring one would expect the shock velocity and stress to change with time and, therefore, the results to differ somewhat as it appears they do. Only with a large number of identical experiments in the region of interest (as is the case for PMMA) can one be sure that a subtle change is really there.

Polysulfone impactors were used in six experiments and the Hugoniot data obtained are plotted in Figure 4 along with Hugoniot data of Carter and Marsh[8] and recent polarization data measured by Graham.[10] Although there is a deviation of the data in our study from the LASL data points above 1.8 GPa, there is no cusp apparent in the Hugoniot. Stress histories at the impact interface are shown for the six shots in Figure 5. Relaxation starts with the 1.76 GPa shot and is observed in all the higher stress shots Relaxation times decrease from about 0.4 µs at 1.76 GPa to less than 0.2 µs at 3.16 GPa indicating that some type of reaction is taking place. According to Figure 4 the beginning of shock-induced polarization occurs at about the same compression (or stress) that relaxation is observed, a good indication that the two observations relate to the same phenomena.

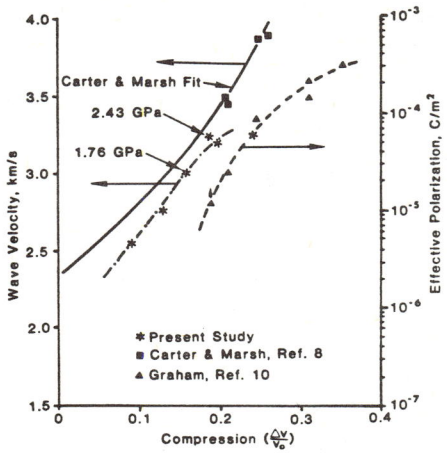

Fig. 4. Comparison of Hugoniot data and shock-induced polarization for polysulfone.

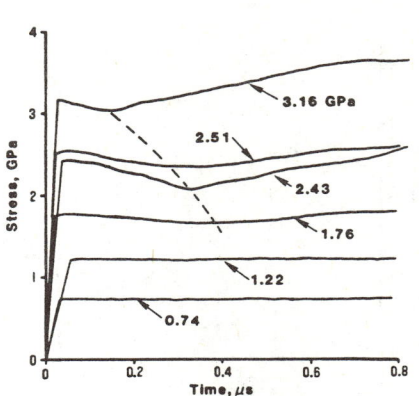

Figure 5. Quartz gauge impact plane stress histories for polysulfone.

Hayes[12] measured stress relaxation at the impact plane using quartz gauges in front surface impact experiments designed to study the 2.1 GPa phase transition in potassium chloride. He found it imperative to control tilt and surface finish on the specimen in order to obtain meaningful data from the records. We believe that the impact states we measured could very well be lower in stress than if the impact had been more precisely controlled. This may be the reason for the particularly large deviation from the Hugoniot curve at the highest stress shot for polysulfone. Although we have observed relaxation in PMMA and Vespel as well as polysulfone, results are not consistent enough to warrant further analysis of the details of the relaxation. However, we do believe the stress relaxation is real and presents another very important indication that a reaction is occurring.

CONCLUSIONS

It has been shown that Vespel and polysulfone exhibit low pressure Hugoniot behavior similar to that observed in PMMA. Vespel has only a slight Hugoniot cusp and polysulfone does not appear to have one at all, but has other abnormalities in the data. Impact plane stress histories indicate that polysulfone has a reaction that occurs above 1.8 GPa. In both of these materials the beginning of the unusual behavior coincides with the threshold of detectable shock-induced polarization.

Relaxation at the impact plane appears to be another

very useful method for observing the results and details of a reaction taking place in polymers as we have shown in polysulfone beginning at 1.8 GPa.

The present and earlier work on several polymers (PMMA, Vespel, and polysulfone) show indications that a shock-induced reaction is taking place at about 2 GPa. This may be some type of a threshold above which shock-induced bond scission and subsequent chemical reaction can be expected.

REFERENCES

1. T. P. Liddiard, Jr., Fourth Symposium (International) on Detonation, Office of Naval Research Report ACR-126, pp.214, 230 (1965).
2. L. M. Barker and R. E. Hollenback, J. Appl. Phys. $\underline{41}$, 4208 (1970).
3. K. W. Schuler, J. Mech. Phys. Solids, $\underline{18}$, 272 (1970).
4. S. A. Sheffield and D. D. Bloomquist, Bull. Am. Phys. S., $\underline{25}$, 567 (1980).
5. D. D. Bloomquist and S. A. Sheffield, J. Appl. Phys. $\underline{51}$, 5260 (1980).
6. D. D. Bloomquist and S. A. Sheffield, "Shock Compression Temperature Rise Determined from Resistivity of Embedded Metal Foils", this conference.
7. R. A. Graham, J. Phys. Chem. $\underline{83}$, 3048 (1979).
8. W. J. Carter and S. P. Marsh, "Hugoniot Equations of State of Polymers", Los Alamos Scientific Laboratory Report LA-UR-77-2062 (1977).
9. G. E. Ingram and R. A. Graham, Fifth Symposium (International) on Detonation, Office of Naval Research Report ACR-184, p. 369 (1970).
10. R. A. Graham, "The Electrical-to-Chemical Connection", this conference.
11. S. A. Sheffiend, unpublished data on an early sample of Vespel obtained in 1974.
12. D. B. Hayes, J. Appl. Phys $\underline{45}$, 1208 (1974).

AN EXPLORATORY STUDY OF REACTIVITY IN ORGANIC
COMPOUNDS SUBJECTED TO SHOCK LOADING*

B. W. Dodson
Sandia National Laboratories,[†] Albuquerque, New Mexico 87185

ABSTRACT

An exploratory study of chemical reactions occurring in organic compounds under shock loading has been carried out. Early results on shock reactivity of the organic compounds acrylamide, adamantane, hexamethylenetetramine, naphthalene, and 1,6-diphenyl-2,4-hexadiyne have established two points: 1) organic reactions occur under shock loading; and 2) chemical structure strongly influences shock reactivity.

INTRODUCTION

An extensive search, translation, and evaluation of Soviet literature has revealed a broad effort on chemical reactions observed in organic compounds subjected to shock loading and subsequently recovered for analysis.[1,2] In order to use this prior work as a foundation for more detailed studies of organic reactivity under shock loading, as well as to develop sample recovery techniques, replication of some of these results is necessary. In addition, other compounds must be studied in an exploratory mode to address specific questions about chemical reactivity under shock loading.

ACRYLAMIDE POLYMERIZATION

Acrylamide ($H_2C=CH-CONH_2$) is a model example of a material which under the conventional "benign shock concept"[3] would be considered non-reactive but which nevertheless undergoes chemical reactions under shock.[4] It is a particularly appropriate object of study in that the reaction yields are large, and the pressure-volume relation has been investigated.[5,6]

An explosive recovery system as reported by Davison, Webb, and Graham[7] was used in this work and in that of subsequent sections. The material under study was pressed to nearly theoretical density (>98%) in a copper capsule. The sample volume was 1 cm^3 or less. All planar interfaces were optically flat and accurately parallel to insure reproducible and calculable shock conditions. The recovery capsule was shocked with an explosive assembly. An explosive plane-wave generator initiated a 12 mm thick pad of explosive. Baratol and Comp-B were the explosives used, giving a shock pressure in the copper of about 13 GPa and 20 GPa respectively.

The acrylamide samples studied were 19 mm ⌀ and either 1 mm or 2.5 mm thick.[8] The pulse duration is sufficiently long that a <u>1 mm sample rings up</u> to the stress in the copper, whereas a 2.5 mm

*This work sponsored by the U.S. DOE under Cont. DE-AC04-76-DP00789.
[†]A U.S. DOE facility.

sample undergoes considerable focusing of the shock wave, resulting in a non-uniform pressure distribution.[7] We have therefore concentrated on studying the thinner samples.

Three samples of acrylamide were shocked and successfully recovered for analysis. These were A) a 2.5 mm sample shocked with Comp-B; B) a 1 mm sample shocked with Baratol; and C) a 1 mm sample shocked with Comp-B. The peak pressures seen by these samples were about 8 GPa for sample A before focusing begins, and about 13 GPa and 20 GPa for samples B and C respectively.

All three samples were found to change from a white pliable disk to a black, crumbly material as a result of the shock loading. The resulting samples were chemically separated using solubility fractionation with methanol and distilled water. The methanol-soluble fraction is the original acrylamide. The water-soluble fraction is a linear polyacrylamide, as confirmed for sample A with FTIR spectra taken by G. W. Arnold of our laboratories. This spectrum appears in Figure 1.

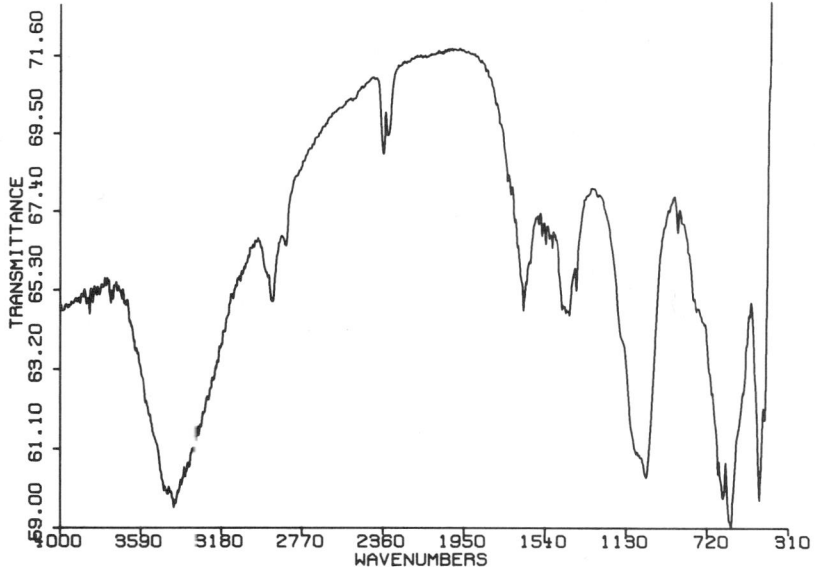

Figure 1. An FTIR absorption spectrum of the water-soluble fraction of shock-loaded acrylamide (sample A in text.)

The absence of absorption between 900 and and 1000 cm^{-1} indicates the lack of the $H_2C=CH-$ functional group. This group is present in the original acrylamide, but absent in the water-soluble fraction. Also, the strong absorption at 1645 cm^{-1} is due to primary amide groups ($-NH_2$). This information, along with the high viscosity of a solution in water (which implies a high molecular weight product), implies the source of the water-soluble fraction is the polymerization reaction

$$n\left[H_2C=CH-C\genfrac{}{}{0pt}{}{\nearrow O}{\searrow NH_2}\right] \xrightarrow{SHOCK} \left[\begin{array}{c}-H_2C-CH-\\|\\C\\NH_2^{\nearrow\,\nwarrow}O\end{array}\right]_n$$

The fraction insoluble in both methanol and water is a cross-linked polymer. These results are consistent with the earlier Soviet work.[4]

The containment of the sample by the recovery fixture is not perfect; accordingly, it is not possible to quantitatively discuss reaction yields. However, the ratio of composition of the material recovered is on the same scale as that reported in the earlier work; about 40% conversion of acrylamide into polyacrylamide, with about 1/2 linear and 1/2 cross-linked polymer at 13 GPa in sample B.

ADAMANTANE AND HEXAMETHYLENETETRAMINE

Adamantane (Tricyclo [3.3.1.1^{3,7}] decane) and hexamethylenetetramine (1,3,5,7 tetraazatricyclo [3.3.1.1^{3,7}] decane) are structural analogs.

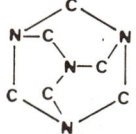

Adamantane   Hexamethylenetetramine

These two substances have the same molecular symmetry (tetrahedral), roughly the same molecular weight and melting temperatures, and both have highly symmetric lattice structures (fcc and bcc, respectively). However, the barriers against molecular motion are different in the two molecules; i.e., the rotational barrier is ~3 kcal/mole in adamantane compared to 19 kcal/mole in hexamethylenetetramine.[8] Because of these very striking similarities and differences, this pair of materials provides a revealing study of the influence of molecular structure on shock-induced chemical reactions.

Using the explosive recovery system described earlier, both 1 mm and 2.5 mm thick samples of adamantane[10] and a 2.5 mm thick sample of hexamethylenetetramine[11] were shocked with a Comp-B explosive pad. The peak pressures seen were similar to those in the acrylamide work.

The adamantane samples showed essentially no sign of reactivity under shock, even in the 1 mm sample, which rings up to 20 GPa. No product with different solubility was found under solubility fractionation using water, methanol, acetone, benzene, cyclohexane, and pentane. Both FTIR and Raman vibrational spectra were identical before and after shock. There is less than 0.5% of amorphous carbon in the 1 mm sample, probably a degradation product. We are currently unable to discount the possible formation of structural isomers and/or other hydrocarbons in small concentrations.

Although the bulk of the hexamethylenetetramine sample was subjected to considerably lower pressures than was the 1mm adamantane sample, elution of the sample using methanol yields several percent of a product. This product is insoluble in normal organic reagents.

The bond strengths and stiffnesses are similar in adamantane and hexamethylenetetramine. As both materials have a tight-packed molecular structure (adamantane has been called a diamond molecule) and the difference in weight between a nitrogen atom and a C-H group is small, the propensity to localize vibrational energy within the molecule should also be similar. We conjecture that the difference in chemical reactivity under shock in these materials is due to a difference in energy barriers against molecular motion within the lattice. If the molecule is free to move as a whole upon application of energies small compared to chemical energies, as in the case of adamantane, the intramolecular degrees of freedom will not couple well to the shock wave, resulting in low chemical reactivity under shock.[2] If the barriers against molecular motion are large, then the coupling of the molecule to the shock wave will be strong, and reactivity will be higher.

AROMATIC COMPOUNDS

Two aromatic compounds, naphthalene and 1,6-diphenyl-2,4-hexa-diyne, have been shocked using the standard explosive recovery fixture. The reaction in a 2.5 mm thick napthalene sample[12] shocked with Comp-B is sufficiently violent that the resulting gas breaches the recovery capsule from within. A 1 mm diphenylhexadiyne sample[13] was shocked with Comp-B and successfully recovered. The sample was totally degraded, being carbonized and greatly reduced in weight.

CONCLUSION

We have confirmed that chemical reactions in selected organic compounds are induced by shock loading. In acrylamide, a polymer forms in large yield by opening of the C=C bond, a result which confirms earlier work.[3] We have explored the relationship between molecular structure and chemical reactivity under shock. Aromatic compounds seem to be most active, while compounds with multiple bonds (acrylamide) give large yields of product. However, materials having only single bonds, such as adamantane and hexamethylenetetramine, despite having rigid molecular structures, are at best slightly reactive under shock loading. In addition, the properties of the crystal lattice, in terms of barriers against molecular motion in the lattice, seem to be important factors in determining chemical reactivity under shock loading.

ACKNOWLEDGEMENTS

The assistance of G. W. Arnold, and of J. A. Shelnutt, both of our laboratories, in providing, respectively, FTIR and Raman spectra of our samples is greatly appreciated.

REFERENCES

1. R. A. Graham and B. W. Dodson, Bibliography on Shock Induced Chemistry, Sandia National Laboratories Report SAND80-1642, Aug. 1980
2. B. W. Dodson and R. A. Graham, present Proceedings.
3. R. A. Graham, Bull. Am. Phys. Soc. $\underline{25}$, 495 (1980).
4. T. N. Ignatovich, I. M. Barkalov, I. N. Dulin, V. N. Zubarev, and P. A. Yampolskii, High Energy Chem. $\underline{4}$, 394 (1970).
5. V. A. Veretennikov, A. N. Dremin, and A. N. Mikhailov, Comb., Expl. and Shock Waves 2, 58 (1966).
6. I. N. Dulin, V. N. Zubarev, A. N. Shuikin, and P. A. Yampol'skii, Russ. J. of Phys. Chem. $\underline{47}$, 475 (1973).
7. L. Davison, D. M. Webb, and R. A. Graham, present Proceedings.
8. 99+% pure Electrophoresis grade acrylamide, Aldrich Chemical Co.
9. H. A. Resing, in Organic Solid State Chemistry, G. Adler Ed., (Gordon and Breach, New York, 1969) p. 101FF.
10. 99+% pure Gold Label adamantane, Aldrich Chemical Co.
11. 99+% pure hexamethylenetetramine, Aldrich Chemical Co.
12. Scintillation grade naphthalene, Eastman Kodak Co.
13. 99% pure 1,6-diphenyl-2,4-hexadiyne, Story Chemical Co.

ANALYSIS OF CAPSULES FOR RECOVERY
OF SHOCK-COMPRESSED MATTER*

Lee Davison, D. M. Webb, and R. A. Graham
Sandia National Laboratories**
Albuquerque, NM 87185

ABSTRACT

There are many circumstances in which it is desirable to recover matter that has been subjected to compression by a strong shock wave. In most cases this necessitates designing an encapsulation device that ensures that the material is processed by the shock in the most uniform way possible and is protected from the aftereffects of the explosion or impact producing the shock. For scientific purposes, it is necessary that the conditions to which the material has been subjected be known. Calculation is necessary for both designing sample recovery capsules and inferring the conditions to which the sample material has been subjected. Means for performing these calculations are discussed, along with their strengths and limitations, and some results for a particular configuration of interest are presented.

INTRODUCTION

In many cases it has proven useful to augment dynamic measurements of shock-wave phenomena with metallographic observation, chemical analysis, and other measurements performed on samples recovered after being subjected to shock compression. Special measures must be taken to ensure that samples are not destroyed during decompression, and further effort is required to determine the sequence of conditions to which the sample has been subjected. Various methods have been developed for recovering material from shock-compression experiments, but the experimental conditions have not often been evaluated carefully, and the work is subject to criticism on this ground[1]. Much can be learned by numerical simulation of recovery experiments, as is exemplified by the work of Skidmore and Lethaby[2] and of Stevens and Jones[3]. In each case these analyses showed that an unfavorable experimental configuration or an interpretation neglecting two-dimensional effects could lead to serious error.

In this paper we present, largely by way of example, some calculations of the performance of an explosively loaded

---

* This work was supported by the U.S. Department of Energy Contract DE-AC94-76-DP00789.

** A U.S. Department of Energy Facility

experimental assembly recently used by Graham and coworkers[4]. We illustrate some features of the assembly itself, and some strengths and weaknesses of the numerical simulation of its performance.

The configuration is as indicated in the cut-away drawing of Fig. 1. It comprises an explosive driver (initiated by a P-22 lens not shown), an iron pulse-forming plate, a copper capsule enclosing the sample, and a large steel wave sink. This assembly, while it does involve some interfaces and screws that are not shown, is quite satisfactorily represented as axially symmetric. Various computer codes are capable of analyzing such configurations. The principal shortcomings of an analysis by any one of them lie with the equations of state used - not only their nominal representation of material behavior, but also their neglect of the very details that the experiment is intended to investigate. These details include, for example, anomalous yield mechanisms or changes in the defect structure or chemical configuration of the sample. It is particularly difficult to evaluate the temperature history in the sample.

In our calculations the explosives were characterized by their JWL equation of state as given by Dobratz[5] (Baratol data[6] were placed in this form as a part of the present work.) The pressure response of the metals was taken to be that given by Thompson[7], while yield and spall strengths were estimated from various literature sources to be: $Y = 0.34$ GPa and $\sigma_s = 2.1$ GPa for copper, $Y = 0.12$ GPa and $\sigma_s = 1.6$ GPa for iron, and $Y = 0.85$ GPa and $\sigma_s = 2.5$ GPa for annealed 4340 steel.

In the examples given, the sample material was plexiglas, and was characterized by the Hugoniot curve and Gruneisen coefficient of McQueen et al.[8], but with a specific heat of 3k per atom.

Plane-wave calculations were made by applying the CHART-D code[9] to the centerline configuration, while the axisymmetric calculations were made using the two-dimensional Eulerian code CSQII[10].

RESULTS

Plane-wave calculations. These calculations were performed to evaluate "nominal" behavior of the assembly and, in particular, the action of the iron pulse-forming plate. This plate was incorporated into the design with the intent of using the 13 GPa phase transformation to convert the triangular pressure pulse in the explosive to a flat-topped pulse. For the material thicknesses of the present configuration, this works very well for such low-pressure explosives as NM and Baratol and satisfactorily but not optimally for more powerful explosives such as composition-B and PBX-9404. The efficacy of the arrangement is indicated in the sequence of pressure distributions of Fig. 2. The explosive considered in this case was

Baratol, the one performing best in the present experimental
configuration and used for most of the experiments.

Two-dimensional calculations. Behavior of the axisymmetric
configuration was simulated for several cases, including various

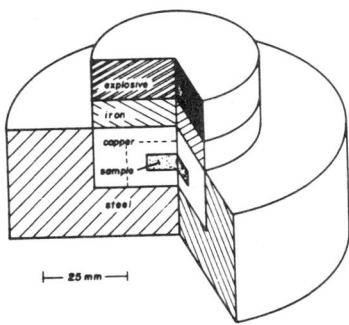

Fig. 1. Cut-away View of the Experimental Assembly.

explosives, two sample (PMMA) thicknesses, and a void sample
cavity. These calculations show that the design performs as
intended, but is in rather delicate balance as regards focusing
the capsule, etc. The sample is compressed by both axially and
radially-propagating waves, and then allowed to expand smoothly and
almost uniformly.

An indication of the wave action that develops when the
configuration of Fig. 1 is loaded by Composition-B is given by the
pressure-contour plots of Fig. 3. The coverage of these plots is
indicated by the dashed line on Fig. 1. In this example, the
compression is best described as taking place in four phases.
First, a plane compression wave is passed from the copper capsule
into the polymer, bringing the latter to a pressure of about 3.5
GPa. This wave then reflects from the high-impedance downstream
boundary of the sample cavity, and returns through the sample
further compressing the part near the centerline and increasing the
pressure in this material to about 7.5 GPa. During this second
phase of the compression process, the peripheral regions of the
sample are being subjected to compression by a radially-converging
disturbance, and this third phase of the loading process eventually
increases the pressure in the material to about 10 GPa. This wave
then reflects from the centerline of the assembly and propagates
outward, leaving material compressed to somewhat over 20 GPa in its
wake. Following this fourth phase of the compression process, the
material is smoothly and uniformly decompressed. The details of the
process vary from one point to another in the sample, but serious
convergence effects are avoided. Calculations on thinner samples
(1 mm) show similar, but somewhat more uniform, processing to about
the same pressure. Examination of the sample cavity contour

(and the positions of Lagrangian tracer points not shown) shows that the sample is subjected to considerable shear near the outer periphery.

Temperatures are evaluated in the course of the computation, but these results are subject to the considerable uncertainty of the specific heat of the compressed material. Nevertheless, we find that the sample is quite uniformly heated to 380 K by the first shock, and to 480 K by the shock reflected from the downstream face of the sample cavity. The radial compression brings the temperature in the central portion of the sample to about 525 K, and the subsequent decompression process leaves the sample at a rather uniform 450 K.

CONCLUSIONS

The calculation reported provides considerable insight into the conditions produced in a specific sample recovery device. Similar calculations show that variations of the configuration, explosive, or sample material can produce pronounced differences in device performance. When the sample chamber is void, for example, focusing

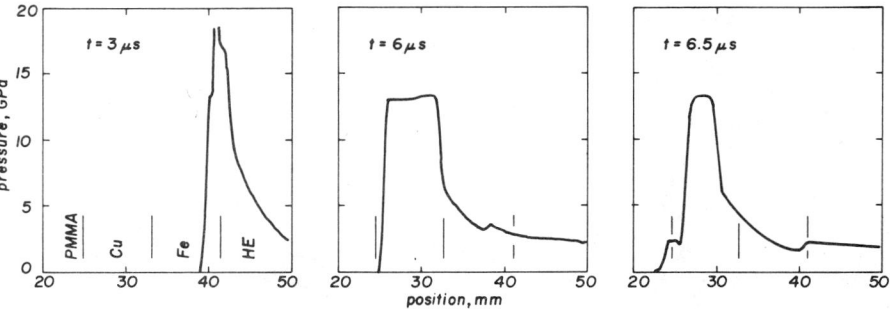

Fig. 2. Action of the iron pulse-forming plate from the time of introduction of the explosive disturbance to the interaction of the wave with a PMMA Sample.

effects lead to pressures well in excess of 100 GPa immediately downstream from the cavity. Spalls are predicted to develop in some designs, and, although serious spallation has not been observed experimentally, its potential effect on both pressure history in the sample and capsule integrity must be considered. The iron pulse-forming plate has been found very effective in tailoring a broad range of explosive loadings to uniform rectangular shape. We note, however, that, although the particular thicknesses of pulse-forming plate considered perform better with the lower-power explosives, the assembly as a whole provides somewhat more regular loading when used with Composition B than with Baratol. A single calculation suggests, in conformity with our expectation, that decreasing the thickness of the sample leads to a more uniform compression process.

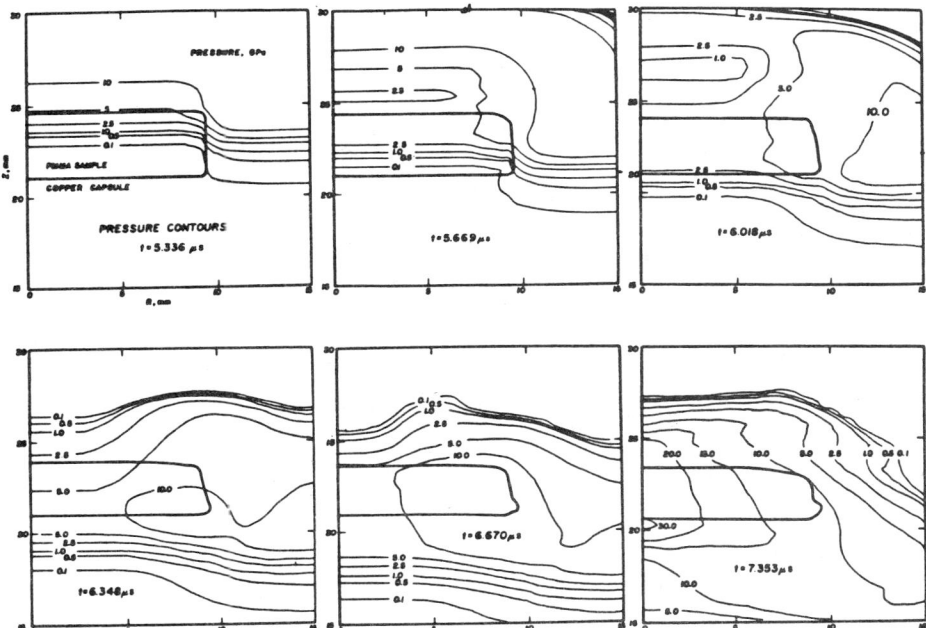

Fig. 3. The nature of the wave interactions in the sample is indicated by a sequence of pressure-contour plots covering the time interval from first encounter to the beginning of a period of smooth decompression. The plots cover the region of the sample within the broken lines on Fig. 1.

REFERENCES

1. L. Davison and R. A. Graham, Phys. Reports 55, 255-379 (1979).
2. I. C. Skidmore and J. W. Lethaby, in Fifth Symposium (Int'l.) on Detonation, ed. by S. J. Jacobs and R. Roberts (U. S. Government Printing Office, Washington, 1970), pp. 573-579.
3. A. L. Stevens and O. E. Jones, J. Appl. Mech. 39, 359-366 (1972).
4. See various papers, this proceedings.
5. B. M. Dobratz, Lawrence Livermore Laboratory report UCRL-51319 Rev. 1, 1974 (unpublished).
6. E. L. Lee, private communication.
7. S. L. Thompson, Sandia Laboratories report SC-RR-71-0714, 1972 (Unpublished).
8. R. G. McQueen, S. P. Marsh, J. W. Taylor, J. N. Fritz, and W. J. Carter, in High-velocity Impact Phenomena, ed. by. R. Kinslow (Academic Press, New York, 1970) pp. 293-417, 515-568.
9. S. L. Thompson, Sandia Laboratories report SLA73-0477, 1973 (unpublished).
10. S. L. Thompson, Sandia Laboratories report SAND77-1339, 1979 (unpublished).

CATALYTIC ACTIVITY OF SHOCK-LOADED TiO$_2$ POWDER*

John Golden and Frank Williams
Department of Chemical Engineering, University of New Mexico

B. Morosin, E. L. Venturini and R. A. Graham
Sandia National Laboratories, Albuquerque, New Mexico 87185†

ABSTRACT

Catalytic activity measurements have been carried out on a pure TiO$_2$ powder which was explosively loaded and preserved for post-shock examination. The oxidation of CO in a flow reactor was used to show an enhancement in catalytic activity of over two orders of magnitude at 720K. The catalytic activity does not correlate with the concentration of two paramagnetic defects as indicated by ESR measurements.

INTRODUCTION

This paper reports measurements of the catalytic activity of a high purity TiO$_2$ powder which has been subjected to shock wave loading and which was preserved for further studies.

This work was motivated by the recent recognition of unique and widespread chemical effects in shock-loaded solids which suggest the importance of shock-induced defect processes.[1,2] The need for better understanding of defect processes has led to increased interest in the detailed studies of defects and shock-induced solid state chemistry studies which reveal significantly enhanced solid state reactivity (see papers in this proceeding). Catalyst materials can show an appreciable increase in activity after quasi-static plastic deformation, which primarily involves the movement and multiplication of dislocations and lattice deformation leading to the formation of large number of line and point defects.[3] Shock loading has been shown to cause orders of magnitude increases in defect concentrations.[4,5] The use of shock waves to enhance catalytic activity of various oxide materials was first systematically studied by Boreskoy et al;[6,7] the first work was carried out by Horiguchi and Nomura.[8] Previous studies on TiO$_2$ shock-loaded to 33 GPa and then employed as a catalyst for the oxidation of carbon monoxide showed a fifty-fold increase in the reaction rate[6,7] at 300K. This increase was attributed to the formation of Ti^{3+} defects in the material.

In the present study, results on TiO$_2$ shock-loaded to 13 and 20 GPa in a copper recovery fixture, and subsequently employed as the catalyst for the oxidation of carbon monoxide reaction are reported. The nature of the induced defects have been examined by ESR and correlation with measured reaction rates was attempted.

*This work sponsored by the U.S. DOE under Cont. DE-AC04-76-DP00789.
†A U.S. DOE facility.

EXPERIMENTAL

A high purity TiO_2 powder was employed in these studies. X-ray diffraction, using $CuK\alpha$ radiation and a 114.5mm Norelco powder camera, showed that this material contains approximately 85% rutile phase and 15% anatase phase. The powder was pressed to 50% density in a copper recovery capsule prior to shock-loading. The fixture was explosively shock-loaded to pressures of about 13 and 20 GPa in the copper.[9]

X-ray diffraction of the recovered material shows broadened high two theta diffraction lines indicating strain-induced lattice defects in large concentrations. The approximate 85% rutile - 15% anatase phase ratio does not appear to be altered in either sample. Diffraction patterns were taken employing the same isochronal temperature schedule as the ESR experiments reported below. The X-ray pattern shows broadened high two-theta lines which persist to anneal temperatures far above those producing a drastic decrease of ESR defects. Temperatures in excess of 1075K are required for hours before these X-ray lines sharpen.

Catalytic activity of the shock loaded materials was determined using the oxidation of carbon monoxide as a test reaction. Reaction rates were measured in a flow reactor as shown in Fig. 1. The reactor and preheat sections are made of pyrex while the rest of the flow

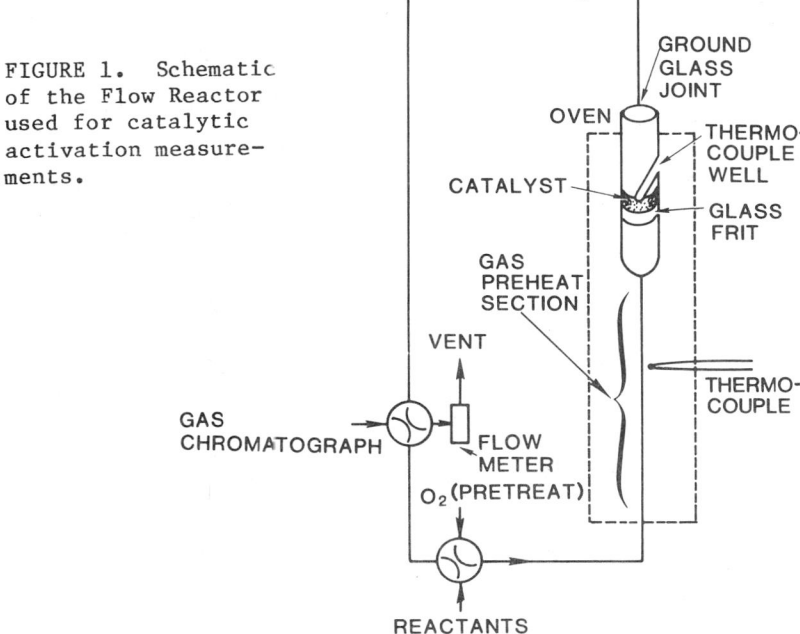

FIGURE 1. Schematic of the Flow Reactor used for catalytic activation measurements.

system is stainless steel. The catalyst temperature was controlled from a thermocouple placed just above the catalyst powder. A ground glass joint was built into the unheated section above the glass frit so that the catalyst powders could be easily removed. The joint was cooled with forced air to minimize any contamination of the catalyst. A stoichiometric mixture of CO and O_2 or pure oxygen could be directed to the catalyst, a soap-film flow meter or a gas chromatograph by means of two four-way valves. The composition and flow rate of gases leaving the reactor can also be determined with this flow arrangement. Reaction rates were calculated from the measured flows and compositions.

Gas chromatographic analysis was carried out with the use of a column that has an annular design. Different packing materials are in the core and the annulus of the column so that CO, O_2 and CO_2 are separated at room temperature in less than 10 minutes. The thermal conductivity detector was calibrated by correlation with GC-MS analysis of gas samples taken from the reactor flow.

The reactor was run without catalyst at the highest temperature used in these experiments. The blank rate was at most ten times less than the lowest rate observed for unshocked TiO_2 powder.

Recovered shocked material was examined by ESR in order to obtain a quantitative measure of the number of paramagnetic defects introduced by the shock process. More detail is reported by Venturini et al.[10] Two paramagnetic defects are observed in the liquid He temperature ESR spectrum, a narrow line isotropic absorption with a g-factor of 2.0029(3) and an absorption with axial symmetry having $g_\parallel = 1.935(5)$ and $g_\perp = 1.971(3)$. The axial defect (concentration $2 \times 10^{25} m^{-3}$) is similar to the Ti^{3+} ion reported in substochiometric TiO_{2-x}) while the isotropic defect (concentration $3 \times 10^{22} m^{-3}$) is believed to be an electron trapped at an oxygen (anion) vacancy (F-center). Both defects are present in TiO_2 powders shocked to 13 and 20 GPa, although their concentrations are considerably lower in the higher shock pressure experiment. The isotropic defect concentration decreased significantly more with pressure than that of the axial defect.

The stability of these defects in the 13 GPa shocked powder was studied by isochronal anneals of one hour at increasing temperatures in 50K steps from 375 to 875K. Approximately 20mg of powder in a sealed quartz sample tube was annealed at a given temperature, followed by recording of the ESR spectrum at liquid He temperature and the X-ray diffraction pattern at room temperature. The axial defects were found to decrease continuously in concentration while the isotropic defect remained roughly constant up to anneal temperatures of 625K at which point a sharp decrease begins. The behavior of the axial defect correlates with the continuous change in sample color from an initial dark gray to light brown and finally pure white at 775K as Ti^{3+} is oxidized to Ti^{4+}.

RESULTS AND DISCUSSION

The catalytic activity of TiO_2 is greatly enhanced by shock-loading. The relative first-order rate constants for the oxidation of carbon monoxide as determined in our flow reactor are shown in Fig. 2. The actual rates were on the order of 2×10^{-3} mole g^{-1} min^{-1} for the shock-loaded sample. The rates for the initial, unshocked TiO_2 powder were so low that measurements were obtained with confidence only at the highest temperature (720K). The 13 and 20 GPa samples give qualitatively the same enhancement compared to unshocked material. The activation energy of 4 Kcal/mole indicated from the catalytic activity measurements appears to be unusually low and it will take further work to determine the mechanisms responsible for such behavior. These results are qualitatively similar to those previously reported[6,7] and confirm the ehnanced catalytic activity induced by shock loading.

Before steady state rates were obtained, the reactivity of the shock treated TiO_2 was observed to slightly increase as the catalyst was maintained at temperature. This is in contrast to the annealing behavior of the axial defects from ESR spectra. Further, the materials maintain their activity even after 7 hours at 720K. Thus, the paramagnetic defects mentioned above, which are annealed during the reaction, evidently are not those responsible for enhanced catalytic activity of the shock-loaded samples.

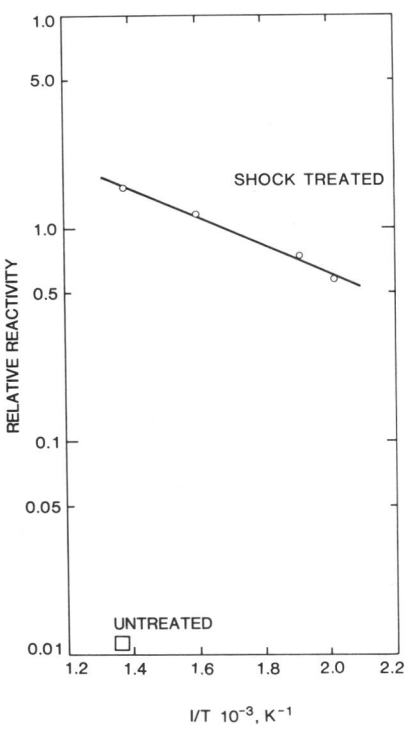

FIGURE 2.

Arrhenius plot for CO oxidation by TiO_2. Relative reactivity is a product of the psuedo first rate constant, k, and the reactor space time, τ.

The TiO_2 material which had been used for catalytic rate measurements was removed from the flow reactor and examined for X-ray line broadening and paramagnetic defect concentration levels. The material shows the same broadened high two-theta X-ray diffraction lines as observed following the 775K anneals described above; however, the ESR spectra show significant differences. The isotropic and axial defects are nearly removed; however, near g = 2.1 a rather wide absorption is found. The detailed nature of this resonance is currently under study.

These ESR measurements of defect concentrations following catalysis experiments suggest a defect other than the Ti^{3+} centers proposed by the earlier investigation[6] as responsible for the enhanced catalytic activity. Though such defects are present, their annealing behavior and catalytic activity of the shocked loaded samples at various temperatures do not correlate.

Studies are underway to identify the nature of the catalytically active defects. It is speculated that surface defects, terminating the lattice strain-induced defects observed by X-ray line broadening, may be responsible and that such defects are not paramagnetically active.

REFERENCES

1. A. N. Dremin and O. N. Breusov, Priroda No. 12, 10 (1971). Translation in Sandia National Laboratories Report SAND 80-6003, March 1980.
2. G. A. Adadurov, V. I. Gol'danskii and P. A. Yampol'skii, Mendeleev. Chemistry Journal 18, 92 (1973).
3. S. Kishimoto and M. Nishioka, J. Phys. Chem. 76, 1907 (1972)
4. R. A. Graham, in Shock Waves and High-Strain Rate Phenomena in Metals - Concepts and Applications, edited by M. A. Meyers and L. E. Murr, Plenum, N.Y. (1981) p. 375 and references therein.
5. L. Davison and R. A. Graham, Phys. Rep. 55, 255 (1979).
6. G. Boreskov, I. Sazonova, N. Keyer, V. Kudinov, G. Gridasova, V. Maly and L. Kefely, in Behavior of Dense Media Under High Dynamic Pressures, edited by J. Berger, Gordon and Breach, N.Y. (1968) p. 389.
7. S. S. Batsanov, G. K. Boreskov, G. U. Gridasova, N. P. Keier, L. M. Kefeli, V. M. Kudinov, V. I. Mali and I. S. Sazonova, Kinetics and Catalysts 8, 1140 (1967).
8. Y. Horiguchi and Y. Nomura, Carbon, 2, 436 (1965).
9. L. Davison, D. Webb and R. A. Graham, this proceedings.
10. E. L. Venturini, R. A. Graham and B. Morosin, this proceedings.

PARAMAGNETIC DEFECTS IN SHOCK-LOADED TiO$_2$*

E. L. Venturini, B. Morosin, and R. A Graham
Sandia National Laboratories,† Albuquerque, New Mexico 87185

ABSTRACT

Electron spin resonance (ESR) has been used to characterize the paramagnetic defects in TiO$_2$ powder samples which have been subjected to explosive loading to enhance catalytic activity and preserved for post-shock study. Two prominent defects are observed in the low temperature ESR powder spectra following peak shock pressures of either 13 or 20 GPa in the copper recovery fixture: an axial defect identified as Ti^{+3} ion, and an isotropic defect tentatively ascribed to an electron trapped at an oxygen vacancy. A series of high temperature isochronal anneals show that both defects are removed rapidly above 670K. Catalytic activity of the shocked TiO$_2$ powder cycled to such temperatures indicates that the paramagnetic defects are not the source of the enhanced catalysis. A study of the x-ray line broadening shows that the strains and dislocations introduced by shock loading can be removed by annealing above 1100K.

INTRODUCTION

Electron spin resonance (ESR) spectra can provide information on the type and concentration of paramagnetic defects in materials. In this paper we report ESR of shock-induced defects in a high purity TiO$_2$ powder and use these measurements in association with catalytic activity data[1] to seek to identify those defects responsible for enhanced catalytic activity in shock-loaded TiO$_2$.

The passage of a shock wave through a solid creates numerous defects and dislocations.[2] Batsanov and coworkers[3,4] report that shock-loaded TiO$_2$ powder exhibits catalytic activity a factor of 50 greater than that of the unshocked material. These authors measured a considerably higher electrical conductivity for the shocked TiO$_2$, and they noted a color change from the initial ivory to a gray-blue. Since similar effects are observed in reduced (oxygen-deficient) TiO$_2$, they attribute the increase in catalytic activity to oxygen vacancies introduced by the shock process.

Linde and DeCarli[5] have examined room temperature ESR spectra of shocked crystals and powders of TiO$_2$ following loading to pressures of 15-100 GPa, which are well above the pressure required to induce a polymorphic phase transition. They report two ESR absorptions, one with a g-factor near 2.008, and the second near 1.987. Unfortunately they do not discuss linewidths or lineshapes, and they show no ESR spectra, making direct comparison with our results rather difficult. By exposing their shocked samples to oxygen or heating to various temperatures in vacuum, they could vary the intensities of these two

---

*This work sponsored by U.S. DOE under Contract DE-AC04-76-DP00789.
†A U.S. Department of Energy facility.

ESR lines, leading to the conclusion that the paramagnetic defects were surface-sensitive. Symons[6] reports ESR spectra at room temperature and 77K for numerous shock-loaded solids. For TiO_2 he observed a doublet ESR absorption with a small axial distortion near g = 2, and he attributes this signal to the presence of a paramagnetic impurity atom such as ^{31}P or ^{19}F. We find no ESR line resembling this defect in our spectra.

EXPERIMENTAL RESULTS

The starting TiO_2 powder for this study was Johnson-Matthey "Puratronic" grade with a total metallic impurity content below 20 ppm by weight. X-ray analysis of this powder showed approximately 85% rutile phase and 15% anatase phase. It was pressed to 50% density in a copper recovery fixture prior to shock loading. Samples subjected to explosive shock-loading of 13 and 20 GPa in the copper were preserved for ESR analysis. These samples were taken from the same shocked powders used in the catalytic activity measurements reported by Golden and coworkers[1] in this proceedings.

X-ray diffraction measurements indicate that the relative amount of each phase present was unchanged in either of the two shock experiments. The TiO_2 powder was shocked in air, and was allowed to remain exposed to the atmosphere in the lab at room temperature for several days after shock-loading. At this point approximately 20 mg of shocked powder was sealed in a quartz capillary tube for subsequent ESR analysis at room and liquid helium temperatures. The quartz tube was not evacuated prior to sealing, and its main purpose

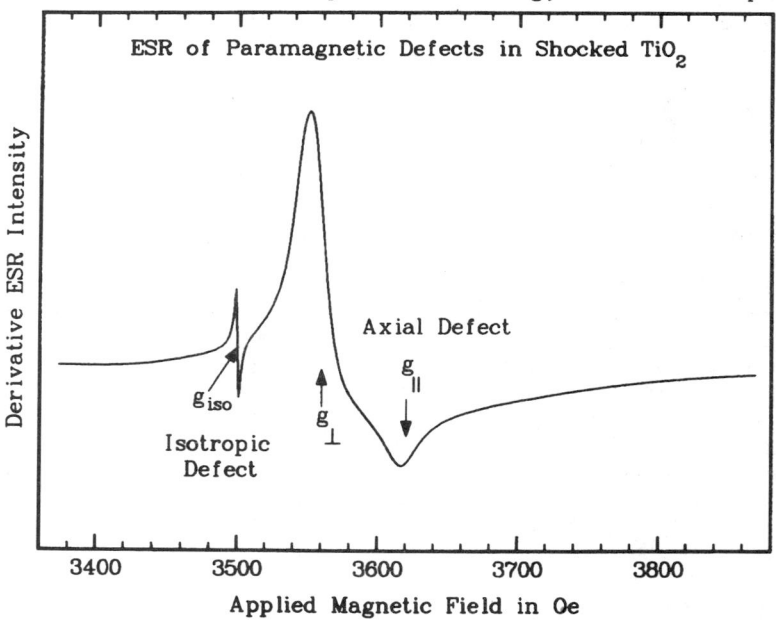

Fig. 1. ESR specrum of 13 GPa sample at 3K and 9.8 GHz.

was to ensure an identical amount of TiO_2 powder was present for successive ESR spectra recorded following various isochronal anneals at high temperatures.

The shocked TiO_2 powder had a dull gray color in contrast to the unshocked white material. A room temperature ESR spectrum for the 13 GPa TiO_2 sample revealed a single isotropic ESR absorption with $g = 2.003(1)$ and a linewidth of 3 Oe. As the sample temperature was lowered below 150K, a second asymmetric ESR absorption appeared at higher magnetic fields.

A typical ESR spectrum recorded at 3K and 9.8 GHz for this material is shown in Fig. 1, where the asymmetric resonance dominates the trace. This asymmetric lineshape can be duplicated by assuming a paramagnetic defect with axial site symmetry with $g_\parallel = 1.937(5)$, $g_\perp = 1.969(3)$, and a isotropic Lorentzian linewidth of 21 Oe. The narrow absorption at lower fields arises from a second paramagnetic defect with an isotropic g-factor = 2.0029(3) and a linewidth of 2.2 Oe, virtually identical with its room temperature position and width.

Comparing a numerical fit to the axial ESR signal in Fig. 1 with a similar fit for standards of known spin concentration allows us to estimate the number of axial defects in the 13 GPa sample as $2 \times 10^{25}/m^3$. These are identified as paramagnetic Ti^{+3} centers, and at this concentration they interact, changing the rhombic symmetry of isolated Ti^{+3} ions to the axial ESR signal observed in the spectrum of Fig. 1.[7] The isotropic defect concentration is $3 \times 10^{22}/m^3$.

In Fig. 2 we present the effect of one-hour isochronal anneals on the two paramagnetic defects using successively higher temperatures. Each point on this figure was determined by annealing the

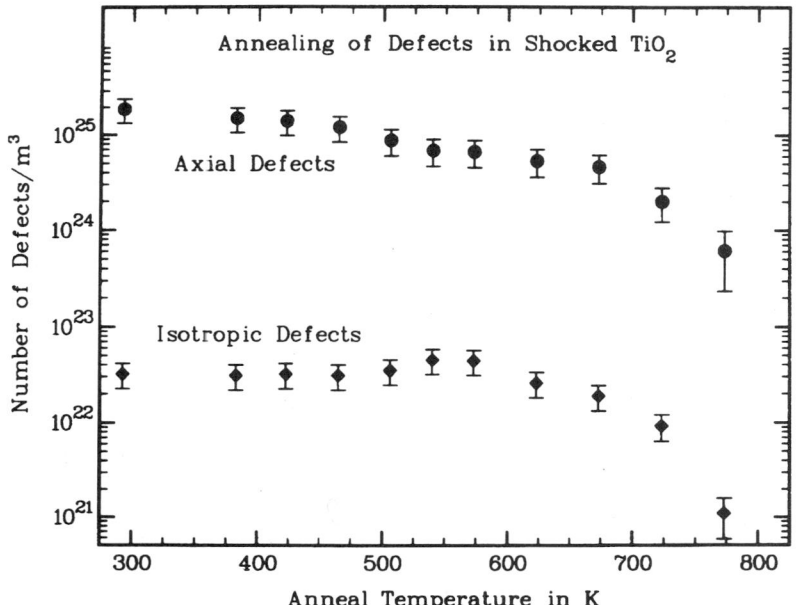

Fig. 2. Annealing effect on paramagnetic defects in 13 GPa sample.

TiO_2 sample in the quartz tube at a given temperature, then recording the ESR spectrum at 3K, followed by numerical analysis to determine defect concentrations. Both defects are reduced drastically following an hour at the highest temperature of 770K, but their behavior differs at lower anneal temperatures. Concentrating first on the more numerous axial Ti^{+3} defects, note the monotonic decrease with increasing anneal temperatures. There is a corresponding change in sample color: the initial dull gray changes to a noticeably lighter gray following the 470K anneal, light brown after the 500K anneal, dirty white after 620K, and pure white (indistinguishable from the unshocked TiO_2 powder) following an hour at 670K. After the latter anneal the Ti^{+3} defect concentration is 1/4 of its initial value.

In contrast, one hour anneals at temperatures up to 620K have little effect on the isotropic defect concentration. The first significant decrease occurs following the 670K anneal. There appears to be no connection between the sample color and the concentration of isotropic defects, suggesting that the gray coloration introduced by shock-loading is due to Ti^{+3} ions.

An ESR examination of the powder which was used in the catalysis studies[1] shows that the two paramagnetic defects which dominate the spectrum shown in Fig. 1 are barely visible above the noise. The ESR spectrum of the powder run in the catalytic reactor contains weak asymmetric absorption with g-values above the isotropic $g = 2.003$ line. Interpretation of this new absorption is not straightforward, and remains a problem. The ESR spectrum for the 20 GPa sample reveals the same defects as the 13 GPa powder, but the concentrations are different.

X-ray diffraction measurements show that the high two-theta line broadening indicative of strains and dislocations in the shock-loaded TiO_2 powder persists to anneal temperatures of 1100K. At these temperatures, the two paramagnetic defects discussed above are below detectable limits, indicating that shock-induced damage involves more than these defects. This point is supported by comparison with the catalytic reactivity measurements[1] of the shocked powder.

DISCUSSION

There have been numerous ESR studies of rutile and anatase phase TiO_2, both as single crystals and powders, particularly reduced (oxygen-deficient) samples.[7,8,9] The gray color of the shocked powder and the asymmetric axial ESR signal have been attributed to Ti^{+3} ions present in oxygen-deficient rutile.[7] There is considerable controversy over where the Ti^{+3} ions reside in the lattice, but it is clear that they are dispersed throughout the bulk.[10,11] Since these defects are introduced by shock-loading, we can rule out the possibility of hydrogen ions being present in the lattice as observed in nonstoichiometric rutile produced by high-temperature hydrogen reduction.[10]

The narrow isotropic resonance at $g = 2.0029$ has been reported in reduced TiO_2 powders prepared by vacuum annealing at high temperatures.[8,9] If the powder is kept under vacuum, the isotropic resonance is seen, but exposure to the atmosphere at room temperature drastically reduces this signal.[9] This resonance has not been reported in reduced single crystals. Based on a series of experiments using illumination and/or various adsorbed gases on vacuum-reduced TiO_2 powder, the isotropic ESR signal at g = 2.003 was identified as due to conduction electrons localized (trapped) on oxygen vacancies at or near the surface.[9]

For our shocked TiO_2 powder sealed in a quartz capillary tube, the isotropic resonance in Fig. 1 is observed despite exposure to air for days (discussed above). This leads to the conclusion that the paramagnetic defects responsible for this signal in shocked TiO_2 are protected from the ambient by their location below the surface.

The present ESR results show that sample temperatures above 670K produce large decreases in the paramagnetic defects due to shock-loading as well as restoring the sample to its original color. Catalytic reactivity measurements[1] on this material using the oxidation of CO indicate that the enhanced reactivity due to shock-loading is improved slightly following an initial heating to 720K, and that this reactivity does not decrease after several hours of operation in a CO/O_2 atmosphere at 720K. The drastic reduction in these defect levels combined with the stable catalytic reactivity of this material does not support the claim that Ti^{+3} ions due to shock-loading are responsible for the enhanced reactivity.[4]

REFERENCES

1. J. Golden, F. Williams, B. Morosin, E. L. Venturini, and R. A. Graham, this conference.
2. For a summary of observations, see L. Davison and R. A. Graham, Phys. Repts. 55, 255 (1979).
3. S. S. Batsanov, G. K. Boreskov, G. V. Gridasova, N. P. Keier, L. M. Kefeli, V. M. Kudinov, V. I. Mali, and I. S. Sazonova, Kinetics and Catalysis 8, 1140 (1967).
4. G. Boreskov, I. Sazonova, N. Keyer, V. Kudinov, G. Gridasova, V. Maly, and L. Kefely, in Behavior of Dense Media Under High Dynamic Pressures, ed. by J. Berger (Gordon and Breach, New York, 1968) p. 389.
5. R. K. Linde and P. S. DeCarli, J. Chem. Phys. 50, 319 (1969).
6. M. C. R. Symons, J. Chem. Soc. A (1971) p. 1648.
7. R. R. Hasiguti in Annual Review of Material Science, Vol 2, ed. by R. A. Huggins, R. H. Bube, and R. W. Roberts (Annual Reviews, Palo Alto, CA, 1972), pp. 69-92, and references therein.
8. P. Meriaudeau, M. Che, P. C. Gravelle, and S. J. Teichner, Bull. Soc. Chim. France (1971), p. 13.
9. E. Serwicka, R. N. Schindler, and R. Schumacher, Ber. Bunsenges. Phys. Chem. 85, 192 (1981); E. Serwicka, M. W. Schlierkamp, and R. N. Schindler, Z. Naturforschung 36a, 226 (1981).
10. L. N. Shen, O. W. Johnson, W. D. Ohlsen, and J. W. DeFord, Phys. Rev. B 10, 1823 (1974).
11. G. V. Chandrashekhar and R. S. Title, J. Electrochem Soc. 123, 392 (1976).

SHOCK-INDUCED CHEMICAL AND STRUCTURAL MODIFICATION OF ZIRCONIA, LEAD OXIDE AND THEIR MIXED POWDERS

J. D. Keck, D. L. Hankey, R. A. Graham and B. Morosin
Sandia National Laboratories
Albuquerque, New Mexico 87185

ABSTRACT

Results of an exploratory study of zirconia, lead oxide and their mixed powders which have been subjected to shock compression and preserved for subsequent study are reported.

INTRODUCTION

Previous investigations of shock-induced modification of lead zirconate titanate systems have shown that shock compression can substantially enhance the solid state reactivity and improve sinterability.[1-6] These prior studies were carried out in attempts to improve properties or decrease production costs of ferroelectric ceramics for piezoelectric applications and are related to a larger effort to improve sinterability of ceramics.[7,8] It is the object of this paper to report results of an exploratory investigation concerned with shock-induced alteration of PbO, ZrO_2, and a mechanical mixture of the two. We have also conducted a single shock-induced modification experiment on a lead, tin, zirconate system with a Nb_2O_5 additive. The present effort is part of a larger program at our laboratory to study shock-induced solid state chemistry.[9]

EXPERIMENTAL PROCEDURE

A. Materials

The powder characteristics (except for PSZT) are tabulated in Table I. The raw material oxides used were 99.9% pure PbO (orthorhombic containing several weight percent tetragonal phase) and 99.6% pure ZrO_2 (monclinic phase). The phases were determined using x-ray diffraction analysis and the purity was determined by spectrographic and loss on ignition analyses. Particle sizes were measured using an x-ray sedigraph and surface areas were determined by a BET method.

A PSZT composition (mixture of predominantly PbO, ZrO_2, and SnO_2, with small amounts of TiO_2 and Nb_2O_5) and equimolar mixtures of PbO:ZrO_2 were homogenized by wet-ball milling the powder in a polyethylene mill using Al_2O_3 grinding media and deionized water. The mixtures were pan dried at 110°C for twenty-four hours and then dry-blended to eliminate any segregation during drying. The orthorhombic PbO was converted to tetragonal PbO during this processing operation (a common occurrence). Therefore, it is important to note that the starting phases (pre-shocked) in the equimolar mixtures were tetragonal PbO and monoclinic ZrO2. Although some of the characterization data such as particle size and surface area do not have any direct bearing on the interpretation of our current experimental results, these parameters will play an important role in relating reactivity changes in shocked samples to defined alterations in the physicochemical properties of the reactant materials.

B. Shock Experiment

The shock compression and sample preservation experiments were carried out in the powder recovery fixture described by other papers in this proceedings.[9]

Samples of the starting materials were pressed in the copper recovery fixtures to the desired densities as defined in Table I. Since the starting powder density has a strong effect on the shock-induced temperature rise and pressure, samples of the mixed powders were compacted to two different densities with the exception of PbO and PSZT. Also, with the exception of PSZT, samples were shocked at low and high pressures (13 and 20 GPa).

Table I Schedule of Shock Compression Experiments and Materials Characterization Data

Mat'l.& Exper. Number	Pressure(a) (Copper) GPa	Density Mg/m^3 (% Crystal)	Mat'l. Properties(b)	
			d_p (μm)	A_s (m^2/kg x 10^3)
PSZT				
8G800	13	4.31(54)	–	–
PbO:ZrO$_2$				
17G800	13	3.92(51)	–	1.85
1G810	13	5.12(67)	–	1.85
6G810	20	4.95(65)	–	1.85
ZrO$_2$(c)				
16G810	13	3.26(56)	0.68	7.56
8G810	20	3.41(59)	0.68	7.56
PbO(d)				
15G810	13	7.38(78)	4.4	0.51
7G810	20	7.49(79)	4.4	0.51

(a) – Quoted pressures are those in the copper recovery capsule, not in the sample. For calculations of pressure see Davison, et al, this proceedings.
(b) – d_p = mean particle size, A_s = surface area.
(c) – Lot Sp 97310A, Teledyne Wah Chang, Albany, Or.
(d) – Lot 7148, Hammond Lead Products, Inc., Hammond, In.

C. Shocked Sample Characterization Procedures

Capsules containing the shocked samples were cut open and the cavities containing the samples were exposed. Visual observations (color, physical structure, etc.) were tabulated and color photographs were taken. The phases of the shock-compressed materials were determined using CuKα radiation and a standard 114.5 mm Norelco powder camera. Several samples were selected from various regions within each recovery capsule for analysis since there were distinct regions with color differences perhaps due to variable pressures and temperature within the capsule. Thermogravimetric analyses were applied to shocked ZrO_2 samples to look for organic contaminants.

RESULTS AND CONCLUSIONS

Before discussing the results on specific samples, some general findings can be presented to avoid repetition. First, lead or lead compounds dominate the diffraction patterns of samples that contain these phases and, therefore, may obscure low concentration of minor phases. Second, the high two-theta lines on all shocked samples are broadened. The diffraction pattern of the shock-compressed ZrO_2 has essentially disappeared at d values below 1.4. Finally, small variations in color of recovered samples may be due to trace quantities of phases not detected by x-ray diffraction or different defect concentration levels in the sample.

Early in our investigation, a PSZT composition was shock loaded to 13 GPa. The capsule partially opened during the shock compression. Analysis of the shocked sample by x-ray diffraction techniques revealed the dominant reactant phases as well as other lines too complex to analyze due to peak overlapping problems. Thus attention was directed towards less complex mixtures.

Equimolar mixtures of PbO and ZrO_2 were pressed to various green densities and shock compressed at 13 GPa and 20 GPa pressure in the copper recovery fixture. The low pressure, low density sample (17G800) provided the most interesting results. The shocked sample had a highly variable color appearance in various regions of the cavity. In the front cavity region, the sample color was grey with a green tint. X-ray analysis concluded that the composition was approximately 50 weight % $PbZrO_3$ and 50 weight % original oxides (orthorhombic PbO and monoclinic ZrO_2). The material at the rear (bottom) of the cavity was a metallic dark grey conductive mass that appeared sintered. The composition of this region in weight percentages was about 80% $PbZrO_3$, 10-15% orthorhombic PbO, and 5-10% Pb. The reason for the conductivity has not been identified at this time. Attempts to relate the conductivity to free lead indicate insufficient free metallic lead to cause conduction. Finally, unreacted phases of the oxides (orthorhombic PbO and monoclinic ZrO_2) surrounded the cavity.

Two other PbO:ZrO_2 samples (1G810 and 6G810) were shock loaded and recovered in well-sealed capsules. Both of these samples appeared uniform in color. The low pressure sample (1G810) was in stratified layers and the high pressure sample (6G810) was in powder form. The high green density sample (1G810) was shocked at the same pressure as was 17G800 but yielded no significant reaction throughout the cavity. Similarly, the high pressure, high green density sample (6G810) gave nearly identical x-ray diffraction patterns as above (orthorhombic PbO and monoclinic ZrO_2) with the presence of a detectable trace of Pb metal (approximately 5%). The starting mixtures contained tetragonal PbO in all PbO:ZrO_2 mixtures. Therefore, the PbO was converted to the orthorhombic phase in all shocked samples. The high degree of reactivity in the low pressure, low green density sample is probably related to the higher temperature attained in the less dense samples.[9]

The oxides were also examined to look for changes in the reactant phases. The recovery capsules of both ZrO_2 samples remained well-sealed during the shock compression. The high pressure shocked sample (8G810) had dark and light-colored regions which yielded identical results in the x-ray analyses. Approximately 20% of the ZrO_2 was converted to the tetragonal phase and the high two-theta lines (below d values of 1.4) were "washed out." The lower pressure sample (16G810) had less conversion to the tetragonal phase (5-10%) but displayed the same "washed-out" effect. Thermogravimetric analyses did not detect the presence of organic substances (sensitivity of 0.3% of sample weight) in ZrO_2 samples.

The sample cavity of the high pressure PbO experiment (7G810) vented open during the shock experiment in a manner which indicated a buildup of internal pressure. Samples from within and external to the cavity were both shown to be orthorhombic PbO with about 10% Pb metal. No tetragonal PbO was detected. The low pressure sample (15G810) remained well-sealed but showed evidence of internal pressure. Three different color regions existed in the cavity: black (center), grey (under and around the black), and greenish-grey (surrounding the core). All were shown to be the orthorhombic PbO phase with traces of Pb metal in the black central region where the internal pressures were highest.

Shock compression has been shown to substantially alter the chemical and structural characteristics of raw materials used in the fabrication of ferroelectric ceramics. Subsequent experiments will be conducted to relate these modifications to enhanced reactivities of the powder mixtures and electromechanical properties of the fired ceramics.

REFERENCES

1. L. R. Zaionts, G. A. Adadurov, R. Ya, Popil'skii, and A. N. Dremin, Elektronnaya Tekhn., Ser. 14, No. 1, p. 139 (1969) in Russian. Translation in Sandia National Laboratories Report, RS3140/81/82, May 1981.
2. A. A. Martynova, I. N. Temnitskii, A. A. Artenova, L. I. Kopaneva and S. S. Batsanov. Inorganic Materials **11**, 626 (1975).
3. I. N. Temnitskii, L. R. Zaionts, V. P. Bokarev, and S. S. Batsanov, in Third All-Union Symposium on Impulsive Pressures, 16-18 October, Moscow, edited by S. S. Batsanov, (1979). Translation in Sandia National Laboratories Report, SAND80-6004, March, 1980, p. 172.

4. A. F. Trudov and S. P. Pisarev, in Third All-Union Symposium on Impulsive Pressures (1979) loc cit, p. 188.
5. A. E. Vitenko, A. V. Popov, V. V. Yablochkin, V. D. Rogozin and N. A. Sadkov in Third All-Union Symposium on Impulsive Pressures (1979) loc cit p. 174.
6. E. A. Atroschenko, E. I. Zharin and E. A. Simonov, in Third All-Union Symposium on Impulsive Pressures, loc cit, p. 214.
7. A. V. Anan'in, O. N. Breusov, A. N. Dremin, V. B. Ivanova, S. V. Pershin, V. F. Tatsii and F. A. Fekhretdinov, in Proceedings, First All-Union Symposium on Shock Pressures, Vol. 2, October 24-26, 1973, Moscow, edited by S. S. Batsanov, Moscow (1974). Translation in Sandia National Laboratories Report, SAND80-6009, April, 1980, p. 28.
8. O. R. Bergmann and J. Barrington, J. Amer. Cer. Soc. $\underline{49}$, 502 (1966).
9. See other papers this proceedings by Graham and coworkers.

SHOCK SYNTHESIS EXPERIMENTS OF Nb-Si SYSTEM

Y. Syono, T. Goto, W. K. Wang* and H. Iwasaki
The Research Institute for Iron, Steel and Other Metals
Tohoku University, Sendai 980, Japan

S. Ohshima and T. Wakiyama
Department of Electronic Engineering, Faculty of Engineering,
Tohoku University, Sendai 980, Japan

ABSTRACT

Shock synthesis experiments have been carried out for Nb-Si system for the pressure range up to 85 GPa, using gun method. A small amount of superconducting component with $T_c \sim 18$ K is observed in the recovery product of arc melted $Nb_{100-x}Si_x$ alloy ($5 \leq x \leq 25$) shocked above about 62 GPa, confirming the results of earlier implosion experiments. A new fcc phase is found to crystallize from amorphous $Nb_{78}Si_{22}$ alloy under shock pressure of 60 GPa.

INTRODUCTION

Considerable attention has been paid to the Nb-Si binary system, motivated with synthesis of a high temperature superconductor of A15 type Nb_3Si. Various techniques such as rapid quenching[1], sputtering[2] chemical vapor deposition[3] or heat pulse technique[4] have been applied to obtain metastable phase of A15 type Nb_3Si. High pressure synthesis, among other things, was considered to be promising, since A15 type Nb_3Si was expected to be denser than the normal pressure phase with the Ti_3P structure[5], on the basis of the predicted value of lattice parameter of the hypothetical A15 type Nb_3Si[6]. Inadaptability of the atomic radius ratio of the Nb-Si pair in the A15 type structure ($R_{Nb}/R_{Si} = 1.14$), which explains instability of A15 Nb_3Si, might be removed by application of high pressure.

Previous implosion experiments[7,8], which used powder mixtures of Nb and Si as starting materials, revealed that a new superconducting phase with critical temperature, $T_c \sim 18$ K, was formed in the products recovered from shock pressure of ~ 100 GPa. Because of the low conversion rate of the high pressure phase, however, the structure of the new phase could not be positively determined, although the A15 structure was suggested to explain high T_c. On the other hand, investigation of the crystallization process of amorphous Nb-Si alloys at static high pressure of 10 GPa resulted in formation of a single-phase A15 $Nb_{81}Si_{19}$[9], although its T_c was found to be rather low[10].

Under these circumstances, reexamination of shock synthesis experiments on Nb-Si alloy system using different starting materials was considered to be meaningful. Two kinds of experiments were made in the present study. One was shock-conversion experiments from arc-melted Nb_3Si specimen with the Ti_3P type structure, and the other was

---

* Present address: Institute of Physics, Chinese Academy of Sciences P. O. Box 603, Beijing, China.

0094-243X/82/780087-05 $3.00 Copyright American Institute of Physics

the crystallization experiments under shock pressure of amorphous Nb-Si alloys prepared by rapid quenching method[11].

EXPERIMENTAL

The starting materials for shock synthesis experiments were prepared either by arc-melting or by rapid quenching technique. The arc-melted specimen was prepared for wide composition range of $Nb_{100-x}Si_x$ (x = 1.5, 5.0, 12.5, 25.0 and 37.5). The specimens with x = 1.5 and x = 37.5 were found contain no Ti_3P phase, while other specimens were mixture of Nb_3Si with the Ti_3P structure and bcc Nb-Si solid solution or Nb_5Si_3 phases. The amorphous specimen prepared by rapid quenching was in a form of thin ribbon with the nominal composition of $Nb_{78}Si_{22}$[11].

Impact experiments for shock synthesis were carried out using a two-stage light gas gun (2SG-TH1)[12] or a single-stage propellant gun (1SG-TH2) for the projectile velocity range of 2.1-3.1 km/s[13]. The specimen assembly (Fig. 1) was impacted by a 2-3 mm thick copper or stainless steel flyer plate glued on the front surface of the projectile. The arc-melted specimen was formed into a disc, 10 mm in diameter and 1 mm in thickness, sandwiched between two niobium discs of the same size, and encapsulated in a copper sample container. The amorphous specimen was placed in a copper cylinder, 10 mm in diameter and 3 mm in height, and tightly pressed into the copper container. The specimen container was protected by a spalling ring and backing plate so as to minimize destructive effects by impact. For the impact velocity range studied, the specimen was completely recovered as embedded in the copper container. The pressure achieved in the specimen was graphically estimated by the impedance match method, using measured projectile velocity and shock impedances of concerned materials.

The specimen thus recovered was examined by x-ray diffraction analysis. The superconducting critical temperature was determined by inductive method which detects susceptibility change associated with the transition to superconducting states.

RESULTS AND DISCUSSION

1. Shock-conversion experiments from arc-melted specimens[14]

Temperature variation of inductive response of the shocked $Nb_{100-x}Si_x$ specimens (x = 1.5, 5.0 and 25.0) to 78 GPa is shown in Fig. 2. Both specimens with x = 5.0 and 25.0 clearly exhibited a transition at 18 K, which is to be ascribed to formation of a new superconducting phase, in addition to the two transitions at 8.5 K and 9.2K due to the bcc Nb-Si solid solution and pure niobium respectively. The new phase in the $Nb_{75}Si_{25}$ specimen was detected for shock pressures above about 62 GPa, but T_c depended little on the shock pressure. The shocked specimens with x = 1.5 and 37.5 which contained no Nb_3Si phase with the Ti_3P structure in the virgin state did not show any high temperature transition. These results suggest that the presence of Nb_3Si with the Ti_3P structure is necessary for formation of the superconducting phase with T_c = 18 K.

X-ray diffraction study indicated some new lines from the sur-

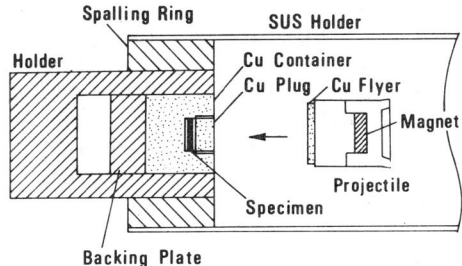

Fig. 1 Experimental set-up for shock recovery

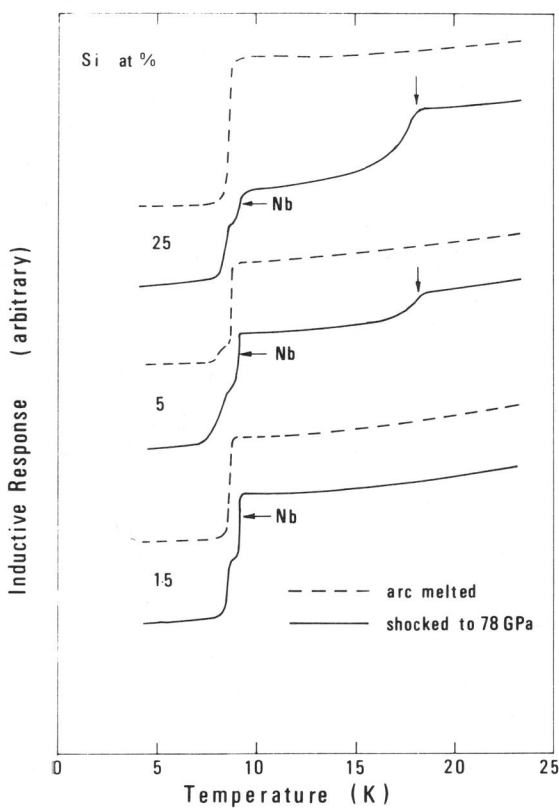

Fig. 2 Temperature variation of inductive response of unshocked and shocked $Nb_{100-x}Si_x$ (x = 1.5, 5 and 25). The arrow indicates onset of superconducting transition.

face layer of the shocked specimen. The new phase could not be conclusively identified as the A15 phase because of low volume conversion and complex x-ray patterns. On the other hand, x-ray diffraction of the ground powder of the shocked specimen consisted solely of the Ti_3P type Nb_3Si and bcc Nb-Si solid solution. Recently, Olinger and Newkirk[15] claimed evidence for formation of the A15 Nb_3Si from the implosion experiments up to 110 GPa, using the same Ti_3P type Nb_3Si as starting materials. Therefore, the new superconducting phase presently observed is most probably due to the A15 Nb_3Si phase which was directly transformed from the Ti_3P structure.

As suggested from the x-ray analysis, the new phase seemed to be formed mostly on the surface layer of the shocked specimen. This was confirmed by the fact that removal of the surface layer of the shocked specimen by successive grinding resulted in disappearance of the high T_c phase[14]. Relatively large inductive signals obtained in the susceptibility measurements compared with low volume conversion might also be explained from the consideration that electromagnetic shielding was achieved by the superconducting surface layer.

The reason why the new superconducting phase was predominantly formed on the surface layer of the shocked specimen might be due to high temperatures locally generated along the surface layer, owing to imperfect contact with sandwiched niobium plates. Low volume conversion could also be explained at least partly. Probably higher shock pressures, accompanying higher temperatures, may promote the conversion and result in higher yield of the high T_c components.

The shocked specimen heated to 600 °C for several ten hours showed no change in the superconducting property, while heat treatment at 700 °C removed high T_c components observed at 18 K. This indicates that high residual temperatures above 700 °C may be harmful for retaining high T_c components in shock synthesis experiments of Nb_3Si. These results suggest that there is some optimum range in shock pressures to increase the yield of the high T_c components.

2. Shock-crystallization of amorphous Nb-Si alloys

Shock-crystallization experiments of amorphous $Nb_{78}Si_{22}$ alloy was performed at 60 and 77 GPa. Shock treatment at 77 GPa resulted in crystallization of the Ti_3P phase of Nb_3Si. On the other hand, the specimen shocked to 60 GPa was found to have a new phase with a face-centered cubic (fcc) structure. Unit cell dimension of the fcc phase was 4.049 ± 0.001 Å. No superstructure lines were observed in the x-ray diffraction pattern, indicating completely disordered arrangements of Nb and Si atoms in the fcc lattice. As the Goldschmidt radius of niobium atom with 12 coordination is 1.47 Å, unit cell dimension of pure niobium with hypothetical fcc structure is estimated to be 4.16 Å. Dissolution of considerable amount of smaller silicon atoms into niobium may reasonably explain the observed unit cell dimension of the fcc $Nb_{78}Si_{22}$ phase. It is noteworthy to mention here that an fcc Nb_3Sn was synthesized at static high pressures[16]. As the fcc lattice is one of the most closely packed structures, it is particularly favorable for stabilization at high pressures. Arguments as to the relative stability of A15 and fcc phases at high pressures, however, might be premature in the present stage.

Details of the experimental results will be reported elsewhere[17].

ACKNOWLEDGMENTS

The authors are grateful to Dr. B. Olinger who kindly informed us of his new results prior to publication. They thank Professor T. Masumoto and Dr. A. Inoue for providing amorphous Nb-Si alloys, and Messrs. N. Sone and M. Saito for assistance in experiments. The work is supported by Grant-in Aid for Scientific Research (No. 546016) given by the Ministry of Education, Science and Culture, Japan.

REFERENCES

1. K. Togano, H. Kumakura and K. Tachikawa, Phys. Lett., 76A, 83 (1980).
2. R. E. Somekh and J. E. Evans, IEEE Trans. Mag., MAG-15, 494 (1975).
3. H. Kawamura and K. Tachikawa, Phys. Lett., 55A, 65 (1975).
4. L. R. Testardi, T. Wakiyama and W. A. Royer, J. Appl. Phys., 48, 2055 (1977).
5. D. K. Deardorf, R. E. Siemens, P. A. Romans and R. A. Cune, J. Less-Common Metals, 18, 11 (1969).
6. S. Geller, Acta Crystallogr., 9, 885 (1956).
7. V. M. Pan, V. P. Alekseevskii, A. G. Popov, Y. I. Belets, L. M. Yupko and V. V. Varosh, JETP Lett., 21, 228 (1975).
8. D. Dew-Hughes and V. D. Linse, J. Appl. Phys., 50, 3500 (1979).
9. C. Suryanarayana, W. K. Wang, H. Iwasaki and T. Masumoto, Solid State Commun., 34, 861 (1980).
10. W. K. Wang, H. Iwasaki, C. Suryanarayana, T. Masumoto and N. Toyoda, to be published.
11. T. Masumoto, A. Inoue, S. Sasaki, H. M. Kimura and A. Hoshi, Trans. Japan Inst. Metals, 21, 40 (1980).
12. Y. Syono and T. Goto, Sci. Rep. Res. Inst. Tohoku Univ., A29, 17 (1980).
13. Y. Syono and T. Goto, This volume.
14. S. Ohshima, N. Sone, T. Wakiyama, T. Goto and Y. Syono, Solid State Commun., (1981), in press.
15. B. Olinger and L. R. Newkirk, Solid State Commun., 37, 613 (1961).
16. J. -M. Leger and H. T. Hall, J. Less-Common Metals, 34, 17 (1974).
17. W. K. Wang, Y. Syono, T. Goto, H. Iwasaki, A. Inoue and T. Masumoto, in preparation.

THEORETICAL STUDIES OF SHOCK DYNAMICS IN TWO-DIMENSIONAL STRUCTURES V. MICROSCOPIC CONSTRAINTS ON SHOCK-INDUCED SIGNALS*

A. M. Karo, F. E. Walker, and W. G. Cunningham
Lawrence Livermore National Laboratory, Livermore, California 94550

J. R. Hardy
University of Nebraska, Lincoln, Nebraska 68588

ABSTRACT

Molecular dynamics calculations are presented that address the extent of microscopic detail that can be deduced from macroscopic gauge measurements of shock propagation in condensed systems. We have simulated large asymmetrically shock-loaded lattices, varying the initial temperature and the strength of shock loading. We have also introduced randomly-placed mass defects into the lattice, and we have studied the degradation of the shock front with the subsequent development of fracture and chunky spall and have compared this with the coherent microscopic spall found for perfect lattices.

INTRODUCTION

An extensive series of molecular dynamics calculations has been performed that addresses the question of the extent to which microscopic detail can, in fact, be deduced from macroscopic measurements of shock propagation in condensed systems. These measurements are those that would typically be taken by Manganin or electromagnetic particle-velocity gauges. In our computer studies we have simulated large asymmetrically shock-loaded lattices, varying both initial temperature and strength of shock loading. Several concentrations of randomly placed mass defects have been introduced into the lattice; the resulting degradation of the shock front, together with the development of fracture and chunky spall, has been studied and compared with the coherent microscopic spall found to occur with perfect lattice structures.

Computer molecular dynamics involves the numerical solution by computer of Newton's equations of motion for all atoms comprising the active region of the assembly. The force acting on each particle is the resultant of all interactions with other atoms in the neighborhood and is obtained as the derivative of an effective many-body potential. The initial positions and velocities of the particles represent the initial conditions of the problem.[1] Thus the coordinates and velocities of the particles are obtained as functions of time.

RESULTS

The present simulations were designed to study the degree to which shocks launched into the same system, but separated in both space and time, maintain their individual integrity. We also wished to determine the influence on this integrity of: (a) the presence of random thermal motion, (b) the size of the specimen, and (c) the presence of mass defects. To this end, two basic lattices were studied: one consisting of 10 columns and 65 rows of atoms and one containing the same number of rows but with 30 columns. In each system the shocks were generated by a triplet of small flying plates, well-separated vertically, initially moving to the right with the same uniform horizontal velocity, and offset to the left by different amounts. This initial situation, illustrated for the thinner lattice by the $t = 0$ configuration in Fig. 1, results in the successive generation of three spatially well-separated shocks. In the present studies the bond parameters are the same for both the plates and the lattice and for both first and second neighbor bonds. Except for defects all masses are equal. (The actual parameters are those for our "model system" discussed in Ref. 2.) The initial situation is shown in Fig. 3 for the thicker lattice. Also shown on this figure are the locations where impurity atoms were introduced.

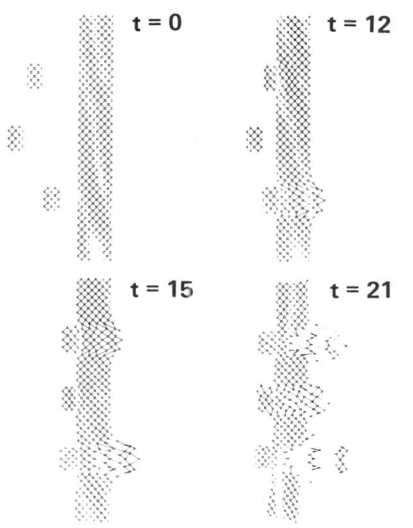

Fig. 1. Configurations of the initially quiescent 10x65 lattice at a sequence of times t as the shocks launched by the triplet of plates, initially moving towards the lattice at $t = 0$ with a velocity of 1.4 units, transit the lattice: at $t = 12$ the first spall is complete, at $t = 15$ the second spall is complete, and at $t = 21$ the third spalled fragment is clearly separated.

In Figs. 1, 2, 4, and 5 we show sequences of configurations from the histories of seven different simulations.

In all of the present calculations carried out on perfect lattices, shock coherence and stability are clearly evident and in qualitative accord with earlier results.[2] Most importantly, in lattices without impurities the lateral transfer of energy is minimal, as shown by the development of three virtually isolated successive spalls in the initially quiescent lattices of Figs. 1 and 4 and also in the thermally highly-excited lattices shown in Fig. 2. A further measure of this lack of lateral energy transfer is the degree to which each spall event is symmetric when the lattice is initially quiescent. It can be seen from Fig. 4 that, even for the thicker lattice, this symmetry is preserved to a surprising degree. In every case a major fraction of the incoming plate energy is carried off by microscopic spall as the associated shock reaches the far

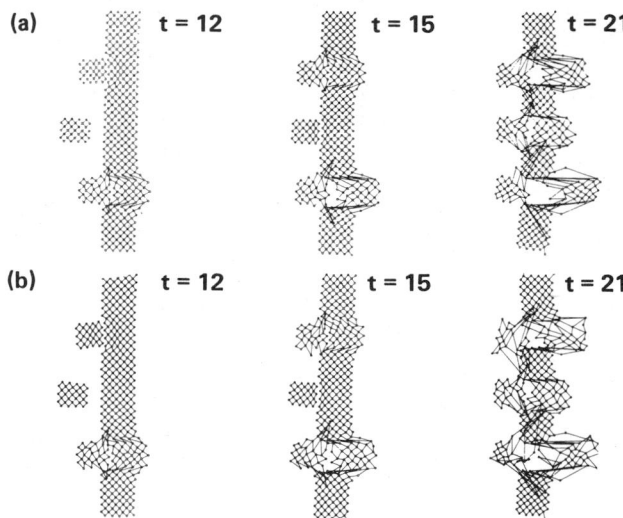

Fig. 2. Configurations corresponding to those in Fig. 1 except for initial thermal motion: (a) thermal motion per bond ~5% of the dissociation energy, and (b) thermal motion per bond ~20% of the dissociation energy. For both cases the spall sequence is strikingly similar to that for the quiescent system.

surface: proportionally, the remainder of the system picks up very little energy. We also observe a very interesting demonstration of what can be referred to as a "memory effect," in that the subsequent history of the shocked system "remembers" the details of the initial loading. Thus, after multiple shock transit through the lattice and the emergence of well-separated spall from the far wall, there is a direct relationship between the pattern of ejected material and that of the initial loading.

The results obtained for shock propagation through imperfect lattices show that coupling of shock energy to defect vibrations occurs readily within the front. Furthermore, these vibrations are initially strongly localized and provide an effective mechanism for lattice disruption near the defect. The manner in which this takes place is qualitatively different for light and heavy impurities. The results shown in Fig. 5 clearly demonstrate this. Heavy impurities appear to pick up more energy from the shock front as it passes, but the resonant vibrations that develop damp out rapidly, producing widespread lattice

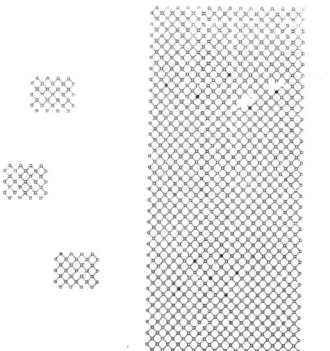

Fig. 3. Configuration of the 30x65 lattice system and associated triplet of loading plates. The sites at which mass defects can be introduced are indicated by circles.

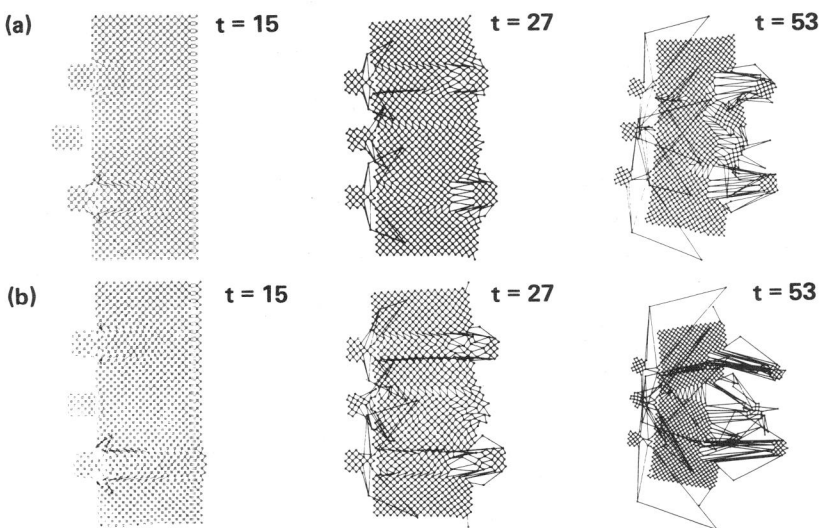

Fig. 4. Configurations of the initially quiescent triple-loaded 30x65 system at a sequence of times t for two different plate velocities: (a) v = 1.2 units and (b) v = 1.4 units. For both systems the first spall is about to commence at t = 15, and at t = 27 double spall is clearly apparent. At t = 53 we see the final state of both systems. Each shows clear double spall: in sequence (b) the central shock has produced a further spall; for sequence (a) this central spall is less definitive.

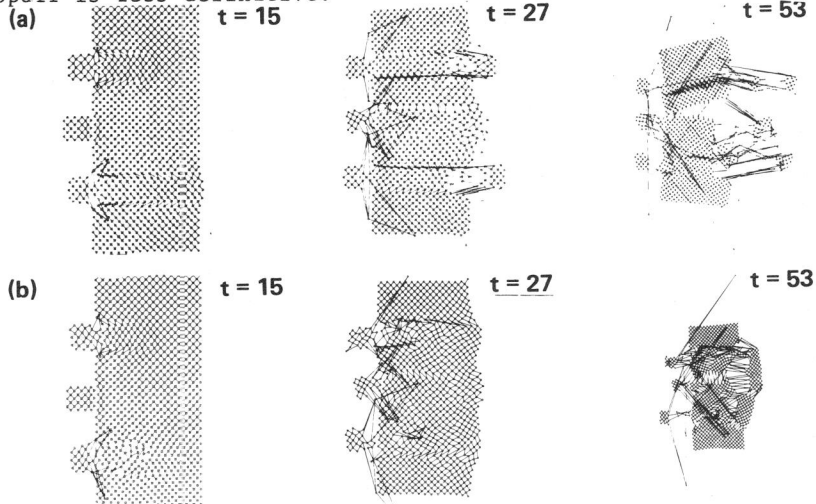

Fig. 5. Configurations for two initially quiescent 30x65 systems containing defects: (a) light mass defects at the sites indicated in Fig. 3, and (b) heavy mass defects at the same sites. Initial plate velocities are 1.4 units. The onset of disruption of both shock and lattice is already apparent at t = 15 when heavy defects are present. For light defects such effects are not present.

fracture and chunky spall, nucleated where the defect concentration is highest. Light impurities tend to take up less vibrational energy, which remains essentially localized: as a result, light impurities have remarkably little effect on the shock. It is only later that we can see evidence of the considerable <u>local</u> distortion and damage anticipated from earlier studies.[2] Again, as expected, this damage is most marked where the impurity concentration is highest.

Finally, in Fig. 6, we show where the present studies lie on the appropriate Hugoniot plot of shock velocity versus particle velocity. This clearly demonstrates that the present shock loading is at the <u>lower limit</u> of those used in experimental studies. Thus, the effects that we have described in the present studies should always be present in any experimental situation, since earlier results[2] show that they are enhanced by heavier shock loading.

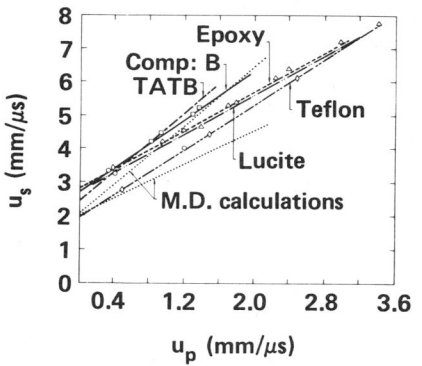

Fig. 6. Theoretical Hugoniots for two different lattice models compared with experimental data. The positions of our systems are indicated by the arrow pointing to the upper M. D. Hugoniot, corresponding to our lattice.

CONCLUSIONS

The present calculations show that the overall pattern of events results from a sequence of microscopic (i.e., atomic or molecular) processes occurring in picoseconds over dimensions of angstroms. It can readily be seen that for shock or detonation waves propagating in solids at 5 to 9 mm/μs, or 5 to 9 A in 10^{-13} s, and with impurities or structural irregularities no larger than 10 μ, an apparent shock rise time of 1 to 2 ns would be the shortest time obtainable. Measurements made at shorter time intervals would only be probing inhomogeneities associated with the intrinsic random defect structure of the material. Even if sample preparation could be improved to the extent that the defects or microcrystalline features in the material would be no more than 0.1 μ and even with subpicosecond instrumentation, the measured, or apparent, rise time would still be about 0.01 to 0.02 ns, i.e., about 100 times the periods associated with phenomena occurring on atomic and molecular scales.

REFERENCES

1. B. J. Alder and T. E. Wainwright, J. Chem. Phys. <u>31</u>, 459 (1959).
2. A. M. Karo and J. R. Hardy, in <u>Proceedings of the NATO Advanced Study Institute on Fast Reactions in Energetic Systems</u>, edited by C. Capellos and R. F. Walker (D. Reidel Publishing Co., 1981).

*Work performed under the auspices of the U.S. Dept. of Energy by the Lawrence Livermore Nat'l. Lab. under contract #W-7405-ENG-48.

EXPERIMENTAL AND THEORETICAL STUDIES ON SHOCK COMPRESSION OF LIQUID CARBON MONOXIDE*

F. H. Ree, W. J. Nellis, M. van Thiel and A. C. Mitchell

University of California, Lawrence Livermore National Laboratory
Livermore, California 94550

ABSTRACT

Dynamic equation-of-state data for liquid CO were measured in the shock pressure range 5-70 GPa (50 - 700 kbar) using a two-stage light-gas gun. The liquid was shocked from initial state near its saturation curves at 77 K for CO. The data were examined by using three theoretical models: (1) chemically nonreactive model, (2) a quasi-chemical-equilibrium model that allows CO to dissociate into gaseous species and graphite, and (3) a chemical-equilibrium model that also includes a dense carbon phase which exists at higher pressures and temperatures than graphite. This dense phase is assumed to be diamond. Our analysis shows that a low pressure, chemical equilibrium takes much longer than a typical shock passage time. As a consequence, the experimental data initially follow the nonreactive Hugoniot to pressures well beyond the chemical dissociation limit. Both the experimental data and the Hugoniot computed with case (3) agree satisfactorily at high pressure. Further consequences of these observations to high-explosive studies are discussed. The experimental and theoretical details concerning this work have been published.[1]

REFERENCE

1. W. J. Nellis, F. H. Ree, M. van Thiel and A. C. Mitchell, J. Chem. Phys. __75__, 3055 (1981).

---

*Work performed under the auspices of the U.S. Department of Energy by Lawrence Livermore National Laboratory under contract #W-7405-Eng-48.

METALLURGICAL EFFECTS OF SHOCK LOADING

D. E. Mikkola and R. N. Wright
Michigan Technological University, Houghton, MI 49931

Impacting a flat specimen with a flat projectile, or flyer, introduces a planar stress wave with a state of one-dimensional strain. The nature of the propagating stress pulse is determined by the loading conditions at the impact surface, as well as by the materials response and geometry of both the specimen and the projectile. Shear strains of tens of percent can be produced with strain rates in excess of 10^6 s^{-1}. For high velocity impacts that create large amplitude planar shock waves, the stress state within the propagating shock pulses is largely hydrostatic, however, the deviator components can be large relative to the common quasistatic yield strengths of materials. The shear response of materials to this deviator is usually assumed to be a simple sum of the elastic and plastic contributions, with the plastic part expressible in terms of dislocation glide, deformation twinning, and in some instances, adiabatic shear.

Most plate impact experiments fall into two general groups:
(1) Time-resolved stress (or velocity) measurements aimed at defining the stress-time profile in detail. These measurements are made at either the impact or back surface, with the target thickness being varied to study changes in stress profile with distance. The shear response of the material is then modeled with some form of plasticity theory incorporating various material constants and adjustable parameters.
(2) Recovery experiments involving the study of the substructures of impacted specimens that have been soft recovered. It is usual to relate the crystal defect densities and arrangements introduced by the shock pulse to the shock pulse parameters. In related experiments the dynamic fracture response of materials can be studied by changing the test geometry to permit the interaction of the release waves from the free surfaces of the specimen and flyer, thereby causing spallation under the action of the resultant tensile stress. The initiation and propagation of the fracture can then be studied and modeled.

Although much has been learned from both of these types of experiments, it is important to note that in many cases the validity of the experimental results can be questioned because of the shock loading procedure and/or the metallurgical condition of the specimens. In the first part of this paper, some of these experimental difficulties will be summarized, particularly those associated with specimen preparation. Against this backdrop of experimental procedures, an attempt will be made to sort out conclusions about metallurgical effects that appear valid based on previous recovery experiments. Progress toward solutions to some of the remaining questions will then be discussed followed by some concluding remarks on the design of future experiments as viewed from the standpoint of metallurgical effects.

EXPERIMENTAL FACTORS

The design of shock wave experiments has been reviewed recently by DeCarli and Meyers[1] and earlier by Orava.[2] In the simplest form of recovery experiments the goal is to develop and propagate a single well-defined stress pulse through the specimen. As a minimum, this requires that the specimen be surrounded at the sides and back by close-fitting, impedance-matched, momentum traps to minimize undesired reflections. In addition, as demonstrated by Stevens and Jones,[3] control of the specimen aspect ratio can be important to provide the lateral constraint over a time span necessary to ensure conditions of uniaxial strain. Employing flyers of specifically designed shapes can also be of advantage.[4] In order to yield conditions of pure one-dimensional strain, the impact of the flyer with the specimen must obviously be highly planar. This is controlled by the experimental setup and can be monitored in a variety of ways. Finally, there are limits to the range of experimental geometries that can be accommodated, for example, the use of very thin specimens (<0.25mm) can make it difficult to achieve the necessary experimental control to assure well-defined conditions of uniaxial strain.

An equally important aspect of shock wave experiments, that has not received adequate attention, is the need for proper specimen preparation and characterization. Once the chemistry of the specimen has been documented, care must be taken to establish the desired metallurgical condition, as well as to ensure that the specimen is homogeneous throughout, including the near-surface regions. Although commonly accepted commercial thermal-mechanical processes can be followed to produce particular microstructures in many materials, it is generally preferred that the microstructure be documented by means of optical and/or electron micrographs together with specification of some statistically significant parameters, such as grain size, volume fractions and sizes of additional phases, and defect concentrations.

It is particularly important to recognize those factors that determine the surface condition of a specimen. Most often the desired specimen geometry is achieved by a combination of procedures involving machining, grinding, and polishing. Of importance here is the fact that all cutting and grinding operations cause large amounts of deformation of the material next to the surface. In the sequence of grinding and polishing each successive abrasive acts to remove the deformed layer left from the coarser stage preceding it, while it in turn leaves a distorted layer of reduced depth. As an important part of the final stage of common metallographic practice,[5] this distorted layer, called the "Bilby layer," can be substantially eliminated by repeatedly etching chemically followed by light repolishing with the finest abrasive. In some cases electrolytic polishing can be used to eliminate the distorted layer, however, unless extreme care is taken, this technique may compromise the planarity of the surfaces.

For a wide variety of materials, fine grinding is generally done with 320, 400, and 600 grit silicon carbide abrasives lubricated with water. The corresponding particle sizes are about 33, 23, and 17 microns, respectively. Rough polishing can be accomplished with a 6 micron particle size diamond dust using a rotating cloth-covered wheel with oil as a lubricant, while final polishing stages can be carried out with 0.3 and 0.05 micron alumina powders on a cloth-covered wheel with water as a lubricant. A variety of other abrasives are available for specific materials and purposes. The specimen is usually etched following final polishing, with the reagent chosen so as to provide a uniform attack for purposes of removing all of the disturbed metal. Typically, these final steps of etching and repolishing are repeated several times. The preparation process can be more complicated for multiphase materials because smearing of more ductile phases can occur, as can deposition or leaching during chemical etching.

With low defect density single crystal specimens, testing can often be done after the preparation of the nearly distortion-free surfaces. On the other hand, it is often advantageous in work with polycrystalline specimens and multiphase materials to carry out the heat treatment to yield the desired microstructure as the last step. For example, the production of a fine-grained polycrystalline specimen might involve cold working the starting stock to the necessary strain, then metallographically preparing a planar distortion-free surface, and finally annealing to give recrystallization and the desired amount of grain growth. Depending on the material it may be necessary to carry out the annealing treatment in a vacuum, or controlled atmosphere, to prevent oxidation, sublimation, or contamination.

The importance of care in specimen preparation cannot be overemphasized regardless of the specimen material. Certainly the introduction of subsurface damage by mechanical abrasion of "plastic" and "semi-brittle" materials is generally accepted. However, it should be noted that so-called "completely brittle" materials, such as alumina, also suffer subsurface damage. This has been demonstrated by Hockey[6] who used transmission electron microscopy to study the subsurface damage in alumina abraded with 0.25 micron diamond, 0.30 micron alumina, or a 325 grit diamond grinding wheel. The near-surface regions of abraded specimens were found to contain high densities of dislocations, and in some cases microtwins. The depth of the easily observable damage varied from less than one micron for the alumina abrasive to nearly ten microns for the grinding wheel.

Specimens of some brittle materials are often prepared by cleavage, but this has the disadvantage of reduced surface planarity, as well as causing the introduction of subsurface dislocations and other near-surface defects.[7]

Finally, there are several ways to examine the metallurgical condition of specimen surfaces. In addition to optical examination of etched and unetched surfaces, x-ray diffraction techniques are quite sensitive to the presence of substructural defects. For example, with single crystals, the smearing of Laue spots, or diffraction topography, can be used to examine the perfection, while with many fine-grained polycrystalline materials a diffraction scan of several peaks to look for peak broadening can be effective.

MICROSTRUCTURAL EFFECTS

Recovery experiments involving the study of the residual effects of shock loading were initiated more than fifty years ago. As a result, numerous reviews of progress in the field have been published.[8-12] The intent here is to summarize some of the more significant results and to comment on those phenomena that are reasonably well understood, as well as to point out areas where a more thorough understanding is needed.

The classic work of Smith and Fowler[13,14] illustrates both the achievements and the shortcomings of many of the early experiments. In these experiments, an explosive plane wave generator was used to accelerate a driver plate, or projectile, to impact the specimens, and the shock pulse amplitude was varied by changes in the driver plate thickness or the amount of high explosive. It was found that the shock pulses caused significant hardening and defect generation with little macroscopic deformation. Also, the shock hardening increased with pulse amplitude (i.e., shock pressure) up to some saturation value, an observation that has been confirmed in many subsequent studies.[15,16] Optical examination established that deformation twinning was a prominent feature of the microstructure of shock-deformed copper, α-brass, and Armco iron, an observation that has also been confirmed in a wide variety of materials.[12,16-23]

Unfortunately, the means by which the shock pressure was varied in these early experiments also caused changes in the shock pulse duration. It is now accepted that the pulse duration can have a significant influence on the shock response of a material.[18,24,25] Also, Smith and Fowler[14] found no twins in certain shocked copper single crystals examined optically; however, DeAngelis and Cohen[26] found with transmission electron microscopy (TEM) and x-ray diffraction (XRD) that there was a high density of twins on a scale too fine to resolve optically. Clearly, care must be exercised in interpreting the results reported for recovery experiments because of limitations in experimental control of the shock process as well as in the resolution capabilities of analysis techniques.

Shock Hardening

Despite the lack of macroscopic plastic deformation, numerous studies have established that shock loading can cause hardening effects equivalent to large plastic strains by conventional deformation methods.[12,27,28] The role of shock amplitude on hardening has been studied most thoroughly, with generally consistent results. As an example, it was found that shock hardening in Cu-8.7Ge increases linearly with increasing shock pressure over the range 5 to 30 GPa when the shock pulse duration was held fixed at various values ranging from 0.02 to 1.5 µsec.[29] The same linear dependence can be found in Murr's summary of results for a variety of materials shocked at 20 to 70 GPa with a pulse duration of 2 µsec.[12] At higher pressures, e.g., 80 GPa with Ni and 304 stainless steel, the hardness saturates and may begin to decrease,[12,30] most likely because of adiabatic heating and dynamic recovery effects.[25] One exception to these general observations on shock hardening is the shock softening reported for beryllium shocked in the range 0.3 to 0.9 GPa.[31]

In contrast with the role of shock pressure, the effect of shock pulse duration has received only limited attention. It has been established that changes in pulse duration can have major effects,[18,24,25,29] however, these effects may be more complex than those associated with pulse amplitude.[25] The hardness of Hadfield steel increased with pulse duration changes from 0.5 to 1.2 μsec,[24] that of copper increased over the range 1.2 to 2.2 μsec at 5 GPa,[18] and that of nickel increased between 4 and 17 μsec at 26.5 GPa.[32] The shock hardening response of Cu-8.7Ge, Fig. 1, was found to vary in a complex manner over a wide range of short pulse durations, from 0.017 to 1.3 μsec at pressures from 10 to 42.5 GPa.[25,33]

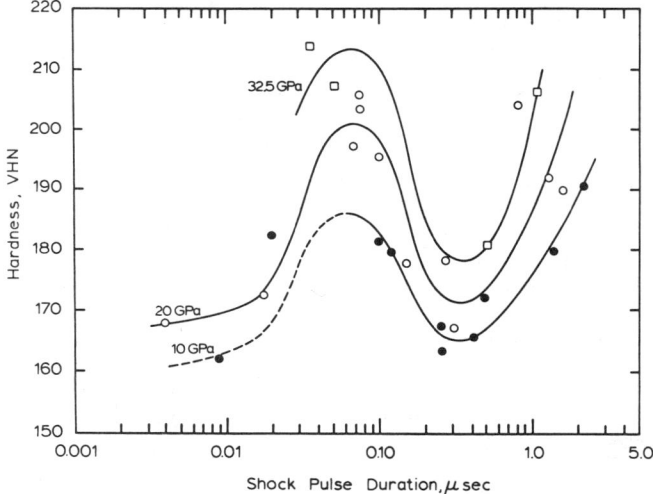

Fig. 1. Variation in hardness of Cu-8.7Ge with shock pulse duration for various shock pressures.[33] Actual pressures for the data points may vary somewhat from the average values given.

The behavior for pulse durations from 0.3 to 1.3 μsec at 20 GPa was similar to that just described; i.e., an increase in hardening with pulse duration. However, the hardness increased on decreasing the pulse duration from 0.3 μsec to 0.07 μsec, reaching a maximum at ∼0.07 μsec with a value near the hardness resulting from the 1.3 μsec pulse. Further decreases in pulse duration from 0.07 μsec caused the hardening to decrease. Substructural studies with XRD and TEM established that this unusual hardening response was caused by the fine structure associated with shock-induced deformation twins. These twinning effects will be discussed in more detail later. The substructural changes in Cu-Ge tend to saturate for pulse durations greater than one μsec.[33,34] This is in agreement with the observation by Murr[35] that the hardening is essentially independent of pulse durations from 0.5 to 6 μsec for a variety of materials.

There are other characteristics of shock pulses that may be important. For example, it has been found that twinning can be activated on different twinning systems during compression and rarefaction.[13,36] Also, Rose et al[15] observed a variation in hardness through the thickness of samples of nickel and concluded that the rarefaction rate, or release rate, was important in hardening response. However, systematic variation of the rarefaction rate with Cu-8.7Ge indicated that it was a second order effect that could be treated as a change in the effective pulse duration.[25]

The shock hardening response of a material is clearly determined by the nature and distribution of defects introduced by the shock pulse. Involved may be dislocations, twins, stacking faults, point defects, or shock-induced second phases, or some combination of these.[8,11,12,15,21,25,37] The remainder of this section will deal with substructural studies relating to all of these, except the introduction of second phases.

Dislocation Substructures

By far the largest number of recovery experiments have been concerned with the behavior of fcc metals. It is generally agreed that, as with conventional deformation, the stacking fault energy of the material determines the nature of the dislocation substructure. Low stacking fault energy materials, such as Hadfield steel,[24] 304 stainless steel,[12] and Cu-8.7Ge,[33] tend to form planar arrays of dislocations. With higher stacking fault energy materials such as copper and nickel, a well-defined cell structure is formed.[12,18,27,38,39] A dramatic exception is aluminum which forms random arrays of heavily jogged dislocations, despite the fact that it has a very high stacking fault energy.[40] This unusual behavior has been explained in terms of the prevention of dislocation cross slip by point defect pinning. While the qualitative features of the dislocation substructures formed by shock loading fcc materials are similar to those formed during conventional deformation, the scale of the substructure is much finer. For example, the dislocation cell size in polycrystalline nickel is almost an order of magnitude smaller than for deformation to a similar strain by cold rolling.[12] As will be discussed later, this can be related to the large deviatoric stresses that are present as the development of the substructure is initiated.

In many cases, increasing the pulse duration has been found to be qualitatively the same as increasing the pulse amplitude. For both nickel and copper the dislocation cell size decreases and the dislocation density increases with increases in either the pulse duration or pulse amplitude.[18,32] Above 40 GPa the dislocation density and cell size in nickel saturate and at very high pressures the dislocation density can actually decrease.[15,30] As mentioned in connection with hardening, this decrease can be related to dynamic recovery and adiabatic heating. Murr and Kuhlmann-Wilsdorf[41] have concluded that the substructure in nickel also saturates for pulse durations greater than a microsecond. The time for saturation was found to be pressure dependent in Cu-8.7Ge shocked at various pressures over a wide range of pulse durations.[34] Rapid increases in dislocation density occurred at short durations, with saturation occurring for times of the order of a microsecond as illustrated in Fig. 2. It has been argued that in many cases the effect of increasing the pulse duration is to permit rearrangement of existing dislocations, rather than to cause additional dislocation generation.[12] This is most likely the case for longer durations; i.e., greater than 2 μsec in most materials.

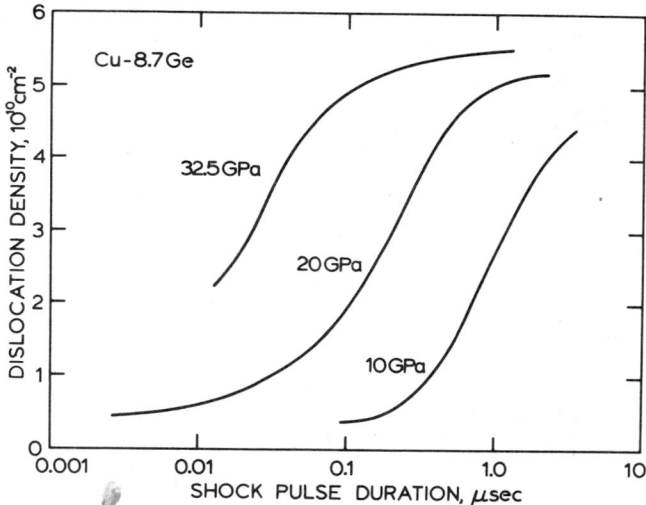

Fig. 2. Schematic showing variation of residual dislocation density with shock pulse duration for Cu-8.7Ge shocked at various pressures.[33,34]

In contrast with fcc metals, the substructures formed in shock loaded bcc metals differ markedly from those formed in conventional deformation. For example, it has been concluded that the shock-induced dislocation substructures in iron[42] and molybdenum[22] resemble those formed by conventional deformation at liquid nitrogen temperature. Iron shock loaded at 7 GPa had a dislocation density of $\sim 10^{10}$ cm^{-2}, which consisted primarily of long screw dislocation segments.[43] This was attributed to the greater mobility of the edge components, which caused loop expansion to leave long screw components behind. The dislocations in molybdenum tend to be arranged randomly,[44] or occasionally in bundles,[22] but with no distinct cell structure; approximately equal numbers of edge and screw dislocations were found.[22]

Planar Defects - Twins and Faults

Shock-induced deformation twinning has been observed in a wide range of metals and alloys including iron,[19,21,45] molybdenum,[22,23] Hadfield steel,[24,46] copper,[13,14,26,38] nickel,[20,47] Cu-8.7Ge,[25,33] Fe30Ni,[48] 304 stainless steel,[12] α-brass,[12] and several alloys based on nickel.[12] Some materials with a high stacking fault energy, such as nickel and copper, had previously been observed to twin only after severe deformation at or below the temperature of liquid nitrogen.[49] In addition to twinning, some materials have also been found to form appreciable numbers of stacking faults.[12,16,25,46]

The incidence of twinning generally increases with increasing shock pressure[12,21,33] and an apparent threshold pressure for twinning has been reported for some materials.[20,21,24] Threshold pressures given for copper range from ∼2 GPa[20] to 34.5 GPa.[50] Johari and Thomas[20] found a threshold pressure of 35 GPa for twinning in nickel specimens that were 76mm in diameter and 6.4mm thick, however, in thin foils of nickel twinning occurred at 3.5 GPa. As was noted previously, in addition to a lack of experimental control, many of the early studies failed to take into account changes in pulse duration. Clearly, the concept of a unique threshold pressure for twinning, that can be separated from experimental factors, remains uncertain.

The effect of shock pulse duration on twinning has been studied for Hadfield steel,[24] 304 stainless steel,[12] and Cu-8.7Ge.[25,34] With the Cu-8.7Ge, increasing the pulse duration from 0.020 to 1.3 μsec, at a constant pulse amplitude of 20 GPa, increased the volume fraction of twins, while the fine structure of the twins also changed. Up to ∼0.7 μsec very thin twins, three to four atom layers thick, formed in groups of four to six making up twin bundles ∼50Å thick. Between 0.07 and 0.30 μsec the thin twins thickened to form more perfect twins ∼50Å thick. These changes caused the maximum in hardening at ∼0.07 μsec shown in Fig. 1. Finally, for longer durations the volume fraction of the ∼50Å thick twins increased.

It has been suggested that there is a minimum pulse duration for twinning.[24,51] However, for molybdenum[22] and Cu-8.7Ge[34] it has been established that twinning and slip occur at very nearly the same time. Also, the fact that twins were observed in Cu-8.7Ge for pulse durations as short as 0.02 μsec,[33,34] (see Fig. 3) indicates that the incubation time in low stacking fault energy materials is very small compared to most experimental pulse durations.

By prestraining molybdenum and iron, Mahajan[24,45] found that shock-induced twinning can be suppressed by as little as 5 percent prior to cold work. It was concluded that twinning occurred when the number of mobile glide dislocations was insufficient to accommodate the imposed strain. The dislocation density was found to be

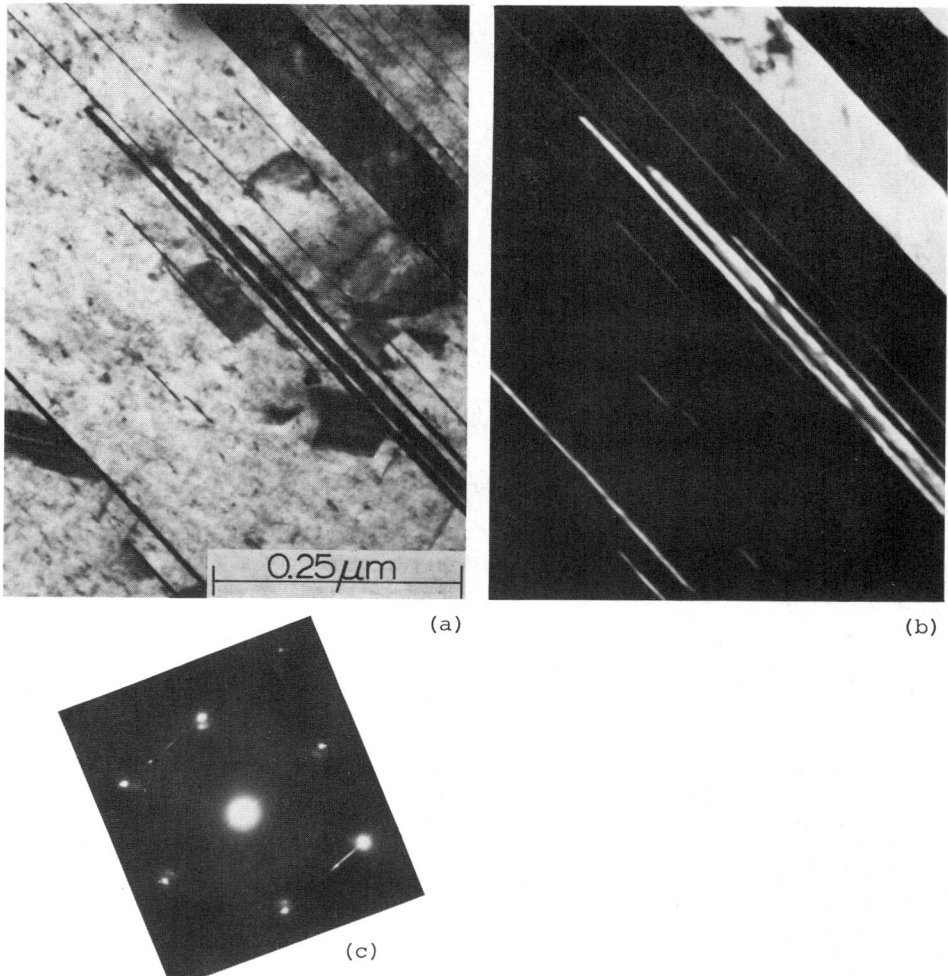

Fig. 3. Bright field (a) and dark field (b) images showing twins in 10 GPa, 0.022 μsec Cu-8.7Ge specimen.[33] Selected area diffraction pattern (c) showing streaking caused by the fine-scale twin structure.

less in the vicinity of deformation twins than in the material as a whole, which was taken as support for these ideas. A similar observation of reduced dislocation densities in the presence of twinning has been made in copper.[20]

Several points should be made concerning these observations on planar defects. First, it must be emphasized that although planar defects constitute an important alternate form of plastic strain, the creation of these defects involves processes based on dislocation interactions and movements. For example, it is generally accepted that dislocation glide (slip) precedes twinning, with the occurrence of twinning being determined by a variety of factors such as: (1) the presence of reactant dislocations to yield an array of partial dislocations to form a twin nucleus, (2) the presence of a heterogeneity or dislocation interaction to aid in nucleation, (3) the surface energy effects associated with the introduction of a twin, and (4) the number of slip systems available and the distribution of glide dislocations among the slip systems. In addition, it should be noted that the relative importance of these factors can vary with experimental conditions, such as the applied stress, temperature, and the metallurgical condition of the specimen. Secondly, because of the nature of twinning mechanisms, the velocities of propagation of twins have the same limitations as those associated with dislocation motion. Considering this and the nature of the twinning shear, the strain rates resulting from twinning are no larger than those arising from normal dislocation motion. Finally, because twinning is accompanied by slip, but not necessarily vice versa, attempts to establish meaningful thresholds in pressure and time for twinning events may not be fruitful. More attention might be given to defining all of the experimental and microstructural conditions leading to twinning in a given material; i.e., creating deformation maps for twinning. In studies of this type, short pulse duration experiments may help to clarify twinning mechanisms in various materials.[34]

Point Defects

The theory of high velocity dislocations predicts that at the velocities achieved in many shock loading experiments, dislocation jogs should be forced to move nonconservatively, thus the formation of vacancies and interstitials should be enhanced over conventional deformation.[52] Experiments on point defect production are difficult to perform because of the fine scale of the defects, or defect aggregates, and because their high mobility can allow them to anneal out at room temperature in many materials. As a result, this is the least understood area in the study of shock-induced crystal defects.

Annealing studies of copper[53,54] and nickel[55] specimens after shocking have shown the presence of appreciable numbers of residual point defects. Interstitial densities of 1.3×10^{-5} and vacancy densities of 6.2×10^{-5} were measured in nickel shocked to 33 GPa at -56°C.[55] A large density of dislocation loops and heavily jogged dislocations have been found by TEM examination of recovered specimens.[40,56] TEM and field ion microscopy study of recovered molybdenum specimens shocked to 25 GPa at room temperature gave dislocation loop densities of 7×10^{14}/cc with a vacancy density of 6.6×10^{-2} per atomic site.[44] It is generally agreed that the concentration of vacancies, and vacancy-type dislocation loops, is considerably higher than the concentration of interstitials and interstitial loops.[40,44,55] However, it should be noted that equal numbers of vacancies and interstitials were found in copper wires shocked at 77°K and kept at that temperature until measurements were made.[54] In a series of carefully controlled experiments, Dick and Styris[57] have used dynamic resistivity methods to measure point defect concentrations created by shock pulses in silver foils. At 10 GPa the vacancy concentration was $\sim 1.5 \times 10^{-3}$ per atomic site. Annealing recovered fragments of the foils indicated that only vacancy clusters remained, with many of the single vacancies having migrated to sinks such as the foil surface.

Although more experiments on point defects are needed, the best experimental approach is uncertain. It may be that low temperature shocking with the use of dynamic resistivity along with annealing studies of the recovered specimens will provide the most information. Positron annihilation has been used effectively recently to study point defects, however, it has the same shortcoming common to other techniques, namely, it is sensitive to the other defects that are created by the shock pulse.[58]

METALLURGICAL EFFECTS IN RELATION TO PARTICULAR EXPERIMENTS

In addition to providing an understanding of metallurgical processes, such as plastic flow, plate impact recovery experiments can yield useful information about shock processes. At the same time, because the initial metallurgical condition of a specimen can influence the results of some experiments, erroneous conclusions about shock effects can result. For this reason in particular, it is necessary that specimen preparation procedures be documented in any published works. The intent of this section is to comment on some specific experiments where metallurgical effects contribute to the understanding of shock processes and material behavior.

Precursor Decay

A problem of defect sources and defect generation?

An area of research that can be affected by metallurgical effects, particularly specimen preparation procedures, is the study of elastic precursor decay under conditions of impact leading to a two wave (elastic and plastic) structure. Typically, these experiments involve measuring the decay of the elastic precursor with propagation distance into the specimen and then modeling the decay process with a decay function based on some form of the Orowan relation, $\dot{\varepsilon}_p = \rho_m v\, b$, that relates the plastic strain rate to the mobile dislocation density, the dislocation velocity, and the slip, or Burgers, vector.[8] The importance that the initial dislocation density can have in determining the nature of the propagating elastic wave was shown clearly by the experiments of Jones and Mote[59] on copper as illustrated in Fig. 4. Although there is general agreement concerning the validity of the precursor decay model, a large number of experiments,[8] particularly those of Duvall and coworkers, have established that in most cases the discrepancies between measured and predicted precursor amplitudes indicate

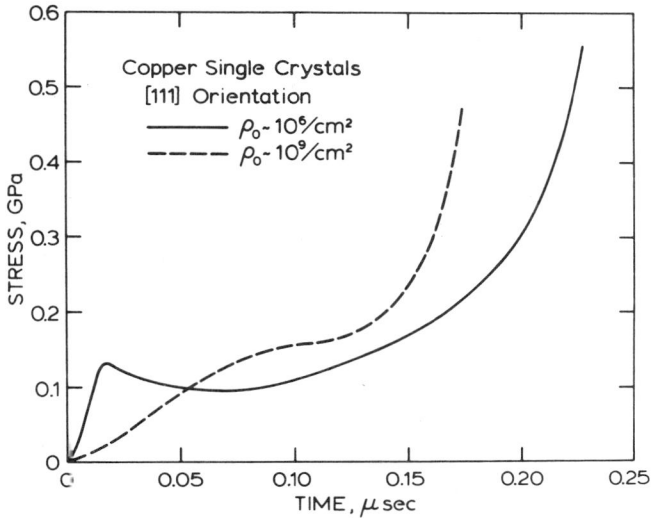

Fig. 4. Schematic showing effect of initial dislocation density on the stress-time wave profile for copper single crystals shocked at 5 GPa along [111]. Additional dislocations introduced by prestraining 3.5% in compression. Original crystal 5.56mm thick, prestrained crystal 4.22mm thick. Prepared from the results of Jones and Mote.[59]

that the initial dislocation density must be increased by two or three orders of magnitude to predict sufficient decay. Based on a recent series of experiments by Clifton[60] in which specimens were recovered to examine the dislocation substructure, it appears that in addition to dislocation generation in the bulk specimen, a contribution to the lower-than-predicted precursor amplitudes can arise from higher dislocation densities formed at the impact surface. These are most likely related to specimen preparation effects of the type discussed earlier and/or to the complexities that must exist on a microscopic scale as the two impacting surfaces meet. This latter point is emphasized by recent observations on α-Cu-Ge specimens that had been carefully prepared and annealed before impact,[33] but which were found to have a very thin (<0.05mm) layer of material at the impact surface that had a higher dislocation density than the bulk.

Whether or not near-surface defect sources make important contributions to precursor decay for a given material may be determined by the microstructure and deformation characteristics of the material. For example, the contribution from surface effects may be small for materials having a small mean free slip distance and a high work hardening rate. It appears that more recovery-type experiments designed to control dislocation generation at the surface through careful choice and preparation of specimen and projectile material, as well as possible use of experimental modifications such as an efficient coupling medium at the impact surface, would be helpful in clarifying precursor decay processes.

Dislocation Generation

Does homogeneous nucleation of dislocations occur? Are there supersonic dislocations?

Conventional deformation experiments have established a linear dependence of residual dislocation density on plastic strain: $\rho = \rho_o + M \varepsilon_p$. Typical values of the multiplication coefficient, M, have been tabulated by Gilman[61] and these have been verified and others added since that time. At the same time the strengthening, $\sigma - \sigma_o$, resulting from the accumulated dislocation substructure has generally been found to obey the relation $\sigma - \sigma_o = \alpha G b \sqrt{\rho}$, where G is the shear modulus and α is a constant. In a typical experiment to study these effects, such as tensile deformation, the imposed average strain rate represents a composite formed by summing over all microscopic elements within the gage section. Obviously, the strain rate for any given element, at any given time, can range from zero to some value that may be orders of magnitude larger than the average strain rate. As a result, little information is available about the time dependence of the processes on a microscopic scale. Plate impact experiments provide a case where each element in the material responds to the imposed shock stress for a very short time that can be controlled in the experiment. Therefore, experiments of this type provide a means of determining lower limits for the dislocation generation rate for various shock amplitudes. As noted by Granato,[62] at high stresses where the dislocation velocity approaches the limiting velocity, the plastic strain rate is approximately proportional to the dislocation generation rate: $\dot{\varepsilon}_p = \rho_m v\, b \cong \dot{\rho} d\, b$, where d is the average distance moved by the dislocations.

There has been considerable debate concerning the processes by which dislocations are generated by a shock pulse. In modeling the stress-time behavior in impact experiments, the strain rate is written in terms of the variation in the elastic and plastic components of strain:

$$\dot{\varepsilon} = \frac{\dot{\tau}}{G} + \frac{3}{2}\dot{\varepsilon}_p$$

The total strain to be accommodated is determined by the impact stress, and for high stresses it is usually assumed that the elastic component is limited to a small fraction of the total strain because of limitations arising from the theoretical strength of the material. However, this approach effectively assumes instantaneous plastic strain, and therefore ignores the time dependence of the plastic processes by which the elastic stresses are relaxed. The

assumption that any process, be it initiation and propagation of
fracture, generation and motion of dislocations, deformation
twinning, or ideal shear (if it exists), can take place instantaneously is physically unreasonable. It is more likely that a
large fraction of the total strain is initially elastic and that
the relaxation of the large elastic shear stresses occurs through
time-dependent plastic flow processes involving dislocation glide
and/or deformation twinning. It is important to note that on the
time scale of impact experiments, the theoretical shear strength
of a material can itself be a time-dependent parameter; i.e., for
times of the order of a nanosecond, or less, the elastic shear
stresses can reach an appreciable fraction of the shear modulus
with subsequent rapid relaxation by plastic flow. Recent experiments in which the residual dislocation density has been measured
as a function of shock pulse duration at constant pulse amplitude
support this argument against instantaneous dislocation generation
within the shock front.[33,34] As shown in Fig. 2, the dislocation
density varies systematically with pulse duration; the generation
rate rises rapidly to a value dependent on the level of shear
stress attained, then decreases as the dislocation density saturates to a value determined by the total strain imposed. If the
dislocations were generated instantaneously within the shock
front, or within the rise time of the pulse, there should be little
dependence of the residual dislocation density on pulse duration.
As with precursor decay, it is suggested that the shear stress
reaches a value determined by the availability of dislocations to
provide plastic flow by slip and/or twinning. Of course, in some
instances processes such as adiabatic shear or phase transformations may intervene. In addition, it should be noted that very
high shock pressures can cause adiabatic heating which can also
provide additional relaxation effects through thermally activated
processes.

Based on the pulse duration dependence of the substructure and
the nature of the substructure, it is suggested that homogeneous
nucleation of dislocations does not contribute importantly to the
plastic flow, nor do the dislocations move at supersonic velocities.
Rather, dislocation generation for a given material occurs at a
rate dependent on the magnitude of the shear stress. At high
stresses a large fraction of available sources and generation
mechanisms are activated, the dislocation velocities are high, and
as a result the generation rate is high. Consequently, the stress
relaxation is rapid causing the generation rate to decrease. In
addition, the creation of the defect substructure, together with
the stress relaxation, leads to a smaller average dislocation
velocity. The variations of ρ, $\dot{\rho}$, and $\bar{v}$ with time for a given
element of material are shown schematically in Fig. 5. Some detailed estimates of the values of $\dot{\rho}$, ε_p, and $\bar{v}$ have been made
recently for shock loaded α-Cu-Ge.[34]

Disordering of Ordered Cu$_3$Au

*Experimental evidence for supercritical shear and homogeneous
nucleation of dislocations?*

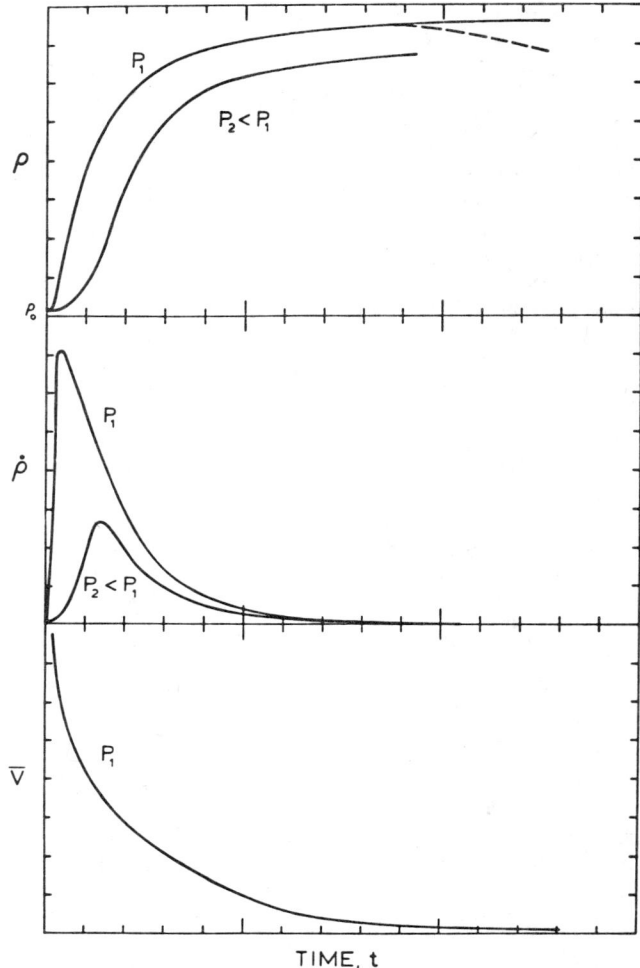

Fig. 5. Schematic representation of the time dependence of the dislocation density, ρ, dislocation generation rate, $\dot{\rho}$, and average dislocation velocity, $\bar{v}$, for Cu-8.7Ge. Dashed line shows expected behavior for a material exhibiting dynamic recovery. Prepared from results given in references 33 and 34.

The early work of Beardmore et al[63] on the effects of shock waves on ordered Cu_3Au has often been cited as evidence for the occurrence of unconventional dislocation behavior during shock loading. In the study, specimens of the alloy in the initially ordered and disordered states were shock loaded at pressures ranging from 16 to 47.5 GPa. The electrical resistivity and stored energy increased rapidly in the range 29 to 37 GPa, Fig. 6, which, based on the calculations of Cowan,[64] was attributed to the creation of supercritical shear stresses at pressures above 29 GPa.

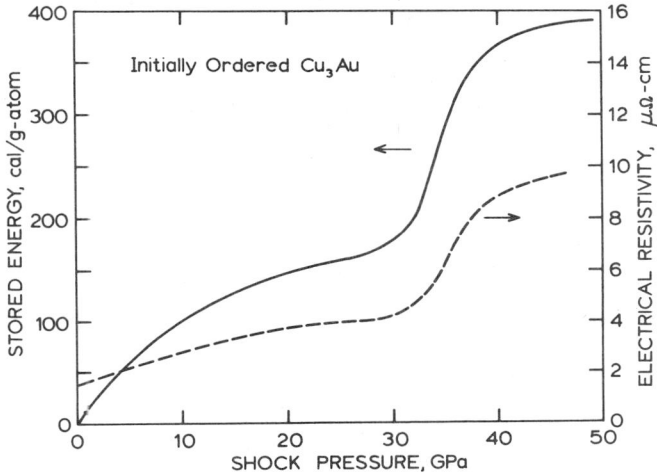

Fig. 6. Summary of results of Beardmore et al[63] showing variation of stored energy and electrical resistivity with shock pressure for initially ordered Cu_3Au. Pulse duration constant at one μsec.

In particular, it was suggested that above 29 GPa dislocations were produced in unconventionally large numbers and on a correspondingly fine scale so as to promote rapid disordering. The possibility of thermal disordering caused by adiabatic heating was dismissed on the basis of both experimental and theoretical considerations. Samples from this study were subsequently examined with electron microscopy and x-ray diffraction.[65] There were no large changes in dislocation density or arrangement over the pressure range where the apparent sharp changes in stored energy and resistivity occurred. Measurements of the degree of order and the antiphase domain boundary density established that most of the disordering resulted from the production of antiphase domain boundaries by the motion of single dislocations. Because the Bravais lattice of the ordered structure is simple cubic, order can only be preserved if the normal $\frac{1}{2}<110>$ fcc dislocations move in pairs called superlattice dislocations. For a variety of reasons this paired movement without disordering involves only a small fraction of the dislocations, so that the preponderant motion of single dislocations causes disordering by cutting the material into ever smaller ordered domains separated by antiphase domain boundary. As the shock pressure is increased giving higher total strains (the pulse durations were of the order of a microsecond), the rate of disordering increases rapidly reaching a maximum and then decreasing as the degree of order becomes small. Calculations of the stored energy from measured antiphase domain boundary densities agreed with the experimentally observed variation. It is interesting to note that the general features of this disordering process are similar to those suggested much earlier by Seeman and Glander.[66] Of importance here is that the disordering process does not require a super-critical stress or homogeneous nucleation of dislocations. Rather, conventional dislocation behavior, coupled with the fine-scale substructure created because of the high shear stresses, leads to a higher rate of disordering per unit strain as compared to conventional deformation.

Spallation

Does the duration of the compressive pulse affect subsequent spallation behavior?

Plate impact experiments have been used to study dynamic fracture under conditions of uniaxial strain at strain rates up to 10^6 s^{-1}. Despite the fact that a stress-strain-strain rate comparison shows wide variations for dynamic and quasi-static fracture tests, the basic microstructural failure processes are apparently the same in all cases.[67] Of interest here is the fact that most plate impact fracture experiments have related the fracture behavior to the amplitude and duration of the tensile stresses causing spallation. In general, the effects of the plastic deformation introduced by the initial compressive shock pulse have been ignored.[68] Because substructural development can be a strong function of pulse duration, it is suggested that the compressive pulse duration can be an important parameter in determining spallation behavior. Unfortunately, changes in the compressive pulse duration can also affect the tensile stress-time variation, however, despite this complication, the complete stress-time sequence affecting a given material element should be able to be established. The argument here is that the effects of the compressive pulse can vary strongly with pulse duration and these effects are not negligible in many cases.

CONCLUDING COMMENTS

Recovery experiments, in which microstructural effects are characterized quantitatively, constitute an important means of increasing the understanding of shock processes. In addition, from a metallurgical viewpoint, these experiments provide information for stress levels and strain rates not generally attainable by other means. As has been emphasized by many authors, particularly recently, it is important to recover and examine specimens from stress-time studies and correspondingly, to characterize the stress-time history of specimens in recovery experiments. It appears that these are the directions being taken currently by many investigators.

The metallurgical design of recovery experiments in the future will undoubtedly exploit those microstructural changes known to have effects on the plastic flow and/or fracture response of materials in conventional deformation. Systematic recovery studies of the effects of such variables as: grain size, temperature, multiphase structures, etc., must be done for a wide variety of materials. Varying the shock pulse duration, particularly in the submicrosecond range, will give additional understanding of the time dependence of atomic-scale plastic flow processes. The re-shock techniques developed recently by Asay and Chhabildas[69] will also provide new information in this area and remain to be exploited in terms of metallurgical effects. In addition, little work has been done on the effects of multiple shocks. An important goal of these recovery experiments should be to provide information that will permit extension of stress-temperature-strain rate maps, the so-called Ashby maps,[70] for a wide variety of materials.

Finally, it may be important to note that with increased understanding of residual microstructures, there exists the possibility for using shock-induced microstructural changes as markers, or gages, to define shock waves in connection with a variety of practical phenomena, such as projectile impacts in structures and in situ rock mechanics experiments.

REFERENCES

1. P. S. DeCarli and M. A. Meyers, Shock Waves and High-Strain-Rate Phenomena in Metals, edited by M. Meyers and L. Murr (Plenum Press, New York, NY, 1981), p. 341.
2. R. N. Orava and R. H. Wittman, Proc. Fifth Int. Conf. High Energy Rate Fabrication (Univ. of Denver, Denver, CO, 1975), p. 1.1.1.
3. A. L. Stevens and O. E. Jones, J. Appl. Mech. 39, 321 (1972).
4. P. Kumar and R. J. Clifton, J. Appl. Phys. 50, 4747 (1979).
5. Metallography, Structures and Phase Diagrams, Vol. 8, Metals Handbook, Eighth Edition (American Society for Metals, Metals Park, OH, 1973).
6. B. J. Hockey, Proc. Brit. Ceram. Soc. 20, 95 (1972); J. Am. Ceram. Soc. 54, 223 (1971).
7. J. J. Gilman, J. Appl. Phys. 30, 1584 (1959).
8. L. Davison and R. A. Graham, Phys. Reports 55, 255 (1979).
9. I. C. Skidmore, Appl. Mat. Res. 4, 131 (1965).
10. E. G. Zukas, Metals Eng. Quart. 6, 1 (1966).
11. W. C. Leslie, Metallurgical Effects at High Strain Rates, edited by R. W. Rohde, B. M. Butcher, J. R. Holland, and C. H. Karnes (Plenum Press, New York, NY, 1973), p. 571.
12. L. E. Murr, Shock Waves and High-Strain-Rate-Phenomena in Metals, edited by M. Meyers and L. Murr (Plenum Press, New York, NY, 1981), p. 607.
13. C. S. Smith, Trans. TMS-AIME 212, 574 (1958).

14. C. S. Smith and C. M. Fowler, *Response of Metals to High Velocity Deformation*, edited by P. G. Shewmon and V. F. Zackay (Interscience Publishers, New York, NY, 1961), p. 309.
15. M. F. Rose, T. L. Berger, and M. C. Inman, Trans. TMS-AIME 239, 1998 (1967).
16. F. I. Grace, M. C. Inman, and L. E. Murr, Brit. J. Appl. Phys. 1, 1437 (1968).
17. D. J. Borich and D. E. Mikkola, *Metallurgical Effects at High Strain Rates*, edited by R. W. Rohde, B. M. Butcher, J. R. Holland, and C. H. Karnes (Plenum Press, New York, NY, 1973), p. 587.
18. A. S. Appleton and J. S. Waddington, Acta Met. 12, 956 (1964).
19. E. Ganin, Y. Komen, and A. Rosen, Mater. Sci. Eng. 33, 1, (1978).
20. O. Johari and G. Thomas, Acta Met. 12, 1153 (1964).
21. J. N. Johnson and R. W. Rohde, J. Appl. Phys. 42, 4171 (1971).
22. S. Mahajan and A. F. Bartlett, Acta Met. 19, 1111 (1971).
23. C. A. Verbraak, *The Science and Technology of Selected Refractory Metals*, edited by N. E. Promisel (Macmillan Co., New York, NY, 1964), p. 219.
24. A. R. Champion and R. W. Rohde, J. Appl. Phys. 41, 2213 (1970).
25. S. LaRouche and D. E. Mikkola, Scripta Met. 12, 543 (1978).
26. R. J. DeAngelis and J. B. Cohen, *Deformation Twinning*, edited by J. P. Hirth and H. C. Rogers (Gordon and Breach Science Publishers, New York, NY, 1964), p. 430.
27. G. T. Higgins, Met. Trans. 2A, 1277 (1971).
28. H. J.Kestenbach and M. A. Meyers, Met. Trans. 7A, 1943 (1976).
29. E. T. Marsh and D. E. Mikkola, Scripta Met. 10, 851 (1976).
30. M. F. Rose and M. C. Inman, Phil. Mag. 14, 925 (1969).
31. J. M. Galbraith and L. E. Murr, J. Mater. Sci. 10, 2025 (1975).
32. A. N. Bekrenev, Z. M. Gelunova, and L. I. Gerasimenko, Phys. Met. Metallography 30, 109 (1970).
33. S. LaRouche, E. T. Marsh, and D. E. Mikkola, Met. Trans. 12A, (1981).
34. R. N. Wright, S. LaRouche, and D. E. Mikkola, *Shock Waves and High-Strain-Rate Phenomena in Metals*, edited by M. Meyers and L. Murr (Plenum Press, New York, NY, 1981), p. 703.
35. L. E. Murr, *Shock Waves and High-Strain-Rate Phenomena in Metals*, edited by M. Meyers and L. Murr (Plenum Press, New York, NY, 1981), p. 753.
36. J. B. Cohen, A. Nelson, and R. J. DeAngelis, Trans. TMS-AIME 236, 133 (1966).
37. T. V. Nordstrom, R. W. Rohde, and D. J. Mottern, Met. Trans. 6A, 1561 (1975).
38. D. C. Brillhart, R. J. DeAngelis, A. G. Preban, J. B. Cohen, and P. Gordon, Trans. TMS-AIME 239, 836 (1967).
39. V. A. Greenhut, P. W. Kingman, and S. Weissman, Microstructural Science 3A, 475 (1975).
40. M. F. Rose and T. L. Berger, Phil. Mag. 17, 1121 (1968).
41. L. E. Murr and D. Kuhlman-Wilsdorf, Acta Met. 26, 847 (1978).

42. W. C. Leslie, E. Hornbogen, and G. E. Dieter, J. Iron and Steel Inst. 200, 622 (1962).
43. E. Hornbogen, Acta Met. 10, 978 (1962).
44. L. E. Murr, O. T. Inal, and A. A. Morales, Acta Met. 24, 261 (1976).
45. S. Mahajan, Phys. Status Solidi (A) 2, 187 (1970).
46. K. Dorph, Scand. J. Metallurgy 6, 38 (1977).
47. R. L. Nolder and G. Thomas, Acta Met. 12, 227 (1964).
48. T. L. Donukis, V. A. Lobodynk, G. I. Savvakin, P. V. Titov, N. P. Fedas, and L. G. Khandros, Phys. Met. Metallography 31, 182 (1971).
49. T. H. Blewitt, R. R. Coltman, and J. K. Redman, J. Appl. Phys. 28, 651 (1957).
50. F. I. Grace, J. Appl. Phys. 40, 2649 (1969).
51. O. E. Jones, Metallurgical Effects at High Strain Rates, edited by R. W. Rohde, B. M. Butcher, J. R. Holland, and C. H. Karnes (Plenum Press, New York, NY, 1973), p. 33.
52. J. Weertman, Response of Metals to High Velocity Deformation, edited by P. G. Shewmon and V. F. Zackay (Interscience Publishers, New York, NY, 1961), p. 205.
53. D. C. Brillhart, A. G. Preban, and P. Gordon, Met. Trans. 1A, 969 (1970).
54. M. A. Mogilevskii, Combustion, Explosion, and Shock Waves 6, 197 (1970).
55. H. Kressel and N. Brown, J. Appl. Phys. 38, 1618 (1967).
56. V. V. Kirichenko and V. N. Rozhanskiy, Phys. Met. Metallography 36, 177 (1973).
57. J. J. Dick and D. L. Styris, J. Appl. Phys. 46, 1602 (1975).
58. G. S. Popov, N. M. Nancheva, and M. R. Minev, Shock Waves and High-Strain-Rate Phenomena in Metals, edited by M. Meyers and L. Murr (Plenum Press, New York, NY, 1981), p. 589.
59. O. E. Jones and J. D. Mote, J. Appl. Phys. 40, 4920 (1969).
60. R. J. Clifton and X. Markenscoff, J. Mech. Phys. Solids 29, (1981).
61. J. J. Gilman, Appl. Mech. Rev. 21, 767 (1968).
62. A. V. Granato, Metallurgical Effects at High Strain Rates, edited by R. W. Rohde, B. M. Butcher, J. R. Holland, and C. H. Karnes (Plenum Press, New York, NY, 1973), p. 255.
63. P. Beardmore, A. H. Holtzman, and M. B. Bever, Trans. TMS-AIME 230, 725 (1964).
64. G. R. Cowan, Trans. TMS-AIME 233, 1120 (1965).
65. D. E. Mikkola and J. B. Cohen, Acta Met. 14, 105 (1966).
66. H. J. Seeman and F. Glander, Z. Metallk. 30, 68 (1938).
67. D. R. Curran, L. Seaman, and D. A. Shockey, Shock Waves and High-Strain-Rate Phenomena in Metals, edited by M. Meyers and L. Murr (Plenum Press, New York, NY, 1981), p. 121.
68. G. L. Moss, U.S. Ballistic Research Laboratory Report 2013, Aberdeen Proving Ground, Maryland, September, 1977.
69. J. R. Asay and L. C. Chhabildas, Shock Waves and High-Strain-Rate Phenomena in Metals, edited by M. Meyers and L. Murr (Plenum Press, New York, NY, 1981), p. 417.
70. M. F. Ashby, Acta Met. 20, 887 (1972).

DEVELOPMENT OF COMPUTATIONAL MODELS FOR MICROSTRUCTURAL FEATURES

L. Seaman
SRI International, Menlo Park, CA 94025

ABSTRACT

Microstructural models have been developed to describe static and dynamic ductile or brittle fracture, wave propagation in composites, phase transformations, and the detonation of explosives. Here an approach to developing such models is presented. The models are continuum models on a macro level, but have additional internal state variables describing the number, size, and other aspects of inclusions, cracks, voids, or hot spots. The models include rules for modifying the internal state variables: these rules are derived from the physical processes of initiation, nucleation, growth, and coalescence. Stress-strain relations are constructed which account for the microstructural features. A ductile facture model is used to illustrate the modeling approach. Advantages of the microstructural approach include the ability to represent very complex, nonlinear processes in a physically reasonable way and the close coordination with experimental results which leads to better understanding of the microprocesses.

INTRODUCTION

Recently, microstructural models have been constructed by several researchers to represent complex material behavior. Typical examples are models for detonation,[1] brittle or ductile fracture,[2-4] compaction of porous materials,[5-6] and wave propagation in composites.[7-8] This paper is an introduction to the development of microstructural models. First, a ductile fracture model is constructed as an example. Then a definition of a microstructural model is given, and the advantages and disadvantages of such models are presented.

DUCTILE FRACTURE MODEL

A model for ductile fracture in metals under impact loading is developed here based on observations of experimental data. First these experiments are examined to determine the microstructural features of interest and the processes these features undergo during fracture. Then the model is outlined to represent these features and processes. Finally the model is tested by combining it with a wave propagation computer program and simulating the experiments under study.

Figure 1, showing the polished cross-section of an impacted target, illustrates the microstructural feature of interest.

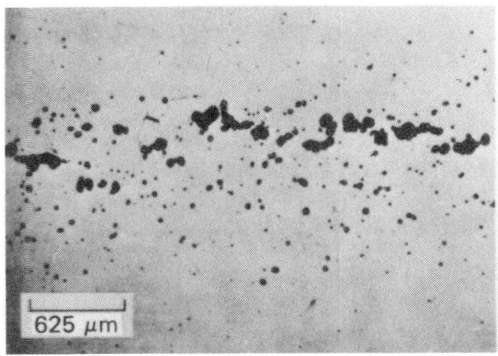

MP-314522-98A

FIGURE 1 SECTIONED TARGET OF 1145 ALUMINUM
IMPACTED AT 145 m/s WHILE AT 400°C

The target is a thin disk of 1145 aluminum impacted on one of its faces by a flyer plate of the same material and half the thickness. Following the impact, compression waves proceed through the thickness and are followed by tensile waves. The duration of the tensile wave is a maximum near the center of the target, and diminishes on either side of the center. From Fig. 1 we see that the damage occurs as circular (spherical) voids. Our interpretations of the processes resulting in this figure are:
- Spherical voids nucleate gradually, so that more voids are seen in regions of longer tensile stress duration.
- Voids grow gradually under the influence of the tensile stress so that larger voids are seen in regions where the tensile stress duration was larger.
- Because the voids are spherical (or circular on the cross-section), the mean stress is the stress component mainly responsible for nucleation and growth.
- At low damage levels, the voids appear to be homogeneously distributed and their growth is essentially independent of the presence of neighbors. Only at high damage is there void interaction and coalescence to form crack-like, anisotropic features.

Having derived this qualitative picture of the microfracture process, we next analyze the data in detail to obtain the processes in a more quantitative form.

The first steps in analyzing the void distribution from one target are illustrated in Figure 2. The left figure shows the sectioned target. For convenience in characterizing the damage, lines have been drawn on the section to separate it into several zones. An average damage in each zone is determined by counting the voids by size and constructing a cumulative size distribution, as shown in the center of Figure 2. Comparison of the distributions for several zones shows that the numbers of voids increase towards the center of the target, and the sizes increase; hence, there appear to be nucleation and growth processes. To generalize these data on the cross-section, we next use Scheil's[9] statistical transformation to compute the probable volume size distribution shown in the right part of Figure 3. Now the voids are represented as a density, number/cm^3: such a quantity can be an internal state variable of the material. For relating these damage distributions to the stress states, we next fit the distributions to the analytical form

$$N_g = N_o \exp(-R/R_1) \qquad (1)$$

where N_g is the number of voids/cm^3 with radii greater than R, N_o is the total number/cm^3 and R_1 is a size parameter.

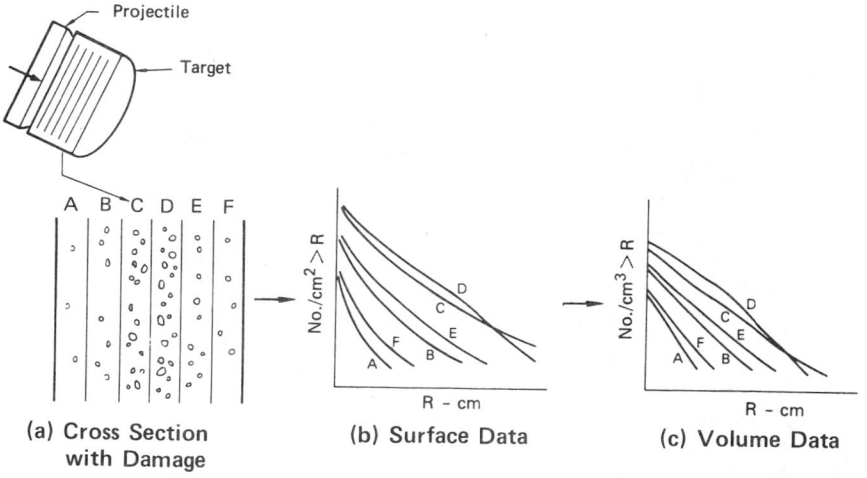

(a) Cross Section with Damage (b) Surface Data (c) Volume Data

JA-314522-12A

FIGURE 2 ACQUISITION AND TRANSFORMATION OF VOID COUNT DATA FROM CROSS SECTIONS OF IMPACTED CYLINDRICAL TARGETS

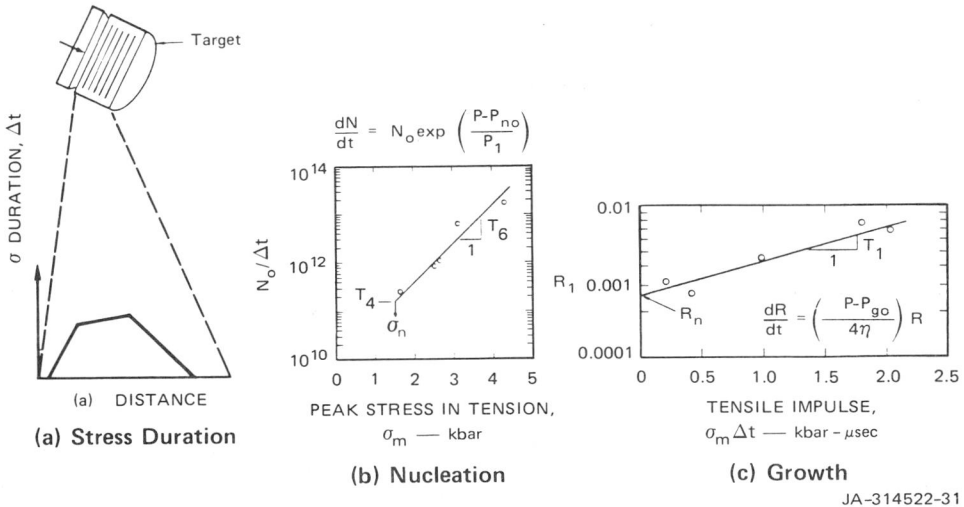

FIGURE 3 COMBINATION OF COMPUTED STRESSES AND DURATIONS WITH OBSERVED DAMAGE TO DETERMINE NUCLEATION AND GROWTH FUNCTIONS FOR A DUCTILE FRACTURE MODEL

Next, the impact is simulated with a wave propagation computer program using a no-damage model for the target material. Stress histories are obtained throughout the target, and especially at the centers of the zones selected for the preceding damage counts. With these histories we can seek relationships between the damage processes and the computed stresses. The nucleation and growth relationships described above are determined quantitatively by forming the plots shown in Figure 3. The coordinates contain combinations of observed damage parameters N_o and R_1, plus the computed quantities P (maximum mean tensile stress) and Δt (duration of the tensile stress). The plotted points are data derived from the planes of maximum damage from a series of impacts at several stress levels (impact velocities) and stress durations (target thicknesses). These plots lead to the following nucleation and growth laws:

$$dN/dt = T_4^* \exp[(P-P_{no})/T_6] \quad (2)$$

$$dR/dt = (P-P_{go})*R/4\eta \quad (3)$$

Here T_4, P_{no}, T_6, η, and P_{go} are taken as new material constants. From the experimental data and the initial simulations we have thus determined the processes and estimates of the material constants.

Besides these empirical relations, we have analytical and other experimental data on the stress-strain relations for use in developing the microstructural model. The stress-strain relations are based on the assumed separation of imposed volume change (ΔV) into components taken by the solid (ΔV_s) and by the voids (ΔV_v):

$$\Delta V = \Delta V_s + \Delta V_v \qquad (4)$$

The mean stress P_s in the solid is computed by the usual Mie-Grüneisen relation:

$$P_s = K_s^* (\rho_s/\rho_o - 1) + \Gamma \rho_s E \qquad (5)$$

where K_s is the bulk modulus, Γ is the Grüneisen ratio, ρ_s and ρ_o are current and initial density and E is the internal energy. Carroll and Holt[10] have given the relation between the average stress P on the section and the solid mean stress P_s:

$$P = P_s \rho/\rho_s \qquad (6)$$

where ρ is the average or gross density, considering both solid and void volumes. We note that this combination of Equations 4 to 6 gives an apparent or gross bulk modulus K which is consistent with the analysis of MacKenzie[11]:

$$K = \frac{K_s}{\alpha + 3K_s^*(\alpha - 1)/4G_s} \qquad (7)$$

where α is the distension ratio and G_s is the shear modulus.

Now the microstructural model can be developed. Until damage begins, there are no microstructural features so the model must provide the usual elastic-plastic response. When the tensile mean stress exceeds the nucleation threshold (P_{no}), damage begins and differential equations 2 and 3 are integrated simultaneously to follow the developing damage. The number and size of voids provide the void volume used in Eq. 4 and the subsequent equations to determine the pressure.

Here we are describing only the material model, not the calculation of the impact and the motion of the material. For solving impact problems, the material model must be coupled with a wave propagation computer program. The program proceeds through the problem with small time steps, providing strains and volume changes to the material model for each material point or computational cell.

The foregoing model was constructed as a computer subroutine and incorporated into one- and two-dimensional wave propagation codes to simulate fracture problems. Having the model we can now verify it by using both the microscopic and macroscopic results. The micro results (the size and number of voids as a function of location in each impacted target) are compared with the observed damage distributions. A sample of such a comparison is shown in Figure 4 for a 14-kbar impact in 1145 aluminum. The micro model was constructed from data of 5 impacts: this figure shows the results for the highest stress case. The measured macro results are stress or particle velocity histories, spall thicknesses, and residual density: these can also be compared directly with the computed results. Cochran and Banner[3] have given samples of this kind of comparison. To accurately define the model parameters and processes, both micro and macro results are compared.

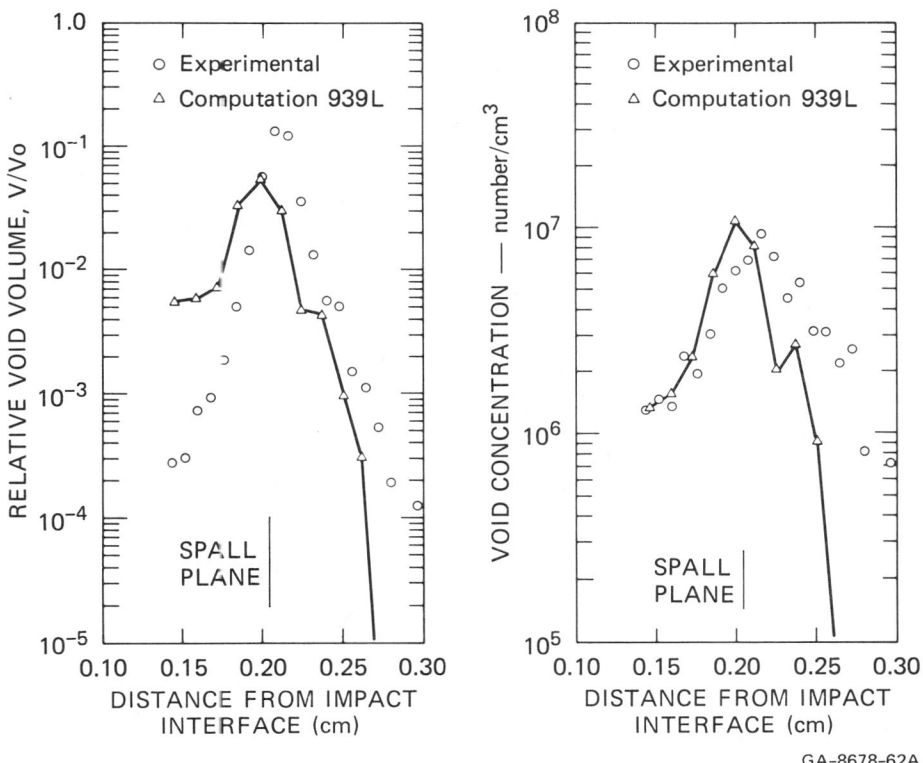

FIGURE 4   COMPARISON OF COMPUTED AND OBSERVED DAMAGE IN A 3.18 mm 1145 ALUMINUM TARGET IMPACTED AT 185.6 m/s

(Impact stress of 14.3 kbar)

The steps in constructing the ductile fracture model were as follows:
- Conduct several experiments with large regions of fairly uniform stress, strain, and damage states. A family of experiments with different load levels and durations are needed.
- Section the targets and observe the damage present. Count the damage, organize it into size distributions as a function of position in the target.
- Simulate the experiments with the best material model available.
- Compare the computed stresses and strains with the observed damage, and assemble the model processes quantitatively.
- Gather theoretical results which may aid in describing the model processes or in the construction of the stress-strain relations.
- Write the microstructural model, including variables describing the microdamage and the rate processes describing the progress of these variables.
- Combine the model with a wave propagation or structural computer program and simulate the series of experiments. Modify the model processes and parameters and repeat the simulations until a satisfactory match is obtained between computational and experimental results.

These steps, or analogous ones, have been used in all of our model development.

OTHER MICROSTRUCTURAL MODELS

Microstructural models have been used in many fields to examine and represent the material behavior. In detonating flow, the microstructural features may be hot spots, voids, shear bands, or grains in the binder. In studies of armor penetration the microdamage features are shear bands, voids, and microcracks. Brittle cracks, layers, and inclusions may be used in rock fracture. In woven composites the fibers, voids produced in forming, and cracks are natural microfeatures.

BRITTLE MICROFRACTURE MODEL

Let us examine a brittle fracture model briefly just to indicate some of the complexities involved. A cross-section of an Armco iron target impacted by an Armco iron flyer to about 40 kbars is shown in Figure 5. There is a high density of cracks and the cracks go in every direction, often intersecting other cracks. So here we have angular distributions as well as size distributions. The size distributions for the Armco iron sample are exhibited in Figure 6: orientation differences are neglected in constructing these distributions for aluminum. But

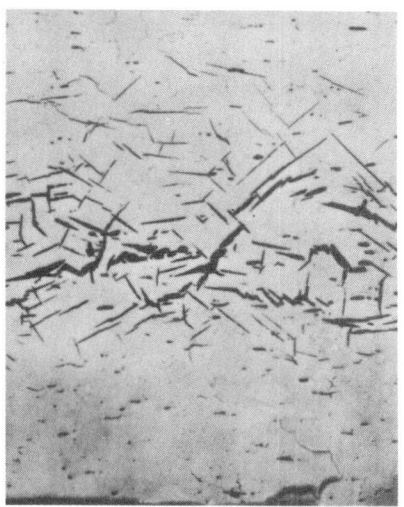

MP-314522-98B

FIGURE 5 CROSS SECTION OF ARMCO
IRON TARGET IMPACTED
AT ROOM TEMPERATURE

Armco iron is quite ductile, although the damage appears as cracks. A crack size distribution for impacted ZnS is shown in Figure 7. Here the cracks tend to be in two groups: large ones of nearly a single size at each location and a distribution of small sizes. The break between the two size groups is probably related to the fracture toughness and the stress level; hence, this material may be brittle enough to follow the patterns of linear fracture mechanics.

The size distributions for brittle fracture must be more complex than those used for the ductile model because the angular distribution and the nonlinearity seen in Figure 7 must be represented. In our brittle fracture model we are using a series of points, instead of an analytical function, to represent the size distribution. Separate distributions are used to represent certain preselected angular orientations. We use 1, 4, 6, or 9 orientations depending on the symmetry and other characteristics of the problem.

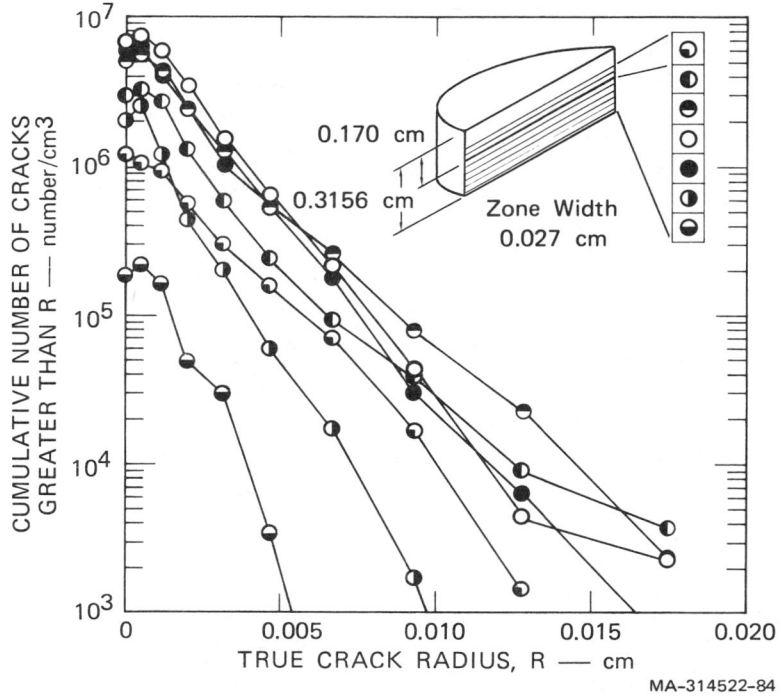

FIGURE 6 CUMULATIVE CRACK SIZE DISTRIBUTIONS AT A SERIES OF LOCATIONS IN A 3.156 mm THICK ARMCO IRON TARGET IMPACTED AT 196 m/s

(Impact stress of 38.4 kbar)

SUMMARY OF MICROSTRUCTURAL MODEL DEVELOPMENT

Having outlined the process of constructing a microstructural model, I'll summarize some of the general features of such models. First, I will give my definition: "A microstructural model is a constitutive model containing internal state variables describing features which are observable in experiments. The model describes one or more processes which modify the internal state variables. Both the micro and macro results of the model can be verified experimentally." This definition is my own. Many current micro models do not fit this definition, usually because the micro-features are not well-defined and therefore not observable or verifiable.

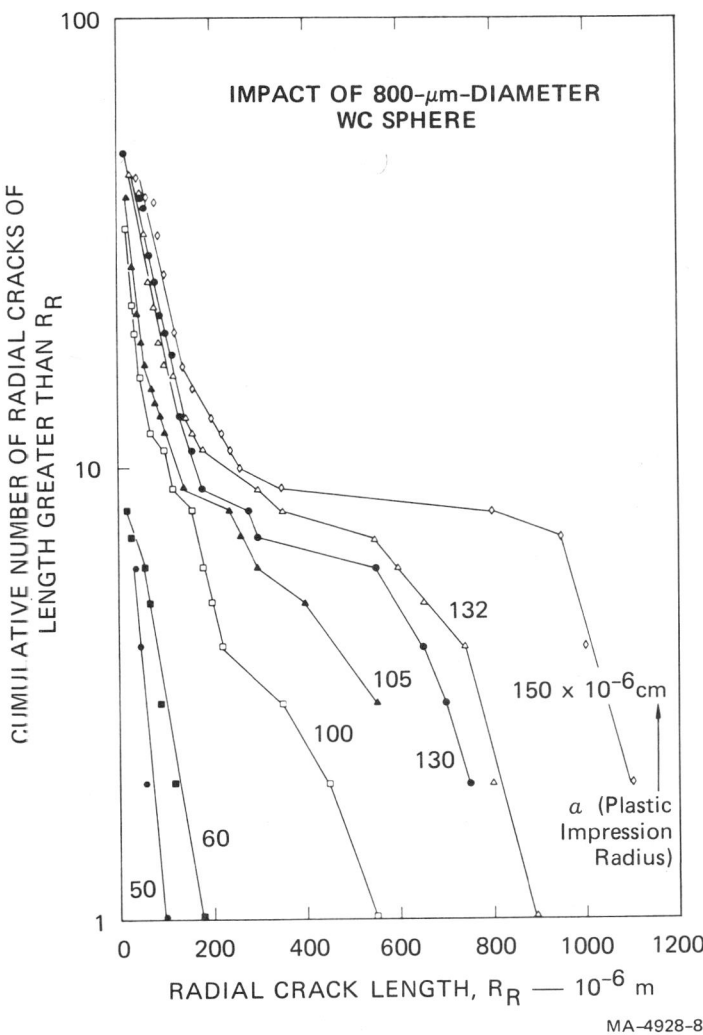

FIGURE 7   MEASURED SIZE DISTRIBUTIONS OF RADIAL CRACKS PRODUCED BY 800-μm-DIAMETER WC SPHERES FIRED AT 6 TO 80 m/s ONTO ZnS TARGETS

There are several difficulties associated with the development of micro models and here are some. First, the experimental observations require closely controlled experiments, painstaking sectioning and counting of microstructural features, and usually transformation of the counted surface features to a volume basis. Hence, there is a lot of work involved in the experimental part of the model development, and there must be an experimental part!

Another problem with microstructural models is the number of added variables per cell required by the model. The ductile fracture model requires two extra variables. But the brittle fracture model usually needs 20 or more extra variables. These numbers of variables may make it difficult to run finite-difference or finite-element programs on large problems with currently available computers. Also, the microstructural models tend to be more complex and require more computer time than a standard elastic-plastic model. In problems I have run the microstructural model calculations often double the computer time.

Uniqueness of the model parameters is the concern of many developers. "With 5 parameters I can match anything" is the conventional wisdom. Uniqueness is a valid concern and I have no proof of uniqueness with which to answer. But I have several points to consider:

- The basic experiments should span the range of stresses, durations, strains states, etc. of interest. Otherwise some aspects of the model are not adequately tested.
- The processes and parameters can be made to represent real, physical processes and properties wherever possible. This requirement greatly limits the flexibility in fitting any given observed result.
- In fitting a model to data, I always do some variation-of-parameters study to determine the parameters. Usually the results of varying each parameter are quite distinct.
- The observable micro and macro results to be fitted by the model are much greater than can be fitted by a few arbitrary parameters, I think. The only possibility for fitting these results is that the model processes approximate the real processes fairly well.

I think the concern for uniqueness is a valid one, but one that can be answered satisfactorily as part of the model development.

There are many advantages to using a microstructural model instead of a continuum model. Such a model gives the analyst the means to use the metallurgical or microscopic observations being provided him by his associates working on the material aspects of the problem. A second advantage is the relative ease with which a micro model can follow very complex stress-strain paths. Complexities encountered are in the representatation of the physical processes rather than in the mathematics. An important advantage is the possibility of using the model outside its range of calibration and in different environments: if the physical processes are reasonably correct, then the model should be extendable in this way. But I think that the most important advantage over continuum models is the material insight which comes from dealing with observable microstructure in the model. The model results deal directly with information required in material design and selection.

The next steps in the work with microstructural models will be to relate the new process parameters to more fundamental material properties. This step will make the models truly predictive.

REFERENCES

1. M. E. Kipp, J. W. Nunziato, R. E. Setchell, and E. K. Walsh, in Seventh Symposium on Detonation, edited by Segmiund J. Jacobs (Annapolis, Maryland, June 1981), Vol. II, p.604.
2. L. Seaman, D. R. Curran, and D. A. Shockey, J. Appl. Phys. 47, 4814 (1976).
3. S. Cochran and D. J. Banner, J. Appl. Phys. 48, #7, 2729, (1977).
4. L. Davison and A. L. Stevens, J. Appl. Phys., 44, 668 (1973).
5. M. M. Carroll and A. C. Holt, H. Appl. Phys., 43, 1626, (1972).
6. J. F. Schatz, in The Effects of Voids on Material Deformation, edited by Stephen Cowin and M. M. Carroll, (Amer. Soc. of Mech. Engin, 1976), p. 141.
7. D. R. Curran, L. Seaman, and M. Austin, J. Comp. Mat., 8, 142 (1974).
8. G. A. Hegemier and G. A. Gurtman, J. Appl. Phys., 45, 4254 (1974).
9. E. Schell, Z. Metallh., 27, 199 (1935).
10. M. M. Carroll and A. C. Holt, J. Appl. Phys., 42, 759 (1972).
11. J. K. MacKenzie, Proc. Phys. Soc., Sect. B, 63, 2 (1950).

TWINNING IN IRON BY LASER GENERATED SHOCK WAVES

F. Cottet and J.P. Romain
Laboratoire d'Energétique et Détonique, L.A. 193
E.N.S.M.A. - C.E.A.T. 86034 POITIERS Cedex - France.

ABSTRACT

Twins are generated in iron targets by shock waves from laser pulses of incident intensity between $5.7\ 10^{11}$ and $1.3\ 10^{14}$ W/cm^2. The successive phases of the compression process : shock formation, increase and decay of shock amplitude are described by a numerical simulation. An interpretation of the observed twin distribution is given in correspondence with the possibility that an elastic precursor, associated with twinning, separates from the shock front at various depth in the target. The penetration depth of the shock wave is estimated from the extent of the twinned zone, depending on the rise-time and amplitude of the applied peak pressure, related to the width and intensity of the laser pulse and to geometry effects. A depth of about 900 μm was found for a 35 J/3.5 ns laser pulse in a 120 μm diameter spot, generating a maximum shock pressure of the order of 8 Mbar.

INTRODUCTION

In a previous work[1], we have shown that iron targets submitted to laser-shocks undergo twinning deformation. This effect is similar to that produced by conventional shocks and is associated with the propagation of an elastic-plastic precursor of the shock wave. The results obtained show that metallurgical observations, together with numerical simulation of the shock compression process, may be used in order to determine approximately the pressure induced in the target by laser pulses.

We have recently developed a more complete model of the formation and decay of laser-generated shock waves[2]. This model is applied in the present work for analysing new experimental results on metallurgical effects of shock waves generated in iron targets by laser pulses of $5.7\ 10^{11}$ to $1.3\ 10^{14}$ W/cm^2 intensity.

The process of generating shocks in a solid target with the use of high power pulsed laser has been largely described[3,4,5]. The shock profile in the target depends on the variation of the pressure applied on the target surface and is governed by the laser pulse shape. Our hydrodynamic model assumes, for simplicity, that the laser pulse is triangular in shape. The successive phases of the compression process, resulting from such initial conditions, are : shock formation, shock amplitude increase and shock decay. For laser pulses in the nanosecond regime, the shock wave formation is achieved within a few microns depth in the target. During the next sequence, the shock amplitude increases and reaches its maximum value at a few tens of microns from the front face of the target. Then rarefaction waves, issued from the pressure drop at the ablation front, overtake the shock front and the shock amplitude decreases.

METALLURGICAL OBSERVATIONS IN LASER-SHOCKED IRON SAMPLES

Experiments were performed with the GRECO ILM[6] neodymium glass laser. Annealed samples of 99,95 % pure iron were irradiated with 3.5 ns (FWHM) laser pulses focused on their front face in 0.1 to 1.5 mm diameter spots. Incident intensities were in the range of 5.7×10^{11} to 1.3×10^{14} W/cm^2.

Figure 1 shows the cross-section of a sample submitted to a pulse of 1.3×10^{14} W/cm^2 intensity, focused in a spot of about 100 µm diameter. The crater depth is 140 µm. Twins appear at about 40 µm distance from the crater bottom. The extent of the twinned zone is about 700 µm. Micrographs 1, 2, 3 and 4 of Fig. 1 show the decay of twins density at increasing distance. This evolution corresponds to the decay of shock amplitude. The spherical limits of the twinned zone are a consequence of the spherical evolution of the shock wave.

The results of the metallurgical observations are reported in table I. A set of samples was irradiated by laser pulses of the same energy and same length, but focused in spots of 0.1 to 1.5 mm diameter corresponding to input intensities of 1.3×10^{14} to 5.7×10^{11} W/cm^2.

DISCUSSION

Assuming that, as in conventional shocks, twins are produced by the elastic-plastic precursor, the zone free of twins observed just under the crater of sample 1 is due to a peak shock pressure exceeding 660 kbar and the disappearance of twins at 870 µm from the front face of the sample is due to a complete attenuation of the shock wave. More information is obtained from the crater depth, the distance at which the shock pressure is high enough to produce the melting of iron. Data of approximate shock pressures necessary for melting are available for several materials[7]. For iron, the value is about 2.5 Mbar. The corresponding point has been reported on

Table 1. Experimental results. Distances are measured from the irradiated surface.

Sample number	Spot diameter (mm)	Input intensity (W/cm^2)	Crater depth (µm)	Front limit of twinned zone (µm)	Back limit of twinned zone (µm)
1	0.1	1.3×10^{14}	140	180	870
2	0.5	5.7×10^{12}	30	30	1130
3	1	1.4×10^{12}	0	0	1050
4	1.5	5.7×10^{11}	0	0	730

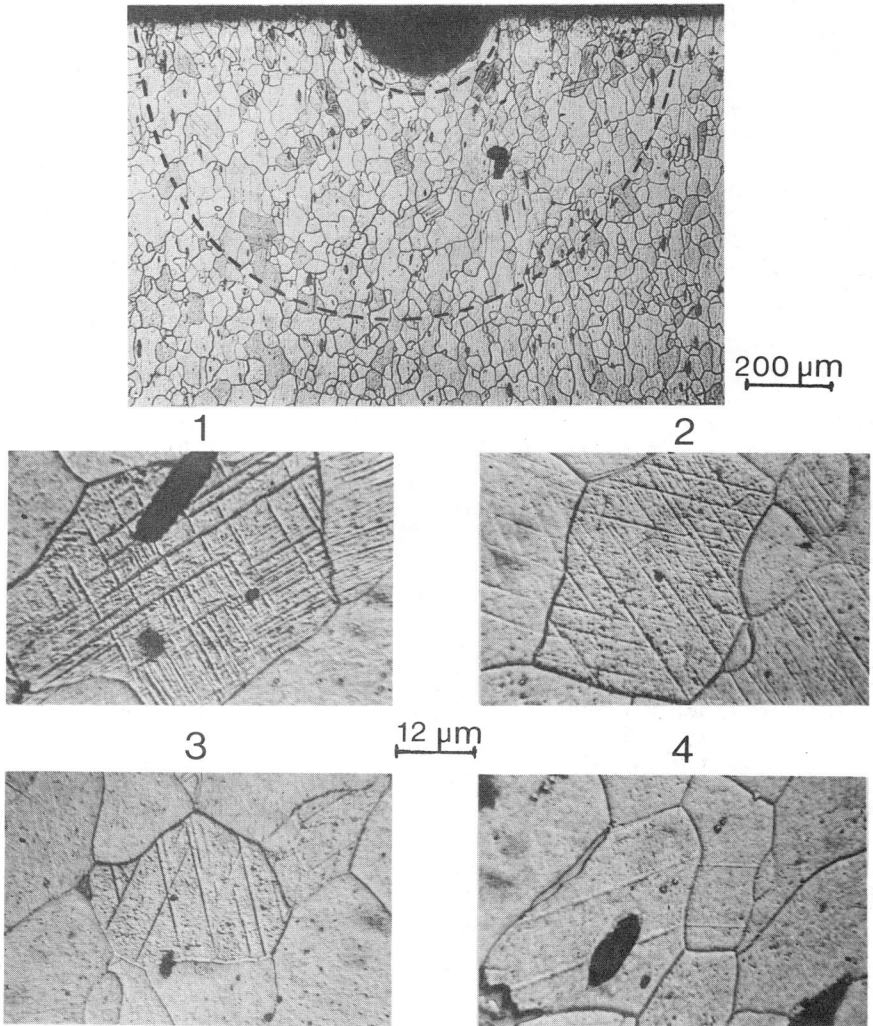

Fig. 1. Cross section of an iron sample irradiated with a
1.3 10^{14} W/cm^2, 3.5 ns laser pulse focused on a spot diameter of
100 µm. Dashed lines represent the approximate limits of the twinned zone. Micrographs 1,2,3 and 4 show the twin-density evolution as a function of distance. 1 : front limit of the twinned zone, 4 : back limit, 2 and 3 : intermediate distances.

the pressure-distance diagram of Fig. 2. The calculated shock amplitude evolution in best agreement with the experimental results shows that the maximum induced pressure should be about 8 Mbar for a laser intensity of $1.3 \; 10^{14}$ W/cm^2. The sequence of shock amplitude increase was calculated in plane geometry. This assumption is justified by the short distance, 70 µm, at which the shock pressure is maximum. Shock decay was calculated in spherical geometry assuming an initial radius of curvature of 50 µm.

Assuming that the shock pressure P is related to the absorbed intensity I by [8]:

$$P \propto I^{2/3}$$

and with 75 % absorption, the maximum shock pressure for experiments 2, 3 and 4 could be evaluated between 0.5 and 1.5 Mbar.

From the experimental results reported in table I, it appears that the shocked zone extent is larger for experiments 2 and 3 than for experiment 1, although the maximum induced pressures are lower. This result is interpreted as a consequence of geometry effects. High pressure shock waves, obtained by focusing the laser pulse on small spot diameters, propagate from a small source region in the form of spherical fronts which attenuate rapidly. With larger focal spots, the incident intensity and consequently the initial pressure are lower, but the planarity of the shock wave is better and the rate of pressure decay is much slower.

It can be also observed that for experiments 2 and 3 there is no zone free of twins under the crater, although the estimated induced presure exceeds 660 kbar. A possible interpretation is given by the diagram of Fig. 3 showing the calculated evolution of the shock profile for maximum pressures of 1 Mbar and 1.5 Mbar in plane geometry. During the increase of the applied pressure on the front face of the target, the elastic wave separates from the shock wave and is overtaken by the shock front if the maximum pressure is high enough (above 1.5 Mbar, diagram 2). When the maximum

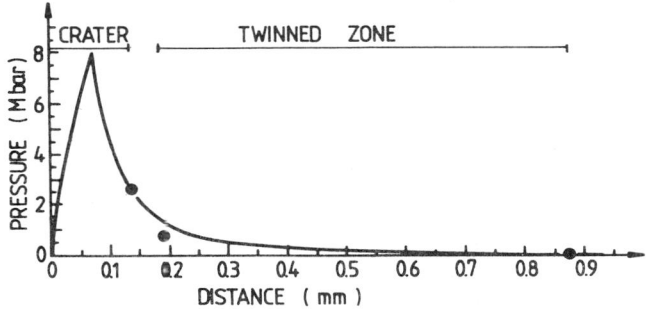

Fig.2. Shock pressure evolution : calculated curve in spherical geometry and experimental results for sample 1. From the comparison of calculated and experimental results, the maximum pressure is evaluated to be about 8 Mbar.

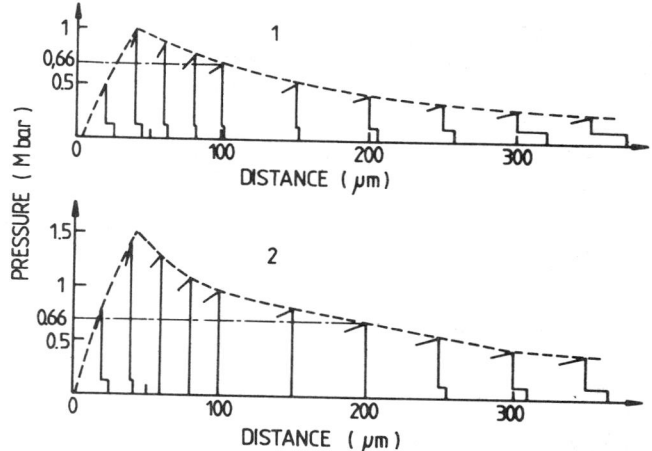

Fig.3. Shock profile evolution calculated in plane geometry.
(1) : maximum shock amplitude 1 Mbar, (2) : maximum shock amplitude 1.5 Mbar. The pressure scale is only valid for the shock wave, the amplitude of the elastic wave is exagerated for a better visualisation.

shock pressure is too low (1 Mbar, diagram 1), the shock wave cannot overtake the elastic wave and consequently twinning occurs throughout the target.

The study of microstructural and mechanical property changes, resulting from laser generated shock waves in solid targets, give information on the induced pressure and the shock profile evolution during its propagation through the target.

REFERENCES

1. J.P. Romain and F. Cottet, in High Pressure Science & Technology (Pergamon Press, Paris, 1980), p. 968.
2. F. Cottet, Thèse n°794, Université de Poitiers, UER-ENSMA,(1981)
3. H. Motz, The Physics of Laser Fusion (Academic Press, London, 1979),p. 145.
4. R.J. Trainor, H.C. Graboske, K.S. Long and J.W. Shaner, Lawrence Livermore Laboratory Report n° UCRL - 52562, (1978).
5. R.J. Trainor, J.W. Shaner, J.M. Auerbach and N.C. Holmes, Phys. Rev. Lett., $\underline{42}$, 1154 (1979).
6. Groupement de Recherches Coordonnées Interaction Laser-Matière, Ecole Polytechnique, 91128 Palaiseau, France.
7. R.C. Shroeder, W.M. Mc Master, Lawrence Livermore Laboratory, Report n° UCRL-51253 (1972).
8. C. Fauquignon, F. Floux, Phys. Fluids, $\underline{13}$, 386 (1970).

STRESS AND TEMPERATURE-DRIVEN NUCLEATION OF MICROSCOPIC VOIDS IN METALS

D. R. Curran
SRI International, Menlo Park, CA 94025

ABSTRACT

Nucleation of microscopic voids in metals obeys similar equations in high-rate plate impact experiments and high-temperature creep experiments. It is postulated that the common mechanism in both cases is vacancy diffusion. Supportive evidence is presented from plate impact data and creep data for aluminum and copper.

I. INTRODUCTION

Microstructural models of void nucleation in solids are basically of two types: (1) tensile stress/temperature-driven and (2) deformation-(plastic strain) driven. Stress/temperature-driven nucleation models are based on the diffusion of vacancies into clusters to form and grow microscopic voids. Such stress/temperature processes are expected to be competitive with deformation-driven nucleation in cases where the temperature is high or where the tensile mean stress is high compared with the deviator (shear) stress or strain. Two such cases at opposite ends of the strain rate spectrum are high-temperature creep rupture of tensile bars and room temperature tensile failure in plate impact experiments.

In the following paragraphs we present the case for stress/temperature-driven vacancy diffusion governing nucleation in both high temperature creep and room temperature plate impact experiments.

II. STRESS/TEMPERATURE-DRIVEN NUCLEATION MODELS

Hull and Rimmer,[1] Raj and Ashby,[2] and Raj[3] worked out the nucleation rates to be expected from tensile stress/temperature-driven vacancy diffusion in grain boundaries at two-, three-, and four-grain junctions and at inclusions in grain boundaries. The threshold initiation condition appears naturally as the attainment of a critical void size, where the criticality condition is a maximum in the free energy function. The initiation condition for a spherical void is

$$\sigma_m d_v = \sigma_m 4\pi v^2 dv > \gamma da = \gamma 8\pi v dv \quad (1)$$

where dv are da are the increase in void volume and areas; σ_m is the mean tensile stress; γ is the void surface energy; and r is the void radius. Thus, the critical void size at nucleation is

$$r_c = 2\gamma/\sigma_m . \quad (2)$$

Raj and Ashby show that this result also holds for non-spherical voids and that the corresponding free energy change is given by

$$\Delta G_c = r_c \sigma_m F_v(\alpha)/2 \quad , \quad (3)$$

where $F_v(\alpha)$ is a geometry factor that depends on the void shape. That is,

$$v = r^3 F_v(\alpha) , \quad (4)$$

where α is the angle formed at the junction of the void and the grain boundary.

The nucleation rate is then derived by Raj and Ashby from the stress/temperature-driven rate theory discussed earlier. First, the number of critical nuclei per unit area in the grain boundary is given by

$$\rho_c = \rho_{max} \exp(-\Delta G_c/kT) , \quad (5)$$

where ρ_{max} is the maximum number of potential nucleation sites in the grain boundary per unit area. Next, the rate P_t of adding vacancies to the critical nucleus is

$$P_t = \frac{4\pi\gamma}{\sigma_m \Omega} \cdot \frac{D_B \delta}{\Omega^{1/3}} \exp(\sigma_m \Omega/kT) , \quad (6)$$

where D_B is the grain boundary self-diffusion coefficient (units: $m^2 s^{-1}$), δ is the grain boundary thickness ($\sim 4 \times 10^{-10}$ m), and Ω is the atomic volume ($1.1 \times 10^{-29} m^3$ for copper).

The nucleation rate is then obtained by forming the product of Eqs. (5) and (6). That is, the nucleation rate is the number of critical nuclei times the frequency that vacancies are added to the critical nucleus:

$$\dot{\rho} = \frac{4\pi\gamma}{\sigma_m \Omega} \cdot \frac{D_B \delta}{\Omega^{1/3}} \rho_{max} \cdot \exp \frac{\sigma_m \Omega - 4\gamma^3 F_v/\sigma_m^2}{kT} \quad (7)$$

In applying this equation Raj and Ashby neglected the $\sigma_m \Omega$ term in the exponential in comparison with the $4\gamma^3 F_v/\sigma_m^2$ term. For high stress applications, it is necessary to retain this term and also convert from $\dot{\rho}$ (number per unit area per unit time) to $\dot{N}$ (number per unit volume per unit time). To do so we multiply $\dot{\rho}$ by one half the ratio of the grain boundary area to the grain volume. The factor of one half arises because each grain boundary shares two grains. For spherical grains, this factor is 3/d, where d is the grain diameter. Thus Eq. (7) is rewritten as

$$\dot{N} = \frac{4\pi\gamma}{\sigma_m \Omega} \cdot \frac{D_B \delta}{\Omega^{1/3}} N_{max} \exp \frac{\sigma_m \Omega - \gamma^3 F_v/\sigma_m^2}{kT} \quad (8)$$

where $N_{max} = (3\rho_{max}/d)$ m^{-3}. The second term in the exponential acts as a stress threshold for creep failure. It is of interest to plot $\dot{N}$ versus σ_m from Eq. (8) over the whole stress range, from the low stresses that apply in creep experiments to the high stresses that apply in impact experiments. This plot has been made for copper in Figure 1 using the following parameters for copper taken from the papers by Raj and Ashby[2] and Raj:[3]

γ = surface energy = 1 J/m^2 .
Ω = atomic volume = 1.2×10^{-29} m^3 .
ρ_{max} = maximum number of nucleation sites per unit area of grain boundary = 10^{17} m^{-2} .
δD_B = boundary width multiplied by grain boundary self-diffusion coefficient.

$\delta D_B = (4)(10^{-15}) \exp\left(\frac{-24.8 \text{ kcal mol}^{-1}}{RT}\right) \text{ m}^3 \text{ s}^{-1}$.

F_V = shape factor for the nucleated void. Two values were chosen for comparison, 10^{-5} and 10^{-3}.

d = grain diameter. In the plate impact experiments on OFHC copper, an appropriate value is 10^{-4} m.

$N_{max} = \frac{3\rho_{max}}{d} = 3 \times 10^{21}$ m^{-3} = concentration of possible nucleation sites.

Figure 1 shows the initial steep rise in nucleation rate as the stress exceeds a threshold, as described by Raj and Ashby. Later, the exponential function of $\sigma_m \Omega / kT$ takes over and causes a rapid rise in $\dot{N}$.

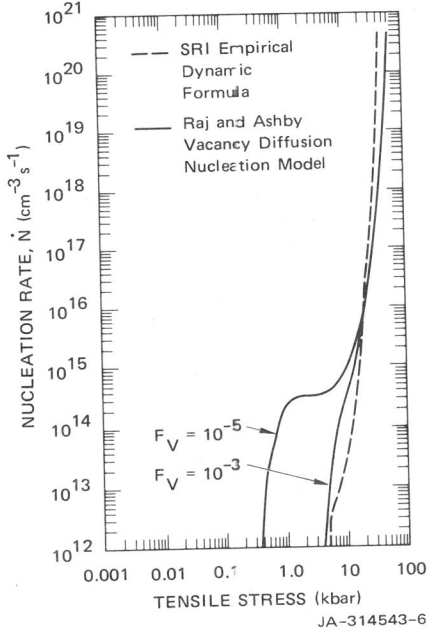

FIGURE 1   NUCLEATION RATE VERSUS STRESS FOR OFHC COPPER (T = 300°K)

The most critical parameter in the low stress region of Raj and Ashby's model is the void shape factor F_V. However, as shown in Figure 1, at high stresses the threshold is no longer important, hence values of $F_V = 10^{-3}$ and 10^{-5} both provide good agreement with the empirical nucleation rate function for OFHC copper obtained in our laboratory plate-impact experiments (Seaman et al[4]). We found that the experimentally observed nucleation rates for 1145 aluminum and OFHC copper were well described by the formula:

$$\dot{N} = \dot{N}_o \exp(\sigma_m - \sigma_{no})/\sigma_1, \quad (9)$$

where $\dot{N}_o$, σ_{no}, and σ_1 are empirically measured material properties. Comparison of Eq. (9) and Eq. (8) shows that at high stresses the plate impact parameter σ_1 must be identical to kT/Ω.

The comparison of σ_1 and kT/Ω for three commercially pure metals studied in our laboratory is given in Table 1.

We see that the agreement is only within an order of magnitude. However, values of σ_1 obtained from plate impact experiments were over a very narrow stress range (1 to 5 GPA) and are therefore subject to uncertainties. Furthermore, as discussed earlier, stress concentrations at impurities could alter the effective value of σ_m.

Another test of the ability of Eq. (8) to describe plate impact results is to examine the predicted temperature dependence of the nucleation rate. Comparison of Eq. (8) and Eq. (9) shows that at

TABLE I Comparison Of σ_1 Derived From
Plate Impact Tests With kT/Ω

Metal	σ_1 (MPa)	kT/Ω (MPa)
1145 Al	40	250
OFHC Cu	200	350
Armco Fe	520	360

high stresses, where
$$\sigma_m \Omega >> 4\gamma^3 F_v/\sigma_m^2 ,$$
σ_1 must equal kT/Ω ; also,
$$\dot{N}_o = \frac{4\pi\gamma}{\sigma_m \Omega} \cdot \frac{D_B \delta}{\Omega^{1/3}} \cdot N_{max} .$$

The most temperature-dependent material parameter in the above equation is D_B, which is proportional to $\exp(-\Delta G/kT)$, where ΔG is the free energy for grain boundary self-diffusion. Thus if plate impact fracture tests were performed at different ambient Kelvin temperatures T_1 and T_2, the ratio of measured values of σ_1 should be given by

$$\frac{\sigma_1(T_2)}{\sigma_1(T_1)} = \frac{T_2}{T_1} , \qquad (10)$$

and

$$\frac{\dot{N}_o(T_2)}{\dot{N}_o(T_1)} \quad D_B(T_2)/D_B(T_1) . \qquad (11)$$

Such tests have been performed on 1145-0 aluminum samples at ambient temperatures of 293K and 673K by Shockey et al. The measured fracture parameters for Eq. (9) are given in Table II.

TABLE II. Dynamic Fracture Properties of 1145-0
Aluminum At Two Temperatures

	293K	673
Nucleation Threshold Stress σ_{no} (MPa)	300	300
Threshold Nucleation Rate $\dot{N}_o$ (n_o/cm^3 sec)	3×10^9	2×10^{12}
Characteristic Stress σ_1 (MPa)	40	96

For the case of σ_1, we see that the ratio $\sigma_1(673K)/\sigma_1(293K) = 2.4$, is in good agreement with the ratio $T_2/T_1 = 2.3$ predicted by Eq. (10). Of course, we recall that the measured absolute value of σ_1 was not in close agreement with kT/Ω (see Table 1); hence, the agreement with Eq. (10) may be fortuitous.

To examine the agreement of the values in Table II with Eq. (11), we must know the temperature dependence of the grain boundary self-diffusion coefficient D_B. Diffusion data have been recently reviewed by Brown and Ashby[6] (1980). Lattice diffusion dominates over grain boundary diffusion at higher temperatures, and between 293K (T/T_m = 0.31) and 673K (T/T_m = 0.72) the diffusion coefficient increases by many orders of magnitude (about 8 for grain boundary diffusion and 15 for lattice diffusion). This can be compared with the measured increase in $\dot{N}_o$ (see Table II) of about three orders of magnitude.

Thus, the increase in $\dot{N}$ with temperature in plate impact data is in the direction predicted by vacancy diffusion, models, but is five orders of magnitude lower than predicted, although incubation time effects, for which there are similarly large uncertainties, might improve the agreement.

In conclusion, vacancy diffusion theories produce nucleation rate functions similar to the function of Eq. (9), which was derived from plate impact data. This suggests that one nucleation function could handle a wide range of stresses and loading rates. However, significant uncertainties in both the experimental data and the theoretical models make conclusions premature. The relative importance of thermal/stress-driven vacancy diffusion processes and deformation-driven nucleation processes must be determined by future work.

ACKNOWLEDGEMENT

Thanks are due to R. Caligiuri for helpful discussions.

REFERENCES

1. Hull, D., and Rimmer, D. E., Phil. Mag., 4, 673, (1969).
2. Raj, R., and Ashby, M. F., Acta Met., 23, 653-666 (1975).
3. Raj, R., Acta Met. 26, 995-1006 (19878).
4. Seaman, L., Curran, D. R., and Shockey, D. A., J. Appl. Phys., 47, 4814-4826 (1976).
5. Shockey, D. A., Seaman, L., Curran, D. R., and Dao, K. C., SRI Final Report, Contract DAAH01-75-1072, for U. S. Army Missile Command, Redstone Arsenal, Alabama (1976).
6. Brown, A. M., and Ashby, M. F., Acta Met., 28, 1085-1101 (1980).

STRUCTURAL DEFORMATION OF EXPERIMENTALLY SHOCK-LOADED PERICLASE (MgO)

H. Schneider and I. Jung

Forschungsinstitut der Feuerfest-Industrie,
An der Elisabethkirche 27, 5300 Bonn 1, Germany
(F.R.G.)

ABSTRACT

The structural deformation of periclase single crystals shock-loaded up to 600 kbar with shock wave direction parallel to one <u>a</u>-axis was investigated by means of X-ray powder (Debye-Scherrer) and single crystal techniques (precession), and by optical microscopy. Slight post-shock cell volume expansion occurs, which is attributed to rarefaction wave expansion. Streaking of hk0, h0l, 0kl, and hhl single crystal reflections is interpreted in terms of rotational glidings parallel (001) producing platelets with a mean size $\gtrsim 1,000$ Å. Probably glidings are due to stress relaxation processes. The same may be true for the planar deformation systems occurring parallel (110).

INTRODUCTION

Periclase (MgO) is an ionic crystal with a NaCl-type structure. The structure can be considered as closely packed cubic arrangement of oxygen ions. The octahedral holes of the oxygen arrangement are occupied by magnesium, whereas the tetrahedral holes are empty.

Ionic crystals of the NaCl-type most easily deform by translation gliding or slip on $\{110\}$ planes. Previous microscopic investigations on periclase, shocked along different crystallographic directions yielded deformation on the {110}, {100}, {111}, and {112} planes along <110>[1].

EXPERIMENTAL

Sample material. Investigations were carried out on clear, cubical, chemically pure synthetic periclase single crystals (MgO: 99.3, CaO: 0.24, SiO_2: 0.34, Fe_2O_3: 0.05, Al_2O_3: 0.06 wt-%).

Shock experiments. Thin single crystal platelets (approx. 0.5 mm thick) cut parallel {100} were used for the shock-recovery experiments. The shock wave direction was parallel to one of the $\underline{a}$-axes ($\underline{a}_3$: arbitrary setting). By a "momentum traps" method [2,3] a peak pressure of about 600 kbar ($\pm$ 4 %) was obtained within the periclase sample.

X-ray studies. The 2θ - angles of at least 5 hkl-reflections (Debye-Scherrer-technique) were carefully measured 6 times using a standard optical measurement device. The cell edges $\underline{a}$ of unshocked and shocked periclase were refined using a least-square fitting. Crystal fragments, approximately 100 μm in size were investigated by conventional precession techniques, using Zr-filtered Mo-radiation. Diffraction photographs were taken from the hk0, h0l, 0kl, and hhl planes.

Microscopic studies. Fragments of shocked periclase single crystals were studied for deformation features with a polarizing microscope.

RESULTS

Various complex deformation modes induced by the shock process do occur in periclase. This paper presents some recent results on the deformation of periclase after shock-treatment at 600 kbar.

X-ray powder data. The X-ray reflections of shock-treated periclase are slightly broadened. The cell edge $\underline{a}$ increases from 4.212 ± 0.001 Å (unshocked) to 4.223 ± 0.002 Å (600 kbar). The slight post-shock cell expansion may be caused by the rarefaction wave travelling behind the shock front and not to shock temperature.

The shocked crystals' X-ray reflections are splitted, the splitting degree increasing with the diffraction angle 2θ. Splitting maxima of the reflections are connected by streaks of lower intensity. The X-ray reflections' splitting is explained by a deformation- induced lowering of the periclases' symmetry.

X-ray single crystal data. The X-ray reflection spots tend to broaden into Debye-Scherrer rings. The peripherical streaks' length is a measure of the magnitude of deformation. Disorientation of lattice blocks or domains with respect to the originally undistorted crystal exhibits maximum reflection streaking for lattice planes perpendicular to deformation planes. Planes of preferred deformation exhibit only weakly streaked reflections. Random glidings and rotations effect symmetrical streaking to both reflection sides. Asymmetrical streaking and the occurrence of intensity maxima may be indicative for an orientated deformation process. According to the X-ray photographs of shocked periclase the following principles can be established:

1. Reflection cuts of the reciprocal hk0 planes show strong streaking, the streaking angle being about 55°. The streaks are asymmetrical. They exhibit two weak and diffuse maxima, both being separated by about 10° (Fig. 1 a).

2. Reflection cuts of the reciprocal h0l and 0kl

planes show only weak reflection streaking (Fig. 1b).

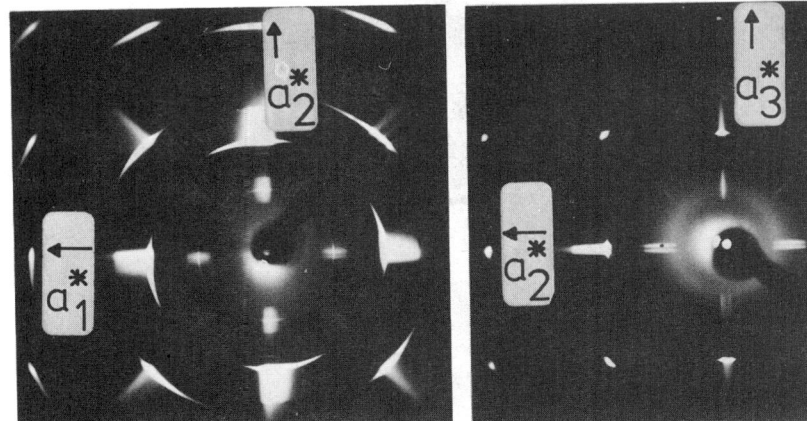

Fig.1 Reciprocal lattice planes of periclase single crystals shock-loaded up to 600 kbar. The shock wave direction was parallel to the a_3-axis (Zr-filtered Mo -radiation, exposure time 48 h). (a) hk0 plane;(b) 0kl plane.

The X-ray reflection spots' morphology implies rotational deformation within (001) around $\underline{a}_3$ (see also [4]).

<u>Microscopic observation.</u> Different deformation modes are visible under the microscope:

1. Microfracturing parallel (100) and (010).

2. Planar deformation systems running parallel (110) (Fig.2). The deformation systems appear as very thin planar discontinuities under the microscope and can optically not be resolved.

According to their similarity to shock-induced deformation systems in quartz[5] and other minerals[6] they should be called "planar elements". Planar elements are believed to be traces of gliding planes. Undulatory extinction occurs along to the planar elements, indicating deformation - produced lattice strain.

Fig.2 Planar elements in periclase shock-loaded up to 600 kbar. The orientation of the elements is parallel (110) and (110), (crossed nicols).

REFERENCES

1. M.J.Klein and J.W.Edington, Phil.Mag. **14**, 21 (1966)
2. D.J.Milton and P.S.Carli, Science **140**, 670 (1963)
3. D. G. Doran and R. K. Linde. in Solid State Physics, edited by F. Seitz and D. Turnbull, 19 (Academic Press, 1966), p. 229.
4. H.Schneider, Phys.Chem.Minerals **4**, 245 (1979)
5. W.v.Engelhardt and W.Bertsch, Contr.Miner.Petrol. **20**, 203 (1969)
6. D.Stöffler, Fortschr.Miner. **49**, 50 (1972)

ULTRAHIGH PRESSURE LASER-DRIVEN SHOCK WAVE EXPERIMENTS*

R. J. Trainor, N. C. Holmes, R. A. Anderson
University of California, Lawrence Livermore National Laboratory
Livermore, California 94550

ABSTRACT

We review recent laser-driven shock wave experiments, with a view toward assessing the prospects of making accurate physical properties measurements at ultrahigh pressures. Recent experimental results on the scaling of shock pressure with laser intensity and wavelength are presented, and preliminary impedance matching experiments are discussed.

INTRODUCTION

We have known for many years that very high pressure shock waves can be produced by intense laser pulses, even though many of the properties of such strong shocks have only been characterized very recently, and some remain poorly understood. A more challenging issue is whether or not it is possible to make accurate and meaningful physical property measurements under the conditions of a strong laser-generated shock compression. To evaluate the prospects of doing this, we need the answers to several key questions. How high are the achievable shock pressures? What are the effects of preheating by suprathermal electrons from the laser plasma and of shock attenuation due to the short lifetime of the laser pulse? How do we deal with them? What are the constraints on beam uniformity? Can we fabricate and characterize targets with the necessary submicron tolerances? What kinds of experiments are possible with present diagnostic capabilities? In this paper we use the results of recent experiments at LLNL as guidance in exploring these questions and speculating about some future directions in laser-generated shock wave research.

STRENGTH OF LASER GENERATED SHOCKS

Simple analytic models developed many years ago[1] predict that the maximum shock pressure P_s is an increasing function of incident laser intensity I_0, varying as I_0^α. This result is substantiated by more detailed calculations[2] using laser-plasma-hydrodynamics codes such as LASNEX.[3] Theoretically, we expect $\alpha = 0.6-0.8$. These predictions have recently been confirmed experimentally for laser intensities between 10^{12} and about 5×10^{14} w/cm^2, as shown in Figure 1. Here we have plotted results

---

*Work performed under the auspices of the U.S. Department of Energy by Lawrence Livermore National Laboratory under contract #W-7405-Eng-48.

for aluminum obtained by Grun et al.,[4] which extend up to about 10^{14} w/cm^2. The solid line is a best fit to the shock pressures they inferred from ion time-of-flight and ballistic pendulum measurements and corresponds to α = 0.8. These are time-integrated measurements and thus probably give a good measure of the peak pressure. We have measured shock velocities produced by 300 ps Gaussian pulses in stepped aluminum targets, using an ultrafast imaging system[5] to record shock transit times across premeasured steps at the Janus laser. The results are used to infer the pressure 20 μm deep inside the target, where it has been hydrodynamically attenuated somewhat; hence the lower pressures than would be predicted by extrapolation of Grun's measurements. We find α = 0.65. Both sets of data are in good agreement with LASNEX simulations.

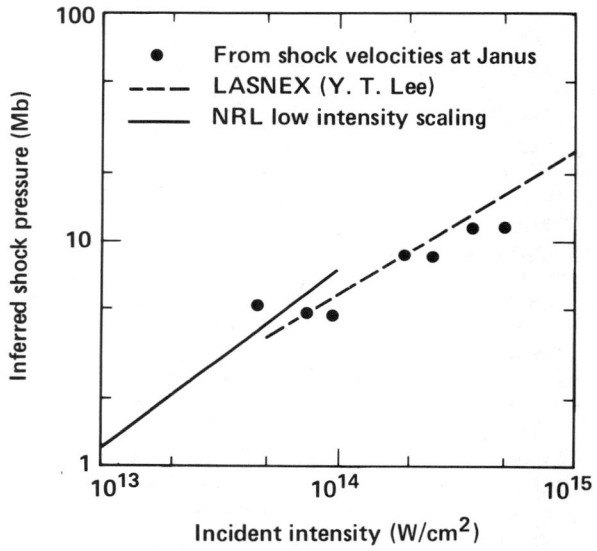

Figure 1. Inferred shock pressure vs incident intensity. The solid line, corresponding to α = 0.8, is a fit to the data of Grun et al.[4] The points, obtained from shock velocity measurements, correspond to α = 0.65.

At higher intensities, there is some evidence that the power-law scaling may break down and that the peak pressure may actually saturate as intensity is increased above $\sim 10^{15}$ W/cm^2.[6] To test the limits of high pressure achievable in a preheat-free, planar laser-driven shock experiment, we have measured the velocity of a strong shock produced in a gold target by ten arms of the Shiva laser. The target consisted of a 22-μm-thick layer of aluminum backed by 32 μm of gold. Atop the gold layer were two 18-μm-high gold steps, one on each side of a 150-μm-wide channel. The roughness of the gold surface was less than 0.3 μm, and the step height was measured to better than 0.2 μm accuracy. The aluminum layer was used to exploit the shock impedance mismatch between aluminum and gold to produce a stronger shock. The thick gold layer was required to assure a preheat-free shock.

The target was irradiated on the aluminum side with a 625 ps FWHM pulse of incident intensity 2.9×10^{15} W/cm^2. The beams were overlapped to produce a spatially uniform spot with intensity variations estimated to be no greater than about 15 percent. The arrival times, measured with respect to the peak of the laser pulse, at the rear of the 32 μm gold layer and at the top of the steps are shown in Figure 2. We obtained a shock velocity across the steps of 17.3 ± 0.3 km/s. The uncertainty of $\pm$ 2% is the best yet achieved in a laser-generated shock experiment and illustrates the advantages of using high laser energy, which permits the use of larger targets. The measured shock velocity corresponds to a pressure of about 3.5 TPa (35 Mbar). Spectrally resolved x-ray emission from the laser plasma was also measured in this experiment, permitting an estimate of suprathermal preheating effects to be made. We estimate that the rear of the target was heated to less than about 500°C, whereas the brightness of the shock signal indicated shock heating of order 5 eV.

Figure 2 also shows the results of LASNEX simulations of this experiment.[7] Two-dimensional (2D) calculations were required to reproduce the experiment due to the two-dimensional character of the plasma blowoff, which reduces shock pressure. However, the shock propagation itself is planar. The two 2D-LASNEX-calculated curves are for slightly differing absorption fractions and are seen to bracket the data. The best curve is for about 30 percent total absorption.

An analysis of the LASNEX calculations which accurately simulate this experiment reveals a modified power-law scaling of peak shock pressure with intensity.[7] However, for intensities below about 3×10^{15} w/cm^2, P_s scales approximately as $I_0^{0.6}$, with a tendency toward saturation only occuring at much higher intensity.

This experiment showed that preheat-free laser-generated shocks in the 3-4 TPa range are possible with present 1.06 μm wavelength lasers. However, we know that the shock was decaying strongly as it entered the step. LASNEX calculates approximately 30 percent attenuation in the pressure during the step traversal. Indeed, a peak pressure of nearly 9 TPa deeper into target (strongly pre-heated, however) was predicted.

EFFECTS OF LASER WAVELENGTH

We know of two promising methods for overcoming the problems of preheat and shock decay. The first approach[2] uses a thin, dense layer imbedded in a slab target to absorb suprathermal electrons. The layer becomes very hot in this process, and its expansion after the termination of the laser pulse helps to maintain the shock pressure at a high value. LASNEX calculations[2] predict that in optimally designed targets of this type the pressure does not decay substantially for several nanoseconds after the termination of the pulse (whose duration was typically 0.3 ns in these calculations).

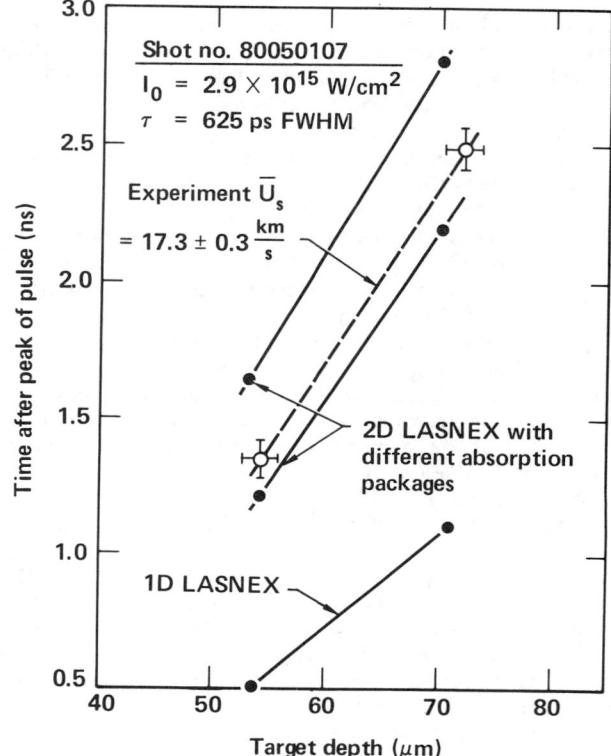

Figure 2. Shock arrival times measured in a thick Al-Au target (see text). The best LASNEX simulation used a total absorption of about 30%. The calculated shock paths are shown as straight lines but actually curve somewhat due to shock decay.

Note that a laser-generated shock in a homogeneous slab target is expected to decay after only about 2τ, where τ is the pulse length. Tests of these predictions are presently underway in our laboratory. The principal drawback to the technique is that it is a relative inefficient use of the laser energy in that for a given intensity somewhat lower peak pressures are generated than in targets without dense layers.

The second approach is the use of laser wavelengths shorter than one micron, which has been predicted to offer considerable advantages to inertial-fusion experiments[8] and shock-wave experiments as well. These advantages include increased absorption of laser energy by the target, reduced suprathermal electron preheat, and possible increased shock pressures. The increased absorption is a direct consequence of the dominance of inverse Bremsstrahlung scattering at short wavelength λ and has been observed in several laboratories.[9] Reduced preheat is a consequence of the reduced importance of resonance absorption at shorter λ. Spectrally resolved x-ray emission measurements indicate dramatic reductions at short λ of the number of suprathermal electrons produced and their characteristic energies.[9]

Several theoretical treatments have shown that P_s should increase with decreasing λ. However, the scaling of P_s with

λ is a subject of some controversy. If we write $P \propto \lambda^{-n}$, various theories predict $n = 0.33-2$, indicating benefits ranging from modest to dramatic when short λ is used. The most sophisticated of these theories[10] predicts $n = 1$, but only for spherical geometry; the case for planar geometry is still being studied and may differ from the case for spherical geometry.[11] In any case, the maximum pressure achievable in a preheat free region of the target should scale much more favorably with λ than is indicated by the models, since the reduced preheat should permit measurements to be made in thinner targets, where shock decay has not significantly attenuated the shock.

We have recently made shock velocity measurements on aluminum targets irradiated by 0.35 μm light produced by frequency-tripling one arm of the Argus laser. The targets[12] were aluminum slabs 25 μm thick. On the irradiated (front) side of the target, Be layers varying between 0-7 μm thick were deposited. A 4 μm-deep, flat-bottomed channel was ion milled into the rear surface of each target. An ultrafast streak camera measured the shock arrival times at the bottom and top of the channel with 20 ps time resolution. Incident laser intensity was nominally $1-2 \times 10^{14}$ w/cm^2, with a few shots being done at higher intensity. Pulse length was typically 700 ps FWHM; absorption was approximately 95 percent. Arrival time data are shown in Fig. 3. These data must be regarded as preliminary in that shot-to-shot corrections for intensity variations about the nominal value have not yet been made; this should have a smoothing effect. For the shots shown in Figure 3, measured shock velocities were about 25 km/s, implying shock pressures of 1-1.2 TPa. These pressures are also inferred by the arrival time data. A simple analysis shows that for a triangular laser pulse with FWHM τ,

$$(t_a - t_o) \simeq \frac{L}{U_s} - \frac{\tau}{(1 + \frac{2}{\alpha})} \tag{1}$$

where t_a is the absolute arrival time, t_o is the time the peak of the pulse arrives at the target, L is the target thickness, U_s is the peak value of shock velocity, and α is the exponent in the intensity scaling law. For $U_s = 25$ km/s, $\tau = 700$ ps, $L = 21$ μm, and $\alpha = 0.7$, Eq. (1) yields $(t_a - t_o) = 300$ ps, in accord with the data. The upward slope of $(t_a - t_o)$ with increasing Be thickness is due to the increased distance which the shock must traverse. A slight dip in this curve at small Be thickness is expected due to the small shock impedance mismatch between Al and Be. The above results are in good agreement with preliminary LASNEX simulations of the experiment by W. C. Mead of LLNL. We also irradiated one of the pure Al targets with a 1.06 μm pulse at 3×10^{14} w/cm^2 incident intensity, keeping all other experimental parameters the same as in the 0.35 μm shots. On this shot, we measured $U_s = 20$ km/s and $(t_a - t_o) = 550$ ps, both corresponding to $P_s \sim 0.6$ TPa. The absorbed intensity was estimated to be about 1.2×10^{14} w/cm^2, comparable to that used

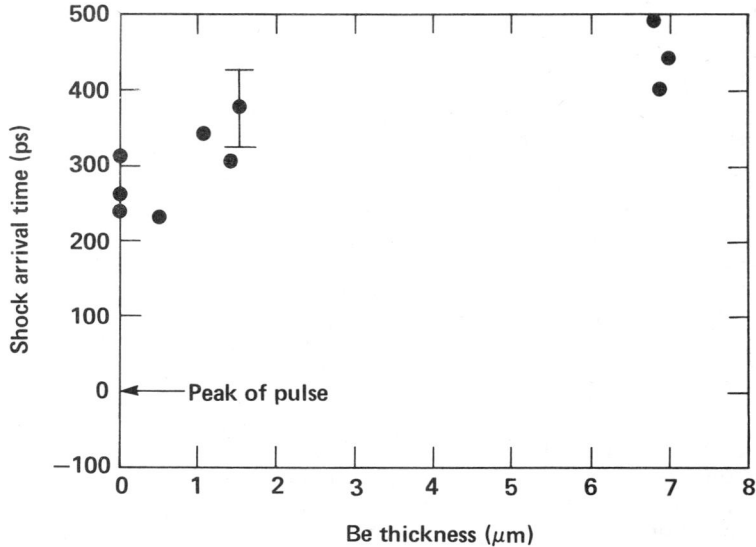

Figure 3. Shock arrival times in Be-coated aluminum targets, produced with 0.35 μm wavelength irradiation and incident intensities of 1-2 x 10^{14} w/cm^2.

in the 0.35 μm experiments. Thus, we have observed a significant enhancement of P_S by a threefold reduction in λ. We must further realize that in the 1.06 μm experiment, the rear surface of the target was preheated to several thousand degrees, whereas the preheat estimated for the short λ shots is negligibly small. In these experiments, shock decay is not believed to have been significant, since the shock was studied at times earlier than 2τ, the approximate onset time for shock decay in Al.

It appears, from our present vantage point, that future ultra-high pressure shock wave experiments will be performed on lasers of high energy and short wavelength.

REQUIREMENTS FOR PHYSICAL PROPERTY MEASUREMENTS

Diverse physical property measurements have been developed for more conventional, lower-pressure experiments, where time resolution on the nanosecond scale is sufficient. Such techniques include measurement of shock and particle velocity (from which the Hugoniot is obtained), sound velocity, temperature, electrical conductivity, x-ray diffraction, etc. Bringing such powerful techniques to bear in laser-shock-compression experiments will require time resolution of 10 ps or better. At present, only shock velocity can be measured with such accuracy (as described above), although particle velocity and temperature measurements appear feasible.

Target requirements for laser-driven shock-wave experiments are also severe. As seen in the discussion of experiments above, targets are typically a few tens of microns thick, and shock velocity measurements are made across parallel planes typically separated by 5-15 µm. A possible target configuration for a copper-aluminum impedance match experiment is shown in Fig. 4. Obviously, surface finishes on such targets must be of the order of 0.1 µm and the interplanar distances measured to about 0.1 µm. Such tolerances are now possible.

Significant progress has been made over the past year toward fabrication and accurate characterization of targets. Reeves and Helm[13] have developed a masking technique for producing smooth, uniform steps in metallic targets. A different technique has been developed by Devine and Evans,[14] using ion milling to produce high quality steps on photo-masked metal substrates. A gold target with two 10 µm-high, ion-milled steps is shown in Fig. 5. Scanning electron microscopy reveals the surface roughness of these stepped surfaces to be about 0.1 µm. Several techniques have been explored for measuring the heights of such steps with high accuracy, including mechanical stylus methods, optical and scanning electron microscopy, and interference microscopy. The interference microscopy technique has been found superior, and the height of a nominal 5-µm-high smooth step can now be measured confidently to about 1%.

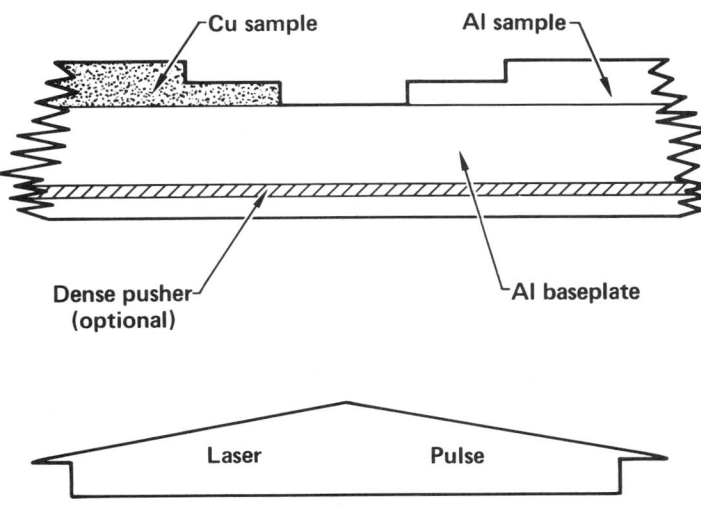

Figure 4. Target configuration for impedance-matching experiments with laser-generated shock waves. Step heights are typically 5-15 µm; total target thickness is typically 20-75 µm.

Figure 5. Scanning electron micrograph of 10 μm high steps ion milled into a gold target. Some surface contamination due to unremoved photoresist residue is seen.

Accurate determination of the initial mass density of the target is a requirement for shockwave experiments in general and is a challenging problem for laser-driven shock experiments, where the total target mass may be only a few mg. Recently, Veeser and Maggiore[15] have developed a technique using low-energy proton backscatter to measure areal density in very small specimens. The technique may provide initial density measurements approaching 1% accuracy. In addition, the method has high spatial resolution (<50 μm), so it can detect density variations in different steps.

Recently, we began a series of shock impedance-matching experiments, using the Janus laser to irradiate copper-aluminum targets. The details of the experiment are discussed elsewhere in these proceedings.[16] Preliminary experimental results are shown in Fig. 6, and are compared to both the present capability to do such experiments using other laboratory methods and to our projected capabilities using upgraded laser sources and realistic diagnostic improvements.

Very accurate data are possible using two-stage, light-gas guns, with shock velocity uncertainities near 1%, up to pressures of about 0.2 TPa in aluminum.[17] Present Janus experiments just overlap the two-stage gun range and extend up to pressures of about 0.6 TPa in aluminum. The overlap, though limited, should be useful since it will allow an accurate check to be made of the low-pressure end of the laser-shock data. The accuracy is about ±10% and is not sufficient to test competing EOS theories. The relatively large scatter is due primarily to laser intensity variations produced by insufficient flatness of the targets, the use of fast irradiating optics, and degraded space/time resolution of the imaging system.[16] These sources of scatter are being eliminated for upcoming experiments, in which we expect the data accuracy to improve to about ±2%. An upgrading of Janus should also extend the range of the data to about 1.2 TPa. It should be mentioned here that in all experiments conducted so far at Janus, the desire to study preheat-free shocks has necessitated target thickness so large that shock decay is inevitable. As mentioned above, the use of imbedded pushers and/or shorter laser wavelengths offer solutions to this dilemma. When Novette is complete in late 1982, the capability will exist to perform high-energy irradiations at short wavelengths (0.53 or 0.35 μm), and accurate shock impedance data up to the 4 TPa range and perhaps beyond, will become possible, thus entering the regime where similar experiments are presently being performed with nuclear-explosives drivers.[18]

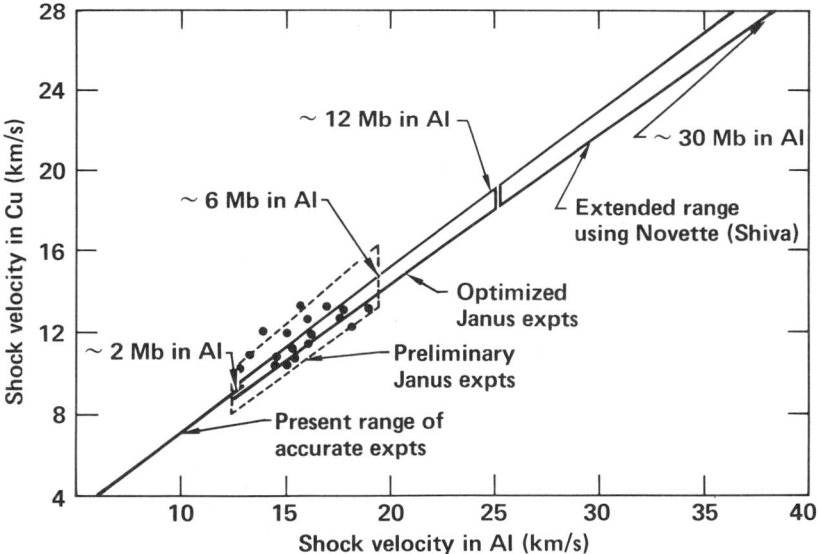

Figure 6. Preliminary results of Al-Cu impedance matching experiments. Dashed lines represent 10% error limits in data. Solid line at pressure below 0.2 TPa (2Mb) in Al is the present range of accurate experiments. Optimized experiments at Janus (see text) should extend to about 1.2 TPa. Novette could produce data at pressures beyond 3-4 TPa.

ACKNOWLEDGMENTS

The work described above involved many collaborators. The Shiva shock velocity measurements were performed in collaboration with D. Banner and D. Phillion. The short wavelength experiments were done in collaboration with E. M. Campbell, W. C. Mead, R. E.

Turner, and F. Ze. The impedance matching experiments were performed jointly with L. Veeser of Los Alamos National Laboratory. Y. T. Lee and R. J. Olness collaborated in the numerical simulations of several of these experiments. We also thank the crews of the Janus, Argus, and Shiva laser facilities for their skillful operation. We also acknowledge W. Hatcher, G. Devine, L. Evans, and G. Reeves (LANL) for fabricating the high quality targets used in these experiments.

REFERENCES

1. R. E. Kidder, Nucl. Fusion $\underline{8}$, 3 (1968).
2. R. M. More, in Laser Interaction and Related Plasma Phenomena, Vol. 5, edited by H. J. Schwarz and H. Hora (Plenum, N.Y., 1981) p. 253.
3. G. B. Zimmerman and W. L. Kruer, Comments Plasma Phys. $\underline{2}$, 51 (1975).
4. J. Grun, R. Decoste, B. M. Ripin, and J. Gardner, Naval Research Laboratory Memorandum Rept. 4410, 1981.
5. L. Veeser and J. Solem, Phys. Rev. Lett. $\underline{40}$, 1391 (1978); R. J. Trainor, J. W. Shaner, J. M. Auerbach, N. C. Holmes, Phys. Rev. Lett. $\underline{42}$, 1154 (1979).
6. M. H. Key, Nature $\underline{283}$, 715 (1980).
7. R. H. Harrach, Y. T. Lee, R. J. Trainor, N. C. Holmes, M. D. Rosen, D. L. Banner, R. J. Olness (these proceedings).
8. C. Max, Laser Program Annual Report, Lawrence Livermore National Laboratory, UCRL-50021-78, p. 3-56 (1979).
9. C. Garban-Labanne, E. Fabre, C. Max, 21st Annual Meeting of Div. of Plasma Physics, American Physical Society, Bull. Am. Phys. Soc. $\underline{25}$, 894 (1980). F. Ze, *et al.*, Laser Program Annual Report - 1979, Lawrence Livermore National Laboratory, UCRL 50021-79, p. 6-46 (1980).
10. C. Max, C. G. McKee, W. C. Mead, Phys. Rev. Lett. $\underline{45}$, 18 (1980).
11. C. Max, R. Fabbro, and E. Fabre, Bull. Am. Phys. Soc., $\underline{25}$, 895 (1980).
12. W. Hatcher and G. Devine of LLNL were responsible for target fabrication.
13. G. A. Reeves, P. J. Helm, CLEOS/ICF 80, Feb. 26-28, 1980, San Diego, Calif.
14. G. Devine and L. Evans, (unpublished).
15. L. Veeser and C. Maggiore, (to be published).
16. N. C. Holmes, R. J. Trainor, R. A. Anderson, L. R. Veeser, and G. A. Reeves (these proceedings).
17. A. C. Mitchell and W. J. Nellis, J. Appl. Physics $\underline{52}$, 3363 (1981).
18. C. E. Ragan, B. C. Diven, M. Rich, W. A. Teasdale, (these proceedings); C. E. Ragan II, Phys. Rev. A $\underline{21}$, 458 (1980).

TIME AND SPACE RESOLVED MEASUREMENTS OF THE
DYNAMICS OF LASER FUSION TARGETS*

Robert H. Price, Mordecai D. Rosen, David L. Banner,
Neil C. Holmes, Marian Kobierecki, James R. Zickuhr,
Harlow G. Ahlstrom
University of California, Lawrence Livermore National Laboratory*

ABSTRACT

A time resolved radiography system has been recently implemented at the Shiva Laser Facility. The radiography system allows measurement of motions in one spatial dimension and one time dimension with a resolution of 4.5 µm and 15 ps respectively, at x-ray energies selected in the range of 0.1 keV to 2 keV. The dynamic range of the system is 10^3. Modifications will extend the capability of the system to a dynamic range of 10^4 for x-ray energies up to 8 keV. The streaked axisymmetric x-ray microscope optical alignment system allows pointing to 5 µm accuracy. Results of recent experiments are described, including the dynamics of the blowoff of laser irradiated disk targets. Measurement of velocities near 10^7 cm/sec for laser irradiated disk targets has been accomplished by use of x-ray backlighting.

INTRODUCTION

There is interest in the shockwave community in extending our understanding of shockwaves and equations of state into the pressure regime beyond that accessible to ordinary laboratory techniques. Laboratory techniques such as high explosives and multistage light gas guns typically reach pressures in the few hundred kilobar and 1 to 5 megabar ranges, respectively.[1] Laser drivers used for inertial confinement fusion offer the possibility of reaching pressures in the 10 to 100 megabar range.
Recent experiments[2,3] have shown that through the use of specially designed and fabricated laser targets, shock velocities can be measured with accuracies comparable to the uncertainties of theory. If particle velocities can also be measured and experimental errors can be improved, useful equation of state data will be obtained in the 10 to 100 megabar pressure range. Spurred by interest in the dynamic behavior of laser-fusion targets, we have developed a streaked x-ray microscope system which will provide the necessary time and space resolution to measure the particle velocity of material during the laser target interaction.

STREAKED X-RAY MICROSCOPE SYSTEM

The streaked x-ray microscope has been installed at the Shiva Laser Facility. This x-ray imaging system couples a large

---

* Work was performed under the auspices of the U.S. Department of Energy by Lawrence Livermore National Laboratory under Contract W-7405-ENG-48.

solid angle Wölter hyperboloidal-ellipsoidal axisymmetric x-ray microscope with a thin photocathode x-ray streak camera and a sophisticated optical alignment camera. The spatial resolution of the streaked x-ray microscope is 4.5 µm, limited by the 22x magnification of the x-ray mirror and the 100 µm resolution of the streak camera. The temporal resolution is 15 ps. The streak camera has an intensity dynamic range greater than 10^3. The streaked microscope system is presently sensitive to x-rays in the 0.1 to 2 keV range, limited on the low end by the streak camera photocathode, and on the high end by the reflectivity cut-off of the grazing incidence x-ray mirror. Mirrors are being fabricated which will extend the sensitivity range to greater than 8 keV.

The optical alignment camera allows visible light alignment of targets relative to the streak camera slit to an accuracy of 5 µm.

To date we have used the streaked x-ray microscope system in several experiments relevant to the measurement of particle velocity in laser fusion targets.

OBSERVATION OF X-RAY SELF EMISSION FROM LASER TARGET INTERACTIONS

We have observed the x-rays emitted from the plasma formed when the laser irradiates small disks of various materials. Figure 1 shows a comparison of the space and time resolved x-ray emission from an aluminum disk, irradiated at 5.8×10^{14} W/cm^2 with a LASNEX-TDG simulation of the experiment.

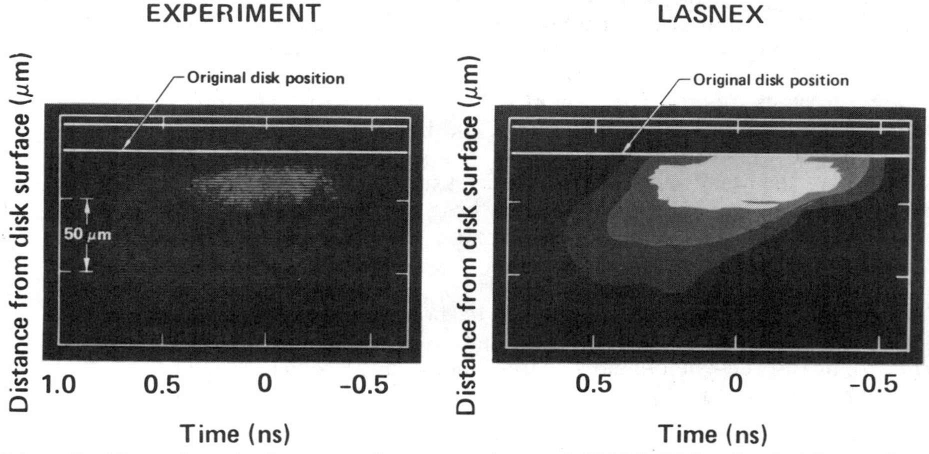

Fig. 1 The streaked x-ray image and the LASNEX-TDG simulation of a 17 µm thick aluminum disk irradiated at 5.8×10^{14} W/cm^2 compare favorably.

The agreement is quite good, and tends to lend confidence to our interpretation of experiments where low Z ablation layers are used on the target. High Z ablation layers, on the other hand, are not as well understood. Figure 2 shows a streaked image of

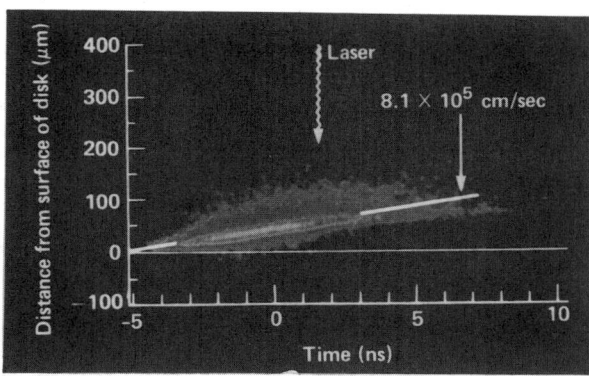

Fig. 2 The streaked image of a gold disk irradiated at 1×10^{14} W/cm^2, 400J, 6.4 ns observed in 270 eV x-rays.

the x-ray emission from the plasma blowoff of a gold disk irradiated at 10^{14} watts/cm^2. There are several striking features in this image which was taken using 270 ev x-rays.

The hot emitting region is very narrow, under 15 μm FWHM, and separates from the disk surface early in time, leaving a cool region between the x-ray emission region and the disk. The velocity of the emission region is nearly constant, independent of the rising and falling laser intensity during the pulse. It would appear that energy transport in the plasma is strongly inhibited for high Z ablator material. LASNEX 1-D simulations have not, to date, been successful in reproducing this image. Thus, if we wish to use an ablator material for EOS targets that is well understood, we can have more confidence in a low Z ablator than in a high Z ablator.

SLAB ACCELERATION EXPERIMENTS

We have also carried out a series of experiments in which streaked x-ray radiography (x-ray backlighting) was used to measure the velocity of a slab of material accelerated by laser driven ablation. The aluminum slab sample (17 μm thick) was mounted on a platform on the target (Figure 3) and irradiated

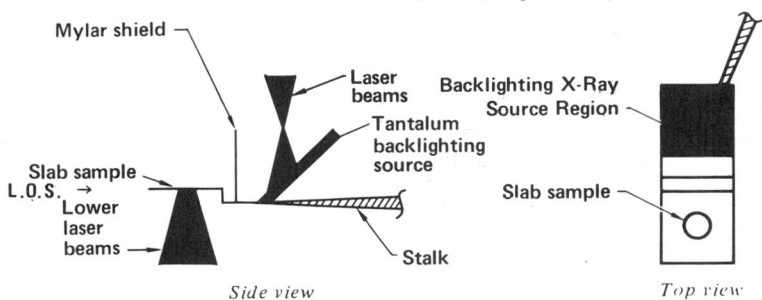

Fig. 3 X-ray backlighting target used for slab acceleration experiments.

with the lower beams of the Shiva Laser. The drive pulse was 6×10^{14} W/cm^2 (110J, 600 ps). The upper beams of Shiva were used to irradiate a tilted tantalum slab to produce a source of 1.9 keV x-rays. These x-rays were used to radiograph the aluminum slab as it accelerated away from the lower laser beams. The image was recorded using the streaked x-ray microscope system.

Figure 4 shows one of the streaked images recorded during this experiment. Both the self emission from the sample and the emission from the tantalum x-ray backlighter are visible. The

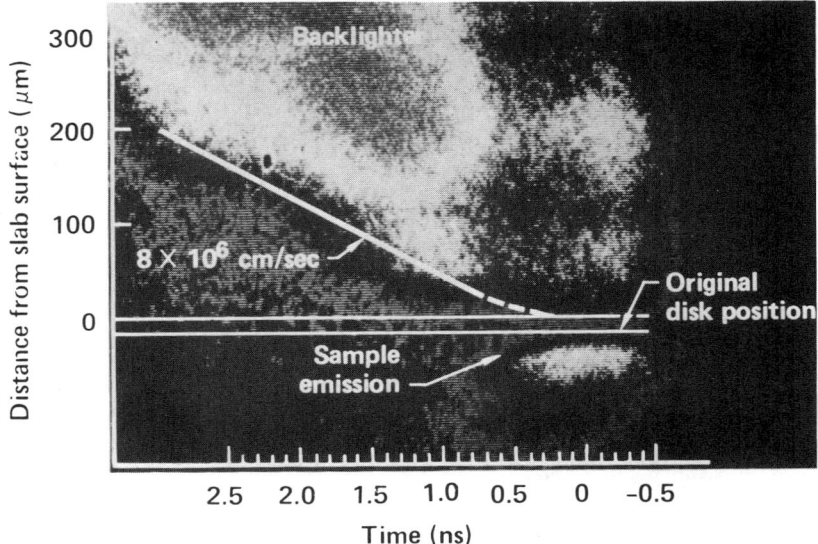

Fig. 4 X-ray backlight streaked x-ray micrograph of a 17 μm thick aluminum slab accelerated to 8×10^6 cm/sec.

shadow sloping up to the left is created by the accelerating aluminum slab and the material ablated from its surface by the laser. The sample reached a velocity of 8×10^6 cm/sec and pressures in the sample are calculated to have reached approximately 15 Mbar.

One problem with this data is that the shadow produced by the slab is not straight after the drive ceases, as one might expect. The reason for this was discovered during 2-D LASNEX simulations of this experiment. The results of the simulation are shown in figure 5. Because of the way the aluminum slab was mounted on the target, laser light was allowed to strike the edge of the disk (figure 5-1). This caused a dense plume to shoot ahead at the edge of the disk, making it appear as if the disk had suddenly accelerated (figure 5-2). After the laser drive pulse had stopped, the plume continued to expand and rarefy (figure 5-3) until the plume became transparent (figure 5-4) and surface of the disk again became visible. This behavior had the

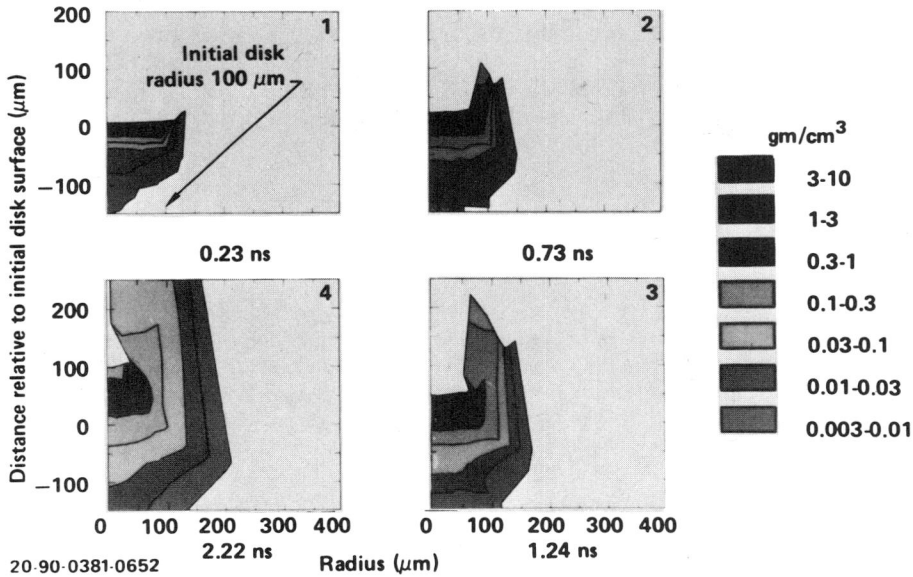

Fig. 5 2-D LASNEX-TDG simulation of an aluminum disk irradiated at 6×10^{14} W/cm^2.

effect of making it appear as if the disk had decelerated late in time.

While it is gratifying to understand this phenomenon, it clearly casts uncertainty on the velocity obtained in the experiment. If we expect to make particle velocity measurements with accuracies on the order of one percent or better, this problem must be solved. Steps will be taken in future experiments to eliminate this problem. First, the target can be built in such a way that the edges of the sample are shielded from irradiation by the laser. Second, higher energy x-rays will be used so the lower density plasmas surrounding the sample will be rendered transparent. Better filtering of the backlighter to reduce extranious radiation and use of shorter wavelength laser light will both reduce superthermal electron preheating of the sample.

SUMMARY

We have demonstrated the use of streaked x-ray microscopy and x-ray backlighting on laser driven targets and have taken the first steps toward realizing its potential for the measurement of equation of state data.

REFERENCES

1. A.C. Mitchell and W.J. Nellis, J. Appl. Phys. 52 3363 (1981).
2. L.R. Veeser and J.C. Solem, Phys. Rev. Lett. 40, 1391 (1978).
3. R.J. Trainor, J.W. Shaner, J.M. Auerbach, and N.C. Holmes, Phys. Rev. Lett. 42, 1154 (1979).

IMPEDANCE-MATCH EXPERIMENTS USING HIGH INTENSITY LASERS*

N. C. Holmes, R. J. Trainor, R. A. Anderson
University of California, Lawrence Livermore National Laboratory
Livermore, California 94550

L. R. Veeser and G. A. Reeves
University of California, Los Alamos National Laboratory
Los Alamos, New Mexico 87544

ABSTRACT

We present the results of a series of impedance-match experiments using copper-aluminum targets irradiated using the Janus Laser Facility. The results are compared to extrapolations of data obtained at lower pressures using impact techniques. The sources of errors are described and evaluated. We discuss the potential of lasers for high accuracy equation-of-state investigations.

DISCUSSION AND RESULTS

High intensity lasers have long been recognized as potentially valuable tools for dynamic equation-of-state (EOS) experiments using ultra-high pressure shockwaves. Recently, a shock pressure inferred to be over 3 TPa (30 megabar) has been generated in a laser experiment.[1] Thus, lasers can be used to generate shock pressures which, until recently, could only be generated in the vicinity of a nuclear explosion.[2] To determine the EOS of a material using shockwaves, the usual method is to measure two dynamic variables (e.g., shock velocity and particle velocity) and use the Rankine-Hugoniot relations[3] to determine an EOS point (the values of pressure, density, and internal energy which describe the state of the material at that point) on the Hugoniot curve. Alternatively, if a material of known EOS is available (a so-called standard material), impedance matching experiments can find the EOS of a test material and only shock velocities in the two materials need be measured.[3] For pressures over about 0.5 TPa, reliable EOS data are very sparse, and no material has a sufficiently well known EOS to serve as a standard. Future experiments using nuclear explosives to generate ultra-high pressure shockwaves will, no doubt, provide accurate EOS points against which these and future laser experiments may be compared. However, the impedance match technique can still be used if many experiments are performed with two materials. Much of the knowledge we now have of the equations of state of many materials at high pressures has been gained using this method in explosively-driven shockwave experiments.

---

*Work performed under the auspices of the U.S. Department of Energy by Lawrence Livermore National Laboratory under contract #W-7405-Eng-48.

We have performed a number of impedance-match experiments using copper-aluminum targets in an effort to evauate the usefulness of laser drivers for EOS research at pressures over 0.5 TPa. Copper and aluminum were chosen because they have been studied extensively at lower pressures, with well known equations of state below about 0.4 TPa and 0.2 TPa, respectively. These two metals are also relatively easy to model theoretically, so we are confident in our extrapolations of previous work into the pressure regime of this work.

The Janus laser facility was used for these experiments. Incident laser intensities were varied between $5 \cdot 10^{13}$ W/cm^2 and $4 \cdot 10^{14}$ W/cm^2, with a pulselength of 300 ps (FWHM) in a gaussian pulse. Laser output energy varied between 30J and 80J at a wavelength of 1.06 µm.

The targets used were thin metallic foils of aluminum, with aluminum and copper steps formed by vapor deposition on the surface of the target opposite to that irradiated by the laser. A typical target is shown schematically in Fig. 1. In most targets, a thin gold layer was included near the laser-incidence side as shown in the figure. The purpose of the layer is two-fold: it acts to shield the stepped region from energetic electrons produced in the laser-absorption region which can preheat the material, and the layer aids in producing a steady shock pressure in the steps.[4]

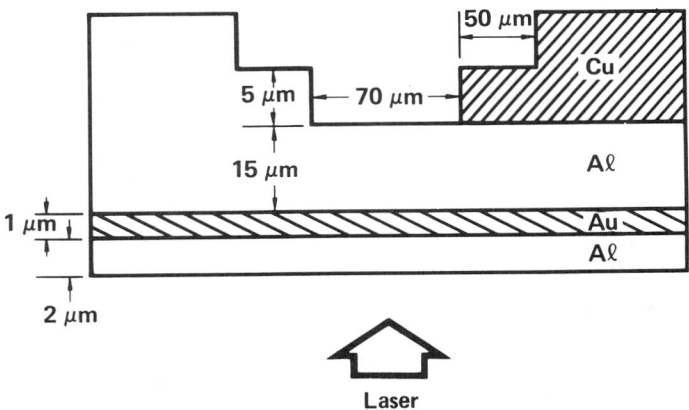

Fig. 1   A typical copper aluminum impedance match target used in our experiments. These targets were prepared by electron-beam vapor deposition.

Shock velocities in the steps were measured using an ultrafast streak camera to record the arrival time of the shock wave as it reached each free surface level. This system has a resolution of roughly 20 ps, and a spatial resolution of about 15 µm. Shock velocities were calculated by dividing the step height, measured with a sensitive stylus gauge to ∼0.1 µm, by the shock transit time across each step.

The measured shock velocity pairs for our experiments are plotted in Fig. 2. The solid line against which they are compared was calculated by extrapolating the EOS data of Mitchell and Nellis[5] for copper and aluminum, approximating the double-shock Hugoniot by a mirror reflection of the principal Hugoniot.[3] Within about 10%, the data are in good agreement with the extrapolation, and the apparently random scatter indicates that systematic errors do not dominate the results. The pressures inferred in these experiments range between 0.2 - 0.6 TPa in aluminum and between 0.4 - 0.8 TPa in copper.

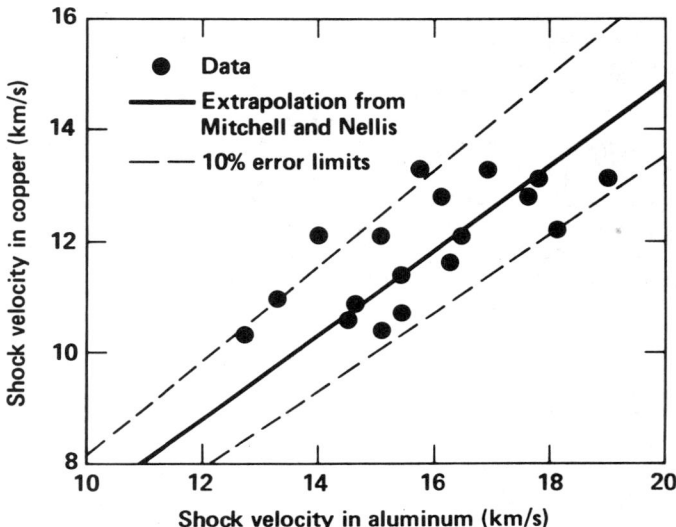

Fig. 2   Impedance-match data for copper-aluminum targets. The data fall within 10% of the bold line calculated from extrapolations of lower-pressure shock wave experiments. The measured shock velocities correspond to pressures of 0.2 - 0.6 TPa in aluminum and 0.4 - 0.8 TPa in copper.

We have identified several causes for the random errors in the experiments. The streak camera time resolution introduces errors of about 5%. The step height measurements we made had errors of roughly 5% as well. We believe the remaining errors are mainly due to non-planarity of the shockwave as it passed into the steps. Laser beam non-uniformity plays a role here. High spatial frequency intensity modulation in the beam can lead to transverse structure in the shock front, but it is difficult to assess its effect on the experiment. It is safe to say, however, that for more accurate experiments, beam modulation should be reduced to a minimum. The laser beam focusing system used to irradiate the targets used f/1 optics. The large convergence angle leads to local intensity variations if the targets are not perfectly flat. Again, these intensity nonuniformities tend to produce non-planar shock fronts. Some streak camera records in this series show evidence that this may have been an important effect.

It now appears that reduction of these error sources is straightforward. Higher time resolution in the recording system will soon be achieved. We are now able to use interferometric microscopy to measure step heights to 0.03 µm accuracy (better than 1%) and we have begun an experiment to measure target density and uniformity with about 1% accuracy using ion backscattering with a spatial resolution of a few micrometers.[6] The technology for producing very uniform laser beams is well known, although it requires much care to eliminate most modulation. Finally, the use of focusing optics of much narrower convergence angle significantly relaxes the requirements on target flatness. We feel that future experiments using laser drivers of relatively modest output energy can provide impedance-match EOS data with roughly 2 - 3% uncertainty, with major impact on our knowledge of material properties at ultrahigh pressures.

The technical assistance provided by A. Susoeff, G. Newman, J. Beard and D. Eckhart during these experiments was invaluable, and we are pleased to acknowledge their contributions to this experimental effort.

REFERENCES

1. R. J. Trainor, N. C. Holmes, Y. T. Lee, D. L. Banner, and D. W. Phillion, Lawrence Livermore National Laboratory, UCID-8574-80-2 (1980)(unpublished).

2. C. E. Ragan III, Phys. Rev. A, 21, 458 (1980).

3. See, for example, Ya. B. Zel'dovich and Yu. P. Raizer, Physics of Shock Waves and High-Temperature Hydrodynamic Phenomena, Vol. 2 (Academic Press, New York, 1966).

4. R. M. More, Proceedings of Fifth Workshop on Laser Interactions with Matter, University of Rochester, 1979 (to be published).

5. A. C. Mitchell and W. J. Nellis, J. Appl. Phys. 52, 3363 (1981).

6. Carl J. Maggiore, "Materials Analysis with a Nuclear Microprobe", edited by Om Johari, (SEM Inc., Chicago, 1980).

CONTRASTS IN ONE- AND TWO-DIMENSIONAL
HYDROCODE CALCULATIONS OF LASER-GENERATED
SHOCKWAVES IN DISK TARGETS

R. J. Harrach, Y. T. Lee, R. J. Trainor, N. C. Holmes,
M. D. Rosen, D. L. Banner, and R. J. Olness
Lawrence Livermore National Laboratory, Livermore, CA 94550

ABSTRACT

A comparison is made between 1D and 2D hydrocode calculations and some recent experimental results on laser-driven planar shockwaves in disk targets. A simple model is described which accounts for the trends seen in the calculations and which gives insight into the regime of laser intensity and pulse duration where two-dimensional effects become significant in these experiments.

INTRODUCTION

At Lawrence Livermore National Lab high power laser systems such as Janus and Shiva have been used to generate ultra-high pressure shockwaves in disk targets. To analyze the experiments we have done both one-dimensional (1D) and two-dimensional (2D) simulations using the Lasnex laser-plasma-hydrodynamics computer code. A very great savings in computer time and expense is realized in the 1D calculations, relative to the 2D ones. However, in cases involving long-duration laser pulses and high laser intensities we find that the 1D results are not sufficiently accurate even when necessary conditions[1] for generating 1D, i.e., planar, shockwaves are met; see, for example, Figure 1. The reason lies in the two-dimensional nature of the blowoff plasma; the laser-generated shock is quasi-1D with regard to its propagation through the disk when the 1D conditions are satisfied, but 2D effects can still be important in its generation.

SIMPLE MODEL

We can understand the need for 2D computations on the basis of a simple model of the laser-target interaction, sketched in Figure 2. The effective cross sectional area A that the target presents to the laser beam is postulated to be, not the laser spot area A_0 at the original surface of the disk, but rather the area of the critical density surface intercepted by the converging laser rays at the time $t = t_{peak}$ of peak laser pulse. A exceeds A_0 by a factor $[1 + (\ell_{ac} \sin\theta / R_s)]^2$ (see the caption of Figure 2 for definition of terms) which depends on the laser pulse duration τ_L and nominal peak intensity I_0 through ℓ_{ac}. The effective value of the laser incident intensity is decreased in the ratio A_0/A compared to I_0, and consequently the peak value of the laser-driven shock pressure, P_s, which scales as I^n, $n \cong 0.7$, is decreased by this two-dimensional effect.

The distance ℓ_{ac} separating the ablation surface and the critical surface is approximately given by the product of the laser pulse duration and the sound speed in the ablating plasma, and thus should vary as $\ell_{ac} \cong \ell_{ac}^{(0)} \tau_L^\beta I_0^\alpha$; from Lasnex calculations for 1.06μm laser light we find $\beta \cong 0.9$, $\alpha \cong 0.3$. The peak value of the laser shock pressure, corrected for 2D effects, is then predicted by this model to be

$$P_s \cong a I_0^n \Big/ \left[1 + \frac{\ell_{ac}^{(0)} \sin\theta}{R_s} \tau_L^{0.9} I_0^{0.3} \right]^{2n} \qquad (1)$$

where $n \cong 0.7$, and the coefficient a has a value of about 8 (depending weakly on τ_L and target composition and more strongly on laser wavelength, which we set to 1.06 μm), $\ell_{ac}^{(0)} \cong 120$ for Al vapor and 1.06 μm light, and the units employed are Mbar (1Mbar = 0.1 TPa) for P_s, 10^{14}W/cm^2 for I_0, ns for τ_L, and μm for ℓ_{ac} and the spot radius R_s.

Equation (1) has not been exhaustively checked against hydrocode calculations covering a wide range of values for I_0, τ_L, and R_s, but we offer the following two comparisons. First, Eq. (1) predicts a ratio 0.7

for the 2D vs 1D calculations pertaining to the Shiva experiment^2 in Figure 1, which is in respectable agreement with the actual value of 90 Mbar/149 Mbar = 0.6. Second, in Figure 3 we show 1D and 2D computations of P_s over a wide range of I_0 values for a set of problems involving 0.6-ns Gaussian-shaped laser pulses incident on 72μm-thick Aℓ disk targets with spot radius R_s = 256 μm and ray convergence angle θ = 22°. Eq. (1) is seen to track the 2D results fairly well over nearly the entire range of intensity. For such a set of problems (having fixed values of R_s and θ) we can use the model to divide the I_0-τ_L plane into regions where the 1D and 2D results agree within a certain tolerance, as shown in Figure 4.

CONCLUSION

In conclusion, the simple model provides physical insight and semi-quantitative information regarding the onset of two-dimensional effects in laser shockwave experiments, and allows one to estimate when 1D calculations suffice. We note in passing that the model provides some justification for the often-used calculational expedient of replacing the cylindrically symmetric disk by a spherical cap of suitably chosen radius of curvature.

REFERENCES

1. These conditions are that the laser spot diameter is large compared to the disk thickness, the distribution of laser intensity throughout the spot area is uniform, and the target surfaces are very smooth and flat.

2. D. L. Banner, R. J. Harrach, N. C. Holmes, Y. T. Lee, D. W. Phillion, M. D. Rosen, and R. J. Trainor, "Ultra-high pressure laser-driven shocks in gold," Laser Program Annual Report-1980, Lawrence Livermore Lab report UCRL-50021-80, 1981.

Work performed under the auspices of the U. S. Department of Energy by the Lawrence Livermore National Laboratory under contract number W-7405-ENG-48.

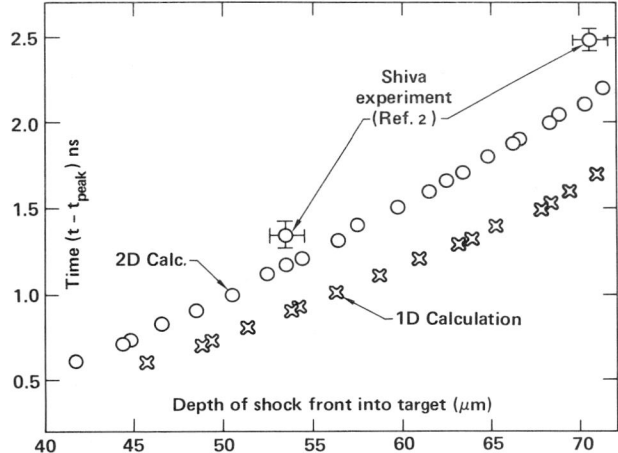

Fig. 1. Comparison of measured (Ref.2) and calculated transit times for laser-driven shockwave in a composite Aℓ/Au gold disk target of thickness (22/50) 72 μm (Aℓ side towards the laser). The peak laser incident intensity and pulse duration are 27 (10^{14} W/cm^2) and 0.60 ns, respectively. The 2D calculation predicts a peak shock pressure of magnitude 90 Mbar just inside the gold layer, decaying to a level of about 32 Mbar after propagating through the target; corresponding numbers for the 1D calculation are 149 Mbar and 54 Mbar. The 2D and 1D calculations required 350 and 29 minutes of CDC 7600 computer time, respectively. By adjusting the amount of laser energy absorbed, the 2D calculation could be brought into close agreement with the two experimental points.

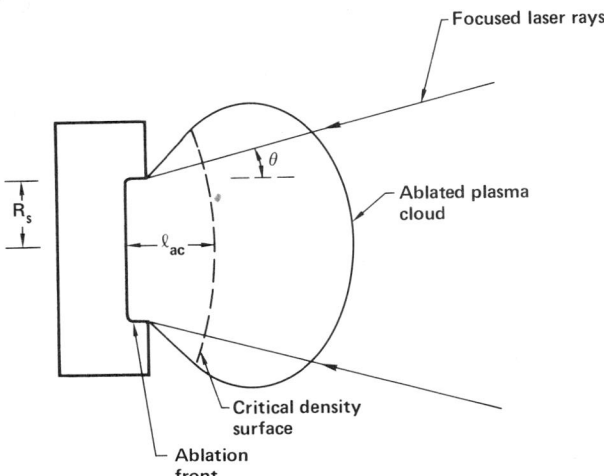

Fig. 2. Schematic diagram of the laser-target interaction, showing the critical surface where the density of free electrons equals the critical density n_c(cm^{-3}) = $10^{21}/[\lambda(\mu m)]^2$, the ablation surface, and the plasma blowing off into the converging incident laser rays. ℓ_{ac} is the distance separating the ablation and critical surfaces. For converging laser rays θ is approximately the ray cone half-angle; for normally incident rays, θ is a more complicated function depending on the amount of beam refraction and the plasma blowoff angle.

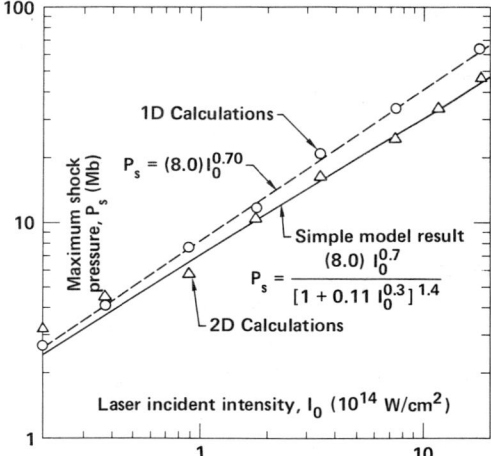

Fig. 3. Comparison of predicted values of the peak shock pressure P_s based on 1D and 2D hydrocode calculations and on the simple model result, Eq. (1), for a set of problems involving 0.6-ns Gaussian-shaped pulses incident on 72-μm-thick Aℓ disk targets with fixed values of the laser spot radius (R_s = 256 μm) and ray convergence half-angle ($\theta = 22°$). The 1D calculations are fit by the dashed line given by $P_s = 8.0\,(I_0)^{0.7}$, with P_s in Mbar and I_0 in units 10^{14} W/cm^2. Our 2D results show large scatter for lower intensities (left side of the graph) for reasons as yet undetermined.

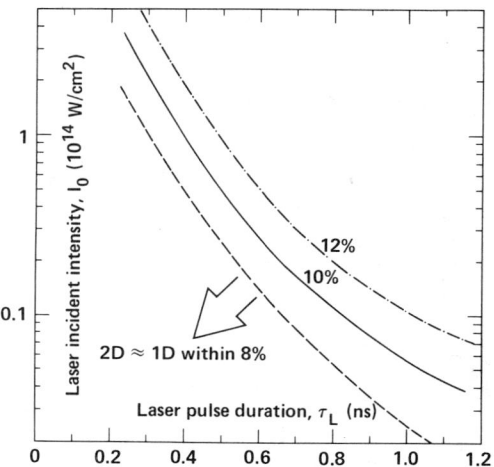

Fig. 4. Diagram of laser incident intensity vs pulse duration corresponding to the results in Figure 3, showing contours within which the 1D and 2D calculations of P_s are expected to agree within a specified amount.

Shock Compression Measurements at Pressures >1 TPa*

C. E. Ragan, B. C. Diven, M. Rich, E. E. Robinson, and W. A. Teasdale
Los Alamos National Laboratory, Los Alamos, NM 87545

ABSTRACT

A nuclear-explosive-generated planar shock has been used to perform impedance-matching experiments relative to a molybdenum standard. Shock velocities were measured with accuracies of 1.5% to 2.5%, thus providing Hugoniot data for samples of aluminum, quartz, iron, molybdenum, and low-density molybdenum at pressures from 2 to 5 TPa.

INTRODUCTION

In the past few years, we have developed several techniques[1-4] for obtaining ultrahigh pressure equation-of-state (EOS) data using underground nuclear explosions. In an absolute measurement at the beginning of the program, we determined[1,2] a Hugoniot point for molybdenum at 2.0 TPa. The calculated Hugoniot based on the SESAME EOS library[5] agreed with the measurement, thus providing increased confidence in the theoretical molybdenum EOS. In a subsequent impedance-matching experiment,[3,4] we used a planar, stable shock to obtain a Hugoniot point for uranium at 6.7 TPa relative to the molybdenum standard; differences from predictions[5] stimulated improved theoretical treatments.[6] The present experiment was fielded to utilize this previously demonstrated[3,4] technique in order to obtain ultrahigh pressure Hugoniot data for a variety of sample materials; a setup similar to that of Ref. 3 with several improvements was used.

EXPERIMENT

The sample arrangement is shown in Fig. 1. The shock passed from the 25-mm-thick lead base plate through the 180-mm-diam by 12-mm-thick molybdenum standard and into 13 samples. The 10-mm-thick samples were arranged in seven stacks as shown in Fig. 1 and consisted of the indicated materials. The horizontal dashed line indicates the effective base of the molybdenum, 10 mm below its upper surface. This package was located ∼3 m from the nuclear explosive, and detailed Monte Carlo calculations were used to optimize the shielding to reduce the neutron and gamma radiations to a level that caused heating of ∼10K.

An array of 75 electrical contact pins[3,4] was used to determine shock-arrival times. Sixteen pins were embedded in the molybdenum standard, three to five pins were embedded in each of the small samples, and four pins were embedded in the lead driver. The pins were separated by 1 to 3 mm in the vertical direction and were positioned horizontally to avoid rarefactions. Five pins were multiplexed onto

---

*Work supported by the U.S. Department of Energy.

each cable, and different decay times were used to provide a unique
signature pulse for each pin. The signals were recorded on sets of
oscilloscopes that provided coverage for 2 to 3 µs along with a
100-MHz sine-wave time base. The signal quality for 25 of the pins
was excellent, and shock-arrival times were determined with ±1-ns
uncertainties. The remaining pins produced lower quality signals
and uncertainties of 3 to 10 ns were assigned to the closure times.

DATA ANALYSIS AND RESULTS

The data analysis procedure involved a sequence of several
hundred least squares fits of the function $t = t(x,y,z)$ to the pin
coordinates and closure times using different functional forms for
t. The results of these fits indicated that the shock velocity was
decreasing slightly with z and that the shock front was curved.
The effective radius of curvature was ∿2 m, and the corresponding
outer-sample tilt angle was ∿1°. Additional fits indicated that
nonplanar effects were purely radial and that asymmetry about the
z-axis introduced <6-ns variation in arrival time along a radius.

These conclusions helped substantiate our procedure for determining shock velocities for different portions of the shock front by
fitting various subsets of the pins. The results of these fits gave
small values (consistent with zero) for the decrease in the shock
velocities. In the previous
experiment,[3,4] the shock velocities
changed by <1% over 10 mm. We
assumed a similar variation in this
experiment and changed the shock
velocities by ±0.5% to determine the
values at the lower and upper surfaces. The resulting shock velocities[7] at the appropriate interface
or sample center are summarized in
Table I, which lists in columns one
and two the sample materials (lower
first) and the experimental shock
velocities. For the molybdenum standard, an overall error of 1.5% that
includes systematic effects has been
assigned to the measured upper surface
value of 27.16 km/s. For the small
samples discussed in this paper
(except iron), the errors in the
shock velocities from the fits were
∿1%; however, overall uncertainties
of ±2% were assigned to each of these
velocities to include systematic
effects. For the iron sample, an
uncertainty of ±2.5% was assigned to
the measured shock velocity.

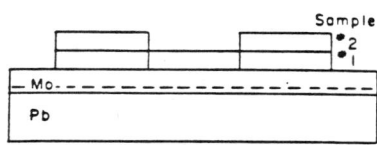

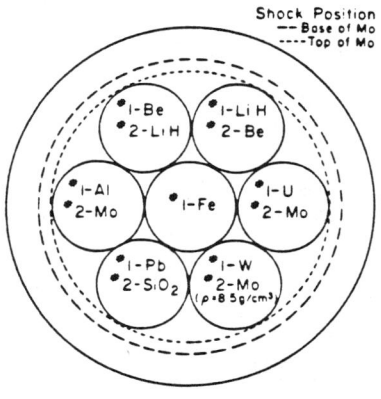

Fig. 1. Schematic drawing
showing the arrangement and
composition of the samples.

Table I. Comparison of Experimental and Calculated Results

Material	Shock Velocity (km/s) Experiment	Difference[b] (%) Old[c]	New[c]	Hugoniot Point[a] Pressure (TPa)	Particle Velocity (km/s)
Mo Std	27.16	---	---	4.900(3.5)	17.67(2.0)
Al	34.39	1.63	0.55	2.226(3.8)	23.89(2.7)
Mo	24.64	-0.13	1.90	4.034(4.5)	16.05(4.3)
Pb	23.79	---	---	4.723(--)	17.51(--)
Quartz	31.95	6.66		1.693(4.3)	24.04(3.0)
W	22.12	---	---	6.351(--)	14.87(--)
Mo[d]	24.28	9.85	9.77	3.693(4.4)	18.35(3.4)
Mo Std	28.02	---	---		
Fe	30.48	-1.92	-3.48	---	---

[a] Based on measured shock velocities and the improved molybdenum EOS; percent errors in parentheses.
[b] $(D_{th} - D_{exp})/D_{exp}$.
[c] Calculated results, "Old" from Ref. 5, "New" from Ref. 6.
[d] $\rho_o = 8.29$ g-cm^{-3}.

These measured velocities were used to obtain Hugoniot points based on impedance-matching analyses for each possible pair of samples. For the lower samples, the measured velocity at the upper surface of the molybdenum standard was used in the analysis. For each upper sample, the corresponding lower sample was treated as the standard material. For the molybdenum atop the aluminum, the measured shock velocity at the upper aluminum surface was used to determine a Hugoniot point. For the quartz and low-density molybdenum, the initial state in the corresponding lower sample was calculated from the pressure in the molybdenum standard. In this calculation, the measured shock velocity of 27.16 km/s was decreased by 1% to account for the decay across the lower sample. These analyses were based on the SESAME EOS library[5] and on both the original and improved[6] molybdenum EOS (when appropriate). In addition, the shock velocity in each sample was calculated based on these theoretical EOSs; columns three and four give the differnces (in percent) between these calculated results and the experimental values.

For the aluminum-molybdenum pair and the iron sample, the differences between calculation and experiment are small; the improved molybdenum EOS gives slightly better agreement, but for either, the EOSs for these samples from the SESAME library[5] are in good agreement with experiment. The calculated and experimental results for the quartz and low-density molybdenum samples differ by more than the experimental uncertainties.

Table I also gives the experimental Hugoniot points and percent errors for the samples. The appropriate shock velocities were used to determine the intersection point in the P-u plane of the straight line $P = (\rho_0 D)u$ with the reflected shock (RS) Hugoniot or the release isentrope (RI) of the lower standard material. The initial state of the standard was determined as described above, and the upper-sample results are based on the assumption that the SESAME EOSs for the lower samples are correct. The errors in parentheses correspond to the quoted uncertainties in the shock velocities. Both the original and improved EOSs for molybdenum were used in this analysis; however, the results given in the table are for the improved molybdenum EOS.

Details of the analysis are illustrated in Fig. 2 for the lead-quartz pair of samples; similar analyses were performed for the other sample stacks. The regions of interest are shown on expanded scales with the calculated[5,6] Hugoniots shown as heavy curves. The initial state in the lead is enclosed in the box in (a) and was calculated using the experimental shock velocity in the molybdenum and its improved[6] EOS. The lead RI from this point is indicated by arrows (→) and intersects the line labeled $P = (\rho_0 D)u$ in (b) to determine a Hugoniot point (circled) for quartz that is 5.1% lower in pressure than the SESAME prediction (large dot). The regions of uncertainty are not indicated in the figure, but uncertainties are given in Table I for the derived Hugoniot points that correspond to the combined shock-velocity uncertainties.

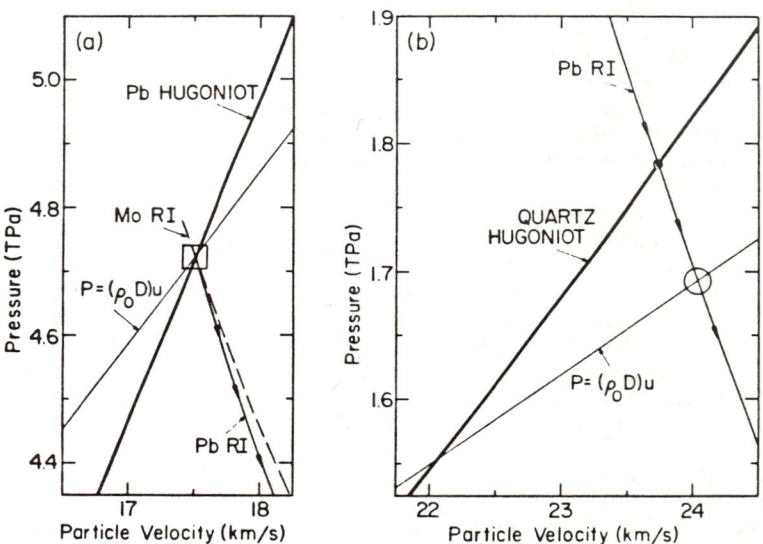

Fig. 2. Illustration of the impedance-matching analysis technique for the lead-quartz pair of samples.

CONCLUSIONS

This experiment provides Hugoniot data that can be used as bench marks for checking the consistency of theoretical EOS calculations. The theoretical EOSs for aluminum, iron, molybdenum, and lead in the SESAME library[5] are in good agreement with these data; the improved[6] molybdenum EOS gives slightly better agreement. The discrepancies for the quartz and low-density molybdenum are larger than the experimental errors and indicate the need for improved theoretical treatments. Such improved theories should provide additional insight into the physics at these extreme conditions.

We are planning an experiment to obtain additional data in this pressure region with ±0.5% shock-velocity uncertainties. Such data with reduced experimental errors for a large number of samples will provide consistency checks for EOS calculations. In addition, we hope to extend this technique to even higher pressures (>30 TPa) to check theoretical predictions at pressures where statistical models are believed to apply.

REFERENCES

1. C. E. Ragan III, M. G. Silbert, and B. C. Diven, J. Appl. Phys. $\underline{48}$, 2860 (1977).
2. C. E. Ragan, III, M. G. Silbert, and B. C. Diven, in High Pressure Science and Technology, edited by K. D. Timerhaus and M. S. Barber (Plenum, New York, 1979), Vol. 2, p. 993.
3. Charles E. Ragan III, Phys. Rev. A $\underline{21}$, 458 (1980).
4. C. E. Ragan III, High Pressure Science and Technology, edited by B. Vodar and Ph. Marteau (Pergamon Press, New York, 1980) Vol. 2, pp. 993-999.
5. B. I. Bennett, J. D. Johnson, G. I. Kerley, and G. T. Rood, Los Alamos National Laboratory Report LA-7130, (1978); "An Invitation to Participate in the LASL Equation of State Library," edited by Necia G. Cooper, Los Alamos National Laboratory Report LASL-79-62 (1979).
6. G. I. Kerley, private communication (1981).
7. Preliminary results given in C. E Ragan, B. C. Diven, E. E. Robinson, M. G. Silbert, and W. A. Teasdale, Bull. Am. Phys. Soc., $\underline{25}$, 513 (1980).

SHOCK HUGONIOT EXPERIMENTS USING AN ELECTRIC GUN

K.E. Froeschner*, H. Chau, G. Dittbenner, R.S. Lee†,
K. Mikkelson, D. Steinberg, R.C. Weingart

Lawrence Livermore National Laboratory, Livermore CA 94550

ABSTRACT

We have made shock Hugoniot measurements in the 0.19 to 0.36 TPa range using an exploding foil electric gun system. Our measurements agree with published data to within our experimental uncertainties. In our most recent experiments these uncertainties are around ±3% in both shock and material velocity and are close to what we expect from an analysis of error propagation in our experimental arrangement. As presently configured the electric gun system can make measurements in several TPa regime. These experiments were done with the system deliberately slowed down so that we could compare our results with those published by other researchers using other methods. We believe that these results establish that valid and accurate high pressure equation of state research can be effectively conducted with electric gun systems.

THE ELECTRIC GUN

In an electric gun, a projectile, usually a thin foil or composite of several foils, is accelerated to high velocity by the expansion of an adjacent metal foil which is electrically exploded by rapid Joule heating. The energy density which may be achieved in such an exploding foil is many times greater than that obtainable in chemical high explosives. In addition, the intense magnetic fields present give rise to substantial Lorentz acceleration. The result is that material may be accelerated to velocities well beyond the reach of other technologies. The major electric gun system in use at LLNL has accelerated 0.300 mm thick Kapton projectiles to about 30 km/s. Performance of this system is discussed in more detail in a companion paper[1] and in other, recent publications.[2]

While simple in concept and generally ideally suited to high pressure equation of state research, there are nonetheless many difficulties which may arise in any real application of electric gun systems in this area. A substantial part of the research conducted over the past year at LLNL was concerned with identifying, understanding and eliminating problems such as hydrodynamic jetting, target preheat and pre-motion, impactor non-planarity and ionization of the residual gas in the barrel.

A number of anomolies seen in early experiments seemed to be associated with hydrodynamic jetting from the interior corner formed at the juncture of the barrel with the impactor foil. This jetting was suppressed by the use of high density metal barrels with a groove called a "clipper" at this position. The function of this "clipper" is not to shear the foil as it may seem, but rather to

---

*Martin, Froeschner & Associates, P O Box 17, Livermore CA 94550
†Kansas State University, Dept of Physics, Manhattan KS 66502

insure parallel non-convergent flow as the foil begins its flight up
the barrel. With clippered metal barrels we are able to accelerate
impactor foils with acceptable smoothness and planarity over most of
the barrel diameter to distances greater than the barrel diameter.
Previously, the interaction of the accelerating impactor foil with
the jet debris limited the useable area to less than a third of the
barrel diameter and the distance to less than half the barrel diameter. The improvement in impactor quality can be seen in
figure 1.

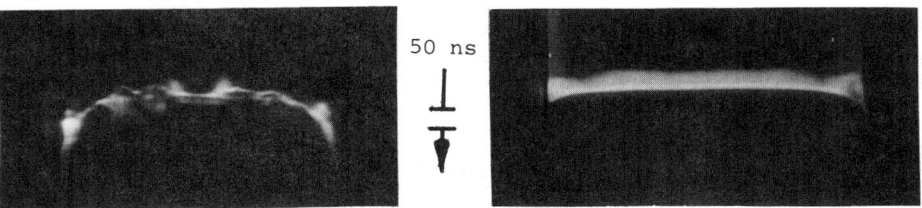

*Figure 1. Comparison of Impact Records Using an Aluminum Barrel
(left) and a Stainless Steel Barrel with "Clipper" (right).*

We demonstrated the absense of target motion prior to impact
of the impactor foil in a recent experiment in which a Fabry-Perot
velocity interferometer was trained on the thin portion of a stepped
Tantalum target. The instrument was sensitive enough to detect
velocities as low as a few meters per second. The record shows no
detectable change from before system trigger until emergence of the
shock produced at impact.

Early experiments had shown apparent preheat of the target,
particularly in the thin region of grooved targets covering the
end of the barrel when the groove was oriented perpendicular to the
direction of the current in the primary exploding foil circuit.
Electrical diagnostics and circuit modeling lead us to believe that
the barrel becomes shorted to the primary exploding foil circuit
shortly after burst. The target in these experiments was thus in a
position to carry substantial directly coupled currents. We have
estimated that this grooved section would be heated to near melting
by such directly coupled currents. Other early experiments with dielectric plastic barrels showed no improvement, perhaps because
even though direct coupling is prevented, inductive coupling is
worse due to absense of the shielding of the metal barrel. In the
experiments reported here, the targets were half-discs oriented so
that they did not bridge the barrel end in the direction of the primary current. This simple precaution seems to have eliminated significant preheat in the target.

EXPERIMENTS

The experiments we have conducted are of the symmetric impact
variety, i.e., the impactor and target are of the same material, thus
the material velocity, u_p, is exactly half the velocity of the impactor measured at the instant of impact. The shock velocity in the

target, u_s, is also directly measured and the other state variables, P, V, E can be derived from the Rankine-Hugoniot relations.

Primary diagnostics are Imacon image converter cameras operated in the streaking mode. One camera has its slit aperature aligned over the barrel diameter and across the steps in the target and a glass flasher. From this record we determine the time of arrival of the impactor at two different distances up the barrel, and the time of shock breakout at two different distances through the target. This record also provides an explicit picture of the shape of the impactor foil at the time of impact. Experiments which show unacceptable degrees of tilt, curvature or surface roughness can be rejected and repeated. A typical record from this primary camera is shown in figure 2, aligned with a cross sectional diagram of the experiment.

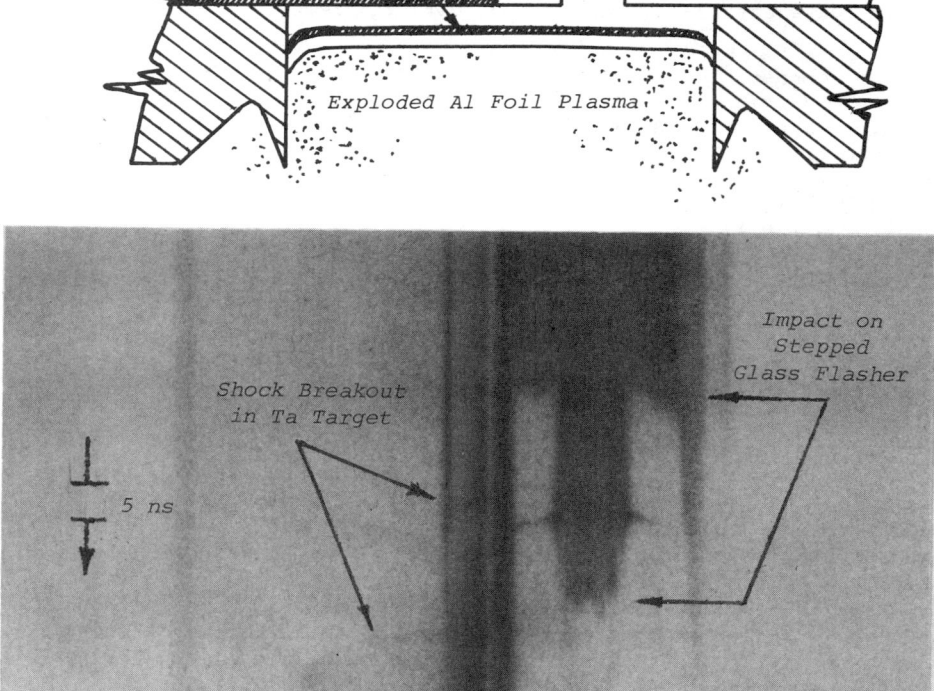

Figure 2. Experimental Record From Impact of a 0.076 mm Ta Foil at 6.4 km/s upon Ta and Glass Stepped Targets. Shot No. 6484. Barrel Evacuated to 10^{-4} Torr.

A second camera is used with a Fabry-Perot velocity interferometer to obtain a continuous history of the impactor velocity. This technique is capable of higher precision than the stepped glass flasher but the second camera has been unavailable. In the experiments described here, impactor velocities were determined by the stepped flasher technique.

RESULTS

The results reported here were obtained using relatively thick Ta impactor foils (0.076 mm) and targets (0.152 to 0.254 mm). This was done to reduce the impact velocities so that our data could be compared with data published by other researchers.[3,4] Our objective was to determine the validity of the electric gun experimental system.

The results we have obtained to date are plotted in figure 3, along with data from the literature on the shock Hugoniot of Ta.

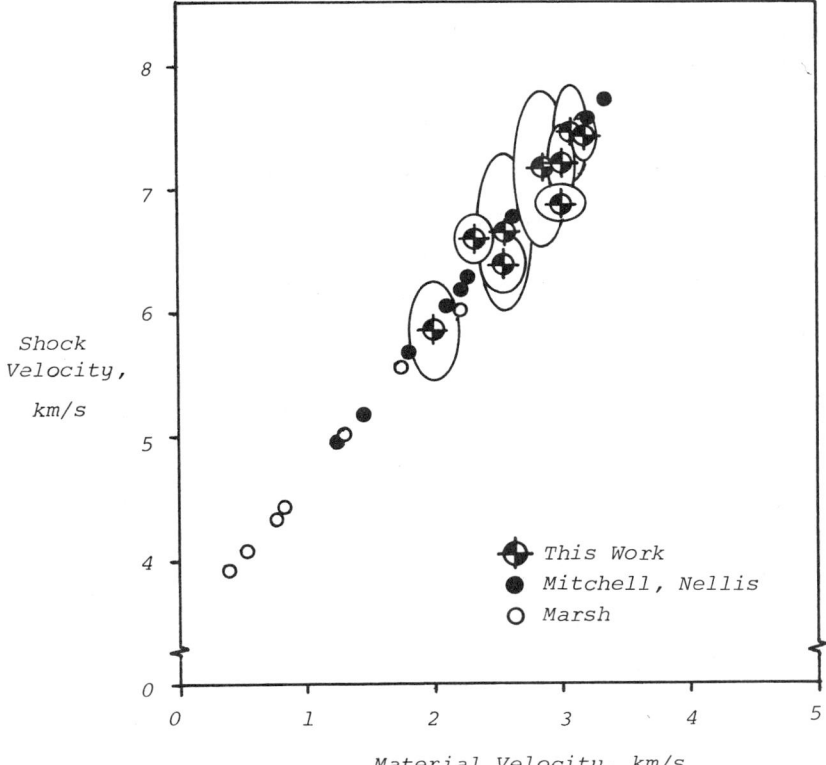

Figure 3. *Electric Gun Measurements of the Shock Hugoniot of Ta Compared to Data Published by Other Researchers.*

The first set of experiments was intended to establish feasibility and develop diagnostic techniques. The uncertainties are relatively large (±5 to 10%) due to uncertainty in the camera writing speed, inadequate step height measurements and poor image quality. Nonetheless, the largest discrepancy when compared to the high precision data of Mitchell and Nellis is less than 4%.

A second series of experiments was recently done with the camera writing speed calibrated to better than 1% with a mode-locked laser, step height metrology precise to about 1% and improved optics and experimental techniques. The uncertainties in these data are more like ±3% and they also agree with the published Hugoniot to within their experimental uncertainties.

An analysis of the ultimate precision attainable in this sort of experiment results in the identification of 8 error terms in the impactor velocity when determined by stepped flasher, and 7 error terms in the measurement of shock velocity in the stepped target. The magnitude of these terms ranges from negligible to a few percent. Assuming Gaussian distribution, we can expect uncertainties of ±2½ to 3% in material velocity and ±2½% in shock velocity. From the Rankine-Hugoniot relations this leads to uncertainties of approximately ±4% in pressure and ±4% in relative volume.

The uncertainties in the data presented here are dominated by the indistinct shock breakout signal in the Ta targets and the apparently fuzzy impact on the glass flashers. The shock breakout signal will be much brighter and easier to read at the higher pressures we intend to explore next. Better surface finish will reduce the inherent risetime of the signal. The indistinct or "fuzzy" character of the flasher impact records is probably due to roughness of the impactor surface. This surface roughness has amplitudes of from 10 to 30 μm at impact time and is thought to result from the presence of rolled-in inclusions of various impurities and oxides in the commercial grade Ta foils used in these experiments or from Taylor-unstable growth of defects in the Kapton surface beneath the Ta impactor foil. Refinements in material quality and preparation aimed at reducing this "fuzz" are in work.

REFERENCES

1. H. Chau, K. E. Froeschner, K. Mikkelson, R. S. Lee, R. C. Weingart, "Performance of a 100 kV Electric Gun System", this proceedings.
2. H. H. Chau, G. Dittbenner, W. W. Hofer, C. A. Honodel, D. J. Steinberg, J. R. Stroud and R. C. Weingart, "The Electric Gun: A Versatile Tool for High Pressure Shock Wave Research", Rev. Sci. Instrum. 51, 0028 (1980).
3. A. C. Mitchell and W. J. Nellis, J. Appl. Phys. 52, 3363 (1981).
4. Los Alamos Shock Hugoniot Data, S. P. Marsh, Compiler (University of California Press, Berkeley, 1979), p. 136.

ACKNOWLEDGEMENT: This work was performed under the auspices of the U S Department of Energy by the Lawrence Livermore National Laboratory under contract number W-7405-ENG-48.

RAILGUNS FOR EQUATION-OF-STATE RESEARCH*

R.S. Hawke, A.L. Brooks, and A.C. Mitchell
Lawrence Livermore National Laboratory, Livermore, CA 94550

C.M. Fowler, D.R. Peterson, and J.W. Shaner
Los Alamos National Laboratory, Los Alamos, NM 87554

ABSTRACT

It appears that a railgun can be used to accelerate an impactor plate to velocities of 10 to 40 km/s, which could generate shock pressures of 1 to 10 TPa. As a first step to determining the potential and limits of railgun accelerators for shock-wave equation-of-state research, a joint team of scientists and engineers at Los Alamos National Laboratory and Lawrence Livermore National Laboratory have initiated railgun research. We have utilized a combined capacitor bank and magnetic flux compression generator to power railguns and have demonstrated the feasibility of accelerating projectiles to ~10 km/s. This paper reports the status of experimental research directed toward launching EOS impactors and achieving higher velocities.

INTRODUCTION

Equation-of-state (EOS) research with shock waves can benefit from two important features of railgun-launched projectiles: high velocity (greater than 10 km/s) and large impactors (1-cm diam by 1-mm thick or larger). Impact velocities of 10 to 40 km/s would result in shock pressures of about 1 to 10 TPa for a tungsten impactor and target. The railgun's ability to accelerate large projectiles will permit fabrication of precision targets and impactors with ordinary machine tools and the use of well-developed diagnostic techniques currently used with two-stage light gas gun and explosive-accelerated impactors. Below we briefly describe the operation of railguns, launcher requirements for EOS research, and the joint Los Alamos National Laboratory and Lawrence Livermore National Laboratory railgun development project.

PRINCIPLE OF OPERATION

The railgun consists of a pair of rigid parallel conductors that carry current to and from an interconnecting movable conductor which functions as an armature (see Fig. 1). The force F on the armature is given by

$$F = \frac{L' I^2}{2}, \qquad (1)$$

---

*Work performed under the auspices of the U.S. Department of Energy by LLNL under contract W-7405-ENG-48, and by LASL under contract W-7405-ENG-36.

where I is the armature and rail current and L' is the inductance per unit length of the rail pair. A plasma arc serves as the armature. Confinement of the plasma arc behind the projectile is provided by the conducting rails on two sides and dielectric rail spacers on the other two sides. For EOS research the impactor is mounted in a sabot to maintain alignment and plasma confinement during acceleration. A variety of power sources can provide the megampere current and millisecond pulse durations needed.
In the past, railguns have been powered by battery,[1] capacitor bank,[2] homopolar generator,[3] and magnetic flux compression generator (FCG).[4,5]

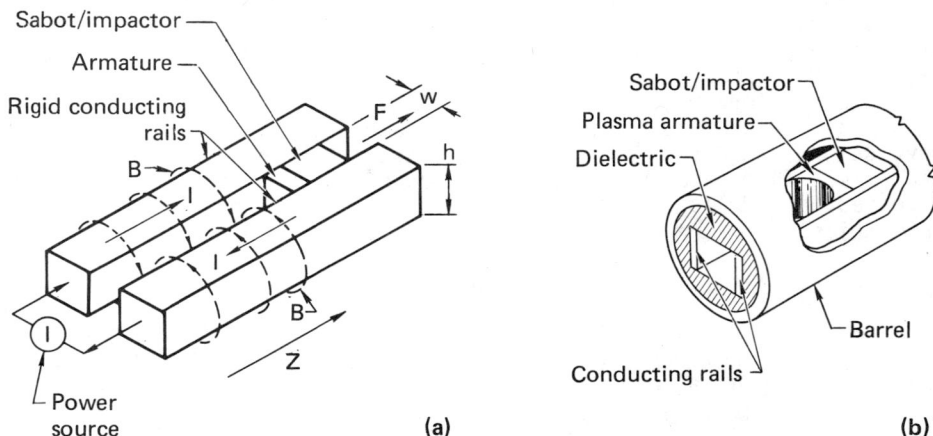

Fig. 1. (a) A railgun accelerator utilizes a power source that supplies the current to generate a magnetic field which, in combination with the armature current, exerts a propulsive force on the backside of the sabot containing the impactor. (b) Cutaway of railgun assembly. The dielectric maintains the rail position and, along with the rails, confines the plasma behind the sabot.

In a drag-free railgun, the projectile momentum is

$$mv = \frac{L'}{2} \int I^2 dt , \qquad (2)$$

where m is the projectile mass, v is the velocity, and t is time. A typical value of L' is about 0.4 µH/m. Hence, 1 MA delivered to a railgun for 1 ms would impart a momentum of 200 N·s (i.e., a 5-g projectile at a velocity of 40 km/s).

REQUIREMENTS FOR AN EOS LAUNCHER

To provide a significant advantage over contemporary shock-wave generation techniques, a railgun must meet a set of requirements that can be summarized as (1) launch velocities in the range of

10 to 40 km/s; (2) impactor size of 10-mm diam by 1-mm thick; (3) impactor flatness sufficient to limit shock velocity measurement uncertainty to 1%; and (4) impactor tilt of less than 1 or 2 degrees.[6] Furthermore, operation of the launcher must not interfere with EOS data acquisition and, in the long run, it is desirable that the cost of maintaining and operating the launcher be comparable to the cost of the EOS experiments.

LAUNCH VELOCITY

Lexan polycarbonate projectiles have been launched intact at velocities up to ~6 km/s and accelerated to velocities of about 10 km/s.[1,7] The operation of railguns has been well enough understood to develop a model[8] which indicates that velocities in the 10 to 40 km/s range might be possible.[9]

IMPACTOR AND SABOT MASS

The mass of a 10-mm-diam by 1-mm-thick impactor will range from 0.145 g for beryllium to 1.51 g for tungsten. A sabot is required to keep the impactor in alignment during acceleration, to electrically insulate the impactor from the rails to prevent current flow through it, and to confine the plasma to its base. A solid 12-mm polycarbonate sabot with an aspect ratio of 2/3 would have a mass of 1.39 g for a square-bore gun and 1.09 g for a round-bore gun. The tungsten impactor and square-bore Lexan sabot result in a total mass of 2.9 g and would have a kinetic energy of 2.3 MJ at 40 km/s.

IMPACTOR FLATNESS AND TILT

In order to limit the uncertainty Δu_s in shock velocity u_s, the uncorrected[10] impact distortion ε must be

$$\varepsilon \leq b \left(\frac{v}{u_s}\right)\left(\frac{\Delta u_s}{u_s}\right), \qquad (3)$$

where b is the thickness of the target. For a 1-mm-thick tungsten impactor impacting a 3-mm-thick tungsten target at 40 km/s, ε must be ±0.04 mm or less in order to limit $\Delta u_s/u_s$ to 1%. Limiting the impactor tilt must be accomplished with sabot and launcher design and precision, particularly at the muzzle of the railgun.

JOINT LOS ALAMOS/LIVERMORE RAILGUN R&D PROJECT

Los Alamos and Livermore are jointly developing railguns for application to EOS research. Experiments at Los Alamos use a 940-kJ (3 mf, 25 kV) capacitor bank coupled with explosively driven FCGs to power railguns[5] designed and fabricated at both Los Alamos and Livermore[11] (see Fig. 2). Experiments at Livermore use a 375-kJ (30 mf, 5 kV) capacitor bank as a power source. The capacitor banks

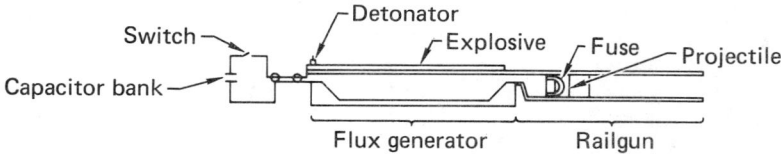

Fig. 2. Components of the FCG-railgun system. The capacitor-bank discharge develops a current in the inductance of the FCG. After the FCG is charged, the detonator ignites the explosive that drives the top conductor toward the bottom one. At this time, the circuit is shorted at the input end of the FCG, trapping the magnetic flux in the FCG-railgun system. Continued implosion of the FCG drives the flux into the railgun and maintains an extended near-constant current pulse in the railgun.

alone can generate railgun currents of about 1 MA. The FCG system can supply inputs ranging from nearly constant megampere currents for about 0.5 ms to shorter duration peak currents of several megamperes. The full-scale, high-energy experiments are done with FCG's (see Table I), while a capacitor bank alone is used for experimentation with similar peak currents, albeit shorter pulse duration.

TABLE I. Summary of results of FCG-powered 12.7-mm-bore railgun experiments.[7]

Accelerator length (m)	0.9	1.8	1.8
Projectile mass (g)	2.9	3.1	3.1
Initial capacitor bank energy (kJ)	70	200	390
Peak current (MA)	0.6	0.8	1.2
Peak acceleration (10^6 "g's")	1.9	3.2	7
Peak stress/elastic limit	5	9	15
Launch velocity (km/s)	2.8	5.5	10
Projectile integrity verified	Yes	Yes	No

We have commenced an effort to launch saboted impactors and implement the use of shielded shorting pins (see Fig. 3). Flatness and tilt measurements will be made in the near future. Sabot design modifications will then be made to reduce tilt and irregularities of the impact if required. Higher energy experiments are expected to achieve higher launch velocities.

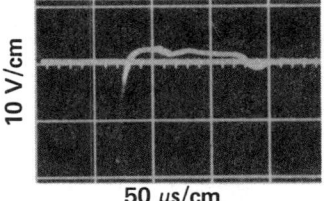

Fig. 3. Pin signal generated by a 1-km/s impact of a 1-g Lexan projectile with an aluminum target.

CONCLUSIONS

In summary, the requirements for a useful EOS railgun launcher are well-defined. The railgun and power supply performance characteristics are fairly well understood up to 10 km/s. We conclude that railguns have the potential to accelerate and launch large impactors at velocities in the 10 to 40 km/s range. The principal areas needing development are (1) techniques to minimize railgun interference with shock-wave diagnostics and (2) the attainment of velocities greater than 10 km/s with flat impacts.

ACKOWLEDGMENTS

We are grateful for the program support and guidance of John W. Kury, and the invaluable technical assistance of C.E. Cummings, B.M. Fox, N.J. Gibson, J.C. King, P. Lorton, and C.D. Wozynski.

REFERENCES

1. J. Hansler cited in Electric Gun and Power Source, Armour Research Foundation Report No. 3 on Project No. 15-391-E (1946).
2. D.E. Brast and D.R. Sawle, "Study of a Rail-Type MHD Hypervelocity Projectile Accelerator," in Proc. 7th Hypervelocity Impact Symp., Martin Co., Orlando, Florida, Vol. 1 (1964), p. 187.
3. S.C. Rashleigh and R.A. Marshall, J. Appl. Phys. $\underline{49}$, 1540 (1978).
4. R.L. Chapman, D.E. Harms, and G.P. Sorenson, "The Magnetohydrodynamic Hypervelocity Gun," in 6th Symp. on Hypervelocity Impact, The Firestone Tire and Rubber Co., Akron, Ohio (1963).
5. C.M. Fowler, D.R. Peterson, R.S. Caird, D.J. Erikson, B.L. Freeman, and J.C. King, "Explosive Flux Compression Generators for Rail Gun Power Sources," Los Alamos National Laboratory, Los Alamos, New Mexico, LA-UR-80-3190 (1980).
6. W.J. Nellis, Lawrence Livermore National Laboratory, private communication (1981).
7. R.S. Hawke, A.L. Brooks, F.J. Deadrick, J.K. Scudder, C.M. Fowler, R.S. Caird, and D.R. Peterson, "Results of Railgun Experiments Powered by Magnetic Flux Compression Generators," Lawrence Livermore National Laboratory, Livermore, California, UCRL-84875 (1981).
8. F.J. Deadrick, R.S. Hawke, and J.K. Scudder, "MAGRAC--A Railgun Simulation Program," Lawrence Livermore National Laboratory, Livermore, California, UCRL-84877 (1980).
9. R.S. Hawke and J.K. Scudder, "Prospects for Generating 1 to 10 TPa Pressures with a Railgun," in High Pressure Science & Technology, B. Vodar and P.H. Marteau, Eds. (Pergamon Press, Oxford and New York, 1980), pp. 979-982.
10. A.C. Mitchell and W.J. Nellis, Rev. Sci. Instr. $\underline{52}$, 347 (1981).
11. A.L. Brooks, R.S. Hawke, J.K. Scudder, and C.D. Wozynski, "Design and Fabrication of Small- and Large-Bore Railguns," Lawrence Livermore National Laboratory, Livermore, California, UCRL-84876 (1980).

ENHANCED PERFORMANCE OF A TWO-STAGE, LIGHT-GAS GUN*

A. C. Mitchell, W. J. Nellis, and B. Monahan

University of California, Lawrence Livermore National Laboratory
Livermore, California 94550

ABSTRACT

We have recently increased the maximum projectile velocity of our two-stage, light-gas gun[1,2] from 7 to 8 km/s by an integrated computational/experimental program. The higher velocity was used to measure Hugoniot data to 480 GPa (4.8 Mbar) in Mo.

EXPERIMENT

The gun is illustrated in Fig. 1. The hot gases from gun powder burned in the breech drive a piston into the pump tube, compressing a light gas, usually H_2. When the gas is compressed to a given pressure, the rupture valve opens and the compressed gas accelerates a projectile along the evacuated barrel. The high velocity is achieved by the increased sound speed of H_2 at the higher temperature in the isentropically compressed state and because of the roughly constant mass flow rate of the gas through the tapered section (the piston mass is very large compared to the gas mass). The physical limits to projectile velocity are the sound speed of the gas and dissipative losses in the flow. The engineering limits are the stresses in the tapered secton and in the breech.

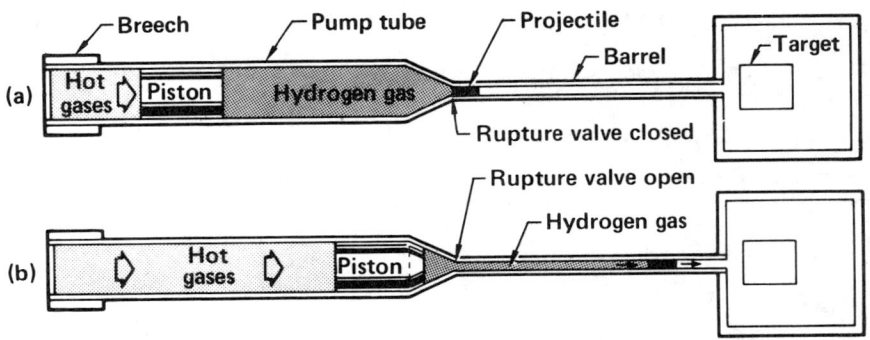

Fig. 1 Schematic of the operation of the two-stage gun. The pump tube is 10 m long with an inner diameter of 90 mm. The barrel is 9 m long with an inner diameter of 28 mm.

*Work performed under the auspices of the U.S. Department of Energy by Lawrence Livermore National Laboratory under contract #W-7405-Eng-48.

The purpose of this study was to optimize performance by using the one-dimensional simulation code GASGUN as a guide. This Lagrangian hydrodynamic code was originally developed at the Naval Ordnance Laboratory[3] and our present version was described recently.[4] The code takes the tapered section into account by a

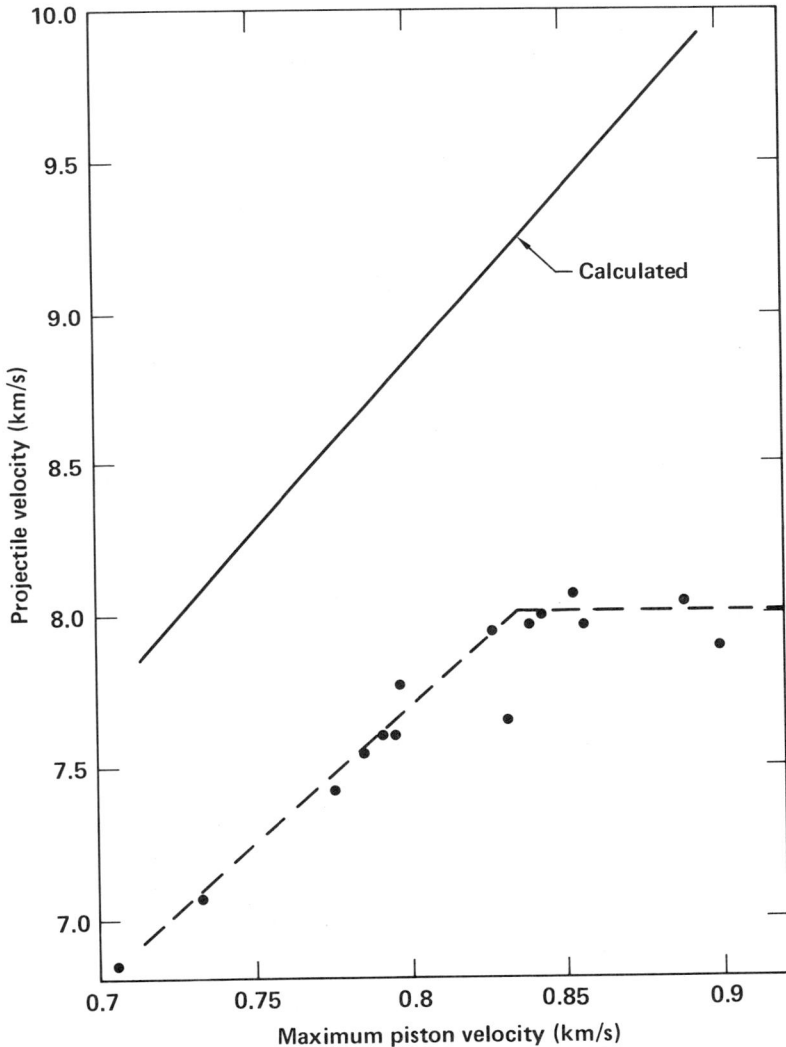

Fig. 2 Measured (dots) and calculated projectile velocities vs maximum piston velocity for an initial H_2 gas pressure of 1.03 MPa (10.3 bar).

flume-type calculation. The driving pulse generated by the gunpowder is calculated by a burn model normalized to measured pressure histories in the breech. All losses are ignored, including friction and turbulence.

It was pointed out to us that the highest velocities are achieved with relativley slow burning gun powder.[5] The effects of replacing our fast-burning powder were then investigated calculationally.[6]. An experimental program guided by the code was initiated in which the piston mass, H_2 fill pressure, powder type and mass, rupture-valve opening pressure, and projectile mass were varied. Piston velocity, projectile velocity, breech pressure, radial growth of the tapered section, and impactor flatness were measured.

RESULTS

The enhanced performance was achieved by using more mass of slower-burning gun powder to achieve a higher driving pressure late in the pulse, more H_2 gas to drive the projectile longer, and a lighter piston which is decelerated more thus reducing stress in the tapered section. In fact, we found growth of the tapered section reduced to practically nothing, which will increase the life of this highly-stressed component.

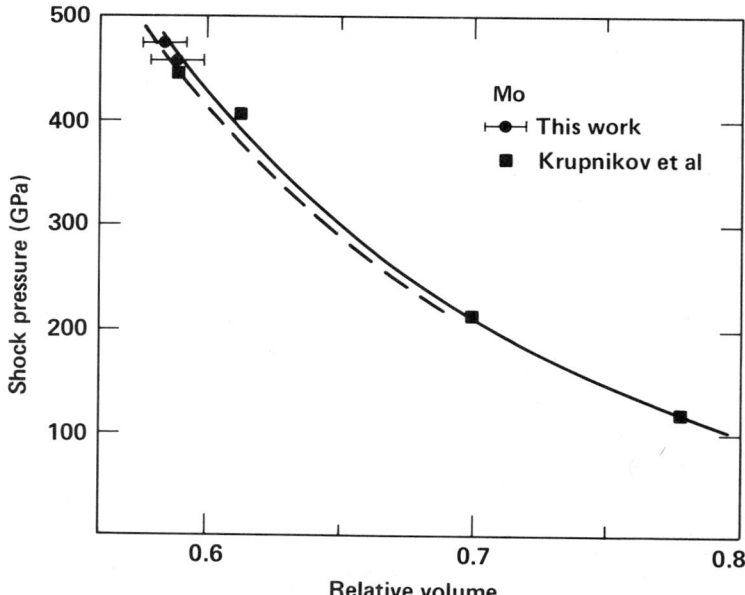

Fig. 3 Shock pressure vs relative volume for Mo. The solid curve is the fit to the data of Krupnikov, et al. (Ref 8). The dashed curve is the extrapolation of the fit to the data of McQueen, et al. (Ref. 9)(100 GPa = 1 Mbar).

The results are summarized in Fig. 2, where projectile velocity is plotted vs maximum piston velocity. The data parallels the calculated velocities up to 8 km/s, which is an experimental limit for the present configuration. The cause is unknown at present.

Two Hugoniot data points were obtained in Mo at 480 GPa by impacting with Ta at 8 km/s and are plotted in Fig. 3. The technique was the same one used previously,[7] except length scales were reduced to achieve a lighter, faster projectile. Our data are in good agreement with previous Soviet results,[8] as well as with extrapolations of Los Alamos data.[9]

REFERENCES

1. A. H. Jones, W. M. Isbell, and C. J. Maiden, J. Appl. Phys. 37, 3493 (1966).
2. A. C. Mitchell and W. J. Nellis, Rev. Sci. Instrum. 52, 347 (1981).
3. R. Piacesi, D. F. Gates, and A. E. Seigel, U.S. Naval Ordnance Laboratory, White Oak, Maryland, NOLTR 62-87 (1963).
4. G. R. Gathers, Lawrence Livermore National Laboratory, Livermore, California, UCID-18574-80-3 (1980).
5. K. Boutwell, General Motors Corp., Goleta, California, private communication (1979).
6. B. Monahan and M. van Thiel, Lawrence Livermore National Laboratory, Livermore, California, UCID-18574-80-1 (1980).
7. A. C. Mitchell and W. J. Nellis, J. Appl. Phys. 52, 3363 (1981).
8. K. K. Krupnikov, A. A. Bakanova, M. I. Brazhnik, and R. F. Trunin, Sov. Phys. Doklady 8, 205 (1963).
9. R. G. McQueen, S. P. Marsh, J. W. Taylor, J. N. Fritz, and W. J. Carter, in High-Velocity Impact Phenomena, edited by R. Kinslow (Academic, New York, 1970), p. 542.

SHOCK EFFECTS IN PARTICLE BEAM FUSION TARGETS

M. A. Sweeney, F. C. Perry, J. R. Asay, and M. M. Widner
Sandia National Laboratories, Albuquerque, NM 87185

ABSTRACT

At Sandia National Laboratories we are assessing the response of fusion target materials to shock loading with the particle beam accelerators HYDRA and PROTO I and the gas gun facility. Nonlinear shock-accelerated unstable growth of fabrication irregularities has been demonstrated, and jetting is found to occur in imploding targets because of asymmetric beam deposition. Cylindrical ion targets display an instability due either to beam or target nonuniformity. However, the data suggest targets with aspect ratios of 30 may implode stably. The first time- and space-resolved measurements of shock-induced vaporization have been made. A homogeneous mixed phase EOS model cannot adequately explain the results because of the kinetic effects of vapor formation and expansion.

INTRODUCTION

The feasibility of inertial confinement fusion (ICF) for commercial power production is dependent upon the extent to which symmetry, stability, and hydrodynamic response can be controlled in an imploding target. At Sandia National Laboratories we are developing techniques to address these questions by studying the shock loading response of a variety of cylindrical and slab targets. The facilities presently in use, the two-stage light-gas gun and the charged particle beam accelerators HYDRA and PROTO I, provide shock wave amplitudes in the 1 - 10 Mbar range, ablation pressures of $\sim$1 Mbar, and stagnation pressures (in cylindrical implosions) of $\sim$40 Mbars. Here we present experimental data and numerical simulations of shock loading response obtained on these facilities. The data, coupled with numerical simulations, should provide a useful link in scaling the required beam uniformity and target quality to thermonuclear conditions. Although our specific goal is to estimate the light ion target performance expected on the modular, multiterawatt accelerator PBFA at pressures of > 100 Mbars, these studies are relevant to the determination of the thermodynamic state, material properties, and hydrodynamic response of matter at high pressures.

SHOCK-INDUCED INSTABILITY STUDIES ON HYDRA

In experiments on the HYDRA accelerator, ablatively-driven double-shell cylinders with and without initial perturbations at the outer edge of the pusher have been imploded using a single electron beam.[1] Here, as on PROTO I and the gas gun facility, four-pulse holographic shadowgraphy has been used to obtain spatially and temporally resolved images of the implosions. Data are obtained by viewing the collapse of the cylinder along its axis.

Figure 1 shows four holograms, spaced 26 ns apart, from one such experiment. The target has a copper ablator, a polyethylene buffer, and a gold pusher. Well-defined initial perturbations at the buffer-pusher interface are provided by 36 regular scorings of the outer surface of the pusher produced by diamond milling.[2] The holograms of the inner surface reveal that perturbations develop on this surface after the shock breaks through. These free-surface irregularities grow and, later in time, the growth saturates due to the interaction of adjacent irregularities. The holograms also indicate that, because a single electron beam is used, material on the side of the cylinder receiving the higher deposition ablates off faster and implodes inward faster.

We have compared this experimental data on the free surface motion with 2-D planar geometry numerical calculations from the materials response code CSQ.[3] In Figure 2 we can see the "bubble and spike" pattern characteristic of nonlinear Rayleigh-Taylor instabilities. The instability is actually a special type of Taylor instability (studied by Richtmyer[4]) due to impulsive acceleration by a shock wave traveling from a light fluid into a heavier fluid. There is also some evidence of deformation of the bubbles and spikes due to Helmholtz instability. At late time the experimental free surface amplitude assumes a random character. This may also be indicative of a turbulent stage of the unstable hydrodynamic motion.[5] The calculation also indicates that fluid-like behavior dominates; i.e., fracture is unimportant and elastic-plastic effects do not control the onset of the unstable growth.[6] Convergence effects are important, however, and contribute to a faster instability growth in the experiment and, consequently, an earlier and more rapid rate of decrease in the free surface perturbation amplitude.

Even without deliberate initial perturbations, the inner surface of the gold as seen in holographic data is not completely smooth. For example, jet formation is sometimes evident at a late phase of the implosion (Figure 3). The jet results from deformation in the implosion cavity due to the presence of a wall during the early part of the collapse. The "wall" is the lower deposition side of the cylinder. Such jet formation is analogous to the collapse of an initially spherical vapor cavity in the neighborhood of a solid boundary.[7]

DEPOSITION AND BEAM AND TARGET STABILITY STUDIES ON PROTO I

In experiments[8] on PROTO I with the radial (applied B) ion diode,[9] we have been studying the ablative acceleration and implosion of aluminum cylinders[2] using holographic shadowgraphy. The multiframe holographic data have been interpreted with CHARTD[10], the 1-D counterpart of CSQ, using an ion-beam ray-trace deposition model.[11] For 3 mm-diameter, 50 μm-thick aluminum cylinders, peak proton intensities of 1 - 2 TW/cm^2, ∼ 1 Mbar ablation pressures, and ∼ 40 Mbar stagnation pressures are inferred. Since the beam pulse duration is much greater than the shock transit time through the shell, the acceleration is much like a rocket rather than a single shock unloading.

The experimental data are encouraging in that the overall implosion symmetry was good for several shots; i.e., $\Delta v/v$ asymmetries were less than 10% over > 50% of the target circumference for some 50 μm-thick shell targets with aspect ratios of 30. Nevertheless, nonuniformities in the implosion are seen in most of the holograms (Figure 4) and, in one instance, what appears to be perturbation growth is seen. Large targets (6 or 9 mm-diameter) with aspect ratios exceeding 30 exhibit asymmetries with a dominant wavelength $\lambda = 2 - 4$ mm. Symmetry improves and the observed perturbation shifts to smaller wavelengths ($\lambda \simeq 0.1 - 1$ mm) if target size--and hence aspect ratio--is decreased.

Detailed microscopic examination of typical targets reveals no significant initial perturbations which can be associated with the observed nonuniformities. If the nonuniformities arise from target irregularities, the growth of these irregularities must be very large (∼ two orders of magnitude). Instead, the nonuniform implosions may be due to beam perturbations: 1) anode plasma nonuniformities, 2) instabilities in the electron space charge cloud, or 3) beam filamentation in the drift region. To establish the validity of these conjectures, calculations and other diagnostic techniques are needed.

SHOCK-INDUCED VAPORIZATION STUDIES ON THE TWO-STAGE LIGHT GAS GUN

On the gas gun facility at Sandia National Laboratories,[12] shock amplitudes to 6 Mbars are attainable. These shocks are sufficient to induce melt and even vaporization in some materials. The unique features of the gun technique as compared to techniques employing particle beam accelerators are well-characterized uniaxial shocks and reproducibility of shock loading to 1% on a shot-to-shot basis. The gas gun can therefore easily provide a data base for the kinetics of melt and vaporization due to shock compression.

In recent experiments, vaporization in lead disks[2] has been studied for shocks up to 2.9 Mbar. The phase diagram for lead indicates that at shock pressures of ∼1.8 Mbar incipient vaporization on the free surface occurs. At pressures of ∼ 3 Mbar, vaporization should be complete. A VISAR diagnostic foil (steel or tantalum) is used to measure continuous mass-velocity distributions. These data, coupled with holographic shadowgraphy, represent the first time- and space-resolved measurements of shock-induced vaporization. In the future, flash radiography will also be used to study the effluent from the shocked lead surface.

Some conclusions from a preliminary analysis of the data are as follows. The shape of the foil velocity profiles (Figure 5) suggests that adiabatic expansion of the vapor ahead of the liquid dominates the early response to shock unloading. At the highest pressures, the VISAR foil is accelerated in a relatively "shockless" manner to a velocity approaching 10 km/s. Vaporization effects dominate over the lower background of surface defect ejecta seen at lower shock pressures. The fringe curvature in the holographic data is 180° out of phase when a low pressure shot is compared with a high pressure shot; this may be indicative of attaining a vapor state.

A CHARTD[10] simulation of one experiment gives good agreement with the observed peak foil velocity but disagrees with the velocity risetime. In the experiment, the risetime of the velocity (ignoring the low velocity foot) is about 100 ns; in the numerical simulation, it is about 10 ns. We conjecture that the simple homogeneous mixed phase model used in the code is inappropriate and that the vapor is actually expanding ahead of the liquid. (The 100 ns risetime is consistent with the sound transit time for the vapor portion of the liquid/vapor mixture in the calculation.) The observed profile shapes are probably related to the rate of vapor formation, whereas in the CHARTD model incipient vaporization occurs instantaneously once a sufficient activation energy is reached.

REFERENCES

1. M. A. Sweeney and F. C. Perry, J. Appl. Phys. $\underline{52}$, (1981).
2. Targets were fabricated by Group CMB6 at Los Alamos National Laboratories.
3. S. L. Thompson, Sandia Laboratories Report SAND77-1339, January 1979.
4. R. D. Richtmyer, Comm. Pure and Appl. Math $\underline{13}$, 297 (1960).
5. N. A. Inogamov, Sov. Tech. Phys. Lett. $\underline{4}$, 299 (1978).
6. D. C. Drucker, Mechanics Today, Vol. 5, ed. by S. Nemat-Nasser (Northwestern University, Evanston, IL, 1979), p. 37.
7. M. S. Plesset and R. B. Chapman, J. Fluid Mech. $\underline{47}$, 283 (1971).
8. F. C. Perry, Bull. Amer. Phys. Soc. $\underline{26}$, 663 (1981).
9. D. J. Johnson et. al., Phys. Rev. Lett. $\underline{42}$, 610 (1979).
10. S. L. Thompson and H. S. Lauson, Sandia Laboratories Report SC-RR-71-0713, February 1972.
11. M. M. Widner, T. A. Mehlhorn, and G. R. Montry, Bull. Amer. Phys. Soc. $\underline{25}$, 944 (1980).
12. L. C. Chhabildas, Bull. Amer. Phys. Soc. $\underline{26}$, 663 (1981).

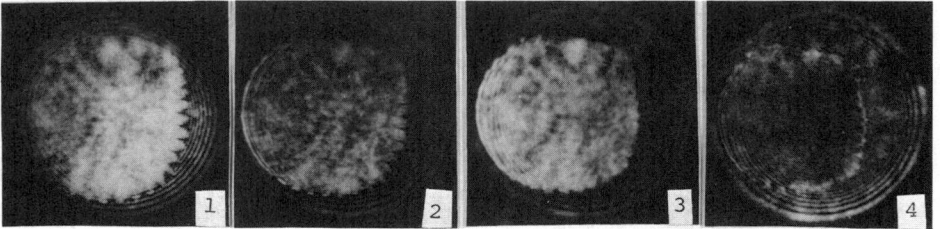

Fig. 1. Holograms of the initially perturbed Cu/CH$_2$/Au cylinder, driven by a single electron beam, at 26 ns intervals.

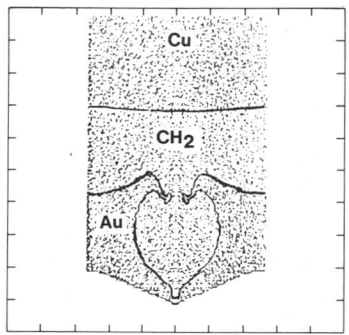

Fig. 2. Imploded state of CSQ calculation showing unstable growth pattern.

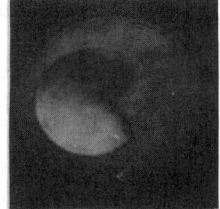

Fig. 3. Hologram showing jet formation in an electron target at a late phase of the cylindrical implosion.

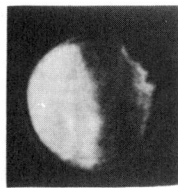

Fig. 4. Hologram of ion target showing nonuniform implosion.

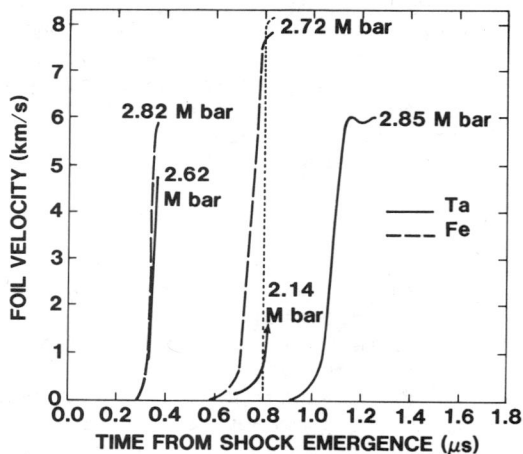

Fig. 5. Foil velocity profiles in shock vaporization studies. Dotted line denotes CHARTD calculation.

CHAPTER IV: Equation of State

SOME TECHNIQUES AND RESULTS FROM HIGH-PRESSURE SHOCK-WAVE EXPERIMENTS UTILIZING THE RADIATION FROM SHOCKED TRANSPARENT MATERIALS

R. G. McQueen and J. N. Fritz
Los Alamos National Laboratory

ABSTRACT

It has been known for many years that some transparent materials emit radiation when shocked to high pressures. We have used this property to determine the temperature of shocked fused and crystal quartz, which in turn allowed us to calculate the thermal expansion of SiO_2 at high pressure and also the specific heat. Once the radiative energy as a function of pressure is known for one material we show how this can be used to determine the temperature of other transparent materials. By the nature of the experiments very accurate shock velocities can be measured and hence high quality equation of state data obtained. Some techniques and results are presented on measuring sound velocities from symmetrical impact of nontransparent materials using radiation emitting transparent analyzers, and on nonsymetrical impact experiments on transparent materials. Because of special requirements in the latter experiments, techniques were developed that lead to very high-precision shock-wave data. Preliminary results using these techniques are presented for making estimates of the melting region and the yield strength of some metals under strong shock conditions.

INTRODUCTION

Los Alamos has been using the radiation from shocked gases for many years to obtain equation of state data for other materials by the flash gap technique developed by J. M. Walsh.[1] Most of those experiments were performed on materials where no radiation could be expected to be observed, and the few where it might be were usually shielded by shims. Probably the first to utilze the radiation emanating from the shock front in materials pertinent to this report was the work of Kormer et al.[2] reported in 1965. They used this effect to infer the temperature of some shocked solids. Until the last few years this powerful tool was not exploited.

During the following years considerable effort was made to measure the sound velocity at high pressures; by ourselves and others, Al'tshuler et al.[3] in particular. Such efforts, if successful, could yield the longitudinal and bulk sound velocities and hence the shear moduli and the quest of the last decade or so; the elusive Grüneisen parameter. These early measurements were basically of the x-t type and hence not time resolved. The early time resolved measurements were limited to low pressure experiments. The ASM probe[4] and VISAR[5] opened a new range of high pressure time resolved experimention.

The techniques described here were developed for two reasons: one to measure the radiation temperature of shocked transparent materials; and two to measure the location of overtaking rarefaction waves behind strong shocks and hence determine the velocity of sound waves at high pressure. In addition observation of the radiation from the shock wave has led to the development of several very accurate shock velocity measuring systems.

In 1968 we performed several experiments to determine the overtaking wave velocity in fused quartz impacted by a stainless steel driver plate. The fused quartz target assembly consisted of a stack of plates held together by double stick tape so that a small ~.05 mm gap separated the plates in the central region. This area was viewed through a small slit with a sweeping image camera. It was felt then, and now, that by observing the shock arrival at the various gaps through the same slit that the shock velocities could be measured very accurately through each layer. These records clearly show the shock arrival at the various levels. In Fig. 1-a the calculated shock velocity vs thickness shows pretty much where the rarefaction overtakes the shock, but clearly more or higher quality data would be needed to locate accurately the exact overtake position. Undoubtedly the most important feature of these records was ignored; that of the intense film darkening at the beginning of the record and subsequent decrease. This record shows that fused quartz radiates copiously and that this could be used to determine its temperature. Moreover it showed that these measurements offer a sensitive method of determining overtake wave velocities. It was over ten years later, while we were now in our present endeavors that these old experiments were recalled. A microdensitometer scan was made of one of the records and plotted 1-b. This shows the radiation from the flash gaps as well as the decrease caused by the pressure

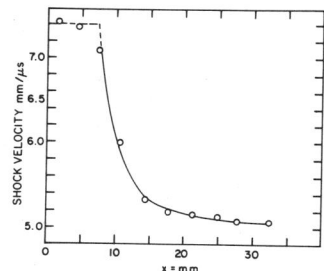

Fig. 1-a. Shock velocity vs thickness through a stack of fused quartz plates.

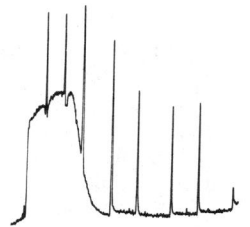

Fig. 1-b. Microdensitometer scan of the photographic record. The spikes are from the flash gaps.

release due to the flash gaps. The most salient feature of course is the decrease in radiation caused by the overtaking release wave. The electronic equipment to exploit these phenomena were available at the time. For example we are using photomultipliers developed even earlier. This should have been a most shining example of serendipity but unfortunately we were looking in a different direction. It is also of interest that the decrease in radiation due to the rarefaction overtake seen by Kormer et al.[2] was also disregarded as an experimental technique.

High quality shock velocity measurements can be made by monitoring changes in radiation levels as the shock passes various interfaces. Since these interfaces can be made to have negligible thickness one has in effect zero-perturbation time markers. The application for making shock velocity measurements will be noted where appropriate since they are usually required as part of either the temperature or overtake measurement.

RADIATION TEMPERATURE MEASUREMENTS

There are several ways to obtain temperatures from radiation measurements. The distinction is basically the difference from looking at the radiation from the following sources: narrow band widths centered on a few wave lengths, so called color temperatures; many small discrete intervals, spectral analysis,[6] and over a broad spectral range, the brightness temperature. We have used brightness temperatures here primarily because of the work done by W. Davis of this laboratory. He determined the brightness temperature of detonating nitromethane by photographic techniques. Since we have comparable photographic capabilities here, we decided to use his work to calibrate the relative radiation of shocked materials to detonating nitromethane.

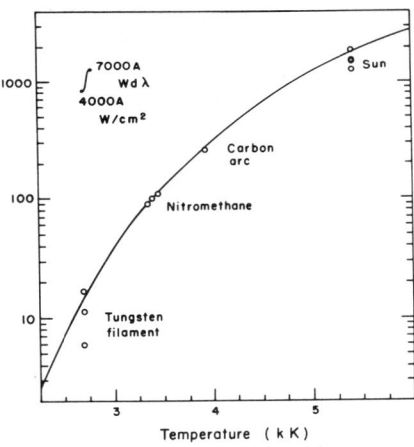

Fig. 2. W. Davis's nitromethane calibration function.

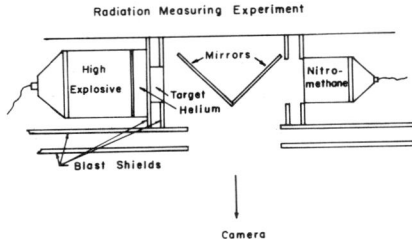

Fig. 3. High explosive system to view both the nitromethane standard and the unknown.

To establish his standards he calibrated to a tungsten filament, the carbon arc and the sun. Assuming that the photographic film responds uniformly to the radiation between 400 and 700 nm he calculated the curve (Fig. 2) of the energy radiated as a function of temperature. The calibration points are plotted on this curve at their estimated temperature with their converted measured brightness to the radiated energy scale, along with the measured nitromethane points. In these experiments we compare the relative intensity of shocked quartz to that of detonating nitromethane by viewing each through discrete openings of various widths placed on the cover plate. The same plate with a range of four in widths was used for many of the nitromethane standards. A suite of cover plates; usually four steps with a range of two were used for the unknown. By suitable combinations over two orders of magnitude in intensities could be compared. An explosive system used on some experiments is shown in Fig. 3 and a reproduction of a photographic record in Fig. 4. A microdensitometer scan of the record is reproduced in Fig. 5.

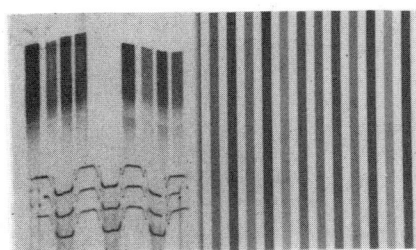

Fig. 4. Photographic record. The unknowns are on the left. The bands on the right are the standards.

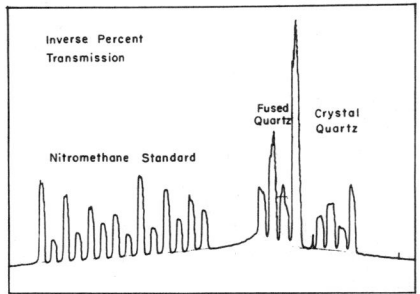

Fig. 5. Microdensitometer scan of a photographic record.

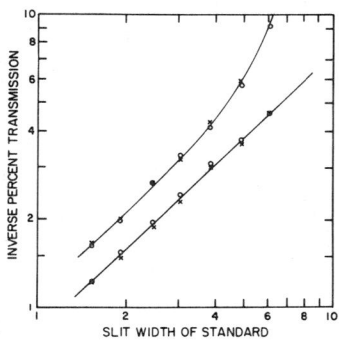

Fig. 6. Inverse percent transmission vs slit width on the nitromethane.

Both film densities and inverse percent transmission of the records have been used to determine relative intensities. A plot of inverse percent transmission of the standard (Fig. 6) gives an indication of the inherent precision of these measurements. The o's and x's are readings from openings of the same width but different location. The two curves on the plot are readings obtained from different settings of the controls of the microdensitometer. Comparing the measured values of the quartz with these curves gives the relative intensities of the unknown to nitromethane.

The results of these measurements are given in Fig. 7 where we have plotted the relative intensities for both fused and crystal quartz as a function of shock particle velocity or what is equivalent to the square root of the internal energy. Below 4.0 km/s both fused and crystal quartz lie on the same curve and except for the region around 4.2 km/s, are approximately on the same but different curve in the higher pressure regime. Clearly in the region from 4.0-4.2 km/s a phase change has occurred. This was first reported at the fall AGU Meeting in 1979. In fused quartz shocked above 4.2 km/s the photographic records show that as the rarefaction from the HE side of the driver overtakes the shock in the quartz that there is first a decrease in radiation and then an increase which on occasion rises to higher levels than the original radiation. This can be seen in Fig. 8-a where a reproduction of an oscilloscope record shows the time history of fused quartz shocked to a particle velocity of approximately 4.2 km/s. There is a fairly rapid increase in radiation, ~ 10 ns, followed by a near constant level and then an increase and subsequent decrease. Examination of the region just before the increase shows a slight decrease. What has happened is that the rarefaction has just overtaken the shock front and with

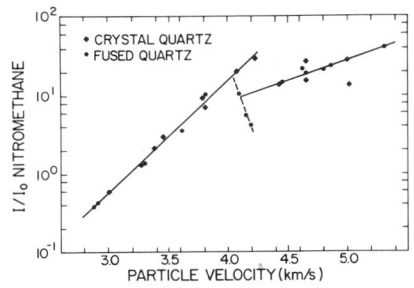

Fig. 7. Internal energy vs relative radiation energy for SiO_2.

further decrease in pressure the radiation increases until it reaches the maximum of the first phase. Assuming that the recording system is linear we find the increase to be approximately a factor of three indicating that perhaps the line drawn for the high pressure phase should have a higher slope going closer to the two low intensity points. Another feature of interest is the crystal quartz data at about 4.2 km/s. It can be seen that the intensity is greater than the maximum of the low pressure fused quartz data. This is not an indication of an error since at a given particle velocity the pressure will be considerably higher. This effect can be seen in Fig. 8-b where crystal quartz (with a piece of fused quartz placed on it) was impacted into this pressure regime. The decrease in intensity is due to the shock entering the high pressure phase of the fused quartz. The rarefaction eventually catches the shock and the pressure decreases until it reaches the phase line where the radiation behaves as in Fig. 8-a. We also note that the four high pressure crystal quartz points show a lot of scatter. These four points actually are from only two shots. It appears that in this pressure regime the quartz has an initial level of radiation that falls off fairly rapidly to a lower but constant level. An oscilloscope record showing this is reproduced in Fig. 8-c.

Fig. 8. Oscilloscope records. (a) Shock running through fused quartz showing the increase in radiation just as the shock wave is degraded by the rarefaction wave. (b) Shock into crystal quartz below the phase change showing sharp decrease in intensity when it reaches fused quartz. Fused quartz is shocked into the high pressure phase. The rarefaction in the fused quartz shows the same features as 8-a. (c) Relaxation seen in other experiments. The radiation increases when it enters the fused quartz. Rarefaction behavior is the same but modified because of the higher pressure.

This would appear to be a manifestation of a metastable state relaxing to some equilibrium value. There are still some interesting features of the behavior of shocked SiO_2 that additional experiments might resolve. In particular are the oscillations in fused quartz when shocked directly into the transition region. This phenomena was reported in the 1979 AGU Meeting. We have since observed this four times, twice with photomultiplier systems. The data of Fig. 7 have been transformed to temperature vs pressure via Fig. 2. We have used those P-T curves to determine the specific heat at constant volume and the thermal expansion at constant pressure. In this work reciprocity of the film response was tested and found to be good over a couple orders of magnitude as well as the black body assumption.

RADIATION MEASUREMENTS USING OTHER DETECTORS

Probably the most desirable feature of the measuring technique just described is having a calibrated standard on each experiment. In addition the logarithmic response of the film, while losing sensitivity, makes it almost impossible not to obtain some data on an experiment. The disadvantages are problems associated with scattered light, nonuniform background and the loss of sensitivity just mentioned. It is well known that photodiodes and photomultipliers have a linear response to radiation over a fairly large range of intensity. This makes them logical choices for measuring radiation temperatures. One weakness in using these is that the system must be calibrated, usually with a tungsten filament. While we have a black body furnace that can be used to almost 3000 K, the most difficult problem is calibrating the system so that the static and dynamic outputs are exactly the same. Accounting for window surfaces is nontrivial and putting a detonating nitromethane radiation pulse on the records through the same optical system is probably impossible. What we have done is something very close to the impedance match technique developed by Walsh.[7] In fact for complete analysis of the experiment one must make such a calculation. We simply sandwich the material to be investigated between two pieces of a "standard." The standard should be characterized both hydrodynamically, so the impedance match calculation can be made, and radiatively-wise so that the radiation from the unknown can be compared with it. It is desirable to match the radiation levels from the standard and the unknown as well as possible for maximum precision. It is also desirable to match their shock impedances so that the reflected shock or release wave from the unknown sample can be used to estimate the Grüneisen parameter after its Hugoniot has been measured. An example of this type of experiment is shown in Fig. 9. Extremely accurate shock velocities can be obtained with this technique.

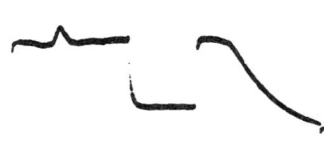

Fig. 9. Radiation from a quartz sandwich experiment done on a 2024 base plate. The radiation begins when the shock enters the first fused quartz plate and decreases when it enters the crystal plate and rises again when it enters the fused quartz cover plate. Relative intensities as well as shock velocities are measured.

If the radiation from the two materials are considerably different, filters can be used at one interface to match the radiation, but the shock velocity measurements will suffer accordingly as will the inherent precision in determining pressures in the three shock states from the radiation levels.

To abstract the temperature from these measurements one must calculate a radiation-temperature curve similar to Davis's, using the spectral response of the system over the appropriate frequency range.

The temperature of the standard for the experiment must be used to establish the relative intensity level for each measurement.

These experiments can be done using blocking filters so that windows with the greatest sensitivity can be employed. Time resolved spectral measurements can also be made. They cannot be performed accurately on multichannel devices that use a time integrated signal. However, we should make some spectral measurements so that the black body assumption can be checked, and to see if there are any regions of high spectral emission on absorption. To this end we are assembling a time resolved spectrometer to sample about ten spectral bands and a time-integrated resolved spectrometer to sample at 512 wave lengths.

RADIATION MEASUREMENTS FOR DETERMINING OVERTAKE VELOCITIES IN SYMMETRICAL IMPACTS

The rarefaction waves to be measured were generated with the widely used method of impacting a relatively thin driver plate onto a target plate. A drawing of the idealized shock-rarefaction process is shown in Fig. 10. This figure and others have been taken from a paper submitted to RSI where the experimental techniques are described in considerably more detail than is done here. In this schematic the interaction of the rarefaction waves from the HE-Driver interface, referred to here as the back side, with the shock wave in the target decreases its velocity in a nonlinear manner. The first degradation comes from what has been designated as the lead characteristic, which travels at the longitudinal velocity in this example of an elastic-plastic solid. The bulk rarefaction wave further decreases the shock velocity to that governed by the tail characteristic. This state is determined by the pressure build up when the HE gases impinge on the back side of the driver when it is decelerated by the initial shock. We have indicated by the small inset drawing the basic difference between the type of experiments described here and those done with an insitu gauge or those that monitor velocities at an interface. If pressure and time were measured both gauges would record the same pressure levels of the different interactions. However, the time scales would not be the same and a linear transformation would not suffice to make them the same. In all the experiments described here we are monitoring the radiation emitted from the shock front.

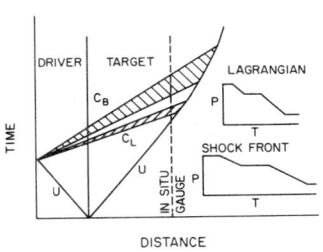

Fig. 10. Schematic of shock and rarefaction from a symmetrical impact of an ideal elastic-plastic solid.

If we restrict our attention to the lead characteristic it can be seen that by equating the time for the shock to go through the driver plus the time when the rarefaction through the driver and target catches the shock in the driver with that shock transit time then the Lagrangian sound velocity, C^L, is given by the following equation:

$$C^L = U_s(R+1)/(R-1) \qquad (1)$$

Here U_s is the shock velocity and R the ratio of target to driver thickness where the catch up occurs. The sound velocity, C, in the shocked material is given by:

$$C = C^L (\rho_0/\rho) \qquad (2)$$

where ρ_0/ρ is the ratio of the initial and final density. In this case the sound speed would correspond to the longitudinal component which we designate as C^L. The velocity of the second lead characteristic is referred to as the bulk velocity, C_B. Thus if R can be measured, the sound velocities can be determined.

In the preceeding sections it was shown that from some transparent materials copious radiation is emitted from the shock front and that the amount is sensitive to pressure. Thus if we made the target thinner, so that the rarefaction wave has not yet overtaken the shock, and placed a piece of fused quartz on its front surface and observed the radiation emitted we might expect to see records like that in the lower insert (Fig. 10). It would be distorted in the P direction since we would be measuring relative light intensity, not pressure. Clearly we should be able to determine the time when the radiation begins to decrease, but also it is obvious that unless one knows almost everything about quartz that the sound velocity in the other material can not be determined. However, if several measurements were made on the same experiment at various target plate thicknesses the times, Δt, for the shock to be overtaken in the transparent material can be used to determine when the rarefaction would have overtaken the shock in the target. Since in this regime all the characteristics are linear it is obvious that these Δt's are a linear function of the target thickness and when extrapolated to zero determines the position where the rarefaction wave would have overtaken the shock in the target. This position is independent of the properties of the transparent material, hence referred to as the analyzer. In Fig. 11 we have drawn the x-t solution for an impact experiment with the location and subsequent interactions of three analyzers. In this figure the left going characteristic from the target-analyzer interaction have been drawn with the same slope, a case that exists if the analyzer has the higher shock impedance. This of course is immaterial if only the leading wave is considered.

In all the experiments the radiation is viewed through small (typically 1 mm diam) apertures placed as close as possible to the analyzers by light pipes ~15 mm away. Thus the signals are averaged over an area ~2 mm in diameter. Two baffles between the light pipes and apertures prevent most of the unwanted radiation in the system from entering the

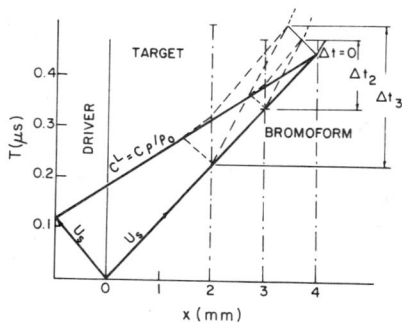

Fig. 11. x-t plot showing how by measuring the time for the rarefaction to overtake the shock in the analyzer that its velocity in the target can be determined.

light pipes. For explosive systems the target assembly is much like that shown in Fig. 12. Here five levels are indicated but sometimes ten or more have been used. The lower part of the assembly is used to measure the differential shock driver velocity, U_B, and hence establishes shock strength. These records also give a measure of bow and tilt so these effects on shock rise times can be accounted for. An assembly used for two-stage gun experiments is shown in Fig. 13. In this particular assembly the steps are placed on a circular array on the impact side of the target. This keeps light coming from strong interactions at corners from complicating the records. Differential velocity measurements can also be made with this assembly by taking PM signals from thick and thin areas of the target and feeding them into the same oscilloscope with one set inverted. The rise and fall represent shock arrivals at the target analyzer interface and when coupled with the measured projectile velocity determine a Hugoniot point. A reproduction of a photographic record used to determine the shock pressure is reproduced in Fig. 14 and the type of oscilloscope record to measure shock velocities on the two-stage gun in Fig. 15. From Fig. 14 it is found that the average time smear caused by bow and tilt over a two mm diameter area is about one ns in the region used to determine the overtake velocities. Records obtained in the central region have essentially no loss of time resolution due to tilt.

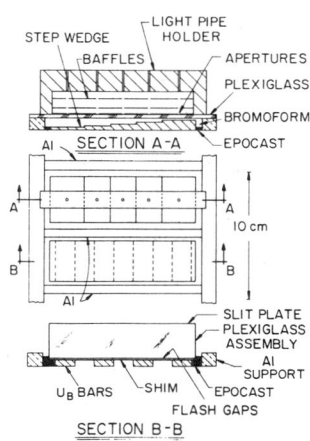

Fig. 12. Target assembly for high explosive experiments.

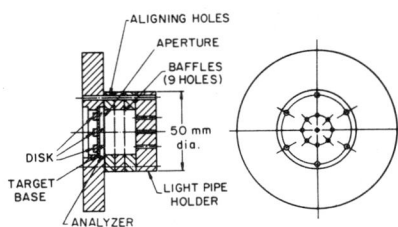

Fig. 13. Target Assembly for two stage gun experiments. Small disk on the impact side give the desired variable target thicknesses.

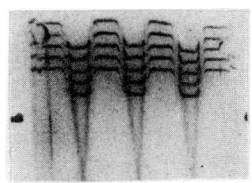

Fig. 14. A reproduction of a photographic record to determine the differential $(U_s - U_D)$ velocity and hence pressure.

We have used three materials as analyzers: fused quartz; a high density glass, $\rho \sim 5$ gm/cm^3; and bromoform, $Br_3HC, \rho \sim 3$ gm/cm^3. To demonstrate that the detection of the first overtake is independent of the analyzer, experiments were performed using all three on the same shot. Reproductions of the records are shown in Fig. 16 and the derived Δt vs thickness data in Fig. 17. The results of an experiment on 2024 Al designed to measure the first wave

Fig. 15. An oscilloscope record showing how the differential velocity can be measured on the two stage gun.

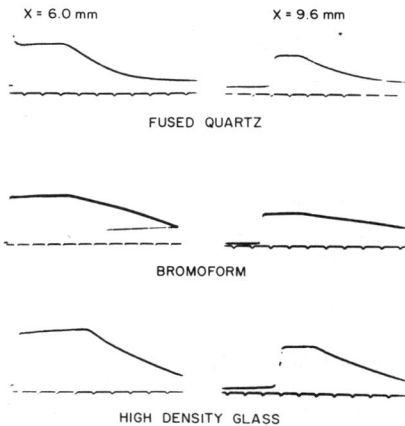

Fig. 16. Overtaking waves at two thickness seen by three different analyzers. The time marks are 0.1 μs.

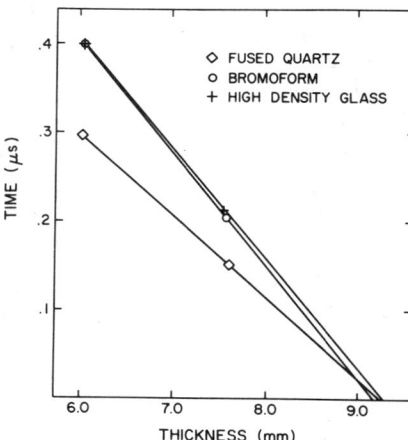

Fig. 17. The measured overtake time vs target thickness from the records of Fig. 16.

arrival are plotted in Fig. 18. The standard deviation for the catch up position determined from least square analysis of the data was 0.3%. Since the catch up ratio is approximately four for most materials these errors are reduced by approximately one half for the derived sound velocities.

Since the sound velocity is proportional to the shock velocity any errors in determining the shock strength are reflected directly in the derived sound speeds. However, this is really only very important if some phenomenon is occurring, such as a phase change where it is desirable to know the absolute value of the pressure. This is because these errors move the sound velocity along a curve that is nearly parallel to the true one. There is a source of error in the explosive experiments that is nontrivial. This is the change in driver plate thickness due to plate stretching while being accelerated. This is not a constant and varies from one explosive driver system to another. We have framing camera records that show that this can be as large as one percent in some geometries. Thin plate accelerated for short distances appear to have very little stretching. For example work done by Brown and McQueen[9] on iron show that explosive experiments done with thin drivers gave the same results as those performed on the two stage gun, which are not bothered by this effect.

So far we have restricted ourselves to measuring the lead characteristic. There is more information on the records but considerably less well defined and a lot less straight forward. A set of records (Fig. 19) for 2024-Al gives an indication of what can be seen in regards to the elastic-plastic behavior of this material under strong shock conditions. In the three records on the left the first break in the traces is due to the longitudinal component of the release wave. The

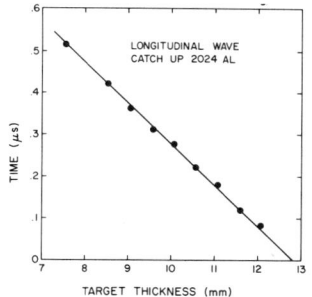

Fig. 18. The results from a very high quality set of records for determining overtake locations.

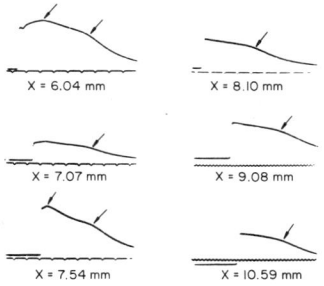

Fig. 19. Records showing the change in slope caused by the arrival of the bulk sound wave.

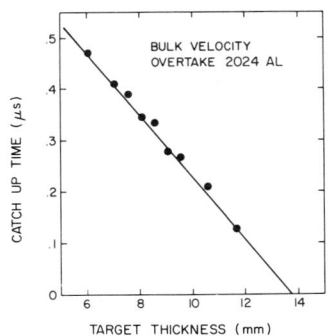

Fig. 20. Catch up time in the analyzer from records of Fig. 19.

second break much less well defined is due to the bulk velocity release. In the records on the right the longitudinal component has overtaken the shock wave in the target before reaching the analyzer and only the break from the bulk is present. The times for these breaks have been measured and are plotted in Fig. 20. In this set of records the second break was reasonably well defined. This is not always the case. The intercept gives an approximate value for the bulk sound velocity using the previous analysis. However, for an accurate value, the interaction of the analyzer (bromoform was used here) with the target must be accounted for as well as the shock degradation in the target due to the finite yield strength exhibited in the records. From the records it can be seen that the shock wave has traveled ~ 8 mm before being degraded by the longitudinal wave. The decrease in shock velocity for the remaining 5.5 mm of run can be accounted for in a reasonable manner. Another correction is needed to account for the decrease in pressure due to the longitudinal release. The bulk release velocity is actually centered on a somewhat lower pressure than that measured for the initial shock wave. A first order correction for this can be readily made. From our many measurements using bromoform as an analyzer we have determined the relative radiation intensity as a function of pressure in the 2024 Al target. Hence the pressure in the target can be estimated by the decrease in intensity of the bromoform. The equation for this is

$$P = P_o - 47 \log (I_o/I) \qquad (3)$$

where P is the pressure in GPa and I_o/I the ratio of intensities. From the records it appears that there is about 10 GPa pressure release before the bulk wave is seen. Even though this looks like a large correction the fact that the R values change very little with pressure implies that even considerable errors in making this correction will have but a small effect on the bulk sound velocity vs pressure locus. This measurement can be used to estimate the yield strength but there will be fairly large errors since this is a differential measurement. It should be noted that the records on the right have one less perturbation, that caused by the interaction of the longitudinal release wave with the analyzer, which makes the system somewhat less complicated. Using bromoform for an analyzer also has the advantage that its behavior should be ideally perfectly plastic.

The Grüneisen ratio, γ_g can be obtained from these measurements from the following equation[9]

$$\gamma = \frac{[(dP/dV)_{HUG} - (dP/dV)_s]2V}{P_H + (dP/dV)_{HUG} (V_o - V_H)} \quad (4)$$

where the standard notation has been used and

$$(dP/dV)_s = -(\rho C_B)^2 \quad (5)$$

Bulk sound velocities calculated so far are in good agreement with those calculated with $\rho\gamma$ = constant.

Work in progress on Cu and 2024 Al indicates that melting does not begin to occur until around 100 GPa in 2024 Al and 160 GPa in Cu. Calculations[10] done earlier indicated that the Cu Hugoniot would cross the melting phase line at about 133 GPa. A tentative explanation for this discrepancy is that the electronic contribution to the specific heat was not used in those calculations. This would have very little effect on the calculated phase line since it was neglected for both phases. However, it would cause the calculated temperatures along the Hugoniot to be too high causing the Hugoniot to cross the phase line at too low a pressure. More data are needed to determine these points more precisely and also the pressures where the materials appear to be completely melted.

OVERTAKING VELOCITIES IN UNSYMMETRICAL IMPACTS

It is unfortunate that symmetrical impacts cannot be used with all the materials that one would like to investigate. This is especially true with the explosive experiments where it is just not possible to accelerate some materials to the velocities desired without breaking them. If the driver has been well characterized it is possible to use it to impact other materials to determine these rarefaction wave velocities. Again if we equate the appropriate times for the catch-up equation but distinguish the driver and target with the appropriate subscripts (D and T) we obtain the following:

$$\frac{R}{U_T} = \frac{1}{U_o} + \frac{1}{C_D^L} + \frac{R}{C_T^L} \quad (6)$$

so

$$\frac{1}{C_T^L} = \frac{1}{U_T} - \frac{1}{R}\left\{\frac{1}{U_o} + \frac{1}{C_D^L}\right\} \quad (7)$$

but
$$C^L = R*U \tag{8}$$

where
$$R^* = (R+1)/(R-1) \tag{9}$$

then
$$\frac{1}{R^*_T} = 1 - \frac{1}{R}\frac{U_T}{U_D}\left\{1 + \frac{1}{R^*_D}\right\} \doteq 1 - F/R \tag{10}$$

and hence
$$C_T = (U_s - U_p)_T \, R^*_T \tag{11}$$

As would be expected, C is more sensitive to errors in measusring R, which is the actual catch up ratio of the experiment, than in the symmetrical experiments.

The results of these measurements, as in the symmetric impact ones, are quite insensitive to errors made in determining pressure. Moreover they are reasonably insensitive to errors in the values of R used for the driver. In our work on 2024 Al it was found to be necessary to use iron drivers if we were to go completely through the melting region. Using R* vs P determined from the data from Ref. 8 below the transition region, we find that the value of F in the above equations changed almost linearly from 2.29 at 80 GPa to 2.346 at 160 GPa. The other pressure dependent term, $(U_s - U_p)$, changes only ten percent in that pressure regime. It is felt that it will be almost impossible to extract any useful knowledge of the elastic-plastic behavior if a material like iron is used for the driver. The bulk velocities will also be difficult, if not impossible, to measure accurately. Since there are some materials, minerals for example, for which it would be desirable to know both the longitudinal and bulk wave velocities, we are planning to characterize a lead alloy, so that elastic-plastic behavior caused by the driver will not be present. This material can then be used as impactors on the gun experiment.

It is probably obvious, but in all these experiments the driver should impact the material of interest directly without any intervening base plate.

If the material to be investigated is transparent it can be used as its own analyzer. By using the equation of state of the driver and target materials to determine the shock velocity in the target and by measuring the time when the radiation begins to decrease, the catch up ratio can be calculated. If the target is made of several layers which have had a thin film of aluminum deposited on them, so that light transmission is reduced by ten percent or so, then it should be possible to measure the shock velocity in the target directly. Fig. 21 illustrates this. Oscilloscope traces are reproduced in Fig. 22 to illustrate the type of measurements that can be made. These are all bonded layers of fused quartz impacted by an iron driver. Epocast is used to bond the layers together because its setting time is long. Pieces are thoroughly cleaned and put in a press with a small drop of epocast in each layer. A plastic cushion is used to maintain pressure and to avoid breaking samples. In the

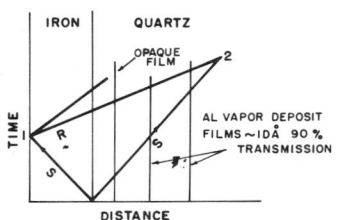

Fig. 21. Schematic for measuring overtake and shock velocities in transparent targets. The opaque film shields unwanted light.

record on the left small discontinuities in the trace can be seen. In the originals these are seen to be small decreases in the radiation caused by glue on the interfaces. This was because insufficient pressure was used in the assembly. The other records show no indication of this, but do show a finite rise time as the shock front passes through the interfaces. It is felt that the increase in radiation is indeed an almost real replica of the shock front. That is, these windows are in a sense acting like insitu gauges. The slow rise time (ten ns is typical) observed is probably real. This could be due to the fact that quartz has a very large phase change and this must be done in the shock front. These records as well as the others are read with an optical comparator that digitizes the record. These are put in the comparator, which displays the records on a screen. Selected areas, for example two adjacent interfaces, are magnified and normalized in amplitude. They can then be matched and the time between them measured. Thus the whole trace in the region of interest is used to obtain velocity measurements.

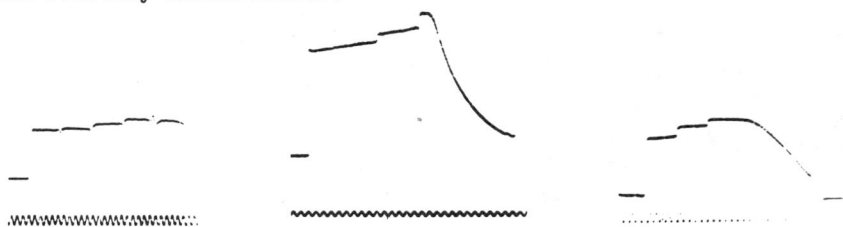

Fig. 22 Records showing shock wave arrivals in fused quartz.

It occurred to us that if we could introduce a small perturbation in the target causing a small short pressure pulse, it might just be possible to see it reflected back from the driver-target interface. If this could be measured the sound velocity could be measured independent of the driver properties. There was also the possibility that this perturbation might travel at bulk sound velocity if the target was on its upper yield surface. Quite thick shims were used in the first experiments so that we could be sure to see something. It didn't take much to put huge perturbations on the records. Later records using thinner shims are reproduced in Fig. 23. The perturbation of interest in the records generally has a decrease followed by a rise. The last record on the right is what was expected. The small decreases probably were caused by glue joints. The perturbators should be deposited rather than glued in place. The smallest aluminum shim stock available here, ~ seven microns produced the small pulse seen in the third figure. If

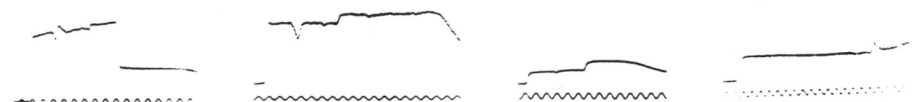

Fig. 23. Four records showing the reflection from a small perturbation in the target reflecting back from the higher impedance driver and eventually overtakes the shock wave.

a represents the distance of the perturbation from the impact interface and b the distance from it when it overtakes the shock. then the Lagrangian velocity is given by

$$C^L = U_s(2a + b)/b \tag{12}$$

These experiments can be done with the overtake experiments just described.

SUMMARY

Several applications utilizing the radiation from shocked transparent materials have been described. These include making temperature measurements on transparent materials, determining the longitudinal sound velocity at pressure and for some materials the bulk velocity, which then can be used to calculate the Grüneisen parameter and the shear modulus at pressure. Estimates of the yield strength can also be made. These experiments have led to techniques for making very accurate shock velocity measurements.

ACKNOWLEDGMENTS

This work was performed under the auspices of the United States Department of Energy.

REFERENCES

1. J. M. Walsh and R. H. Christian, Phys. Rev. <u>97</u>, 1544 (1955).
2. S. B. Kormer, M. V. Sinitzyn, G. A. Kirillov and V. O. Urlin, Sov. Phys. JETP <u>21</u>, 689 (1965).
3. L. V. Al'tshuler, M. I. Brozhnik, and G. I. S. Telegin, J. Appl. Mech. Tech. Phys. <u>12</u>, 921-926, (1971).
4. J. N. Fritz and J. A. Morgan, Rev. Sci. Instrum. <u>44</u>, 2 (1973).
5. L. M. Barker, Proceedings of 1967 IUTAM Symposium 1967 (Gordon and Breach, N. Y., 1968), p. 483.
6. G. A. Lyzenga and T. J. Ahrens, Geophys. Let. 7, 141, (1980).
7. J. M. Walsh, M. H. Rice, R. G. McQueen, and F. L. Yarger, Phys. Rev. <u>108</u>, 196 (1957).
8. J. M. Brown and R. G. McQueen, Geophys. Res. Let. <u>7</u> 533, (1980).
9. R. G. McQueen, Metallurgical Society Conference <u>Vol. 22</u>, (Gordon and Breach, N. Y., 1964), p. 45.
10. R. G. McQueen, W. J. Carter, J. N. Fritz and S. P. Marsh, NBS Spec. Publ. 326, edited by E. C. Lloyd (U. S. Government Printing Office, Washington, D. C., 1968), p. 219.

THEORETICAL EQUATIONS OF STATE FOR METALS

G. I. Kerley*
Los Alamos National Laboratory, Los Alamos, NM 87545

ABSTRACT

Equation of state (EOS) calculations for xenon, iron, and beryllium are described and shown to give good agreement with shock wave experiments. The theory includes models for both solid and fluid phases, calculation of the melting curve, and effects of atomic shell structure on thermal electronic contributions to the EOS.

DESCRIPTION OF THE MODELS

Two computer codes were used in our calculations. The PANDA code[1] computes the pressure, internal energy, and Helmholtz free energy, as functions of density and temperature, for both solid and fluid phases; the melting line is located by matching the pressures and Gibbs free energies of the two phases as a function of temperature. The INFERNO code of Liberman[2] computes the thermal electronic contributions to the EOS, which are input to PANDA.

The solid EOS consists of three terms. For example, the pressure is given by

$$P_s(\rho,T) = P_c(\rho) + P_1(\rho,T) + P_e(\rho,T) \qquad (1)$$

where ρ and T are the density and temperature, respectively. P_c is the zero Kelvin isotherm (cold curve), which we construct from band theoretical calculations and experimental data. P_1, the lattice vibrational term, is given by the Debye model. P_e is the thermal electronic term from INFERNO, discussed below.

The fluid EOS consists of two terms. The pressure is written

$$P_f(\rho,T) = P_n(\rho,T) + P_e(\rho,T). \qquad (2)$$

P_n includes contributions from the ground electronic state and from the motion of the nuclei. This term is computed from hard sphere perturbation theory, using the CRIS model.[3] In this model, the energy of a fluid atom in the cage formed by its neighbors is determined from the cold curve of the solid. Therefore, it is not necessary to specify the interatomic potentials or any other parameters in order to calculate the fluid EOS. Contributions from the electronic and nuclear degrees of freedom are strongly coupled; therefore, we do not separate P_n into a cold curve and a thermal term as we do for the solid.

---

*Supported by the U.S. Department of Energy, Office of Basic Energy Sciences

0094-243X/82/780208-05 $3.00 Copyright 1982 American Institute of Physics

The INFERNO model of Liberman[2] is used to compute the thermal electronic contributions to the EOS for both solid and fluid phases. In this theory, the Dirac equation is solved to obtain wave functions and thermodynamic properties for an average atom, as a function of both density and temperature. An important feature of the theory is that it includes effects due to atomic shell structure, that are smeared out in simpler models such as Thomas-Fermi-Dirac (TFD) theory.[4]

XENON

Like all materials, xenon is expected to become metallic at high densities. Zero temperature band theoretical calculations of Ross and McMahan[5] predict the energy gap between the filled 5p band and the empty conduction band to close at a density of 12 g/cc. In shock wave experiments, where there is thermal electronic excitation, effects due to this insulator-metal transition can be observed at lower densities.

The thermal electronic pressure calculated using the INFERNO model is shown in Fig. 1. At densities in the range 2-10 g/cc, narrowing of the band gap allows increased electronic excitation and also results in a negative contribution to the pressure. This result agrees with Ross's model[5] for the rare gases.

In this work, the cold curve was taken from the band calculations of Ross and McMahan,[5] shown as circles in Fig. 2. This result, together with the CRIS model and the INFERNO model, completely define the fluid EOS a priori. Our predictions of the Hugoniot are compared with experiment[6] in Fig. 2. If no electronic excitation is allowed, the calculations deviate sharply from experiment at high pressures. When the TFD model is used to describe the electronic excitations, the results are better but still not satisfactory. Calculations using the INFERNO model are in excellent agreement with the experimental data. INFERNO predicts metallization to occur at about 10 g/cc, in fair agreement with the band theoretical calculations.

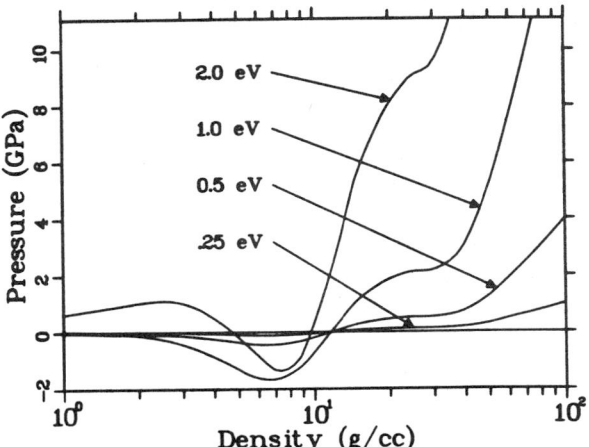

Fig. 1. Thermal electronic pressure for xenon, as a function of density, at several temperatures.

IRON

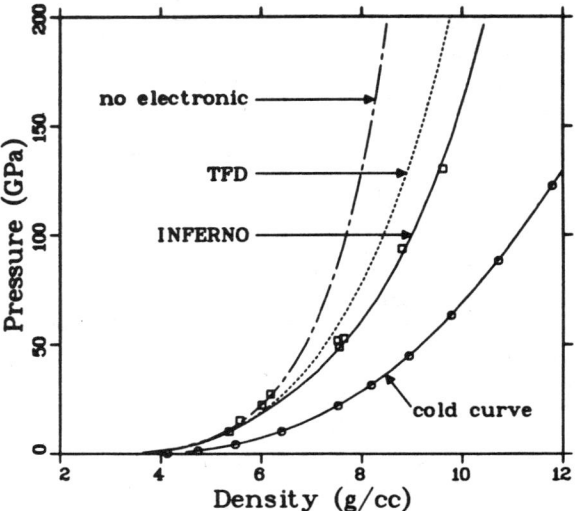

Fig. 2. Cold curve and Hugoniot for xenon.

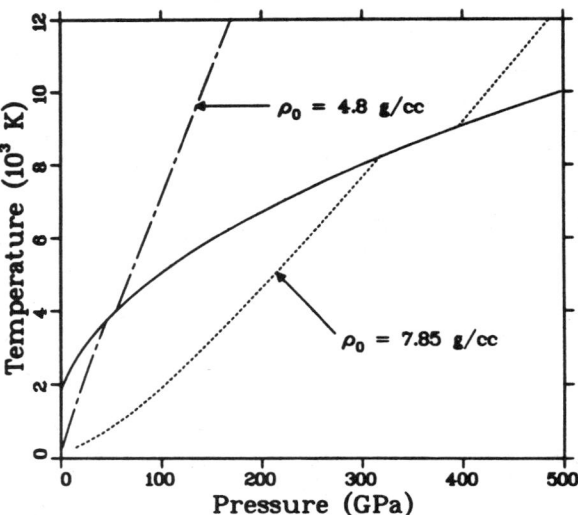

Fig. 3. Iron melting curve and Hugoniots for two initial densities.

Application of the theory to iron is complicated by the existence of several solid phases. In this work, we considered only the close packed ε-phase, which is stable at pressures above 13 GPa. The cold curve was taken from the static high pressure experimental data of Mao and Bell.[7]

Our theoretical melting curve for iron is shown in Fig. 3. In this calculation, we forced agreement with the experimental melting point at zero pressure by subtracting an empirically-determined constant from the free energy of the fluid. (This correction was about 4% of the solid binding energy.) Calculated Hugoniots for two initial densities are also shown in Fig. 3. Alpha-phase iron, having a density of 7.85 g/cc, is predicted to begin melting at about 300 GPa, in fair agreement with the value of 250 GPa obtained by Brown and McQueen.[8] Porous α-phase iron, having a density of 4.8 g/cc, is predicted to begin melting at 45 GPa.

Shock velocity vs. particle velocity curves for iron of various initial densities are shown in Fig. 4. Agreement between the theory and the experimental data [8-10] is very good for both solid and fluid phases over the entire range, which extends to 1000 GPa. More detail can be seen in Fig. 5, showing the shock data [8,10] for an initial density of 4.8 g/cc. In the calculated Hugoniot, the mixed phase region is shown by a dashed line. Agreement with the experimental data is excellent except at the lowest pressures, for which the shocked state is in the α-phase.

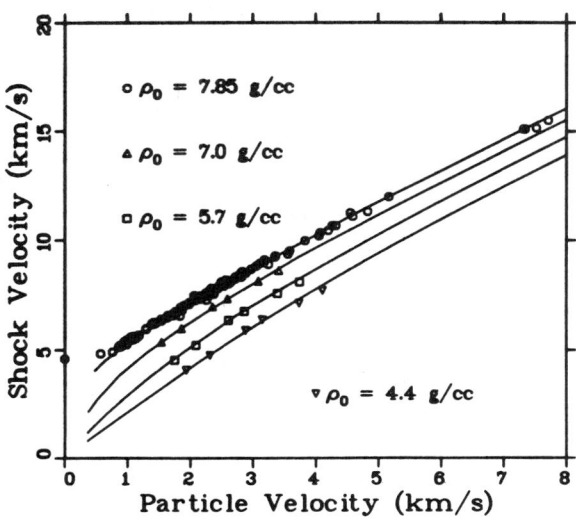

Fig. 4. Shock velocity vs. particle velocity for iron at several initial densities.

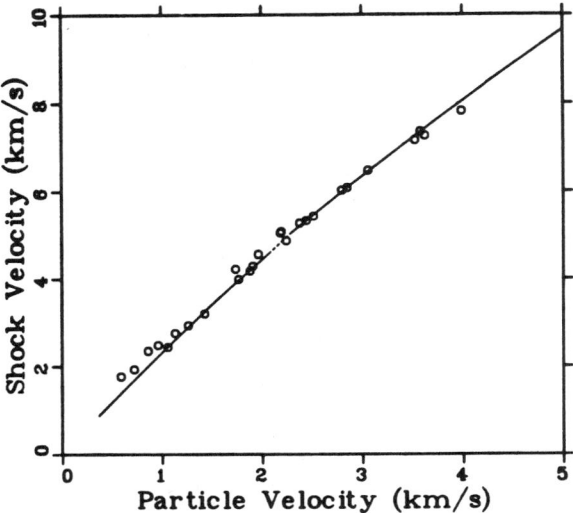

Fig. 5. Hugoniot for porous iron having an initial density of 4.8 g/cc.

BERYLLIUM

The cold curve for beryllium was taken from the band theoretical calculations of Perrot,[11] shown as circles in Fig. 6. The solid EOS is fairly sensitive to the form of the Grüneisen paramater. We used the expression

$$\gamma = 1.3/\rho + 2/3 \qquad (3)$$

which agrees with both thermodynamic data and Neal's shock wave measurements [12] to within experimental error.

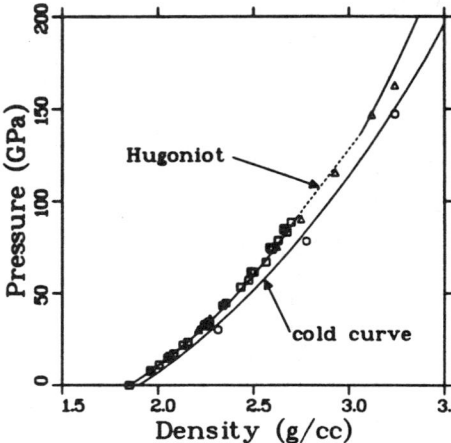

Fig. 6. Cold curve and Hugoniot for beryllium.

The calculated melting point at zero pressure is 1545 K, in good agreement with experiment. The theoretical melting curve has a small negative slope over the pressure range 10-160 GPa, and melting is predicted to begin at about 90 GPa, under shock loading.

The theoretical Hugoniot is compared with experiment [10,13] in Fig. 6. The mixed phase region is depicted by a dashed line. Our theory predicts a significant softening of the Hugoniot due to melting. These calculations are in very good agreement with the data of Isbell, et. al.,[13] shown by triangles. New measurements in this high pressure region would be useful.

REFERENCES

1. G. I. Kerley, "Users Manual for PANDA: A Computer Code for Calculating Equations of State," Los Alamos National Laboratory report LA-8833-M (1981).
2. D. A. Liberman, Phys. Rev. B20, 4891 (1979).
3. G. I. Kerley, J. Chem. Phys. 73, 469 (1980); 73, 478 (1980).
4. R. D. Cowan and J. Ashkin, Phys. Rev. 105, 144 (1957).
5. M. Ross and A. K. McMahan, Phys. Rev. B21, 1658 (1980).
6. W. Nellis, M. van Thiel, A. C. Mitchell, and M. Ross, "The Equation of State of D_2 and Xe in the Megabar Region," presented at the Eighth International Thermophysical Properties Conference, National Bureau of Standards, Wash. D.C., 1981.
7. H. K. Mao and P. M. Bell, J. Geophys. Res. 84, 4533 (1977).
8. J. M. Brown, and R. G. McQueen, J. Geophys. Res., to be published.
9. L. Y. Al'tshuler, N. N. Kalitkin, L. V. Kuz'mina, and B. S. Chekin, Sov. Phys. JETP 45, 167 (1977).
10. S. P. Marsh, LASL Shock Hugoniot Data (University of California, Berkeley, 1980).
11. F. Perrot, Phys. Rev. B21, 3167 (1980).
12. T. Neal, "Determination of the Grüneisen γ for Beryllium at 1.2 to 1.9 Times Standard Density," in High Pressure Science and Technology, Sixth AIRAPT Conference, edited by K. D. Timmerhaus and M. S. Barber (Plenum Press, NY, 1979) pg. 80.
13. W. M. Isbell, F. H. Shipman, and A. H. Jones, "Hugoniot Equation of State of Eleven Elements to Five Mbars," Materials Science Laboratory report MSL-68-13.

A NEW FULL-RANGE EQUATION OF STATE FOR COPPER*

K. S. Long, D. Young and F. H. Ree
University of California, Lawrence Livermore National Laboratory
Livermore, California 94550

ABSTRACT

We present a new global equation of state (EOS) for copper which extends from 10^{-3} g/cm^3 to 10^3 g/cm^3 and from room temperature to 5×10^4 eV. The EOS itself is a table in which pressure and energy are given at each point of a rectangular grid of temperature and density. The pressures and energies of the full-range EOS were calculated with six different theoretical models, each of which is valid in a different regime.

RESULTS

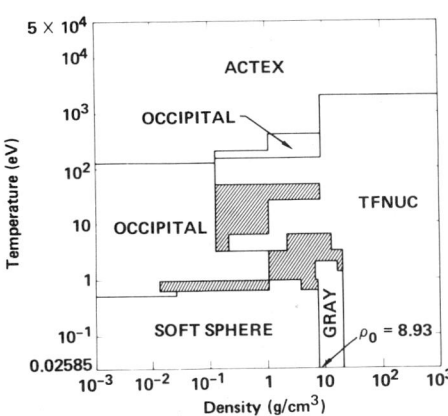

Fig. 1 Theoretical subregions of the new EOS for copper.

Figure 1 shows where each of the physics models was used. (The names refer to the computer codes which embody the theories.) Above 1 eV in expansion, copper is in ionization equilibrium, and we must use physics models which treat that regime. At high temperature, Rogers' rigorous quantum-statistical-mechanical model ACTEX[1] calculated the EOS. This theory, based on a perturbation expansion, is valid in the presence of moderate to strong plasma coupling. As the code which embodies the ACTEX theory moves into lower temperature or higher density, it begins to have convergence problems, so we finished the ionization equilibrium EOS with the OCCIPITAL model developed by Rouse.[2]

This model calculates the statistical mechanics of partially ionized plasmas using the Saha method modified with first order Debye-Hückel Coulomb corrections and Planck's partition function.

At low temperature below liquid density, Young calculated the copper EOS with his soft sphere model[3] for liquid metals. This is a semi-empirical model; that is, certain input parameters are adjusted until experimental data in the form of isobars of enthalpy, volume, and sound speed vs temperature are reproduced.

*Work performed under the auspices of the U.S. Dept. of Energy by Lawrence Livermore Natl. Lab. under contract #W-7405-Eng-48.

The most complicated area in any EOS is the so-called multiphase region, which includes the melt and liquid-vapor transitions. For the solid-melt-liquid area, we used the GRAY model,[4] developed by Grover, Royce and Young and which is also semi-empirical. GRAY treats the solid EOS with the Dugdale-McDonald form of Grüneisen-Debye theory, and the liquid is based on a scaling law; that is, the solid Grüneisen EOS at a given temperature and density is corrected with terms depending on the ratio of the temperature to the melting temperature at that density. The melt transition is calculated according to the Lindemann Law. The experimental data directly input into GRAY are normal-state data (solid density, sound velocity, Grüneisen gamma and the electronic specific heat coefficient), the melting temperature at 1 atm, and the cohesive energy. Also input are analytic fits for Grüneisen gamma as a function of density, and shock velocity as a function of particle velocity.

The rest of the compression EOS was generated by McMahan with APW band theory[5] and with the TFNUC model. APW (augmented plane wave) calculated the zero degree isotherm (identical to the room temperature isotherm for all practical purposes) between 20 g/cm^3 and 60 g/cm^3. TFNUC, used for the balance of the compression EOS, is a combination of two models. First, an electronic EOS is calculated using Thomas-Fermi theory with the Kirzhnits form of quantum and exchange corrections,[6] then a nuclear correction[7] is added to this basis. The ionic corrections are based on one-component-plasma theory at high temperature and on Grüneisen-like theory at low temperature, with an interpolation scheme in between. TFNUC also includes a Lindemann Law melt consistent with the GRAY melting line.

The composite pressure and energy surfaces were constructed by joining the separate EOS subsurfaces as shown in Fig. 1. At high temperature, the models merged smoothly into one another. At low temperature, however, the physics is more complex (for example, pressure ionization and phase transitions occur), and there were in some cases substantial mismatches between adjacent theories. In this case, portions of each EOS subsurface on either side of a boundary had to be replaced with numerical interpolation (indicated by the shaded regions in Fig. 1) to insure a smooth and continuous join. During the joining procedure, we demanded that
  (1) all EOS functions be continuous (except at a phase transition),
  (2) the isothermal bulk modulus be positive, and
  (3) the heat capacity be positive.

Figure 2 shows the resultant composite EOS in the form of energy vs density along constant temperatures. Multiphase physics is illustrated by the solid-vapor and liquid-vapor transitions. Shell-structure ionization effects dominate the surface between 100 eV and 1000 eV. The L-shell strips off between 100 and 300 eV, followed by a region where the Cu^{+27} plasma is stable; finally K-shell ionization sets in and is complete at T = 1600 eV. M-shell ionization (not indicated in Fig. 2) is also evident between 25 eV and 100 eV, but the ionization of other shells at lower temperatures is obscured due to shell overlap effects.

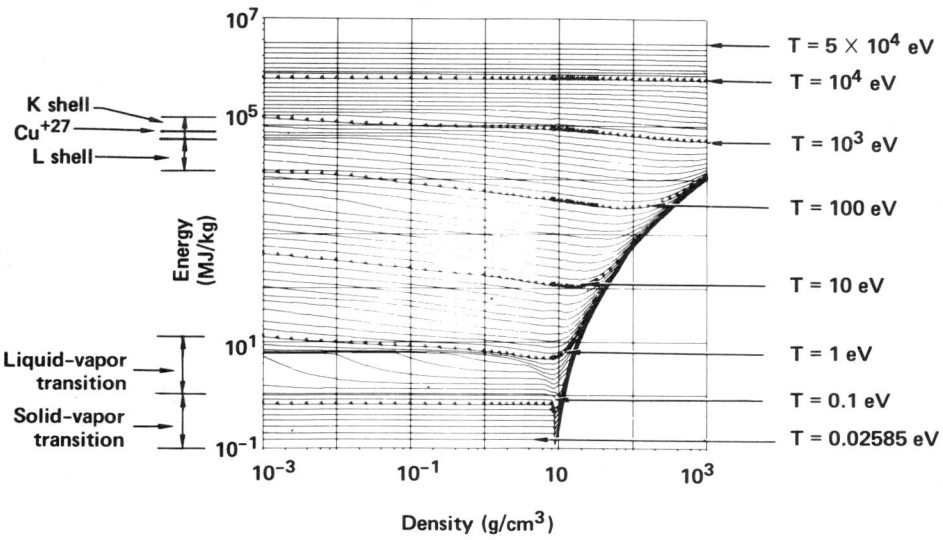

Fig. 2   Energy vs density isotherms of the new global copper EOS.

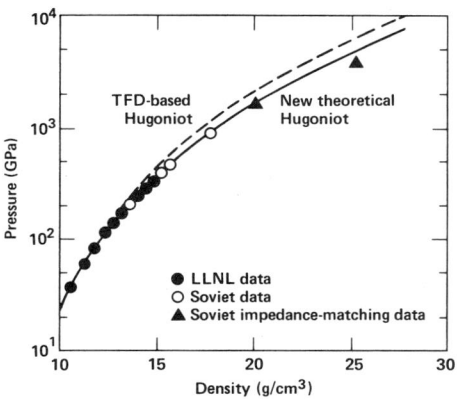

Fig. 3 Comparison of the new theoretical Hugoniot (solid line) with experimental data and with the Thomas-Fermi-Dirac-based theoretical Hugoniot (dashed line).

The true test of any theoretical EOS is its agreement with experimental data. Figure 3 shows a principal Hugoniot comparison (in pressure-density space) between the new theoretical Hugoniot (solid line) and experimental data[8,9,10] (points). Agreement is excellent. It is no surprise that the LLNL Hugoniot experiments (which are accurate to 1%) are reproduced well, since the GRAY calculations were normalized to an analytic fit of those data. However, the Soviet high-pressure, impedance-matching data, not used in our normalization, also lie nearly on the calculated Hugoniot--even the 4-TPa (40 Mbar) point. Also superimposed on this plot is the theoretical Hugoniot from the Thomas-Fermi-Dirac EOS.

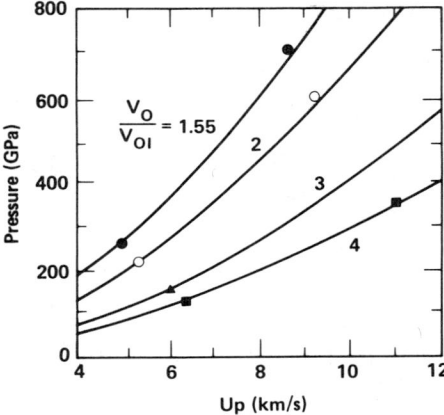

Fig. 4   Pressure vs particle velocity. Theoretical calculations (solid lines) are compared with experiments performed by Kormer, et al. (Ref. 11). The initial densities of the porous copper samples were 1.55-, 2-, 3-, and 4-fold expanded.

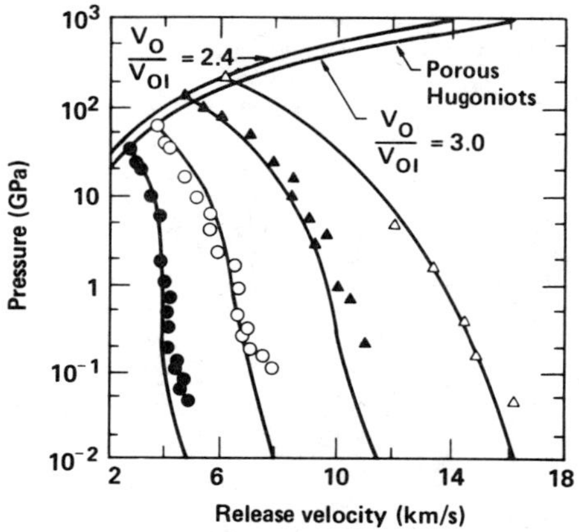

Fig. 5   Pressure vs release velocity comparison of theoretical release isentropes with experiments performed by Al'tshuler, et al. (Ref. 12).

Off-Hugoniot states of the EOS in compression can be accessed by means of porous Hugoniot experiments. That is, higher pressures and internal energies (at a given density) can be achieved upon shock compression for an initially porous sample than for a sample shocked from solid density. Kormer, et al., have done Hugoniot measurements[11] for porous copper specimens starting from initial densities that are 1.55-, 2-, 3-, and 4-fold expanded. These are compared in Fig. 4 with theoretical calculations using the new EOS for copper.

The furthest experimental probes into the expansion EOS are an excellent set of Soviet experiments which have mapped out the release isentropes of shock-compressed porous samples of copper.[12] Figure 5 is a pressure vs release velocity comparison of the data with theoretically-computed release adiabats. The match is quite good, especially above 1 kbar. The lowest points on these isentropic curves reach into the critical region.

REFERENCES

1. F. J. Rogers and H. E. DeWitt, Phys. Rev. A$\underline{8}$, 1061 (1973); and F. J. Rogers, Phys. Rev. A$\underline{10}$, 2441 (1974).

2. C. A. Rouse, Astrophys. J. $\underline{136}$, 636 (1962).

3. D. A. Young, "A Soft-Sphere Model for Liquid Metals", Lawrence Livermore National Laboratory, Livermore, CA, UCRL-52352 (1977).

4. R. Grover, High-Temperature Equation of State for Simple Metals, Proceedings of the Seventh Symposium on Thermophysical Properties, ed. by A. Cezairliyan (The American Society of Mechanical Engineers, New York, 1977), p. 67.

5. Description of the APW method is given in T. L. Loucks, Augmented Plane Wave Method (Benjamin, New York, 1967); and in L. F. Mattheiss, J. H. Wood, and A. C. Switendick, in Methods in Computational Physics, edited by B. Alder, S. Fernback, and M. Rotenberg (Academic, New York, 1968), Vol. 8, p. 63. The manner in which the method is made self-consistent is described by J. C. Slater and P. DeCicco, Solid State and Molecular Theory Group, MIT Quarterly Progress Report No. 50 (1963) (unpublished), p. 46.

6. S. L. McCarthy, The Kirzhnits Correction to the Thomas-Fermi Equation of State, Lawrence Livermore National Laboratory, Livermore, CA, UCRL-14362 (1965).

7. F. Ree, D. Daniel, and G. Haggin, NUC--A Code to Calculate Ionic Contributions to P and E, Lawrence Livermore National Laboratory, Livermore, CA, Internal Document HTN-291 (1978).

8. A. C. Mitchell and W. J. Nellis, J. Appl. Phys. $\underline{52}$, 3363 (1981).

9. L. V. Al'tshuler, S. B. Kormer, A. A. Bakanova, and R. F. Turnin, Zh. Eksp. Teor. Fiz. $\underline{38}$, 790 (1960) [Sov. Phys. JETP $\underline{11}$, 573 (1960)].

10. L. V. Al'tshuler, N. N. Kalitkin, L. V. Kuz'mina, and B. S. Chekin, Zh. Eksp. Teor. Fiz. $\underline{72}$, 317 (1977) [Sov. Phys. JETP $\underline{45}$, 167 (1977)].

11. S. B. Kormer, A. I. Funtikov, V. D. Urlin, and A. N. Kolesnikova, Zh. Eksp. Teor. Fiz. $\underline{42}$, 686 (1962) [Sov. Phys. JETP $\underline{15}$, 477 (1962)].

12. L. V. Al'tshuler, A. V. Bushman, M. V. Zhernokletov, V. N. Zubarev, A. A. Leont'ev, and V. E. Fortov, Zh. Eksp. Teor. Fiz. $\underline{78}$, 741 (1980) [Sov. Phys. JETP $\underline{51}$ (2), 373 (1980)].

ACCURATE SELF-CONSISTENT-FIELD ISOTHERMS FOR NaCl TO 30 GPa

M.S.T. Bukowinski
University of California, Berkeley, CA 94720

ABSTRACT

Results of a self-consistent-field calculation of the equation of state of NaCl are presented. The predicted zero temperature lattice constant, isothermal bulk modulus and its pressure derivative are in excellent agreement with available data. The room-temperature isotherm agrees with Decker's equation and experimental isotherms. The bulk modulus agrees with data while its pressure derivative is more consistent with static compression measurements than ultrasonic measurements.

INTRODUCTION

The equation of state of NaCl is of considerable interest in high pressure research. Its history as an internal pressure standard was recently reviewed by Chhabildas and Ruoff[1]. Existing data on its equation of state, elastic constant and velocities were critically analyzed by Birch[2].

Sodium chloride's role as a pressure standard is largely due to the theoretical isotherms derived by Decker[3,4] from a Born-Mayer potential. Decker's equation of state is in very good agreement with pressure values determined by the nonlinear ruby scale[5] and MgO secondary calibration curve based on the same ruby scale[6]. Since the ruby scale still contains some uncertainty, it is not known how accurate Decker's equation is above 20 GPa; although apparently successful, its theoretical foundation is weak[7]. It is thus of some interest to attempt a more rigorous calculation.

We report here a self-consistent field calculation of the zero-temperature isotherm of NaCl. Thermal corrections are estimated to allow a comparison with finite temperature measurements.

THEORY

The total pressure at volume V and temperature T is given by

$$P(V,T) = P_o(V,0) + \frac{1}{V} \int_0^T \gamma(V,t) C_v(V,t) dt \quad (1)$$

The first term on the right hand side is the ground state pressure, and the second term is the correction due to thermal vibrations, expressed in terms of the thermal Grüneisen parameter,

$$\gamma(V,T) \equiv \left(\frac{\partial P}{\partial T}\right)_V \frac{V}{C_v} \quad (2)$$

and the heat capacity at constant volume, $C_V(V,T)$.

The ground state pressure is the sum of a static lattice pressure, P_S, and the zero-point vibrations pressures, P_{ovib}. The latter may be computed with the convenient approximation of Domb and Salter[8]

$$P_{ovib}(V,0) = -\frac{\partial}{\partial V} \frac{9}{8} nk\theta_\infty = \frac{9}{8} \gamma \frac{nk\theta_\infty}{V}, \quad (3)$$

where n is the number of atoms, k is the Boltzmann constant and θ_∞ is the high temperature limit of the Debye temperature. Errors due to the use of eqn (3) for the zero-point pressure are of the order of one percent[9].

The volume dependence of θ_∞ was modeled by using its elastic definition and assuming that all the elastic velocities have the same volume dependence as the bulk velocity. This leads to

$$\theta_\infty = 290 \left(\frac{K}{K_o}\right)^{1/2} \left(\frac{V}{V_o}\right)^{1/6}, \quad (4)$$

where quantities with a zero subscript or superscript refer to zero pressure, K is the bulk modulus and 290^K is the value of θ_∞ at zero pressure[9]. Use of $\gamma = -\partial \ln\theta/\partial \ln V$ with $\gamma = \gamma_o(V/V_o)$ leads to results that differ at most by one kilobar at the highest compression studied. We prefer eqn (4) because it is better constrained.

The static lattice pressure was computed from a self-consistent band structure calculation. We used the symmetrized-Augmented-Plane-Wave method[10] to obtain the energy spectra and charge densities of NaCl for several values of the lattice constant. The exchange-correlation potential was approximated by the Hedin-Lundqvist model[11,12] and the "muffin-tin"[10] approximation was used for the charge density and potential. The resulting expression for the ground state pressure is:

$$P_s(V,0) = \frac{1}{3V}\left[2E_K + E_{coul} - 3E_{xc} + 3\int \rho(\vec{r}) v_{xc}(\vec{r}) d^3\vec{r}\right], \quad (8)$$

where E_K is the electronic kinetic energy, E_{coul} is the total Coulomb energy, E_{xc} is the total exchange-correlation energy, v_{xc} is the corresponding potential and $\rho(\vec{r})$ is the electronic charge density. The "muffin-tin" approximation may lead to total energy errors of the order of a few hundredth's of a Rydberg and, therefore, pressure errors that can be as large as several GPa at large compressions. The first order perturbation theory method of Danese and Connolly[15] was used to include the effect of the intersphere non-muffin-tin charge density and potential.

The thermal pressure was estimated by integrating low temperature data of Meincke and Graham[16]. It was assumed that $\gamma(v,T) = \gamma(V_o,T) (V/V_o)^A$. All calculations were done with $A = 1$; use of other values caused only minor differences. The $C_V(V,T)$ data were fitted with the Debye model and values for compressed densities were estimated by letting the Debye temperature vary according to eqn (4).

DISCUSSION OF RESULTS

The computed room temperature equation of state is an excellent agreement with Decker's equation and with available high pressure experimental data. As shown in Figure 1, differences between the pre------ computed here and those of Decker are nowhere larger than   the theoretical uncertainty.

The estimated uncertainty in the computed pressures is of the order of 0.5 GPa at 30 GPa. This does not include the possibly significant, but unknown, errors implicit in the form of the Hamilitonian use. Because of the good agreement, however, among the independent curves shown on Figure 1, it is probable that the present equation, or Decker's equation, is sufficiently accurate for purposes of calibration to 30 GPa.

Table 1 shows values of the calculated zero pressure lattice constant a, bulk modulus K_o, its pressure derivation K_o', and second pressure derivative K_o''. All parameters were determined from a 4th order Eulerian finite strain fit. The calculated K_o' is more consistent with static measurements (except those of Chhabildas and Ruoff) than acoustic measurements. Following a suggestion by Fritz[17], four additional P-V points below 0.1 GPa were computed. A fit to 5 points below 0.1 GPa yielded K_o = 24.0 ± 0.3 GPa and K_o' = 5.2 ± 0.3 at room temperature. Although pressures below 0.1 GPa do not constrain K_o' very well, the result does suggest that there is a rapid initial change in the curvature of the NaCl equation of state, which may explain the apparent discrepancy between static and acoustic data. Some very careful static or acoustic work to 5 GPa would be helpful in settling this issue.

Table I Comparison of some theoretical and experimental parameters

T (°K)		a (Å)	K_o (GPa)	K_o'	$K_o K_o''$
0	th.	5.603 ± .01	26.7 ± .5	4.67 ± .3	-5.45
	exp.	5.596[a]	26.6[b], 27.4[c]	4.88[d]	
298	th.	5.645 ± .01	23.85 ± .5	4.78 ± .3	-5.77
	exp.	5.640 ± .001[d]	ultrasonic		-16.21[d]
			23.30 - 23.87[f]	5.02 - 6.0[f]	-4.77 to
			static		-23.84[g]
			23.40 - 26.4[f]	3.90 - 4.92	
			23.85[h]	5.73[h]	

[a] Reference 16. [d] Reference 20. [g] Reference 1.
[b] Reference 18. [e] Reference 21. [h] Reference 22.
[c] Reference 19. [f] Reference 2.

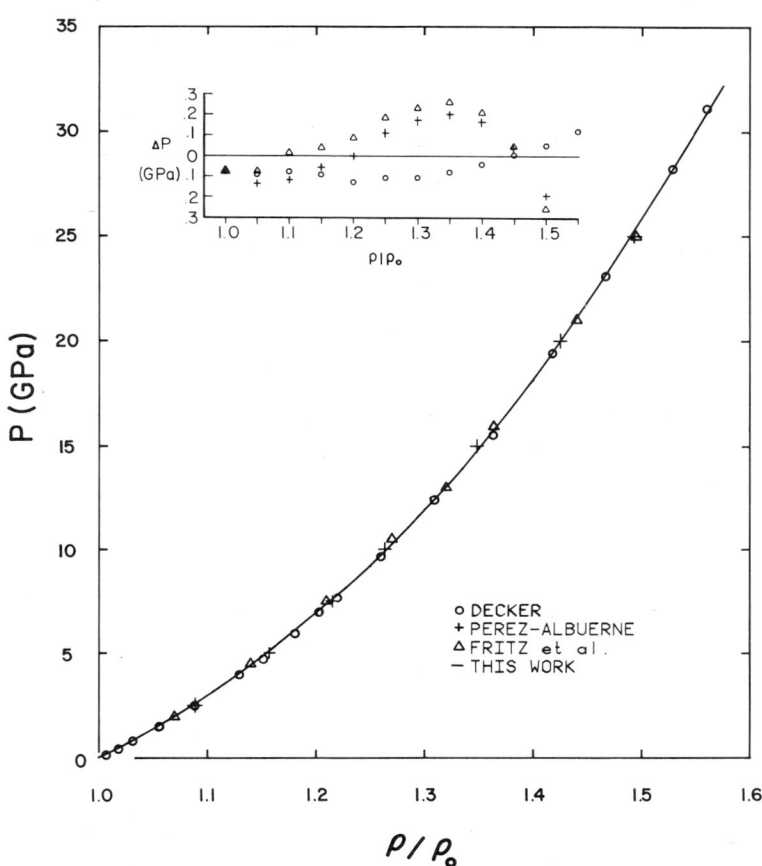

Fig. 1. Theoretical and experimental isotherms for NaCl. Insert shows values of other isotherms minus our theoretical isotherm. ρ is here the crystal density and ρ_0 is its experimental zero-pressure values. P is the pressure. Data is that of Decker (Ref. 3), Perez-Albuerne (Ref. 13), and Fritz et al. (Ref. 14).

ACKNOWLEDGEMENTS

M. John Aidun provided many hours of patient help in the numerical calculations. Korda Cordes patiently and skillfully typed several versions of the manuscript. I benefited from a stimulating conversation with J. Fritz. The research was supported by NSF grants EAR-78-12943 and EAR-80-08214 and funds provided by the Institute of Geophysics and Planetary Physics, University of California.

REFERENCES

1. L.C. Chhabildas and A.L. Ruoff, J. Appl. Phys. $\underline{47}$, 4182 (1976).
2. F. Birch, J. Geoph. Res. $\underline{83}$, 1257 (1978).
3. D.L. Decker, J. Applied Phys. $\underline{42}$, 3239 (1971).
4. D.L. Decker, J. Applied Phys. $\underline{37}$, 5012 (1966).
5. T. Yagi, Carnegie Inst. Washington Yearbook $\underline{76}$, 528 (1977).
6. H.K. Mao, P.M. Bell, J. Shaner and D. Steinberg, in High Pressure Science and Technology, edited by K.D. Timmerhaus and M.S. Barber (Plenum, New York, 1979), p. 739.
7. L. Thomsen, J. Phys. Chem. Solids $\underline{31}$, 2003 (1970).
8. C. Domb and L. Salter, Phil. Mag. $\underline{43}$, 1083 (1952).
9. T.H.K. Barron, W.T. Berg and J.A. Morrison, Proc. Roy. Soc. London $\underline{A242}$, 478 (1957).
10. L.F. Mattheiss, J.H. Wood and A.C. Swittendick, Methods in Computational Physics $\underline{8}$, 63 (1968).
11. L. Hedin and B.I. Lundqvist, J. Phys. C.; Solid St. Phys. $\underline{4}$, 2064 (1971).
12. B.I. Lundqvist and S. Lundqvist, Computational Solid State Physics, edited by F. Herman, N.R. Dalton and T. Koehler (Plenum, New York, 219, 1972).
13. E.A. Perez-Alberne and H.G. Drickamer, J. Chem. Phys. $\underline{43}$, 1381 (1965)
14. J.N. Fritz, S.P. Marsh, W.D. Carter and R.G. McQueen, Accurate Characterization of the High Pressure Environment, edited by E.C. Lloyd (NBS Special Publ. 326, Washington, D.C., 1979) p. 201.
15. J.B. Danese and J.W.D. Connolly, J. Chem. Phys. $\underline{61}$, 3063 (1974).
16. P.P.M. Meincke and G.M. Graham, Proc. 8th Intnl. Conf. on Low Temp. Phys., London, Butterworths $\underline{401}$ (1963).
17. J. Fritz, Personal communication.
18. J.T. Lewis, A. Lehoczky and C.V. Briscoe, Phys. Rev. $\underline{161}$, 877 (1967).
19. R.Q. Fugate and D.E. Schuele, J. Phys. Chem. Solids $\underline{27}$, 493 (1966).
20. H. Spetzler, C.G. Sammis and R.J. O'Connell, J. Phys. Chem. Solids $\underline{33}$, 1727 (1972).
21. Natl. Bur. Std. Circular No. 539 (U.S. GPO, Washington, D.C., 1957), vol. 2, pp. 41 and 44.
22. Average of values in reference 1.

THE SHOCK COMPRESSION OF LIQUID H_2 TO 10 GPa (100 kbar)*

W. J. Nellis, M. Ross, M. van Thiel,
A. C. Mitchell and G. J. Devine
University of California, Lawrence Livermore National Laboratory
Livermore, California 94550

N. Brown
Specialty Enginnering Associates
Santa Clara, California 94505

ABSTRACT

Hugoniot data for liquid H_2 to 10 GPa was used to test the H_2 intermolecular potential.

INTRODUCTION

The principal Hugoniot of H_2 initially in the liquid state was measured between 2 - 10 GPa (20-100 kbar). At maximum pressure the liquid is compressed a factor of 3.2 over the initial density and the calculated shock temperature is about 3250 K. The experiment is intended to provide data for fluid, molecular H_2 at high densities and temperatures from which an intermolecular potential may be obtained. Fluid theory can then be used to generate a reliable equation of state for applictions such as modeling the giant planets and refining the prediction of high pressure phase transitions in hydrogen.

EXPERIMENT

Shock pressures were generated with a two-stage, light-gas gun.[1,2] The diagnostic system was described previously.[3] The technique used to perform Hugoniot measurements in cryogenic liquids has also been described.[4,5] The coolant was liquid H_2. The cryogenic design was modified somewhat because the heat of vaporization per unit volume is significantly less for liquid H_2 than for the coolants used previously. The details of the cryogenic modifications will be published in a later paper. The initial state was very close to the saturation curve: 20.2 K and a sample pressure of ~800 torr. Both self-shorting and piezoelectric crystal pins were used as shock wave detectors.[3]

RESULTS

The Hugoniot data for liquid H_2 are plotted in Fig. 1 as shock pressure versus molar volume. Also plotted are the point of van Thiel and Alder[6] and the data of Dick and Kerley.[7]

*Work performed under the auspices of the U.S. Department of Energy by Lawrence Livermore National Laboratory under contract #W-7405-Eng-48.

DISCUSSION

A theoretical Hugoniot was computed using fluid perturbation theory.[8] The internal partition function was assumed to be the same as for H_2 gas. The theory is plotted as the solid line in Fig. 1. The intermolecular potential was approximated as pair-wise additive and spherically symmetric. It is a modified form of the Silvera-Goldman potential[9] adjusted at small interatomic separations to predict our new, unpublished D_2 Hugoniot data. This potential was constrained to fit the 4.2 K isotherm of Anderson and Swenson to 2.5 GPa,[10] the isotherms of Mills, et al.[11] to 2.0 GPa and 300 K, and the melting line of Diatschenko and Chu[12] to 300 K and 5.4 GPa. Because of the high temperature the potential was tested down to intermolecular separations of 1.7 Å. The results for both single and double-shock experiments and theory for liquid D_2 will be published in a later paper.

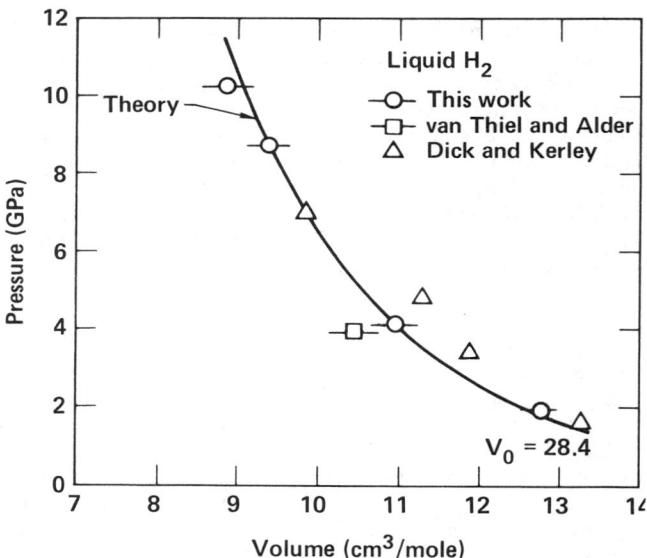

Fig. 1. Shock pressure vs molar volume for liquid H_2. V_o is the initial molar volume of the unshocked liquid (10 GPa = 100 kbar).

REFERENCES

1. A. C. Mitchell, W. J. Nellis, and B. Monahan, this proceedings.
2. A. C. Mitchell, W. J. Nellis, and R. J. Trainor, this proceedings.

3. A. C. Mitchell and W. J. Nellis, Rev. Sci. Instrum. **52**, 347 (1981).
4. W. J. Nellis and A. C. Mitchell, J. Chem. Phys. **73**, 6137 (1980).
5. W. J. Nellis, F. H Ree, M. van Thiel, and A. C. Mitchell, J. Chem. Phys. **75**, 3055 (1981).
6. M. van Thiel and B. J. Alder, Mol. Phys. **10**, 427 (1966).
7. R. D. Dick and G. I. Kerley, J. Chem. Phys. **73**, 5264 (1980).
8. M. Ross, J. Chem. Phys. **73**, 4445 (1980).
9. I. F. Silvera and V. V. Goldman, J. Chem. Phys. **69**, 4209 (1978).
10. M. S. Anderson and C. A. Swenson, Phys. Rev. **B10**, 5184 (1974).
11. R. L. Mills, D. H. Liebenberg, J C. Bronson, and L. C. Schmidt, J. Chem. Phys. **66**, 3076 (1977).
12. V. Diatschenko and C. W. Chu, Science **212**, 1393 (1981).

EQUATION OF STATE EXPERIMENTS AND THEORY RELEVANT TO MODELING THE MAJOR PLANETS*

Marvin Ross and William J. Nellis
University of California, Lawrence Livermore National Laboratory
Livermore, California 94550

ABSTRACT

Shock compression measurements and theoretical studies on molecular fluids of interest to structural models of the major planets have been carried out at LLNL for H_2, CH_4, H_2O, NH_3 with work on He in progress.

INTRODUCTION

The giant planets are thought to consist of three layers: an outer layer of molecular hydrogen and helium; a middle layer either of metallic hydrogen and helium (for Jupiter and Saturn) or of icy ammonia, methane, and water (for Uranus and Neptune), and a rocky core of iron, nickel, silicon, and magnesium oxides.[1,2] Hydrogen and helium are subjected to a very wide range of pressures (e.g., from 1 bar to 45 Mbars in Jupiter). To construct models that predict the observed characteristics of the giant planets, requires equations of state of the constituent materials. Most of the recent shock-wave data used to model the giant planets were obtained with the LLNL two-stage gas gun. This device can accelerate a 20-g metal projectile against a target at velocities up to 7 km/s.[3] To achieve conditions comparable to those in the planetary interiors, H_2, D_2, CH_4, NH_3, and H_2O must be shocked from the liquid phase. The first four have boiling points below room temperature, and thus the targets are actually small cryostats.

PROPERTIES OF HYDROGEN AND HELIUM

To illustrate the significance of our recent liquid hydrogen shock-wave data to planetary modeling, we have in Fig. 1 compared pressure-temperature plots for computed isentropes of the current models of Jupiter and Saturn with these experimental results. Typically, when a Hugoniot and an isentrope originate from the same initial conditions, they progress along very different pressure-temperature paths. However, because the planetary isentropes originate at very low (atmospheric) densities and low temperatures, they follow an extended compression range and reach pressures and temperatures comparable to those obtained by the high-density Hugoniot. Thus we can use single-shock and

---

*Work performed under the auspices of the U.S. Department of Energy by Lawrence Livermore National Laboratory under contract #W-7405-Eng-48.

reflected-shock experiments to sample a part of the pressure-temperature region of interest to planetary models. This contrasts the situation found in mantle materials where terrestrial processes and shock experiments start with a solid at 1 bar with the shock process driving the temperature up much more rapidly.

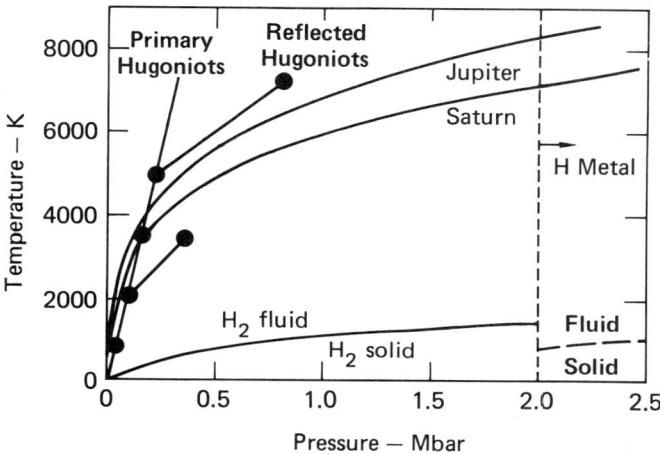

Fig. 1   Experimental Hugoniots compared with Saturn and Jupiter isentropes. The metal is believed to exist above 2 Mbars.

We constructed a comprehensive equation of state for hydrogen by determining an intermolecular potential which would correctly predict the measured high pressure properties of the solid, fluid and melting curve as well as the shock-wave data.

Although no shock-wave data exists for helium there are static high pressure measurements of the solid, fluid and melting curve. The helium equation of state was extended to higher pressure by calculating the 0 K isotherm to 200 Mbars using electron band theory.[4] Our calculations predict that helium becomes metallic at 112 Mbars. An intermolecular pair potential for helium was determined by fitting all these data and was used to predict a liquid Hugoniot. We intend to carry out shock-wave experiments on liquid helium to check this theory.

THE ICES (H_2O, CH_4, NH_3)

H_2O, CH_4 and NH_3 (the ices) are believed to constitute the middle layer of Uranus and Neptune.[2] The estimated pressures and temperatures of this layer ranges from about 6 Mbars and 7000 K at the inner core/ice boundary, to about 0.2 Mbars and 2200 K at the outer ice/hydrogen-helium boundary.

The shockwave data on water and ammonia, including electrical conductivities, have been measured over part of the Uranus and

Neptune pressure-temperature range.[5] Figure 2 compares some shock-wave data for these molecules with an isentrope calculated for Uranus (Neptune is similar). The electrical conductivity data,[3] shown

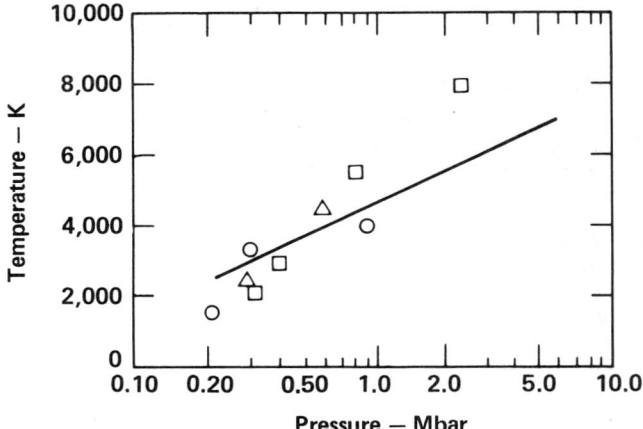

Fig. 2 Uranus (or Neptune) isentrope; solid line and experimental shock-wave results. ☐ Water, o methane, and △ ammonia.

in Fig. 3, suggests that above 0.2 Mbars and about 2000 K the conductivities for H_2O become constant at about 20 to 30 mho/cm, as if the processes leading to ionization have become saturated. It would appear that water has become fully ionized. For ammonia similar results are observed. Shock compression studies on CH_4 show an apparent deviation from predicted molecular behavior that can be explained by the tendency for CH_4 and hydrocarbons above 0.20 Mbars and 2000 K to dissociate into elemental carbon and molecular hydrogen.[6] These results are consistent with earlier work by Ree[7] who showed that shock-wave data on hydrocarbons above these conditions could be interpreted in terms of a conversion to carbon and molecular hydrogen. Grover[8] recently proposed a carbon phase diagram which predicts that below 3000 K and 0.4 Mbar, carbon will exist in the well-known four coordinated diamond form and above these conditions it will be in a metallic solid phase. Consequently we conclude that in the Uranus and Neptune "ice" layer H_2O and NH_3 are not molecular but ionic, and that the carbon in methane has been converted to a solid phase.

THE ROCKY CORE

All of the giant planets are believed to contain a rocky core of iron, silicon, nickel, and magnesium and their oxides. Most planetary structural studies use theoretical models to compute the thermodynamic properties of the rock mixture. New high-pressure shock-wave data for iron, nickel, magnesium oxide, and silicon oxide have been used to develop revised empirical equation-of-state models for rock mixtures.[9] These studies show that the experimentally based equation-of-state models are in much better

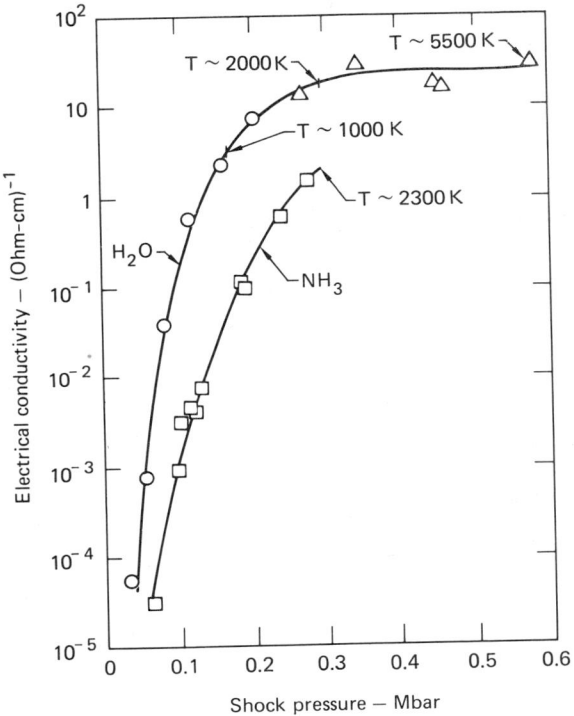

Fig. 3 Plot of the electrical conductivities of water and ammonia indicating that above 2000 K and 0.2 Mbar, the conductivity of water (and probably of ammonia) has reached its maximum and that these materials are thus almost completely dissociated and ionized.

agreement with the planetary structural models than are the theoretical equation-of-state models.

SUMMARY

Using new equations of state based on shock-wave experiments, the internal structure of Jupiter can be modeled in such a way as to explain all of its observed properties, while retaining the same proportions of constituent elements as existed in the sun when it was formed.[9] The characteristics of Uranus and Neptune can also be reconciled under the assumption of a solar composition of ice and rocky components. Only Saturn remains anomalous, and investigations of the thermodynamics and solubility of hydrogen-helium mixtures may provide the solution to this discrepancy.[10]

REFERENCES

1. W. L. Slattery, Icarus 32, 58 (1977).

2. W. B. Hubbard and J. J. MacFarlane, J. Geophys. Res. 85, 225 (1980).

3. A. C. Mitchell and W. J. Nellis, Rev. Sci. Instrum. 52, 347 (1981).

4. D. A. Young, A. K. McMahan, and M. Ross, "Equation of State and Melting Curve of Helium to Very High Pressure", Lawrence Livermore National Laboratory Rept. UCRL-85788, Livermore, CA. Accepted for publication in the Physical Review.

5. A. C. Mitchell, M. I. Kovel, W. J. Nellis, and R. N. Keeler, in High Pressure Science and Technology, Vol. II (B. Vodar and P. Marteau, eds.) (Pergamon, Press, Oxford, 1980) p. 1048.

6. M. Ross and F. H. Ree, J. Chem. Phys. 73, 6146 (1980).

7. F. H. Ree, J. Chem. Phys. 70, 974 (1979).

8. R. Grover, J. Chem. Phys. 71, 3824 (1979).

9. A. S. Grossman, J. B. Pollack, R. T. Reynolds, A. L. Summers and H. C. Graboske, Jr., Icarus 42, 358 (1980).

10. D. J. Stevenson, J. Phys. F: Metal Phys. 9, 791 (1979).

ONE-DIMENSIONAL ISENTROPIC COMPRESSION

Gregory A. Lyzenga* and Thomas J. Ahrens
Seismological Laboratory, California Institute of Technology
Pasadena, California 91125

ABSTRACT

The generation of nearly isentropic pressure-density states in a molecular fluid sample, e.g. H_2O is examined by a series of one-dimensional finite difference calculations. We employ a series of buffer materials of increasing shock impedance (Lexan, Al, Fe, W) behind the sample and impact it with a composite flyer plate of the same series of materials. In the case of H_2O impacted at 2.5 km/sec, three-fold nearly isentropic compression to a pressure of 70 GPa is achieved in 10 μsec with a 3 cm thick composite impactor.

INTRODUCTION

Dynamic pressure-density isentropic compression data can usefully supplement shock and high pressure ultrasonic, x-ray diffraction and high pressure Raman and Brillouin scattering data because they provide thermodynamic information along a unique thermodynamic path (Fig. 1). Such data may be used to develop inter-atomic and molecular potential functions for molecular media such as H_2O, CO_2, H_2 and NH_3 as well as study phase transitions.

All previous experiments and experimental concepts have achieved isentropic compression of non-electrical conductors by utilizing electrically or explosively driven converging cylindrical geometries and in many cases magnetic fields to obtain ultra high pressures within cylindrical volumes[2,3,4]. Pressures up to 800 GPa have been reported in H_2 with this configuration. The density of cylindrical samples can be obtained via flash x-ray radiography to precisions which vary from 10 to 40%. The pressure must be calculated using an assumed equation of state, or inferred from magnetic field intensity measurements[3,4].

In 1972 Kompaneets et al.[5] proposed an experimental configuration to achieve an isentropic compression in one-dimensional planar flow which utilized a shock wave driven into a medium of initial variable porosity with a massive explosive charge. Motivated by this paper, we carried out an extensive series of one-dimensional finite-difference flow calculations[6] using an initial variable density samples. The phenomenon of a shock wave gradually transforming to a dispersed isentropic compression wave, predicted[5] was verified to a limited extent. However, because of the greater complexity of

*Present address: Jet Propulsion Laboratory, Pasadena, California

carrying out such experiments we chose to rather study flows induced by symmetric impact of layered geometries (Fig. 2).

CALCULATIONS

With the goal of eventually carrying out an experimentation using a propellant gun and initially carrying out experiments on water, we utilized the simplified equations of state indicated in Table 1 for our calculations. A composite impactor and target materials in the configuration of Fig. 2 comprised a sequence of materials whose impedances steadily increase from that of H_2O.

When H_2O is impacted by a composite flyer of Fig. 2 a relatively smooth build-up of pressure in the central H_2O layer results.

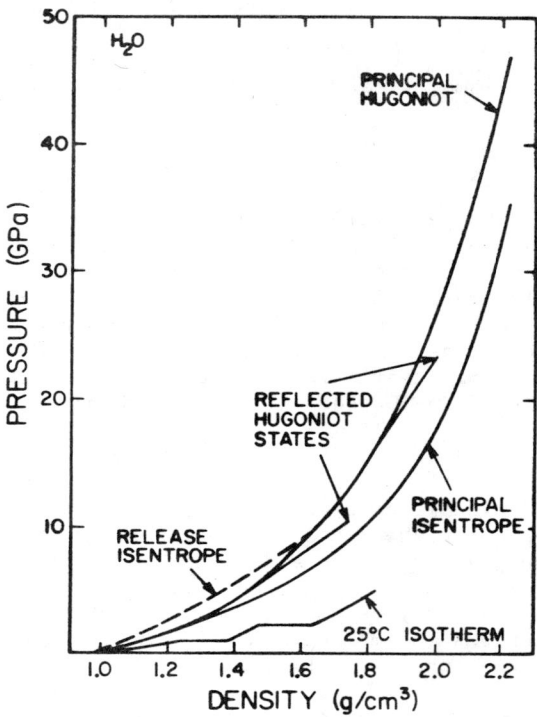

Fig. 1. Pressure-density relations for water. Horizontal portions of isotherms at 1.0 and 2.2 GPa correspond to phase transitions. Representative reflected shock state shown are typical of data now available to 220 GPa[1].

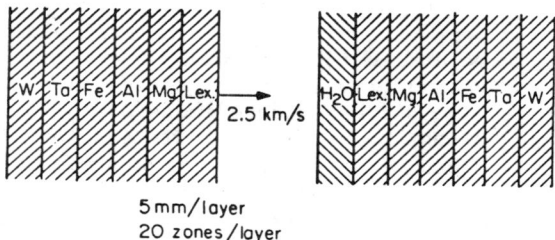

Fig. 2. Symmetric impact configuration for nearly isentropic compression of water.

Table I Assumed Hugoniot Parameters for Impact Calculations[a]

Material	ρ_0 (g/cm^3)	C_0 (km/s)	s	γ_0
W	19.30	4.005	1.268	1.20
Ta	16.66	3.423	1.214	1.69
Fe	7.86	3.768	1.655	1.30
Al	2.70	5.355	1.345	2.13
Mg	1.78	4.650	1.200	1.46
Lexan	1.196	2.796	1.258	2.00
H_2O	1.00	3.111	1.160	2.00

(a) data source, Reference 6

Figure 3 shows the pressure in a particular H_2O zone as a function of time from the impact. The time scale of the pressure rise is determined by the width of the material layers.

The solid curves show the theoretical Hugoniot curve and isentrope of H_2O obtained from the assumed equation-of-state properties (Table I). The H_2O pressure rises monotonically for about 10 μsec following a P-V path which is near the theoretical isentrope. In terms of pressure, the deviation from the isentrope is negligible although some shock heating is apparent from entropy calculations.

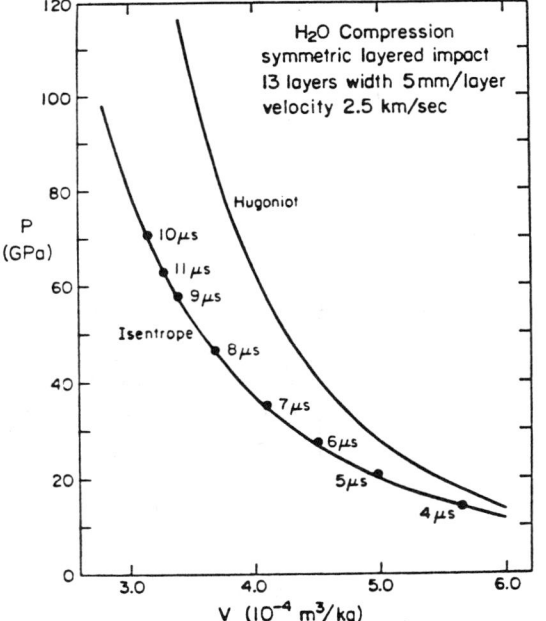

Fig. 3. Computed results of 13-layer symmetric impact. Points indicate computed pressure and specific volume within H_2O layer at the indicated times after impact. Theoretical Hugoniot and isentrope for H_2O are shown for comparison.

The peak H_2O pressure is ~70 GPa. This is below the 135 GPa limiting pressure possible with infinitely thick tungsten impacting tungsten but above the 15 GPa maximum obtainable via a single shock with the same impact velocity. This result indicates that the method illustrated by Fig. 2 is generally feasible for producing isentropic compression in planar geometries. In order to examine a simpler configuration the flow in a 8-layer experiment utilizing a Lexan, Al, Fe and W projectile impacting also a water Al, Fe, and W assembly was calculated. The entropy generated in the water sample, Δs, from standard conditions in going to an arbitrary state (P_i, V_i, E_i) may be expressed in terms of the temperatures in state i and on the isentrope at the same volume. This is

$$\Delta s = \int_{T_s}^{T_i} \frac{C_v dT}{T} . \qquad (1)$$

The specific internal energy difference between these same states is

$$\Delta E = E_i - E_s = \int_{T_s}^{T_i} C_v dT. \qquad (2)$$

Assuming C_v = constant, then the above integrals yield

$$\Delta s = C_v \ln(T_i/T_s) = C_v \ln\left(\frac{\Delta E}{3RT_s} + 1\right). \qquad (3)$$

Since ΔE is obtainable from the finite difference calculations and the state, E_i, the theoretical isentrope may be calculated, T_s is calculated from Mie-Grüneisen theory, and thus, Δs evaluated.

Figure 4 shows the entropies calculated in this manner for the two symmetrical impact configurations and illustrates the result that the entropy production is small and relatively constant up to very high pressures. Additionally, this entropy production is apparently not strongly influenced by the number of stacked plates.

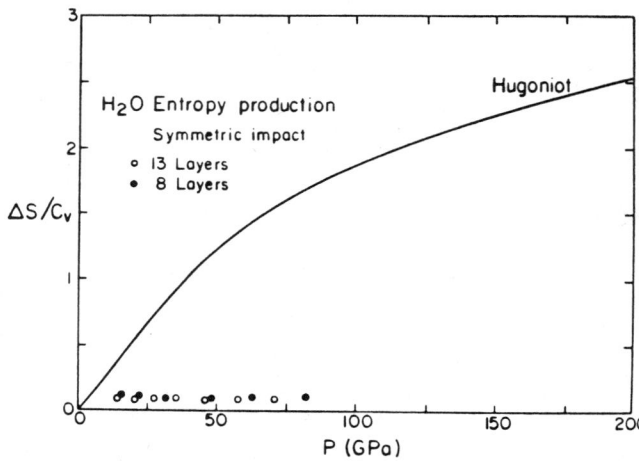

Fig. 4. Specific entropy rise calculated from results of symmetric impact calculations.

CONCLUSIONS

We conclude that nearly isentropic compression may be achieved in one-dimensional flow and the density and pressure may be measured by techniques described elsewhere in this volume[8]. The entropy which is generated is largely the result of the first shock traversing the investigated layer. In the present symmetric impact configuration the first shock in the H$_2$O layer is caused by the impacting lexan layer and has an amplitude of ∿5 GPa. Interestingly, the observed entropy production is very nearly the Hugoniot entropy at just this first shock pressure of 5 GPa. Evidently, the bulk of the entropy contribution comes from the initial shock, with essentially negligible contribution from succeeding disturbances which elevate the sample to the final isentropic pressure.

Contribution No. 3653, Division of Geological and Planetary Sciences, California Institute of Technology, Pasadena, California. This research was supported under NASA grant NAGW 205.

REFERENCES

1. A. C. Mitchell and W. J. Nellis, in High Pressure Science and Technology, Vol. 1, edited by K. D. Timmerhaus and M. S. Barber (Plenum, N.Y., 1979), p. 428.
2. C. M. Fowler, Science 180, 261 (1973).
3. F. V. Grigor-ev, S. B. Kormer, O. L. Mikhailova, A. P. Tolochko and V. D. Urlin, JETP Letters 16, 201 (1972).
4. R. S. Hawke, D. E. Duene, J. G. Huebel, R. H. Keeler and H. Klapper, Phys. Earth Planet. Interiors 6, 44 (1972).
5. A. S. Kompaneets, V. I. Romanova and P. A. Yompol'skii, JETP Letters 16, 183 (1972).
6. R. J. Lawrence and D. S. Mason, Sandia Laboratories, Albuquerque, N.M., SC-RR-710284 (1975).
7. M. van Thiel, Univ. of Calif. Lawrence Livermore National Laboratory, UCRLA 50108 Rev. 1 (1977).
8. M. B. Boslough and T. J. Ahrens, this proceedings.

A METHOD OF DETERMINING POINTS ON THE PRINCIPAL
ISENTROPES OF MOLECULAR LIQUIDS

M. B. Boslough and T. J. Ahrens
Seismological Laboratory, California Institute of Technology
Pasadena, California 91125

ABSTRACT

We have examined the feasibility of using a large-diameter, projectile-target impact to carry out one-dimensional, isentropic compression experiments on molecular fluids. By employing a three-layered target geometry, with a thin foam driver layer and a thick, high-impedance anvil layer, liquid H_2O can be compressed to a state within 0.1% of its principal isentrope at pressures up to about 30 GPa. The pressure and density of the state achieved can be determined from electromagnetic particle velocity gauges imbedded on the interfaces bounding the sample.

INTRODUCTION

Because of the relative simplicity of shock wave experimental techniques, most high pressure data obtained for dynamically compressed materials have been Hugoniot data. However, there are many problems--particularly involving low impedance materials such as molecular fluids--for which knowledge of the principal isentrope would be of great value. Several methods have been conceived for the production of large isentropic compressions in one dimension,[1-5] none of which seem to be directly amenable to measurement. The most likely configuration for an experimental attempt is a simplified version of the symmetric impact experiment proposed by Lyzenga[4] and Lyzenga and Ahrens[5] (Fig. 1).

This experiment involves the use of several layers with large shock-impedance contrast arranged in a way such that the sample layer is compressed by a series of small shocks, as opposed to a single large shock wave as in a conventional Hugoniot experiment. This idea is based on the principle that a compression will be isentropic in the limit of small shocks. Because it has been found empirically that in such a configuration, the net entropy gain is controlled by the first shock wave,[4,5] a low density foam layer (Fig. 1, Material I) is used to break up the initial shock wave into small shocks before it enters the sample layer (Material II). These two layers then reverberate together up to some peak pressure dictated by the high impedance of the flyer and anvil layers (Material III in Fig. 1, but not necessarily the same material).

Another attractive property of using foam for the first layer is that it can be chosen so that above its crush-up pressure it is a good impedance match to the sample layer, so rarefactions and recompressions, which increase the final entropy, are avoided.

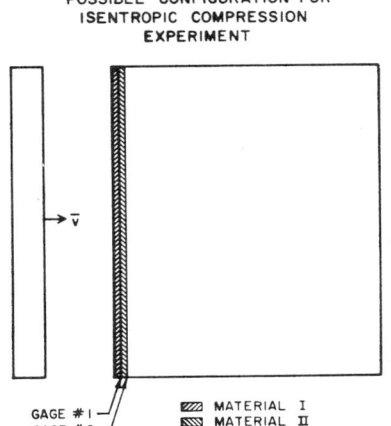

Fig. 1. Three-layer target for producing isentropic compressions is impacted by flyer plate at velocity $\bar{v}$. Shock impedance increases from material I to III.

However, it takes about ten shock wave transits to reverberate up to peak pressure. As a result, the flyer and anvil must be about twenty times as thick as the other layers combined in order to prevent rarefactions from the free surfaces from entering the sample layer. This constraint, along with the constraints imposed by edge effects, requires us to consider a sample layer with a thickness on the order of only a few millimeters for a realistic laboratory experiment.

COMPUTER SIMULATIONS

Several variations of this three-layered target geometry were tested using a one-dimensional finite difference code.[6] A simplified equation of state (EOS) was assumed for H_2O, with a linear u_s-u_p relationship, where $c_0 = 2.335$ mm/μsec and $s = 1.359$ are taken from a fit to published data.[7,8] For the foam layer, we used parameters approximating a polyurethane reference EOS,[9-11] with an initial distension of 10.0, a foam yield stress of 0.02 GPa and a crush-up stress of 0.2 GPa. The important aspects of the simulation were insensitive to variation of the foam parameters.

A greater effort was made to use accurate parameters for the anvil layer, because these dictate the final sample pressure. A high impedance insulator is necessary; high impedance to achieve high peak pressures, and insulating to allow use of electromagnetic gauges. We used parameters for polycrystalline Al_2O_3, which has been studied in detail up to 16 GPa.[12] The constitutive equation for this layer includes an elastic limit and strain rate dependence.

RESULTS AND DISCUSSION

Several simulations were run in the configuration shown in Fig. 1, varying parameters such as velocity, layer thickness, and foam constitutive equation. In the case shown in Fig. 2, the foam layer was 0.5mm, the H_2O layer was 1.0mm and the Al_2O_3 layer was 40mm thick. The "infinite impedance" flyer velocity was 0.75mm/μsec. This is roughly (but not exactly) equivalent to an Al_2O_3 flyer with a velocity of 1.50mm/μsec.

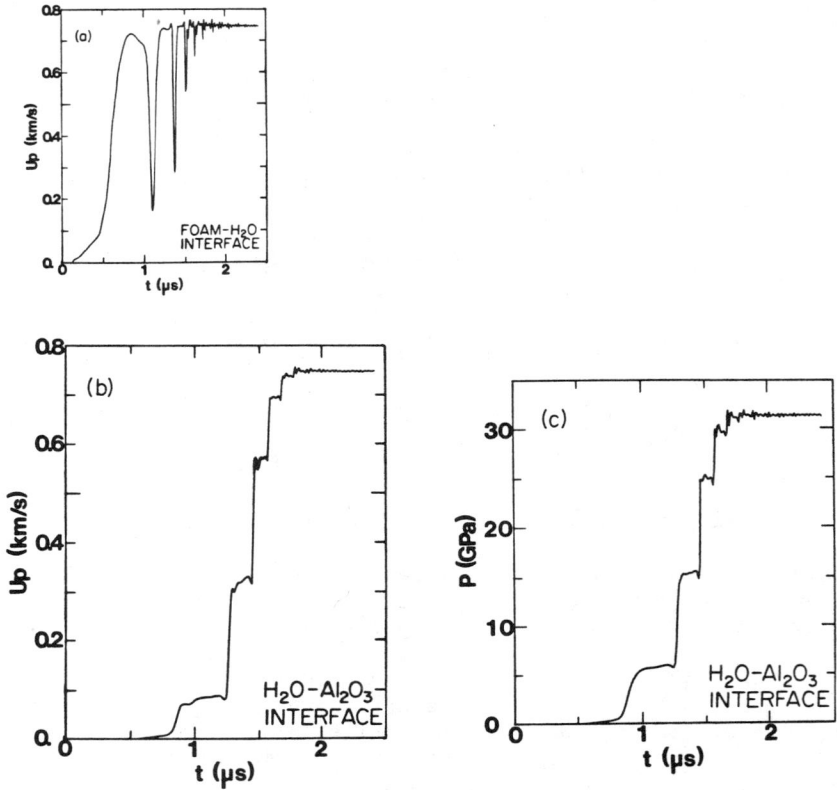

Fig. 2. Particle velocity (Up) and pressure (P) histories calculated by finite difference code for a simulated impact on a three-layer target. (a) $u_p(t)$ at left boundary of sample layer. (b) $u_p(t)$ at right boundary of sample layer. (c) $P(t)$ at right hand boundary of sample layer.

The particle velocity histories at the H_2O layer interfaces are shown in Figs. 2a and 2b, and the pressure history at the $H_2O-Al_2O_3$ interface is shown in Fig. 2c. These illustrate the effect of the foam in breaking up the first shock wave. The first disturbance is effectively a shallow ramp wave up to about 1 GPa. After a few reflections, the ramp steepens into a shock, but by then the shock waves are not as dissipative, or entropy-producing. By 2.0 μsec, the right-hand interface of the H_2O has reached 31.4 GPa, and a density of 2.18 g/cm^3. The calculated pressure on the isentrope of H_2O at this density is 31.39 GPa, while the Hugoniot pressure is 42.76 GPa. Thus the non-isentropicity in this experiment is clearly negligible. The difference in pressure across the H_2O layer after 2 μsec averages about 0.1 GPa, well less than 1% of the total pressure. For realistic measurement precisions, effects of non-isentropicity and non-uniformity are therefore not important.

The density of the H_2O layer can be determined by integrating the difference of the particle velocities of its interfaces over time, by the equation

$$\eta(\tau) = \left(1 - \frac{\rho_o}{\rho}\right) = \frac{1}{x_o} \int_0^\tau (u_{p1}(t) - u_{p2}(t)) dt \quad (1)$$

where $u_{p1}(t)$ and $u_{p2}(t)$ are the velocities of the left- and right-hand interfaces, respectively, and x_o is the initial thickness of the layer. The particle velocities at these interfaces can be measured by electromagnetic gauges,[13] which have an experimental uncertainty of about 1%.[14] This uncertainty combined with a 1% oscilloscope uncertainty leads to a net uncertainty of about 2% in the final density measurement. At times prior to 2 μsec in our simulation example, the density determined in this manner is meaningless, because of the non-uniformity of the layer (a shock wave being in transit); however, it may be possible to determine the intermediate states at the moment the shock wave is reflecting from one of the boundaries.

There are several possible methods of determining the pressure, the simplest being an impedance match solution for the final (peak) state. This method would not give pressures of the intermediate states, however, and is limited by knowledge of the outer-layer parameters. If the anvil is a well-studied material, as the polycrystalline Al_2O_3 is below 16 GPa, the particle velocity at the left boundary of the anvil can be used with the anvil constitutive equation to obtain a stress history at that boundary, providing no rarefaction has arrived from the anvil free surface. If the anvil material has a strain-rate dependence, the velocity of the interface can be used as a boundary condition along with the material parameters in the one-dimensional code to obtain the stress. If there is no rate dependence, the Riemann integral can be inverted to obtain $P(u_p)$.[15] In cases where the material properties are not well known, it may be necessary to use an inclined electromagnetic gauge,[16] or Manganin stress gauge. By one or a combination of these it may be possible to determine the peak pressure to an uncertainty of about 2%.

Contribution No. 3654, Division of Geological and Planetary Sciences, California Institute of Technology, Pasadena, California. This research was supported under NASA grant NSG-7129.

REFERENCES

1. A. S. Kompaneets, V. I. Romanova and P. A. Yompol'skii, JETP Letters 16, 183 (1972).
2. G. A. Adadurov, V. V. Gustov, V. S. Zhuchenko, M. Yu. Kosygin and P. A. Yampolskii, Fizika Goreniya i Vzryva 9, 576 (1973).

3. A. S. Kompaneets, V. I. Romanova and P. A. Yompol'skii, Fizika Goreniya i Vzryva 11, 807 (1974).
4. G. A. Lyzenga, PhD Thesis, Calif. Inst. of Tech., Pasadena, CA (1980).
5. G. A. Lyzenga and T. J. Ahrens (1981) this volume.
6. R. J. Lawrence and D. S. Mason, Sandia National Laboratories, Albuquerque, N.M. SC-RR-710284, (1975).
7. J. M. Walsh and M. H. Rice, J. Chem. Phys. 26, 815 (1957).
8. A. C. Mitchell and W. J. Nellis, in High Pressure Science and Technology, Vol. 1, edited by K. D. Timmerhaus and M. S. Barber, (Plenum Press, 1979), p. 428.
9. R. G. McQueen, S. P. Marsh, J. W. Taylor, J. N. Fritz, and W. J. Carter, in High-Velocity Impact Phenomena, edited by R. Kinslow (Academic, New York, 1970), p. 555.
10. W. Herrmann, Sandia National Laboratories, Albuquerque, N.M. SC-DC-71 4134, (1972).
11. W. J. Carter and S. P. Marsh, unpublished data.
12. D. E. Munson and R. J. Lawrence, J. Appl. Phys. 50, 6272 (1979).
13. A. N. Dremin and G. A. Adadurov, Sov. Phys.-Solid State 6, 1379 (1964).
14. D. B. Larson and G. D. Anderson, J. Geophys. Res. 84, 4592 (1979).
15. J. W. Enig, J. Appl. Phys. 34, 746 (1963).
16. C. Young, R. Fowles and R. P. Swift "Shock Waves and the Mechanical Properties of Solids" edited by J. J. Burke and V. Weiss (Syracuse University Press, Syracuse, 1971).

MOLECULAR DYNAMICS STUDY OF SODIUM
USING A MODEL PSEUDOPOTENTIAL*

Richard E. Swanson,[†] Galen K. Straub, and Brad Lee Holian
Los Alamos National Laboratory, Los Alamos, NM 87545

ABSTRACT

The dynamics of sodium is investigated using the coulomb and Born-Mayer interaction augmented by a model pseudopotential to represent the electron interactions including screening, exchange, and correlation. The model parameters were previously determined and have been shown to accurately reproduce experimental equation-of-state, lattice vibration, and crystal phase properties of sodium in the harmonic limit. In this paper the equation-of-state and structural properties are examined in molecular dynamics calculations. The long range effects of the potential are included. Typically, each particle interacts with about 500 neighbors. The calculated equation of state of sodium in the hcp, bcc, and liquid structures is discussed.

INTRODUCTION

This paper presents the main results[1] of molecular dynamics (MD) calculations for sodium using an interaction potential ϕ derived from pseudopotential theory. ϕ consists of an indirect interaction, V_{IND}, between pairs of ions, including the effect of electron gas screening, exchange, and correlation, along with the coulomb and Born-Mayer repulsion terms:

$$\phi(r) = \frac{z^2 e^2}{r} + \alpha e^{-\gamma r} + V_{IND}(r) \ . \tag{1}$$

The indirect interaction is

$$V_{IND}(r) = \frac{\Omega_0}{\pi^2} \int_0^\infty F(q) \frac{\sin qr}{qr} dq \ , \tag{2}$$

where the three empirical parameters in $F(q)$ and the α and γ in the Born-Mayer term of Eq. (1) were determined by fitting to zero temperature and pressure data and harmonic frequencies of sodium.[2] Figure 1 is a plot of $\phi(r)$.

The average kinetic and potential energies of the MD system of ions interacting by way of the effective interaction, $\phi(r)$, are not the total system energy. We denote these energies as E_s, which depends on the detailed structure of the ions. However, the major contribution to the total system energy is given by the volume-dependent

---
*Work supported by the U.S. Air Force and the U.S. Department of Energy.
[†]U.S. Air Force Academy, Co. 80840.

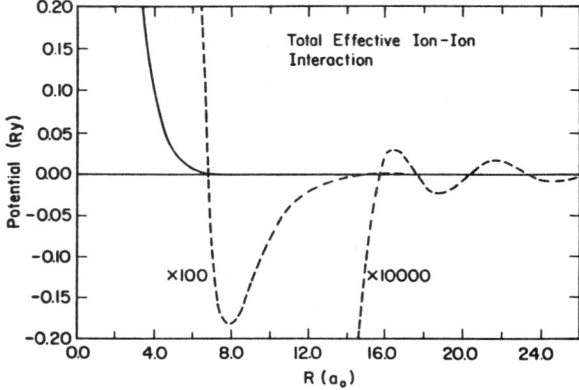

Fig. 1. The total effective ion-ion interaction potential, $\phi(r)$.

terms, E_V. E_V includes the average electron kinetic energy, exchange, and correlation energy, and q = 0 terms of the pseudopotential. The total system energy is given by

$$E_{TOT} = E_s + E_v .$$

SOLID SODIUM

The hexagonal close-packed (hcp) and body-centered cubic (bcc) crystal structures of sodium were simulated by 672-particle MD systems. For each structure we calculated the static (T = 0 K) crystal potential as a function of volume for the range of atomic volumes (Ω_0) from 230 to 280 a_0^3. The results were fit to the equation

$$\phi_0(\Omega_0) = P_1 + P_2\Omega_0 + P_3\Omega_0^2 + P_4\Omega_0^3 .$$

The coefficients are given in Table I. The calculated bcc static crystal energy is -0.475 Ry, the zero pressure atomic volume is 255.1 a_0^3, and the bulk modulus is 5.08 x 10^{-4} Ry/a_0^3. These values agree within 1% of the observed values. We calculated finite-temperature energies from time averages of the MD kinetic and potential energies to obtain E_s, to which was added the volume-dependent terms, E_v.

Table I. Coefficients of the cubic fit of the calculated static crystal potential as a function of atomic volume

Term	bcc	hcp
P_1 (Ry)	-0.310428 ± 0.011	-0.297263 ± 0.035
P_2 (Ry/a_0^3)	$(-1.676211 \pm 0.13) \times 10^{-3}$	$(-1.828394 \pm 0.42) \times 10^{-3}$
P_3 (Ry/a_0^6)	$(5.575895 \pm 0.51) \times 10^{-6}$	$(6.161591 \pm 1.6) \times 10^{-6}$
P_4 (Ry/a_0^9)	$(-5.985510 \pm 0.66) \times 10^{-9}$	$(-6.744432 \pm 2.1) \times 10^{-9}$

Typical resulting equation-of-state points are shown for bcc and hcp sodium at an atomic volume of 256 a_0^3 in Fig. 2. We notice in this figure the deviation from the harmonic result (the straight lines). The anharmonic contributions to the energy, $f(\Omega_0,T)$, defined by

$$E(\Omega_0,T) = \Phi_0(\Omega_0) + 3kT + f(\Omega_0,T) , \qquad (3)$$

was calculated for hcp and bcc sodium for volumes from 232 to 270 a_0^3 (10% compression to 10% expansion) and temperatures from 0 to 400 K. The data was fitted by (see Table II)

$$f(\Omega_0,T) = C(\Omega_0)T^2 \qquad (4)$$

LIQUID SODIUM

The molecular dynamics system melts when the velocities of the particles are great enough to allow them to diffuse away from the perfect lattice positions and form a more random structure. When this liquid state is cooled by artificially removing kinetic energy from the system, it remains in a metastable glassy state to low temperatures. This glassy state may be studied in a manner similar to that described above for the solid to determine its equation of state.[1]

Figure 3 is a plot of the equation-of-state points calculated for bcc and the glassy state of sodium at an atomic volume of 256 a_0^3. The circles

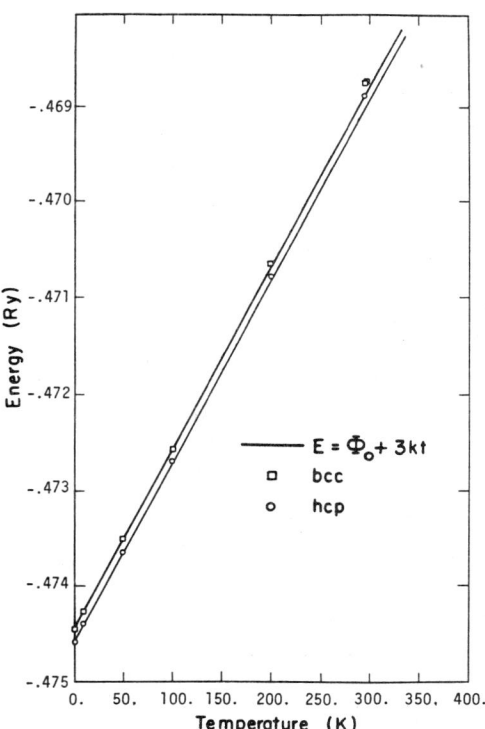

Fig. 2. Total system energy vs temperature for hcp and bcc sodium at an atomic volume of 256 a_0^3.

Table II. Values of the Coefficient $C(\Omega_0)$

Ω_0 (a_0^3)	bcc (× 10^{-9} Ry/k^2)	hcp (× 10^{-9} Ry/k^2)
232	0.908 ± 0.64	1.301 ± 0.23
250	1.012 ± 0.52	1.475 ± 0.50
256	1.250 ± 0.10	1.563 ± 0.25
270	1.350 ± 0.51	2.057 ± 0.33

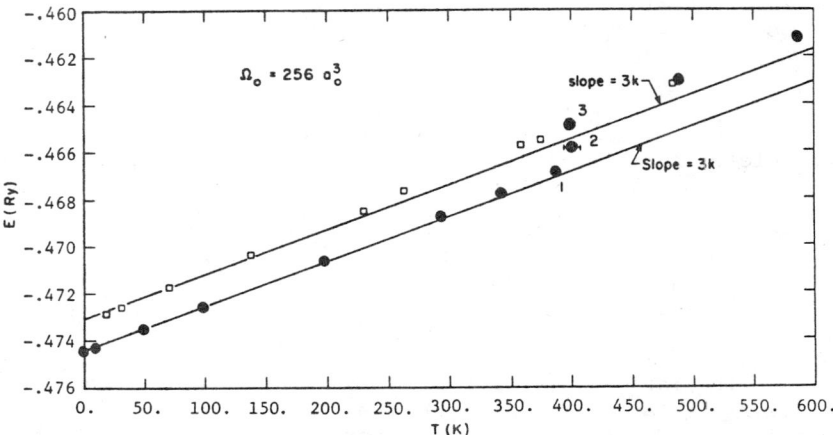

Fig. 3. Equation-of-state points for bcc and liquid sodium. The circles denote points started in the bcc configuration. The squares denote points started in the glassy solid configuration.

that lie on the upper, or liquid, curve represent calculations for which the system melted from the bcc structure.

The points marked 1, 2, and 3 in Fig. 3 represent calculations that had initial temperatures of 400, 450, and 500 K, respectively. Points 1 and 3 have equilibrated to about the same temperature. Point 1 remained on the solid line, point 3 melted and equilibrated at the liquid line, and point 2 formed a partially melted state. We calculated the average mean square displacements from the initial positions, $<\Delta r^2>$, for these points. The results are plotted in Fig. 4. After about 400 time units of equilibration, the curve for point 3 attained a constant slope which yielded a diffusion coefficient of 3.7×10^{-5} cm^2/s. The experimental value reported by Faber[3] is 4.2×10^{-5} cm^2/s. The curve for point 1 shows that the particles are not diffusing. The energy difference between the solid and liquid curves of Fig. 3 at 400 K is 1.7×10^{-3} Ry/ion, which compares favorably with the experimental latent heat of fusion for sodium of 2.32×10^{-3} Ry/ion.

DYNAMIC PHASE CHANGE

We have simulated with MD a bcc-to-hcp martensitic phase change in sodium. Both the hcp and bcc MD systems contained 672 particles, with 12 close-packed planes normal to the z-axis, in the same volume but with differently-shaped periodic calculational boxes. We began a calculation of bcc sodium at 50 K and allowed the system to equilibrate for 150 time units. At this point we changed the shape of the periodic box to make it appropriate to the hcp structure. This had the effect of "pushing" the system over a potential hill, forming a face-centered tetragonal (fct) system. The system then relaxed spontaneously to the hcp structure by a slipping of close-packed planes

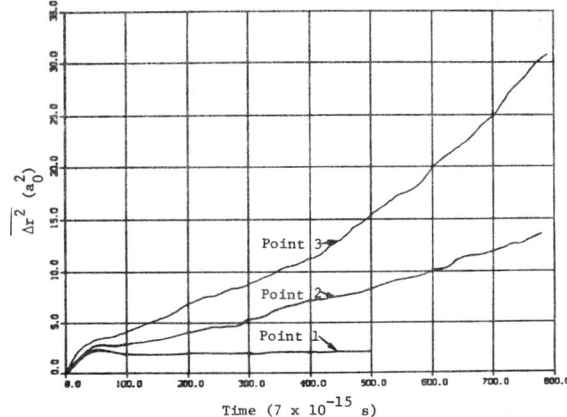

Fig. 4. The mean square displacements vs time for the calculations that produced points 1, 2, and 3 of Fig. 3.

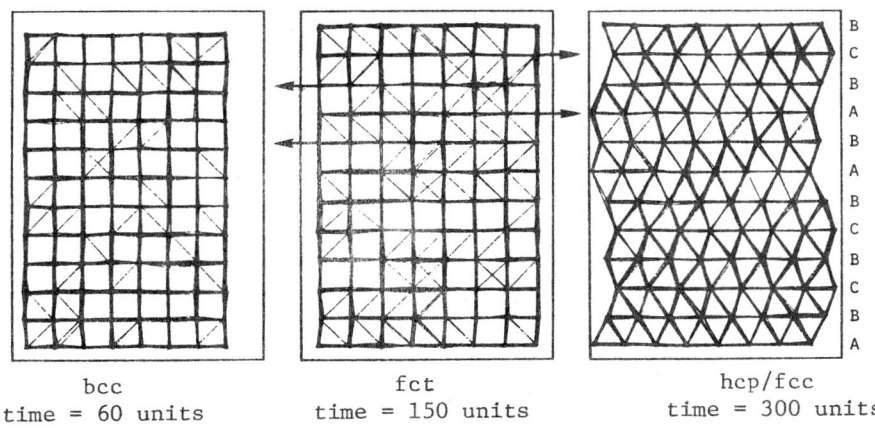

Fig. 5. The molecular dynamics system at three different times during the dynamic phase change. Near neighbors are connected by lines.

relative to one another, as is characteristic of martensitic phase transitions. Fig. 5 shows the molecular dynamics system viewed down rows of atoms with near neighbors connected by lines for the bcc, fct, and hcp systems. The final equilibrated state of the system is not perfect hcp because stacking faults exist as indicated by the standard A, B, C designations of the relative positions of close-packed planes.

REFERENCES

1. Richard E. Swanson, Los Alamos report LA-8877-T Los Alamos National Laboratories, Los Alamos, NM 1981.

2. Duane C. Wallace, Phys. Rev. 176, 832 (1968).

3. T. E. Faber, Introduction to the Theory of Liquid Metals (University Press, Cambridge, England, 1972).

CHARACTERIZATION OF THERMOMECHANICAL RESPONSE OF POROUS VANADIUM

R. E. Tokheim and A. B. Lutze
SRI International, Menlo Park, CA 94025

ABSTRACT

Constitutive relations and equation-of-state parameters for plasma-sprayed vanadium were determined from physical, thermal and mechanical data. Also, electron-beam data taken on the Physics International Owl II machine were used for characterizing thermomechanical porous and liquid-vapor material response under rapid heating conditions.

INTRODUCTION

Our goal was to develop an equation-of-state (EOS) model for porous and fully dense vanadium. Consider the pressure-volume-energy EOS surface shown in Figure 1. The porous EOS surface is superimposed onto the fully dense EOS surface comprising solid, liquid, gas, and mixed-phase regions. The advantage of using a porous material in studying the EOS surface is that, with in-depth radiative heating, we obtain a liquid and vapor material response without response from the solid and solid-liquid regions. We also obtain a porous material response, as indicated in Figure 1 at $t = t_1$ and later at $t = t_2$, where the faster traveling liquid stress wave has partially overtaken the slower porous stress wave. We establish the EOS model by comparing computed transmitted stress histories with gage records.

DATA BASE

For plasma-sprayed vanadium ($\rho_{po} = 5.35$ g/cm^3), we obtained acoustic and static confined compression data from Lawrence Livermore National Laboratory.[1] Isobaric thermodynamic data,[2] yield-strength temperature data,[3] Hugoniot data,[4] and the solid Grüneisen ratio[4] were available from the literature for isotropic solid vanadium. The e-beam experiments on the 1-MeV Owl II machine at Physics International (PI) used the small area beam (typically 400 cal/cm^2) with a 44-mm target aperture. The effective range in the porous vanadium was about 1 mm. The dose-depth profile needed for stress wave computations was based on target current, estimated by matching diode conditions with a Faraday cup data base and on an effective e-beam angle obtained from filtered multiple Faraday cup data as input to the PI Monte Carlo code.

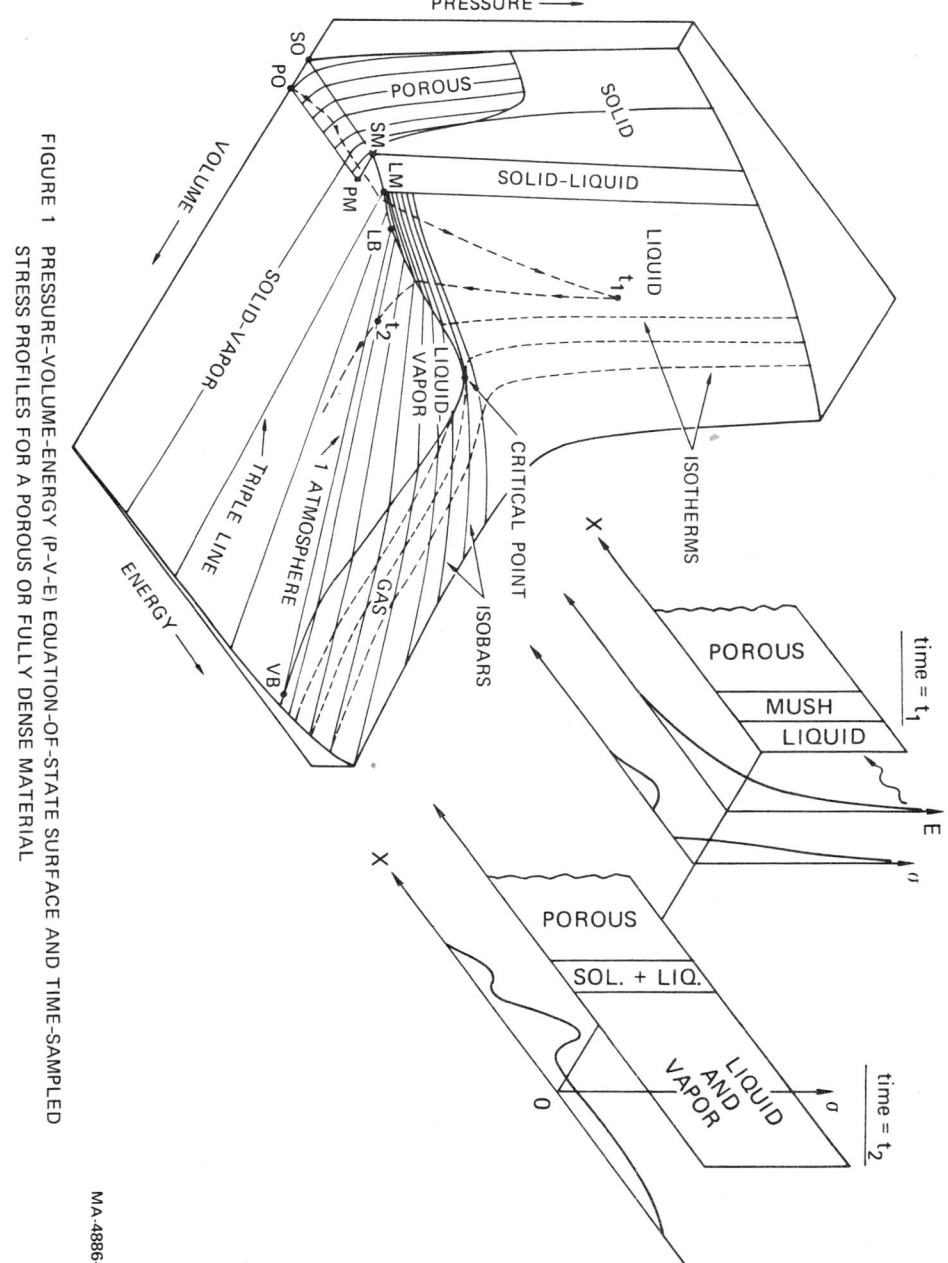

FIGURE 1 PRESSURE-VOLUME-ENERGY (P-V-E) EQUATION-OF-STATE SURFACE AND TIME-SAMPLED STRESS PROFILES FOR A POROUS OR FULLY DENSE MATERIAL

MODEL

Plasma-sprayed vanadium is anisotropic because plasma spraying flattens molten particles on a cooled substrate, producing porous vanadium with mechanical transverse isotropy. We used acoustic and static data to determine the elastic constants (C_{33}, C_{11}, C_{12}, C_{13}, C_{44}) (C_{11} and C_{12} were obtained by scaling from another plasma-sprayed material.) and then the initial porous moduli ($K_3^* = 51.5$ GPa, $E_3^* = C_{33} = 105$ GPa; starred quantities are anisotropic parameters) corresponding to uniaxial strain. The shape of the pressure surface for zero energy was estimated using the Herrmann p-α model.[5] The energy-dependent pressure surface is defined in terms of the density referred to zero energy and a thermal function F(e) based on static yield strength versus temperature data:

$$p = p \left(\frac{\rho}{1 + \Gamma_{so}^* \rho_{so} e/C^*} \right) F(e) \qquad (1)$$

The porous Grüneisen ratio becomes $\Gamma_p^* = (K_3^* \rho_{so} F(e)/(C\rho)) \Gamma_{so}^*$ where solid Grüneisen ratio $\Gamma_{so}^* = (K_3^*/K) \Gamma_{so} = 1.1$ is scaled from the isotropic value Γ_{so} and the porous bulk modulus K. A yield strength of 40 MPa was inferred from static data. The fully dense model for $\rho \geq \rho_{so}$ was the Mie-Grüneisen model used with anisotropic modulus and Grüneisen ratio. For $\rho \leq \rho_{so}$ the following expansion model was used

$$p = \rho \Gamma_e^* \left[e - e_{subl} \left(1 - \exp\left(\frac{C^*}{\Gamma_{so}^* \rho} \left(1 - \frac{\rho_{so}}{\rho} \right) \right) \right) \right] \qquad (2)$$

with an effective Grüneisen ratio

$$\Gamma_e^* = H^* + (\Gamma_{so}^* - H^*)(\rho/\rho_{so})^{0.5} \qquad (3)$$

The most sensitive modeling parameters are K_3^*, $F(e)$ and Γ_{so}^*.

RESULTS AND CONCLUSION

Figure 2 shows the time-dependent dose-depth profiles for the two e-beam shots obtained from Monte Carlo calculations. EB4479 was a normal shot. However, because of diode flashing, EB4473 was a low fluence shot and required a special inductance correction to obtain the accelerating voltage input for Monte Carlo computations.

Figure 3 shows transmitted stress histories compared with quartz gage records. The low fluence shot gave primarily a porous material response, whereas the high fluence shot was dominated by the fully dense material response. Uncertainties from gage (g), e-beam diagnostics (e), and from combined diagnostics and modeling parameters (e + m) are so indicated. These correlations with e-beam data show that (1) our modeling is within diagnostic and modeling uncertainties for stress peaks, (2) the acoustic data were appropriate for determining initial moduli, (3) no strain rate effects are apparent, and (4) model unloading behavior in the liquid-vapor region probably needs improvement, although the correlation would be better using the upper limit in the fluence uncertainty.

We have found that the electron-beam machine is an essential tool for constructing an energy-dependent EOS of a material. Also, correlations with uncertainties less than 20% for porous peaks and 25% for fully dense peaks are possible, provided that the e-beam diagnostics and data analysis are performed with sufficient care.

REFERENCES

1. H. C. Heard, private communications (1978).
2. D. R. Stull and G. C. Sinke, Thermodynamic Properties of the Elements (American Chemical Society, Washington, D.C., 1956).
3. W. Rostoker, The Metallurgy of Vanadium (Wiley, N.Y., 1958).
4. K. A. Gschneidner, Jr., Solid State Physics, Vol. 16 (Academic Press, N. Y., 1964), p. 275.
5. W. Herrmann, J. Appl. Phys. $\underline{40}$, p. 2490 (1969).

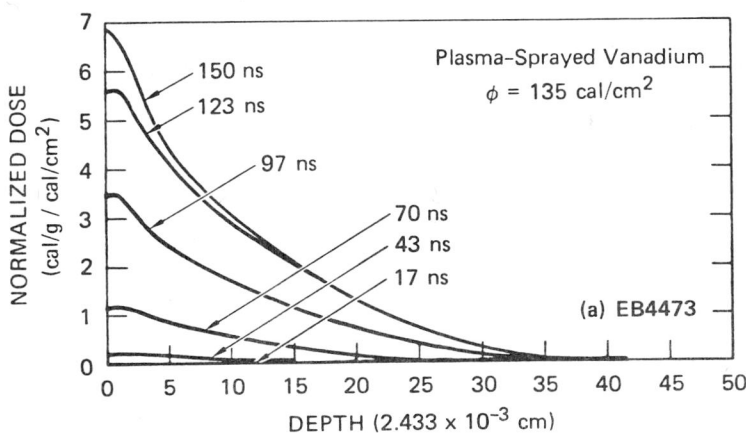

FIGURE 2 TIME-DEPENDENT DOSE-DEPTH PROFILES FOR OWL II e-BEAM SHOTS

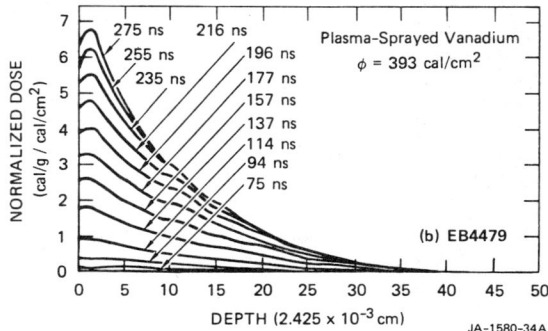

FIGURE 2 TIME-DEPENDENT DOSE-DEPTH PROFILES FOR OWL II e-BEAM SHOTS (Concluded)

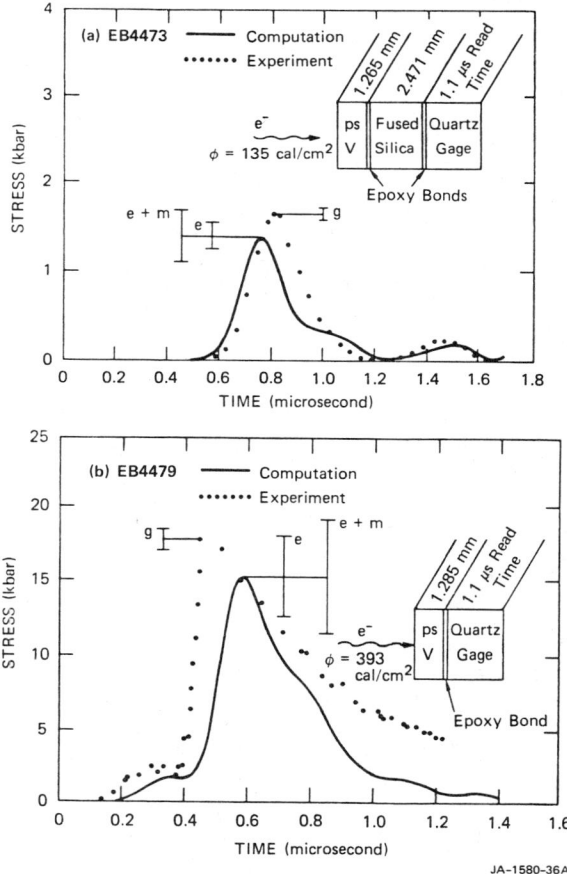

FIGURE 3 STRESS WAVE COMPUTATION FOR PLASMA-SPRAYED VANADIUM COMPARED WITH TRANSMITTED STRESS HISTORY FOR OWL II e-BEAM SHOTS

CHAPTER V: Electrical and Optical Properties

BEHAVIOR OF FERROELECTRIC CERAMICS AND PVF$_2$ POLYMERS UNDER SHOCK LOADING

F. BAUER

Institut Franco-Allemand de Recherches de Saint-Louis (ISL)
12 rue de l'Industrie, 68301 Saint-Louis, France

ABSTRACT

Electric energy can be released under shock action by the rapid depolarization of a prepolarized ferroelectric material or piezoelectric polymer of the PVF$_2$ type. For the behavior of various PZT compositions undergoing shock loading, it is well known that high energy densities can only be attained with ceramics having both a high remanent polarization and low permittivity. The energy is released via a F (ferroelectric) → AF (antiferroelectric) phase transition. A few "hard" PZT ceramics have these properties. Some ternary compositions are seen to exhibit this F → AF transition.

PVF$_2$ (polyvinylidene fluoride) is well suited for use in the mechanoelectric energy conversion induced by a plane shock wave. PVF$_2$ presents a strong piezoelectric activity under specific conditions of polarization which will be discussed (at room temperature remanent polarization attains 8 at 11 µC/cm^2 for PVF$_2$ films). The study of shock loaded electric energy released from PVF$_2$ films will be reported. Also shown will be the Hugoniot curve of isotropic PVF$_2$ as well as that of poled PVF$_2$ one-dimensionally stretched between 5 and 25 kbar. The results achieved will be discussed.

INTRODUCTION

Prepoled and rapidly depoled ferroelectric materials and piezoelectric polymers can be regarded as power or energy sources which are particularly suitable for specific applications, e.g. the production of intense pulse currents with high voltages or energies needed to initiate gases or explosives. When a material is traversed by a shock wave, its physical properties and mechanical properties (internal energy, density etc...) undergo a discontinuity of the first order at the instant of passage of the shock wave. If the material used is ferroelectric or piezoelectric, its electrical energy can be released into a resistor circuit.

Pioneering work in USA was carried out by NEILSON[1], REYNOLDS and SEAY[2], DORAN[3], and HALPIN[4,5] who performed early investigations of both electrical and mechanical behavior of ferroelectric ceramics...

Pioneering work in France was done by VOLLRATH and STENZEL[6,7]. In 1953 they carried out early investigations of barium titanate ceramics and electrets used in shaped-charge projectiles as an energy source for the initiation of electrical detonators with primary explosives. They showed that poled barium titanate ceramics or electrets subjected to the impact of a projectile can adiabatically release their energy into a resistor circuit.

The theoretical model[8] of this adiabatic depolarization using ferroelectric materials shows the maximum energy density of the recoverable potential energy in a mechanoelectrical adiabatic conversion process to be equal to $P_r^2/2\varepsilon$ where P_r is the remanent polarization of the material and ε the average dielectric permittivity during the depolarization. High energy will therefore be obtained with a material having a high P_r and low ε.

Suitable doped materials of the $Pb(Zr_{1-x}Ti_x)O_3$ type and of the $Pb(Hf_{1-y}Zr_y)_{1-x}Ti_xO_3$ satisfy the above criteria under specific conditions.

Polyvinylidene fluoride (PVF_2) seems actually to be a strong candidate for the investigation of the mechanical electrical energy conversion under shock loading. PVF_2 presents a remanent polarization, a strong piezoelectric activity, a low dielectric permittivity, and the mechanical properties of a flexible film combined with those of a piezoelectric element.

Also the quest of materials capable of storing and releasing high electrical energy densities under shock loading leads to the comparative experimental investigation of ferroelectrics and piezoelectrics and of their behavior under shock loading. The results reported here are limited to ferroelectric ceramics as well as to PVF_2 piezoelectric polymers.

Section I of this paper will provide general information on interesting ferroelectric compositions. Moreover ferroelectric energy conversion and shock compression will be recalled.

Section II will provide in more detail general information on PVF_2 polymer and on its "ferroelectric" character. Shock compression behavior of unpoled and poled PVF_2 sheets will be described as well as the first results achieved in adiabatic energy conversion processes.

I. FERROELECTRIC ENERGY CONVERSION UNDER SHOCK WAVE ACTION[9,10]

As outlined above high electrical energy will be obtained with a shock loaded ferroelectric material having a high remanent polarization P_r and a low dielectric permittivity ε. Lead zirconate titanate ceramics $Pb(Zr_{1-x}Ti_x)O_3$ were experimentally studied under shock wave action. These compositions[8] were produced and made available by the "Laboratoire de Ferroélectricité" of Lyon.

Table I: Characteristics of Compositions Investigated

PZT composition	T (^0C)	ε_r	P_r ($\mu C/cm^2$)	ρ_0' ($\Omega \cdot cm$)	d_{33} ($10^{-12} C/N$)	k_p
96.5/3.5+1 mol% Nb_2O_5	221	315	33	$1.2 \cdot 10^{14}$	65	0.15
85 / 15+1 mol% Nb_2O_5	290	330	37.8	10^{14}	100	.23
75 / 25+1 mol% Nb_2O_5	332	380	34.4	$5 \cdot 10^{12}$	125	.29
65 / 35+1 mol% Nb_2O_5	358	470	33.8	$4.5 \cdot 10^{13}$	200	.39
53 / 47+1 mol% Nb_2O_5	420	800	33.2	$2 \cdot 10^{13}$	260	.48
95 /5+0.8 mol% WO_3	236	303	36.4		63	

Two groups of compositions were prepared. The first group containing increasing concentrations of titania and 1 mol% Nb_2O_5, was formulated to investigate the influence of the Ti concentration. The following compositions were prepared:

$PbZr_{0.85}Ti_{0.15}O_3$ + 1 mol% Nb_2O_5 (PZT 85/15)
$PbZr_{0.75}Ti_{0.25}O_3$ + 1 mol% Nb_2O_5 (PZT 75/25)
$PbZr_{0.65}Ti_{0.35}O_3$ + 1 mol% Nb_2O_5 (PZT 65/35)
$PbZr_{0.53}Ti_{0.47}O_3$ + 1 mol% Nb_2O_5 (PZT 53/47)

The second group containing

$PbZr_{0.95}Ti_{0.05}O_3$ + 0.8 mol% WO_3 (PZT 95/05, 0.8 mol% WO_3)
$PbZr_{0.965}Ti_{0.035}O_3$ + 1 mol% Nb_2O_5 (PZT 96.5/03.5, 1 mol% Nb_2O_5)

was prepared to study the influence of the dopants on two metastable ferroelectric ceramics. (For characteristics see Table I.) These two compositions have an AF (antiferroelectric) state at room temperature, but the F (ferroelectric) and AF states are fairly close (maximum ferroelectric coupling and minimum antiferroelectric coupling in a model with two sublattices) and a metastable ferroelectric phase occurs when an electric field is applied[11]. The ceramic then keeps its remanent polarization over a wide range of temperature.

We have studied the total irreversible adiabatic depolarization of these ceramics under shock compression[10]. In our experiments each disk of poled ceramic is traversed axially by a shock wave generated by impact of a projectile and the recoverable electrical energy is measured via a resistor circuit[9].

The experimental results achieved with various compositions of the $Pb(Zr_{1-x}Ti_x)O_3$ type with $0.03 < x < 0.47$ should be briefly recalled here. At pressures ranging from 10 to 20 kbar, the influence of the composition on both the liberated energy and electric load is investigated. The electrical energy delivered on a linear load resistance is found to be maximum in the case of a finite value of the resistance of the discharge circuit. PZT 96.5/3.5 is seen to release 3 times as much electrical energy (1.8 J/cm^3) as materials having a composition close to morphotropic transformation (about 0.7 J/cm^3). The power gain is markedly higher, i.e. on the order of 10 to 15. The optimum shock-pressure range is derived (12 to 20 kbar) over which the conversion efficiency of these "hard" materials is best. Changes generated in shock pressure and calculations allowed to demonstrate the decrease of internal resistivity[12] (e.g. energy conversion efficiency) of a ferroelectric ceramic at pressures above 20 kbar.

The sintering process has a large effect on the internal resistance of the shock loaded material[9]. "Hot pressed" PZT 96.5/03.5+1% Nb_2O_5 releases a maximum electric energy of 2.5 J/cm^3. This high density of electric energy released can only be explained by the presence of a phase change. The experimental determination of the Hugoniot curve shows that under the action of shock compression, the PZT 96.5/03.5 releases its electric energy via a phase change $F \rightarrow AF$[10]. The results demonstrate the advantages of the conversion process achieved with PZT ceramics via a phase change $F \rightarrow AF$.

Also materials suitable for this type of conversion must be antiferroelectric and polarizable at room temperature with a stable induced ferroelectric phase. Their composition must be close to a morphotropic phase transition between antiferroelectric and ferroelectric states. Furthermore they must present a low dielectric permittivity and a high remanent polarization.

These criteria can also be applied to $Pb(Hf_{1-y}Ti_y)_{1-x}Ti_xO_3$ called PHZT for low values of x. Studies[13] of the temperature composition phase diagram of this composition as well as the use of the Goldschmidt factor allow to find a solid solution $Pb(Hf_{0.3}Zr_{0.7})_{0.915}Ti_{0.085}O_3 + 1\% La_2O_3$ which exhibits a stable ferroelectric phase at room temperature. This material delivers its electrical energy by F → AF transition under shock wave action. But the energy densities are limited to 1.2 J/cm^3 because electrical conduction occurs in the shock loaded material. This problem remains to be solved.

II. PVF$_2$ POLYMER AND ITS BEHAVIOR UNDER SHOCK LOADING

Polyvinylidene fluoride also called PVF$_2$ or PVDF is a semicrystalline polymer whose monomer is $-CH_2-CF_2-$. In 1969 KAWAI[14] found out that PVDF films became strongly piezoelectric after having undergone mechanical stretching and subsequent action of an electric field.

This element is of particular interest. In effect it combines the characteristics of a plastic material with those of a piezoelectric element. Some years ago KUREHA Chemical Company was the only firm to produce those polymers. PENNWALT Corporation, RHONE POULENC, and THOMSON LCR, France, are presently engaged in producing piezoelectric PVF$_2$ polymers of high quality. For this reason it was felt worth-while to initiate a research program in which we tried to investigate the physical characteristics of those polymers when undergoing the action of shock loading.

The studies of PVF$_2$ described herein cover four regions of interest:
- General remarks and demonstration of the ferroelectric character of PVF$_2$
- Determination of the PVF$_2$ Hugoniot curve
- Study of the electric energy released from PVF$_2$ films and its use for mechanoelectric energy conversion
- Behavior at very high pressure levels: examples for possible use.

II.1 PREPARATION OF THE POLAR PVF$_2$ POLYMER - DETERMINATION OF ITS PHYSICAL CHARACTERISTICS

As reported from KEPLER[15] we know that polyvinylidene fluoride is a crystalline polymer which exists in three stable polymorphs. In phase I the polar form, the molecules are in a planar zig-zag conformation and the unit cell is orthorhombic and there are two polymer chains per unit cell[19]. Theories[15] and calculations[17,18] of the induced polarization could be equal to a maximum value of

13.2 µC/cm². BROADHURST's model[19] proposes a maximum value for polarization which is equal to 22 µC/cm². Properties of other phases will not be considered here.

The polar material can then be poled when a strong electric field is applied. As knowledge of the electric energy stored in those materials is of fundamental interest, the polarization of biaxially stretched thin PVF₂ film was experimentally investigated.

II.1.1 POLARIZATION OF PVF₂: FERROELECTRIC CHARACTERISTICS

We have run hysteresis loops at room temperature on polyvinylidene fluoride using electrical fields up to $5 \cdot 10^8$ V/m. The polymer film used 23 µm thick and biaxially oriented film from RHONE POULENC Films. Circular aluminium electrodes were evaporated of 1 cm² area. The sample was placed between spherical electrodes in a dielectric liquid. A novel experimental arrangement was designed and built for measuring directly the polarization of the material under investigation (Figure 1). A high-voltage triangular

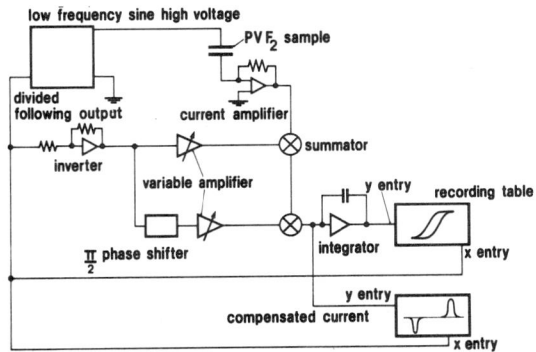

Fig.1. Cycling hysteresis device

signal was applied to the specimen at a very low frequency and the current traversing the PVF₂ sample was measured. Since D is equal to ε E + P, the intensity i passing through the polar dielectric is given by

$$i = i_D + i_R = \varepsilon \frac{dE}{dt} + \frac{dP}{dt} + E/R \quad \text{where}$$

i_D = displacement current, i_R : current into the resistance R ;
D = electric induction ; P = polarization of the material ;
R = internal resistance of the material.

From the intensity i measured, the resistive component E/R as well as the capacitive component ($\varepsilon \frac{dE}{dt}$) are subtracted. We obtain a cycle for i such that $i = \frac{dP}{dt}$.

By integrating the current we obtain the curve which indicates the total polarization of the material as a function of the electrical field applied. This polarization appears as a hysteresis loop between the temperature of the melting point of PVF_2, $T_f = 170°C$, and the temperature of the vitreous point, $T_g = -40°C$. Typical experimental results for a bi-axially oriented sample are shown in Figure 2. The hysteresis loop was traversed many times at room

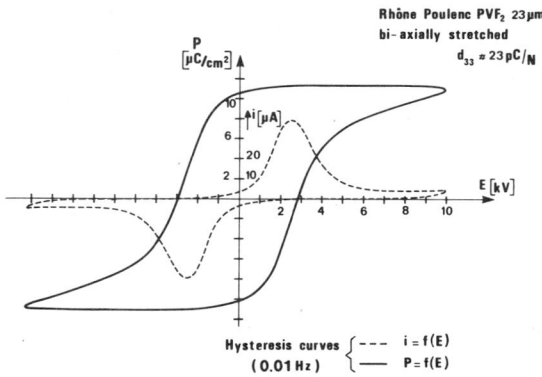

Fig.2. Hysteresis curves for bi-axially stretched PVF_2

temperature at a maximum value of electric field just below that for which electrical breakdown can occur, in order to drain the ions and local space charges. PVF_2 showing this hysteresis loop is therefore "ferroelectric". The loops thus obtained are drawn as isothermal curves at a very low frequency (0.01 Hz). The remanent polarization attains 10 $\mu C/cm^2$ on the samples of 23 μm made available by RHONE POULENC. At room temperature the coercive field equals 1MV/cm. The piezoelectric coefficient d_{33} attains 22 pC/N.

In order to confirm this last value, we have measured the amount of charge released by each sample of RHONE POULENC when it was suddenly heated to above its melting point = 170°C. For the RHONE POULENC PVF_2 we obtained an average value of 7.7 $\mu C/cm^2$ of delivered electric charge. The ratio of the polarization measured in the thermal depoling experiment to that measured from the hysteresis loop was 0.86. The discrepancy increases with the thickness of the PVF_2 sample. Our measurement disagrees with the prediction of the vacuum moments model. If the samples were 100% crystalline, the saturation polarization would be of the order of 12.1 $\mu C/cm^2$. Since polyvinylidene fluoride is typically 50% crystalline, the saturation polarization should be on the order of 6 $\mu C/cm^2$.

Our results seem to be in better agreement with BROADHURST's model[19]. In the latter case P_r may attain 11 $\mu C/cm^2$ for a polymer containing 50% of crystallites.

The polarization of the material is stable between -40°C and +100°C.

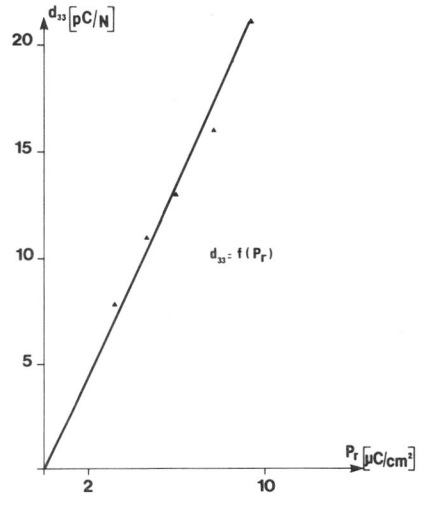

Fig.3. d_{33} piezoelectric coefficient versus remanent polarization

It is verified that the coefficient d_{33} (measured with d_{33} Berlincourt piezometer) is proportional in a satisfactory manner to the remanent polarization [17,18] (Figure 3) and that the polarity of the material under investigation is inverse to the high voltage applied, e.g.

$$d_{33} = - \frac{P_r}{V} \frac{dV}{dx_3} = -\gamma P_r$$

P_r remanent polarization

$\frac{dV}{V}$ compressibility

γ Young modulus

x_3 stress applied.

II.1.2 INVESTIGATION CONDUCTED UNDER UNIAXIAL COMPRESSION

An experimental arrangement has been designed and built[23] which allows the cycles giving the electric charge released in a short-circuit for a PVF_2 specimen of 23 µm in thickness to be drawn as a function of the uniaxial pressure. The piezoelectric activity of PVF_2 is practically reversible up to 9 kbar (Figure 4).

II.2 DYNAMIC ADIABAT OF PVF_2

The investigation of both the physical behavior and electric data of PVF_2 subjected to the action of a shock wave requires its dynamic adiabatic to be known. The experiments performed will be described and the dynamic adiabatic of non polar as well as of poled PVF_2 samples will be given for pressures ranging from 5 to 30 kbar.

II.2.1 MEASURING METHOD AND EXPERIMENTAL TECHNIQUE USED

It should be briefly recalled that the natural shock data (pressure p, relative specific mass ρ_0/ρ, shock velocity U, particle velocity u) can be determined from measurements of both pressure p and velocity U and from the use of conventional formulae applying to strong shocks:

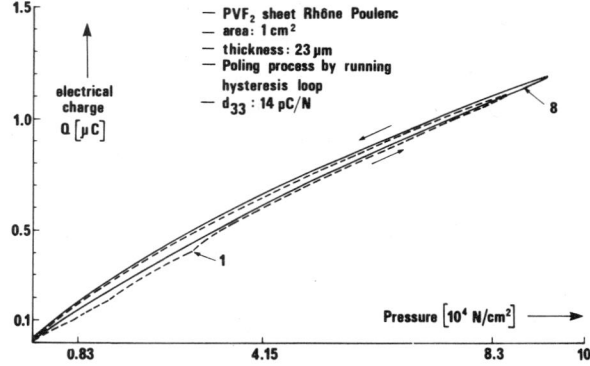

Fig.4. Electrical charge versus static compression

$$p = \rho_0 \cdot U \cdot u$$ where ρ_0 = the initial density and

$$\rho_0/\rho = 1 - \frac{u}{U} = \frac{U-u}{U} . \qquad (1)$$

The adiabatic curve for a short-circuited PVF_2 unpoled or poled was determined using the Sandia Shorted quartz gage method proposed by R.A. GRAHAM[20,21].

Table II

Trial	Origin PVF_2	Characteristics	Thickness [µm]	p [kbar]	U [m/s]	u [m/s]	ρ_0/ρ
1	KYNAR	isotropic	900	20	3000	370	0,876
2	KYNAR	"	900	7	2300	169	0,926
3	Kureha	"	1450	19	2940	350	0,877
4	"	"	1450	16,5	2790	330	0,879
5	"	"	920	15,4	2660	321	0,88
6	"	"	1450	13,5	2540	295	0,883
7	"	"	920	13,4	2520	295	0,883
8	"	"	920	10,8	2450	244	0,9
9	KYNAR	poled	550	12,5	2570	270	0,895
10	"	"	"	10,9	2500	242	0,903
11	"	"	"	17,8	2722	363	0,866
12	"	"	"	24,6	3022	452	0,850
13	"	"	"	8,5	2606	181	0,930
14	Thomson LCR	"	830	26	3074	469	0,847
15	"	"	"	9,3	2643	195	0,926
16	"	"	"	21,5	2804	425	0,849
17	"	"	870	22	2960	413	0,860

II.2.2 RESULTS

In the above table II are shown the results achieved with isotropic and unpoled PVF_2 samples and with poled samples, respectively.

Figure 5 exhibits the dynamic adiabat of isotropic PVF_2 as well as that of poled PVF_2. The shock pressure versus particle velocity is plotted in Figure 6.

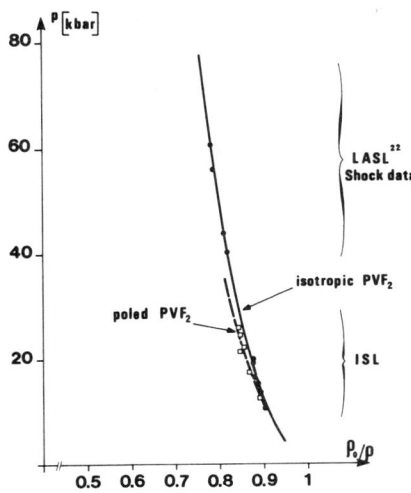

Fig.5. Shock pressure versus compression of poled PVF_2 and isotropic PVF_2

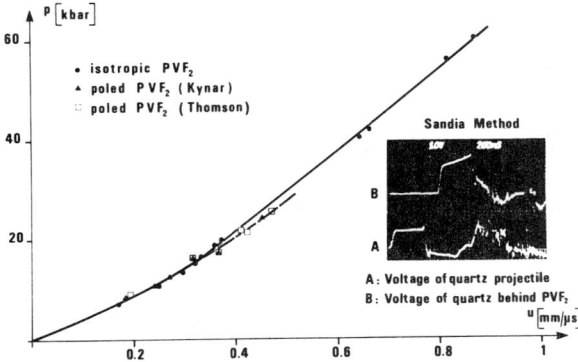

Fig.6. Shock pressure versus particle velocity of poled and isotropic PVF_2

If the dynamic adiabat of isotropic PVF_2 is an extension of the curve drawn by LASL[22], the dynamic adiabat of poled PVF_2 is seen to differ slightly from that of isotropic PVF_2. It can be assumed that polar and polarized PVF_2 is more compressible than isotropic PVF_2. This higher compressibility results from the high anisotropy of the material, which is due to the action of piezo-crystalline fields appearing in the polarization process.

II.3 INVESTIGATION OF PVF_2 UNDERGOING MECHANOELECTRIC ENERGY CONVERSION

Let us apply to PVF_2 the theoretical computation of the density of potential electric energy which can be recovered via mechanoelectric energy conversion when a shock wave is applied. For this we use the following relationship:

$$W_{elec.} = \frac{1}{2} P_r^2 / \varepsilon \qquad \begin{array}{l} P_r \text{ remanent polarization} \\ \varepsilon \text{ permittivity or } \varepsilon_0 \varepsilon_r \end{array}$$

and we obtain about 38 J/cm^3.

From the above it is clearly seen that those materials are promising candidates for shock induced energy conversion.

II.3.1 EXPERIMENTAL TECHNIQUE

Experiments are performed on circular disks of poled PVF_2 materials in a holder which is schematically shown in Figure 7. The

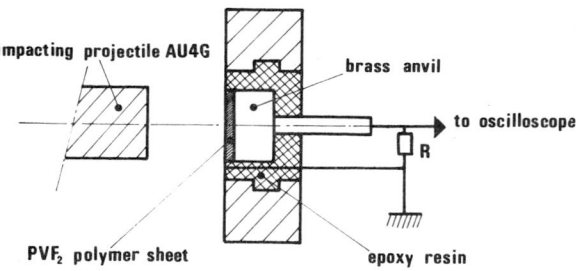

Fig.7. Experimental arrangement for shock studies

shock wave is generated by the impact of duraluminium or copper cylinders and is made to propagate along the disk axis in the same direction to that of polarization.

Taking into account the small thickness of the disks (23 µm), both projectile and anvil are polished optically. The tests are performed in a vacuum. The recoverable electric energy is measured through a resistor circuit of negligible inductance.

The maximum of releasable electric energy, i^2R, is established in assigning several values to the resistance of the discharge circuit for a given shock pressure. Thereafter the influence of the shock pressure induced is investigated in increasing or in decreasing the intensity of the latter. The shock pressure is computed with the aid of shock data of both projectile and PVF_2 and of the measured projectile velocity[20].

II.3.2 EXPERIMENTAL RESULTS

PVF_2 polymers with following given characteristics were investigated.

Table III

PVF_2	Thickness [µm]	P_r [µC/cm^2]	d_{33} [10^{-12}C/N]	ε_r
Rhône Poulenc Films	23	10	20 − 22	10 − 12
Thomson LCR	370 830		22 − 25	9 − 13
Pennwalt Corporation	370 550		24 − 28	11 − 12

a) Electric Characteristics of PVF_2 Samples Subjected to Shock Loading

Poled PVF_2 samples of 830 µm in thickness made available by THOMSON LCR[27] were equipped with (silver painted) electrodes in "Sandia guard ring configuration". The d_{33} coefficients range from 17 to 22 pC/N. In order to investigate the electric response of the PVF_2 disks, the latter were connected to a 10 Ω resistor and subjected to sustained shock loading of different intensities. The shape of the recorded voltage is characteristic and exhibits a shock wave profile which is analogous to that given by a quartz X cut in the same "Sandia guard ring configuration"(Figure 8). This comparison should not be continued. In effect the more the shock intensity increases the more the velocity of the shock wave induced in the PVF_2 sample increases while the pulse length decreases. Except for the experimental errors involved the previously measured shock velocities are found again (section II.2.2).

If the curve indicating the amount of electric energy released in a short-circuit is plotted as a function of the shock pressure induced, the shape of the curve is seen to be practically linear (Figure 8). It is verified that the released charge Q is approximately proportional to the applied stress σ (Q = d_{33} σ). The deviation of the results achieved can be correlated with the dispersion of the previously measured d_{33} coefficients. Simultaneously is drawn in Figure 8 the value of the average electric voltage

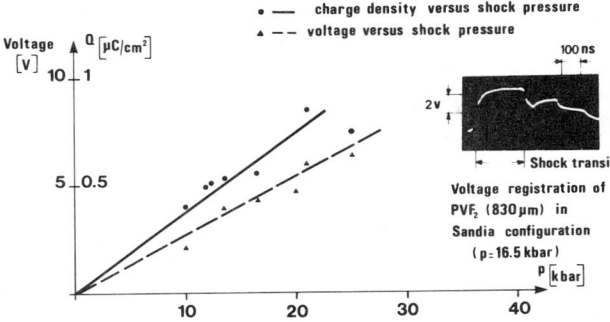

Fig.8. Electric charge and average voltage versus shock pressure of PVF$_2$ gauge in Sandia guard ring configuration

measured as a function of the shock pressure induced. The curve is practically linear over the pressure range considered. From these experimental findings it can be concluded that the "ferroelectric" PVF$_2$ shows a behavior analogous to that of a piezoelectric material although it is not linearly elastic.

b) Investigation of PVF$_2$ Samples Undergoing Electromechanical Energy Conversion Between 8 and 26 kbar

The first tests carried out in a depolarization regime on a resistive charge were done with PVF$_2$ disks (active area: 1 cm^2) of different thicknesses: 20 μm, 23 μm, 300 μm, 350 μm. The shock wave

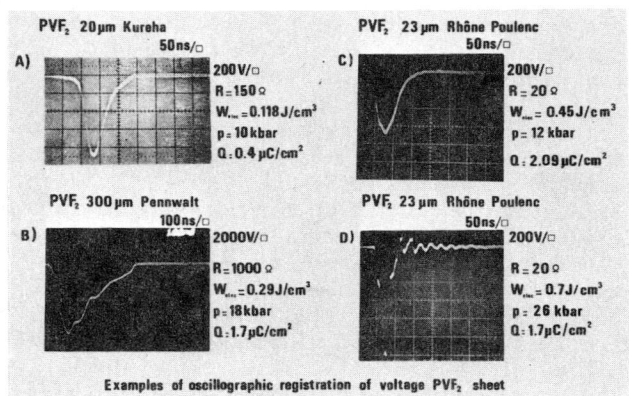

Fig.9. Examples of oscillographic registrations

generated is a sustained one. Figure 9 exhibits a few characteristic recordings.

- At a pressure of 10 kbar, the density of the electric energy released equals 0.1 J/cm^3 (Figure 9A).
- At still higher pressures acting on Rhône Poulenc samples poled by running hysteresis loop (P_r = 10 µC/cm^2) we attain 0.45 J/cm^3 at 12 kbar and 0.7 J/cm^3 at 26 kbar (peak voltage: 1200 V on 20 Ω) (Figures 9C,9D). The energy densities attained are limited by dielectric breakdowns occurring in test running which this shortcoming generated at the instant of impact has a marked effect. Furthermore some tests performed with PVF$_2$ samples of 23 µm in thickness placed between the brass anvil and an insulator have allowed to attain 1.19 J/cm^3.
- In all Figures the voltage is seen to increase stepwise: the first step corresponds to the transit time of shock wave while the second step corresponds to the compression generated as the wave is reflected from the brass anvil. Depolarization tests performed with poled PVF$_2$ samples during a very short time period (< 2 minutes) show that the signal recorded presents anomalies which must be due to space charges trapped and released at the instant of shock wave transit.

Figure 9B exhibits high voltages attained with thick films. In all Figures the rise time corresponds to the transit time of the shock wave propagating in the material.

The quantities of released charges represent only 20% of the remanent polarization of the material under investigation. At low pressures PVF$_2$ is depolarized very slightly only and its internal resistance remains at a high level even under shock loading.

c) Tests Performed at High Pressures by BOUCHU[+] at CEA

BOUCHU[25] made tests in order to observe the signals delivered by a miniature probe. The probe consists of an active PVF$_2$ specimen made available by Rhône Poulenc and characterized by ISL (diameter: 0.75 mm, thickness: 23 µm). The PVF$_2$ specimen was placed in a coaxial configuration (Figure 10B) and located along a line closed on

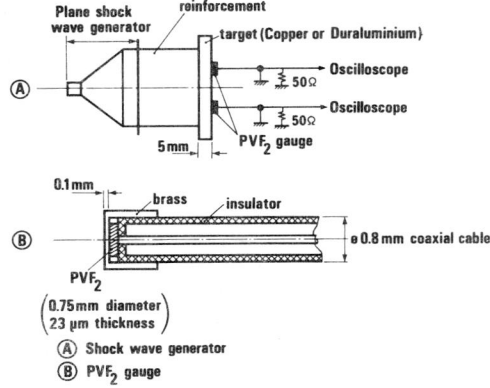

Fig.10.

Experimental arrangement A - Scheme of the PVF$_2$ gauge B -

[+]Mr. BOUCHU, CEA, 93 Sevran (France)

its characteristic impedance equalling 50 Ω. The experimental arrangement is depicted in Figure 10A. It includes a plane wave generator consisting of conventional explosives. For each trial 6 or 8 probes were used which were in contact with the target and distributed along a circumference of 40 mm in diameter.

The pressures attained in the target are known and reproducible: p = 96 kbar in an AU4G target
p = 560 kbar)
p = 1060 kbar) in a copper target

The results achieved are listed in the following table[25]:

Table IV

Target Material	Pressure attained in the target [kbar]	Signal Amplitude [V]	Transit time of the shock wave	U PVF$_2$ [m/s]	p in PVF$_2$ specimen [kbar]	p in KYNAR [kbar]
AU4G	96	$\bar{V}_{average}$=25	T = 5,5 ns	4180	44	40
Cu	560	$\bar{V}_{average}$=64	T = 4,4 ns	5230	172	176
Cu	1060	$\bar{V}_{average}$=39	T = 3,4 ns	6760	330	328

The transit time of the shock wave in the PVF$_2$ specimen is given by (t_2-t_1) (Figure 11). Knowlegde of the shock polar of the

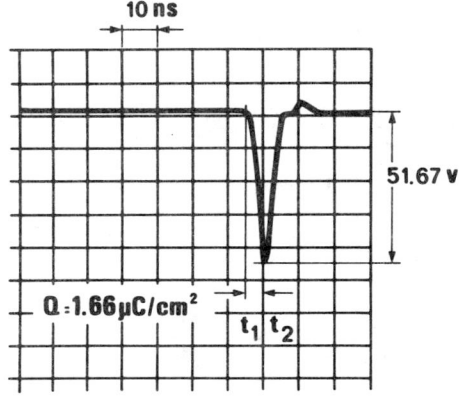

Fig.11. Example of oscillographic registration (p=560kbar in copper)

target material as well as the determination of the straight line p/u with the aid of approximation U = A+Bu allow three points of the shock polar p = f(u) of the PVF$_2$ sample to be drawn (Figure 12).

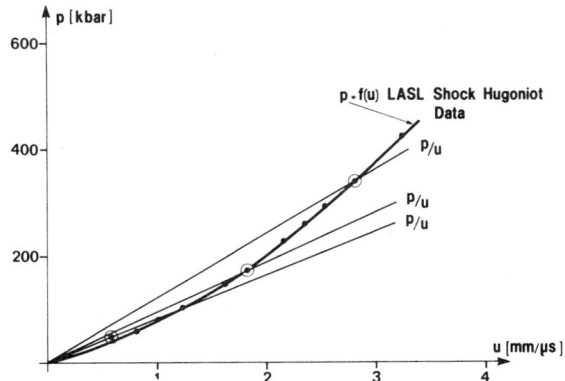

Fig.12. Shock pressure versus particle velocity

There is a good agreement between these three points and the shock polar of KYNAR[22]. This confirms the choice of parameters used for evaluating the shock pressure intensity in the active element. It must be noted, however, that the signal amplitudes are somewhat scattered at the same shock pressure. This dispersion is due to the anisotropy of the piezoelectric characteristics of the PVF_2 samples cut without taking particular care. PVF_2 remains insulating even at high pressures.

d) Behavior of PVF_2 Placed Inside an Explosive[26,27]

Some highly interesting experiments should be reported here. These tests were performed (in the framework of a DRET contract) by PFEFFER[+] with 30 μm thick PVF_2 samples of 3 mm in diameter developed by MICHERON[++]. The PVF_2 sample is placed inside the explosive. At the instant of passage of the shock wave PFEFFER records on 50 Ω a voltage pulse which attains 1500 V in 20 ns. The quantities of electric charges released were found to be of the order of 8 μC/cm².

CONCLUSION

Some methods and experimental arrangements were developed in order to investigate semicrystalline PVF_2 polymers. In the framework of these investigations the piezoelectric activity of these polymers, their ferroelectric characteristics and their behavior under shock wave action were studied in particular. Experimental observations could be reported with respect to the polarization of the material and its electric characteristics could be established when undergoing the influence of a plane shock wave.

[+] Mr. PFEFFER, Thomson Brandt Armement, 45 La Ferté St-Aubin, France
[++] Mr. MICHERON, Thomson LCR, Centre de Corberville, 91 Orsay, France

The electric energy densities recoverable via shock loading attain high values. They correspond to those of the best ceramic materials. PVF_2 appears to be a strong candidate for being used as an energy converter under shock loading in ballistics and detonics in view of chronometric recording of strong shock waves. PVF_2 material offers a variety of possible applications in which the ferroelectric characteristics of this material may find a widespread use.

ACKNOWLEDGEMENT

The author wishes to thank for helpful advice Prof. L. Eyraud I.N.S.A. of Lyon, R.A. Graham Sandia Laboratories, Dr. M. Eyraud of Rhône Poulenc France, Dr. Ph. Bloomfield Pennwalt Corporation, Dr. Pfeffer and Dr. Micheron Thomson France, especially. M. Bouchu of CEA France who provides the results obtained at high shock pressure level, and T. Wingler ISL for having contributed to the English translation.

REFERENCES

1. F.W. Neilson, Bull. Am. Phys. Soc. 2, 302 (1957)
2. C.E. Reynolds & G.E. Seay, J. Appl. Phys. 32, 1401-1402 (1961)
3. D.G. Doran, J. Appl. Phys. 40, 40-47 (1968)
4. W.J. Halpin, J. Appl. Phys. 37, 153-163 (1966)
5. W.J. Halpin, J. Appl. Phys. 39, 3821-3826 (1968)
6. K. Vollrath, A. Stenzel, Report ISL 19/53, unpublished
7. K. Vollrath, A. Stenzel, Report ISL 20/53, unpublished
8. M. Troccaz, L. Eyraud, Y. Fetiveau, P. Gonnard, J. Paletto, C.R. Acad. Sc. Ser. B 276, 13, 547-550 (1973)
9. F. Bauer, L. Eyraud, Y. Fetiveau, K. Vollrath, J. Am. Ceram. Soc., 5-6, 63 (1980)
10. F. Bauer, Thesis Dr. Sc. University Lyon, Report ISL 131/76 (1977)
11. M. Troccaz, P. Gonnard, Y. Fetiveau, G. Grange, Ferroelectrics 14, 1-2 (1976)
12. P.C. Lysne, J. Appl. Phys. 44, 2, 577-82 (1973)
13. G. Grange, M. Troccaz, L. Eyraud, F. Bauer, Appl. Phys. 23, 289-293 (1980)
14. H. Kawaï, Jpn. J. Appl. Phys. 8, 975 (1969)
15. R.G. Kepler, Org. Coatings and Plastics Chemistry 38.278 (1978) (Reprints for ACS Meeting Anaheim C.A. March 1978)
16. R.G. Kepler, R.A. Anderson, J. Appl. Phys. 49, 3, 1232-35 (1978)
17. F. Micheron, G. Bichon, C. Lenonon, H. Facotti, M. Royer, Pro. Inter. Workshop, Kyoto Oct. 1978
18. M. Royer, F. Micheron, C.R. Acad. Sci. Serie B, 287, 6, 145-7 (1978)
19. M.G. Broadhurst, G.T. Davis, J.E. McKinney, R.E. Collins, J. Appl. Phys. 49, 10, 4992-7 (1978)
20. R.A. Graham, F.W. Neilson, W.B. Benedick, J. Appl. Phys. 36, 5, 1775-83 (1965)
21. R.A. Graham, J. Appl. Phys. 46, 1901-09 (1975)
22. LASL Shock date, St. P. Marsh, University of Califor. Press (1980)
23. F. Bauer, Report ISL Co 215/80, unpublished
24. M. Bouchu, CEA Report DO-0047 (1976) unpublished
25. M. Bouchu, ME/EMP, 524/81, CEA Report (1981) unpublished
26. Pfeffer, private communication
27. F. Micheron, private communication
28. Ph.E. Bloomfield, NBSIR 75-724 [R] (1975)
29. Ph.E. Bloomfield et al., Naval. Research Rev., May (1978)
30. R.A. Graham, J. Phys. Chem. 83, 3048 (1979)
31. L. Davison, R.A. Graham, Physics Reports 55, 255 (1980)

OPTICAL PYROMETRY AT HIGH SHOCK PRESSURES AND ITS INTERPRETATION

G. A. Lyzenga
Jet Propulsion Laboratory, Pasadena, Ca. 91109

ABSTRACT

The utility of multi-color pyrometry has long been recognized in the study and elucidation of the equations of state of condensed matter subjected to shock pressures in the range of several tens to a few hundreds of GPa. The widespread application of this technique is hampered somewhat by ambiguity in the interpretation of the results of experiments on certain materials and in certain pressure ranges. The nature of shock-generated thermal radiation is discussed within the framework of models of transient electronic structure near the shock front. These models explain the emission of equilibrium radiation at the Hugoniot temperature for intermediate pressures, and such data can then be useful for characterizing thermal equations of state and phase transitions. The particular case of SiO_2 is discussed, and the detection of the stishovite melting transition and phase disequilibrium phenomena are presented.

INTRODUCTION

Since the early 1960's, optical pyrometry of shock compressed solids and liquids has been an accepted tool for the study of thermal equations of state (EOS) and phase transitions. The pioneering work of Kormer[1], served to demonstrate some of the utility as well as the limitations of this experimental technique.

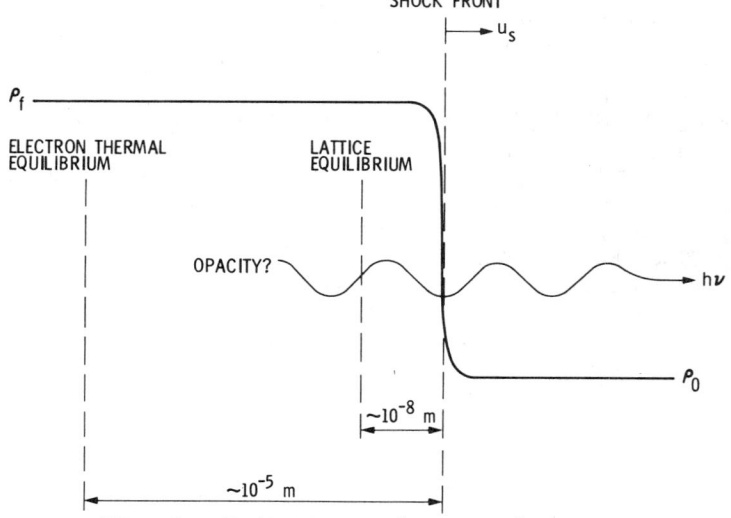

Fig. 1. Radiation and shock structure

The fundamental measurement of the shock pyrometry experiment depends upon having a transparent material medium ahead of the propagating front, so that thermal radiation from the shocked region may be remotely detected. Under these conditions, the compressed and heated region behind the front as illustrated schematically in Figure 1 gives rise to radiation. The volume of material giving rise to the detected signal (and temperature) depends upon the opacity of this region.

Kormer pointed out that while thermal equilibrium among the ions or "cores" comprising the lattice is achieved within a few hundred interatomic spacings, the thermal equilibrium of associated free electrons may be considerably delayed. In the case that electrons ordinarily populating the valence band must be ionized into conduction band states, Kormer finds that 10^{-9}s is required for these electrons to assume the lattice temperature. It is for this reason that the spatial distribution and density of free electrons giving rise to radiation and opacity in the shock front are of importance in determining the effective temperature measured by shock pyrometry.

Prior to experiments with shock pyrometry in solids and liquids, considerable work has been done in measuring the brightness temperature of gas shocks. As discussed by Zel'dovich and Raizer[2], these experiments give reasonable results up to a threshold of some tens of thousands of degrees K. Above this however, the apparent temperatures for stronger shocks decline and saturate at some lower value. Figure 2 shows that in condensed matter (CsBr in this case) completely analogous behavior is observed. While the cause in this case cannot be the radiative preheating which causes opacity and

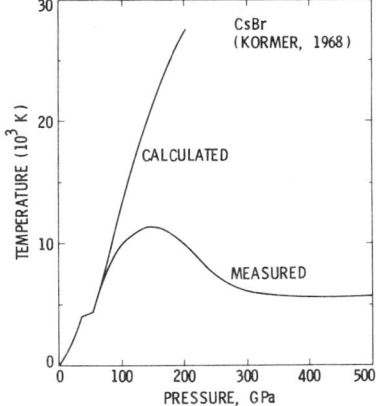

Fig. 2. Brightness temperature in shocked CsBr

"screening" in the gas case, this serves to illustrate that the approach to thermal equilibrium behind the shock front can be important in determining the results of pyrometry. In this case, at very high pressures, an optically thick electron layer is achieved in the early part of the front, prior to equilibration of the electron temperature.

The current research reported here and elsewhere[3,4] was carried out at Lawrence Livermore National Laboratory, employing a two-stage light gas gun for shock generation and a six-wavelength, few-ns-time-resolution optical pyrometer. This paper describes the phenomenological aspects of shock pyrometry in these experiments, and illustrates its application to the particular case of shocked SiO_2.

PYROMETRY EXPERIMENTS

The experimental technique employed here records the spectral radiance of the shocked sample in each of six visible wavelength bands (9 nm effective width) as a function of time, using high speed oscilloscopes. An alternate approach such as that reported by Sugiura et al.[5] records an essentially continuous spectrum, but at the expense of detailed time dependence information.

Figure 3 shows the experimental oscilloscope records from two experiments on NaCl shocked to different pressures. These records illustrate the features typical of records obtained in most solids and liquids in the general range of pressures near 100 GPa. The first arrival of the shock front in the transparent sample is accompanied by a very rapid increase in brightness, which remains approximately constant until the shock encounters the sample free surface, at which point the measured radiation abruptly decreases.

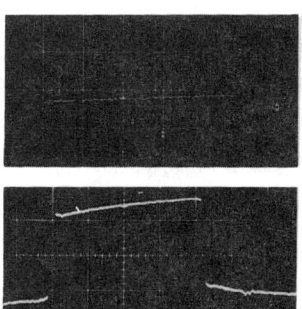

Fig. 3. Radiance records for NaCl (100 ns per div.)

The apparent brightness is not precisely constant throughout the shock transit period, but increases with downward concavity, approaching an apparent limit.

The magnitude of this time dependence depends weakly upon pressure, being smallest at low pressure, and increasing to a 10-15% effect at the highest observed pressures. The wavelength dependence of the brightness increase is quite weak, with only a slight (few per cent) relative enhancement of the blue wavelengths at later times. Generally, the temperatures obtained from these records are constant in time within the uncertainties of fitting and the brightness increase is generally ascribable to a uniform increase in apparent emissivity at all wavelengths, with little change in color temperature.

Values of radiative temperature are typically obtained through a least squares fit of the spectral radiance data to a temperature and emissivity model. This procedure combines the sensitivity of the brightness temperature method to absolute intensity and the sensitivity of the color temperature method to spectral shape. In the current work, a single-parameter, constant-emissivity model is assumed. This approach works well as long as the radiating material does not display strong departures from "gray emissivity".

Figure 4 shows the results of an experiment with a well constrained temperature fit. The spectrum is well fit by the exponential blue cutoff of the Planck blackbody distribution, so that the shock temperature, in this case for liquid water, is determined with ~ 100 K uncertainty. ϵ denotes the apparent graybody emissivity, and has the maximum physical value of unity for a perfect blackbody.

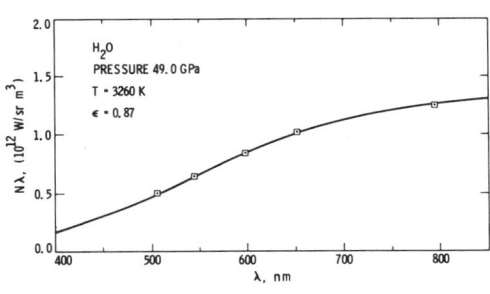

Fig. 4. H_2O spectrum. The solid line is the fit to a graybody spectrum.

Given such high quality unambiguous temperature determinations, use of these values in evaluation of various equations of state would be straightforward. Unfortunately, such unambiguous interpretation of the spectra are not always possible. While typically good temperature fits are obtained in the lower range of investigated pressures, at moderately high pressures (> 100 GPa) some materials display anomalous spectra. In particular, the materials SiO_2, NaCl, and CaO have shown spectra not well fit by a Planck distribution, principally because of an apparent deficiency in the intensity at blue wavelengths below ~ 500 nm. This non-thermal spectrum is well illustrated in Figure 5 for the case of crystalline α-quartz. The absolute brightness and blue cutoff profile cannot be

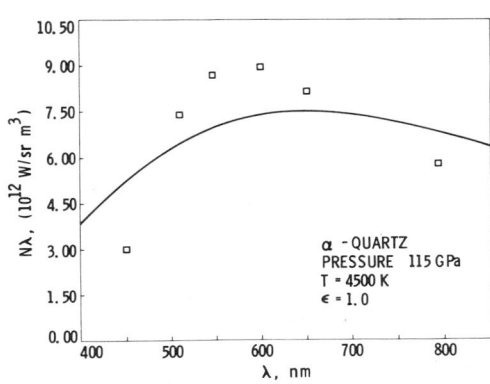

Fig. 5. SiO_2 spectrum.

simultaneously fit with physically reasonable values of ϵ, and T. Shock compressed H_2O has not been observed to display this effect up to 80 GPa pressure. A satisfactory theory of radiative emission in the shock compressed state should explain this blue extinction at high pressures (or temperatures).

This selective attenuation of blue wavelengths generally leads to a systematic underestimate of shock temperatures. Since the effect seems to set in progressively at higher pressures, the net effect may be to lower the apparent rate of temperature increase with shock pressure. Since in both SiO_2 and NaCl, high pressure P-T gradients have been observed which are lower than anticipated from theoretical EOS, this may be an important effect. Such systematic errors, however, cannot go completely undetected, since the spectral signal of blue attenuation is in all cases fairly obvious.

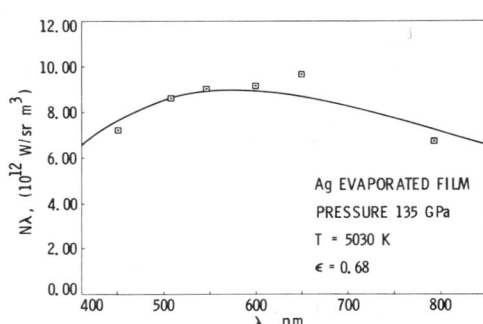

Fig. 6. Ag Spectrum.

Figure 6 shows the detected spectrum of a metallic Ag interface, in contact with a transparent sapphire crystal. The shocked metal shows no discernable blue cutoff at this pressure, and this result suggests that the blue cutoff seen in insulators is related to the peculiar nature of radiative transfer in these transparent materials under shock.

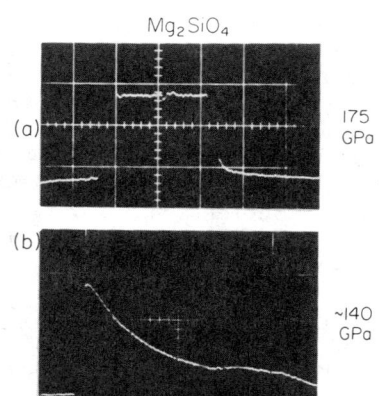

Fig. 7. Mg_2SiO_4 radiance records (100 ns/div.).

One further aspect of the radiative behavior of shocked solids and liquids has to do with the low shock amplitude limit. It has been shown that the optical pyrometry method is not generally applicable to arbitrarily high pressures, presumably because of limitations imposed by the electronic radiative mechanism. It turns out that related problems may limit the usefulness of pyrometry in some materials at low pressure. Figure 7 shows the intensity records resulting from two different experiments on pure single crystal forsterite (Mg_2SiO_4). This material experiences relatively modest shock heating, and therefore displays the low-

temperature radiative limit at rather higher pressures than other more compressible and strongly heated dielectrics. The first frame (a) displays the outcome of an experiment at a shock pressure of 175 GPa. This record shows the expected shock luminescence during the period of shock wave run in the sample. At this pressure, the radiation is constant in time and the spectral agreement with a blackbody shape is good, yielding a temperature value of 4950K.

The second frame (b) shows the quite different result of the same experiment carried out at approximately 140 GPa Hugoniot pressure. Instead of showing a constant intensity profile which begins and ends abruptly with shock arrival at the sample faces, this shot yields a decaying radiation intensity profile in which the shock free surface arrival appears only subtly. Such a record is interpreted as indicating that the sample remains transparent and non-radiating behind the shock front. Only the decaying surface temperature of the hot metallic base plate (in this case copper) in contact with the relatively cool sample is seen by the pyrometer.

Evidently, at some pressure below 150 GPa in forsterite, the density of shock-induced free electrons and/or ionized donor states drops below the value required for significant optical opacity. While this failure of the radiative mechanism at low pressure complicates the problem of shock temperature measurement in certain materials, the observation provides valuable insight into the nature of the thermal radiation source. Furthermore, the transparency of certain crystals under shock can be exploited to investigate the shock temperatures of non-transparent materials. Sapphire (Al_2O_3) is another incompressible insulator which shows transparency up to large shock pressures. This material has been used for that reason as a transparent "anvil" to measure the temperature of shock compressed metallic silver films.

INTERPRETATION OF SHOCK TEMPERATURES

In the range of pressures that shock pyrometry yields reliable temperatures, valuable insights into EOS, including phase changes are obtained. Indeed, the early results of Kormer revealed melting of the alkali halide crystals under shock, while conventional Hugoniot EOS measurements could not detect the effect. The current experimental work has succeeded in confirming and extending these results for NaCl. In this paper the results for experiments on quartz (SiO_2) are briefly summarized, since they illustrate many of the aspects of interpretation of shock pyrometry results.

Figure 8 shows the measured shock temperature as a function of pressure for experiments on synthetic quartz. Both single crystal (α-quartz) and amorphous (fused quartz) samples were investigated, in order to investigate as large a region of the SiO_2 phase diagram as possible. Apparent in the plot are large temperature declines along the Hugoniot, with minima near 4500 K. The range of investigated pressures is entirely above the transition to the six-fold coordinated SiO_2 polymorph stishovite, and evidently these temperature drops signal the occurrence of a major phase change from stishovite to a new form of SiO_2.

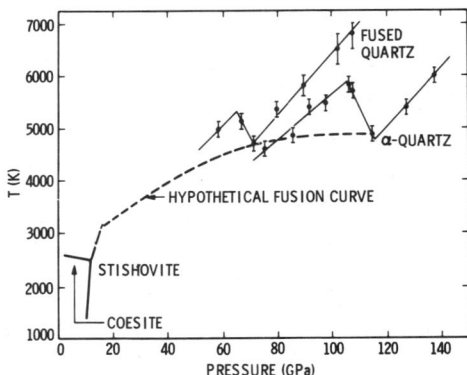

Fig. 8. Shock temperatures in SiO$_2$

From a consideration of relative densities and energetics discussed elsewhere, the transition has been identified as melting of stishovite. As indicated in the figure, the solid-liquid phase boundary is tentatively drawn through the temperature minima of the Hugoniots, and the result is a revised phase diagram for SiO$_2$, of importance to problems of the earth's interior. These results have allowed an estimate for the latent heat of fusion and volume change to be obtained using previously available Hugoniot data. At 70 GPa pressure, these values are 3.0 MJ/kg and 3.4% respectively. These values imply a Clapeyron phase line slope of 11 K/GPa, which is consistent with that drawn in Figure 8.

It should be noted that the magnitudes of the temperatures quoted here may be somewhat uncertain, due to systematic errors which compromise somewhat the fidelity of the determined spectra. While additional experiments will refine these errors, the current results, as presented in Figure 8, do reflect correct relative relationships, and serve to indicate underlying physical processes.

While the melting interpretation is self-consistent, Figure 8 implies that the pyrometer "sees" the temperature of superheated solid stishovite throughout the mixed phase region, rather than the equilibrium value along the phase boundary. Apparently, the explanation of this phenomenon again requires invoking the structure of the radiating front, and its relation to the rate of phase transformation behind the front. The conclusion of this paper deals with the radiative mechanism and explanations of this and the above described anomalous behavior.

CONCLUSIONS

If the free electron emissivity model proposed by Kormer accurately describes processes occurring behind the shock front, it should successfully account for the above described time and spectral dependences of the shock luminescence at various pressures. In this picture, it is postulated that immediately behind the strong shock front, transient defects and vacancies introduced into the distorted lattice may give rise to donor level concentrations upwards of 10^{19} cm^{-3}.

Figure 9 illustrates schematically the variation of free electron concentration along the profile of the front. In each of the three pressure regimes indicated, electron density exceeds a "critical value" at some distance into the wave which shortens with increasing amplitude. Furthermore, the depth of shocked material from which significant radiation escapes to infinity depends upon the gradient in electron concentration, and is indicated by cross-hatching.

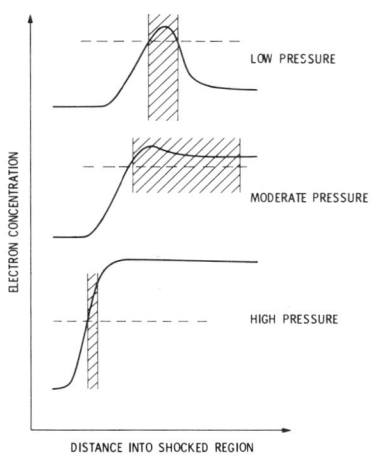

Fig. 9. Electron density and opacity on shock front

The observed radiation time dependence is explained by the optical depth effect. At the lowest pressures for which critical density is exceeded, only a fairly narrow band contributes, while behind this region the defects are "annealed" and free charge density drops. Thus the detected radiation is essentially time independent as this profile propagates. At somewhat higher pressures, the persistent electron density becomes appreciable, and the observed intensity and emissivity should increase with propagation distance following the initial "spike". Finally, at the highest pressures in which large optical depths are achieved early in the wave, this model predicts that the time independent blackbody limit should be approached, and this prediction may be checked experimentally using new very high pressure techniques.

The spectral departures from blackbody may be due to either anomalous emission or absorption. The former possibility requires the free electron gas to become a poor emitter/absorber at blue wavelengths, such as from a displacement of the plasma frequency into the visible range. A weakness of this model is that it predicts a shift of the edge to higher frequencies with increasing electron concentration, in apparent contradiction of observation. Absorptive mechanisms require selective blue attenuation in the non-radiating toe of the shock. Band gap closure, or more likely, absorption by defect bands such as color centers could account for the blue absorption edge.

Finally, the superheating phenomenon seen in SiO_2 shocked above the melting line is explained if within the mixed phase region, 10^{-9}s or longer is required for thermal equilibration of the bulk mixed phase assemblage. In this case, the observed radiation temperature remains at the pre-transition superheated value nearer the wave front, until the pressure of 100% melting is approached.

ACKNOWLEDGMENTS

The contributions of T. J. Ahrens, M. B. Boslough and K. Kondo of California Institute of Technology and W. J. Nellis and A. C. Mitchell of Lawrence Livermore National Laboratory to the theoretical and experimental results presented herein are gratefully acknowledged. This work was supported by NSF Grant EAR78-12942, and under DOE contract number W-7405-ENG-48, and also in part by NASA/JPL.

REFERENCES

1. S. B. Kormer, Sov. Phys. Usp. <u>11</u>, 229 (1968).
2. Ya B. Zel'dovich and Yu. P. Raizer, Physics of Shock Waves and High-Temperature Hydrodynamic Phenomena, (Academic, N.Y., 1967), p. 598.
3. G. A. Lyzenga and T. J. Ahrens, Rev. Sci. Instrum. <u>50</u>, 1421 (1979).
4. G. A. Lyzenga, T. J. Ahrens, and A. C. Mitchell, Shock Temperatures and Equations of State of SiO_2, submitted for publication (1981).
5. H. Sugiura, K. Kondo, and A. Sawaoka, Rev. Sci. Instrum. <u>51</u>, 750 (1980).

MICROWAVE DIELECTRIC CONSTANT OF SHOCK-LOADED LITHIUM NIOBATE*

J. K. Hartman, J. L. Wise, R. A. Graham,
R. O. Johnson, G. E. Clark, and T. J. Burns
Sandia National Laboratories,† Albuquerque, N. M. 87185

ABSTRACT

The microwave dielectric constant of single-crystal, Z-cut $LiNbO_3$ has been measured in a sample subjected to a 35 GPa impact loading. Interpretation of the measurements indicates that the average dielectric constant behind the shock wave is a factor of 3 larger than the initial value.

INTRODUCTION

Although recent publications concerning the interaction of microwave radiation with shock-compressed materials are few, the history of such experiments is quite long. Microwaves were used twenty-five years ago to study detonation phenomena in explosives.[1,2] Non-reactive shocks were first studied in 1965 by Johnson.[3] More recently, the dielectric constants and conductivities of shock-compressed fluids and solids were measured using microwaves and employing an equivalent transmission line model by Hawke, et al.[4,5] The present work utilized a double-balanced mixer to detect Doppler-shifted frequencies, resulting in a determination of the elastic wave velocity, the shock velocity, and the average dielectric constant behind the shock front.

THE EXPERIMENT

As part of a series of exploratory experiments, a Z-cut $LiNbO_3$ crystal (56 mm diameter, 24.12 mm thick, 4.641 Mg/m^3 density) was impacted by a 12.62 mm thick copper impactor at 2.034 km/s. During the impact process, the $LiNbO_3$ sample was irradiated by 17.5 GHz microwaves from a circular horn antenna mounted 10.2 mm behind the free surface. The impact conditions resulted in an initial shock pressure of approximately 35 GPa in the $LiNbO_3$. The face of the crystal toward the impactor was coated with a 5000Å film of vapor-deposited aluminum to prevent pre-impact reflection of the microwaves from the approaching projectile. The reflected signal was heterodyned in a double-balanced mixer with part of the original signal generated by the oscillator as shown in Figure 1. The output of the mixer was recorded on three Tektronix transient digitizers. Standard velocity and tilt pins were included in the experimental

---
*This work was supported by the U. S. Department of Energy under Contract DE-AC04-76-DP00789.
†A U. S. Department of Energy Facility

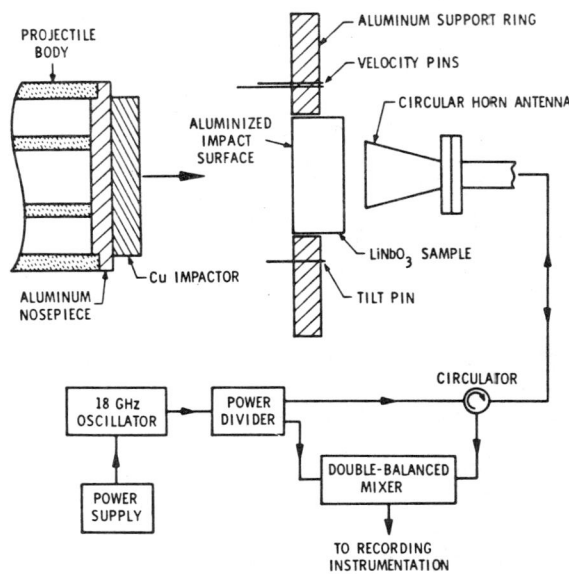

Fig. 1. Experimental Configuration

set-up to determine the projectile velocity and orientation.

Figure 2 shows the output of the mixer as a function of time compared with a computer-generated position vs. time diagram of stress-wave and particle trajectories in the sample and impactor. The copper impactor impacts the aluminized face of the $LiNbO_3$ at time t_0 as indicated by the sudden breakaway of the measured signal. Next, the recorded trace displays a series of oscillations, created by multiple internal reflection of the microwaves between the impact interface, the shock front, and free surface, and possibly the elastic precursor, lasting until time t_1. At time t_1 the elastic wave reaches the free surface and sets it in motion. The microwave signal gives $t_1 - t_0$ to be 3.41 μs, which corresponds to a velocity of 7.07 km/s compared to a velocity interferometer measurement of 7.11 km/s. Both measurements compare favorably with the 7.28 km/s value determined by Stanton and Graham.[6] After t_1, the excursion of the trace becomes larger since it is now primarily affected by the motion of the free surface. The corresponding reflected signal is quite strong compared to the attenuated signal still being reflected from the interfaces interior to the crystal.

The next time of interest, t_2, corresponds to the arrival of the plastic shock wave at the free surface. The measured transit time of 3.77 μs indicates a shock velocity of 6.40 km/s, which should tend to be a lower bound since interaction effects with the reflected elastic wave have been ignored. The shock velocity calculated from the Hugoniot relation[6] for $LiNbO_3$ is 6.25 km/s for the present impact conditions.

ANALYSIS

If the frequency of the reflected microwave signal were fixed, problem analysis would be straightforward. However, there are three interfaces at which the microwaves may be reflected, if the presence of the elastic precursor is ignored. Two of these, the impact surface and the shock front, are moving at high velocity. In addition, multiple reflections occur, resulting in a variety of different path lengths. If changes in the path lengths as a function of time are taken into account, the Doppler frequency shifts are automatically included in the model. The Fresnel equations,

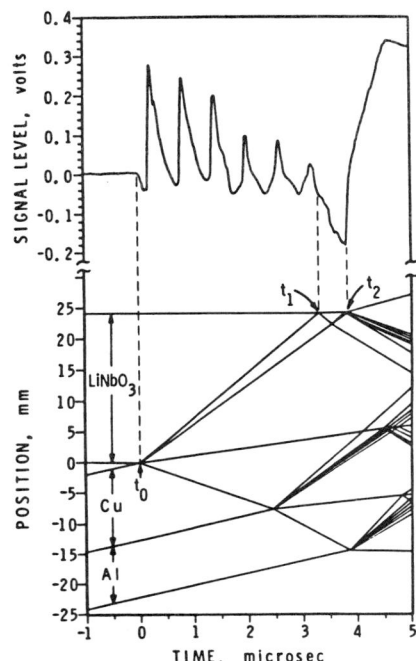

Fig. 2. Mixer Output Compared to Position-Time Diagram

$$\rho_R = (\sqrt{K_1} - \sqrt{K_2})/(\sqrt{K_1} + \sqrt{K_2}) \text{ and}$$

$$\rho_T = 2\sqrt{K_1}/(\sqrt{K_1} + \sqrt{K_2}) \text{ , where}$$

K_i are dielectric constants, give the reflection and transmission coefficients for a plane electromagnetic wave as the wave travels from medium 1 into medium 2 at normal incidence. The reflection coefficient for a perfectly conducting metal surface is given by $\rho_{RM} = -1$. Each individual contribution to the electric field takes the form

$$E_i = \rho_i A e^{-\beta_i} \cos\left(2\pi f_o t + \phi_i\right) , \quad \text{where}$$

$$\rho_i = \prod_j \rho_{ij} \quad , \quad \beta_i = n_i \alpha_2 d_2 + m_i \alpha_3 d_3 \quad , \quad \phi_i = \frac{2\pi}{\varepsilon_o} \sum_j \sqrt{K_j}\, d_{ij} f_o$$

The ϕ_i are the phase shifts introduced by the different path lengths followed by various waves, and the β_i represent the attenuation-distance product for the various paths. In the present problem medium 1 is the vacuum, medium 2 is the unshocked LiNbO$_3$ and medium 3 is the shocked LiNbO$_3$. A computer program has been used to calculate the electric field in region 1 to the sixth order in the ρ_{ij} product, where only the $\rho_{ij} < 1$ contribute to the order of the product.

For the present experimental configuration, it is approximately true that

$$V_{LO} \propto A\cos(2\pi f_o t + \phi_o) \text{ and } V_{RF} \propto \sum_i E_i ,$$

where V_{RF} and V_{LO} are the voltages that appear at the input ports of the double-balanced mixer, f_o is the oscillator frequency, and ϕ_o is a phase shift associated with all the unmeasured paths in the instrumentation. The mixer output is given by

$$V_{OUT} \propto V_{RF} V_{LO} .$$

Due to bandwidth limitations, this output will not follow the high frequency terms, such as given by the sum of the input frequencies, but will only give the difference in frequencies and phase shifts. Hence,

$$V_{OUT} \propto \sum_i \rho_i e^{-\beta_i} \cos(\phi_i - \phi_o) .$$

Consequently, the output is proportional to the Doppler shifts (i.e., the part of the phase shift that varies with time) and the difference in path lengths.

This output has been calculated for various values of the dielectric constant and attenuation parameter to obtain a fit with the experimental data. While the fine structure differs somewhat from the experimental curve, the primary frequency signal has been constrained to have the same period as the experimental curve by adjusting the dielectric constant. The curve is quite sensitive to the value chosen for the dielectric constant; in this case, the selected value was 125. If one were to calculate the dielectric constant from a straight Doppler shift off the impact interface, the value obtained would be 137. Thus, it is seen that interactions with the shock front do affect the value chosen for the dielectric constant. It was not deemed worthwhile to attempt to obtain a better fit for the attenuation parameter with the experimental data, as the poor experimental geometry allowed many extraneous factors (e.g., edge effects) to affect the signal amplitude.

DISCUSSION

While this experiment was designed with specific engineering objectives in mind, it demonstrates the usefulness of the information available in the Doppler-shifted microwave signal. The ability to track the impact interface, the shock front, the elastic wave, and the free surface, as well as material changes, offers potential for future studies. In particular, from the original experimental trace it appears that one might be able to determine the arrival times of the elastic wave as it reverberates between the free surface and the shock front.

A digital Fourier spectral analysis of the recorded signal reveals the necessity of including waves that are reflected more than once. In addition, parts of the spectrum seem to indicate that interactions with the elastic wave must be included for an accurate description of the situation. Even higher-order terms need to be retained in the summation for the electric field so that steeper rise times may be attained on the model output.

Lastly, the question of the dielectric constant of 125 behind the shock front versus the initial value of 45 must be addressed. Using the isentropic shock compression model, calculations show that the density behind the shock front has increased to 5.78 Mg/m^3 and the bulk temperature has risen 92°C above ambient. Neither of these estimates can explain the observation that the average dielectric constant increases in value by a factor of nearly 3, even considering that LiNbO$_3$ is a ferroelectric. This leads to the suggestion that heterogeneous yielding is occurring, as was first suggested by Stanton and Graham.[6] Localized heating along slip bands can drive portions of the ferroelectric far up the dielectric constant versus temperature curve, leading to very high values for the dielectric constant at these hot spots. The average effect seen by the microwaves (λ = 2.6mm in LiNbO$_3$) results in an effective dielectric constant of 125. This effect may be investigated further by repeating the experiment with a ferroelectric having a lower Curie temperature or by studying lithium niobate over a wider pressure range.

REFERENCES

1. M. A. Cook, R. L. Doran, and G. J. Morris, J. Appl. Phys. **26**, 426 (1955).
2. G. F. Caswey, J. L. Farrands, and S. Thomas, Proc. Roy. Soc. (London) **248**, Series A, 499 (1958).
3. E. G. Johnson, Rohm & Haas Co. Redstone Arsenal Research Div., Report No. S-87, Contract #DA-01-021 ORD-11909(Z) (1965).
4. R. S. Hawke, A. C. Mitchell, and R. N. Keeler, Rev. Sci. Inst. **40**, 632 (1969).
5. R. S. Hawke, R. N. Keeler, and A. C. Mitchell, Appl. Phys. Lett. **14**, 229 (1969).
6. P. L. Stanton and R. A. Graham, J. Appl. Phys. **50**, 6892 (1969).

EFFECT OF SHOCK WAVES ON THE ABSORPTION SPECTRUM OF RUBY

R. S. Hixson
Los Alamos National Laboratory, Los Alamos, NM 87545

P.M. Bellamy, G.E. Duvall, C.R. Wilson
Washington State University, Pullman, WA 99164

ABSTRACT

The effect of shock loading upon the unpolarized absorption spectrum of ruby has been measured. Experiments were performed both above and below the Hugoniot elastic limit (HEL), with the experiment above the elastic limit failing due to extinction of the light upon impact. The experiments below the elastic limit were both done at about the same pressure ($\sim$10 GPa), and show a shift of both absorption bands in the visible region toward shorter wavelength, the shifts agreeing well with those measured at a comparable hydrostatic pressure. The magnitudes of the shifts below the HEL complement shifts recently measured by Goto et al.[1]

INTRODUCTION

Absorption spectrum measurements on minerals under pressure can provide information on the compressed crystal structure, and so are of geophysical interest. Measurements of changes in the absorption spectrum of single crystals of various kinds have been performed by several workers[1-6] using both static and shock compression techniques.

The spectral properties of ruby have been studied extensively as a result of laser applications. The chromium doping results in ruby having two strong absorption bands in the visible range, one at 400 nm (25000 cm^{-1}), and one at 550 nm (18000 cm^{-1}), making it an excellent choice for experiments using a spectrometer tuned to the visible light region. The band at 400 nm corresponds to the electronic transitions $^4A_2 \rightarrow \, ^4T_1$ and the 550 nm band to the $^4A_2 \rightarrow \, ^4T_2$ transitions. In addition to the experimental information available there have also been very complete theoretical studies of the spectrum of Ruby[7-8].

EXPERIMENTAL TECHNIQUES

The experimental arrangement was that shown in Fig. 1. The crystallographic C-axis of the ruby sample was oriented at 60° with respect to the impact surface.

0094-243X/82/780282-05 $3.00 Copyright 1982 American Institute of Physics

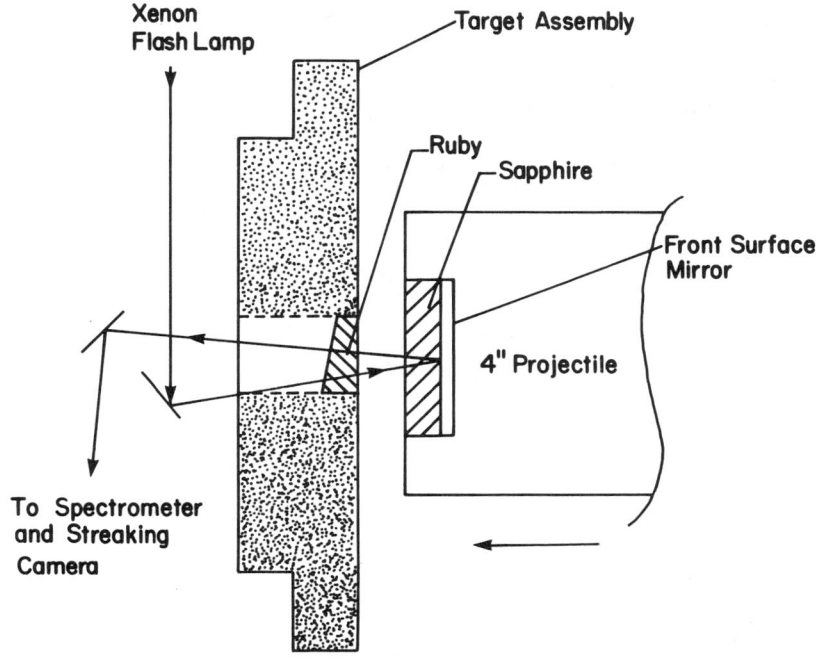

Fig. 1. Experimental details.

A 10.2-cm-diameter aluminum projectile was accelerated by means of a light gas gun to velocities between 0.5 and 1.0 km/sec. A sapphire (Al_2O_3) plate backed by a front surface mirror was embedded in the front of the projectile, the sapphire positioned to strike the ruby (0.05% chromium doped) target producing a symmetric impact. The Hugoniot data of Graham and Brooks[9] was used to find the pressure corresponding to the known particle velocity in the ruby sample.

A coaxial pin in the target was shorted by the projectile impact triggering a xenon flashlamp and the oscilloscopes used to monitor a photomultiplier tube on the streaking camera. The light from the flashlamp was focused onto the target assembly, the reflection from the beveled first surface of the ruby sample being sent off axis so as not to add to background light on the film. The light then passed through the ruby and sapphire, was reflected by the mirror, and directed onto the entrance slit of a diffraction grating spectrometer with grating blazed to 500 nm. The spectrum produced by the spectrometer was directed onto the rotating mirror of a streaking camera (Beckman & Whitely 339) and projected onto Kodak #2475 spectrographic film, thus producing a time resolved spectrum. A photograph of the spectrum for a 10.4 GPa experiment is shown in Fig. 2.

Fig. 2. Time resolved absorption spectrum of ruby under shock loading. The xenon emission lines occur at the start of the streak, and the absorption shifts under compression can clearly be seen.

The film is sensitive to wavelengths between 300 nm and 700 nm, and the maximum writing rate of the camera is 9 km/sec. The writing rate in Fig. 2 is about 7.2 km/sec.

EXPERIMENTAL RESULTS

The results of the three experiments are shown in Table I.

Table I Results of Ruby Experiments. Absorption band shifts are denoted by $\Delta\lambda$.

Shot	P (GPa)	U_p (km/s)	$\Delta\lambda$ (550 nm) (nm)	$\Delta\lambda$ (400 nm) (nm)
80-014	20.0	0.512	--	--
80-015	10.4	0.234	181±10	70±10
80-016	9.8	0.227	--	

Shot 80-016 was performed with an electronic streaking camera, with a wavelength window of approximately 120 nm, and was tuned to follow the $^4A_2 \rightarrow {}^4T_2$ transition. The results from the experiment using the electronic camera were unsatisfactory.

The data from shot 80-015 were digitized and stored on magnetic tape. A fixed-time densitometric scan of these data in the unshocked region is shown in figure 3, and the two absorption bands can clearly be seen.

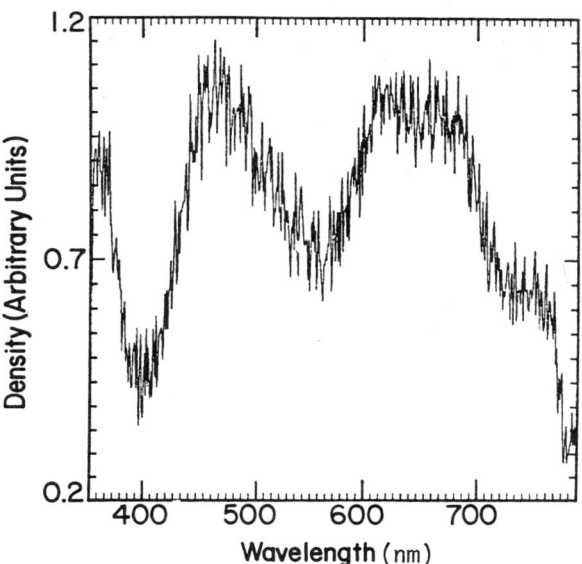

Fig. 3. Absorption spectrum of ruby at 1 atm. at fixed time.

The digital information is contained within a matrix of 512 x 512 "pixels"; i.e. with a total of about a quarter of a million points. A wavelength calibration is achieved by looking at the xenon emission lines at the start of the streak, and noting the pixel location at which they occur.

DISCUSSION

In addition to the ruby shot performed above the elastic limit, one was also done using a sapphire plate as a target. Both of these shots showed similar behavior upon impact; the light was no longer able to pass through the target/impactor combination. This behavior may be related to the phenomenon of plastic yielding, and so possibly due to the stress deviator collapse seen previously in sapphire[9] and the mechanism of heterogeneous melting. Grady[10] has recently discussed a possible mechanism in which a brittle solid may lose bulk shear strength due to a heterogeneous yielding process and the formation of shear bands. Further evidence that this loss of light intensity is due to plastic yielding is provided by the experiment below the elastic limit in which no noticeable loss of intensity upon impact is seen.

The crystal field parameter 10Dq, and the Racah parameter B can be calculated from the results of shot 80-015, and the results compare well with the shifts observed by Stephens and

Drickamer[3] at an equivalent hydrostatic pressure. These parameters are calculated from the equations of Tanabe and Sugano:[7]

(1) $E(^4A_2 \rightarrow {}^4T_2) = 10\,Dq \equiv \Delta$

(2) $E(^4A_2 \rightarrow {}^4T_1) = 3\Delta/2 + 15B/2 - 1/2[(\Delta - 9B)^2 + 144B^2]^{1/2}$,

from the absorption band energy shifts. The results of this calculation are $\Delta = 18625$ cm^{-1}, and $B = 665$ cm^{-1}.

*This work was supported by ONR Contract No. N00014-77-C-0232.

REFERENCES

1. T. Goto, T.J. Ahrens, and G. R. Rossman, Phys. Chem. Minerals 4, 253 (1979).
2. E. S. Gaffney and T. J. Ahrens, J. Geophys. Res. 78, 5942 (1973).
3. D. R. Stephens and H. G. Drickamer, J. Chem. Phys. 35, 427 (1961).
4. S. Minomura and H. G. Drickamer, J. Chem. Phys. 35 (1961).
5. D. R. Stephens and H. G. Drickamer, J. Chem. Phys. 34 (1961).
6. R. M. Abu-Eid, The Physics and Chemistry of Minerals and Rocks, edited by R. G. Streus (John Wiley, N.Y., 1976).
7. Y. Tanabe and S. Sugano, J. Phys. Soc. Japan 9, 753 (1954).
8. S. Sugano and Y. Tanabe, J. Phys. Soc. Japan 13, 880 (1958).
9. R. A. Graham and W. P. Brooks, J. Phys. Chem. Solids 32, 2311 (1971).
10. D. E. Grady, J. Geophys. Res. 85, 913 (1980).

TEMPERATURE MEASUREMENTS OF SHOCKED TRANSLUCENT MATERIALS BY TIME-RESOLVED INFRARED RADIOMETRY*

William G. Von Holle
Lawrence Livermore National Laboratory
Livermore, California 94550

ABSTRACT

Infrared emission in the range 2-5.5 µm has been used to measure temperatures in shock-compressed states of nitromethane, cyclohexane and benzene and in polycrystalline KBr. Polymethylmethacrylate shows anomolous emission probably associated with some heterogeneity.

INTRODUCTION

Early work on the time-resolved measurement of infrared emission from shock heated explosives proved valuble in the experimental investigation of "hot spots" produced in pressed explosives by low amplitude shocks.[1] It became necessary to extend the investigation of "mock" explosives and inert materials reported at that time. Except for the work of Raikes and Ahrens[2] on silicate minerals, very little other work has been done on temperature measurements by radiation pyrometry in non-metals shocked to low pressures. This paper presents the highlights of the work in shocked inert materials showing the contrasting responses of organic liquids, a polymer and a polycrystalline solid.

EXPERIMENTAL

Space does not permit a complete description of the experiment, which can be found in a recent paper.[3] Plane shock waves were introduced into the samples by 102 mm plane projectile impact in a gun facility. Two radiometers were used to measure the infrared radiance of shocked samples in two overlapping bands, 2-5.5 µm (Band #1) and 4-5.5 µm (Band #2), with a response time of about 0.4 microsecond.

All liquids were analytical reagent grade except nitromethane, which was spectrochemical grade, containing at least 98.7% CH_3NO_2. 7 mm-OFHC copper flyer plates impacted 3.2 mm thick x 51 mm dia stainless steel driver plates polished to a mirror finish in contact with the liquid. Liquid path lengths were approximately 21 mm, followed by a 2 mm sapphire window.

---

*Work performed under auspices of the U. S. Department of Energy by Lawrence Livermore National Laboratory under contract No. W-7405-Eng-48.

The polymethylmethacrylate (PMMA) samples were 6.4 mm thick by 51 mm diameter and were backed by 6 mm thick KCl crystals with Kel F-90 grease between to prevent gaps and transmit infrared. The impact face was covered with either a 3.2 mm 2024 Aluminum driver plate without grease or .08 mm aluminum foil and silicone grease.

The polycrystalline potassium bromide (KBr) targets were prepared by pressing infrared quality KBr powder onto 1.5 mm stainless steel disks at 0.2GPa (30,000 psi).

RESULTS

The time-resolved radiance from the neat liquids resulted in signals which increase exponentially until the sapphire window is reached. Figure 1 shows the brightness temperatures from one band for the three liquids where the arrows indicate the approximate time of arrival of the shock at the sapphire which causes an impedance mismatch. Figure 2 is a typical logarithmic plot of the observed signal. Peak brightness temperatures (at sapphire arrival) for single shocks in nitromethane are plotted in Figure 3 and compared with a calculated curve.[4]

The solid polymer PMMA was examined for comparison to the liquid results. Figure 4 shows the Tektronix 7903 oscilloscope traces for the 2-5.5 μm band for two experiments with identical Al flyer plate velocities of 0.98 mm/μs. The only difference between the two was that the lower trace resulted when a film of silicone grease was used between the driver plate and the PMMA surface. The lower trace is similar to that resulting from a 2.0 GPa shock into the PMMA (not shown). The radiance at PMMA/KCl breakout corresponds to a brightness temperature of about 100°C for about a 3.8 GPa first shock into PMMA.

The infrared response from shocked polycrystalline KBr is illustrated in Figure 5. This nearly linear increase in radiance with time starting from impact contrasts with the above results.

DISCUSSION

Solution of the radiation transfer equation[5] for a planar system with only normal emission is given by equation (1)

$$S = \int_{\Delta\lambda} (1 - e^{-\alpha_1 d}) e^{-\alpha_2 x} W^{bb}(\lambda, T) d\lambda \qquad (1)$$

where S is the signal; d is the thickness of shocked material with absorption coefficient α_1; x is the thickness of remaining unshocked material of absorption coefficient α_2; $W^{bb}(\lambda,T)$ is the Plank black body function into which is incorporated the responsivity of the detector.

The case of the shocked liquid is approximated by assuming that $\alpha_1 d$ becomes large and that the wavelength-dependent absorption coefficient, α_2, can be replaced with an average quantity, $\bar{\alpha}_2$, over the wavelength range $\Delta\lambda$. Equation (2) then

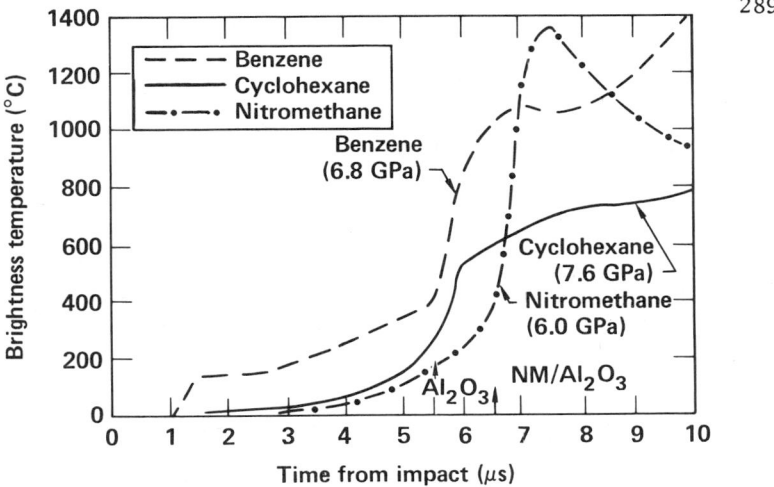

Fig. 1. Brightness temperatures for 3 liquids for the 2-5.5 μm band. Arrows indicate shock arrival at sapphire interface.

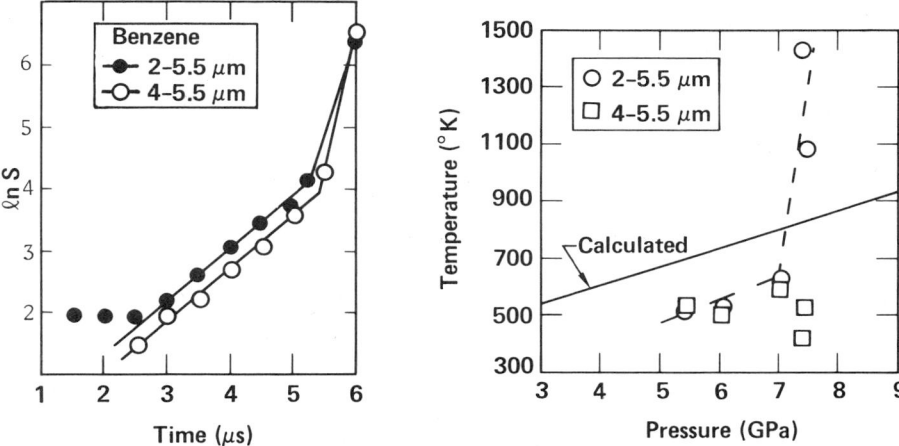

Fig. 2. Logrithmic plots for benzene signals.

Fig. 3. Shocked nitromethane brightness temperatures. Points are experimental.

describes the resulting radiance as a function of time, t, where d_o is the unshocked path length.

$$S(t) = \exp\left\{\bar{\alpha}_2 \, (U_s t - d_o)\right\} \int_{\Delta\lambda} W^{bb}(\lambda, T) \, d\lambda \qquad (2)$$

Linear plots of the logarithm of the signal vs time are expected and observed for all the liquids studied (see Figure 2). The brightness temperature can be derived from these plots independently of the peak measurement. Although there is more scatter

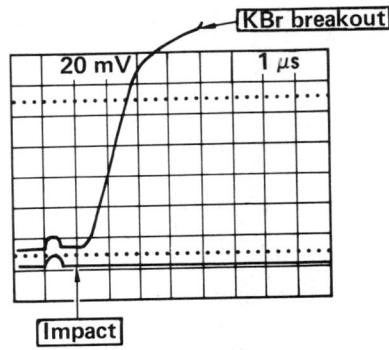

Fig. 5. Pressed KBr radiance history (4-5.5 μm band) for 10.9 GPA.

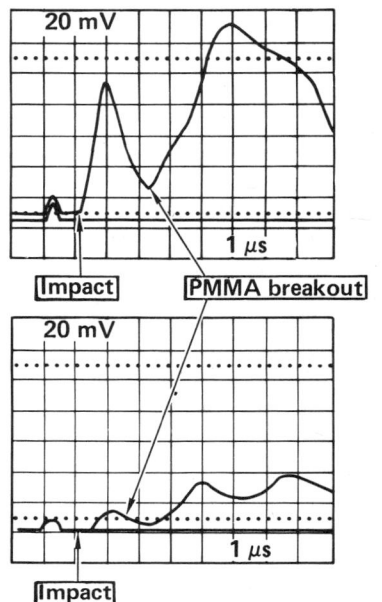

Fig. 4. Oscilloscope traces from two PMMA experiments (2-5.5 μm band)

in the data, results are generally in agreement with the peak temperatures. This indicates that our assumption that $\alpha_1 d \rightarrow \infty$ is correct and that our derived temperatures are the true temperatures within the assumptions made.

For nitromethane shocked to less than 7.0 GPa the near coincidence of the results for Bands 1 and 2 supports the above conclusion. For the 7.4 GPa experiments, non-resonant emission from reaction products is postulated to explain the sharp divergence of the two brightness temperatures.[3]

The results for polycrystalline KBr typified by Figure 5 can be explained by assuming that $\alpha_1 d$ is very small (optically thin) and α_2 is nearly zero. Again, if we expand the exponential and use an average value $\alpha_1 = \bar{\alpha_1}$, under the assumptions under which equation (1) is valid, equation (3) describes this case for the early part of shock travel.

$$S(t) = \alpha_1 d \int_{\Delta\lambda} W^{bb} (\lambda, T) \, d\lambda \qquad (3)$$

where

$$d = (U_s - U_p)t.$$

One must explain the origin of the optical radiation in KBr, which is perfectly transparent in this wavelength range. The results are consistent with the emission from impurities situated near inhomogeneities which are heated to higher temperatures than the bulk KBr.

Finally we turn our attention to the PMMA results of Figure 4. It is not surprising that the two adjacent 600 grit polished surfaces result in a large radiance spike near impact due to intense heating, but elimination of this gap by using a film of silicone grease also greatly reduces the radiance following the PMMA-KCl breakout. For these experiments the time of PMMA-KCl breakout corresponds approximately to the arrival of the rarefaction at the Al-PMMA interface. Therefore, the large signal increase commencing at PMMA breakout time in the upper part of Figure 4 could be the result of some phenomena associated with this rarefaction in material damaged by the initial impact. Imbedded thermocouple[6] and copper foil[7] temperature measurements on shocked PMMA exhibit large changes in the 2 to 4 GPa range which the authors ascribe to chemical reaction in the shock-compressed material. There is no indication of reaction from the infared emission histories. Exothermic chemical reaction would most probably result in large signals before the shock breaks out of the PMMA, as in the case of nitromethane above 7.0 GPa. Although peliminary, these data indicate the extreme care to be taken in data interpretation when introducing inhomogeneities even at moderate shock amplitudes.

REFERENCES

1. William G. Von Holle and E. L. Lee, in Behavior of Dense Media Under High Dynamic Pressures, (Commissariat a l'Energie Atomique, Paris, 1978), p. 425.
2. S. A. Raikes and Thomas J. Ahrens, Geophys. J. R. Astr. Soc. 58, 717 (1979).
3. William G. Von Holle, and C. M. Tarver, to be published in proceedings of the 7th Detonation Symposium, June, 1981, Annapolis, Maryland.
4. P. C. Lysne and D. R. Hardesty, J. Chem. Phys. 59, 6512 (1973).
5. Zeldovitch and Raizer, Physics of Shock Waves and High-Temperature Hydrodynamic Phenomena, Vol. I (Academic Press, New York, 1966). pg. 128
6. D. D. Bloomquist and S. A. Sheffield, J. Appl. Phys. 51, 5260 (1980).
7. D. D. Bloomquist and S. A. Sheffield, Appl. Phys. Lett. 38, 185 (1981).

TIME RESOLVED SPECTROSCOPY OF SHOCK COMPRESSED LIQUIDS

K. Ogilvie and G. E. Duvall
Department of Physics
Washington State University, Pullman, WA 99164

ABSTRACT

An experimental procedure has been developed for using a rotating mirror camera to record time-resolved absorption spectra of liquids undergoing shock compression. Experimental records have been obtained for cells containing liquid carbon disulfide shocked, through reverberation, to peak pressures of 55, 80, 100 and 120 kbar. Experiments have been performed using both reflected and transmitted light. Time and spectral resolution were limited to approximately 30 nsec and 30 Å; spectral range was from 4000 to 2500 Å. This initial work on carbon disulfide shows it to become highly absorptive when shocked to low pressures of 8 to 14 kbar, and to progressively become a better broadband reflector as the pressure in a thin layer rings up to the final value. A decay in the reflectivity after reaching peak pressure in the 120 kbar experiment may indicate chemical decomposition. This is in accord with earlier results of S. A. Sheffield based on measurement of flow parameters.

INTRODUCTION

Techniques for recording time resolved absorption spectra of liquids have been developed. The sample is contained in a sapphire cell and compressed by a reverberating shock wave to pressures that can be as high as 130 kilobars. The spectral range available runs from the visible to 2500 angstroms. Resolutions of 30 angstroms and 30 nanoseconds have been achieved. The techniques have been applied to carbon disulfide at pressures up to 123 kilobars in the spectral region between 4000 and 2500 angstroms. The results show carbon disulfide to become an excellent broadband reflector starting at pressures below 10 kilobars and becoming a progressively better reflector as the pressure increases. In the highest pressure experiments the reflectivity reaches a peak, then decays. The decay may be due to chemical decomposition of the carbon disulfide.

EXPERIMENTAL DETAILS

The sample cell consists of two pieces of Z-cut sapphire; each 31.75 mm in diameter. The front or impact side of the cell is 2.0 mm thick and the back is about 12.7 mm thick. The cell cavity is varied according to the individual experiment. Cavities as thin as 0.0008 mm have been constructed and they may be as thick as several millimeters.

0094-243X/82/780292-04 $3.00 Copyright 1982 American Institute of Physics

The sapphires are mounted in brass rings by inserting the sapphire into the hot rings then letting them cool and contract tightly about the sapphire. A layer of solder between the brass and sapphire acts as a gasket.

The thinnest cells were measured by forcing the cell front against the back then counting interference fringes as the force was removed. The cell is impacted by a 14.7 mm thick sapphire impactor mounted on a projectile for a 10 cm gas gun.

Two main versions of the experiments are distinguished by the location of the xenon flashlamp that is used as the spectral light source. In the first setup the flashlamp was located outside the target tank and the light was focused into the back of the cell by a lens and mirror. The light passed through the cell and impactor and was reflected by an aluminum layer on the back of the impactor. The light then passed through the impactor and cell a second time. In the newer setup the flashlamp is mounted in the projectile behind the impactor so that the light makes only one pass through the cell. The flashlamp receives its power from contacts on the target face which mate with brass strips on the front of the projectile. A ring of insulating rubber surrounds the contacts in the target face. After contact is made, this rubber ring insulates the high voltage conductors from the vacuum in the target tank, preventing unwanted discharges. High voltage to the contacts is turned on by a spark-gap after contact is made. This occurs about four microseconds before the impactor strikes the cell so the flashlamp has time to reach peak brightness for the experiment.

Commercial flashlamps with a 3 mm bore, 2.54 cm arc length and 450 torr xenon fill were used in the original setup outside the target tank. These same lamps were tried in the projectile but proved unreliable. Their failure is believed due to the electrodes breaking during acceleration in the gun. The electrodes had an unsupported projection of about 17 mm into the lamp cavity normal to the direction of acceleration. Because of this problem we manufactured our own lamps by using high vacuum epoxy to seal 3.2 mm thoriated tungsten welding rods into each end of a quartz tube. This tube has a 3 mm bore in a 13 mm long center section with a 15 mm long, 7 mm bore section fused to each end. A 1.6 mm copper tube is epoxied in with one of the electrodes. The tube is used for evaculating and filling the lamp with xenon to 450 torr. The electrodes project less than 2 mm beyond the supporting epoxy into the lamp. The epoxy is scorched, contaminating the interior when the lamp is fired so they are strictly a one shot design.

After the light leaves the cell it is reflected toward a telescope, which focuses it into the slit in a Beckman and Whitley model 1500 dispersion unit. The dispersed spectrum then enters a Beckman and Whitley model 339B continuous-writing streak camera. The record produced has wavelength dispersed across the 35 mm film and time dispersed along the film. A 300 line per millimeter grating yields a dispersion of 80 angstroms per millimeter and the streak camera gives a temporal dispersion of about 8.7 millimeters per microsecond. The 0.50 mm by 0.27 mm slit usually used yields resolutions of 40 angstroms and 30 nanoseconds. The film used is Kodak 2475 recording film.

RESULTS

Four successful experiments were conducted using the original setup in which light from an external flashlamp was focused into the back of the cell and reflected. Three of these were carbon disulfide filled cells about 0.1 mm thick. The ultimate pressures in these shots were 83, 99, and 123 kilobars. All three records show the carbon disulfide spectrum prior to impact and a slight decrease in intensity at the time of impact. After the transit time for the shock through the front sapphire, the cells became abruptly absorbing for a period of time about equal to the transit time for the first shock through the carbon disulfide (30 to 40 nsec). Following this the cells become progressively more reflective as multiple shocks ring up the pressure in the carbon disulfide. Further experiments to be discussed shortly make it apparent that the reflection is occurring at the carbon disulfide layer, not at the mirror behind the impactor. The reflection is broad band; having no spectral characteristics attributable to the carbon disulfide. In the highest pressure experiment the reflectivity reaches a peak then decays to about 0.7 of its peak value in 800 nanoseconds. This decay may be due to chemical decomposition of the carbon disulfide. This is in accord with earlier results of S. A. Sheffield based on measurement of flow parameters.[1] Reflectivity is constant after reaching its peak in the two lower pressure experiments.

The fourth experiment of this type used carbon disulfide diluted to 0.1 of its pure concentration by hexane. When the shock reached the solution in this cell, it rapidly became completely absorptive well before the pressure in the solution reached its maximum.

A fifth experiment, similar to the above three pure carbon disulfide experiments, was modified by replacing the mirror on the impactor by a black absorbing cavity. This shot produced a record surprisingly similar to those shots with the mirror. Reflections from the various sapphire-vacuum interfaces provided enough light to show a weak carbon disulfide spectrum prior to impact. After the shock reached the carbon disulfide, the record showed the reflectivity of carbon disulfide to increase then decay, just as it had in the 123 kbar experiment. This led us to believe that the light was actually being reflected from the carbon disulfide layer. This conclusion is supported by the results of flashlamp-in-the-projectile-type experiments which are described below.

Three successful experiments have been conducted with the flashlamp mounted in the projectile. Cell thicknesses for these experiments were 0.0008 mm, 0.01 mm, 0.14 mm. All three experiments had peak pressures in the range of 50 to 60 kilobars. In the thickest cell the absorption bands were observed to extend toward longer wavelength as the first few shocks rang through the carbon disulfide. By the time the third shock traversed the carbon disulfide the transmission was completely cut off. The other two experiments were too thin to resolve the shock transit times; the light cut off abruptly when the shock reached the carbon disulfide. This strong cutoff of the light even in very thin cells confirms that the records obtained in the original setup were produced by light reflected from the carbon disulfide layer.

The possibility that the high reflectivity of the carbon disulfide is due to a transition to a highly conductive state is being investigated and will be reported on separately.

This work is supported by ONR Contract No. N00014-77-C-0232.

REFERENCES

1. S. A. Sheffield and G. E. Duvall in *Proceedings of the Symposium on Behavior of Dense Media under High Dynamic Pressures*, (Commissariat à l'Energie Atomique, Paris, 1978), pp. 381-389.

THE RESISTIVITY OF LIQUID CARBON DISULFIDE DURING SHOCK COMPRESSION

C. R. Wilson, G. E. Duvall, and K. Ogilvie
Department of Physics
Washington State University,* Pullman, WA 99164

ABSTRACT

Changes in the resistivity of CS_2 under shock compression (60 kbar) have been measured. Also reported is the reflectivity of shocked CS_2 (120 kbar) for light in the 2100 to 3100 Å region. Carbon disulfide appears to become reflective at pressures as low as 6 kbar. The magnitude of the reflection seems to imply a greater conductivity than is indicated by the resistance measurements. The minimum resistivity of the shocked CS_2 was 5×10^3 ohm-m.

INTRODUCTION

The transformation of liquid CS_2 under static and dynamic compression has been studied for over forty years. Under static compression of $\simeq 40$ kbar it forms a stable black polymeric solid with semiconductor properties.[1-3] The nature of the reaction products for dynamic compression of CS_2 is less certain. Hugoniot data for shocked CS_2 indicates a phase transition at 62 kbar.[4] Most workers believe the transition is decomposition to the elements C and S since CS_2 is thermodynamically unstable with respect to its elements and a decreased density phase is favored at high pressures.[5]

In an effort to further understand this phase change we have been studying time resolved ultraviolet absorption spectroscopy of shocked CS_2. Initial experiments in which the light was transmitted twice through the shocked CS_2 (by a reflector in the projectile) showed a substantial increase in light during the shock process. A subsequent experiment in which a light absorber was placed in the projectile ruled out an increase in transparency as a source of the increased light. Thus CS_2 appears to reflect light during shock compression.

In order to understand the nature of the increased reflectivity we have carried out resistivity experiments on CS_2. This paper presents our experimental methods, an estimate of the reflection coefficient of shocked CS_2 and the time resolved resistance changes of shocked CS_2.

EXPERIMENTAL TECHNIQUES

In our reflectivity experiment a target cell consisting of a 0.09 mm layer of CS_2 contained between two 3.175 cm sapphire discs was impacted by a 10.2 cm-diameter Al projectile traveling at 5.6 mm/μsec. The projectile head consisted of a 3.175 cm sapphire disc epoxied in the Al body. Light from a Xe flashlamp (ILC lamp #L3958) which passed through the cell and the sapphire disc on the projectile was absorbed by the flat black coated interior of the projectile. The Xe flashlamp was turned on by the projectile shorting a charged coaxial cable imbedded in the target assemble. A pair of mirrors carried the light to the cell and then up to the entrance slit of a diffraction grating spectrometer. The grating had 600 lines/mm blazed at 3000 Å. The spectrum was then recorded by a rotating mirror streak camera (Beckman and Whitley 339) on Kodak 2475 film. The region covered was $\simeq$ 2100 to 3100 Å. The total light output was monitored by a photmultiplier tube (RCA 4831) mounted on the camera. The voltage output from the photomultiplier was recorded by oscilloscopes whose sweep was also initiated by the charged coaxial cable.

The resistivity experiment involved the same cell configuration with a 0.05 mm layer of CS_2. The projectile traveled at a speed of 0.29 mm/μsec. Circular electrodes 9 mm in diameter were deposited on the sapphire discs. These electrodes were obtained by vacuum depositing Nichrome V wire onto the sapphire followed by a coating of Al. The film thickness was 1-2 μm. Resistivity changes during the experiment were computed by discharging a 100 μF capacitor (charged to 45 volts) through the cell and recording the voltage appearing across a 50 ohm load resistor which is in series with the cell. Prior to the experiment the discharge circuit was calibrated by shorting across the cell with known resistors and recording the discharge voltage across the load resistor. Thus the output for the experiment consisted of a time resolved voltage which was converted to resistance using the calibration graph of resistance vs load resistance voltages.

Pressures obtained in the shocked CS_2 were determined by the impedence matching method using Hugoniots for sapphire and CS_2.

EXPERIMENTAL RESULTS

The streak camera film record of the reflectivity experiment provided an estimation of the reflection coefficient of shocked CS_2. A fixed wavelength scan (along the 2700 Å band pass) of the film record by a densitometer provided the incident light intensity and the reflected light intensity of the shocked CS_2. Seven percent of the incident light intensity is subtracted from both intensities because of the reflection from the air sapphire interface. The ratio of the reflected light to incident light was 0.8 ± 0.2. Variability is caused by the time variation in the Xe flash intensity. A value obtained from the photomultiplier record, which integrates over all the wavelengths, was of similar magnitude. The pressure attained in this experiment was 120 kbar.

In the resistivity experiment we found a decrease in resistance followed by an increase. The resistance fell to 9000 ohms in 40 nsec after the shock passed into CS_2. The pressure at this time was 6 kbar. The minimum resistance of 4000 ohms occurred 200 nsec after the shock entered the CS_2. The pressure at this time was 63 kbars. The resistance then increased to 82,000 ohm in another 200 nsec. Using our cell dimensions we calculate a minimum resistivity of 5×10^3 ohm-m.

DISCUSSION

Some investigators have observed a loss in transparency of shocked CS_2 and have partially attributed it to light scattering from C in the decomposition products.[6] We feel that scattering is not the source of the increase in light based on a scattering angle of 2π radians and a distance of over 3 meters to the first objective mirror in the telescope. While our resistivity measurements do not indicate as great a conductivity as implied by the reflection coefficient, it may be that CS_2 becomes more conductive at higher pressures. Dick has indicated the conductivity of CS_2 increases significantly at pressures greater than 80 kbar.[4] We plan to make resistance measurements at higher pressures.

*This work was supported by ONR Contract No. 00014-77-C-0232.

REFERENCES

1. P. W. Bridgeman, Proc. Am. Acad. Arts and Sci. 74, 399 (1941).
2. E. Whalley, Can. J. Chem. 38, 2105 (1960).
3. E. G. Butcher, M. Alsop, J. A. Weston, and H. A. Gebbie, Nature 199, 756 (1963).
4. R. D. Dick, J. Chem. Phys. 52, 6021 (1970).
5. S. A. Sheffield, "Shock Induced Reaction in Carbon Disulfide," Ph.D. Thesis, Washington State University, Pullman, WA (1978).
6. O. B. Yakusheva, V. V. Yakushev, A. N. Dremin, Russian J. Phys. Chem. 51, 973 (1977).

ELECTRICAL AND OPTICAL MEASUREMENTS ON FUSED QUARTZ UNDER SHOCK COMPRESSION

K. Kondo and T. J. Ahrens
Seismological Laboratory, California Institute of Technology
Pasadena, California 91125

A. Sawaoka
Research Laboratory of Engineering Materials
Tokyo Institute of Technology, Midori, Yokohama 227, Japan

ABSTRACT

The resistivities of specimens of SiO_2 (fused quartz) singly and doubly shocked in the 10 to 45 and 27 to 90 GPa range, respectively, demonstrate a marked decrease from values of $\sim 10 \Omega \cdot m$ to $\sim 0.1 \Omega \cdot m$ at single shock pressure of ~ 40 and double shock pressure of ~ 74 GPa. The shock-induced polarization profiles also show a sudden change of electrical properties of the material at those pressures. The rapid decrease in resistivity suggests a further transformation to an unknown phase or production of an electron bound level.

INTRODUCTION

The shock-wave equations of state of SiO_2 for various initial forms have been studied over a wide range of dynamic pressure[1-4]. However, other measurements on physical properties of SiO_2 under shock compression are necessary to identify the high pressure and high temperature state because a small change of volume or pressure can not be detected by the EOS experiments. We have performed measurements of electrical resistivity, shock-induced electromotive-force, and shock-induced radiation spectra of fused quartz specimens. The resulting data are discussed in light of our knowledge of the electrical properties of dielectrics and their relation to the previous equation of state studies. The data are also discussed with the recent observation of shock-induced radiation.

EXPERIMENTAL

Both resistivity and shock-induced polarization measurements were carried out in longitudinal geometry by using a charged capacitor method[5]. The disc-shaped specimens were impacted by a copper or tungsten flyer plate at speeds from 1.1 to 3.2 mm/μsec. The impedance match method was used to determine both the initial and reflected shock states in the specimen using the Hugoniot data of McQueen et al.[6] and Wackerle[1], for copper, tungsten, and fused quartz. Both reflected Hugoniot and isentropic states were approximated using the principal Hugoniot curve. Shock temperatures were estimated on the basis of standard thermodynamic formulae.

Experimental details are described elsewhere[5].

The basic element of apparatus for measureing shock induced radiation of matter was the use of a EG&G Princeton Applied Research optical multichannel analyzer (OMA2). The light from the shock-compressed state via the uncompressed material was focused onto an entrance slit of a spectrometer by a collecting lens system. The incident light was dispersed into spectrum in the region between 400 and 820 nm, and recorded by a sensitive vidicon detector with a silicon diodes target. The gating time was about 300 nsec, which was during shock compression. The experiments were carried out by using 40 mm propellant gun of California Institute of Technology. Details will be described elsewhere.

RESULTS AND DISCUSSION

Resistivity measurements are summarized in Fig. 1. It is clearly seen from Fig. 1 that the resistivity of fused quartz under shock loading gradually decreases with the first shock pressure up to 36 GPa and rapidly decreases by about two orders of magnitude at 36∼40 GPa. Results of the reflected shock-compression state also show that the rapid decrease of resistivity occurs at about 70 GPa.

Polarization signals at pressure below 37.5 GPa were too small and unclear to be recorded. However, at 43 GPa, a positive voltage of about 1V is gradually generated following shock propagation and the sign of the voltage changes upon a reflection of the shock wave. This suggests that changes of the electrical

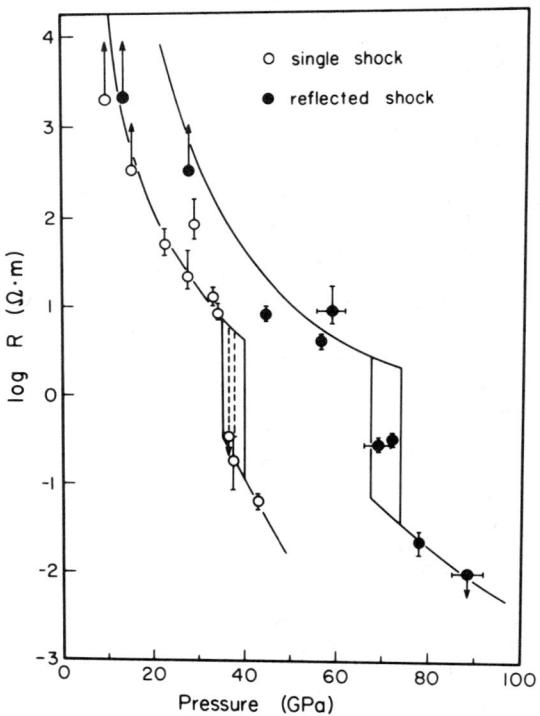

Fig. 1. Logarithm of electrical resistivity (R) as a function of shock pressure for fused quartz.

properties at shock pressures of 36.5 to 37.5 GPa depend on time and that the true transition in the electrical state of fused quartz occurs nearer 40 GPa and 3250K. The polarization signal also suggests that an effective positive charge is produced at the shock front as the conductivity increases.

The resistivity versus reciprocal shock temperature (Fig. 2) clearly demonstrates two regimes of electrical resistivity behavior. The temperature dependence of resistivity for fused quartz at atmospheric pressure has been reported[7] and is also shown in Fig. 2. This is approximately represented by the equation, $R = 0.251 \times \exp(0.88 /kT)$, where R and kT are in $\Omega \cdot m$ and eV, respectively. An extrapolation of this line agrees well with shock-wave data with larger resistivity than 1 $\Omega \cdot m$. In the regime over which the high temperature and high pressure data agree, we believe that fused quartz has partially transformed to stishovite. Therefore, we infer no significant difference exists between ionic transport processes in normal fused quartz and the presumed shock-induced high pressure phase, stishovite.

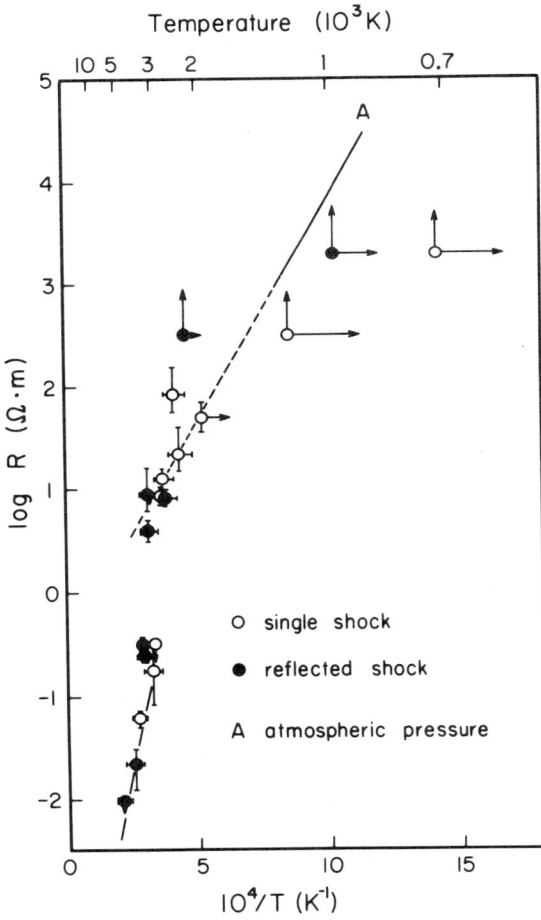

Fig. 2. Logarithm of electrical resistivity (R) as a function of reciprocal shock temperature for fused quartz. Solid line A shows temperature dependence of resistivity for fused quartz (ref. 7).

However, at the higher temperature, the activation energy suddenly changes and the resistivity is approximately represented as, $R = 2 \times 10^{-5} \times \exp(2.4 \pm 0.5 /kT)$. This sudden change of activation energy suggests the occurrence of structural change in

SiO_2 into another high-temperature and high-pressure phase.

However, no appreciable increase of density is observed in the Hugoniot data for fused quartz above ∼40 GPa. Some evidence has been found for the occurrence of phase change of SiO_2 with a similar or slightly higher density than stishovite[8-10]. If a four- or six-coordinated glass still exists under shock conditions because of the reconstructive phase transition to stishovite, glass-transition or crystallization may probably occur at that pressure and temperature. Although a resistivity decrease of about two orders of magnitude occurs on melting in some minerals[11,12], this does not appear to be consistent with the shock temperature measurements for fused and crystalline quartz, which suggest that the onset of melting stishovite under loading occurs at higher pressure and temperature than those for the change in electrical properties[13].

Preliminary observation of shock-induced radiation spectrum within the mixed phase region at 22 GPa shown in Fig. 3 suggest a few broad lines in the spectrum, which are superimposed into lattice-temperature radiation. The strongest emission band appears at 509 nm with half height width of 20 nm, and the second one at 589 nm with 20 nm. These wave lengths correspond to 2.44 and 2.10 eV.

Molecular orbital calculations on the electronic structure of fused or amorphous quartz have been carried out using a variety of cluster-models, and show that the width of valence band and energy gap are of the order of 9 and ∼13 eV, respectively, while the experimental value of the gap is 11 eV[14-16]. The defect electronic states are due to oxygen vacancies and proton interstitials. The latter and Na^+ are the main impurities in commercial fused quartz. They give rise to electron levels in the gap which

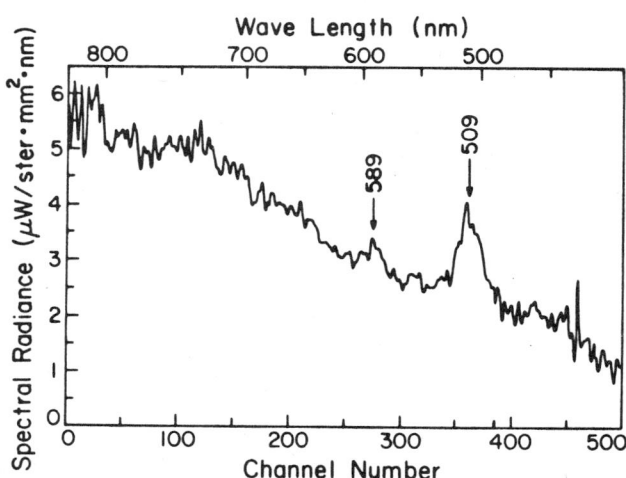

Fig. 3. Shock-induced radiation spectrum of fused quartz at 22 GPa. Collecting time is 300 nsec.

are dependent on the model cluster size[16]. The oxygen vacancy is responsible for some of the ultraviolet absorption bands. If we assume that the electron level of 2.4 ± 0.5 eV in the gap is produced during shock-compression, it should be an acceptor bound level or an electron trap because the polarization result suggests a positive carrier. It is likely that disruptive effects of shock front produce this defect as an electron trap similar to the luminous defects introduced by mechanical damage of semiconductor. It is also likely that activated free-atomic state exists between quartz and stishovite in the mixed phase region. When a change in phase is taking place, such a condition may exist in the high pressure region because of the reconstructive and slow reaction to stishovite. However, to explain the sudden change of the resistivity, the above model needs a sudden change of carrier concentration or increase of the number of traps because the thermal activation of electron increases continuously with temperature.

Contribution No. 3652, Division of Geological and Planetary Science, California Institute of Technology, Pasadena, CA 91125.

REFERENCES

1. J. Wackerle, J. Appl. Phys. 33, 922 (1962).
2. L. V. Al'tshuler, R. F. Trunin, and G. V. Simakov, Izv. Acad. Sci. USSR, Phys. Solid Earth 10, 657 (1965).
3. R. F. Trunin, G. V. Simakov, and M. A. Podurets, Izv. Acad. Sci. USSR, Phys. Solid Earth 2, 102 (1971).
4. H. Sugiura, K. Kondo, and A. Sawaoka, J. Appl. Phys. submitted.
5. K. Kondo, A. Sawaoka, and T. J. Ahrens, J. Appl. Phys. 52, 5084, 1981.
6. R. G. McQueen, S. P. Marsh, J. W. Taylor, J. N. Fritz, and W. J. Carter, in *High Velocity Impact Phenomena*, edited by R. *Kinslow* (Academic Press, New York, 1970) p. 244.
7. R. W. Wallace and E. Ruh, J. Am. Ceram. Soc. 50, 358 (1967).
8. L. Liu, W. A. Bassett and J. Sharry, J. Geophys. Res. 83, 2301 (1978).
9. V. N. German, M. A. Podurets and R. F. Trunin, Sov. Phys. JETP 37, 107 (1973).
10. V. N. German, N. N. Orlova, L. A. Tarasova and R. F. Trunin, Izv. Acad. Sci. USSR, Phys. Solid Earth 11, 431 (1975).
11. T. Murase and A. R. McBirney, Geol. Soc. Amer. Bull. 84, 3563 (1973).
12. T. J. Shankland, Rev. Geophys. Space Phys. 17, 792 (1979).
13. G. A. Lyzenga, T. J. Ahrens, and A. C. Mitchell, J. Geophys. Res. (to be published).
14. K. L. Yip and W. B. Fowler, Phys. Rev. B10, 1400 (1974).
15. A. G. Revesz, Phys. Rev. Lett. 27, 1578 (1971).
16. A. J. Bennett and L. M. Roth, J. Phys. Chem. Solids 32, 1251 (1971).

SHOCK-COMPRESSION TEMPERATURE RISE DETERMINED FROM RESISTIVITY OF EMBEDDED METAL FOILS*

D. D. Bloomquist and S. A. Sheffield
Sandia National Laboratories, Albuquerque, New Mexico 87185[†]

ABSTRACT

The temperature rise induced by shock compression of polymethyl methacrylate (PMMA) was determined from measurements of the electrical resistivity of embedded copper foils. The temperature of the copper was determined from the observed foil resistance and known values of the change in copper resistivity with temperature and shock compression. Temperature values obtained over a stress range from 0.9 to 6.0 GPa are in good agreement with thermocouple measurements reported previously.

INTRODUCTION

The need for an accurate means of measuring temperature in the shock-compressed state of condensed materials has been recognized for many years and is of timely importance because of an increasing interest in chemical activity under shock loading conditions.[1] Efforts to develop a technique for measuring temperature in shock wave experiments have been centered primarily on detection of infrared radiation and the use of embedded thermocouples. With each of these methods, unique limitations and obstacles must be overcome; neither method has been completely successful or widely applied.[2]

The first experiments in which temperature rise in a shock compressed material was inferred from a change in resistivity were reported by Dremin et al.[3] In that study, temperature was inferred by comparing a calculated time dependence of temperature as a function of foil thickness for embedded metal foils to the results of resistivity measurements using nickel foils that varied in thickness from 6.5 to 100 μm. The results are somewhat ambiguous because of neglected demagnetization effects that accompany shock compression of nickel, the use of only two leads in the resistance measurements, and the reportedly large uncertainties involved in the analysis.

The present study was undertaken to determine the feasibility of using the resistance of metal foils to measure shock temperatures and to confirm the earlier thermocouple measurements in PMMA.

EXPERIMENTAL PROCEDURE

The present experiments consisted of shock loading a sample of PMMA and observing the resistivity of a 5 μm-thick embedded copper

---

*This work sponsored by the U.S. Department of Energy under Contract DE-AC04-76-DP00789.
[†]A U. S. Department of Energy facility.

foil using a π voltage probe. The experimental configuration is illustrated in Fig. 1. The copper foil was sandwiched between two flat, polished surfaces using a thin film of solvent (dichloroethane) or Hysol R8-2038 epoxy to bond the pieces together. Care was taken in the case of the glue bonds to ensure a thin bond which resulted in a separation between the polished surfaces approximately equal to the foil thickness of 5 μm. Shock loading was achieved by projectile impact as indicated in Fig. 1, and the shock wave propagated in a direction parallel to the signal leads. The probe was positioned with the resistive element between 1 and 3 mm from the impact plane.

The resistance of the copper foil was measured using a standard four-point resistance measurement technique, so that effects of contact resistance and the changing resistance in the shocked leads were eliminated. A constant currrent power supply[4] designed for pulse operation was connected across the outer leads and triggered about 20 μs before impact. The current remained constant (after a 1 μs settling time) to within 0.02% per μs. The voltage was

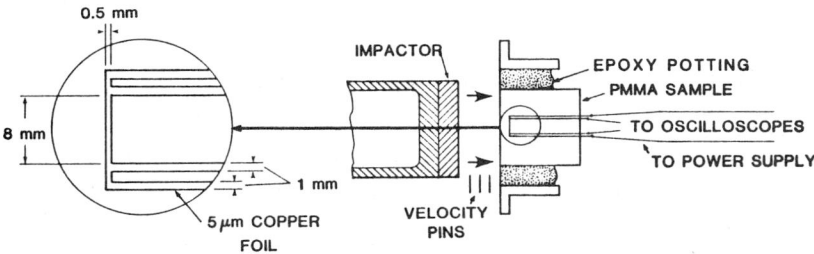

Fig. 1 Experimental Configuration in which shock temperature rise is determined from observed changes in electrical resistivity of an embedded copper foil.

monitored across the inner leads using a Tektronix 7A13 differential preamplifier in a Textronix 7903 oscilloscope. The change in resistance as the shock wave traverses the foil position was observed directly as a change in voltage.

THEORY

To determine the temperature of a metal foil from the observed ratio α of the electrical resistivity of shocked material to the preshock value, it is assumed that the effects of temperature rise and volume compression in a shock wave are independent[5] and of the form

$$\frac{\rho(\eta_c, T)}{\rho_o(0, T_o)} \equiv \alpha(\eta_c, T) = \alpha(0, T_o) + \beta(T - T_o) + \gamma\eta_c + \delta\eta_c^2, \quad (1)$$

where ρ is the Lagrangian electrical resistivity (i.e., relative to the undeformed cross section), T is absolute temperature, η_c is volume compression of the copper defined as $1 - v/v_0$, v is specific volume, $\alpha(0,T_0)$ has the defined value of 1, and the subscript zero designates values just prior to shock compression. In the Bloch-Gruneisen theory of electrical resistivity of metals, which holds for copper above room temperature,[6] the rate of change of resistivity with temperature is constant and independent of the defect or impurity concentrations. The coefficient β represents the rate of change of resistivity with temperature and can be determined from standard resistivity-temperature tables[7] for copper as $\beta = 4.38 \times 10^{-3}$ K^{-1}. The constants γ and δ, which represent the rate of change of electrical resistivity in copper with shock compression, can be determined from data reported by Keeler[8] in which the resistivity of copper was observed as a function of shock stress. These data can be corrected to eliminate the effects of the inherent shock heating using calculated values of temperature rise on the Hugoniot[9] and the zero-pressure dependence of resistivity on temperature. These corrected data have been fit to the expression

$$\alpha(\eta_H, T_0) - 1 = -3.14\ \eta_H - 32.8\ \eta_H^2, \quad \eta < 0.1, \tag{2}$$

by a least squares method with resulting standard error of estimate of less than 1%. In this expression the subscript H indicates values on the Hugoniot of copper. The constants γ and δ can be determined by comparison of Eq. (2) with Eq. (1) as -3.14 and -32.8, respectively.

The ratio α is determined from the observed voltages as

$$\alpha(\eta_c, T) = \rho/\rho_0 = V/V_0, \tag{3}$$

where V is the voltage across the inner leads and η_c is the compression of the copper at the stress of the PMMA. The temperature rise in the resistive element can be determined from Eq.(1) in the form

$$\Delta T = [\alpha(\eta_c, T) - 1 - \gamma\eta_c - \delta\eta_c^2]/\beta. \tag{4}$$

In interpreting the data, we assumed that the volume compression in the copper was that which would have been achieved in a single shock to the final stress state of the PMMA, thus neglecting the effects of the rather complicated compression process in the foil.

This is not expected to lead to a significant error at the low pressures achieved in these experiments.

RESULTS AND DISCUSSION

The observed temperature histories consisted of a rapid rise in temperature when the shock wave reached the resistive element followed by a relatively constant plateau. The results of experiment RT-9 at 3.3 GPa are shown in Fig. 2 along with temperature histories predicted by thermal diffusion calculations[10,11] of the temperature at the center of the 5 μm foil assuming an instantaneous change in the surrounding temperature as the shock wave pressurizes the foil. If it is assumed that the thermal equilibration process of the foil-host-material system determines the shape of the rise, the thermal diffusivity of the PMMA appears to be as much as four orders of magnitude larger than the zero pressure value of $D_o = 1.3 \times 10^{-7}$ m^2/s. This is consistent with earlier observations using thermocouples.[3]

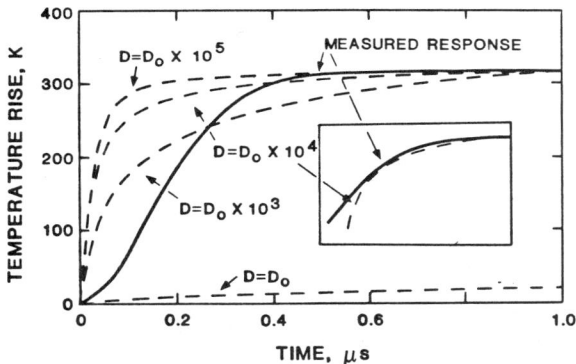

Fig. 2. Temperature history for experiment RT-9 at 3.3 GPa compared with calculated temperature histories based on thermal constants of the unshocked material (D=Do), and on thermal diffusivities enhanced by up to five orders of magnitude.

Results of experiments over the pressure range from 0.9 to 6.0 GPa are plotted in Fig. 3 along with earlier data obtained from thermocouple experiments in PMMA[10] and a calculation based on zero pressure thermal constants.[9] Data from two previously reported experiments using the present techniques are also included[12] in Fig. 3. Temperature measurements obtained in the present experiments are consistent with the earlier results and provide independent confirmation of the previous conclusions that the rate of thermal equilibration is greatly enhanced under shock loading and that the temperature in PMMA rises very rapidly with stress for shock stresses greater than 2.0 GPa. The observation of anomalously high thermal equilibration rates, which was also reported by Dremin et al.,[3] suggest that a fast heat transfer mechanism is active behind a shock wave in this type of material which was previously unknown and is anomalous in terms of conventional shock wave phenomenology. These present results also indicate that observing the resistivity change in embedded metal foils is a viable way to measure temperature in shock-wave experiments. This technique may be applicable in situations where thermocouples or infrared detection techniques are unsuitable.

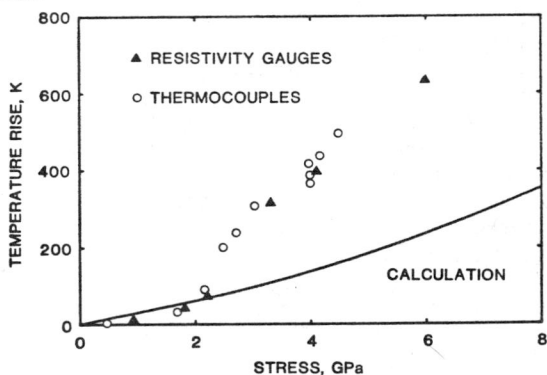

Fig. 3. Plot of measured temperature rise as a function of shock stress obtained from resistivity gauges and thermocouple.

REFERENCES

1. R. A. Graham, Bull. Am. Phys. Soc. 25, 495 (1980).
2. Lee Davison and R. A. Graham, Phys. Rep. 55, 255 (1979).
3. A. N. Dremin, B. P. Ivanov, and A. N. Mikhailov, Combustion Explos. Shock Waves 9, 784 (1973).
4. Pulsar model 301 Manganin gage power supply, Pulsar Instruments, 817 Burlingame Avenue, Redwood City, CA 94063.
5. J. J. Dick and D. L. Styris, J. Appl. Phys. 46, 1602 (1975).
6. N. F. Mott and H. Jones, The Theory of the Properties of Metals and Alloys, 2nd edition (Dover, New York, 1958).
7. W. F. Roeser and H. T. Wensel, Temperature, Its Measurement and Control in Science and Industry (Reinhold, New York, 1941).
8. R. N. Keeler, in Proceedings of the International School of Physics Enrico Fermi, Course XLVII, Physics of High Energy Density (Academic Press, New York, 1970), pp106-125.
9. R. G. McQueen, S. P. Marsh, J. W. Taylor, J. N. Fritz, and W. J. Carter, in High-Velocity Impact Phenomena, edited by Ray Kinslow (Academic Press, New York, 1970), pp. 293-418.
10. D. D. Bloomquist and S. A. Sheffield, J. Appl. Phys. 51, 5260 (1980).
11. R. Grover and P. A. Urtiew, J. Appl. Phys. 45, 146 (1974).
12. D. D. Bloomquist and S. A. Sheffield, Appl. Phys. Lett. 38, 185 (1981).

REFLECTION OF A LASER-GENERATED OPTICAL SIGNAL FROM A SHOCK FRONT IN WATER

P. Harris
U. S. Army Armament R&D Command, Dover, New Jersey 07801

H. N. Presles
University of Poitiers, 86034 Poitiers, France

ABSTRACT

The reflectivity of a 0.58 GPa (5.8 kbar) shock front in water has been measured with a 0.5145 µm laser generated optical signal. Optical polarizations both parallel and perpendicular to the plane of incidence were employed. Measurements were carried out at large angles of incidence. Concurrently a closed form expression was derived which predicts reflectivity for all angles, optical polarizations, and shock front thicknesses. Comparison between theory and experiment shows that the shock front thickness is less than or equal to 4×10^{-2} µm. A detailed description of this work has been published.[1]

1. P. Harris and H. N. Presles, J. Chem. Phys. <u>74</u>, 6864 (1981).

HIGH TEMPERATURE PHASE TRANSFORMATION IN THE
TITANIUM ALLOY Ti-6Al-4V*

J. E. Shrader and M. D. Bjorkman
Boeing Aerospace Company, Seattle, WA 98124

ABSTRACT

A test program using both gas gun plate impact and pulsed electron beam techniques was conducted to explore the effect of polymorphic phase changes on the mechanical and thermal properties of Ti-6Al-4V at temperatures above 800 K. Shock waves were used to probe material properties either side of the phase line. The measured data are consistent with the equilibrium thermodynamics of a phase line between 900 K to 1370 K, across which the adiabatic bulk modulus abruptly increased 10%, the Gruneisen coefficient abruptly increased 40% and the dynamic yield strength decreased 50% near the phase line.

INTRODUCTION

Plate impact and pulsed electron beam techniques were used to probe the equation of state surface of Ti-6Al-4V at elevated temperature and stress. Effects from the completion of the polymorphic phase transformation (hexagonal close packed α-phase to body centered cubic β-phase) in Ti-6Al-4V were measured by shock loading specimens heated to an initial state either side of the phase line. No attempts were made to observe shock wave-induced phase transformation effects.

METHODOLOGY

Ti-6Al-4V specimens with two microstructures were studied: specimens heated above the β-transis to produce an acicular grain structure and specimens heated in the $\alpha + \beta$ region to produce an equiaxed α-grain structure. Both the $\alpha + \beta$ and the β-processed alloys contained $6 \pm 0.5\%$ wt. aluminum, 4 ± 0.5 wt. vanadium, <0.3% wt. iron, <0.9% wt. total impurities and the balance titanium.

The Boeing 2.5-inch bore light gas gun was used for the plate impact experiments. The specimen was mounted in a holding fixture which allowed for simultaneous heating by passing current through the specimen, and pretensioning by pulling the specimen using a hydraulic piston. The temperature was monitored using a thermocouple spot welded to the specimen, impact stress was measured using an impacting X-cut quartz gage mounted in the projectile, and the rear surface velocity history was measured with a VISAR.

The electron beam tests were made with the Boeing FX-75 electron beam machine, which was operated with a mean beam enery between 2.0 and 2.7 MeV, and a shine time of 54 ns. Fluence was varied by adjusting charge voltage, tank pressure, and by moving the target closer to or further away from the anode. Test specimens were radiatively heated from a carbon sheet in close proximity to the specimen. The carbon sheet was heated resistively. The heating time was approximately 30 s, and temperature was measured using a thermocouple spot-welded to the specimen surface. The dose was measured using a thin foil calorimeter, and the rear surface velocity of the specimen was measured using a combined VISAR (with a lensed delay leg) and displacement interferometer.

RESULTS

Figure 1 shows some of the rear free surface velocity histories from impacted β-processed specimens with elevated initial temperatures. Note that dynamic yield strength decreases with temperature. Wave profiles from specimens with initial temperatures of 291 K and 903 K are characteristic of viscoplastic flow. The wave profile from the specimen with an initial temperature of 1371 K (which is above the β-transis at atmospheric pressure) shows a drop in particle velocity behind the elastic shock, while the two lower initial temperature shots show no drop. This may indicate a change in flow mechanism with temperature or phase.

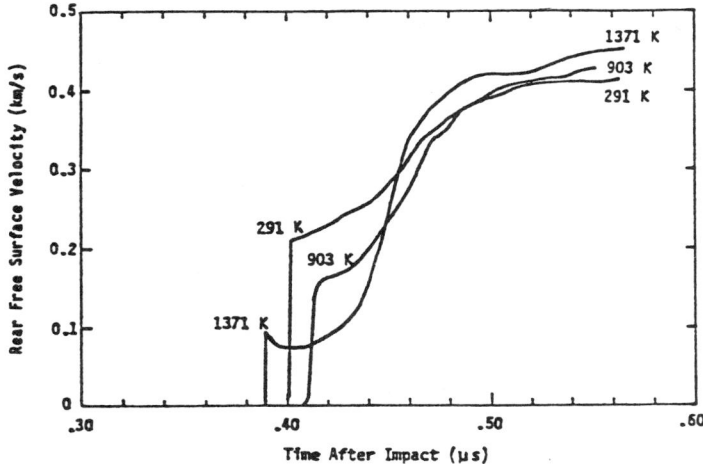

Figure 1. Some of the rear free surface velocity histories for β-processed Ti-6Al-4V.

Figure 2 is a plot of the temperature dependence of the elastic precursor amplitude. The curve connecting the points was drawn freehand. Shown on the temperature axis are the β-transis temperature at atmospheric pressure and the temperature at which 90% of the titanium is α-phase and 10% is β-phase, when in equilibrium, at atmospheric pressure. Shock velocity data from electron beam tests and gas gun tests are shown in Fig. 3.

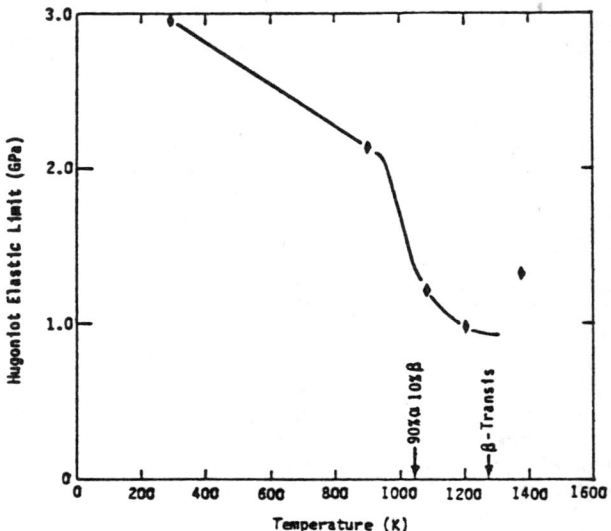

Figure 2. Temperature dependence of the elastic precursor amplitude at the rear free surface, of β-processed Ti-6Al-4V.

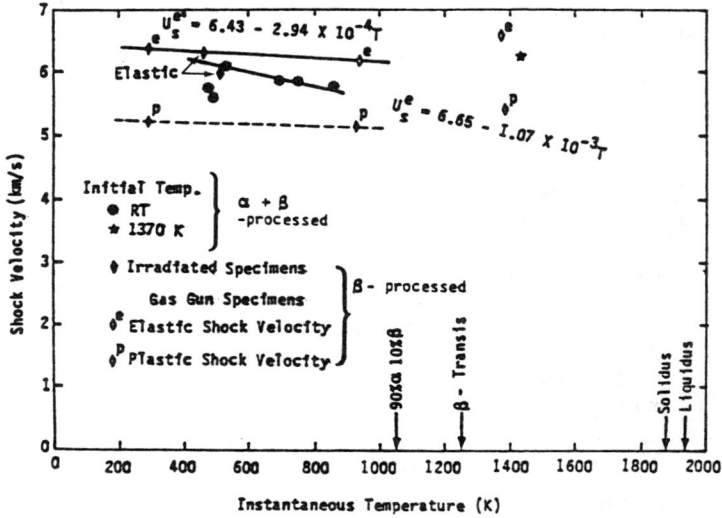

Figure 3. Shock velocity from transit time in irradiated specimens of β-processed and α + β-processed Ti-6Al-4V.

Note in Fig. 3 that for both the electron beam and plate impact tests, the elastic shock velocity data increase in the vicinity of the β-transis.

Figure 4 gives the stress-particle velocity Hugoniot data for β-processed Ti-6Al-4V measured in plate impact experiments. The σ_x - u_p Hugoniot curve for α + β-processed alloy measured by L. Seaman et al.[1] is shown for comparison.

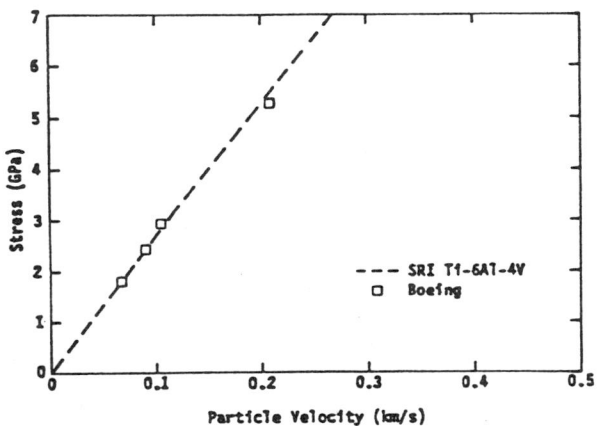

Figure 4. σ_x - u_p Hugoniot for β-processed Ti-6Al-4V.

Gruneisen coefficient was calculated from the electron beam tests of α + β-processed Ti-6Al-4V. These results are given in Fig. 5. Scatter in the data make it impossible to tell conclusively whether there is a jump in Gruneisen coefficient at the phase line or not. If the two suspect points are deleted from Fig. 5, then the remaining data suggests that the Gruneisen coefficient increases when crossing the phase line.

DISCUSSION

The consistency of the measured data with equilibrium thermodynamics was checked with the procedure developed by K. A. Holsapple[2] for calculation of the jump in thermomechanical property at the phase line of the alloy. Using this procedure, the relationships in Table I are obtained across the boundary for Ti-6Al-4V.

The shock wave measurements show that the adiabatic bulk modulus increased approximately 10% at the phase line, and the Gruneisen coefficient increased approximately 40%, which compares favorably with the calculations.

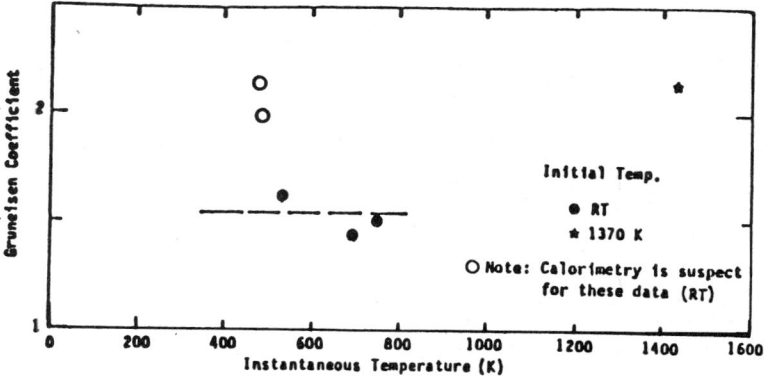

Figure 5. Temperature dependence of Gruneisen coefficient for α + β-processed Ti-6Al-4V.

Table I - Jump in thermomechanical properties across phase boundary

$\dfrac{\gamma_2}{\gamma_1}$	$\dfrac{K_2^T}{K_1^T}$	$\dfrac{C_{V_2}}{C_{V_1}}$	$\dfrac{K_2^S}{K_1^S}$
1.37	1.02	0.876	1.05

where subscript 1 = below phase line
subscript 2 = above phase line
γ = Gruneisen parameter
K^T = isothermal bulk modulus
C_V = specific heat at constant volume
K^S = adiabatic bulk modulus

REFERENCES

*Work performed under the auspices of DNA, Contract DNA001-79-0149.

1. L. Seaman, R. F. Wiliams, J. T. Rosenberg, D. C. Erlich and R. K. Linde Air Force Weapons Laboratory Tech. Report No. AFWL-TR-69-96, 1969, (AD865505).
2. K. A. Holsapple, Boeing Aerospace Co. Doc. No. D180-19075-1, 1975. See also K. A. Holsapple and R. M. Schmidt, J. Appl. Phys. 49 (11), 5493-5501 (1978).

GLASS TRANSITION IN SHOCK LOADED CERAMICS

Y. Horie
U.S. Army Research Office and North Carolina State University
Raleigh, N.C. 27650

The initiation of the diaplectic glass in shock-loaded ceramics is examined in terms of the excess Gibbs free energy of dislocations. The novelty is an attempt to model the dislocation interaction energy that is often neglected, but which is considered important by Edward and Warner. The critical temperatures calculated for MgO and Al_2O_3 are not inconsistent with the values found in the discussions of heterogeneous yielding in those materials, but are well below their melting temperatures.

INTRODUCTION

Heterogeneous yielding in shock loaded oxides like quartz, sapphire, and magnesium oxide is thought to be a yield process in which high local temperatures appear as a result of adiabatic shear yielding similar to the shear banding process observed in metallic systems.[1] This paper is an attempt to model the consequence of such a yielding to explain the appearance of diaplectic glass in the recovered specimens of shock loaded ceramics.[2,3] The basic idea is that in such a hot spot, the crystal may become unstable with respect to the state with a large number of dislocations. The supporting materials for this idea are: (1) although melting may be suspected, there is little atomic diffusion during shock-loading, (2) dislocation theory of amorphous solids, resembling a random packing model, predicts pair distribution functions which are in good agreement with experiments,[4,5] (3) mechanical shearing is known to transform a crystalline solid into an amorphous material.[6] Therefore, it seems reasonable to suggest that the observed diaplectic glass is a remnant of the transition triggered by a sudden growth of dislocations. But our analysis confines itself to the calculation of the excess free energy of dislocations.

THE GIBBS FREE ENERGY OF DISLOCATIONS

The excess free energy of dislocations relative to a perfect solid is $U - TS + Pv$ where U is the strain energy of dislocations, T is the absolute temperature of the solid, S is the change in vibrational entropy of the solid due to dislocations (the configurational entropy is known to be negligibly small[8]), P is the hydrostatic pressure in the solid, and v is the volume dilatation due to dislocations. The internal strain energy consists of three parts: the self energy U_1, the dilatation energy U_2, and the interaction energy U_3. We adopt the following approximate expressions for U_1 and U_2: U_1 (per unit length) = $\alpha \mu b^2 / 4\pi$, and

U_2 (per unit atomic volume) = $1/2 \ B \ (a^2\rho)^2$ where μ the shear modulus, a a lattice constant such that $a^3 = v_o$ = atomic volume, ρ the dislocation density (the length of the dislocation line per unit atomic volume).[7,8]

The evaluation of the interaction energy of dislocations, which are arranged in complex, tangled, curved arrays can become very complicated.[7] So, rather than attempt a general analysis of the complex interactions, we present a simple physical model based upon the observation[5] that the simulated dislocation configuration in a highly faulted solid resembles the network built by Mizushima[9] in his dislocation theory of melting. This network is built by randomly assembling the lattice cube whose edge length is given by $1/\rho^{1/2}$ and contain at their center a straight dislocation line element in one of the three [100] axes. The continuity of the network is established by inserting additional line segments between the center of the cube and the open end of dislocation line elements. The intersecting center called a node may be associated from three to six line segments. Assuming that these nodes are formed to establish the minimum-energy configuration, the interaction energy may be estimated by the energy associated with such a node. For an intersecting corner with a right angle, the energy reduction is known to be of the order, $\mu b^3/4\pi$[10]. Then, the total reduction of energy resulting from the formation of the network is

$$U_3 (\text{per unit mass}) = (N \ a^3 \ \rho^{3/2})(\beta\mu b^3/4\pi)$$

where N is the number of atoms per unit mass and β a constant related to the number of line elements associated with a node.

For the change in vibrational entropy, we shall use the approximation,[11,12] $(Nkv_o\rho \ln 1/\lambda/b)$ (per unit mass) in which λ is the constant that expresses phonon frequency shift due to dislocations. It is estimated that for oxides $\ln(1/\lambda) \approx 2$.

Upon assembling the preceding expressions for U, S, and v, one obtains the excess Gibbs energy.

$$G/Nv_o = g = (\alpha\mu/4\pi) \ C_d + 1/2 \ B \ C_d^2 - (\beta\mu/4\pi) \ C_d^{3/2}$$
$$+ P \ C_d - (NkT/V_o) \ \ln(1/\lambda) \ C_d \qquad (1)$$

where $V_o = Na^3$, $C_d = a^2\rho$, and it is assumed that $b = a$.

DISLOCATION DENSITY IN SHOCK LOADED MATERIALS

An order-of-magnitude estimate of the dislocation density in shock loaded solids may be made through the use of the total transient plastic strain ε introduced by Holtzman and Cowan[13] and a polynomial relation between the dislocation density and ε[14]

$$V/V_o = \exp(-.75\varepsilon), \qquad (2)$$

$$\rho = \rho_o + M\varepsilon^m \simeq M\varepsilon^m \quad \text{for } \rho \gg \rho_o, \qquad (3)$$

where V is the specific volume under compression by a shock pressure P_h, and M and m are constants. The shock pressure and the associated temperature are calculated by use of the empirical linear relation between shock velocity u_s and the particle velocity.

$$P_h = c^2(V_o-V)/(V_o-s(V_o-V))^2 \qquad (4)$$

$$T = T_o + VP_h/\gamma C_v \qquad (5)$$

where $u_s = c + su_p$, γ the Gruneisen constant, and C_v the specific heat.

EXAMPLES AND DISCUSSIONS

Fig. 1 shows the excess free energy g for MgO and Al_2O_3 along the Hugoniot calculated through the use of Eqs. (1)-(5). Their physical properties are summarized in Table 1. The constant α and β are assumed to be 1 and 3, respectively. The latter is the minimum number of branches a node can have in the dislocation network. It is also assumed that m = 1 and a^2M = 0.05.

The excess free energy is a monotonically increasing function of pressure, and no transition can be expected in homogeneous shock compression. The reason for this monotonicity appears to be the dominant influence of the pressure term in Eq. (1). This can be seen in Fig. 1 by juxtaposing g at the adiabatically decompressed state with the same dislocation density. Now there appears a critical pressure at which the excess energy disappears as a result of shock heating. Although it is suggestive of the instability of a solid, this disappearance cannot be used to infer such a possibility, because the density is fixed relative to the Hugoniot pressure.

To discuss the instability of the solid with respect to the state with a large number of dislocations, we need to examine the variation of g in terms of the dislocation density. The results are shown in Figs. (2) and (3). In view of the observed heterogeneous yielding which causes a local hot spot, we varied the temperature at a fixed pressure. All the results show a critical value of T at which the dislocation density can suddenly jump from a small number to a large one. In Ref. 7, the appearance of a minimum is attributed to the interaction energy that is proportional to $\rho^{3/2}$. If a is about 2.5×10^{-6} m, then C_d of 0.05 corresponds to ρ of $8 \times 10^{17}/m^2$.

We are not aware of the experimental data that are directly comparable with those predicted, but the magnitude of the critical temperatures is not inconsistent with those discussed in relation to heterogeneous yielding.[15] The density is also consistent with those reported for MgO.[16] Therefore, shock lamella observed in ceramics[2,3,15] may have its origin in the disordering process discussed above.

REFERENCES

1. L. Davison and R. A. Graham, Physics Report $\underline{55}$, 255 (1979).
2. B. M. French and N. M. Short (eds.), Shock Metamorphism of Natural Materials (Mono Book, Baltimore, 1968).
3. D. Stöffler and U. Hornemann, Meteoritics $\underline{7}$, 371 (1972).
4. H. Koizumi and T. Ninomiya, J. Phys. Soc. Jpn. $\underline{49}$, 1022 (1980).
5. E. J. Jensen, W. D. Kristensen and R. M. J. Cotterill, Phil. Mag. $\underline{27}$, 623 (1973).
6. R. Roy, J. Non. Cryst. Solids $\underline{3}$, 33 (1970).
7. S. F. Edward and M. Warner, Phil. Mag. A$\underline{40}$, 257 (1979).
8. D. Kuhlman-Wilsdorf, Phys. Rev. $\underline{140}$, 1599 (1965).
9. S. Mizushima, J. Phys. Soc. Jpn. $\underline{15}$, 70 (1960).
10. J. P. Hirth and J. Lothe, Theory of Dislocations (McGraw-Hill, New York, 1968), p. 139.
11. A. H. Cottrell, Dislocations and Plastic Flows in Crystals (Oxford Univ. Press).
12. T. Ninomiya, J. Phys. Soc. Jpn. $\underline{44}$, 263 (1978).
13. H. M. Otto and J. R. Holland, in Lattice Defects and Their Interactions (ed. by R. R. Hasiguti, Gordon and Breach, 1967), p. 1089.
14. J. J. Gilman, Micromechanics of Flow in Solids (McGraw-Hill, New York, 1967).
15. D. E. Grady, in High-Pressure Research (ed. by M. H. Manghani and S. Akimoto, Academic Press, 1977), p. 389.
16. M. J. Klein and J. W. Edington, Phil. Mag. $\underline{14}$, 21 (1966).

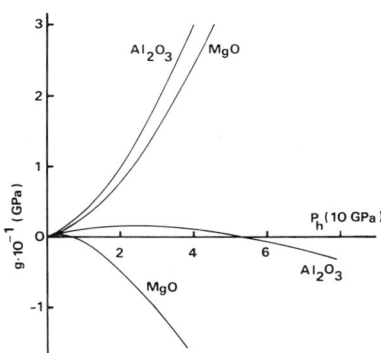

Fig. 1 The excess free energy of the Hugoniot state (the upper curves) and of the decompressed state.

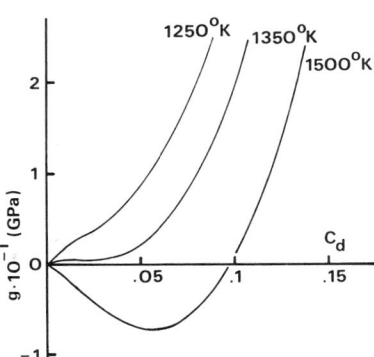

Fig. 2 The excess free energy of MgO at 10 GPa.

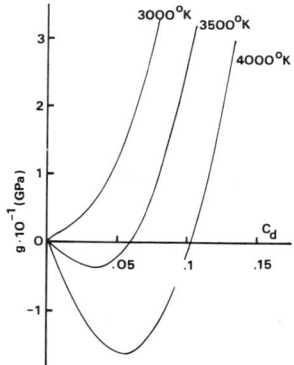

Fig. 3 The excess free energy of Al_2O_3 at 10 GPa.

Table 1 Properties of MgO and Al_2O_3*

	MgO	Al_2O_3
ρ_o (Kg × $10^3/m^3$)	3.58	4.0
C (m × 10^{-2}/μsec)	0.68	0.86
S	1.25	1.1
γ	1.4	1.9
β (100 GPa)	1.65	2.3
μ (100 GPa)	1.17	1.24
C_v at room temp. (J/g °k)	1.1	1.1
Nk (J/g °k)	0.415	0.166

* The source of the data are (1) R. G. McQueen, J. N. Fritz, and S. P. Marsh, J. Geophys. Res. 69, 2947 (1964); (2) V. N. Zharkov and V. K. Kalinin, Equations of State for Solids at High Pressures and Temperatures (Consultants Bureau, New York, 1971).

SHOCK-INDUCED PHASE TRANSITION IN GaP

Tsuneaki Goto and Yasuhiko Syono
The Research Institute for Iron, Steel and Other Metals,
Tohoku University, Sendai 980, Japan

ABSTRACT

Shock compression and adiabatic release experiments have been performed on single crystal GaP oriented parallel to the (221) plane up to about 140 GPa using a two-stage light-gas gun. Inclined mirror experiments reveal the three-wave structure, yielding the Hugoniot elastic limit and apparent phase transition pressure to be 7.2 and 22.9 GPa, respectively. The relative volume discontinuity due to the phase transition is almost the same as those of Si, Ge and Sn. The release adiabat originated from the shock-induced high-pressure phase is determined by means of the buffer technique. The volume at 0 GPa on the release adiabat is smaller by about 20 % than that of the initial low pressure phase, indicating that the high pressure phase is almost retained within a submicrosecond time scale after pressure is released. A double-shocked state is measured at about 140 GPa. Gruneisen parameter of the high pressure phase estimated from this result is about four times greater than that of the normal pressure phase. Resistivity measurements under shock loading indicate that the shock-induced high-pressure phase is metallic.

INTRODUCTION

A semiconductor-metal transition in a typical III-V compound GaP was first observed under static loading by Onodera et al.[1] There after, many investigators examined the transition pressure using various types of high-pressure apparatus to adopt as a fixed-point of the pressure standard above 10 GPa, and indicated that the transition pressure was assigned at 23 ± 1 GPa.[2-6] This value is very close to the calculated transition pressure by Van Vechten[7] on the basis of the quantum dielectric theory of electronegativity.

Preliminary shock experiments of GaP was also carried out by means of the argon flash gap method.[8,9] A contraction of volume due to the phase transition was observed at around 26 GPa. However, the transition point was not determined explicitly because this technique could not detect a multiple shock structure produced by the shock-induced phase transition in the sample.

In the present study we have redetermined precise Hugoniot states of GaP to about 90 GPa by means of the inclined mirror method with a two-stage light-gas gun. Adiabatic release- and twofold-compression experiments have also been performed to clarify the nature of the transition and high-pressure phase. Moreover, the electrical resistivity has been measured under shock loading to make sure whether the shock-induced high-pressure phase is metallic or not.

EXPERIMENTAL

High-purity single crystal discs of GaP, 2 to 3 mm in thickness, used in the experiments are oriented with their faces parallel to the (221) crystal plane. The bulk density determined by the Archimedean method is 4.125 g/cm^3, which agrees well with the x-ray density of 4.129 g/cm^3. Platelet specimens for Hugoniot measurements, typically 10 × 14 mm^2 in lateral extent, were cut and carefully polished on both surfaces to a flatness and parallelism of ±2 μm.

A two-stage light-gas gun (2SG-TH1)[10] was used to produce plane shock waves. The plastic projectile-bearing flyer plate, 20 mm in diameter, was accelerated in the velocity range of 1.6 to 4.1 km/s. Driving pressures in the samples were controlled by changing the impact velocity and the material of flyer and target. The impact velocity was precisely measured with the magnetoflyer method[11] to within ±0.2 %.

The inclined mirror method was adopted to detect the multiple shock wave produced by the dynamic yielding or phase transition. A sample assembly for this method is shown in Fig. 1. The setting angle α of the inclined mirror was 1 to 3°, measured to an accuracy of ±1 %. The light reflected from the mirrors was recorded with a continuous access streak camera with writing rate of about 10 mm/μs.[12]

The shock state of the Hugoniot elastic limit or phase transition point was analyzed on the basis of the free-surface approximation. The final shock state was determined by the impedance match solution centered at the preceding shock state using the measured impact velocity, shock parameters of the flyer and target material, and shock velocity of the final state.

Partially released states centered at the Hugoniot states in the high-pressure phase were determined with the buffer technique.[12] Glass, fused silica, or PMMA was used for buffer material. The released state at zero pressure was estimated from the final free-surface velocity measured by the inclined mirror method. The twofold compression state was realized by the shock reflection technique,[12] and determined by measuring the shock velocity in a standard material (copper) with higher shock impedance, placed on the sample.

The electrical resistivity under shock compression was observed with the charged-capacitor circuit.

RESULTS AND DISCUSSION

The inclined mirror run of driving pressure between 33 GPa and 42 GPa caught successfully the three successive shock waves produced by the elastic-plastic transition and phase change in the sample: elastic precursor and plastic I and II waves, as shown in Fig. 1. The observed values of shock (U_s) and particle velocities (u_p) are shown in Fig. 2, together with ultrasonic data[14] of longitudinal and bulk sound velocities. The velocity of the elastic shock wave is consistent with the longitudinal sound velocity. Almost linear relations hold between the shock and particle velocities for the

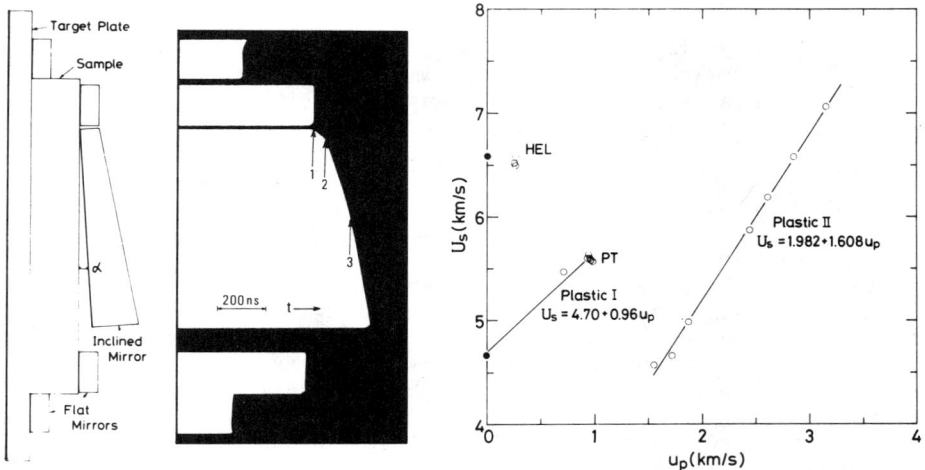

Fig. 1　Sample assembly and streak photograph for the inclined mirror method. Arrows 1, 2 and 3 indicate arrival times of three successive shock wave on the free-surface of the sample

Fig. 2　$U_s - u_p$ Hugoniot for GaP

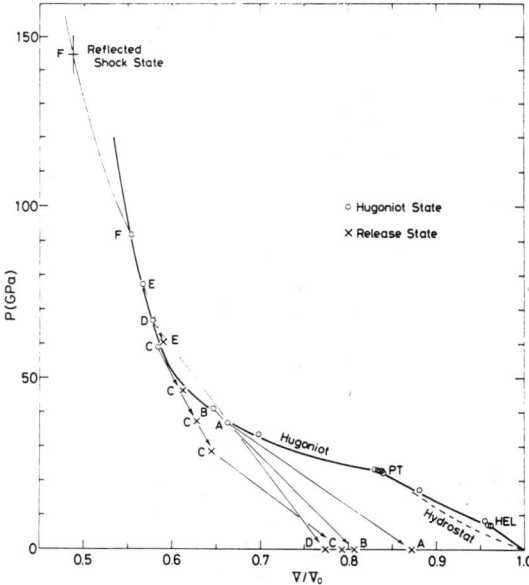

Fig. 3　Shock compression curve of GaP. Adiabatic release states and reflected shock state are shown

plastic I and II waves. For the plastic I wave, shock data including the bulk sound velocity are well described by the linear least squares fit

$$U_S = 4.70 + 0.96 \, u_p \quad (r = 0.988) \, , \quad (1)$$

where r is the correlation coefficient. For the plastic II waves, a linear $U_S - u_p$ trend is represented by

$$U_S = 1.982 + 1.608 \, u_p \quad (r = 0.998) \, , \quad (2)$$

Figure 3 shows the shock compression curve of GaP computed from the measured shock and particle velocities. The observed elastic limit (HEL) lies within the range 6.7 to 8.4 GPa, which is comparable to those of Si[15] and Ge.[16] The shock compression curve of the low-pressure phase approaches rapidly to the hydrostatic compression curve estimated from the ultrasonic data (K_S = 89.8 GPa[14], K_S' = 4.76[17]) with increasing pressure above HEL, suggesting that shear stresses in the sample are substantially lost above HEL. The phase transition point of GaP under shock loading is determined to be P = 22.9 ± 0.4 GPa and V/V_0 = 0.836 ± 0.004. The transition pressure is in good agreement with that under static loading.[2-6] The relative volume discontinuity is estimated to be about 20 %, which is almost the same as those of Si, Ge and Sn.[18] This implies that the high-pressure form of GaP is of the β-Sn structure.

Data of the release- and reflected-shock states are plotted in the P-V plane (Fig. 3). The final volume upon release to zero pressure from the Hugoniot state in the high-pressure phase is smaller by about 20 % than the zero pressure volume of the original low-pressure phase. This value is very close to the volume change due to the shock-induced phase transition. The final volume released from the Hugoniot state in the mixed phase region is also smaller than the original zero-pressure volume. These results suggest that the shock-induced high-pressure phase is retained at least within a submicrosecond time scale after the pressure is released to zero.

The state compressed twofold by direct and reflected shocks is different from the Hugoniot state at the same volume, because the increase in internal energy due to the twofold compression is smaller than that due to the direct compression. Both states are connected by the equation

$$P_2 = \frac{P_H - (\gamma_2/V_2)[(P_H - P_1)(V_0 - V_2)/2]}{1 - (\gamma_2/V_2)(V_1 - V_2)/2} \quad (3)$$

with the Grüneisen parameter γ, where the subscript 2 refers to the reflected shock state, and P_H is the pressure on the original Hugoniot at the volume V_2. The point P_1, V_1 is the centering point for the reflected shock state on the Hugoniot. The locus of reflected shock states centered at the state F on the Hugoniot in Fig. 3 is calculated from equation 3 with γ/V = 12.5 g/cm³ on the assumption that the Hugoniot state in the high-pressure phase is represented by the linear relation 2 even at higher pressures. The observed reflected shock state is just on this locus. The value of γ/V estimated in the high-pressure phase is about 4 times greater than that of the low-pressure phase evaluated at zero pressure ($\gamma/V = 3.31$ g/cm^3). This compares favorably with the case of Sn, where the value of γ/V increases from 3.9 to 16.3 g/cm^3 by the α-β phase transition.

The sample used in the resistivity measurement is n-type semiconductor with a resistivity of 43 Ωcm. Electrical measurements under shock loading indicate that the resistivity decreases to 9.4 Ωcm at 41 GPa in the mixed phase region and to less than 0.1 Ωcm at 61.5 GPa in the high-pressure phase. This suggests that the shock-induced high-pressure phase is also metallic.

ACKNOWLEDGEMENT

The authors wish to thank Professor Y. Nakagawa for warm encouragements and continuous support. They are grateful Messrs. J. Sato and H. Moriya for technical assistance.

REFERENCES

1. A. Onodera, K. Kawai, K. Ishizaki and I. L. Spain, Solid State Commun. 14, 803 (1974).
2. G. J. Piermarini and S. Block, Rev. Sci. Instrum. 46, 973 (1975).
3. C. G. Homan, D. P. Kendall, T. E. Davidson and J. Frankel, Solid State Commun. 17, 831 (1975).
4. F. P. Bundy, Rev. Sci. Instrum. 46, 1318 (1975).
5. S. C. Yu, I. L. Spain and E. F. Skelton, Solid State Commun. 25, 49 (1978).
6. A. Onodera and A. Ohtani, J. Appl. Phys. 51, 2581 (1980).
7. J. A. Van Vechten, Phys. Rev. B7, 1479 (1973).
8. T. Goto, Y. Syono, J. Nakai and Y. Nakagawa, Solid State Commun. 18, 1607 (1976).
9. Y. Syono, T. Goto and Y. Nakagawa, in High-Pressure Research: Applications in Geophysics, edited by M. H. Manghnani and A. Akimoto (Academic, New York, 1977), p. 463.
10. Y. Syono and T. Goto, Sci. Rep. RITU, A29, 17 (1980).
11. K. Kondo, A. Sawaoka and S. Saito, Rev. Sci. Instrum. 48, 1581 (1971).
12. T. Goto and Y. Syono, Sci. Rep. RITU, A29, 32 (1980).
13. T. J. Ahrens, J. Appl. Phys. 37, 2532 (1966).
14. W. F. Boyle and R. J. Sladek, Phys. Rev. B11, 2933 (1975).
15. W. H. Gust and E. B. Royce, J. Appl. Phys. 42, 1897 (1971).
16. W. H. Gust and E. B. Royce, ibid 43, 4437 (1972).
17. Y. K. Yogurtcu, A. E. Abey, A. J. Miller and G. A. Saunders, in High Pressure Science and Technology, edited by B. Vodar and Ph. Marteau (Pergamon, Paris, 1980), Vol. 1, p. 302.
18. J. C. Jamieson, Science 139, 762 (1963).

SHOCK COMPRESSION AND PHASE TRANSFORMATION
OF AlN and BP

K. Kondo, A. Sawaoka, K. Sato and M. Ando
Research Laboratory of Engineering Materials
Tokyo Institute of Technology, Midori, Yokohama 227, Japan

ABSTRACT

High density samples of AlN and BP were prepared for shock wave experiments by a high-pressure sintering method. The shock velocity of AlN decreases linearly with increasing particle velocity up to 0.73 km/sec; at this point, a drop, followed by a linear shock-velocity increase occurs. The Hugoniot compression curve shows a phase transformation at 21 ± 1 GPa, accompanying a volume change larger than 20%. The structural transition from wurzite to sodium chloride is inferred by analogy. Although the BP Hugoniot is considerably ambiguous, a kind of phase transition at around 40 GPa is suggested.

INTRODUCTION

Various III-V and II-VI compounds undergo semiconductor-to-metal phase transitions at elevated pressures. These transition pressures have been determined by many researchers for their possible application as fixed points at high pressure[1]. Van Vechten has developed the theoretical predictions for many kinds of such materials by analogy of IV group transitions[2]. Recent developments of various high pressure techniques have produced many data on such phase transitions, most of which are consistent with his predictions[1]. However, some kinds of materials show a different type of transition, for example, InSb[3]. It is well known that several II-VI compounds transform into NaCl structure at high pressure as compared to β-Sn or its diatomic equivalent in case of IV or III-V compounds[1,4,5]. In any case, it is possible to suppose that all materials with open structure transform into a denser phase. Aluminium nitride (AlN) and boron phosphide have wurzite and sphalerite structures, respectively. A phase transition for both materials has not been found yet.

SAMPLE PREPARATION

A high-pressure sintering technique is convenient to prepare these materials, which are very hard and unstable at high temperature. A static high pressure apparatus was a hexahedral anvil type with 16 x 16 mm² anvils. The cross section of sample assembly used is shown in Fig. 1. AlN powder of 99.9% purity and 2 μm particle size was shaped into a pellet by cold compression at 0.5 GPa, put into the sample assembly, and then kept at 6.5 GPa and 1500°C for 15 minutes. Apparent density of the sintered body of AlN was 3.26 Mg/m³, which agreed with the theoretical X-ray density within an experimental error. Vickers hardness was about 1700 kg/mm^2, which was higher than any other published values of AlN sintered body.

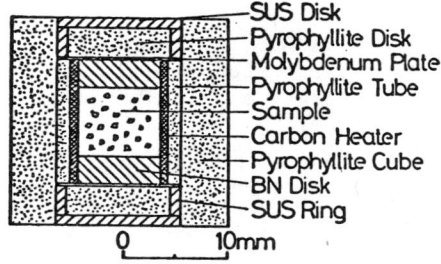

Although the sintered body of BP was prepared by the same way, the apparent density of it was 97-99% of the theoretical density of 2.97 Mg/m^3. Vickers hardness was 1500-2000 kg/mm^2. The X-ray diffraction

Fig. 1. Cross section of sample assembly for high-pressure sintering.

patterns for both materials were the same as the starting powders, respectively. The sintered materials were about 7.5 mm in diameter and 6 mm in thickness. These were cut into two pellets of 2-2.5 mm thickness and both flat surfaces were polished in parallel.

SHOCK EXPERIMENTS

Shock experiments were performed using a two-stage light-gas gun (HS-2 and HS-3B, Tokyo Institute of Technology [6,7]. Three kinds of experiment were carried out for AlN; (1) shock wave velocity measurement using a self-shorting pin, (2) particle and shock wave velocities by electromagnetic induction gauges[8], and (3) electrical resistivity measurement by a simple longitudinal configuration[9]. For BP, experiments (1) and (3) were carried out.

The particle-velocity gauge (Up-gauge), consisted of 10 μm thick copper foil backed by polyimide film, 12.5 μm thick, of which the active portion for the electromotive force was 4 x 0.5 mm^2. Since the sample was smaller than the previous particle-velocity experiments[8], the front gauge was inserted between a sapphire driver plate and the sample. The smaller gauge, of which the active portion was 3 x 0.5 mm^2, was directly assembled as the rear gauge without a gauge base. The emf output voltage of both gauges were calibrated for particle velocity in the same configuration as AlN experiment by using sapphire crystal[10]. The sapphire crystal was also used as a flyer plate.

Metallic driver and flyer plates (tungsten, copper and 2024 Al) were used for the other experiments[10]. The flyer velocity was measured by a magnetoflyer[6] or a dc X-ray technique[7]. It was assumed except for the U_p-gauge experiments that the expansion isentropes of the driver plate materials were identical with the mirror images of their shock adiabats. The impedance match method was used to determine the particle velocity in the experiments (1) and (3).

RESULTS AND DISCUSSIONS

All the results obtained for AlN are presented graphically in Fig. 2, using shock velocity (U_s)-particle velocity (U_p) coordinates. The negative slope of U_s with U_p is due to the ramp wave precursor of the U_p-gauge results. This precursor wave continued to a plateau at particle velocity of 0.65 to 0.75 mm/µs. The lowest edge of the ramp wave corresponded to the velocity of 10.13 to 10.93 mm/µs, which was in good agreement with the longitudinal sound velocity of AlN, 10.0∼10.4 mm/µs. The shock wave arrived at the rear gauge with a considerably slow velocity after the plateau.

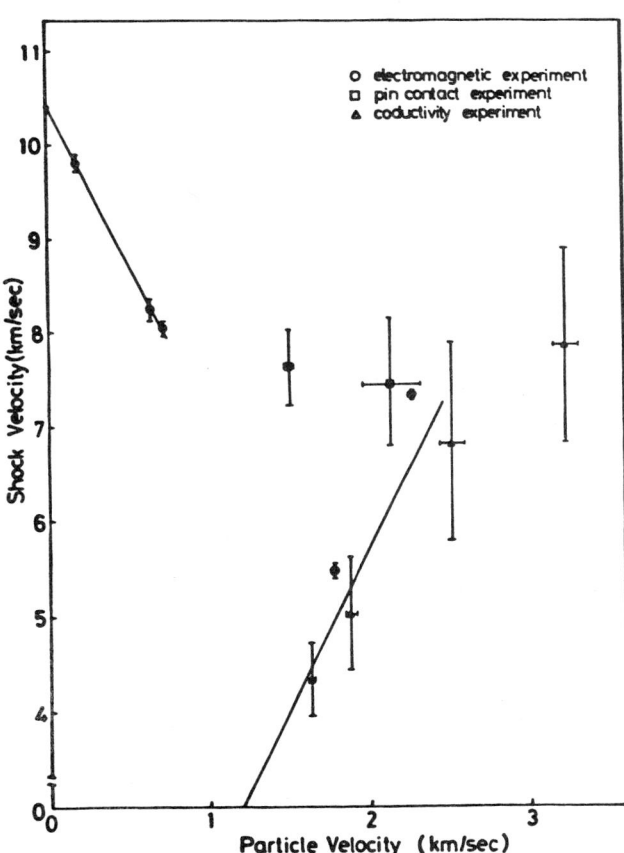

Fig. 2. Shock velocity-particle velocity Hugoniot of AlN.

Since the self-shorting pins results are constant with particle velocity, they seem to be activated at the nearly maximum pressure of the ramp wave. On the other hand, the shock velocities estimated by the resistivity measurements correspond to the final shock state, which is consistent with the U_p-gauge experiments. It is clearly seen that the value of U_s extrapolated to zero particle-velocity of AlN is not consistent with the bulk sound velocity. This Hugoniot, therefore, suggests that a kind of phase transition occurs at pressure about 20 GPa, and that the volume difference between low pressure and high pressure phases is approximately 22%. Since the resistivity of

AlN at the highest pressure measured is 2 Ω·m, the high pressure phase is not metallic.

It is known that wurtzite (ZnS) transforms to sphalerite below 4.5 GPa[11], and that sphalerite undergoes a transition at about 24 GPa[12]. These transitions are very sluggish and strongly depend on shear stress. Recent experiments[1] show that the high pressure phase is reasonably identified as sodium chloride structure, and that the transition pressure is 15.0 ± 0.5 GPa. The structural transition from wurtzite to sodium chloride is inferred for the AlN result by analogy.

The pressure derivatives of elastic constant of CdS and ZnS are positive except particular shear modes [13,14].

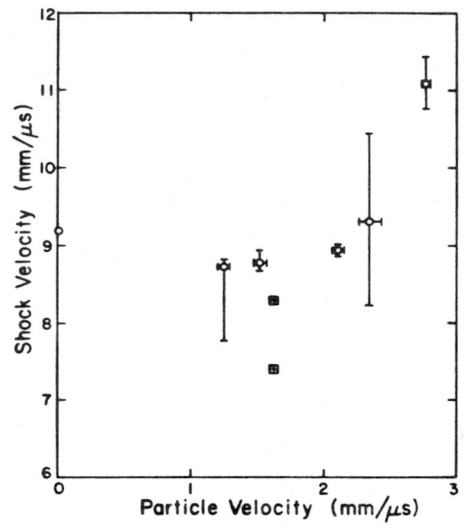

Fig. 3. Shock velocity particle velocity Hugoniot of BP. The open circle denotes shorting-pins results, and the other the polarization analysis.

Only for a few materials, such a ramp precursor wave has been observed, whose pressure derivatives of both longitudinal and shear moduli are negative with pressure [15,16]. Although the ramp precursor wave might be influenced by the HEL of the sapphire driver plate, the negative slope of U_s-U_p relationship is possibly related with the sluggish transition mentioned above. The thermodynamical stability between the sphalerite and wurzite phase must be also related with this result observed for AlN.

The U_s-U_p Hugoniot of BP is shown in Fig. 3. The shock velocities obtained by self-shorting pins are constant up to 2.1 mm/μs, which are slightly lower than the longitudinal sound velocity at ambient condition, but suddenly increase at U_p over 2.5 mm/μs. The feasible two velocities obtained from the polarization measurement are considerably slow. Since the shock velocities obtained from the electrical resistivity measurements are very ambiguous, they are not plotted in Fig. 3. However, the resistivity of the sample is at least less than 10^{-2} Ω·m at U_p over 2.1 mm/μs which is the detectable minimum value. It is likely that a kind of discontinuity exists at particle velocity around 2.1 mm/μs, which may correspond to the pressure from 30 to 45 GPa.

REFERENCES

1. S. C. Yu, I. L. Spain, and E. F. Skelton, Solid State Commum. $\underline{25}$, 49 (1978).
2. J. A. Van Vechten, Phys. Rev. $\underline{78}$, 1479 (1973).
3. M. D. Banus and M. C. Lavine, J. Appl. Phys. $\underline{40}$, 409 (1969).
4. S. Minomura and H. G. Drickamer, Phys. Chem. Solids $\underline{23}$, 451 (1962).
5. G. J. Piermarimi and S. Block, Rev. Sci. Instrum. $\underline{46}$, 973 (1975).
6. K. Kondo, A. Sawaoka and S. Saito, Rev. Sci. Instrum. $\underline{48}$, 1581 (1977).
7. H. Sugiura, K. Kondo, A. Sawaoka, Rev. Sci. Instrum. $\underline{51}$, 750 (1980).
8. K. Kondo, Y. Yasumoto, H. Sugiura, and A. Sawaoka, J. Appl. Phys. $\underline{52}$, 772 (1981).
9. K. Kondo, T. Mashimo, and A. Sawaoka, J. Geophys. Res. $\underline{85}$, 977 (1980).
10. R. G. McQueen, S. P. Marsh, J. W. Taylor, J. N. Fritz, and W. J. Carter, in High-Velocity Impact Phenomena, edited by R. Kinslow (Academic, New York, 1972), pp 293.
11. C. F. Cline and D. R. Stephens, J. Appl. Phys. $\underline{36}$, 2869 (1965).
12. G. A. Samara and H. G. Drickamer, J. Phys. Chem. Solids, $\underline{23}$, 457 (1962).
13. J. A. Corll, Phys. Rev. $\underline{157}$, 623 (1967).
14. E. Chang and G. R. Barsch, J. Phys. Chem. Solids, $\underline{34}$, 1543 (1973).
15. I. J. Fritz, J. Appl. Phys. $\underline{49}$, 4423 (1978).
16. K. Kondo, S. Iio, and A. Sawaoka, J. Appl. Phys. $\underline{52}$, 2826 (1981).

CRYSTALLOGRAPHIC PROPERTIES OF LITHIUM NIOBATE SHOCK-LOADED FROM 3.7 to 20 GPa AND PRESERVED FOR POST-SHOCK STUDY*

B. Morosin and R. A. Graham
Sandia National Laboratories, Albuquerque, New Mexico 87185[†]

ABSTRACT

Optical and X-ray diffraction observations are reported for lithium niobate crystals which have been subjected to shock loading along the c axis and recovered for post-shock study. At pressures of 3.5 and 5.5 GPa, no unusual X-ray diffraction results are observed. Optical observations on these samples show planar features similar to prior observations in other crystals. At 9.5 and 19 GPa, the X-ray observations indicate that the samples are "recrystallized" into micron size crystallites with well coordinated crystallographic alignment changes from the original crystal axis. The observed effects are thought to be initiated by localized heating accompanying heterogeneous yielding.

INTRODUCTION

It is the object of this paper to report X-ray diffraction observations on single crystal samples of lithium niobate which have been subjected to high pressure shock wave loading and preserved for post-shock study.

The present work was largely motivated by an attempt to clarify prior studies of the shock compression behavior of lithium niobate which revealed a succession of unusual compressional features. These effects are thought to be a manifestation of a heterogeneous yielding phenomena in which localized slip along selected crystallographic directions leads to temperatures of thousands of degrees which persist for a few microseconds due to the low thermal conductivity of lithium niobate.

EXPERIMENTAL

The crystals used in the present work are from the same source and prepared in the same manner as in prior work which studied compressional[1] and piezoelectric[2,3] behavior of lithium niobate under shock loading and static high pressure.[4]

Shock loading was carried out along the c axis with both impact and explosive techniques. The schedule of experiments is shown in Table I. The explosive loading experiments were carried out in a recovery fixture holder described by Kiker[5] and used for studies on metals.[6] All surfaces were lapped flat to several sodium light wave bands to assure intimate contact. The experiment numbered 35G796 was carried out with the sample precisely placed in an Invar metal

*This work sponsored by the U.S. DOE under Cont. DE-AC04-76-DP00789.
[†]A U.S. DOE facility.

ring. At the higher explosive pressure the sample, significantly larger in diameter than in the other cases, was placed in a steel ring.

The impact loading experiments were carried out on a high velocity compressed gas gun.[7] Each sample is precisely encapsulated in a copper recovery fixture developed by Graham and Browning which has provision for both lateral and longitudinal momentum traps to minimize uncontrolled releases of pressure.[8]

TABLE I Schedule of Shock-Loading Experiments for z-cut LiNbO$_3$

Experiment Number	Peak Pressure (d) GPa	Impact Velocity km/s	Sample Dimensions(e)
1S780 (a)	19	--	32mmϕx6.4mm
35G796 (b)	9.5	--	18.7mmϕx2.5mm
1716 (c)	5.5	0.378	18.7mmϕx2.5mm
1712 (c)	3.5	0.229	18.7mmϕx2.5mm

(a) Shock-loading with a 56mm diameter plane wave lens, 12.5mm thick PBX9404 explosive pad and 12.5mm thick mild steel wave shaper. This experiment was carried out by W. B. Benedick and P. L. Stanton. (b) Explosive-loading as in (a) except with 12.5mm thick baratol explosive pad. (c) Planar impact studies on a high velocity compressed gas gun. (d) For calculations of explosively driven pressure pulses produced in the copper see Davison, Webb and Graham, these proceedings. Pressures in the lithium niobate were calculated by impedance mismatch methods. (e) Dimensions shown are diameter and thickness.

CHARACTERISTICS OF SHOCKED SAMPLES

(a) 3.5 GPa Sample: Optical microscopy on this specimen was carried out on crystallites carefully separated with an index oil using a Leitz Silux Pol petrographic microscope and showed mosaic type crystallite blocks ranging in sizes from 2 to 50 μm. The larger crystallites showed optical planar features with similar densities as reported by Muller and Hornemann on shock-loaded olivine.[9] Precession photography (MoKα radiation) on various pieces, chosen with well presented surfaces so that the original c axis orientation is known, show that these mosaic blocks are mostly well aligned with respect to the original axes. The occasional misorientation observed might have resulted from specimen mounting and appears not to be structure related. Powder diffraction patterns (CuKα using a 114.55 mm Norelco camera) show sharp high two-theta lines with values identical to unshocked material from the same boule.

(b) 5.5 GPa Sample: Similar large variation in crystallite sizes as observed in the 3.5 GPa sample were found. A higher density of optical planar features was also observed, though optically clear tiny crystallites were clearly in abundance. Precession photography also revealed reasonably well aligned mosaic blocks. The spread in diffraction spots suggest an 8-9 degree misorientation. Sharp high two-theta diffraction lines with values identical to unshocked material are observed.

(c) 9.5 GPa Sample: Optical examination revealed a rather surprising observation on this shocked specimen: that of rather uniform size for the tiny crystallites. These were consistently 2-3μm in size, with very few smaller or larger in size. Under crossed polaroids, each crystallite shows sharp extinction. There were no optically observed planar features or faults. No evidence for any glassy or amorphous particles was found. X-ray powder diffraction showed sharp high two-theta lines with d-values corresponding to the unshocked material. Precession photographs also revealed an unusual, surprising and interesting result. Both 110 and 104 Bragg peaks are found aligned along the same direction, obviously from different crystallites. Examination of the mounted specimen along other directions revealed that the 006 and 202 were also coincident. Yet the spread for misalignment is rather small (~ 9 degrees) and necessary when different mosaic or crystallite blocks are made coincident with the above mentioned directions.

(d) 19 GPa Sample: The results for this specimen are essentially identical to that for the 9.5 GPa sample. The small differences concerned the precession photography which, in addition to the unusual recrystallization showed a slightly wider spot spread. This was attributed to our specimen mounting procedure. This sample is shock-loaded above the suspected 13.9 GPa transition. No evidence was found for a dense phase.

DISCUSSION

The X-ray observations on recovered, shock-loaded $LiNbO_3$ are not those typically reported in the literature for an open structure type like quartz or felspar in which both glassy-like material and altered lattice constants are found.[10] The present results are more like those for a dense and compact structure type. For example, single crystals of andalusite, $AlSi_2O_5$, subjected to 40 GPa shock compression results in a mosaic type of crystallite blocks (internally fragmented domains) of μm size and it appears that such fragmentation is highly structure controlled.[11,12] Such mosaic blocks remain together with the crystal axes in an orientation which is not far from those for the original crystal. Similar results are found at 33 GPa in ilmenite[13] shocked along the c axis, though in this case sufficient anisotropy due to the structure yields a relative rotation of the mosaic blocks which is smaller within the c plane than that containing c as well as some a component. The ilmenite results are very similar to those observed in our 3.5 and 5.5 GPa samples; a somewhat better alignment is observed for our samples on directions not confined to the c plane.

Optical planar features observed in larger fragments in our experiments are probably regions of localized deformation as expected from heterogeneous yielding. But the concentration level is sufficiently low that no effect is seen on the high two-theta X-ray diffraction lines. Most other shock-loaded materials examined in our laboratory (see other papers in this proceedings) show evidence for lattice strain in the X-ray patterns.

The 9.5 and 19 GPa samples yielded similar, though unusual results and, hence, will be discussed as if they are identical samples. These unusual observations include the rather uniform 2-3 μm crystallite size, optically clear crystallites with no glassy component, sharp high two-theta diffraction lines with identical d-values as unshocked material, and well-aligned, differently oriented crystallites within small samples selected for precession photography. Since the X-ray beam is of the order of 1mm in size (diameter) and the mounted specimen varied by less than a factor of 3 smaller, the number of crystallites contained within such a mounted specimen is near 10^8. To find only a few orientations well aligned is rather unusual. Further, the d values of the pairs of coincident directions are within 10% of each other; i.e., d_{110} and d_{104} have d-values of 2.58 and 2.74, respectively while those for d_{006} and d_{202} are 2.31 and 2.12 respectively. The lack of variation of intensity corresponding to these different crystallite orientations when different mounted specimens were examined suggests a uniform distribution throughout rather than an accidental "twinning" or sample selection.

An examination of the recently published precession photographs on ilmenite[13] subjected to 55 GPa appear to show similar aligned crystallites, particularly the 110 and 104 peaks, though the authors did not discuss this fact. Other orientations reported by these authors appeared to show a much more pronounced spot spread than that found in the present study.

The crystal structures of ilmenite and lithium niobate, as well as Al_2O_3, consist of a hexagonal close packing of oxide ions in which the smaller metallic ions occupy some of the interstices in slightly different, ordered arrangements. Mossbauer spectrum of shock-loaded ilmenite was shown to compare favorably with that for disordered ilmenite, synthesized at static high pressure.[13] This is primarily evidenced by a broadening of the line profile which is a function of the peak loading pressure. Even at the highest loading the shocked material did not attain complete disorder. Our present diffraction intensities on $LiNbO_3$ do not show variations required for such cation disorder; however, our peak-loading pressures are significantly lower than those employed in the ilmenite study.

We can currently only speculate how the final state was attained in our 9.5 and 19 GPa $LiNbO_3$ specimens. It is clear that the shock-loading has placed the crystals in a highly unusual state. A homogeneous recrystallization from the melt is not a credible mechanism since a temperature calculation based on benign shock concepts gives an increase in temperature of only 90K at 13 GPa. The melt temperature at atmospheric pressure is about 1525K. A rapid quenching process with passage of the shock wave is a credible mechanism for the growth of small crystallites of uniform size. On the other hand, these crystallites need to be ordered and aligned with respect to each other. We have carefully examined the atomic arrangement in the crystal structure to attempt to postulate a mechanism for a solid state transformation. An extremely orderly shearing with a sizeable atomic motion is necessary. We can only speculate that perhaps the approximately 10% differences in interplanar spacings mentioned above became small at these shock pressures in $LiNbO_3$ and that in the shock-loaded environment sufficient cation motion can occur. Such motion together with a "recrystallization-like" transformation might sweep defects to the crystallite boundaries, removing evidence for strain-broadening and optical planar features.

Even though full description of the process leading to the observed formation of reoriented crystallites cannot be determined from information available, it appears necessary to invoke heterogeneous yielding concepts to obtain conditions necessary for recrystallization. Within the local regions subject to slip, temperatures are expected to be thousands of degrees,[14] and the heated, likely melted, regions would quench rapidly. Nevertheless, the recrystallization is observed in virtually the entire sample volume and it is not clear how the slipped regions can reach such a high density. The influence of compressibility change, specific heat change, volumetric distortion and latent heat due to ferroelectric to paraelectric transition and its inversion, as well as electrostrictive effects are all complicating factors which could possibly lead to a different effect than in electrically inactive materials.

REFERENCES

1. P. L. Stanton and R. A. Graham, J. Appl. Phys. 50, 6892 (1979).
2. P. L. Stanton and R. A. Graham, Appl. Phys. Lett. 31, 723 (1977).
3. R. A. Graham, J. Appl. Phys. 48, 2153 (1977).
4. R. A. Graham, Ferroelectrics, 10, 65 (1976).
5. J. L. Kiker, J. Sci. Instrum. 43, 269 (1966).
6. A. R. Champion and R. W. Rohde, J. Appl. Phys. 41, 2213 (1970).
7. S. Thunborg, G. E. Ingram and R. A. Graham, Rev. Sci. Instr. 35, 11 (1964).
8. R. A. Graham and J. R. Browning, to be published.
9. W. F. Muller and U. Hornemann, Earth and Planetary Science Letters, 19, 251 (1969).
10. D. Stoffer, Fortsehr. Mineral 49, 50 (1972).
11. H. Schneider, Phys. Chem. Mineral 4, 245 (1979).
12. H. Schnieder and U. Hornemann, Phys. Chem. Minerals 1, 257 (1977).
13. Y. Syono, H. Takei, T. Goto & A. Ito, Phys. Chem. Minerals 7, 82 (1981).
14. D. E. Grady, in *High-Pressure: Applications in Geophysics*, edited by M. H. Manghnani and S. Akimoto, Academic, New York.

SHOCK-INDUCED DEFECTS IN CADMIUM SULFIDE*

B. W. Dodson and E. L. Venturini
Sandia National Laboratories,[†] Albuquerque, New Mexico 87185

ABSTRACT

We report on the explosive shock loading of CdS powders pressed to 50% of theoretical density in a copper recovery capsule, to pressures in the copper of about 3, 13, and 20 GPa, and preservation of the samples for post-shock study. No residual structural changes were detected via X-ray diffraction, but large densities ($\sim 10^{16}/cm^3$) of two types of defects were detected using ESR. One defect had an isotropic spectrum, with g = 2.0028, and the other had an axial spectrum with $g_\parallel$ = 1.790 and $g_\perp$ = 1.775. The axial defect seems to be connected with the wurtzite-rocksalt phase transition in CdS.

INTRODUCTION

Our work on shock loading and recovery of CdS was motivated by publications on Homan, et al[1] and Brown, et al[2] reporting magnetic flux exclusion at 77°K, with the suggestion of high temperature superconductivity, in samples of CdS powder which had been "pressure-quenched". The pressure-quenching process involves static loading of the powder to pressures about 4 GPa, and then releasing the pressure at a rate of 10^{10} Pa/sec. Such samples exhibited a large diamagnetic effect, in some cases amounting to 25% flux exclusion.[2]

The pressure quenched samples which exhibit the anomalous magnetic effects changed structure from the original wurtzite (hexagonal) structure to a mixture of two cubic phases (sphalerite and rock salt). Under hydrostatic pressure, single crystal CdS undergoes a transformation from the wurtzite structure to that of rocksalt with a volume reduction of $\sim$ 21% at a pressure of 2.3 $\pm$ 0.1 GPa.[3] Upon release of pressure, the recovered material generally is reported to contain a mixture of the wurtzite and sphalerite structures, or just the wurtzitic phase.[4] The structural effects of rapid pressure-quenching are apparently not equivalent to those of quasi-static pressure cycles.

In connection with a broad, exploratory effort investigating the chemical effects of shock loading in condensed media, we attempted to determine if shock loading and recovery of CdS powder would produce changes similar to those caused by the pressure-quenching treatment. There are, of course, significant differences. There is considerable evidence that application of hydrostatic pressure is not equivalent to shock loading.[5] Furthermore, the loading and unloading rates in the shock experiment will be 5 to 10 orders of magnitude faster than in the pressure-quenching treatment.

In this paper, we will describe the shock recovery technique and characterization of the recovered samples, with special emphasis

---
*This work sponsored by the U.S. DOE under Cont. DE-AC04-76-DP00789.
†A U.S. DOE facility.

on the nature of shock-induced defects as determined using ESR measurements. In the conclusion we shall discuss the relation between the experimental data and the "catastrophic shock concept".

EXPERIMENTAL CONFIGURATION

An explosive recovery system whose shock compression properties are described in detail by Davison, et al[6] was used in this work. CdS powder[7] was pressed into a copper capsule to 50% of the crystalline density. The final dimensions of the sample were 19 mm $\emptyset$ x 2.5 mm thick. The sample was contained in a series of copper plugs and plates designed so that deformation would give a tightly sealed capsule. All surfaces were optically flat and accurately parallel to give reproducible and calculable shock conditions.

The recovery capsule was shocked with an explosive assembly consisting of an explosive plane-wave generator with 56 mm diameter which initiates an explosive pad with dimensions 56 mm $\emptyset$ x 12.5 mm thick. The explosive pads used were Baratol and Comp-B. The loading pulse was shaped by transmission through a mild steel (SAE 1018) disk. The resulting pressure pulses have durations on the order of microseconds and a peak pressure of 13 GPa (Baratol) and 20 GPa (Comp-B).[6] In two experiments this wave was propagated through 5 mm of copper and then into the sample. In another experiment the wave passed through 5 mm of PMMA, and then into the copper capsule, giving a stress of about 3 GPa in the copper.

The lack of information concerning the pressure-volume relation for pressed CdS powder makes the determination of pressures in the sample difficult. However, estimates based on the P-V relation for low-density sintered aluminum[8] indicate that 3 GPa in the copper capsule corresponds to about 0.6 GPa in the CdS sample, well below the pressure of the phase transition. Similar estimates for 13 GPa in copper indicate a pressure of about 3 GPa in the sample; thus, this sample is shocked in the vicinity of the transition. The sample shocked to 20 GPa in the copper will probably still be in the mixed-phase region of the transition because of the large volume reduction associated with the wurtzite to rocksalt transition.[9].

SAMPLE CHARACTERIZATION

X-ray diffraction measurements by B. Morosin of our laboratories show no trace of either sphalerite or rocksalt phase in the recovered samples. Accordingly, the recovered material is different from that subjected to pressure quenching.

Although measurements on the samples by J. E. Schirber of our laboratory failed to indicate any magnetic anomalies even in samples cooled in liquid nitrogen within a few minutes of shock loading, modified optical properties indicate significant changes in the material. The initial powder is a bright yellow, whereas the recovered sample is a gray-orange. Such substantial color changes with no corresponding permanent structural change imply the formation of significant densities of defects. FTIR spectroscopy performed by G. W. Arnold of our laboratories indicated a qualitative similarity, both in absorption coefficients and in spectral shape, between the 13 GPa (copper) sample and a radiation-damaged CdS single crystal.

ESR SPECTRA OF PARAMAGNETIC DEFECTS IN SHOCK-LOADED CdS POWDER

The recovered CdS powder samples were examined using electron spin resonance (ESR). The ESR spectrum for a given material provides a quantitative measure of the type and concentration of paramagnetic defects. By comparing the ESR spectrum of the unshocked CdS powder with that for the shocked samples, one can distinguish magnetic centers produced by the shock loading. We found no paramagnetic defects in the CdS powder in sufficient concentration to give a detectable ESR signal prior to shock compression.

In Figure 1 we show the ESR spectra for the recovered samples. The spectra are derivative ESR absorption versus applied magnetic field, recorded at 3K and 9.8 GHz. There are two prominent signals:

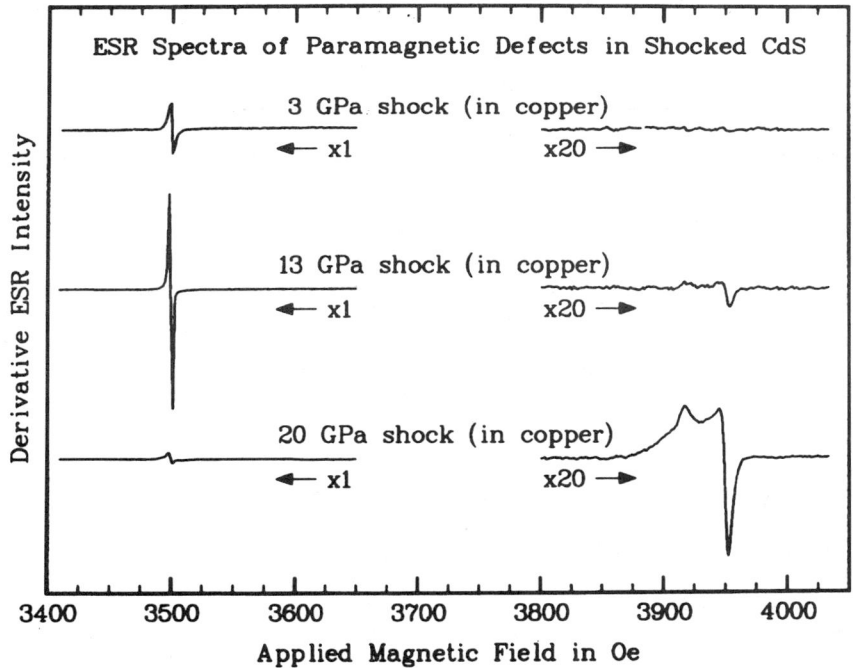

Figure 1. Comparison of ESR spectra for CdS powder shock-loaded to three different pressures. An isotropic defect appears on the left, and an axial defect on the right. All spectra were acquired at 3K and 9.8 GHz.

One near 3500 Oe has an isotropic g-factor of 2.0028(3), and the other near 3925 Oe has an axial g-factor with $g_{\parallel}$ = 1.790(2) and $g_{\perp}$ = 1.775(1). The paramagnetic defect concentrations have been determined by numerical analysis of the spectra in Fig. 1, using a known amount of diphenylpicryl hydrazyl as a reference. The isotropic defect concentration is approximately $2 \times 10^{16}/cm^3$ for samples from the 3 and 13 GPa experiments, and $7 \times 10^{15}/cm^3$ for the 20 GPa experiment. The axial defect is below the noise level for the 3 GPa sample, and its concentration is approximately $7 \times 10^{15}/cm^3$ for the 13 GPa sample and $7 \times 10^{16}/cm^3$ for the 20 GPa sample.

Although there are no reports of ESR spectra for shock-loaded CdS, there have been several studies of CdS. The isotropic ESR signal has been observed in CdS heated in vacuum above 450°C, and was attributed to a sulfur vacancy or F-center[10], possibly at the surface[11]. The axial defect has also been reported[12,13,14] as due to a singly charged sulfur vacancy or a donor impurity such as chlorine. Since this latter defect was not observed prior to shock-loading, and since its concentration increases with increasing pressure, it is unlikely that impurities are responsible.

DISCUSSION

CdS powder samples which have been shock-loaded and recovered for post-shock analysis show the presence of large concentrations of residual defects. It appears that the presence of the axial defect is related to that portion of the sample which is cycled through the phase transition. Hence the 3 GPa (copper) sample, which does not undergo the phase transition, shows no sign of the axial defects. However, the 20 GPa (copper) sample, which is shocked well into the mixed-phase region, shows an order of magnitude more axial defects than does the 13 GPa (copper) sample, which is near the lower boundary of the mixed-phase region.

In an attempt to shed some light on the origin of this defect, we have examined the ESR spectra of two CdS single crystals as grown, one with low initial resistivity (0.5 ohm-cm) and one with high resistivity (5×10^{10} ohm-cm). No paramagnetic centers were seen in the high-resistivity crystal, while the low resistivity sample had a concentration of $2 \times 10^{16}/cm^3$ of the same axial defect as seen in the shock loaded CdS powder. The low-resistivity crystal was hydrostatically loaded to 2.5 GPa, above the structural phase transition, and then recovered for analysis. Upon reexamination, the axial ESR defect was no longer observed. We can come to three conclusions: 1) Since the axial defect concentration was similar for a single crystal and the finely divided powders, it is probably a bulk effect; 2) the axial defect is related to the structural phase transition in some manner; and 3) since the defect disappears on quasi-static pressure cycling through the transition, it is not due to some macroscopic nonstoichiometry, but rather must be a collection of local chemical defects.

It is more difficult to discuss the origin of the isotropic defect, beyond noting that there is no apparent change in the residual defect density with shock pressure. Accordingly, this defect is not linked with the phase transition.

The production of large densities of defects does not reconcile naturally with "benign shock concepts", but is seen as the basis for enhanced solid-state reactivity in the "catastrophic shock concept" of Graham.[5] It is important to extend the current view of shock loading to include the microscopic processes which produce effects seen as anomalous under the benign shock concept. Post-recovery study of the nature and systematics of production of shock-induced defects is a step towards this goal.

ACKNOWLEDGEMENTS

In addition to those mentioned in the text, we would like to thank T. H. Geballe of Stanford University for initially suggesting the study of CdS under shock loading, B. E. Hammons of Sandia National Laboratories for assistance with the single-crystal work, and R. A. Graham of Sandia National Laboratories for much advice and encouragement.

REFERENCES

1. C. G. Homan, D. P. Kendall, and R. K. MacCrone, Solid State Commun. 32, 521 (1979).
2. E. Brown, C. G. Homan, and R. K. MacCrone, Phys. Rev. Lett. 45, 478 (1980).
3. G. A. Samara and A. A. Giardini, Phys. Rev. 140, 1 (1965).
4. R. T. Johnson and B. Morosin, High Temp-High Press. 8, 31 (1976).
5. R. A. Graham, Bull. Am. Phys. Soc. 25, 495 (1980).
6. L. Davison, D. M. Webb, and R. A. Graham, present Proceedings.
7. 3 μ luminescent grade CdS powder from the General Electric Co.
8. LASL Shock Hugoniot Data, edited by S. P. Marsh, University of California Press, Berkeley, (1980).
9. J. D. Kennedy and W. B. Benedick, J. Phys. Chem. Solids 27, 125 (1966).
10. G. D. Sootha, G. K. Padam and S. K. Gupta, Phys. Stat. Solidi (a) 52 K125 (1979).
11. T. Arizumi, T. Mizutani and K. Shimakawa, Japan. J. Appl. Phys. 8 1411 (1969).
12. R. S. Title in Physics and Chemistry of II-VI Compounds, ed. by M. Aven and J. S. Prener (North-Holland, New York, 1967), Ch. 6.
13. K. Morigaki and T. Hoshina, J. Phys. Soc. Japan 24, 120 (1968).
14. A. L. Taylor, G. Filipovich, and G. K. Lindeburg, Solid State Commun. 9, 945 (1971).

THE α-β TRANSITION IN T=0 HIGH PRESSURE BERYLLIUM*

A. K. McMahan
University of California, Lawrence Livermore National Laboratory
Livermore, California 94550

ABSTRACT

Linear-muffin-tin-orbital calculations are reported for the static lattice energy of α (hcp) and β (bcc) Be under pressure. These results indicate a T=0 α-β transition below 2 Mbar with a volume change between one and two percent. The bulk moduli of the two structures are found to be essentially identical, so that any change in the lattice Grüneisen parameter at the transition must arise from structural dependence in other combinations of the elastic constants, most likely Poisson's ratio.

INTRODUCTION

Beryllium is known to undergo an hcp-bcc (α-β) phase transition at 1523 K and ambient pressure.[1] There has been some interest recently in the question of whether or not the α-β transition also occurs at high pressure but low temperature.[2] To investigate this question, self-consistent linear-muffin-tin-orbital[3] (LMTO) calculations have been carried out for the static lattice energy of T=0 Be in the two structures, and also in the fcc structure for comparison. The results do indicate an α-β transition. However, estimates of the zero-point energy are more than twice the static lattice energy difference between the two phases. As it is the total energy, static lattice plus zero-point, which determines the location of the transition at zero temperature, a careful calculation of the lattice vibrational properties of Be will be needed to accurately determine the transition pressure. This is a difficult problem, and beyond the scope of the present work. The present work does suggest, however, that the transition will occur below about 2 Mbar, and that the volume change at the transition will be between 1 and 2 percent. Furthermore, these calculations indicate that any significant change in the lattice Grüneisen parameter at the transition, as has been predicted,[2] will arise most likely from differences in Poisson's ratio between the two phases.

---

*Work performed under the auspices of the U.S. Department of Energy by Lawrence Livermore National Laboratory under contract #W-7405-Eng-48.

COMPUTATIONAL DETAILS

The LMTO technique has been described at length elsewhere.[3,4] The present calculations used the von Barth-Hedin exchange-correlation potential,[5] and treated all electrons self-consistently, and as bands. The 1s states were sampled with relatively few points in the Brillouin zone. The conduction states were sampled with 150 (hcp) and ∿500 (bcc and fcc) points in the irreducible wedge to insure convergence of the total energy to better than 0.002 eV. The experimentally measured c/a ratio, 1.568, was taken for the hcp phase.[1] The combined correction term of Andersen[3] was included, but the muffin-tin correction of Glötzel and Andersen[6] was not. The latter affects the α-β energy differences reported here by less than 10%. The calculated pressure was found to be too low by 60 kbar at the measured equilibrium volume. Since the bulk modulus is 1.10 Mbar,[7] this represents 5.5% agreement between calculated and measured zero-pressure volumes, which is within the limits expected from local density band structure calculations.

RESULTS

The static lattice energy differences obtained from the LMTO calculations are shown in Fig. 1, relative to the energy of the hcp phase. The pressure scale at the top of the figure includes estimates of the hcp zero-point pressure, as well as a constant 60 kbar

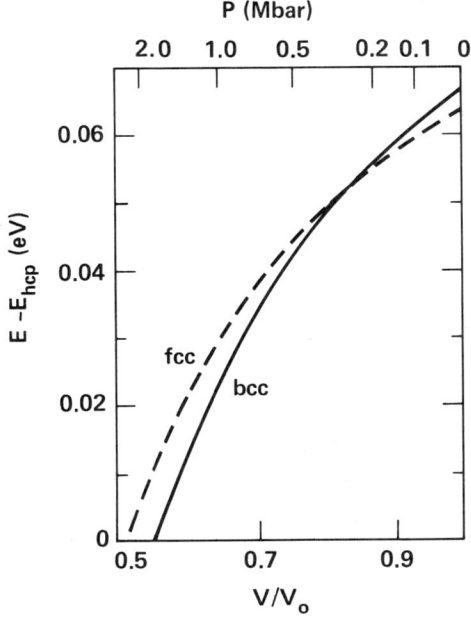

Fig. 1 Static lattice energy differences for T=0 Be vs. relative volume

shift to bring agreement with the measured zero-pressure volume, V_o. It can be seen in the figure that the bcc static lattice energy becomes lower than that of the hcp phase at about 2 Mbar. The zero pressure Debye temperature for hcp Be is known to be 1440 K,[8] so that with reasonable estimates for the Grüneisen parameter, the zero-point contribution to the energy of the hcp phase will range from 0.14 to 0.29 eV for $V/V_o = 1 - 0.5$. This is more than twice the bcc-hcp energy difference shown in the figure. The Debye temperature for hcp Be is larger than that for any other metal, with most metals having values smaller by a factor of two or more.[8] Thus it is not unreasonable to assume that the bcc phase will have a smaller Debye temperature, and that the α-β transition will therefore occur below 2 Mbar at zero temperature. According to the present results, the transition would occur at 200 kbar if the Debye temperature for the bcc phase were 1070 K, and that for the fcc phase larger than 1080 K, at this pressure. The estimated hcp Debye temperature at this pressure is 1680 K.

The volume offset in the static lattice pressure curves between the hcp and bcc results is slightly less than 1% over the whole range 0 - 5 Mbar. It is unlikely that the zero-point contribution could increase this offset by more than a percent, so that the zero temperature volume change at the α-β transition should be less than 2%.

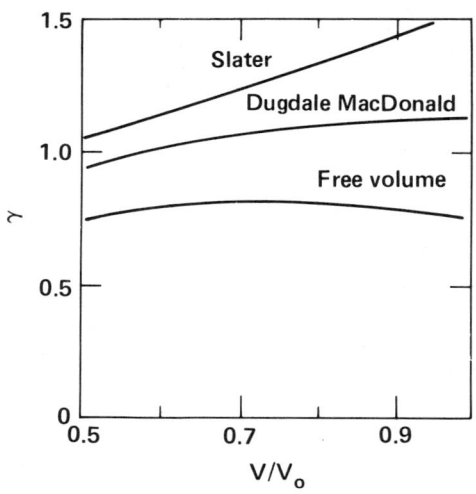

Fig. 2. Lattice Grüneisen parameter for T=0 Be vs. relative volume.

Estimates of the lattice Grüneisen parameter, γ, are shown in Fig. 2. The Slater ($\eta=2/3$), Dugdale-MacDonald ($\eta=2/3$), and free volume ($\eta=4/3$) models[9] have been used where

$$\gamma_\eta = -\frac{1}{6} - \frac{1}{2}\frac{d\ln B/d\ln V + \eta}{1 - \eta P/B} \; , \tag{1}$$

and where P and B are the calculated static lattice pressure and bulk modulus, respectively. Some insight into Eq. (1) is gained by noting that for an isotropic material, the high temperature limit of the Debye model yields[9]

$$\gamma = -\frac{1}{6} - \frac{1}{2}\frac{d\ln B}{d\ln V} + \frac{V\nu'(4-5\nu)}{3(1-\nu^2)(1-2\nu)} \tag{2}$$

where ν is Poisson's ratio, and ν' is its volume derivative. Different choices for η correspond to different assumptions about the behavior of Poisson's ratio. The important point to be made is that the bulk moduli for all three structures investigated in this work are essentially identical. Thus according to Eq. (2), any significant change in γ at the α-β transition must arise from structural dependence of Poisson's ratio. Since hcp Be is quite anisotropic, one should actually say that changes in γ must arise from structural dependence in combinations of the elastic constants other than the bulk modulus. Nevertheless, Poisson's ratio is a likely candidate. The value[7] for hcp Be is 0.02, more than an order of magnitude smaller than values near 0.3 found for most metals.[10] If bcc Be were in some sense a more normal metal, then one might indeed get a significant change in γ at the transition. It is tempting to view the spread in Fig. 2 as indicative of the possible extent of such a change in γ. Unfortunately, the Dugdale-MacDonald model seems to be in the best agreement with measurements for most metals,[11] and thus the most likely guess for the bcc phase. Since the normal density value[10] of γ for hcp Be is 1.18, and already in agreement with the Dugdale-MacDonald model, this does not suggest a dramatic change in γ at the transition.

DISCUSSION

Hcp Be is considered to have a significant covalent character to its bonding. The electron density of states shows a deep minimum at the Fermi level suggestive of splitting between bonding and anti-bonding states characteristic of covalency.[12] It is believed that this feature leads to the low superconducting transition temerature in this material.[13] The present calculations show $N(\varepsilon_F)/N(\varepsilon_F)_{fe}$ = 0.25, 0.38, and 0.54 for the hcp, fcc and bcc structures at normal density, where $N(\varepsilon_F)_{fe}$ is the density of states for a free electron gas of equal electron density. According to these numbers, the bcc structure appears to be the least covalent of the three structures. One might view the α-β transition in compressed Be as a transition from a more covalent to a less covalent state of this material. Electron density contours suggest a bond in hcp Be which is parallel to the c axis,[12] and whose natural length may well have caused the rather non-ideal c/a ratio found in this phase. As the solid is compressed, contraction of this bond may prove energetically unfavorable, thus eventually favoring the less covalent and more metallic-like bonding in the bcc phase.

ACKNOWLEDGMENTS

I am grateful to Drs. P. B. Allen and D. A. Young for conversations, and Dr. H. L. Skriver for his LMTO computer program.

REFERENCES

1. J. Donohue, The Structures of the Elements (John Wiley, New York, 1974), p. 37.
2. H. C. Graboske, R. Grover and K. Long, Bull. Am. Phys. Soc. $\underline{24}$, 724 (1979); Lawrence Livermore National Laboratory Report No. UCRL-50048-78-3 (1978, unpublished), p. 9; T. Neal in High-Pressure Science and Technology, edited by K. D. Timmerhaus and M. S. Barber (Plenum Press, New York 1979), Vol. 1, p. 80.
3. O. K. Andersen, Phys. Rev. B$\underline{12}$, 3060 (1975); O. K. Andersen and O. Jepsen, Physica $\underline{91B}$, 317 (1977).
4. The LMTO computer program used in this work is a modification of a program provided by H. L. Skriver (see, e.g., H. L. Skriver and J.-P. Jan, Phys. Rev. B$\underline{21}$, 1489 (1980)) and is described in A. K. McMahan, H. L. Skriver, and B. Johansson, Phys. Rev. B$\underline{23}$, 5016 (1981).
5. U. von Barth and L. Hedin, J. Phys. C$\underline{5}$, 1629 (1972).
6. D. Glötzel and O. K. Andersen, J. Phys. F, in press. This correction is the same as the Ewald Correction of E. Esposito, A. E. Carlsson, D. D. Ling, H. Ehrenreich and C. D. Gelatt, Jr., Phil. Mag. A$\underline{41}$, 251 (1980).
7. D. J. Silversmith and B. L. Averbach, Phys. Rev. B$\underline{1}$, 567 (1970).
8. American Institute of Physics Handbook, 3rd. ed. (McGraw-Hill, New York, 1972), Table 4e-10.
9. E. B. Royce, in Physics of High Energy Density, edited by P. Caldirola and H. Knopfel (Academic, New York, 1971), p. 84.
10. K. A. Gschneider, Jr., in Solid State Physics, edited by H. Ehrenreich, F. Seitz, and D. Turnbull (Academic Press, New York, 1964), Vol. 16, ps. 292 and 412.
11. M. H. Rice, R. G. McQueen, and J. M. Walsh, in Solid State Physics, edited by H. Ehrenreich, F. Seitz, and D. Turnbull (Academic Press, New York, 1958) Vol. 6, p. 1.
12. S. T. Inoue and J. Yamashita, J. Phys. Soc. Jpn. $\underline{35}$, 677 (1973).
13. P. B. Allen and P. C. Dynes, Phys. Rev. B$\underline{12}$, 905 (1975).

BASIC FEATURES OF DYNAMIC PHASE TRANSITION IN FINITE SAMPLES*

N. Salansky, H. Mar, D. Hawken, Y. Kleiman
3M Canada Inc.

ABSTRACT

In a system comprised of a sample "locked into" a sample holder through which a shock wave is propagating, the boundary effects due to shock impedance mismatch at the interfaces has been taken into account in determining the spatial distributions of the pressure (P) and temperature (T) in the sample. Various energy loss mechanisms were considered in evaluating the contributions to the T distribution. Dynamic diagrams in P(t), T(t) coordinates at different cross-sections of the sample were constructed based on the time (t) dependence of the distributions. Superposition of the dynamic diagrams on the static phase-transition diagram for graphite now makes it possible to determine the spatial distributions where conditions necessary for the phase transition to occur exist in the sample. The technique has been applied to the graphite-diamond dynamic phase transition[1] and results compared with experimental data.

REFERENCES

1. J. Kleiman, H. Marr, N. Salansky, CAP Congress June 1980, McMaster University.

*Supported by NRC of Canada.

ON CONSTITUTIVE MODELLING FOR THE SHOCK PHYSICIST*

W. Herrmann
Sandia National Laboratories,† Albuquerque, New Mexico 87185

ABSTRACT

Despite the fact that the theory of constitutive equations has been well developed in continuum mechanics, the literature continues to blossom with constitutive equations fabricated from many ad-hoc assumptions, some of which may be inconsistent or implicit, and are often very restricted in the types of motion to which they are applicable. Many of these constitutive equations in fact are special cases of general constitutive theories which have been studied in detail, and for which many general properties are known. The methodology of continuum constitutive theory will be reviewed, with some examples drawn from viscous, viscoelastic, viscoplastic and mixture theories which have recently been applied to describing certain features of shock wave propagation.

INTRODUCTION

Shock wave research has two major objectives: the investigation of material behavior under extreme conditions, and the solution of technological problems involving impacts, explosions, etc. The latter usually involves the former in the construction of constitutive equations for use in the numerical solution of initial/boundary value problems. The former also frequently involves the latter in interpreting experiments, so that constitutive theory lies at the heart of shock wave research.

Many current attempts at the construction of constitutive equations still begin with the assumption of a number of very specific functions relating stress, strain, temperature, etc., which might be suggested on empirical or theoretical grounds. The resultant equations are often very restricted in their applicability, for example to one-dimensional motions or infinitesimal displacements, but even so have the potential for kinematic or thermodynamic inconsistencies and instabilities. Their proper generalization to arbitrary motions is by no means trivial.

Fortunately, constitutive theory has been developed intensively over the past few decades. The kinematics of arbitrary deformations was completed during the 1950's following earlier work on finite strains. During the 1960's it was shown that the second law of thermodynamics is intrinsically a stability postulate which places certain restrictions on the constitutive relations. A number of very general constitutive theories have been developed in detail, which

---

*This work was supported by the U.S. Department of Energy under Contract DE-AC04-76-DP00789.
†A U.S. Dept of Energy facility.

satisfy kinematical and thermodynamic constraints. It is remarkable that many aspects of wave propagation behavior predicted by these constitutive theories can be worked out in detail without inserting any very specific functions. Much of the behavior observed experimentally can be described by a specialization of one or other of these constitutive theories. There is an enormous advantage in using an existing theory, if possible, since all of the previous work which has been done to establish consistency, stability, isotropy, approximations, general properties etc. then carries over directly, and need not be established anew.

There are some other considerations regarding constitutive theories which are receiving considerable attention at this time. It is evident that many materials exhibit physical instabilities which must be described correctly. For example, many materials exhibit strain softening under certain conditions. The resultant material instability does not necessarily lead to globally unbounded solutions, i.e. disaster, but may merely lead to strain localization in shear bands, slip bands or faults. Chemically reactive systems offer a host of other examples of material instabilities. The onset of instabilities must be recognized and, for many purposes, means of obtaining the proper "post-failure" solution are needed.

Finally, means of obtaining solutions to specific initial/boundary value problems are needed. In most cases this involves numerical solution methods. Since the constitutive equations frequently introduce additional terms into the partial differential equations, it does not suffice to force ad-hoc finite difference analogs of these terms into a numerical solution method designed to solve the wave equation, adding liberal amounts of artificial viscosity if necessary to prevent numerical problems. It is clearly necessary to develop a numerical method for the complete set of partial differential equations and establish its convergence properties.

It is clearly desirable to investigate the properties of the differential equations and of their numerical representation in some generality, rather than for each special case in which specific expressions are inserted for each function in the constitutive equations. This is an area of considerable interest at present, although results are scarce due to the difficulty of the mathematical theory.

In order to illustrate some of the above points, a simple and familiar example will first be discussed, that of a single phase pure fluid. This will then be generalized to the more interesting case of a multiphase chemically reactive mixture. The analogous equations for solid materials with internal state variables undergoing arbitrarily large deformations will then be discussed. Many phenomena observed in shock wave experiments can be described by these simple theories, but by no means all. Some indication will be given of other theories which are receiving continued attention for use in shock wave research.

SINGLE PHASE PURE FLUID

Consider, as a first example, the familiar constitutive equations for a single phase pure fluid. The configuration in a fluid is fully described by its specific volume υ. In almost all non-trivial problems involving shock waves, it is necessary to include thermodynamic effects. The thermodynamic state can be characterized by the absolute temperature T. The present response of the material is assumed to be a function of the present values of υ and T and their past histories. Various properties are conferred, depending on the assumed history dependence. For this example, we choose to express the entire history dependence in the present rate of change of the specific volume. This will introduce a bulk viscosity. The basic assumption then is that the material response, which might be characterized in terms of the total pressure P, the Helmholtz free energy A and entropy S, depends on υ, $\dot{\upsilon}$ and T.

$$A = \hat{A}(\upsilon,\dot{\upsilon},T) \qquad P = \hat{P}(\upsilon,\dot{\upsilon},T) \qquad S = \hat{S}(\upsilon,\dot{\upsilon},T) \qquad (1)$$

These constitutive equations are still too general. For any arbitrary motion, the first and second laws of thermodynamics may be violated by a material governed by them. Restrictions arise on (1) if such violation is to be avoided. The arguments are now so familiar that only the results will be given. It is found that the Helmholtz free energy function $\hat{A}$ must be independent of $\dot{\upsilon}$ and that the total pressure P may be expressed as a sum of an equilibrium thermodynamic pressure p and a viscous pressure q such that

$$A = \hat{A}(\upsilon,T) \qquad p = -\frac{\partial}{\partial \upsilon}\hat{A}(\upsilon,T) \qquad S = -\frac{\partial}{\partial T}\hat{A}(\upsilon,T)$$

$$q = \hat{q}(\upsilon,\dot{\upsilon},T) \qquad \delta = -\hat{q}(\upsilon,\dot{\upsilon},T)\,\dot{\upsilon} \geq 0 \qquad (2)$$

$$\hat{q}(\upsilon,0,T) = 0 \qquad \eta = -\upsilon\frac{\partial}{\partial \dot{\upsilon}}\hat{q}(\upsilon,0,T) \geq 0$$

The first three equations establish $\hat{A}$ as a thermodynamic potential function while the last three define properties of the viscosity function $\hat{q}$, i.e. $\hat{q}$ is such that the dissipation δ, representing the rate of viscous stress work, is non-negative, which further implies that the viscous pressure is zero and the viscosity coefficient η is non-negative in mechanical equilibrium defined by $\dot{\upsilon} = 0$.

There are some further restrictions on the constitutive functions which arise from stability considerations. In this simple example they are Gibb's inequalities, which guarantee stability of equilibrium states of an isolated finite body of fluid[1]

$$\frac{\partial^2 \hat{A}}{\partial \upsilon^2} > 0 \qquad \frac{\partial^2 \hat{A}}{\partial T^2} < 0 \qquad (3)$$

The first and second of these imply that the bulk modulus and specific heat at constant volume are positive. Together they provide convexity conditions on the Helmholtz free energy function $\hat{A}$ which admit the application of Legendre transformations,[2] resulting in alternate descriptions in terms of the internal energy $\hat{\mathcal{E}}(\upsilon,S)$, enthalpy $\hat{H}(p,S)$ and Gibb's free enthalpy $\hat{G}(p,T)$. Note that while the consequences of the second law (2) are local restrictions at a material particle, Gibb's inequalities arise from considerations of equilibrium of an isolated finite body of fluid.

In solving problems, the constitutive equations (2) are used, together with the equations of mass, momentum and energy conservation, to define the governing partial differential equations. The equation expressing conservation of internal energy $\mathcal{E}$ is

$$\dot{\mathcal{E}} = -(p+q)\dot{\upsilon} \qquad (4)$$

The constitutive equations are best introduced in differential forms. The Helmholtz free energy A is related to the internal energy $\mathcal{E}$ by

$$A = \mathcal{E} - TS \qquad (5)$$

Differentiating $(2)_1$, $(2)_2$, $(2)_3$ and (5) with respect to time, the resulting equations together with (4) may be solved for $\dot{p}$ and $\dot{T}$ in terms of υ to give differential equations for the pressure and temperature

$$\dot{p} = -\frac{1}{\upsilon}\left(K_T + \frac{\upsilon T \varphi_T^2}{C_v}\right)\dot{\upsilon} - \frac{1}{\upsilon}\left(\frac{\upsilon \varphi_T}{C_v}\right)q\dot{\upsilon}$$

$$\dot{T} = -\left(\frac{T\varphi_T}{C_v}\right)\dot{\upsilon} - \left(\frac{1}{C_v}\right)q\dot{\upsilon} \qquad (6)$$

where K_T, C_v and φ_T are the isothermal bulk modulus, specific heat at constant volume, and pressure-temperature coefficient respectively, which are all functions of (υ,T) defined by

$$K_T = \upsilon\frac{\partial^2 \hat{A}}{\partial \upsilon^2} \qquad C_v = -T\frac{\partial^2 \hat{A}}{\partial T^2} \qquad \varphi_T = -\frac{\partial^2 \hat{A}}{\partial \upsilon \partial T} \qquad (7)$$

while q is a function of $(\upsilon,\dot{\upsilon},T)$ by (2). The first coefficient in brackets in $(6)_1$ turns out to be the isentropic bulk modulus K_s, while the second is just the Grueneisen ratio γ. The first coefficient in $(6)_2$ is the pressure-entropy coefficient φ_s. The forms (7) have been retained in (6) to emphasize their derivation from the Helmholtz free energy function $\hat{A}$, which is often convenient since it may be obtained directly from partition functions for the material. However, second derivatives of all the thermodynamic potential functions are related, so that it is also possible to work in terms of $\hat{\mathcal{E}}$, $\hat{G}$, or $\hat{H}$.

Note that the internal energy has been eliminated from explicit consideration in (6). Instead of using the temperature as an explicit variable, $(6)_2$ may be replaced by (4) so that the internal energy is reintroduced. The coefficients in $(6)_1$, that is the isentropic bulk modulus and Grueneisen's ratio must now be expressed as functions of the internal energy and specific volume, but this can always be done if any one of the potential functions is known. The temperature description $(6)_2$ is to be preferred, since it is often easier to infer the temperature dependence of material properties than their internal energy dependence.

While there are many numerical solution strategies for the fluid flow equations, the equations (6) together with the equations of conservation of mass and momentum provide a particularly convenient basis for a general solution method. Note that (6) do not involve spatial gradients. Gradients are confined to the mass and momentum conservation equations, where they may be handled by conventional finite-difference or finite-element schemes. In fact, (6) are ordinary differential equations (ODE) in time, for given average values of $\dot{\upsilon}$ and q over a finite time step, and thus can be integrated over a time step by standard ODE methods. Note that the thermodynamic potential function itself need not be known explicitly. Only second derivatives of the potential function must be known functions of (υ,T). The forms (6) also lend themselves to investigation of stability and convergence of the numerical methods.

It is possible, without inserting specific functions for $\hat{A}$ and $\hat{q}$, to investigate the wave propagation behavior of materials governed by the constitutive equations of the general form (2). The simplest involve plane waves, for which one-dimensional specializations of the governing equations suffice. It is easily shown that discontinuous shock waves, on which υ suffers a jump discontinuity, and acceleration waves, on which $\dot{\upsilon}$ suffers a jump discontinuity, cannot exist. Plane steady propagating waves, that is, plane waves with a continuous profile that does not change in time, can exist. Their velocity and profile are related to the constitutive functions. Damped plane acoustic (infinitesimal sinusoidal) waves may also propagate, their speed and damping being related to the constitutive functions. No real material, of course, exhibits purely a bulk viscosity. Its introduction is useful chiefly to introduce artificial damping into finite-element or finite-difference methods which are otherwise unable to handle discontinuous shock and acceleration waves.

REACTIVE FLUID MIXTURE

We turn now to a more interesting example, that of a mixture of m reactive fluid components, which may be applied equally well to explosives, chemical decomposition, phase changes and dissociation or ionization. The quantity of each component present in the mixture is characterized by its molar concentration ω_a, $a = 1, \cdots m$. The constitutive equations for a mixture are taken to depend on the molar concentrations of all of the components, as well as on $\upsilon, \dot{\upsilon}$ and T. It is more convenient to use the distances of the molar concentration

from their values in some reference state, i.e. the extents of
reaction, expressed by components of an m-dimensional vector $\underset{\sim}{\xi}$.

$$A = \hat{A}(\upsilon,\dot{\upsilon},T,\underset{\sim}{\xi}) \qquad P = \hat{P}(\upsilon,\dot{\upsilon},T,\underset{\sim}{\xi}) \qquad S = \hat{S}(\upsilon,\dot{\upsilon},T,\underset{\sim}{\xi}) \qquad (8)_1$$

The $\underset{\sim}{\xi}$ change by chemical reactions, governed by rate equations of the form

$$\underset{\sim}{\dot{\xi}} = \underset{\sim}{f}(\upsilon,\dot{\upsilon},T,\underset{\sim}{\xi}) \qquad (8)_2$$

The first and second laws of thermodynamics again place limitations on the constitutive functions in (8). In particular $\hat{A}$ is again found to be a thermodynamic potential function independent of $\dot{\upsilon}$ such that

$$A = \hat{A}(\upsilon,T,\underset{\sim}{\xi}) \qquad p = -\frac{\partial \hat{A}}{\partial \upsilon} \qquad S = -\frac{\partial \hat{A}}{\partial T} \qquad \underset{\sim}{\zeta} = -\frac{\partial \hat{A}}{\partial \underset{\sim}{\xi}} \qquad (9)$$

where the components of the vector $\underset{\sim}{\zeta}$ are chemical affinities corresponding to the components of $\underset{\sim}{\xi}$. The condition $(2)_5$ now is found to take the form

$$\delta = \underset{\sim}{\zeta} \cdot \underset{\sim}{\dot{\xi}} - q\,\dot{\upsilon} \geq 0 \qquad (10)$$

which, in general, does not provide quite as specific restrictions on the form of $\hat{q}(\upsilon,\dot{\upsilon},T,\underset{\sim}{\xi})$ as in the previous example. However, the conditions $(2)_5$, $(2)_6$ and $(2)_7$ are regained if q is independent of $\underset{\sim}{\xi}$. Analogous conditions are obtained for $\underset{\sim}{f}$ if $\underset{\sim}{f}$ is independent of $\dot{\upsilon}$, and we will henceforth make this assumption.

The constitutive equations (9) now depend on the extents of reaction, i.e. the amounts of each constituent present. The behavior of the mixture is determined by the kind of dependence of $\hat{A}$ on $\underset{\sim}{\xi}$ and the forms of the reaction rate functions $\underset{\sim}{f}$. Very complicated behavior is possible, with rich possibilities for material instabilities which have been incompletely explored, but many of which are physically interesting. Only one specially simple case will be considered. The functions $\underset{\sim}{f}$ may be such that for each (υ,T) the equations

$$\underset{\sim}{f}(\upsilon,T,\underset{\sim}{\xi}) = \underset{\sim}{0} \qquad (11)$$

have unique solutions $\underset{\sim}{\xi}^*$. Then to each (υ,T) there corresponds an equilibrium composition $\underset{\sim}{\xi}^*(\upsilon,T)$. Inserting this into (9) it is seen that there exists an equilibrium equation of state identical in form with that for a single pure fluid (2). At a fixed (υ,T), $(8)_2$ are then a set of ODE of form

$$\underset{\sim}{\dot{\xi}} = f(\underset{\sim}{\xi}) \qquad (12)$$

These may have interesting stability problems. If they are such that for any initial composition $\underset{\sim}{\xi}_0$ (holding υ,T constant) the solutions of (12) approach the equilibrium state ξ^* as time proceeds, then the material is asymptotically stable, and no material instabilities can occur. More often, less stringent conditions apply to $\underline{f}$ and material instabilities are possible.

It is again convenient to work with the constitutive equations in differential form. We now differentiate $(9)_1$, $(9)_2$ and $(9)_3$ with respect to time, and use (4) and the derivative of (5) to obtain expressions for $\dot{p}$ and $\dot{T}$

$$\dot{p} = -\frac{K_s}{\upsilon}\dot{\upsilon} - \frac{\gamma}{\upsilon}q\dot{\upsilon} + \left(\frac{\gamma}{\upsilon}\underline{\zeta} + \underline{\Lambda}_T - \frac{\gamma T}{\upsilon}\underline{I}_v\right)\cdot\dot{\underline{\xi}}$$

$$\dot{T} = -\varphi_s\dot{\upsilon} - \frac{1}{C_v}q\dot{\upsilon} + \left(\frac{1}{C_v}\underline{\zeta} - \frac{T}{C_v}\underline{I}_v\right)\cdot\dot{\underline{\xi}} \tag{13}$$

where K_s, γ, C_v and φ_s are defined as before, and

$$\underline{\Lambda}_T = -\frac{\partial^2 \hat{A}}{\partial \upsilon \partial \underline{\xi}} \qquad \underline{I}_v = -\frac{\partial^2 \hat{A}}{\partial T \partial \underline{\xi}} \tag{14}$$

The first two terms on the right of each of (13) are the same as those in (6), while the last terms in each case are inner products over all the reactions, which, because of $(8)_2$, (9), (14) and the assumption that the $\underline{f}$ are independent of $\dot{\upsilon}$, are functions of $(\upsilon,T,\underline{\xi})$. Recapitulating the constitutive equations

$$\dot{p} = -\frac{1}{\upsilon}K_s(\upsilon,T,\underline{\xi})\dot{\upsilon} - \frac{1}{\upsilon}\gamma(\upsilon,T,\underline{\xi})q\dot{\upsilon} + g(\upsilon,T,\underline{\xi})$$

$$\dot{T} = -\varphi_s(\upsilon,T,\underline{\xi})\dot{\upsilon} - \frac{1}{C_v(\upsilon,T,\underline{\xi})}q\dot{\upsilon} - h(\upsilon,T,\underline{\xi}) \tag{15}$$

$$\dot{\underline{\xi}} = \underline{f}(\upsilon,T,\underline{\xi})$$

where g and h may be termed relaxation functions. We again have a set of m+2 ODE, for fixed average values of $\dot{\upsilon}$ and q over a finite time step. This set is frequently stiff, that is, different reactions may occur on vastly different time scales, and special care must be taken with their integration.

An important special case occurs when the motion is such that reactions can keep up with the changing volume and temperature so that the mixture is effectively in equilibrium at all times. The last of (15) is then replaced by the equilibrium condition (11), and g and h vanish, but the coefficients of $(15)_1$ and $(15)_2$ continue to depend on the current equilibrium value of $\underline{\xi}$. A very convenient numerical solution method in this case is to integrate the ODE $(15)_1$

and $(15)_2$ subject to the conditions (11) which are treated as implicit constraints.

The various types of wave propagation which may be obtained in a material governed by the above constitutive equations can again be discussed in considerable detail without inserting specific functions in (15), but merely by observing certain inequalities and restrictions on their forms. The discussion will be deferred until these equations are further generalized to solid materials in the next section.

SOLIDS WITH INTERNAL STATE VARIABLES

While the constitutive theory in the last section was motivated by a consideration of a chemically reacting mixture of fluids, the internal state variables $\underline{\xi}$ may be interpreted as applying to material characteristics other than extents of reactions for other applications. For example, the equilibrium theory of porous materials with negligible strength of Herrmann[3], and its non-equilibrium extension by Butcher[4] have constitutive equations which may be put into the same general forms if the porosity α is interpreted as an internal state variable. There is an obvious advantage to recognizing this fact, since all of the results which have been obtained for the general theory then apply directly.

Most useful generalizations apply to solids, rather than to fluids. Since the geometric complexities of arbitrary deformations in solids can be handled very compactly in tensor notation, the equations of the previous section will be translated into forms suitable for solids before proceeding.

The configuration in a solid must be described in terms of a strain from some reference configuration, which may, but need not necessarily be the initial configuration. Figure 1 shows schematically a portion of material in its current deformed configuration and the same portion of material in its fixed reference configuration. A material line element dX_I ($I = 1,2,3$) in the reference configuration corresponds to the line element dx_i ($i = 1,2,3$) in the current deformed configuration. Strains are related to changes in length and relative angle of pairs of line elements between the reference and current configurations. Two common finite strain measures

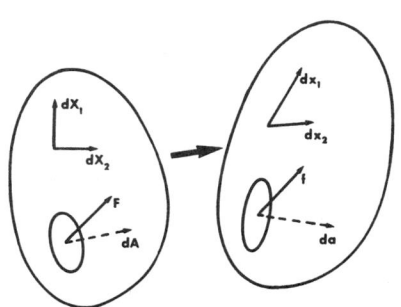

Fig. 1. Reference and Current Configurations

are Green's strain tensor $\underline{E}$ and Almansi's strain tensor $\underline{e}$ defined by

$$E_{KM} = \frac{1}{2}\left(\frac{\partial x_i}{\partial X_K}\frac{\partial x_i}{\partial X_M} - \delta_{KM}\right) \qquad e_{km} = \frac{1}{2}\left(\delta_{km} - \frac{\partial X_J}{\partial x_k}\frac{\partial X_J}{\partial x_m}\right) \qquad (16)$$

where the usual summation convention on repeated indices applies. Green's strain is particularly convenient for constitutive purposes, since it refers to elements in the reference configuration, which is stationary. Consequently its components are unaffected by rigid body motions of the current configuration.

The pressure in a fluid is replaced by the stress in a solid, which is related to the contact force vector $\underline{f}$ acting on an element of area $d\underline{a}$ in the current deformed configuration. Since $\underline{f}$ and $d\underline{a}$ have counterparts $\underline{F}$ and $d\underline{A}$ in the stationary reference configuration, two stress measures may be defined: Piola's stress tensor $\underline{\Sigma}$ and Cauchy's stress tensor $\underline{\sigma}$

$$F_M = \Sigma_{MN} \, dA_N \qquad f_i = \sigma_{ij} \, da_j \qquad (17)$$

Again, Piola's stress is particularly convenient, if less familiar, since it refers to force vectors and elements in the reference configuration which is stationary, and its components too are unaffected by rigid body motions.

The constitutive theory paralleling that of the last section can be given quite generally, by assuming that the response of the material, expressed in terms of its Helmholtz free energy function A, stress $\underline{\Sigma}$ and entropy S depends on its configuration expressed by $\underline{E}$, its history expressed almost trivially by $\underline{\dot{E}}$, its temperature T and a number of internal state variables $\underline{\xi}$ whose meaning will be determined later

$$A = \hat{A}(\underline{E},\underline{\dot{E}},T,\underline{\xi}) \qquad \underline{\Sigma} = \hat{\underline{\Sigma}}(\underline{E},\underline{\dot{E}},T,\underline{\xi}) \qquad S = \hat{S}(\underline{E},\underline{\dot{E}},T,\underline{\xi}) \qquad (18a)$$

The change of internal state variables is assumed to be governed by evolution equations of the form

$$\underline{\dot{\xi}} = \underline{f}(\underline{E},\underline{\dot{E}},T,\underline{\xi}) \qquad (18b)$$

The development parallels that of the previous section. The consequences of the first and second laws of thermodynamics are found to be[5]

$$A = \hat{A}(\underline{E},T,\underline{\xi}) \qquad \underline{\Sigma}_e = \rho_R \frac{\partial \hat{A}}{\partial \underline{E}} \qquad S = -\frac{\partial \hat{A}}{\partial T} \qquad \underline{\zeta} = -\frac{\partial \hat{A}}{\partial \underline{\xi}} \qquad (19)$$

where $\underline{\Sigma}_e$ is the equilibrium part of the stress and ρ_R is the density in the reference configuration. The viscous part of the stress is given by a constitutive equation of form

$$\underline{\Sigma}_v = \hat{\underline{\Sigma}}_v(\underline{E},\underline{\dot{E}},T,\underline{\xi}) \qquad (20)$$

The dissipation inequality, paralleling (10) becomes*

$$\delta = \underset{\sim}{\zeta} \cdot \dot{\underset{\sim}{\xi}} + \frac{1}{\rho_R} (\underset{\sim}{\Sigma}_v \cdot \dot{\underset{\sim}{E}}) \geq 0 \qquad (21)$$

When $\underset{\sim}{f}$ is independent of $\dot{\underset{\sim}{E}}$ the same conditions are obtained on the form of $\underset{\sim}{f}$ as before. If $\underset{\sim}{\Sigma}_v$ is independent of $\underset{\sim}{\xi}$, then $\underset{\sim}{\Sigma}_v$ must vanish in mechanical equilibrium $\dot{\underset{\sim}{E}} = \underset{\sim}{0}$, and the fourth-order viscosity tensor

$$\underset{\sim}{M} = \frac{\partial \underset{\sim}{\Sigma}_v}{\partial \dot{\underset{\sim}{E}}} \qquad (22)$$

must be positive semi-definite there, paralleling $(2)_6$ and $(2)_7$.

The second derivatives of $\hat{A}$ are now the fourth-order isothermal elasticity tensor $\underset{\sim}{L}_T$, second-order stress-temperature tensor $\underset{\sim}{\Phi}_T$ and scalar specific heat at constant strain C_E defined by

$$\underset{\sim}{L}_T = \frac{\partial \underset{\sim}{\Sigma}_e}{\partial \underset{\sim}{E}} = \rho_R \frac{\partial^2 \hat{A}}{\partial \underset{\sim}{E} \partial \underset{\sim}{E}} \qquad C_E = -T \frac{\partial \hat{S}}{\partial T} = -T \frac{\partial^2 \hat{A}}{\partial T^2}$$

$$\underset{\sim}{\Phi}_T = \frac{\partial \underset{\sim}{\Sigma}_e}{\partial T} = \rho_R \frac{\partial^2 \hat{A}}{\partial T \partial \underset{\sim}{E}} \qquad (23)$$

A major question remains concerning the proper generalization of Gibb's inequalities (in the absence of internal state variables). Coleman and Noll have postulated that $\underset{\sim}{L}_T$ is positive semi-definite, generalizing $(3)_1$.

The constitutive equations can again be put into differential forms by noting that the equation expressing conservation of energy is, in tensor form

$$\dot{e} = \frac{1}{\rho_R} (\underset{\sim}{\Sigma} \cdot \dot{\underset{\sim}{E}}) \qquad (24)$$

Differentiating $(19)_1$, $(19)_2$, $(19)_3$ and (5), and solving for $\dot{\underset{\sim}{\Sigma}}_e$ and $\dot{T}$ by steps paralleling those leading to (15) yields**

$$\dot{\underset{\sim}{\Sigma}}_e = \underset{\sim}{L}_S\{\dot{\underset{\sim}{E}}\} + \Gamma(\underset{\sim}{\Sigma}_v \cdot \dot{\underset{\sim}{E}}) - \underset{\sim}{G}$$

$$\dot{T} = \frac{1}{\rho_R} (\underset{\sim}{\Phi}_S \cdot \dot{\underset{\sim}{E}}) + \frac{1}{\rho_R C_E} (\underset{\sim}{\Sigma}_v \cdot \dot{\underset{\sim}{E}}) - h \qquad (25)$$

$$\dot{\underset{\sim}{\xi}} = \underset{\sim}{f}(\underset{\sim}{E}, T, \underset{\sim}{\xi})$$

---

* We use the dot notation for the product $A_{ij} B_{ij}$ of second order tensors $\underset{\sim}{A}$ and $\underset{\sim}{B}$ i.e. $\underset{\sim}{A} \cdot \underset{\sim}{B} = \text{tr}(\underset{\sim}{A}^T \underset{\sim}{B})$
** The fourth-order tensor operation $L_{ijkl} \dot{E}_{kl}$ is denoted by the mapping $\underset{\sim}{L}\{\dot{\underset{\sim}{E}}\}$

where $\underline{L}_s$ is the fourth-order isentropic elasticity tensor, $\underline{\Gamma}$ is the second-order Grueneisen tensor, and $\underline{\Phi}_s$ is the second-order stress-entropy tensor which are related directly to the quantities defined in (23) in much the same way as before, and are therefore functions of $(\underline{E}, T, \underline{\xi})$. The second-order stress relaxation function $\underline{G}$ and scalar temperature relaxation function h are also functions of $(\underline{E}, T, \underline{\xi})$ if the dependence of $\underline{f}$ on $\underline{\dot{E}}$ is omitted. The equations (25) closely parallel their scalar counterparts (15) for a fluid.

The remarks made in the previous section regarding equilibrium, stability and numerical solution methods carry over to the present case.

The equations of this section apply in general to an anisotropic material. In most cases, the material is likely to exhibit certain symmetries in its properties if the reference configuration is chosen to be an undistorted configuration. Material symmetries result in further restrictions on the forms of the constitutive functions in (25). For example, if the material is isotropic in its reference configuration, then

$$\underline{L}_s\{\underline{E}\} = \lambda \, \text{tr}(\underline{E})\underline{1} + 2\mu\underline{E} \qquad \underline{\Gamma} = \gamma\underline{1}$$
$$\underline{G} = g_0\underline{1} + g_1\underline{E} + g_2\underline{E}^2 \qquad (26)$$

where the scalar Lamé elastic coefficients λ and μ, the Grueneisen ratio γ, and the relaxation coefficients g_0, g_1 and g_2 may depend on $\underline{E}$ through its invariants as well as on $(T, \underline{\xi})$. An advantage of the present referential description in terms of $\underline{E}$ and $\underline{\Sigma}$ is that these symmetries are preserved in arbitrary deformations, since the reference configuration remains fixed, although the material in the current deformed configuration will exhibit strain-induced anisotropies. Representations of tensor functions are available for all the point symmetry groups.

DISCUSSION

The theory of viscous fluids and the theory of chemically reacting fluid mixtures, discussed in earlier sections, are obviously special cases of the more general theory of materials with internal state variables discussed in the previous section. Other special cases include fluids or fluid mixtures with tensor viscosity, and non-linear elastic solids without internal state variables. When internal state variables are included, a wide variety of phenomena can be represented, depending on the meaning which is assigned to these variables, and the specific formulation of the constitutive functions $\underline{f}$ and $\hat{A}$ (or rather its second derivatives). A few of the applications which have been made will be indicated here.

Nunziato et al.[6] have used the above theory of materials with internal state variables (but zero viscosity except for an artificial viscosity introduced for numerical purposes) to describe viscoelastic polymers. A number of internal state variables was used each of which corresponded to terms in a discrete relaxation

spectrum of the material. This represents a relatively simple case
in that the evolution equations for the internal state variables
exhibit asymptotic stability of equilibrium states. Since most wave
propagation experiments involve uniaxial strain, one-dimensional
specializations of the governing equations suffice. Properties of
one-dimensional plane acceleration waves, shock waves, steady propagating waves and acoustic (infinitesimal sinusoidal) waves for
this case have been investigated in considerable generality, and
have been reviewed by Herrmann[7] where references to the original
works may be found. These properties, in turn, serve to connect
observable features, such as the growth or decay of shock and
acceleration waves, the shape of steady propagating waves, and the
dispersion and attenuation of acoustic waves to the constitutive
functions, allowing direct evaluation of certain features of the
constitutive functions from experiments. Uniaxial strain experiments, of course, do not permit complete evaluation of the general
three-dimensional constitutive functions. However, the existence
of the general theory, together with assumptions of material
symmetries, provides guidance for the generalization of the one-
dimensional theory, and may suggest how any additional information
which is required in the three-dimensional theory may be obtained.

Viscoplasticity may be described by a theory which is almost
identical to the viscoelastic theory described above, if the relaxation functions are chosen so that stress relaxation is rapid when
the state is far from equilibrium, but within some "elastic" region
about the equilibrium state the relaxation is extremely slow so
that, for all practical purposes, anelastic deformation ceases.
The so-called unified creep plasticity theories[8] are of this type,
in which the introduction of discrete yield surfaces is avoided.
Elastic-viscoplastic theories arise when there is a finite region
in strain space, for given values of $(T,\underline{\xi})$, upon which the equilibrium condition (11) is satisfied. For motions in which the state
remains within this region, the $\underline{\xi}$ do not change, and the material is
thermoelastic. In the equilibrium region, there is a one-to-one
correspondence of stress and strain (at a given temperature) through
$(19)_2$ so that the equilibrium region in strain space maps into a
region in stress space, bounded by a yield surface. The shape and
position of the equilibrium region, and hence the yield surface, may
be taken to depend on certain internal state variables. A number of
questions remain in the general theory; however, these do not arise
in its one-dimensional specialization. A review of one-dimensional
wave propagation in elastic-viscoplastic materials has been given by
Herrmann[9], where further references are given. The behavior of
acceleration, shock and steady waves closely parallels that of viscoelastic materials, and these results may be used similarly in the
deduction of constitutive functions from experiment.

Another application of the theory is to solid materials undergoing damage. For example, solids undergoing spallation may do so
by the growth of finely distributed voids (in ductile materials) or
microcracks (in brittle materials). The macroscopic behavior may be

described in terms of one or more internal state variables representing the extent of void growth or cracking.[10,11] The constitutive functions are so formulated that the stress reduces as the damage increases. Consequently it is to be expected that physical instabilities should occur in this case. These are expected in the form of localized shear banding. The consequences have not been studied in detail.

The theory of phase changes of Hayes[12] is another direct application of the present general theory, and most of the general properties of waves cited above carry over to this case. Another example is the theory of detonation of homogeneous explosives of Kipp and Nunziato[13] which has been very successful in predicting detonation failure of liquid nitromethane in various cylindrical symmetric two-dimensional configurations, although all of the constitutive functions were evaluated from one-dimensional wave propagation experiments.

Many other potential applications exist in shock wave physics for the internal state variable theory discussed in this paper. Internal state variables may be used to describe many different types of history effects arising from various physical processes occurring in the material. It has been shown that the theory and the proposed numerical solution method are applicable both to the rate dependent theory, and to its rate independent equilibrium approximation (as well as to the simpler cases of elastic solids or fluids without history effects). While kinematical and thermodynamical consistency have been fully explored, and the restrictions arising therefrom on the constitutive functions have been rendered exlicit, questions of stability have not yet been studied in detail. The theory admits many cases of instability, nonuniqueness and bifurcation, many of which are, in fact, expected on physical grounds. Similarly, convergence of numerical solution methods remains to be established, especially in cases where physical instabilites are expected. When instabilities are not present, general properties of plane acceleration, shock, steady and acoustic waves have been established, and these results are of great value in evaluating constitutive functions from wave propagation experiments. Clearly, it is very advantageous to use this, or other, existing constitutive theory if possible, since all of the known results then apply, and need not be reestablished anew.

There are, of course, many phenomena which cannot be described by the theory of materials with internal state variables, outlined in this paper. A classical example is a mixture of finely divided gas bubbles in a liquid. If it is assumed that the gas and liquid do not diffuse relative to each other, and that the pressure and temperature of the gas and liquid are equal at all times, then the theory discussed above applies. However, this is the limiting case of the more usual situation in which these conditions do not pertain. The gas bubbles do not usually move at the same velocity as the liquid in a mixture in motion, since the motion of the liquid is transmitted to the gas bubbles through a drag force which requires relative motion to act. Expansion of the gas bubbles relative to

the liquid requires that the pressure be higher in the gas. Such
expansion entails an inertia of the gas-fluid system which is not
reflected in translational inertia of the mixture. Similarly
heating of the gas by the fluid requires a temperature difference
to effect the heat transfer in a motion involving changing temperatures. New conservation laws appear in such theories to account
for the added microstructural degrees of freedom. Again, the concepts can be generalized to cover many diverse physical phenomena.
A general theory has been given by Drumheller and Bedford[14] with
applications to mixtures which are being discussed elsewhere in this
conference. Applications may include not only mixtures of different
components as illustrated by the simple example above, but situations
where a single material may undergo heterogeneous processes, such as
hot spot initiation of explosives or possibly heterogeneous initiation of phase changes or heterogeneous deformation by shear banding,
which are also being discussed elsewhere in this conference.

REFERENCES

1. B. D. Coleman and J. M. Greenberg, Arch. Rational Mech. Anal. $\underline{25}$, 321 (1967).

2. H. B. Callen, Thermodynamics,(Wiley 1963).

3. W. Herrmann in Appl. Mech. Aspects of Nuclear Effects in Materials, ed. C. C. Wan, (ASME 1971).

4. B. M. Butcher in Shock Waves, ed. J. J. Burke and V. Weiss,(Syracuse University Press 1971).

5. B. D. Coleman and M. E. Gurtin, Phys. Fluids $\underline{10}$, 1454 (1967).

6. J. W. Nunziato, K. W. Schuler and D. B. Hayes, Proc. Intl. Conf. on Comp. Methods in Nonlinear Mechanics, Austin, TX, Sept. (1974).

7. W. Herrmann, 7th U.S. Natl. Cong. Appl. Mech., Boulder, CO, June (1974).

8. R. D. Krieg, Proc. 4th Conf. on Structural Mechanics in Reactor Technology, San Francisco, CA (1977).

9. W. Herrmann in Propagation of Shock Waves in Solids, ed. E. Varley,(ASME 1976).

10. L. W. Davison and A. L. Stevens, J. Appl. Phys. $\underline{44}$, 668 (1973).

11. D. E. Grady and M. E. Kipp, Intl. J. Rock Mech. and Mining Sci. $\underline{17}$, 147 (1980).

12. D. B. Hayes, J. Appl. Phys. $\underline{46}$, 3438 (1975).

13. M. E. Kipp and J. W. Nunziato, 7th Symp. on Detonation, Annapolis, MD, June (1981).

14. D. S. Drumheller and A. Bedford, Arch. Rational Mech. Anal. $\underline{73}$, 257 (1980).

DYNAMIC STRESS-STRAIN CURVES AT PLASTIC SHEAR STRAIN RATES OF 10^5 s^{-1}*

C. H. Li** and R. J. Clifton***
Brown University, Providence, RI 02912

ABSTRACT

Pressure-shear impact of thin (0.2-0.4 mm) aluminum plates sandwiched between hard steel plates has been used to impose plastic shear strain rates up to 2×10^5 s^{-1} and to sustain these rates for a duration of 10^{-6} s. The particle velocity and shear traction at the interface between the aluminum specimen and the steel target plate are obtained from measurements of the normal and transverse components of particle velocity at the free surface of the steel target plate. At times after nominally homogeneous states of stress are obtained in the specimen the shear strain rate, averaged through the specimen thickness, and the average shear stress can be obtained directly from the transmitted wave. Results for commercially pure aluminum specimens indicate that the flow stress (at 15% strain) at shear strain rates of 2×10^5 s^{-1} is approximately 140 MPa compared to reported values of 60 MPa at shear strain rates of 10^3 s^{-1}. This large increase in flow stress is viewed as indicating that a transition in the rate controlling mechanism for plastic flow occurs between strain rates of 10^3 s^{-1} and 10^5 s^{-1}.

INTRODUCTION

As an important first step in measuring the inelastic response of materials at strain rates beyond those accessible in conventional testing machines Kolsky[1] introduced the so-called split-Hopkinson bar technique. In this technique a thin, cylindrical specimen, sandwiched between two long elastic bars, is subjected to a compressive stress pulse with a duration of the order of 10^{-4} seconds. By monitoring the incident and reflected waves in one of the elastic bars, and the transmitted wave in the other, one can determine the time history of the average stress and strain-rate in the specimen. Once the strain is obtained by integration of the strain-rate with respect to time, a dynamic stress-strain curve can be obtained. Such a curve is understood to be a satisfactory representation of the mechanical response of the specimen after a sufficient number of longitudinal wave reflections have occurred for nominally homogeneous states of deformation to be set up in the specimen. The technique has been widely used for obtaining dynamic stress-strain curves of metals at strain rates of approximately 10^3 s^{-1}. Modifi-

---

*This work has been supported by the Army Research Office and by the NSF Materials Research Laboratory at Brown University.
**Graduate Student, Division of Engineering
***Professor of Engineering

cation of the technique to use torsional waves instead of longitudinal waves has removed uncertainty regarding the effects of lateral inertia and of friction between the specimen and the adjoining bars[2].

Extension of the technique to higher strain rates is possible by using high yield strength elastic bars and by reducing the length of the specimen. In this way, strain rates of 10^4 s^{-1} have been obtained under reasonably well controlled conditions[3,4]. Extension to still higher strain rates by further reductions in configuration dimensions does not appear to be an attractive option. Instead, the need for understanding mechanical response at strain rates of the order of 10^5 s^{-1} appears to call for a new approach. Such an approach is introduced here, based on a synthesis of concepts from split-Hopkinson bar experiments and pressure-shear plate impact experiments[5].

EXPERIMENTAL PROCEDURES

A schematic of the experiment is shown in Fig. 1. The specimen

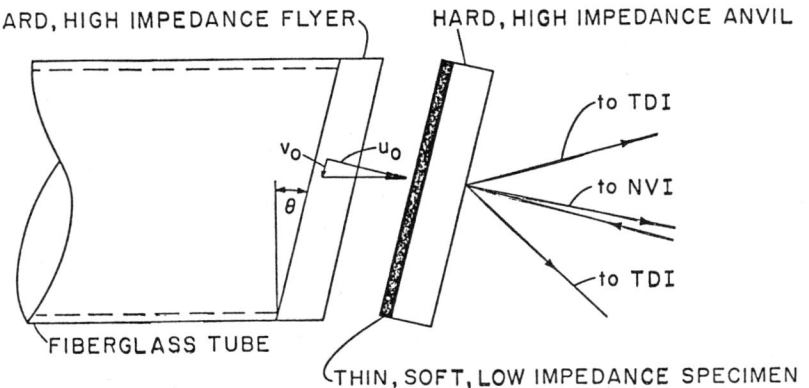

Fig. 1. Schematic of High Strain-Rate Experiment

is subjected to pressure-shear loading by impact of flat, parallel plates inclined relative to their direction of approach. The plates on each side of the specimen have sufficiently high yield strengths to remain elastic under the impact conditions. Consequently, elastic wave theory can be used to infer the particle velocity at the back of the specimen from the particle velocity at the back of the anvil plate. The latter particle velocity is monitored by means of a normal velocity interferometer (NVI)[6] and a transverse displacement interferometer (TDI)[7]. The bounding plates have higher acoustic impedance than the specimen in order to insure that unloading of the specimen does not occur during reflection of the waves from the bounding plates. Specimen thicknesses are large relative to grain sizes, but small relative to the thicknesses of the bounding plates.

The experiments are conducted using a 63.5 mm single-stage gas gun with a 2.5 m long barrel. With compressed nitrogen gas this gun can be used to launch projectiles at velocities up to 0.3 mm/ sec. A keyway in the barrel and a key in the projectile prevents rotation of the projectile during its acceleration. The target, consisting of a thin aluminum specimen attached to a high strength steel anvil, is mounted in a vacuum chamber at the muzzle end of the gun. The chamber is evacuated to a pressure of 25 μm to 70 μm Hg before the flyer is launched in order to reduce the possibility of an air cushion that would allow sliding between the flyer and the specimen. More details on the gun, the pressure-shear impact configuration, and the alignment procedure are given elsewhere[5,8].

For the experiments reported here both the flyer and the anvil were made from Stentor steel produced by the Carpentor Steel Co. After being machined to the final dimensions, both pieces were heat treated to the maximum strength. This state usually has a tensile yield strength of 17 to 20 kbar. After heat treatment, both pieces were ground and lapped to a flatness of better than 0.5 μm over the surface. Then the rear surface of the anvil was mirrorized and a ruled glass grating was used to copy a 200 lines/mm diffraction grating onto the surface by means of a photo-resist process, using Kodak photo-resist type 3.

The specimens were made from 0.625 mm thick sheets of 1100-H14 and 6061-T6 aluminum. The sheet was first punched to 45 mm diameter discs. These discs were bonded together with epoxy resin and then machined to 37.5 mm diameter with four 4.8 mm holes equally spaced on a 25.4 mm diameter circle. The specimens were separated by dissolving the resin. After the 1100-H14 specimens were separated, they were annealed at 750°F for one hour and cooled in the furnace for twenty-four hours. Because of this annealing the specimens are designated as 1100-0 aluminum. The 6061-T6 aluminum pieces were not annealed. Because all specimens were thin and easily bent, they were attached with Buehler wax to steel pieces of the same diameter before lapping to final thickness and flatness. The specimens were then lapped until the flatness of the aluminum pieces was better than 0.5 μm over the whole surface. This flat surface of an aluminum specimen was attached to a high strength steel anvil of the same flatness by applying epoxy at the perimeter of the two pieces. A little pressure was applied first to insure that no epoxy could get into the specimen-anvil interface. After the epoxy cured, the assembly was put on a hot plate to melt the wax and separate the specimen and anvil from the other steel piece. Then, four 3 mm diameter copper pins (which were used to trigger the oscilloscopes and to detect the tilt between two impact faces) were glued into the centers of the four holes in the specimen. Afterwards the unpolished surface of the aluminum specimen was lapped to a flatness of better than 0.5 μm over its diameter. This surface became the impact face, which was aligned parallel to the flyer plate by means of the technique[8] except that rear surface mirrors were attached temporarily to the impact faces to eliminate the need for highly polished impact faces. The latter modification was introduced to reduce the possibility of relative sliding across the impact plane.

The normal and transverse motion of the rear surface of the anvil was monitored by means of an NVI and a TDI as described by Kim and Clifton[5,7]. From elastic wave theory the normal and transverse components of surface traction and particle velocity at the interface between the specimen and the anvil are, respectively,

$$\sigma = -\frac{\rho c_1}{2} u_{fs} \qquad \tau = -\frac{\rho c_2}{2} v_{fs} \qquad (1a,b)$$

$$u = \frac{1}{2} u_{fs} \qquad v = \frac{1}{2} v_{fs} \qquad (2a,b)$$

where ρ is the mass density of the anvil, c_1 and c_2 are the elastic longitudinal and shear wave speeds for the anvil and (u_{fs}, v_{fs}) are the normal and transverse components of the particle velocity at the free surface of the anvil. In equations (1) and (2) the times at which σ, τ, u, v are computed are earlier, by the corresponding elastic wave transit times, than the times at which u_{fs} and v_{fs} are measured. Equations (1) and (2) apply before the respective waves reflected at the free surface of the anvil return to the specimen-anvil interface. After two or three reverberations of elastic longitudinal waves through the specimen thickness the normal stress σ and normal velocity u become essentially constant and equal to $-\rho c_1 u_0/2$ and $u_0/2$, respectively, where u_0 is the normal component of the initial flyer plate velocity. After enough reverberations of elastic shear waves through the thickness h of the specimen to allow nominally homogeneous states of stress and deformation to be established, the nominal shear strain rate in the specimen is

$$\dot{\gamma} = \frac{v_0 - 2v}{h} \qquad (3)$$

where v_0 is the transverse component of the initial flyer plate velocity. Plots of the shear stress obtained from (1b) versus the nominal shear strain obtained from the integration of (3) with respect to time are the dynamic stress-strain curves obtained from the experiments.

EXPERIMENTAL RESULTS

Dynamic stress-strain curves for the 1100-0 aluminum specimens are shown in Fig. 2. Three curves are shown for each of two values of the angle of inclination θ (see Fig. 1). The values $\theta = 14°$ and $\theta = 26.6°$ correspond to ratios u_0/v_0 of 4 and 2, respectively. A range of strain rates was obtained by varying the angle θ, the projectile velocity, and the specimen thickness. Fourteen tests were conducted; the six curves in Fig. 2 are representative of the best of the fourteen tests.

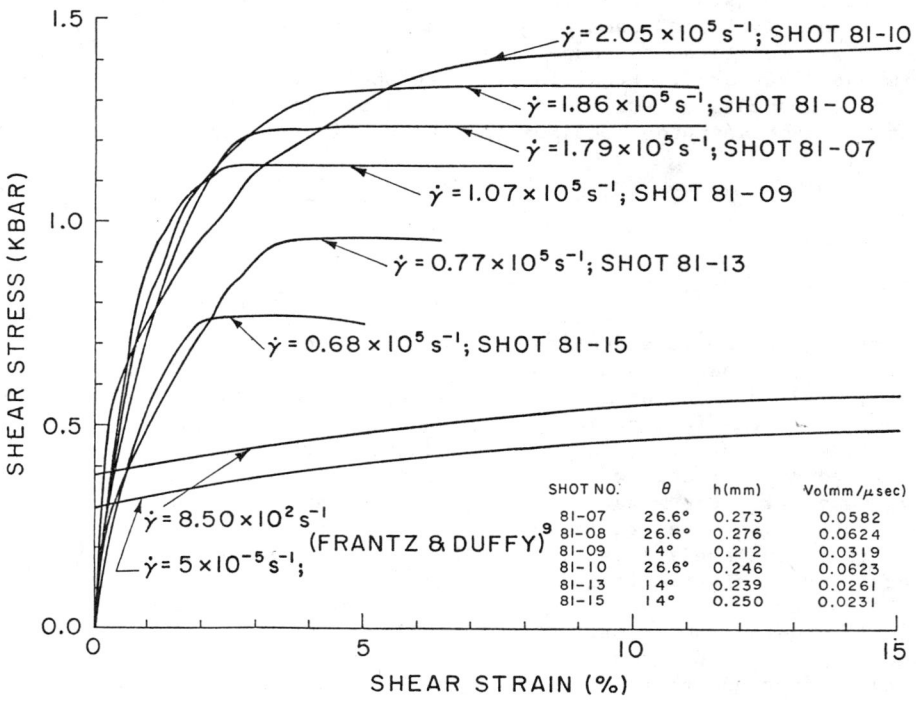

Fig. 2 Dynamic Stress-Strain Curves for 1100-0 Aluminum

For each test the shear stress increases rapidly initially, and then becomes essentially constant. As in split-Hopkinson bar type experiments the steeply rising part of the curve should be disregarded because the stress and deformation are nonhomogeneous initially. The existence of a stress plateau suggests that a nominally uniform state of stress has been established in the specimen and that strain hardening is relatively unimportant at the strains and strain rates shown in Fig. 2.

For comparison, two stress-strain curves are shown from tests on 1100-0 aluminum by Frantz and Duffy[9] at shear strain rates of 5×10^{-5} s^{-1} and 8.5×10^2 s^{-1}. These curves have been shifted along the strain axis to allow direct comparison with results of the pressure-shear experiments in which pre-compression due to longitudinal compressive waves produces plastic strains corresponding to an equivalent plastic shear strain of approximately 2.5%. Consequently, the curves from Frantz and Duffy[9] have been shifted so that, for example, a shear strain of 5% in Fig. 2 corresponds to a shear strain of 7.5% in their paper.

The plateau stress levels show a marked dependence of flow stress on strain rate over the strain rate range from 0.68×10^5 s^{-1}

to 2.05×10^5 s^{-1}. The contrast between the relative strain rate insensitivity of the flow stress over the strain rate range from 5×10^{-5} s^{-1} to 8.5×10^2 s^{-1} and the pronounced strain rate sensitivity above strain rates of 6.8×10^5 s^{-1} suggest a change in the rate controlling mechanism for plastic deformation. At strain rates up to 10^3 s^{-1} a linear relationship between τ and $\log \dot{\gamma}$ is compatible with the rate controlling mechanism being the thermally activated motion of dislocations past obstacles. At strain rates of approximately 10^5 s^{-1} the flow stress increases more nearly linearly with $\dot{\gamma}$ than with $\log \dot{\gamma}$, suggesting that at these strain rates viscous drag on dislocations is the rate controlling mechanism.

Results of two pilot experiments on 6061-T6 aluminum specimens are shown in Fig. 3, along with a reference stress-strain curve at

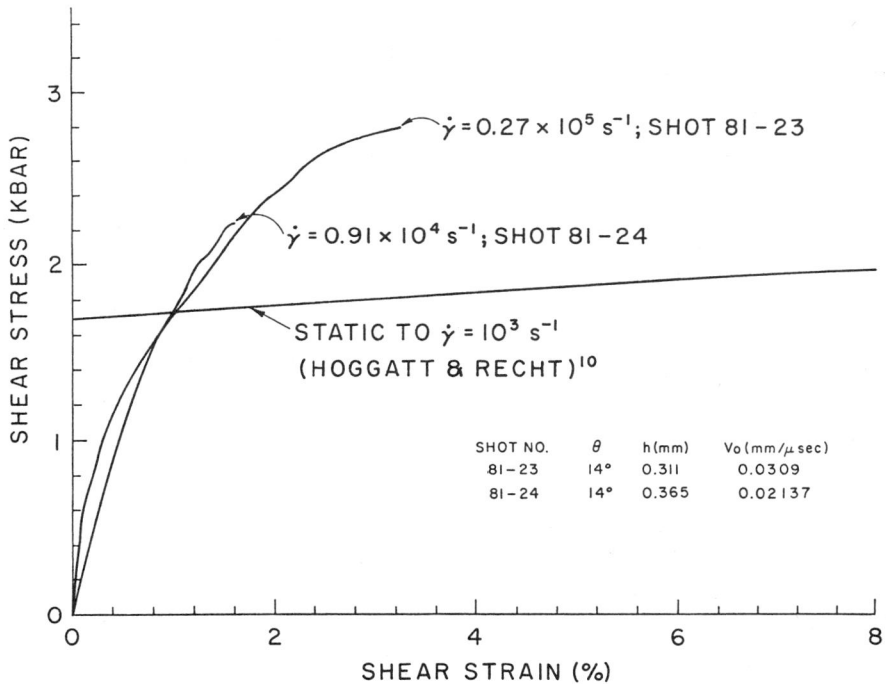

Fig. 3 Dynamic Stress-Strain Curves for 6061-T6 Aluminum

lower strain rates. The latter curve has also been shifted as described for the reference curves in Fig. 2. Interpretation of the dynamic stress-strain curves in Fig. 3 is less straightforward than for those in Fig. 2 because of the lack of stress plateaus in Fig. 3. Consequently, comparison of flow stress levels requires consideration of the plastic strain, which is determined with less certainty. Nevertheless it is clear that there is an appreciable strain rate sensitivity of 6061-T6 aluminum at strain rates of 10^4 s^{-1} and higher. Further tests are required at larger values of θ so that larger strain rates and more nearly constant levels of flow stress are obtained.

CONCLUSIONS

Pressure-shear impact of thin specimens sandwiched between hard, high impedance plates offers a means for obtaining the dynamic response of materials under well characterized conditions and at strain rates that are higher than those accessible by other means. By using flyer and anvil plates made from such high strength, high impedance materials as tungsten carbide it should be possible to determine the flow stress of most metals at strain rates of approximately 10^5 s^{-1}.

REFERENCES

1. H. Kolsky, Proc. Phys. Soc. B62; 676 (1949).
2. J. Duffy, J. D. Campbell and R. H. Hawley, J. of Appl. Mech. 38, 83 (1971).
3. D. A. Gorham, Mechanical Properties at High Rates of Strain edited by J. Harding (The Institute of Physics, Bristol and London, 1979), p. 16.
4. U. S. Lindholm, IUTAM Symposium on High Velocity Deformation of Solids, edited by K. Kawata and J. Shioiri (Springer-Verlag, Berlin, Heidelberg and N.Y., 1978), p. 26.
5. K. S. Kim and R. J. Clifton, J. of Appl. Mech. 47, 11 (1980).
6. L. M. Barker and R. E. Hollenbach, J. Appl. Phys. 41, 4208 (1970).
7. K. S. Kim, R. J. Clifton and P. Kumar, J. Appl. Phys. 48, 4132 (1977).
8. P. Kumar and R. J. Clifton, J. Appl. Phys. 48, 1366 (1977).
9. R. A. Frantz and J. Duffy, J. of Appl. Mech. 39, 939 (1972).
10. C. R. Hoggatt and R. E. Recht, Exp. Mech. 9, 441 (1969).

INTERPRETATION OF SHOCK-WAVE DATA FOR BERYLLIUM AND URANIUM WITH AN ELASTIC-VISCOPLASTIC CONSTITUTIVE MODEL

Daniel J. Steinberg and Richard W. Sharp, Jr.
Lawrence Livermore National Laboratory, Livermore, California 94550

ABSTRACT

We have developed a new elastic-viscoplastic constitutive model for metals which can easily be incorporated into a one-dimensional Lagrangian hydrodynamic computer code. Rather than the usual constant relaxation time, we have used one which is a function of temperature (i.e., thermal energy) and strain rate. This allows us to successfully calculate shock wave profiles in beryllium and uranium over a wide range of stress, temperature, strain and strain rate.

INTRODUCTION

In a previous paper, Steinberg, Cochran and Guinan[1] described an elastic-plastic constitutive model for metals applicable at high strain rate. In this paper, we will generalize this model to include the effect of strain rate, in particular shock-smearing, which occurs when a material is shock loaded to a few times its yield strength.

In classic elastic-plastic theory, when a flyer strikes a target, there are only two waves initiated at the interface; the elastic and plastic waves. In actual fact, a whole spectrum of waves are launched or perhaps one wave of varying velocity (see the dashed curves in the figures). Thus, there is really no clear distinction between elastic and plastic waves, and simple elastic-plastic theory must be modified to include a "time-constant", τ, which is indicative of a dispersion of the plastic wave from the longitudinal wave velocity to the velocity characteristic of the peak stress.

As the stress loading is increased, the experimentally observed shock front becomes sharper and more distinct, and at strain rates of $\geq 10^5$ s^{-1} shock-smearing becomes insignificant. To model this observation, we need a τ that decreases rapidly as the stress increases. In liquids, rate-dependent effects such as viscosity appear to decrease exponentially with temperature.[2] Therefore, as increasing stress implies increasing temperature, T, we chose a τ that decreased exponentially with T.

Even this modification was not sufficient, and τ had to be made a function of the strain rate as well. It is the purpose of this paper to show how a τ which is a function of temperature and strain rate can significantly improve the agreement with experimental data.

THE STRAIN-RATE DEPENDENT CONSTITUTIVE MODEL

We represent the strain-rate dependent stress deviator S by what is commonly referred to as the standard linear solid model; i.e.,

$$\dot{S} = 2G\dot{\varepsilon} - \frac{S - S_{eq}}{\tau}, \qquad (1)$$

where ε is the distortional strain, G is the shear modulus, dots over S and ε denote Lagrangian derivatives with respect to time, and S_{eq} is the equilibrium or strain-rate independent stress deviator.

$$\dot{S}_{eq} = 2G\dot{\varepsilon} \text{ when } |S_{eq}| < \frac{2}{3} Y, \qquad (2)$$

where Y is the yield strength; otherwise $|S_{eq}| = \frac{2}{3} Y$. We write τ as

$$\tau = A_1 \exp[Q/E_{in}], \qquad (3)$$

where the "instantaneous thermal energy", E_{in} is

$$E_{in} = (E - E_c) + A_2 |S\dot{\varepsilon}|. \qquad (4)$$

E and E_c are respectively the total and zero-Kelvin-compression energies, and A_1, A_2 and Q are material constants.

Our calculations are not very sensitive to the choice of Q. We have set Q equal to the activation energies in other rate-dependent effects in solids such as creep or self-diffusion.

Neither A_1 nor A_2 were determined independently of our shock-wave data, but A_2 can be estimated to about a factor of two from the average grain size, X, and diffusion coefficient, D, of the sample materials; i.e. $A_2 = X^2/D$. This is based on the idea that A_2 is the time it takes an average grain to reach thermal equilibrium when E_{in} is initially greatest at the grain boundary. Some of this uncertainty in pre-determining A_2 is due to the factor-of-two variation in grain size in the materials. It follows then that A_1 is the only completely free parameter in our strain-rate model.

In a strain-rate independent calculation, when S reaches 2/3 Y, it remains equal to 2/3 Y for the duration of shock loading. With the addition of the strain-rate dependence in Eq. 1 - 4, S may exceed 2/3 Y for a time duration related to τ. The qualitative effect of our form for τ on the stress deviator is demonstrated in Figs. 1a and 1b. In these figures the deviator is compared with the yield strength for two different shock profiles -- one at low strain rate (Fig. 1a), and the other at high strain rate (Fig. 1b).

In the low strain rate example, typical of substantial shock smearing, $\dot{\varepsilon}$ only varies from 2 to 5 x $10^4 s^{-1}$. This small variation, coupled with the low maximum $\dot{\varepsilon}$, means that the second term in Eq. 4 will not only be small compared to the thermal energy $E-E_c$, but will be nearly constant. Therefore, only the thermal energy will have an effect in shaping the shock wave profile. Fig. 1a shows that after the initial overshoot, the stress deviator gradually relaxes back to the yield surface.

In the high strain rate case (i.e., for a distinct well-formed shock), $\dot{\varepsilon}$ varies from 8 x 10^4 to 5 x 10^6 s^{-1}. At the shock front, where $\dot{\varepsilon}$ is a maximum, the second term in Eq. 4 now completely dominates E_{in}. This term can be thought of as a distortional energy, large in the immediate vicinity of the shock. This in turn makes τ decrease so rapidly that the strain rate dependence of S essentially vanishes. This causes S to drop back abruptly to the yield surface.

Thus, Eq. 1 - 4, when added to an elastic-plastic constitutive model, allow the qualitative features of shock-wave profiles at quite different stress levels to be properly accounted for.

COMPARISON OF THE MODEL WITH EXPERIMENTAL DATA

Using the model outlined in the previous section, we have calculated several shock-wave experiments with S-200E beryllium reported by Christman and Feistmann.[3] In these experiments, fused quartz flyers impacted beryllium targets with fused quartz windows and a velocity interferometer recorded the motion of the Be-quartz interface. We changed two of the parameters in our constitutive model[1] to conform with the data of Christman and Feistmann on their particular material; the initial shear modulus became 145.3 GPa, and the exponent in our work-hardening function became 0.44.

For the hydrodynamic equation of state, we used a Gruneisen form with a shock and particle velocity (U_s-U_p) equation $U_s = C + SU_p$ = 7.76 + 1.4 U_p mm/μs and a γ equal to $1.11-1.0(1-1/\eta)$.[1] The parameter C is the bulk sound velocity determined by Christman and Feistmann, and S was determined from an ultrasonic measurement of the isentropic pressure derivative of the bulk modulus.[4] The compression, η, is the ratio of the density to the initial density.

The two parameters for the Bauschinger model[1] were determined from the uniaxial stress data of Christman and Feistmann. Called G_1 and G_2 respectively in Ref. 1, they are here equal to 61 GPa and 0.

It is important to emphasize that up to now, no information from the shock wave experiments has been used to normalize the calculations. However, there are no data to completely predetermine two of the parameters in the strain-rate model. Therefore, we turn to some of these shock-wave experiments to determine A_1 and A_2 (see fig. 2b). The best fit to the data was with $A_1 = 1.0 \times 10^{-14}$ μs and $A_2 = 2.0$ μs, the value of Q = 214.3 kJ/mole having been predetermined from data in Refs. 5 and 6.

An independent test of the entire constitutive model was then done by calculating the shock-wave profiles shown in Figs. 2a and 3a. The agreement between experiment and calculation is excellent.

Fig. 3b shows that both the Bauschinger and the strain-rate models are needed to correctly reproduce the shape of the measured wave profiles.

Cochran and Banner[7] have reported capacitor free-surface velocity gauge measurements on shocked uranium. We have compared calculations of these experiments with and without our strain-rate model.

For the hydrodynamic equation of state, we used a Gruneisen form with $U_s = 2.48 + 1.53\ U_p$ mm/μs[8] and $\gamma = 2.32-1.8\ (1-1/\eta)$.[1] The initial shear modulus is 88 GPa,[8] and $G_1 = 15$ GPa and $G_2 = 0$.[9] The remaining information needed for our constitutive model can be found in Ref. 1. As with beryllium, A_1 and A_2 are the only parameters which had to be determined by a best fit to at least one of the experiments. We found $A_1 = 4 \times 10^{-14}$ μs and $A_2 = 0.75$ μs. Q was determined from a scaling rule using the melt temperature[6] and is 218.7 kJ/mole. The comparison of the calculations with the data for two of these experiments are shown in figs. 4 and 5.

It is apparent that we must reproduce the shock-loading profile correctly if we are to reproduce the spall signal. For a complete description of the work, including the integration of our strain-rate model with the spall model of Cochran and Banner[7], see ref. 10.

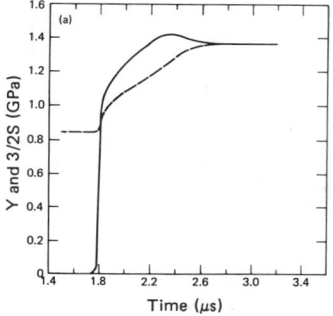 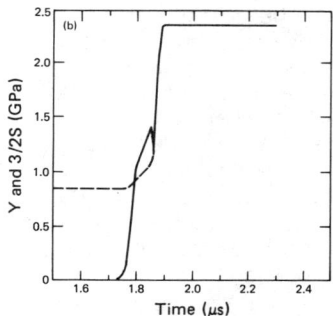

Fig. 1. Calculated comparison of the yield strength (dashed line) and 3/2 times the stress deviator (solid line) vs. time for a maximum strain-rate of (a) $5 \times 10^4 s^{-1}$ and (b) $5 \times 10^6 s^{-1}$.

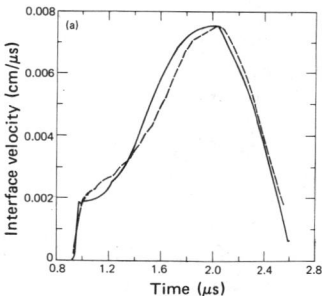 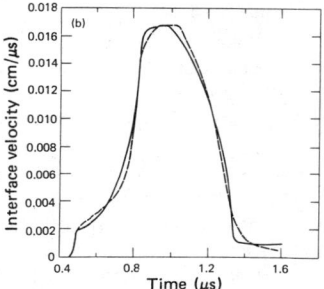

Fig. 2. Quartz-Be interface velocity vs. time. Solid line is the calculation; dashed line is the experiment. (a) 1.1 GPa peak stress (b) 2.4 GPa peak stress; target is 4.1 times thicker than the flyer.

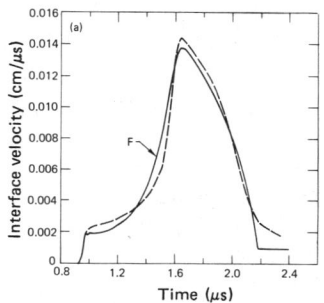 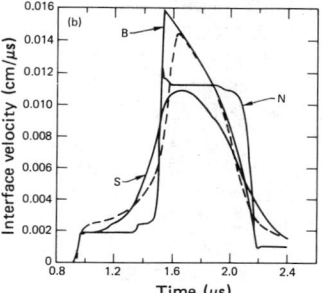

Fig. 3. Quartz-Be interface velocity vs. time. 2.4 GPa peak stress; target is 7.4 times thicker than the flyer. The dashed line is the experiment. The solid lines are calculations with F) full model, S) strain-rate model only, B) Bauschinger model only, N) neither model.

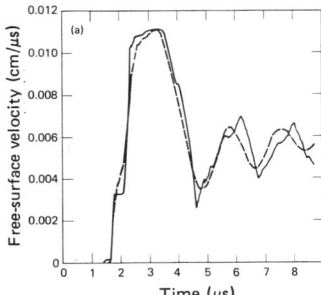 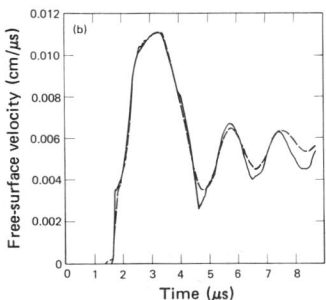

Fig. 4. Free-surface velocity vs. time for a 6-mm-thick uranium target shocked to 3.2 GPa. The dashed line is the data; the solid line is the calculation without (a) and with (b) the strain-rate model.

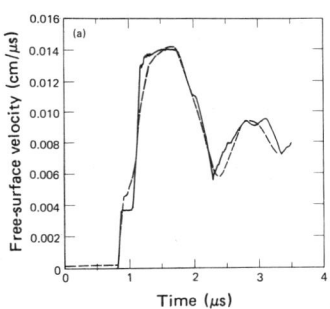 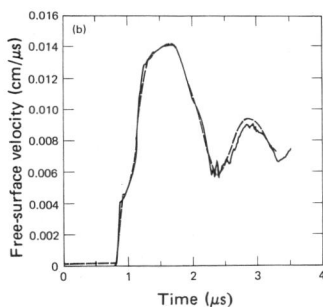

Fig. 5. Free-surface velocity vs. time for a 3-mm-thick uranium target shocked to 3.9 GPa. The dashed line is the data; the solid line is the calculation without (a) and with (b) the strain-rate model.

REFERENCES

1. D. J. Steinberg, S. G. Cochran, and M. W. Guinan, J. Appl. Phys. 51, 1498 (1980).
2. E. N. da C. Andrade, Philos. Mag. Suppl. 17, 497 (1934).
3. D. R. Christman and F. J. Feistmann, Materials and Structures Laboratory, General Motors Tech. Center Rept. MSL 71-23 (1972).
4. D. Silversmith and B. Auerbach, Phys. Rev. B. 1, 567 (1970).
5. F. A. McClintock and A. S. Argon, Eds., Mechanical Behavior of Materials (Addison-Wesley, Reading, MA, 1966), p. 631.
6. D. Lazarus, in Energetics in Metallurgical Phenomena, 1, edited by W. M. Mueller (Gordon and Breach, New York, 1965), pp 51-52.
7. S. Cochran and D. Banner, J. Appl. Phys. 48, 2729 (1977).
8. Compendium of Shock Wave Data, M. Van Thiel ed., Lawrence Livermore National Laboratory Report UCRL-50108, Rev. I, 1977, p. 288.
9. S. Cochran and M. Guinan, Lawrence Livermore National Laboratory Internal Report UCID-17105, 1976 (unpublished).
10. D. Steinberg and R. Sharp, to be published in J. Appl. Phys.

ON A CRITERION FOR THERMO-PLASTIC SHEAR INSTABILITY*

T. J. Burns, D. E. Grady and L. S. Costin
Sandia National Laboratories,† Albuquerque, NM 87185

ABSTRACT

Dynamic torsional Kolsky (split-Hopkinson) bar experiments on thin-walled tubes of 1018 cold-rolled and 1020 hot-rolled steel are modeled using a deformation plasticity theory which incorporates a specific constitutive model for the shear stress in terms of strain, strain-rate, and temperature into a system of differential equations. The exact time-dependent homogeneous flow solution of the equations is found and used to derive a special case of a generally accepted instability criterion. For given material parameters, this criterion predicts a critical strain at which a homogeneous deformation can bifurcate into a localized deformation, i.e., a shear band, at constant strain-rate. Stability diagrams of strain-rate vs. strain can be constructed for the two types of steel using the criterion. The Kolsky bar data is shown to be consistent with this analysis, and an explanation for the instability criterion is given which assumes that small perturbations on the non-steady homogeneous flow are isentropic to first order.

INTRODUCTION

The dynamic flow stress of a metal is frequently modeled[1] as a function not only of the strain but also of the strain-rate and temperature. Thus, within this framework, for a simple shearing deformation, the flow stress satisfies a rheological equation of the form

$$\tau = f(\gamma, \dot{\gamma}, \theta) \quad , \qquad (1)$$

where τ is the shear stress, γ and $\dot{\gamma}$ are the engineering strain and strain-rate, and θ is the temperature. There is a great deal of experimental evidence[2,3] which indicates that homogeneous shearing deformations at high strain rates becomes unstable to small perturbations, so that shear bands can form, when the condition

$$d\tau = Qd\gamma + Rd\dot{\gamma} - Pd\theta = 0 \qquad (2)$$

is satisfied, where $Q = \partial\tau/\partial\gamma|_{\dot{\gamma},\theta}$ is the work-hardening, $R = \partial\tau/\partial\dot{\gamma}|_{\gamma,\theta}$ is the strain-rate hardening, and $P = -\partial\tau/\partial\theta|_{\gamma,\dot{\gamma}}$ is the thermal softening; generally, Q, R, and P are positive, so that the flow stress increases with increasing strain and strain-rate and decreases with increasing temperature. Because of the strain-rate and temperature dependence in (1), the stress and temperature in a homogeneous

---

*This work was supported by the U. S. Department of Energy under Contract DE-AC04-76-DP00789.
†A U. S. Department of Energy facility.

shearing deformation of a material governed by (1) is highly time-dependent. It is well known from the theory of hydrodynamic stability[4] that this non-steadiness of the homogeneous flow must be taken into account in studying the stability of the flow with respect to small perturbations. The purpose of this paper is to present a derivation of (2) which takes into account the non-steadiness of the homogeneous flow, for the specific constitutive equation[5]

$$\tau = c[1 - a(\theta - \theta_o)](1 + b\dot{\gamma})^m \gamma^n , \qquad (3)$$

where the parameters have been determined by Costin et al.[6] for 1018 cold-rolled steel and 1020 hot-rolled steel, and to show that (2) is consistent with the experimental results obtained by Costin et al. The derivation of (2) assumes that small perturbations on the homogeneous flow are isentropic to first-order; however, the method of analysis used here can also be applied without this assumption.

DERIVATION OF CRITERION

Under the assumption that the flow is adiabatic, incompressible, rigid-thermoplastic, and simple shear, the equations describing momentum and energy balance are given by

$$\rho_o \frac{\partial^2 u}{\partial t^2} = \frac{\partial \tau}{\partial x} ,$$

$$\rho_o C_v \frac{\partial \theta}{\partial t} = \tau \frac{\partial^2 u}{\partial x \partial t} , \qquad (4)$$

where $u(x,t)$ is the displacement in the y direction, ρ_o and θ are the density and temperature, and C_v is the specific heat at constant volume of the material. The constitutive law (3), with $\gamma = \partial u/\partial x$ and $\dot{\gamma} = \partial^2 u/\partial x \partial t$, and the initial and boundary data

$$u(x,0) = 0 , \quad \theta(x,0) = \theta_o ,$$

$$\frac{\partial u}{\partial t}(0,t) = 0 , \quad \frac{\partial u}{\partial t}(d,t) = v_o , \qquad (5)$$

$$\frac{\partial \theta}{\partial x}(0,t) = 0 , \quad \frac{\partial \theta}{\partial x}(d,t) = 0 ,$$

where d is the sample thickness and v_o is the initial velocity of the upper boundary of the sample, complete the specification of the problem. System (4) and (5) has the simple non-steady homogenous solution,

$$u_h(x,t) = \dot{\gamma}_o xt , \quad \dot{\gamma}_o = \frac{v_o}{d} ,$$

$$\theta_h(t) = \theta_o + \frac{1}{a}\left\{1 - \exp\left[-\frac{ac(1 + b\dot{\gamma}_o)^m}{\rho_o C_v(n+1)}(\dot{\gamma}_o t)^{n+1}\right]\right\} \qquad (6)$$

Letting $\gamma_h = \dot{\gamma}_o t$, the corresponding stress is given by $\tau_h(\gamma_h(t), \dot{\gamma}_o, \theta_h(t))$. This stress changes from increasing to decreasing with time when $d\tau_h/dt = 0$, which corresponds to (2) for the constitutive equation (3). In terms of γ_h instead of time (2) corresponds to the condition that $d\tau_h/d\gamma_h = 0$, which occurs at the critical homogeneous strain of

$$\gamma_{crit} = \left[\frac{\rho_o C_v n}{ac(1+b\dot{\gamma}_o)^m} \right]^{\frac{1}{n+1}} \tag{7}$$

Using the parameters of Costin et al., equation (7) predicts critical strains, at which shear bands can form, of 0.18 at $\dot{\gamma}_o = 500/\text{sec}$ for 1018 cold-rolled steel and of 0.85 at $\dot{\gamma}_o = 1000/\text{sec}$ for 1020 hot-rolled steel. This is consistent with the experimental results of Costin et al. at the same constant strain rates, in which shear bands were found to occur after strains of 0.1 to 0.2 in the cold-rolled steel and no heterogeneous deformations were found in the hot-rolled steel after strains of up to 0.7. Equation (7) can thus be used to plot stability diagrams of γ_{crit} vs. $\dot{\gamma}_o$ for the two types of steel.

NON-STEADY STABILITY ANALYSIS

The assumption that any small perturbations are isentropic to first order is equivalent to assuming that the perturbations satisfy the linear variational equation of the nonlinear system

$$\rho_o \frac{\partial^2 u}{\partial t^2} = \frac{\partial \tau}{\partial x} \quad , \quad \tau = \tau_h\left(\frac{\partial u}{\partial x}\right) \quad , \tag{8}$$

with respect to the homogeneous solution $u_h(x,t) = \dot{\gamma}_o x t$. This variational equation is obtained by linearizing (8) about the homogeneous solution, and thus it is given by

$$\frac{\partial^2 \delta u}{\partial t^2} = F(t) \frac{\partial^2 \delta u}{\partial x^2} \quad ,$$
$$\delta u(0,t) = \delta u(d,t) = 0 \quad , \quad \delta u(x,0) = \mu(x) \quad , \tag{9}$$

where

$$F(t) = \frac{1}{\rho_o}\left(\frac{d\tau_h}{d\gamma}\right) \quad \gamma = \dot{\gamma}_o t \quad ,$$

and $\mu(x)$ is the small initial perturbation amplitude. It is standard to decompose the initial perturbation $\mu(x)$ into normal Fourier modes and to analyze the stability of each mode, given by $\beta = 2\pi k/d$, where $k = 1, 2, \ldots$. When this is done, it is clear that (9) does <u>not</u> have any solution of the form $e^{\alpha t + i\beta x}$, due to the time dependence of the squared "sound speed" $c^2 = F(t)$. However, it is easy to see that (9) does have solutions of the form $\delta u(x,t) = U_\beta(t) \sin \beta x$, which leads to the non-steady stability equation for the growth of Fourier modes,

$$\frac{d^2 U}{dt^2} + \beta^2 F(t) U = 0 , \qquad (10)$$

where the subscript β has been dropped. For any stress-strain relationship $\tau_h(\gamma_h)$ which increases for strains less than the critical strain and decreases otherwise, $F(t)$ is positive for $t < \gamma_{crit}/\dot{\gamma}_o$ and negative otherwise, and it can be shown that in this case, any solution of (10) must remain bounded and oscillatory for $F(t)$ positive and grow exponentially for $F(t)$ negative. This may explain the observations that shear bands form after strains greater than the critical strain in metals. This analysis is also consistent with the numerical calculations of Olson et al.,[7] who used an isentropic stress-strain relationship of the form (8) and found that shear bands formed in a type of steel after the critical strain γ_{crit} was exceeded in pure shear.

The approach of this section can also be applied to non-isentropic variational systems derived from (4), (5), (6), but the analysis becomes considerably more difficult, so that numerical or other approximate methods appear to be necessary for the stability analysis. It should be emphasized, however, that the method of computing eigenvalues, which is commonly used in studying the stability of steady flows, can lead to incorrect conclusions about the stability of non-steady flows.

ACKNOWLEDGMENT

The authors would like to thank S. T. Montgomery and T. G. Trucano for reviewing the manuscript.

REFERENCES

1. Y. L. Bai, in Shock Waves and High Strain-Rate Phenomena in Metals, Edited by W. A. Meyers and L. E. Murr, (Plenum, NY, 1981) p. 277.

2. R. S. Culver, in Metallurgical Effects at High Strain Rates, Edited by R. W. Rohde et al., (Plenum, NY, 1973) p. 519.

3. M. R. Staker, Acta Metallurgica 29, 683 (1981).

4. P. G. Drazin and W. H. Reid, Hydrodynamic Stability, (Cambridge Univ. Press, UK, 1981) p. 353.

5. J. Litonski, Bull. de l'Academie Polonaise des Sci. 25, 7 (1977).

6. L. S. Costin et al., Inst. Phys. Conf. Ser. No. 47, Chapter 1, p. 90 (1979).

7. G. B. Olson et al., in Shock Waves and High Strain-Rate Phenomena in Metals, Edited by W. A. Meyers & L. E. Murr, (Plenum, NY, 1981) p. 221.

DYNAMIC COMPACTION OF ELASTIC-VISCO-PLASTIC POROUS MATERIALS UNDER SHOCK

K. Kim
Naval Surface Weapons Center
White Oak, Silver Spring, Maryland 20910

S. I. Oh
Battelle Columbus Laboratories
Columbus, Ohio 43201

ABSTRACT

When organic crystals such as RDX are compacted by a strong shock, their material behavior near pores may be mathematically described as elastic-visco-plastic. The elastic contribution is from elastic deformation of small sub-particles while the visco-plastic contribution is from shear banding and friction between the sub-particles. The hollow sphere model of Carroll and Holt[1] has been applied to this problem to obtain an expression describing dynamic compaction.

INTRODUCTION

The development of constitutive equations to describe the dynamic compaction behavior of porous materials is a topic of considerable interest. This paper identifies friction, particle rearrangement without much friction, shear banding, and plastic deformation as various forms of mechanical deformation during compaction of porous organic crystals. Their deformation behavior is put together and expressed in the context of the hollow sphere model[1,2] as elastic-viscoplastic. The hollow sphere model, explained in brief, assumes that the only relevant aspects of pore geometry during compaction of porous systems are the initial porosity and typical pore diameter. Then one needs only to consider an alternative system: a hollow sphere of the matrix material whose inner and outer radii are such that the pore diameter and the overall porosity are those of the porous material.

DESCRIPTION OF MATERIAL PROPERTIES

A phenomenological description of the compaction process of porous systems is as follows. If the porosity is relatively high, then the particles try to rearrange themselves, like a particle finding its way into a hole which is larger than or approximately equal to its size, without requiring too much outside force applied to the system. As the porosity becomes smaller, the rearrangement of the original size particles becomes difficult, and a significant amount of friction exists between the particles. Then, in rubbery materials, plastic (either visco- or non-visco-) deformation takes place to fill voids further. In brittle materials, dislocation planes will be generated across the particles to accomodate the required deformation. Depending on material properties, the dislocation planes can either combine together creating thick shear bands or can break the particles. In the latter case, the broken particles will try to rearrange themselves (with or without significant friction) or further break down into still smaller particles.

In compaction of organic crystals (most of the explosive materials are organic crystals), all of the processes mentioned are assumed to occur. Evidence of the brittle behavior can be seen in Reference 3, where an RDX crystal surface was examined after indentation. Many slip (or dislocation) planes crisscrossing the surface can be observed near the indented area. Evidence of the rubbery behavior can also be observed[4] where RDX crystals were subjected to pressures of up to 30 Kbars and showed the gradual disappearance of their boundaries with neighboring particles and gradual loss of crystallinity.

Of course, the study of every shear band, frictional surface between every particle, or rearrangement motion is not possible. One wonders, then, if there is a way to predict a global behavior of the deforming material using a continuum approach assuming a statistical randomness of the irregularities in a given control volume small enough to be part of a single organic crystal, but sufficiently large to contain many smaller irregularities so as to warrant a statistical average. This approach forms the basis of this paper.

It is recognized that in this control volume, global stress would be dependent upon both global strain and strain rate. It is also recognized that there is a global yield strength of the material which may be close to the original yield strength of the solid material. For the sake of elucidating the salient features of compaction of a shocked porous medium, a linear elastic-viscoplastic material description is given to this control volume.

One still needs further simplification to study the three-dimensional compaction process. This is done by adapting the hollow sphere model discussed in the Introduction. Details of the model are shown in Figure 1. The goal of the study is now reduced to calculating u, σ_r, and σ_θ when P_0 is applied at the time origin at the outer surface of the hollow sphere whose outer and inner radii are r_0 and r_i, respectively.

ANALYSIS

The set of equations describing the above system, neglecting inertial effects and bulk compressibility, is:

$$\frac{\partial \sigma_r}{\partial r} + 2 \frac{\sigma_r - \sigma_\theta}{r} = 0 \tag{1}$$

$$\frac{\partial v}{\partial r} - \frac{v}{r} = \sqrt{3} \, \gamma \left[\frac{\sigma_r - \sigma_\theta}{\sqrt{3} \, k} - 1 \right] + \frac{1}{2G} \frac{\partial}{\partial t} (\sigma_r - \sigma_\theta) \tag{2}$$

$$\frac{\partial v}{\partial r} + 2 \frac{v}{r} = 0 \tag{3}$$

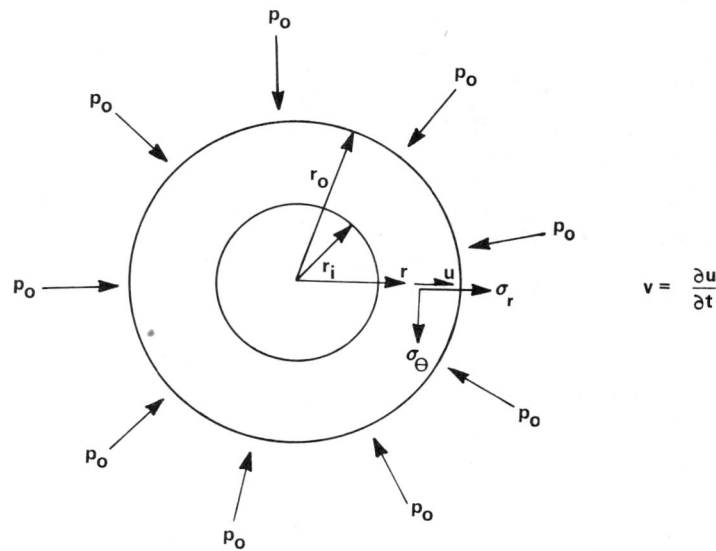

Fig. 1. Hollow sphere model. r_i is the pore radius; r_o is the solid radius; r is the space coordinate; u is the displacement; σ_r is the radial stress; σ_θ is the tangential stress; p_o is the applied stress; and τ is the time.

with initial and boundary conditions,

$$\sigma_r(r_i, t) = 0, \quad \sigma_r(r_o, t) = -P_o \qquad (4), (5)$$

$$\sigma_r(r, 0) = 0, \quad \sigma_\theta(r, 0) = 0 \qquad (6), (7)$$

where $v = \frac{\partial u}{\partial t}$, γ represents a proportionality between stress and strain rate, k is shear yield strength, G is shear modulus, and t is time. Equation (1) describes the force balance of the system, Equation (2) describes the linear-viscoplasticity, and Equation (3) relates volume conservation of the solid material. The negative sign in Equation (5) indicates that P_o is a compressive stress.

The solution for the set of equations can be obtainable in a closed form. For example, v can be expressed as:

$$v(r, t) = \delta\left\{\frac{P_o}{4G(r_o^{-3} - r_i^{-3})r^2}\right\} + \frac{P_o - 2\sqrt{3}\, k\, \ln(r_o/r_i)}{2(r_o^{-3} - r_i^{-3})r^2 \eta} \qquad (8)$$

where $\eta = \frac{k}{\gamma}$ and $\delta\{\}$ represents a delta function at t = 0. Integration of v over time would give compaction behavior of the hollow sphere, and, therefore, pore collapse behavior of the original porous materials. A typical calculation result is shown in Figure 2.

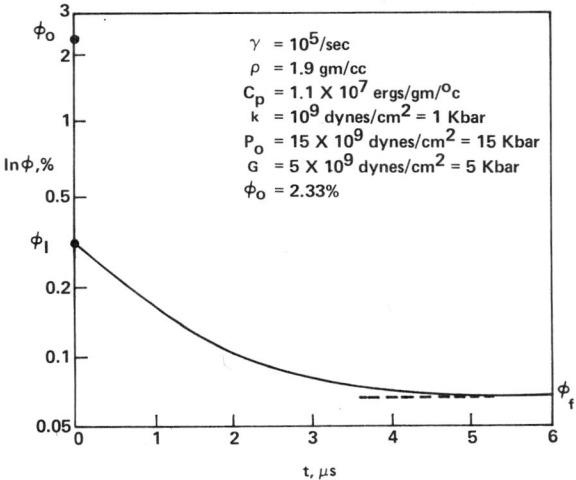

Fig. 2. Typical dynamic compaction process

COMPARISON OF RESULTS WITH EXPERIMENTAL DATA

It is difficult to compare the results directly with experimental observation of dynamic compaction. It is because exact values of γ and G are not accurately known.

Asymptotic and indirect comparisons, however, are possible. First, the final compaction state can be compared against the empirical static compaction law in the literature.[5] The empirical law is shown as:

$$P = \alpha k \ln \frac{1}{\phi} - \beta , \qquad (9)$$

where $\phi = \left(\frac{r_i}{r_o}\right)^3$ is porosity. This compares favorably with the final stage of the dynamic compaction obtainable by taking the limiting value of v equal to zero as time approaches infinity in Equation (8), as follows:

$$P = 2 \frac{k}{\sqrt{3}} \ln \frac{1}{\phi} . \qquad (10)$$

Equation (10) is identical to the final compaction state obtained by Carroll and Holt.[1]

Indirect comparisons come from examination of shock sensitivity data of porous explosives.[6] If one assumes that the shock sensitivity of porous explosives is directly related to the maximum temperature obtainable from the elastic-viscoplastic compaction process discussed above, the following observation can be explained qualitatively. For details, see Reference 6.

(A) Initial porosity effect on sensitivity: It has been observed that increased initial porosity enhances the shock sensitivity of porous explosives.

(B) Ramp wave effect on sensitivity: A ramp wave whose rise time is beyond a certain finite value is less effective in initiation of explosives than a step stress, even if the final plateau stress after the rise time is the same as the step stress.

(C) Observation of P-τ relationship in sensitivity: A square pulse stress of magnitude, P, is required to have a certain finite duration, τ, for initiation of given explosives. A typical example is $P^2\tau$ = const.

SUMMARY

A phenomenological description of dynamic compaction of shocked, porous, organic crystals has been given with a use of a hollow sphere model. A closed form expression of time-dependent compaction behavior was obtained for linear-elastic-viscoplastic materials. The result has been compared indirectly against existing experimental observations. The comparison is favorable.

REFERENCES

1. M. M. Carroll and A. C. Holt, J. Appl. Phys. 43, 4 (1972).
2. A. C. Holt, M. M. Carroll, and B. M. Butcher, Pore Structure and Properties of Materials (Academia, Prague) (1974).
3. W. L. Elban and R. W. Armstrong, 7th Symp. on Det. (1981).
4. P. Miller, private communications (1981).
5. R. W. Heckel, Trans. Metallur. Soc. of AIME, 221 (1961).
6. K. Kim, Proc. JANNAF Propulsion Systems Hazards Subcommittee Mtg., Monterey, (1980).

NUMERICAL ANALYSIS OF HOPKINSON BAR EXPERIMENTS ON DISLOCATION DYNAMICS*

Minao Kamegai
University of California, Lawrence Livermore National Laboratory
Livermore, California 94550

ABSTRACT

We have simulated with a two-dimensional hydrocode, HEMP, the Hopkinson bar experiments on dislocation dynamics in pure aluminum polycrystals. The experiments were performed by Dharan and Hauser[1] using a modified Kolsky thin-wafer technique. The flow stress in aluminum was measured at strain rates, ranging from 10^3 to 10^5 sec^{-1}. Dharan and Hauser analyzed these results in terms of dislocation damping. The HEMP simulation utilizes a phenomenological model of work-hardening applied to the aluminum strength. The computation gives a clear picture of stress relaxation in the aluminum specimens. A comparative study was made on the dislocation model and the work-hardening model. An essential difference between the two models is scalability. Based on the scaling experiments on aluminum conducted by Wilkins and Guinan[2], an argument is presented that the dynamics of aluminum deformation is a work-hardening process rather than dislocation damping.

REFERENCES

1. C.K.H. Dharan and F. E. Hauser, Journal of Applied Physics **44**, 1468 (1973).
2. M. L. Wilkins and M. W. Guinan, Journal of Applied Physics **44**, 1200 (1973).

*Work performed under the auspices of the U.S. Department of Energy by Lawrence Livermore National Laboratory under contract #W-7405-Eng-48.

QUASI-ELASTIC HIGH-PRESSURE WAVES IN 2024 Al AND COPPER*

C. E. Morris, J. N. Fritz and Brad Lee Holian
Los Alamos National Laboratory, Los Alamos, NM 87545

ABSTRACT

Release waves from the back of a plate slap experiment are used to estimate the longitudinal modulus, bulk modulus and shear strength of the metal in the state produced by a symmetric collision. The velocity of the interface between the metal target and a window material is measured by the axially symmetric magnetic (ASM) probe. Wave profiles for initial states up to 90 GPa for 2024 Al and up to 150 GPa for Cu have been obtained. Elastic perfectly-plastic (EPP) theory cannot account for the results. A relatively simple quasi-elastic plastic (QEP) model can.

INTRODUCTION

States of solids along shock loci have stresses more complicated than the simple fluid pressure that has been assumed for simplicity in the high pressure regime. In the past decade three techniques have become available that can record continuous wave profiles at very high pressures which can be used to study elastic-plastic flow in solids at these extreme conditions. They are the VISAR[1], the ASM probe[2], and the use of radiation from shock fronts in transparent materials[3]. In this brief communication we shall give some preliminary results obtained from use of the ASM probe on plate slap experiments where the driver has been accelerated by high explosive systems. Asay and Chhabildas[1], in their work on 6061Al, describe similar experiments and analysis. Because of the brevity of this communication, we rely heavily on their paper for discussion of the concepts involved and references to previous work.

EXPERIMENTAL, SIMPLISTIC RESULTS

Figure 1 shows the ASM probe assembly and driver before impact. Fringing lines from the magnet are pinned to the front face of the target. Motion of the front face is taken on by the magnetic field lines, resulting in a loss of flux in the coil. The induced signal in the coil can be analyzed to give u(t) of the target front face.

Figure 2 shows the interactions of interest in our experimental system. Impact at 1 and outgoing shock waves, release at 2 and 3, the extended interaction of these release waves 4567, and the extended interaction of the forward moving release with the window 8910. The probe records the velocity of the interface, i.e., along the path 3, 8, 10..... An ideal window material would match the metal in impedance everywhere, have low (and independent of pressure) wave velocities, and have a sufficiently low conductivity to be transparent to a diffusing magnetic field. Such does not exist. In these high pressure experiments it was necessary to design in a large

---

* Work supported by the US Department of Energy.

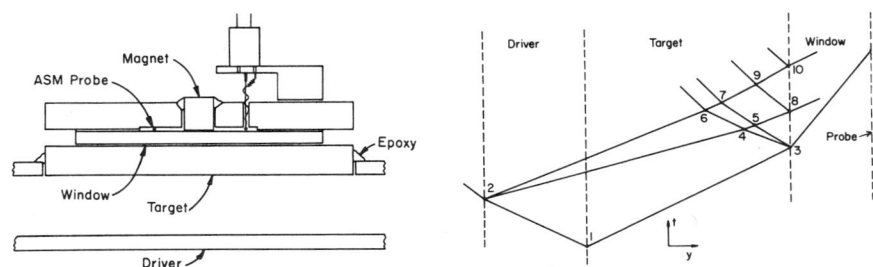

Fig. 1 Experimental Assembly   Fig. 2 Lagrangian y-t Diagram

plateau time, t_8-t_3, so that the wave from the interaction 8 would not overtake the initial shock in the window, interact and come screaming back to confuse the information recorded between t_8 and later. This causes the 4567 interaction to be buried more deeply in the target, which makes the change in slope of the characteristics of the foward facing release a non-neglectable correction. If the window does not match the metal in impedance (and it is the match of small waves around the high pressure states that we must be concerned with) a further untangling of the 8910 interaction must be done. The goal of this analysis is the $\sigma_n(\eta)$ path ($\eta = 1-\rho_0 V$) in a simple wave (also $\sigma_n(u)$ for impedance matching). Given the symmetric impact, time independent flow, uniqueness of the release path (i.e. experimental design is such that no hysteresis occurs), and a complete EOS of the window material, a characteristic code analysis could be done that would give $\sigma_n(\eta)$ from the measured u(t). It would not be a simple forward analysis nor a simple backward one, since unscrambling 4567 depends on the information at 8910, which in turn depends on 4567. A characteristics code would be useful whose elementary steps are solved by integral rather than difference equations. It is however, a determined problem (aside from a necessary extrapolation of the $\sigma_n(\eta)$ curve) and amendable to an iterative analysis. In lieu of such an analysis we have utilized concepts from EPP theory to analyze the results. This amounts to treating the characteristics in Fig. 2 as "shocks" carrying discontinuous waves, the earliest being the elastic release, and later, the bulk release, or an appropriate fraction of it. This results in the interaction diagram shown in Fig. 3. The striking feature of this is the large fraction of the release that is accomplished "elastically".

A typical analyzed experimental record is shown in Fig. 4 for a mid range 2024 Al shot. The u_p of the plateau is used to determine a point on the window Hugoniot and, with an impedance match, the initial metal state. The time difference, t_8-t_3, primarily determines the longitudinal velocity in this metal state and hence the initial slopes in the σ_n-u diagram. The velocity difference, u_3-u_{10} determines the size of the structure in the interaction diagram, and hence determines the change in shear stress[1], $\tau_0+\tau_c$, that the material would support. Complete use of EPP concepts in this scheme would result in a shear strength too large by a factor of

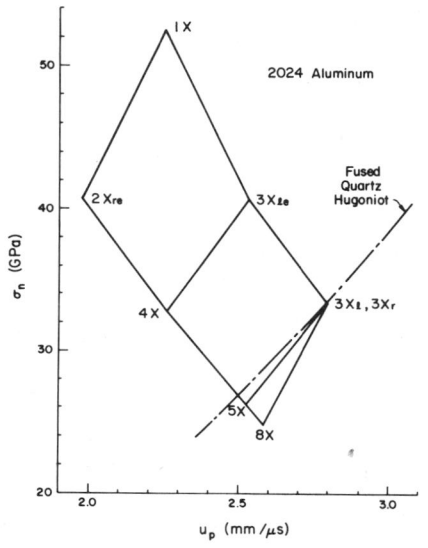
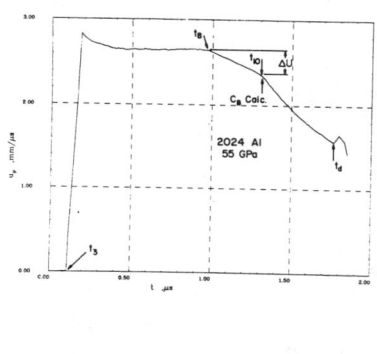

Fig. 3 Simplified EPP σ-u Diagram   Fig. 4 Exp. u(t) Record

two. Sufficient dispersion in the wave velocities (curvature in the $\sigma_n(\eta)$ plane) was included to match the ramp in the u(t) curve. The knowledgeable reader will recognize the corrections and interactions necessary in these procedures. Figure 3 was for a fused quartz window while Fig. 4 was for a teflon window. They are close enough for illustrative purposes, but the data we shall now present was taken using teflon as a window material. Further characterization of fused quartz and a high density leaded glass needs to be done before we can rely on our results for these windows.

Figures 5 and 6 show C_L for the two metals studied here. We

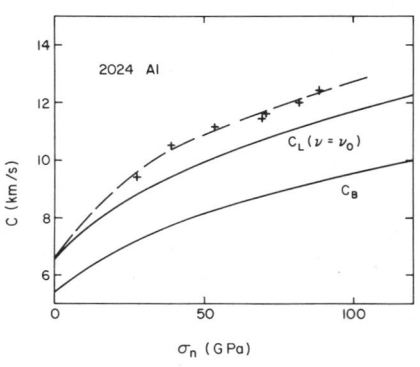
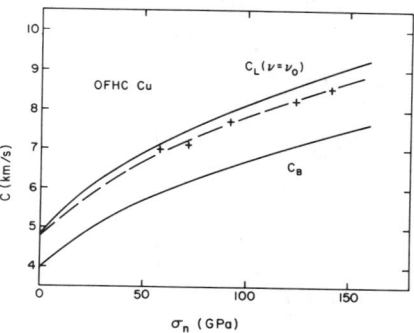

Fig. 5 Sound Speed, 2024 Al   Fig. 6 Sound Speed, OFHC Cu

observe that the longitudinal sound speed lies above and below the curve where Poisson's ratio equals a constant. This behavior for these two metals agrees with the systematics inferred by Roman et.al.[4] from lower pressure data. The disappearance of the longitudinal velocity indicating the onset of melting has not been observed at the highest pressures we have obtained with our explosively driven drivers.

Figure 7 shows the shear strengths we have obtained for these two metals. We have clearly gone past the maximum value for both of these metals. It is tempting to use an extrapolation of these curves to estimate the melting transition, but it is not clear that a 'reasonable extrapolation' of these curves would necessarily give even an upper bound to the melting transition.

DISCUSSION

We must add several caveats to the data as we have presented them here. Most of the uncertainty arises from explosively driven drivers. Wave traversal time through the driver is quite important in our experiments and hence a stretching of the driver could cause us to overestimate C_L. A 1% thinning of the driver could cause a 1.8% (in 2024 at 55 GPa) and a 1.1% (in Cu at 70 GPa) overestimate. Our experiments have a self calibrating feature in that the time from initial plate motion to magnet destruction gives the shock velocity through the window material. Since these two events are not at the same radius, bow in the driver would cause this shock velocity to be overestimated. We have used the window shock velocity to estimate the pressure of the interaction. This has yielded pressures 2-10% higher than those obtained from the plateau u(t) associated with the interface. Elimination of a positive bow effect and a suspected ~2-3% error in magnet calibration due to a probe misalignment would bring these pressures into agreement. Framing camera studies on drivers have been performed but are as yet unanalyzed. Many of these problems will be eliminated and higher pressure will be achieved by transferring these experiments to our two-stage light gas gun. A better knowledge of the 8910 target-window interaction must be

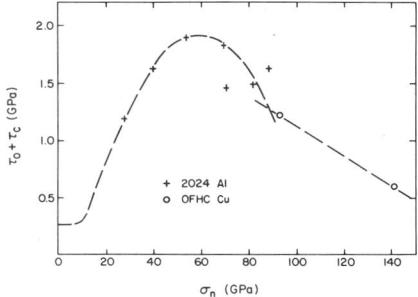

Fig. 7 Shear Strength vs Initial σ_n in the Metal. These data are not in agreement with Al'tshuler, et al.[5]

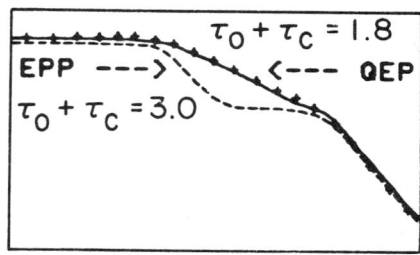
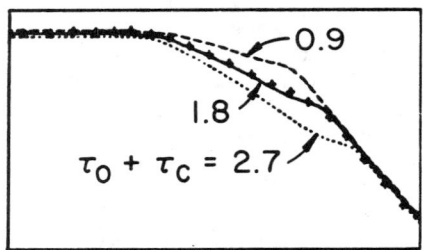

Fig. 8 EPP-QEP Comparison   Fig. 9 QEP with Various $\tau_o+\tau_c$

obtained. A series of experiments investigating release waves in suitable window materials is underway.

In order to verify our simplistic method of calculating C_L and $\tau_o+\tau_c$ we have performed some 1D Lagrangian hydrodynamic calculations for the 55 GPa 2024 Al record. Figures 8 and 9 show these calculations. The EPP model requires 3.0 GPa for $\tau_o+\tau_c$ and does not fit the data. Our QEP local model is characterized by the following expression for the rigidity modulus:

$$\mu(\sigma_n) = \mu_o + \mu_o' \ (\sigma_n - \tfrac{4}{3}\tau) \ (\tau_c+\tau)/(\tau_c+\tau_o)$$

The dimensionless quantity μ_o' was chosen to give the right C_L at pressure and is in excellent agreement with the ultrasonic data of Thomas[6]. A global model such as that of Steinberg et.al.[7] would incorporate the dependence of μ_o' on pressure and temperature to a greater extent. From Fig. 9 one can see that $\tau_o+\tau_c$ is fairly well determined within the assumptions of the model and agrees with our earlier simplistic method of estimating it. Nothing profound is implied by this model other than it gives a $\sigma_n(\eta)$ curve close to the experimental one.

For this analysis to be complete, reloading waves need to be introduced into the driver-target. This is almost impossible to do with explosively driven flyers. This will also be done on the gas gun.

REFERENCES

1. J. R. Asay and L. C. Chhabildas, in Shock Waves and High Strain Rate Phenomena in Metals, edited by M. A. Meyers and L. E. Murr (Plenum Press, New York, 1981).
2. J. N. Fritz and J. A. Morgan, Rev. Sci. Instr. 44, 215 (1973).
3. R. G. McQueen, these proceedings.
4. J. P. Romain, A. Migault, and J. Jacquesson, J. Phys. Chem. Solids 37, 1159 (1976).
5. L. V. Al'tshuler, M I. Brazhnik, and G. S. Telegin, J. Appl. Mech. Tech. Phys. 12, 921 (1971) (English Translation).
6. J. F. Thomas, Jr. Phys. Rev. 175, 955 (1968).
7. D. J. Steinberg, S. G. Cochran, and M. W. Grunan, J. Appl. Phys. 51, 1498 (1979).

SHOCK RELEASE OF 2024-T351 ALUMINUM IN THE 10 GPa RANGE

Y. Partom, D. Yaziv and Z. Rosenberg
A.D.A., P.O. Box 2250, Haifa, Israel.

ABSTRACT

We conducted plane-impact experiments to measure release adiabats of 2024-T351 aluminum. We used our 2.5" laboratory powder gun to accelerate aluminum and copper impactors at the aluminum sample targets. The target discs were backed by a lower impedance material to cause the desired amount of unloading. Stress histories were measured by manganin gauges embedded in the target. The results are not predicted by the classical elastoplastic model. Release adiabats from different shock levels do not coincide, and release and shock adiabats are not symmetrical with respect to the hydrostatic adiabat. We were able to account for the experimental results by means of our "grain model". We adjusted the grain parameters to match the experimental shock adiabat, and then used them to predict release adiabats. We obtained a good agreement with the experimental data.

INTRODUCTION

Measurements were designed to accurately determine the release adiabats in 2024-T351 Al. Shock data in this aluminum have not been published. Fowles[1] and other[2,3] have measured the shock adiabat in 2024 Al. Barker[4] inferred the release adiabat of 6061 Al from a free-surface velocity history. Kusubov and Van Thiel integrated manganin stress gauge histories in 5050B Al and 2024-T4 Al to obtain release adiabats, but the accuracy seemed poor. We concluded that to determine release adiabats accurately we need to measure stress and particle velocity along the release path, and we designed our experiments accordingly.

EXPERIMENTAL

Our samples were made of 2024-T351 aluminum which contains 4.4% Cu, 1.5% Mg, and 0.6% Mn, and has a mass density of 2.774 gr/cm^3. The treatment T351 is described in Ref. 7.

The plane impact experiments were conducted in our terminal ballistics laboratory with a 2.5" powder gun described in Ref. 8. We record stress histories in the target by means of manganin stress gauges emplaced as described in Refs. 9 and 10. Our gauge calibration is similar to that in Ref. 11.

Release adiabats of 2024 aluminum were obtained in five plane impact experiments. The target consists of two 100 mm diameter and 5 mm thick aluminum discs, and a backing disc of PMMA, magnesium, or water. Manganin gauges were placed between the two aluminum discs and occasionally between the aluminum and the backing disc. A schematic gauge output is shown in Fig. 1.

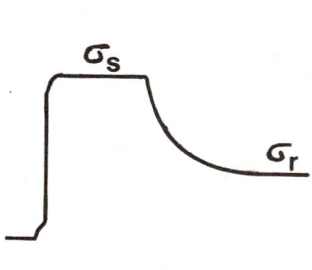

Fig. 1. Schematic gauge output.

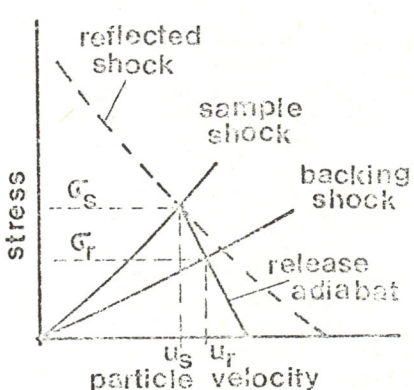

Fig. 2 Impedance mismatch for obtaining u_r.

From the measured stress levels σ_s and σ_r, a point on the release adiabat in the stress-particle velocity (σ, u) plane is inferred through the impedance mismatch shown in Fig. 2. The Hugoniot curves for Al and Mg needed for the mismatch computation were taken from Ref. 3, that for PMMA from Ref. 12, and that for water from Ref. 2. The results are in Table I.

The release points from Table I are plotted in Fig. 3 together with the shock adiabat (Hugoniot curve) and theoretical curves from the model to be presented. For each shock level σ_s there is a different release curve, rather than a common curve as predicted by the elastic-perfectly-plastic (EPP) theory. A model to represent these release curves is presented next.

Table I. Shock release Experiments

Impactor[a] Velocity m/sec	σ_s/σ_r GPa	$u_r^* = 2u_s - U_r$ m/sec
806[b]	6.58/4.52	296
797	6.53/2.55	186
476	3.75/1.45	105
728[c]	5.90/1.65	130
378[d]	10.50/4.34	294

a Al impactor and PMMA backing plate except as noted.
b Mg backing plate
c Water backing material
d Cu impactor

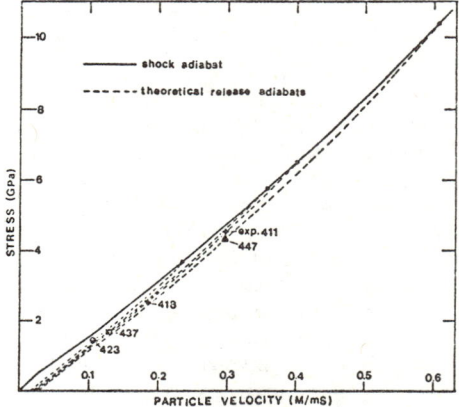

Fig. 3. Release data points compared to grain model theoretical predictions in the $(\sigma-u^*)$ plane.

GRAIN MODEL

Grain models, such as those of Duwez[13] and Taylor,[14] represent polycrystals made of many randomly oriented single crystals. Non-self-consistent (NSC) grain models require displacement or stress continuity: in ours[15,16] all grains have the same strain, the macroscopic strain. In this model each grain is isotropically elastic and has one slip plane which can slip when the shear stress reaches a critical value.

The macroscopic stress state in a planar impact (uniaxial strain) is characterized by the principal stresses σ_x, σ_y, and σ_y in an xyz coordinate system with flow in the x-direction. We also assume that this is the stress state in each grain.

To obtain the shear stress on the slip plane, define a local coordinate system pqr, where pq is the slip plane and p the slip direction. With the direction cosines μ_{ij}, we obtain the shear stress as

$$\tau = \sigma_{pr} = (\sigma_x - \sigma_y)\mu_{xp}\mu_{xr} = \frac{3}{2}\sigma'_x\mu_{xp}\mu_{xr} = f\sigma'_x \quad (1)$$

where σ'_x is the deviator stress. In the model the continuum of slip plane orientation is divided into N discrete, equally spaced orientations.

Macroscopic stress is computed by imposing uniaxial strain ($\varepsilon = 1 - \rho_0/\rho$) in a series of increments $\Delta\varepsilon$ where ρ is density. The pressure P is computed from the experimental adiabatic hydrostat of Syassen, et al:[16]

$$P = 75.9\varepsilon + 201.2\varepsilon^2 + 368.3\varepsilon^3 \quad \text{(GPa)} \quad (2)$$

The deviator stress is computed elastically from

$$\Delta\sigma'_x = (4/3) G\Delta\varepsilon \quad (3)$$

where the measured[17] variation of G with ε is used:

$$G = 29.0 + 233\varepsilon + 938\varepsilon^2 \quad \text{(GPa)} \quad (4)$$

The shear stress τ_i on the ith grain is obtained either from the elastic relation (3) or, if yielding has occurred, from the yield function τ_y, which was fitted to the experimental shock adiabat

$$\tau_y = .03 - 0.19\varepsilon + 9.5\varepsilon^2 \quad \text{(GPa)} \quad (5)$$

The macroscopic stress is the average over all grains:

$$\sigma'_x = \frac{1}{N}\sum_{i=1}^{N}\tau_i/f_i \qquad \sigma_x = P + \sigma'_x \quad (6)$$

where f_i is the orientation factor defined in Eq. 1.

RESULTS AND DISCUSSION

With the grain model we computed four release adiabats corresponding to the four stress (σ_s) levels in Table 1 and plotted these curves in Fig. 3. The reflected particle velocity u* is computed by:

$$\Delta u^* = \Delta\sigma_x/(\rho_o \bar{C}_L) \quad \text{where} \quad \bar{C}_L^2 = \Delta\sigma_x/(\rho_o \Delta\varepsilon) \tag{7}$$

We see from the figure that the agreement is good. In Fig. 4 we plot the shock adiabat and five computed release adiabats. Also shown is the hydrostatic adiabat given by Eq. (2). We conclude that release adiabats for different shock levels follow different paths, but for shocks stronger than about 8 GPa the lower parts of the release curves practically coincide. Generally, release adiabats and the shock adiabat are not symmetric about the hydrostatic adiabat. This type of behavior is usually referred to as a Bauschinger effect. The results of Fig. 4 are thus clearly not according to the predictions of the classical EPP model.

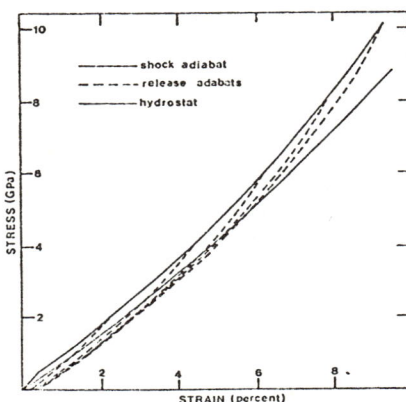

Fig. 4. Computed release adiabats.

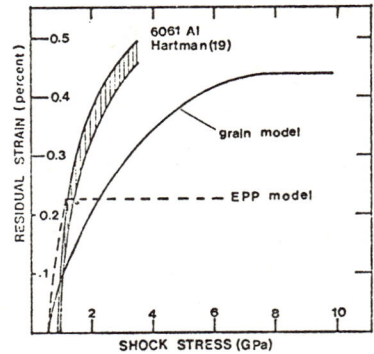

Fig. 5. Residual strain in 2024 aluminum samples that were shocked and then unloaded to zero stress.

Upon complete unloading we get from our grain model a residual strain ε_{res} that increases with σ_s as shown in Fig. 5. By contrast the EPP model predicts a constant value of ε corresponding to the residual pressure $P = 2Y/3$. Experimental values of ε_{res} for 6061 aluminum by Hartman[18] are also shown. His data curve has a shape similar to the grain model prediction.

REFERENCES

1. G.R. Fowles, J. Appl. Phys. $\underline{32}$, 1475 (1961).
2. M. van Thiel ed., Compendium of Shock Wave Data, UCRL-50108(1966).
3. R.G. McQueen et al., High Velocity Impact Phenomena, Edited by R. Kinslow, p.293, Academic Press (1970).
4. L.M. Barker, HDP Symposium, p.483, Gordon and Breach (1968).
5. A.S. Kusubov and M. van Thiel, J. Appl. Phys. 40, 3776 (1969).
6. M. van Thiel and A. Kusubov, Accurate Characterizations of High Pressure Environment, edited by E.E. Lloyd, (Nat. Bur. Stand. U.S. 1971), p.125.
7. Aluminum Standard and Data, Aluminum Association U.S. (1968).
8. Y. Oved, G. Luttwak and Z. Rosenberg, J. Comp. Mat. $\underline{12}$, 84 (1978).
9. Z. Rosenberg, D. Yaziv and Y. Partom, J. Appl. Phys. $\underline{51}$, 3702 (1980).
10. D. Yaziv, Z. Rosenberg and Y. Partom, J. Appl. Phys. $\underline{51}$, 6055 (1980).
11. E. Barsis, E. Williams and C. Skoog, J. Appl. Phys. $\underline{41}$, 5155 (1970).
12. L.M. Barker and R.E. Hollenbach, J. Appl. Phys. $\underline{45}$, 4872 (1974).
13. P. Duwez, Phys. Rev. $\underline{47}$, 494 (1935).
14. G.I. Taylor, J. Inst. Metals $\underline{62}$, 307 (1938).
15. Y. Partom, D. Yaziv and Z. Rosenberg, J. Appl. Phys. $\underline{52}$, to be published (1981).
16. K. Syassen and W.B. Holzapfer, J. Appl. Phys. $\underline{49}$, 4427 (1978).
17. D. Yaziv, Y. Partom and Z. Rosenberg, J. Appl. Phys. $\underline{52}$, to be published (1981).
18. W.F. Hartman, J. Appl. Phys. $\underline{35}$, 2090 (1964).

SHOCK-WAVE COMPRESSION OF A BOROSILICATE GLASS UP TO 170 KBAR*

J. Cagnoux
Centre d'Etudes de Gramat, 46500 Gramat, France

ABSTRACT

Strain mechanism of a borosilicate glass was investigated using manganin gauges in two types of shock-wave experimental assemblies. A method is proposed to obtain the pressure-specific volume and the deviatoric stress-specific volume planes. Above the HEL (∼75kbar), the borosilicate undergoes a significant, but not complete, loss of shear strength, and the material shows a viscoplastic behavior. The two-wave structure, obtained above the HEL, seems consistent with the densification of glass.

INTRODUCTION

The response of materials shocked to high pressures is often assumed hydrodynamic. However, strength effects cannot be ignored at low stresses. Recent shock-wave studies on some non-metallic materials suggest that substantial strength reductions occur during loading. That is the case in quartz rock,[1] polycrystalline tungstene [2] In a previous published paper,[3] we observed, in glass, a constant deviatoric stress above the Hugoniot elastic limit (H.E.L.). We conducted this present study to specify and to attempt to explain this behavior.

EXPERIMENTAL TECHNIQUE AND RESULTS

The glass studied is a borosilicate with a density of 2.226 and longitudinal sound velocity of 5450m/s. Two types of experiments were performed with a gas gun technique and explosive systems.

A. Lagrangian analysis experiments (AL) : Target samples are 80mm diameter. Each experimental assembly was constructed with four parallel glass plates. Stress gauges (300mΩ) were mounted between these slabs, two gauges in each of the three planes.

B. Deviatoric stress experiments (D) : A grid gauge (1Ω) is perpendicularly located to the shock propagation direction, and a wire gauge (300mΩ) is placed in a perpendicular plan to the other gauge. Details of this technique and its validity are described by Chartagnac in Ref 4.

The loading and unloading calibration coefficients of gauge material was provided by Chartagnac.[5]

Figures 1 and 2 show experimental profiles recorded from shots AL5 and D6.

---

* Work supported by French Ministry of Defense

0094-243X/82/780392-05$3.00 Copyright 1982 American Institute of Physics

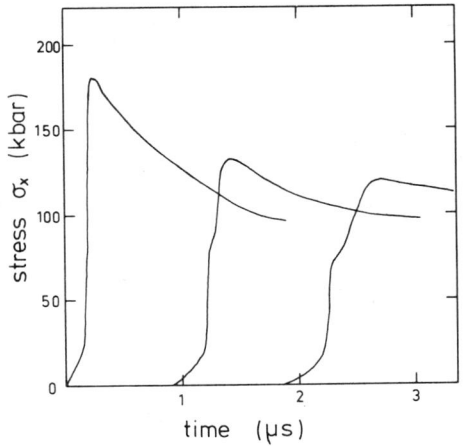

Fig.1.Representative stress-time profiles as obtained in borosilicate glass(shot AL5),for propagation distances 10,15,20mm.

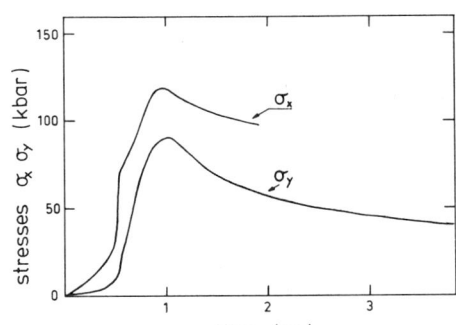

Fig.2.Representative stress-time profiles as obtained in borosilicate glass(shot D6), for propagation distance 10mm. The higher amplitude is from the gauge parallel to the shock front, and the lower amplitude is from the gauge perpendicular to the shock-front.

ANALYSIS OF EXPERIMENTAL RESULTS

Results of shots AL were analyzed with the method described by Seaman.[6] Results of shots D were analyzed with the hypothesis of uniaxial strain.[4] We performed AL experiments and D experiments with the same initial conditions. Thus, an original analysis method can be proposed. This method is only valid if the stress profiles $\sigma_x(t)$ obtained from the two experiments (AL and D) are superposable within the experimental error. The combination method is described on Figure 3. Then, we observed the deviatoric stress S_x and the pressure P ($\sigma_x = P + S_x$) in the stress-specific volume plane. Results obtained are presented on Figures 4, 5 and 6.

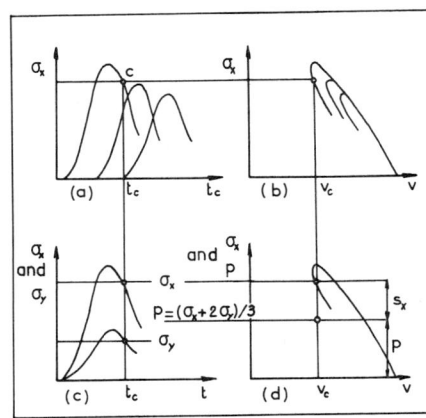

Fig.3. Technique used to obtain the pressure-specific volume and the deviatoric stress-specific volume curves. The plane(b) is obtained by the lagrangian analysis of plane(a). The plane (c) is the experimental plane obtained by the deviatoric stress experiment,performed with the same initial conditions as (a). The deviatoric part S_x and the pressure P appear in the plane (d). The construction is here realised for a point c of the unloading curve.

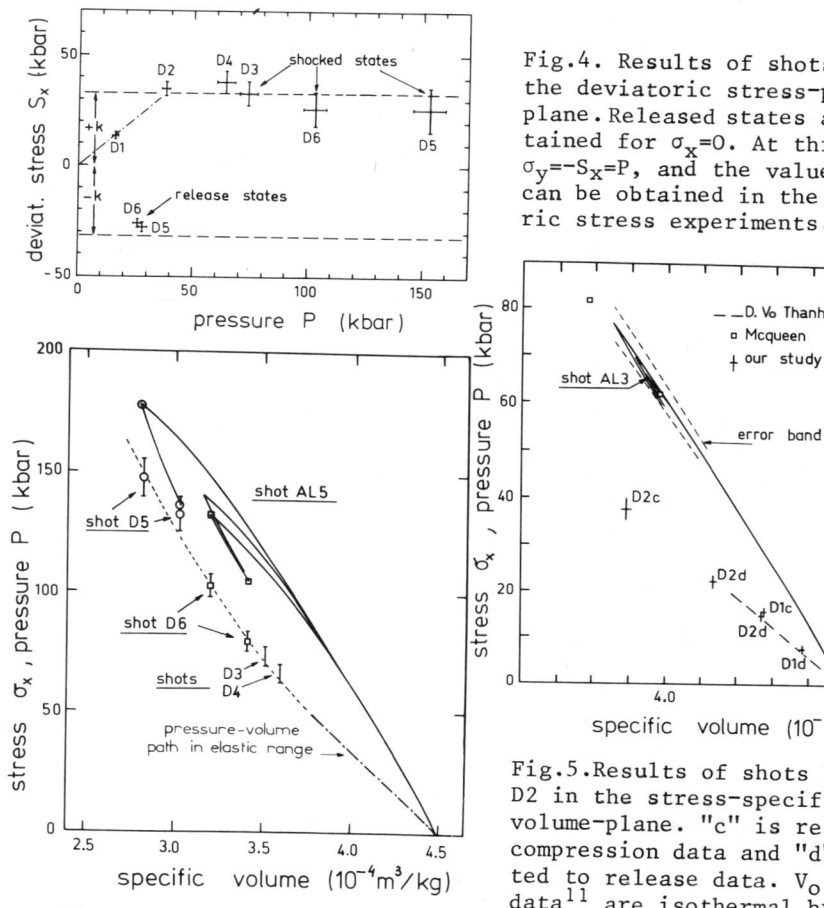

Fig.4. Results of shots D in the deviatoric stress-pressure plane. Released states are obtained for $\sigma_x=0$. At this limit $\sigma_y=-S_x=P$, and the value of σ_y can be obtained in the deviatoric stress experiments.

Fig.5. Results of shots D1 and D2 in the stress-specific volume-plane. "c" is related to compression data and "d" is related to release data. V_o Thanh's data[11] are isothermal hydrostatic compression of an identical borosilicate glass.

Fig.6. Results of shots D3, D4 D5 and D6 in the stress-specific volume plane.

DISCUSSION

For shots AL1 and AL2, the glass clearly exhibits a reversible loading-unloading path. For AL3 (Figure 5), there is a very weak hysteresis, but the unloading path is located within the experimental error band. On the other hand, the loading-unloading path observed with the shot AL5 (and AL4) exhibits a large hysteresis. From these investigations, an elastic response of the glass between 0 and about 75 kbar can be assumed. Above this limit, the wave propagation in borosilicate glass shows a typical two-wave structure. The precursor wave has been described as elastic, with an apparent Hugoniot elastic limit increasing with driving stress. The second wave shows dispersive spreading that gives rise to plastic-like permanent deformation.

A. Elastic response.

In the range 0-75kbar, the borosilicate exhibits two different behaviors. First, a non-linear elastic response between 0 and 30 kbar, and after an approximately linear behavior from 30 kbar to 75 kbar. We can establish an analogy between the dynamic behavior of fused silica and borosilicate glass up to the HEL. Results of deviatoric stress experiments permit to calculate the Poisson's ratio. It is 0.22 ± 0.02, that is, the static value.

B. Yield and permanent deformation flow.

- Yield condition : We assume that points of maximum compression reported on Figure 4 are related to equilibrium states. Above the HEL, these states lie on the straight line $S_x = k \approx 32$kbar, inside the experimental error band. This observation justifies describing the glass behavior above the HEL by a Von Mises or a Tresca yield condition.

- Plastic flow rule : On Figures 5 and 6, the deviatoric stress S_x appears as the difference between $\sigma_x(V)$ and $P(V)$. On Figure 7, we reported S_x versus the specific volume V. The observed overstepping of the limit surface results from a viscoplastic behavior, which explains the fact that the apparent HEL increases with driving stress.

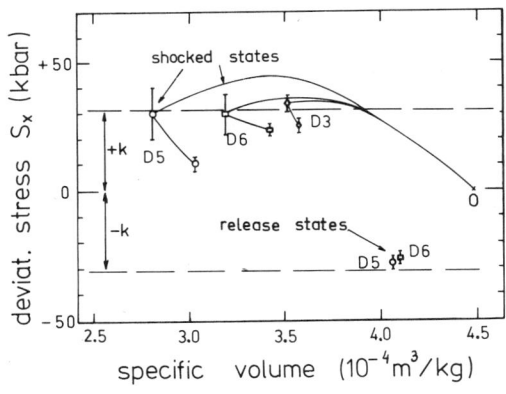

Fig.7. Deviatoric stress versus specific volume during loading and unloading of borosilicate glass.

MICROMECHANICAL ASPECTS

A complete loss of shear strength was not observed above the HEL, so, results of the previous study[3] are here confirmed. This work does not permit assuming that there is crushing of the material during the shock wave compression, as concluded by Kanel.[7] To confirm this interpretation, we intend to perform shots with steady waves and spall-strength measurements.

The two-shock wave structure, obtained above the HEL, is consistent with the densification of glass.[8,9] In an unpublished study,[10] we measured a limit permanent deformation (for $\sigma_x=0$) $(\Delta V/V)_R = 10.9\%$ $\pm 1\%$, for a shock stress level of 170kbar. For a maximum compression stress σ_x greater than 170kbar, we observed[10] a permanent deformation smaller than 10.9%, and the material recovered the initial specific volume for a maximum compression stress $\sigma_x=240$kbar. This behavior is consistent with an annealing phenomenon occuring above

σ_x=170kbar. Under this value, the densification increases with the maximum compression stress σ_x, and the model previously defined is applicable.

CONCLUSION

The main conclusions of this study are as follows :

i) The two experimental methods described ("lagrangian analysis experiment" and "deviatoric stress experiments") used together represent a powerful tool for determining the behavior of solid materials under shock-wave compression.

ii) On their elastic range (0-75kbar), behaviors of borosilicate glass and fused silica are identical.

iii) Above the HEL, borosilicate glass undergoes a significant, but not complete, loss of shear strength. This behavior seems consistent with a densification of glass.

iiii) Above the HEL and up to 170kbar, the response is viscoplastic, with a Von Mises or a Tresca yield condition.

REFERENCES

1. D. E. Grady, W. J. Murri, and P. S. De Carli, J. Geophys. Res. 80, 4857(1975).
2. J. R. Asay, L. C. Chhabildas, and D. P. Dandekar, J. Appl. Phys. 51, 4774(1980).
3. J. Cagnoux, in High Pressure Science and Technology, Proc. VIIth Int. AIRAPT Conference, Le Creusot, France, 1979, Vol.2, p. 1022.
4. P. Chartagnac, 1981 Topical Conference on Shock Waves in Condensed Matter, SRI International, Menlo Park, CA 94025 (to be published).
5. P. Chartagnac, unpublished work.
6. L. Seaman, J. Appl. Phys. 45, 4303 (1974).
7. G. I. Kanel, A. M. Molodets and A. N. Dremin, Combustion Explos. Shock Waves 13, 772 (1977).
8. R. V. Gibbons and T. J. Ahrens, J. Geophys. Res. 76, 5489 (1971)
9. J. Arndt, U. Hornemann and W. F. Müller, Physics and Chemistry of Glasses 12, 1 (1971).
10. J. Cagnoux, unpublished work.
11. D. Vo Thanh and A. Lacam, High Temperatures High Pressures 10, 655 (1978).
12. R. G. McQueen, S. P. Marsh, J. W. Taylor, J. N. Fritz and W. J. Carter, p 369, in High Velocity Impact Phenomena, edited by Ray Kinslow, Academic Press, New York and London, (1970).

DETERMINATION OF MEAN AND DEVIATORIC STRESSES IN SHOCK LOADED SOLIDS

P.F. CHARTAGNAC
Centre d'Etudes de Gramat, 46500 Gramat, France

ABSTRACT

We analyse the methods used to obtain the mean and deviatoric shock stresses in solids and emphasis is placed on the method of "piezoresistive deviatoric gauges" that we have been using since 1973.

INTRODUCTION

The general purpose of our work concerns the study of the methods used to obtain the equation of state and the plastic yield surface of shock loaded solids. The data required can be provided by the well established technique of plane shock wave experiments. The strain (ε_x) in the direction of wave propagation (x direction) is uniaxial, so, those two relations can be written as follows :

Equation of state : $\quad p = f(\varepsilon_x) \quad (1)$
Plastic yield curve : $\quad \sigma'_x = g(p) \quad (2)$

where p is the mean stress and σ'_x the deviatoric part of the longitudinal stress (σ_x).
In this paper we discuss the way to determine p and σ'_x in a shock loaded solid.

STATE OF ART

A. TABLE I. Methods which do not use piezoresistive measurements of stresses.

Methods	Assumptions	Facilities	Remarks
Hugoniot + Hydrostat	Mean shock stress = hydrostatic pressure	Shock wave generators static loader + hydrostatic device	Reasonnable for metals and inorganic crystalline solids
Hugoniot + Von Mises model	Deviatoric stress constant beyond H.E.L	Shock wave generators	Fits most metals
Combined compression and shear loading (GUPTA[1])	Use of elastic E.O.S Shear wave transmitted through sample bonds	Laboratory gun (slotted barrel) IMPS device	Appropriate to non ferromagnetic materials
Unloading and reloading waves (ASAY[2])	Yield curve determined by tangents from unloading and reloading paths	Laboratory gun VISAR	Describes work-hardening Appropriate to non viscous materials

B. Method of the "piezoresistive deviatoric gauges".

In 1973, we began to use the "piezoresistive deviatoric gauges" and published the results obtained in limestone in 1976.[3] A similar method was published in 1977 by Kanel et al.[4] In fact, the first attempts for measuring the lateral stress were those of Bernstein.[5]

1. Principle.

In plane shock wave measurements in isotropic materials, the uniaxial strain condition (in the x direction) provides an easy calculation of the mean and deviatoric stresses :

$$p = \frac{1}{3} (\sigma_x + 2\sigma_y)$$
$$\sigma'_x = \frac{2}{3} (\sigma_x - \sigma_y)$$

Both σ_x and σ_y are measured with piezoresistive foil-like gauges[6] from the same batch (Fig.1)

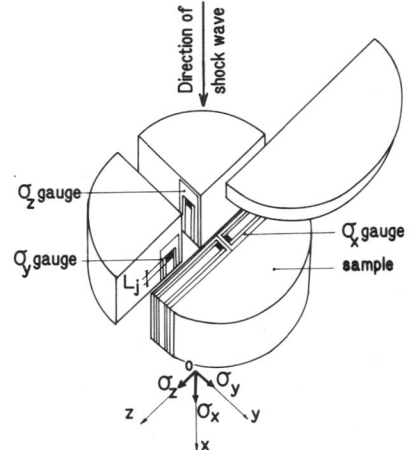

Fig.1. Sample arrangement

The yield curve $\sigma'_x(p)$ corresponds to an equilibrium plastic surface, so we generated steady shock waves (Fig.2). Let us notice about Fig.2 that the rise time of lateral transducers[6] mainly depends upon the width LJ of the piezoresistive element (2.5mm). The criterion of choice of the gauges is precision but not rise time which does not interfere with the steady stress measurements.

The principle of σ_x piezoresistive measurement is now well known and does not need to be explained. On the contrary, the way the σ_y gauges work must be explained. The following demonstration is valid if the thickness of the solid submitted to the steady shock is several times bigger than LJ (this condition is satisfied for all tests reported in this paper). If the encapsulation[6] of the gauge behaves hydrodynamically,[7] it is possible to deduce the three following consequences :

- The shear stresses are not transmitted to the piezoresistive element of the gauge.
- When the transducer is in a mechanical equilibrium with the solid, that is to say when the signal from the gauge is constant, then the hydrostatic pressure in the planar encapsulation slab is equal to σ_y in the solid.
- The calibration of the transducers established for the same encapsulation, with longitudinal stresses, is valid for σ_y.

But is this assumption of an hydrodynamic behavior of the encapsulation correct, and specifically for small σ_y stresses ?

The encapsulations are made of Kapton or polyimide; their shock behaviors are not well known: that is why we have developed a test at a moderate stress (DPL5-Fig.4) where the gauge was imbedded in a 0.18mm thick film of silicone oil, which is assumed to behave like a fluid under steady shock stress. Fig. 4 shows that the value of σ_y in test DPL5 is equal to the one measured in test DPL1 which is identical except that there is no oil but a classical epoxy bond.

In addition to this experiment there is another proof of the σ_y measure's validity: the results obtained in the elastic range of steel[8] indicate a Poisson's ratio equal to the static value. This agrees with the fact that very little dissipation effects occur in the elastic range of steel.

2. Remarks.

The two main qualities of that method are to include no theoretical assumption and to suit all solids (ferromagnetic or not). Its main shortcoming concerns precision because p and σ'_x result from piezoresistive measurements (precision is ±0.28Kb for the carbon gauges used in the range 0-20Kb and ±5% for the manganin gauges used in the range 0-200Kb).

APPLICATIONS OF THE PIEZORESISTIVE DEVIATORIC GAUGES METHOD

We have applied this method to limestone,[3] steel[8] and PMMA. and J.Cagnoux applied it to a borosilicate glass.[9] The only example reported here is PMMA because data obtained with other methods have been previously published[1,10] and will be compared with our results.

A. Experimental technique for PMMA study.

Plane shock waves were generated with a 110mm smooth bore gas gun (tilt less than 2×10^{-3}rd). The samples were 80mm diameter and 45mm thick (Fig.1). The σ_x gauges plane was located $5 {}^{+0.00}_{-0.05}$ mm from the front face of the sample. The faces receiving the gauges were carefully ground. Tolerances are typically 3×10^{-4}rd in parallelism and ±0.1mm in flatness. After gauging and reassembling, the front face of the sample was ground. In PMMA, we used DYNASEN carbon gauges calibrated by M. Perez.[11] The total thickness of the transducer is 0.11mm.

B. Results.

The purpose of this section does not concern the study of the behavior of PMMA. The results will only be considered as an example of application of the "piezoresistive deviatoric gauges method".

The polymethyl methacrylate(PMMA) used was Altuglas obtained in sheet stock. Its density was measured as $1.18 g/cm^3$.

The σ_x records (Fig.2) show an initial rise time of 50ns, due to the transducer, up to about two thirds of the final stress. This is followed by a 600ns more gradual rounding up to be steady stress, which is to be expected of a rate-sensitive or viscoelastic response.

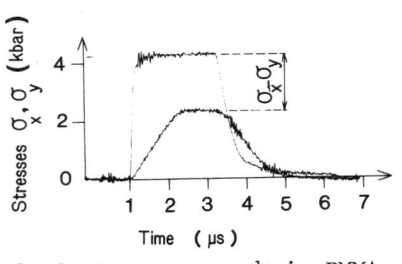

Fig.2. Stress records in PMMA.

In order to obtain the $p(\varepsilon_x)$ curve (Fig.3) we have plotted our σ_x and p data on the $\sigma_x(\varepsilon_x)$ Hugoniot curve from Barker and Hollenbach.[12] The deviatoric stress σ'_x appears as the difference between σ_x and p. One can see that Gupta's[1] $p(\varepsilon_x)$ data (combined compression and shear loading method) agree with our results in the elastic range. Fig.3 also shows that the hydrostatic curve[1] cannot represent the mean shock stress. This probably comes from the viscous dissipations resulting in temperature increase : so the mean shock stress cannot lie on a 20°C isotherm-hydrostat.
The $\sigma'_x(p)$ curve (Fig.4) shows that there is a very good agreement between our data and two measurements recently published by Gupta[10] using the same method of "piezoresistive deviatoric gauges".

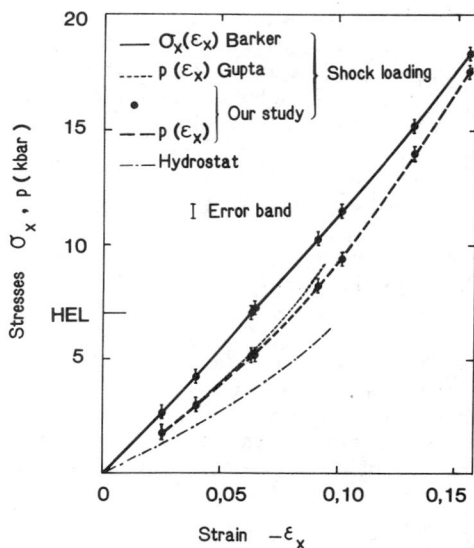

FiG.3. $\sigma_x(\varepsilon_x)$ and $p(\varepsilon_x)$ for PMMA

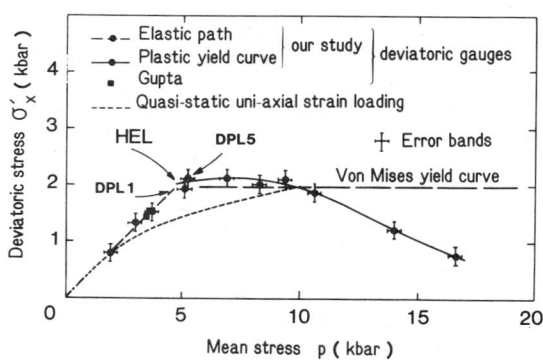

Fig.4. $\sigma'_x(p)$ for PMMA

Beyond the Hugoniot elastic limit(H.E.L) (σ_x=7kbar[12] and p=5kbar), the deviator remains constant ($\sigma'_x \approx 2.1$kbar) indicating a Von Mises or Tresca yield condition ; then when p=8kbar($\sigma_x \approx 10$kbar), the deviatoric stress decreases: shocked PMMA tends to behave hydrodynamically. We did not perform any test beyond 17kbar because Bloomquist and Sheffield[13] report an important increase in temperature above a longitudinal stress of 17-16kbar and we don't have any temperature dependant calibration of the gauges.

CONCLUSIONS

The method of "piezoresistive deviatoric gauges" does not include any theoretical assumption and is suitable for all materials. However it requires a precise calibration of the gauges. The examples cited show that the precision depends upon the material (deviator magnitude) and upon the stress range explored.

Finally, we must not forget that Gupta's and Asay's methods also supply thermodynamic out of equilibrium paths that can be used to introduce rate dependance in the behavior equations. Though this was not the purpose of the present work, we point out that it is possible to obtain the same results by combination of the "piezoresistive deviatoric gauges method" and the lagrangian analysis method.[9]

ACKNOWLEDGEMENTS

B. Jimenez is gratefully acknowledged for all gas gun experiments. J. Cagnoux and M. Perez are thanked for many helpful discussions. This work was supported by the French Ministry of Defense.

REFERENCES

1. Y. M. Gupta, J. Appl. Phys. 51,5352 (1980).
2. J. R. Asay, J. Lipkin, J. Appl. Phys. 49,4242(1978).
3. P. Chartagnac. Thèse Docteur Ingénieur n°528 UPS Toulouse (1976).
4. G. I. Kanel et al, Combustion Explosion and Shock Waves 13,772 (1978).
5. D. Bernstein et al, In Behavior of Dense Media Under High Dynamic Pressures (Gordon and Breach, New York (1968), p.461.
6. "Gauge" means piezoresistive element plus encapsulation. "Transducer" means gauge plus the epoxy bond.
7. We only study the encapsulation behavior because its thickness (0,09mm for carbon gauges to 0,13mm for manganin gauges) is about ten times bigger than the epoxy bond thickness (0,01mm).
8. P. Chartagnac. Determination of Mean and Deviatoric Stresses in XC 38f Steel. (to be published).
9. J. Cagnoux. Shock Wave Compression of a Borosilicate Glass up to 170Kbar, APS Conference SRI (1981).
10. Y. M. Gupta et al, Appl. Phys. Lett. (1980).
11. M. Perez. Thèse Docteur Ingénieur n°714 UPS Toulouse (1980).
12. L. M. Barker, R. E. Hollenbach, J. Appl. Phys. 41,4208 (1970).
13. D. D. Bloomquist, S. A. Sheffield, Bull Am. Phys. Soc. 24,714 (1979).

SYMMETRIC ROD IMPACT TECHNIQUE FOR DYNAMIC YIELD DETERMINATION

D. C. Erlich, D. A. Shockey, L. Seaman
SRI International, Menlo Park, CA 94025

An experimental technique has been designed and tested to determine the dynamic yield curve of a rod in compression at high strain rates ($\sim 10^4 - 10^6$/sec). The technique involves impacting a gas-gun accelerated cylindrical rod onto an identical stationary rod at high velocities, and using a high-speed framing camera to take silhouettes (at $\sim 10^6$ frames/second) of the resulting deformation. A two-dimensional wave propagation code is then used to simulate the experiment, varying the yield parameters until the computational results approximate the experimental silhouettes. This technique is a refinement of the classical Taylor test, and eliminates the uncertainties of the boundary conditions as well as those caused by the Taylor formulae approximations. Results are presented for an experiment using 6061-T6 aluminum.

INTRODUCTION AND BACKGROUND

Knowledge of the yield curve of solid materials at high strain rates ($10^4 - 10^6$/sec) and large strains would be of immense value in a wide range of fields ranging from armor penetration studies to explosive metal forming. Unfortunately, no routine method exists for obtaining such information. Recently, however, we have developed a technique that appears capable of providing reasonably accurate dynamic yield curve determinations. It is based on the rod-on-rigid plate impact technique of Taylor and Whiffin.[1] A brief history of this technique is provided below.

In 1947 Taylor and Whiffin[1] determined dynamic yield strengths of solids at high strain rates by accelerating rods of a specimen material against "rigid" thick plates as indicated in Figure 1. They analyzed the test using one-dimensional rigid-plastic theory and found a relationship between the dynamic yield strength σ_y^{dyn} and the ratio of initial and final rod lengths L_o/L_f. These authors performed experiments on various steels, copper, lead, and paraffin wax and showed that σ_y^{dyn} was independent of impact velocity and specimen geometry.

Six years later Lee and Tupper[2] performed a more careful analysis of the test, using elastic-plastic theory and a one-dimensional characteristics code. They computed the final profiles of the rod after impact and obtained fair agreement with observed shapes. In 1968 Hawkyard, Eaton, and Johnson[3] performed experiments with copper and steel rods at elevated temperatures. These authors applied several one-dimensional analyses to their result and found that none could successfully predict the final shape of the deformed rod.

0094-243X/82/780402-05$3.00 Copyright 1981 American Institute of Physics

Wilkins and Guinan[4] in 1972 accounted for the two-dimensional nature of plastic flow in an impacting rod, and by using a two-dimensional wave propagation code and an elastic-plastic constitutive relation with work hardening were able to compute the observed shape of the rod after impact. Their results showed a high correlation between the dynamic yield stress and L_o/L_f for a wide range of rod velocities and geometries, thus verifying the basic correctness of the original Taylor-Whiffin analysis.

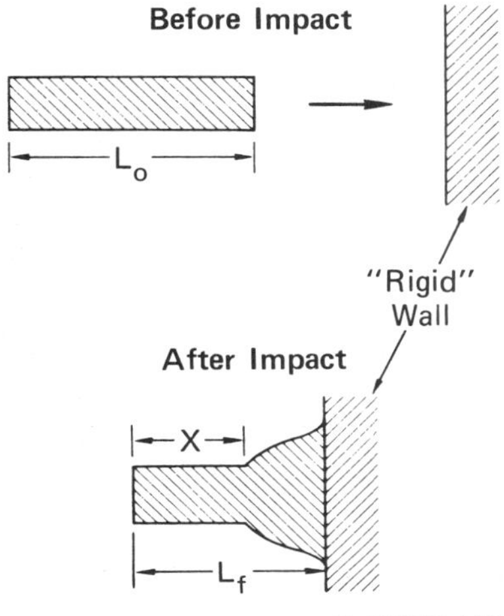

FIGURE 1 ROD AND PLATE IMPACT TEST DEVISED BY TAYLOR[1] TO DETERMINE DYNAMIC YIELD STRENGTH OF SOLIDS AT HIGH STRAIN RATES ($10^4 - 10^6$/sec) AND LARGE STRAINS

EXPERIMENTAL TECHNIQUE

Recently, workers at SRI International attempted to obtain the entire high-strain-rate flow curve of a material instead of only the dynamic yield strength and designed a modified verson of the rod-on-plate-impact technique. In the modified method, the deformation history of the rod is monitored by high speed photography and compared with the computed deformation history at intermediate, as well as final, deformation times. A further modification is the replacement of the plate with a rod of the same material and geometry as the impacting rod. This arrangement allows the impacting bodies to deform symmetrically, thus, eliminating boundary condition uncertainties in the analysis that arise from possible movement of the impact boundary into the "rigid" plate and from the unknown friction conditions at the rod-plate interface.

The technique was first tested using rods of 6061-T6 aluminum 38.1 mm long by 9.5 mm in diameter, machined so that the impact ends were flat and normal to the rod axis. Rod-on-rod impact was accomplished by mounting and carefully aligning one rod at the muzzle of a gas gun, mounting and aligning the second rod on the leading end of a projectile, and accelerating the projectile in the gun barrel by suddenly releasing a volume of compressed helium. The experimental arrangement is shown in Figure 2. The rods are held firmly but lightly in place to minimize perturbation of the wave propagation in the rods. After impact both rods fly into a recovery tank filled with soft, energy-absorbing material to minimize additional damage to the rods.

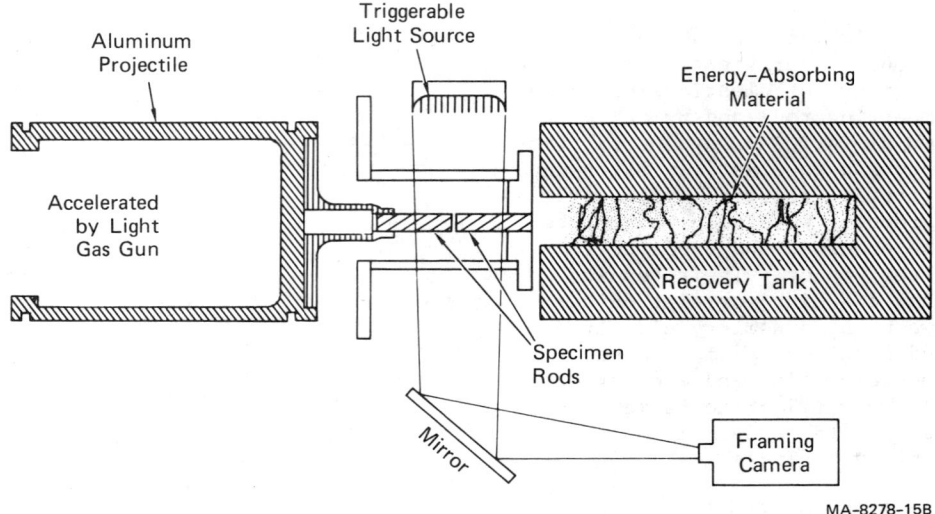

FIGURE 2   ARRANGEMENT FOR MEASURING HIGH RATE CONSTITUTIVE EQUATIONS

Changes in contour of the rods as they deform during impact are monitored by backlighting the rods and photographing their silhouettes with a high-speed framing camera at approximately 10^6 frames per second. A series of exploding bridge wires served as the light source.

RESULTS

The series of high-speed photographs in Figure 3 shows the change with time of the profiles of two aluminum rods that collided at 556 m/s. The profiles were digitized, and then compared with the simulations performed using HEMP, a two-dimensional, finite-difference, wave propagation code,[5] having an elastic-plastic constitutive law with asymptotic work hardening.

Several iterative computations using different values for the dynamic yield strength and the asymptotic limit were required before the computed rod profiles matched the observed profiles. The excellent agreement shown in Figure 4 for three times during the deformation was obtained using an initial yield strength of 3.2 kbar and an asymptotic limit of 4.1 kbar.

Near the impact surfaces, large shear stresses were computed close to the rod periphery and large tnesile stresses were computed in the vicinity of the rod axis. If the conditions are sufficient to cause shear instabilities or internal fracture, the rod profiles will be affected, and attempts to match the profiles in the manner discussed above will be meaningless and yield erroneous constitutive relations. To ascertain whether shear or tensile failure has occurred in these experiments, rods were sectioned after impact on a plane

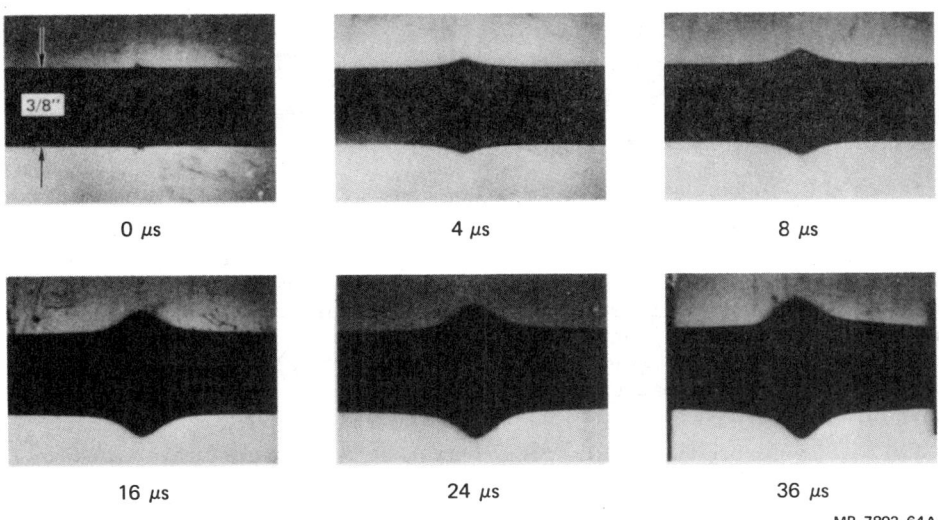

FIGURE 3 DEFORMATION HISTORY FOR TWO ALUMINUM RODS COLLIDING END-ON AT 556 m/s
Elapsed time after impact is indicated.

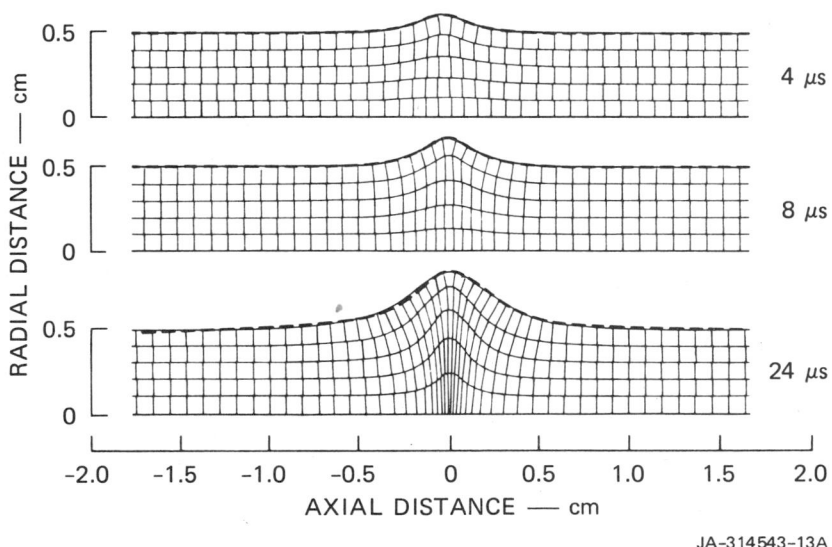

FIGURE 4 COMPARISON OF RECORDED (DASHED LINE) AND COMPUTED ROD PROFILES AT THREE TIMES AFTER IMPACT

containing the rod axis, and examined metallographically. In the experiment described above, no failure was observed. However, at higher velocities, dynamic tensile failure in the form of a population of spherical voids was observed in aluminum.[6] In a titanium alloy we observed the incipient stages of shear failure that would lead to a cone-shaped fragment whose base was the impact surface.

DISCUSSION

The sensitivity of the computed deformation profile to the constitutive relation needs to be assessed to establish the usefulness of this procedure. Computational simulations in which the values of dynamic yield strength and work-hardening rate were systematically and independently varied indicated that constitutive relations accurate to about 10 to 15% could be expected for low work-hardening-rate material whereas the error for high work-hardening-rate materials may be only 2 to 4%.

The computational simulations show that strain rate varies with distance from the impact surface; the material near the impact surface experiences substantially higher rates of strain than material a fraction of a rod radius further removed. Thus, if the material is highly rate sensitive, a more complicated constitutive law in which the yield strength and work-hardening are prescribed functions of strain rate may be required to obtain a good match between computed and observed profiles. In the present example of 6061-T6 aluminum, the rate-independent constitutive law predicted well the rod profile throughout the deformation process.

ACKNOWLEDGMENTS

The computational simulations of the 6061-T6 aluminum experiment were performed by Bryan Scott of the Ballistics Research Laboratory, Aberdeen, MD.

REFERENCES

1. Taylor, G. I. and Whiffin, A. C., Proc. Roy. Soc. A, $\underline{194}$ 289-322 (1947.
2. Lee, E. H. and Tupper, S. J., J. Appl. Mech. $\underline{21}$ 63-70 (1953).
3. Hawkyard, J. B., Eaton, D. and Johnson, W., Mech. Sci. $\underline{10}$ 929-948 (1968).
4. Wilkins, M. L. and Guinan, M. W., J. Appl. Phys., $\underline{44}$, 1200-1206 (1973).
5. Wilkins, M. L., "Calculation of Elastic-Plastic Flow," Lawrence Livermore Laboratory Report No. UCRL-7322 (1969) (unpublished).
6. Kipp, M. E. and Davison, L., "Analysis of Ductile Flow and Fracture in Two Dimensions," in Shock Waves in Condensed Matter (American Physical Society, New York, 1981).

RE-EXAMINATION OF THE PRECURSOR DECAY ANOMALY*

R. J. Clifton
Brown University, Providence, RI 02912

ABSTRACT

Consideration of the elastodynamics of moving dislocations gives a precursor decay relation that is essentially the same as the well known decay relation obtained from an elastic/viscoplastic model of the material. Recent results from plate impact recovery experiments are used to assess the validity of commonly used models for dislocation velocity and mobile dislocation density. Current models for dislocation velocity based on a linear viscous drag resistance at low velocities and a cut-off at the elastic shear wave speed appear to be acceptable. Satisfactory models for mobile dislocation density require consideration of large dislocation densities that are produced at the impact face during specimen preparation and during the impact of rough surfaces. Inclusion of these surface effects and analysis of precursor decay using a finite deformation theory appears to provide a satisfactory framework for explaining the observed precursor decay. Error analysis of precursor decay experiments indicates that determination of the value of the plastic strain rate at the wavefront from precursor decay measurements may be unreliable.

INTRODUCTION

An ordinary differential equation for the decay of the precursor amplitude with distance of propagation in plate impact experiments on isotropic materials has the form (Taylor[1])

$$\frac{D[v_z]}{Dz} = \frac{c_2^2}{c_1^2} \dot{\varepsilon}^p_{zz} \tag{1}$$

where $D[v_z]/Dz$ is the total derivative, along the wavefront, of the jump in particle velocity; $\dot{\varepsilon}^p_{zz}$ is the normal component of plastic strain rate in the z-direction; c_2 and c_1 are, respectively, the speeds of elastic shear waves and elastic longitudinal waves. If the material has slip systems oriented at 45° to the wavefront, then the plastic strain rate $\dot{\varepsilon}^p_{zz}$ due to a shear strain rate $\dot{\gamma}^p$ on one of these families of shear planes is

$$\dot{\varepsilon}^p_{zz} = \frac{1}{2} \dot{\gamma}^p . \tag{2}$$

The plastic shearing rate $\dot{\gamma}^p$ is related to the motion of dislocations by (Orowan[2])

$$\dot{\gamma}^p = bNv_d \tag{3}$$

---
*Supported by NSF GRant ENG 79-10633 and the NSF Materials Research Laboratory at Brown University.

where b is the Burgers vector of the dislocations, N is the dislocation line length per unit volume, and v_d is the average dislocation velocity. Substitution of (2) and (3) into (1) gives

$$\frac{D[v_z]}{Dz} = \frac{c_2^2}{2c_1^2} bNv_d \ . \tag{4}$$

as the expression for precursor decay in terms of the dislocation density and dislocation velocity at the wavefront. Analogous expressions are obtained for anisotropic crystals with slip on several different slip systems (Johnson et al.[3]).

Precursor decay predicted by relations of the form (4), with N taken to be the initial dislocation density in the bulk of the specimen and with reasonable estimates for v_d, is much less than observed in plate impact experiments on single crystals[3-10]. Indeed, discrepancies between measured and predicted amplitudes are so large that the value used for dislocation density in the calculations must be increased by a factor of the order of $10^2 - 10^3$ before sufficient decay is predicted. Potential explanations of these discrepancies have been considered; however, none has received general acceptance. In this paper we review briefly the results of several investigations related to the re-examination of precursor decay and its dependence on the motion of dislocations.

ELASTODYNAMICS OF PRECURSOR DECAY

Clifton and Markenscoff[11] have re-derived the precursor decay relation based on consideration of elastic waves emanating from long, straight dislocations that are set in motion by the precursor. Their result is

$$\frac{D[v_z]}{Dz} = \frac{c_2^2 \, bNv_d}{2c_1^2 (1 - v_d/(\sqrt{2} c_1))} \tag{5}$$

which is the same as (4) except for the term $(1 - v_d/(\sqrt{2} c_1))$ in the denominator. The presence of this term in (5) indicates that dislocations moving towards the wavefront have a greater effect on precursor decay than those moving away from the wavefront. This effect is not large. The maximum value of the ratio of the contribution of forward moving dislocations to the contribution of backward moving dislocations is 2.1, obtained for $v_d/c_1 = \pm 1/2$. Moreover, the cumulative effect of dislocations of opposite sign will reduce the influence of the additional term in the denominator of (5). Overall, (5) is viewed as indicating that (4) is essentially correct. Further elastodynamic analysis by Markenscoff and Clifton[12] for the case of circular dislocation loops indicates that the decay relation is independent of the initial radius of the loops. Thus, as along as the elastic response can be taken to be linear it appears that the precursor decay anomaly can be addressed by examining how well the various quantities in (4) are determined by experiments.

NONLINEAR ELASTICITY AND PRECURSOR DECAY

Nonlinear elasticity effects on precursor decay have been analyzed by Ahrens and Duvall[13]. If the shock wave velocity is denoted by U and the longitudinal sound speed immediately behind the wavefront is denoted by c, then the precursor decay relations can be written as (Clifton[14])

$$\frac{D\sigma}{DX} = \frac{-F + (c^2/U^2 - 1)\left[\frac{\partial \sigma}{\partial t}\right]_X}{U\left(\frac{1}{2} + \frac{3}{2}\frac{c^2}{U^2}\right)} \tag{6}$$

where σ is the normal stress on the wavefront; X is distance from the impact face, measured in the initial configuration; F is a relaxation function (Duvall[15]) that corresponds to the rate of stress relaxation due to inelastic processes; and $(\partial \sigma/\partial t)_X$ is the rate of change of σ with respect to time at fixed X, immediately behind the wavefront. Calculation of the precursor decay corresponding to (6) requires integration of the field equations in the region behind the wavefront, such as was done by Herrmann et al.[16] and Clifton[17]. In these calculations the precursor amplitude at remote positions is affected by inelastic deformation that occurs at finite times after arrival of the front at positions nearer the impact face. Thus, when nonlinear elasticity effects are included the value of $D\sigma/DX$ at any X is influenced by the inelastic deformation at, for example, the impact face.

If the slope $D\sigma/DX$ of the precursor decay curve is measured from a series of experiments on targets of different thicknesses, and the slope $(\partial \sigma/\partial t)_X$ is obtained from the stress-time profiles for each of the experiments, then (6) can be used to compute the value of the relaxation function F at each thickness. The relaxation function F is essentially proportional to the plastic strain rate in the direction of the normal to the wavefront. Consequently, from (2) and (3), a value of mobile dislocation density at the wavefront can be calculated from a value of F obtained from (6) and an assumed value for the dislocation velocity at the wavefront. This procedure can lead to large errors in the estimate of the dislocation density N if the precursor decay slope $D\sigma/DX$ is small and the slope $(\partial \sigma/\partial t)_X$ is large and negative since then F is computed as the difference of two numbers that may be of comparable magnitude. This difficulty is made more troublesome by the relatively large uncertainty in the values of $D\sigma/DX$ and $(\partial \sigma/\partial t)_X$ due to smoothing of the measured wave profile details near the wavefront because of, for example, the bond layer between the specimen and a quartz gage transducer used to measure the stress as well as because of the effects of finite transducer size on the monitoring of waves that are propagating in directions that differ from the direction of the normal to the rear surface of the specimen. Rosenberg and Duvall[18] showed that for impact of LiF in <111> directions the uncertainties of the measured values of $(D\sigma/DX)$ and $(\partial \sigma/\partial t)_X$ are so great that the sign of F is not even determined at distances greater than 0.5 mm.

DISLOCATION VELOCITY AND DISLOCATION DENSITY

Further insight into the modeling of N and v_d in precursor decay relations has been sought through plate impact experiments designed to allow recovery of the specimen for examination of dislocation configurations caused by a known stress pulse. Recovery experiments on LiF crystals by Kumar and Clifton[19] and on MgO crystals by Kim[20] indicate that glide bands of closely spaced dislocations are formed. Two principal conclusions are drawn from the lengths of the bands and their distribution through the thickness of the specimens. First, the average dislocation velocity appears to be essentially proportional to the resolved shear stress for experiments on high-purity annealed LiF crystals at room temperature. Furthermore, the proportionality constant, expressed as a drag coefficient $B = \tau b/v_d$ where τ is the resolved shear stress, has a value of approximately 3×10^{-4} dyn s cm^{-2} which is approximately the value obtained by others[21-23] at much lower stress levels. Thus, at least for high-purity annealed LiF, it appears that a linear viscous drag model is applicable up to one-fourth of the elastic shear wave speed. More limited results on MgO by Kim[20] suggest a similar proportionality of average dislocation velocity and resolved shear stress. Analysis of the energy radiated from supersonic dislocations indicates that the stress required for sustained supersonic motion of dislocations is higher than occurs in most precursor decay experiments.[11] Overall, it appears unlikely that the lack of certainty in the modelling of dislocation velocity is an important factor in the precursor decay anomaly.

The second principal conclusion is that dislocation densities in impacted crystals are much larger near the impact face than they are in the bulk of the specimen[19,20,10]. Part of the increased density near the surface is due to damage induced during preparation of surfaces to obtain the required flatness. Another contribution appears to come from the remaining surface roughness of the impact faces. Even roughnesses of the order of hundredths of microns require the generation of large dislocation densities in order to achieve geometric compatibility of the impacting surfaces. Kim[20] observed dislocation densities of the order of 10^9 cm^{-2} in transmission electron microscope pictures of glide bands near the impact face of an MgO crystal that was impacted at the relatively modest velocity of 0.07 mm/µsec. Such large densities, if present at such early times as the times at which closure of the gap between mismatched rough impact faces occurs, would have a marked effect on precursor decay. Large initial dislocation densities near the impact face would cause large rates of decay near the impact face and the large rate of generation of dislocations would result in wave profiles in which $(\partial\sigma/\partial t)_X$ in (6) is large initially. From (6) such shaping of the wave profile contributes to the rate of precursor decay at all positions X.

CONCLUSIONS

Resolution of the precursor decay anomaly most likely requires thorough consideration of dislocation densities near impact faces and of errors associated with the interpretation of experimental records. Precursor decay relations which include nonlinear elasticity effects appear to provide a satisfactory framework for analyzing precursor decay. Modeling of dislocation velocity by a linear viscous drag law with a relativistic cut-off at the elastic shear wave speed is expected to be satisfactory for high purity crystals. Critical experiments are required which can be used to eliminate incorrect explanations of the anomaly.

REFERENCES

1. J. W. Taylor, J. Appl. Phys. 36, 3146 (1965).
2. E. Orowan, Proc. Phys. Soc. A52, 8 (1940).
3. J. N. Johnson, O. E. Jones, and T. E. Michaels, J. Appl. Phys. 41, 2330 (1970).
4. T. E. Michaels, Ph.D. Thesis, Washington Sate Univ. (1972).
5. P. L. Studt, E. Nidick, F. Uribe, and A. D. Mukherjee, Metallurgical Effects at High Strain Rates, edited by R. W. Rohde, B. M. Butcher, J. R. Holland, and C. H. Karnes (Plenum Press, N.Y., 1973) p. 379.
6. P. P. Gillis, K. G. Hoge, and R. J. Wasley, J. Appl. Phys. 42, 2145 (1971).
7. J. R. Asay, G. R. Fowles, and Y. M. Gupta, J. Appl. Phys. 43, 2132 (1972).
8. Y. M. Gupta, G. E. Duvall, and G. R. Fowles, J. Appl. Phys. 46, 532 (1975).
9. M. D. Bjorkman, Ph.D. Thesis, Washington State Univ. (1979).
10. J. E. Vorthman, Ph.D. Thesis, Washington State Univ. (1979).
11. R. J. Clifton and X. Markenscoff, J. Mechs. Phys. Solids 29, (1981).
12. X. Markenscoff and R. J. Clifton (unpublished manuscript).
13. T. J. Ahrens and G. E. Duvall, J. Geophys. Res. 71, 4349 (1966).
14. R. J. Clifton, Mechanics Today, Vol. 1 edited by S. Nemat-Nasser (Pergamon Press, Elmsford, New York, 1972), p. 102.
15. G. E. Duvall, Stress Waves in Anelastic Solids, edited by H. Kolsky and W. Prager (Springer-Verlag, Berlin, 1964), p. 20.
16. W. Herrmann, D. L. Hicks, and E. G. Young, Shock Waves and the Mechanical Properties of Solids, edited by J. J. Burke and V. Weiss, (Syracuse University Press, 1971), p. 23.
17. R. J. Clifton, Ibid., p. 73.
18. G. Rosenberg and G. E. Duvall, J. Appl. Phys. 51, 319 (1980).
19. P. Kumar and R. J. Clifton, J. Appl. Phys. 50, 4747 (1979).
20. K. S. Kim, Ph.D. Thesis, Brown University (1979).
21. J. E. Flinn and R. F. Tinder, Scripta Met. 8, 689 (1974).
22. F. Fanti, J. Holder, and A. V. Granato, J. Acoust. Soc. Am. 45, 1356 (1969).
23. O. M. Mitchell, J. Appl. Phys. 36, 2083 (1965).

A THERMAL-VISCOUS MODEL FOR HETEROGENEOUS YIELDING IN ALUMINUM*

D. B. Hayes
D. E. Grady
Sandia National Laboratories, Albuquerque, NM 87185

ABSTRACT

The finite rise times of high intensity shock waves in metals are commonly interpreted in terms of a bulk viscosity which is observed to vary inversely as the square root of the strain rate.[1] To interpret this result, a two-temperature thermal model of a heterogeneously-yielding metal has been constructed. The model assumes that the body is composed of thin shear bands in which irreversible energy is dissipated as heat and which are separated by the bulk of the material which undergoes isentropic compressive heating only. A condition of wave steadiness quantifies the irreversible energy dissipated during the passage of the shock. Two important results follow: First, the dependence of the apparent viscosity on the inverse square root of the strain rate stems from the microstructural, heterogeneous nature of the process, and is only expected for steady waves; second, when the model is used to interpret recent experiments,[2] the viscosity calculated in the shear band itself is seen to be temperature independent.

INTRODUCTION

Conventional continuum mechanical descriptions based solely on average stresses, strains, strain rates and temperatures are inadequate to establish the conditions responsible for many shock physics, chemistry, and metallurgical effects. A more general constitutive description with sufficient internal structure to describe the local gradients in temperature, strain and strain rate is necessary. The present study is not that ambitious, since the current state of experimental knowledge cannot support such an approach at this time. Instead, an intermediate step is taken in which a number of simplifying assumptions have been made, allowing construction of a model useful for interpreting experimental results.

The objectives of this work are to explore some of the implications of localized deformation during the shock compression process. This is done in the context of a two-temperature model. The model relies heavily on conventional concepts of temperature production and heat conduction and, in part, relies upon recent experimental evidence regarding trends in shock wave rise times and density of deformation features. Although we make some comparisons with experimental work, the principle objective remains to provide a consistent, although probably simplistic framework, to guide continuing experimental work.

---

* This article sponsored by the U. S. Department of Energy under Contract DE-AC04-76-DP00789.

Two-Temperature Model. The two-temperature model uses the concept of a partitioning of energy in the shock-loaded solid. A series of equi-spaced planar deformation bands (N per unit length) are assumed to pre-exist at sites in the aluminum. All of the irreversible energy which accompanies the shock jump is dissipated in these bands. Dissipation occurs at a constant rate over a period of time equal to the rise time of the shock front. Dissipation by the shearing strain rate is assumed to be Newtonian viscous and leads to a local true viscosity within the bands and an apparent or bulk viscosity for the solid as a whole. Thermal conduction limits peak temperatures and allows finite widths in the shear bands. In the interior of the material, in regions remote from shear bands, the material compresses along the isentrope. The assumption is that deformation arises from the passage of a steady shock wave. This allows use of the Hugoniot energy jump condition to quantify the dissipation. Because of wave steadiness, the magnitude of the viscous stress produced must be just sufficient to bring the stress-volume response in coincidence with the Rayleigh line.

Specific analytical forms are developed to describe the model features. The shock rise time is taken as $\varepsilon/\dot{\varepsilon}$, where ε is final engineering strain in the longitudinal direction behind the shock and $\dot{\varepsilon}$ is the average strain rate in the shock front. The required shear stress between the equilibrium curve and the Rayleigh line is calculated using the first nonvanishing term of a power series expansion of the Hugoniot about the initial state,

$$\tau = \frac{3}{16} \frac{S}{\rho_o C_o^2} P^2 \qquad (1)$$

where ρ_o is the initial material density, P is the shock pressure, and C_o and S are constants in the usual linear shock-velocity/particle-velocity relation. This dissipation creates an energy flux at each lamella,

$$F_o = \frac{1}{6} \rho_o C_o^2 S \varepsilon^2 \frac{\dot{\varepsilon}}{N} , \qquad (2)$$

which penetrates the surrounding material at a rate described by classical thermal conduction. The depth of penetration of the thermal pulse (namely the shear-band width), and the resultant temperature are taken as the values which characterize the entire process; no attempt is made to resolve variations in these during the shock rise time. Combining the above results leads to an equation for the temperature rise in the shear band, where

$$T_H = T_o e^{\Gamma \varepsilon} + \frac{\rho_o C_o^2 S}{3NK} \left(\frac{k}{\pi}\right)^{1/2} \varepsilon^{5/2} \dot{\varepsilon}^{1/2} ,$$

Γ, K and k are the Grüneisen ratio, thermal conductivity and thermal diffusivity, respectively.

We have assumed the mass fraction λ of material subjected to these high temperature, high strain rate conditions is limited by thermal conduction during the shock rise time. An expression for the width of the shear band,

$$\Delta x = \left(\frac{\pi k \varepsilon}{\dot{\varepsilon}}\right)^{1/2}, \qquad (3)$$

combined with the known N deformation bands per unit length, yields

$$\lambda = N \left(\frac{\pi k \varepsilon}{\dot{\varepsilon}}\right)^{1/2}. \qquad (4)$$

Implementation. To implement the model, shock wave properties for aluminum at moderate pressures were used (Table 1). The experimentally observed steady wave relationship between strain and strain rate,

$$\dot{\varepsilon} = 10^{12} \varepsilon^4 \text{ s}^{-1}, \qquad (5)$$

is imposed on the model, eliminating the need for knowledge of the viscosity.

No model is proposed here for the dependence of N, the density of shear bands on the strain and strain rate. Such a model may not even follow from a thermal description alone, but rather a balance between thermal and mechanical processes occurring in the shock front. Therefore, experimental results for the density of shear bands supplement this model,

$$N = 0.56 \, \dot{\varepsilon}^{3/4} \text{ m}^{-1} \qquad (6)$$

These data were obtained in a variety of experimental situations and although in any individual measurement there is considerable uncertainty, there appears to be a consistent trend of increasing shear band density with increasing strain rate.

Applications of the Model. It is useful to apply the model to several situations so that experiments can be designed and interpreted in a consistent framework: Figure 1a shows the temperatures anticipated as a function of the shock pressure. The adiabatic temperature applies to the regions in the interior of the aluminum remote from shearing regions while the local shock temperature corresponds to the temperature in the adiabatic shear bands themselves. Note that nowhere in the material is the average shock temperature achieved. Figure 1b shows the shear band density calculated as a function of shock pressure. The two lower pressure aluminum data from the laboratory were used to specify the model. A significant decrease in band spacing is anticipated with increasing pressure. On the same figure cell sizes observed in shock metallurgy studies on nickel are compared.[3] Figure 1c shows the expected mass fraction of preferentially heated material as a function of pressure.

Again, there are sparse data determined from shock data.[2] As the Hugoniot pressure increases, the mass fraction increases rapidly toward a more homogeneous state. Figure 1d displays the calculated true viscosity in the shear band regions inferred from the model and the specific descriptions chosen for the shear band density N and the strain/strain rate relationships. It is noteworthy that this viscosity does not depend strongly upon the shock pressure, hence, the temperature in the shear banding regions.

TABLE I - Material Properties for Aluminum

Density	ρ_o	2.71×10^3	kg/m^3
Grüneisen ratio	Γ_o	2.11	--
Intercept	C_o	5.25×10^3	m/s
Slope	S	1.5	--
Thermal Diffusivity	k	5.2×10^{-5}	m^2/s
Thermal Conductivity	K	1.34×10^2	W/mK

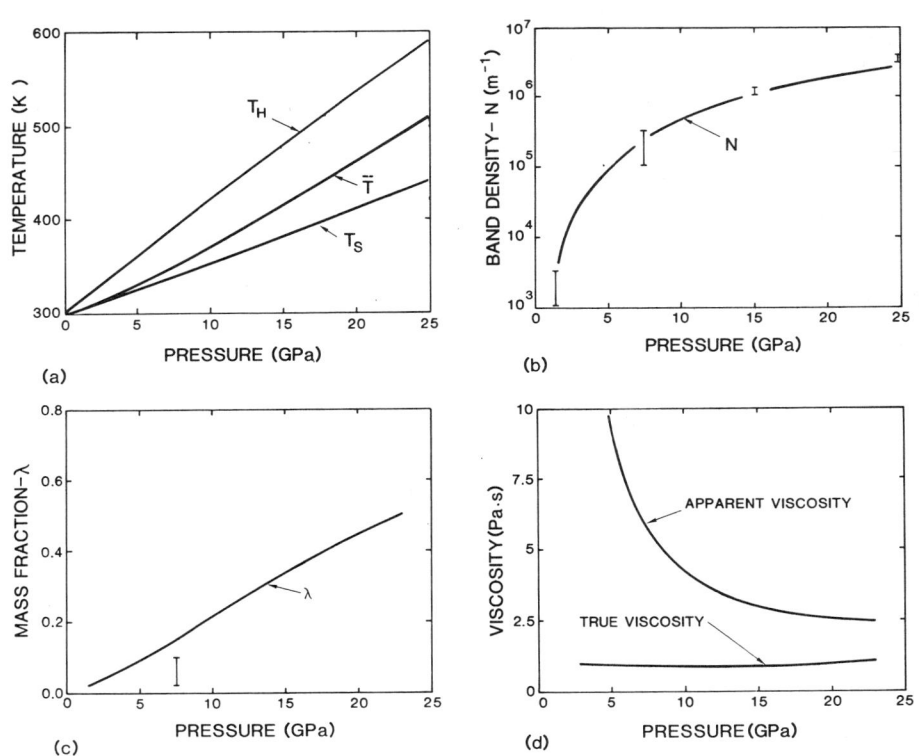

Figure 1. Results from model.

CONCLUSIONS

There are three interesting observations which are derived from this study. The first regards the shear band density where experimental results were used. We noted that over the entire experimental range, physical width of the shock front coincided roughly with the distance spanned by three shear bands. This is partially intuitive since under no circumstances would one expect for the shear bands to be so sparse that the physical width of the shock front was small compared to shear band spacing. Otherwise the entire yielding would have to take place between bands negating the need for heterogeneous yielding. There still remains, however, a gap in our understanding as to why this specific relationship is observed. A second observation is that the calculated true viscosity in the shear bands does not depend strongly upon the experimental shock pressure, hence the temperature in the shear bands. We have no assessment of the fundamental significance of this observation, being unable to distinguish between an athermal viscous process or competing effects of temperature and strain rate which make the true viscosity not vary from experiment to experiment, but which might not be generally true. It seems important then that experiments be conducted under different sets of conditions such as conditions of preheating in order to test the applicability of this observation to other situations. Finally, it is of interest to examine the apparent or bulk viscosity within the context of this model. Apparent viscosity is related to the true viscosity through

$$\eta_A = \eta_T / N\Delta x \quad . \tag{7}$$

Evaluating the denominator in terms of strain rate, and using the observed approximate constancy of η_T, we see that

$$\eta_A \alpha \frac{1}{\dot{\varepsilon}^{3/8}} \, , \tag{8}$$

which demonstrates that the apparent viscosity ought to, within the context of this model, vary as the inverse three-eights power of the strain rate. A widely observed proportionality between viscosity and the inverse one-half power strain rate is seen to approximately follow from the microstructural description and is not an inherent property which is necessarily expected in other than steady shock wave experiments.

REFERENCES

1. D. E. Grady, Appl. Phys. Let., June, (1981).
2. J. R. Asay and L. C. Chhabildas, Proceedings of the International Conference on the Metallurgical Effects of High Strain Rate Deformation and Fabrication, Albuquerque, NM, June 22-26, 417 (1980).
3. F. Greulich and L. E. Murr, Mat. Sci. Eng. 39, 81 (1979).

SHOCK COMPRESSION OF BERYLLIUM*

J. L. Wise, L. C. Chhabildas and J. R. Asay
Sandia National Laboratories,† Albuquerque, New Mexico 87185

ABSTRACT

Shock/reshock and shock/release experiments have been performed on beryllium to obtain wave profile measurements spanning the stress range of 6-35 GPa. Particle velocity histories in the beryllium samples were determined using a velocity interferometer in conjunction with lithium fluoride windows. The shock velocity-particle velocity response of beryllium was obtained from the interferometer signals, which allowed determination of the Hugoniot for Be over the stress range studied.

INTRODUCTION

Previous shock-wave investigations of beryllium have been confined to either time-resolved wave profile measurements to assess rate-dependent effects at low stresses (<10 GPa),[1] or to shock velocity-particle velocity measurements at high pressures (>10 GPa) to determine the dynamic equation of state.[2,3] Various attempts to explain the results of earlier experiments have applied such concepts as strain hardening and polycrystalline modeling to the wave profile work,[1] and have suggested the possibility of an hcp-bcc phase transition in the 6-30 GPa stress range to correlate equation of state studies.[3,4]

In order to resolve questions relating to the dynamic response of beryllium, the present experiments were conducted to produce the first systematic, time-resolved wave profile measurements in the intermediate stress regime (6-35 GPa). The interferometer records obtained from these experiments provide shock velocity-particle velocity Hugoniot data which are reported here. In addition, details of these particle velocity histories have been analyzed to investigate material strength variations and to evaluate the possible existence of a phase transition at elevated pressures (see Ref. 5). Finally, measurements of the risetime of the initial plastic wave permit estimation of the effective viscosity of beryllium for steady shock compression (see Ref. 6).

EXPERIMENTAL PROCEDURE AND RESULTS

Beryllium specimens used in the current investigations were prepared and provided by Lawrence Livermore National Laboratory. The chemical composition by weight was at least 98% beryllium, with maximum impurity levels of 1.5% beryllium oxide and 0.5% other elements. The average grain size was 35 microns or less, with a maximum size of

---

*This work was supported by the U. S. Department of Energy under Contract DE-AC04-76-DP00789.
†A U. S. Department of Energy facility.

125 microns. Measured sample densities ranged from 1.849 to 1.854 Mg/m^3, with a mean of 1.851 Mg/m^3. Ultrasonic sound speed measurements indicated values of 13.13 and 9.03 km/s for the longitudinal and shear wave velocities, respectively, corresponding to a Poisson's ratio of 0.051.

The experimental configuration used for the present studies is illustrated in Fig. 1. Shock loading of a beryllium sample (nominal thickness = 4, 8, or 9 mm) was produced upon impact by a projectile assembly which had been accelerated in a powder gun to a pre-selected velocity. The projectile was faced with an appropriate impactor material and backing plate to produce the desired initial stress state in the sample, followed by subsequent unloading or reloading from the shocked state. For initial shock stresses in the 6-11 GPa range, a Z-cut sapphire impactor was used with either a polymethylmethacrylate (PMMA) backing plate for release experiments, or a tungsten carbide (WC) backing plate for reshock experiments. For initial states in the 16-17 GPa range, beryllium impactors were typically used with backing plates of either PMMA (release) or copper (reshock). Experiments involving impact stresses of 24-25 GPa employed copper impactors with PMMA (release) or WC (reshock) backing plates. The highest initial stresses (34-35 GPa) were produced by WC impactors with PMMA backings (release only). Coaxial velocity pins were used in each experiment to determine the impact velocity to an accuracy of ~ ± 0.3% from a time-of-flight measurement. The average angular misalignment of the impactor relative to the beryllium target was 4.8 mrad, as measured by four equally spaced tilt pins. Table I summarizes the impact experiments and results.

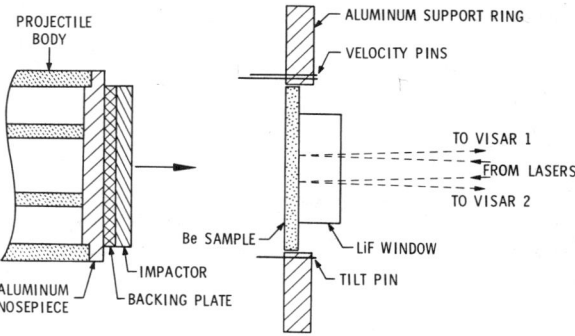

Fig. 1. Experimental configuration for beryllium studies.

In each experiment, a single-crystal lithium fluoride (LiF) window with axis parallel to the <100> crystalline direction was bonded to the rear surface of the beryllium sample. Wave profile data were obtained by using a pair of VISAR velocity interferometers[7] to monitor the particle velocity at two different locations on the sample/window interface. Lithium fluoride affords an excellent impedance match with beryllium. At worst, the shock impedance mismatch between

Table I: Summary of beryllium experiments

Shot No.	Impact Velocity[a,b] (km/s)	Sample Thickness (mm)	Shock Velocity (km/s)	Particle Velocity (km/s)	Impact Stress (GPa)	Impact Strain	Peak Strain Rate (μs^{-1})	Peak Stress Offset[f] (GPa)	Viscosity (Pa-s)
BE18	0.536(S)	8.049	8.38c	0.403	6.44	0.0469	0.9	0.17	189
BE3	0.871(S)	3.946	8.72c,e	0.637e	10.5	0.0721	2.9	0.42	145
BE8	0.899(S)	8.062	8.71d,e	0.662e	10.9	0.0748	3.4	0.46	134
BE19	0.946(S)	8.062	8.88d	0.687	11.4	0.0766	3.2	0.49	154
BE1	1.968(B)	3.954	9.00d	0.984	16.6	0.109	7.4	1.01	134
BE9	1.984(B)	8.063	9.15c	0.992	17.0	0.108	10.1	1.02	100
BE4	2.040(B)	3.896	9.11d,e	1.020e	17.4	0.111	9.0	1.08	119
BE14	1.969(A)	9.019	9.21c,e	1.016e	17.5	0.110	11.6	1.07	92
BE13	1.962(C)	8.052	9.54d	1.378	24.5	0.144	70.4	1.97	28
BE10	1.970(C)	8.050	9.60d	1.402	25.1	0.146	53.2	2.05	38
BE15	1.987(C)	9.027	9.53d,e	1.417e	25.2	0.149	44.5	2.09	46
BE2	1.990(C)	3.947	9.44d,e	1.438e	25.2	0.152	58.6	2.15	36
BE6	2.000(C)	3.881	9.58c	1.433	25.7	0.149	---	---	---
BE16	2.204(T)	4.062	10.01d	1.844	34.2	0.184	---	---	---
BE17	2.210(T)	8.061	10.18d	1.835	34.6	0.180	---	---	---

a Accuracy in impact velocity is ~ ± 0.3%.
b Letter in parenthesis designates impactor material: A = 6061-T6 aluminum; B = beryllium; C = OFHC copper; S = Z-cut sapphire; T = tungsten carbide.
c Measurement based on absolute timing.
d Measurement based on a velocity of 13.5 km/s for the initial breakaway of the elastic precursor ramp.
e Data acquired by single VISAR only.
f Offset determined using U = 7.998 + 1.124u, km/s.

419

Be and LiF is only 6.3% at 35 GPa (based on Be Hugoniot data provided
in Ref. 2 and LiF Hugoniot data from Ref. 8). By correcting the
interface velocity for the Be/LiF impedance mismatch, the particle
velocity behind the incident shock in the beryllium has been determined from the present experiments and is listed in Table I. Unless
otherwise noted, the listed particle velocity is an average of two
separate VISAR measurements. The optical behavior of LiF under shock
loading has been previously characterized,[9] and a 28% correction has
been applied to all present particle velocity data to account for the
variation of refractive index with stress. An estimated 1% uncertainty in the optical correction term, coupled with the observed
agreement of 0.5-1.0% between two independent interferometer measurements, results in an overall accuracy of 1-2% for the particle
velocity data.

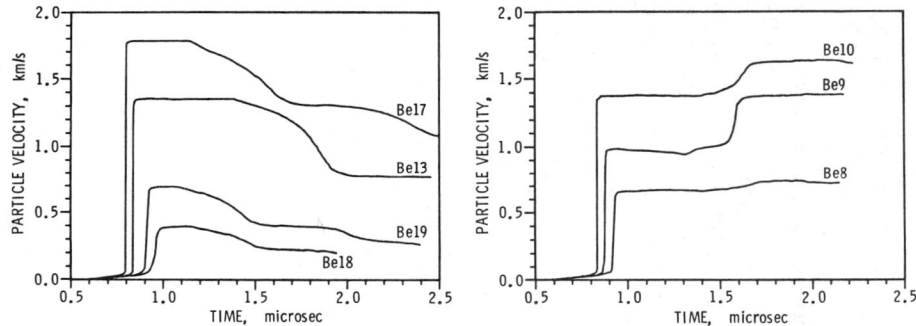

Fig. 2. Interface velocity histories for beryllium samples
of 8-mm nominal thickness.

Representative interface velocity histories are shown in Fig. 2,
where data for shots Be13, Be17, Be18, and Be19 correspond to shock/
release experiments, and data for shots Be8, Be9, and Be10 correspond
to shock/reshock experiments. A ramped elastic precursor with mean
amplitude of ~ 0.3 GPa was detected in all experiments. This ramping
behavior has been observed previously in beryllium and explained by
consideration of the polycrystalline nature of the material.[1] Additional features of the current wave profile data are analyzed in
detail in Refs. 5 and 6. Data for the peak strain rate, peak stress
offset, and viscosity associated with incident bulk shocks in Be are
listed in Table I and discussed in Ref. 6.

When interpreted in conjunction with timing data from velocity
and tilt pins, the interferometer records yield propagation velocities for the initial elastic precursor and plastic shock wave in the
beryllium samples. For those shots where absolute timing was available, the velocity of the leading edge of the precursor ramp averaged
13.5 km/s, with a standard deviation of 0.2 km/s. Experimental
shock velocity data are listed in Table I. Unless otherwise noted,
the reported shock velocity corresponds to the average value obtained

from two separate VISAR measurements. For those experiments where absolute timing was not available, the shock velocity was based on a velocity of 13.5 km/s for the leading edge of the precursor ramp. A plot of shock velocity vs. particle velocity data from the individual VISAR measurements is shown in Fig. 3. The corresponding stress vs. strain data are also plotted in Fig. 3, and average stress and strain values are listed for each experiment in Table I. A linear least-squares fit to the unaveraged shock velocity-particle velocity data results in $U = 7.982 + 1.131u$, km/s, where U = shock velocity, u = particle velocity, and the standard deviations of the intercept and slope of this fit are 0.049 km/s and 0.039, respectively. The standard deviation of the shock velocity data about the fitted line is 0.087 km/s. The present relation agrees excellently with the fit of $U = 7.998 + 1.124u$ reported by McQueen, et al.[2]

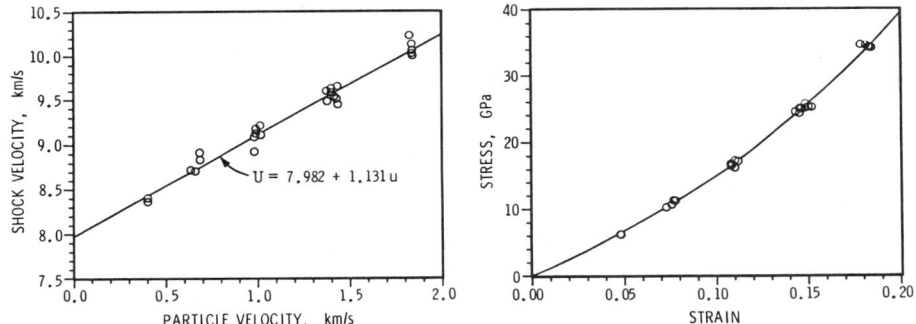

Fig. 3. Experimental data for shock velocity vs. particle velocity and stress vs. strain.

REFERENCES

1. R. E. Swanson, Los Alamos Scientific Laboratory Report LA-5943-MS (1975).
2. R. G. McQueen, S. P. Marsh, J. W. Taylor, J. N. Fritz and W. J. Carter, in High-Velocity Impact Phenomena, ed. by R. Kinslow (Academic Press, NY, 1970).
3. L. W. Davison and J. N. Johnson, Sandia Laboratories Report SC-TM-70-634 (1970).
4. H. C. Graboske, R. Grover and K. S. Long, Lawrence Livermore Laboratory Report UCRL-50028-78-3 (1978).
5. L. C. Chhabildas, J. L. Wise and J. R. Asay, "Reshock and Release Behavior of Beryllium," this proceedings.
6. J. R. Asay, L. C. Chhabildas and J. L. Wise, "Strain Rate Effects in Beryllium under Shock Compression," this proceedings.
7. L. M. Barker and R. E. Hollenbach, J. Appl. Phys. 43, 4669 (1972).
8. W. J. Carter, High Temp. - High Press. 5, 313 (1973).
9. J. L. Wise and L. C. Chhabildas, Bull. Am. Phys. Soc. 25, 566 (1980).

RESHOCK AND RELEASE BEHAVIOR OF BERYLLIUM

L. C. Chhabildas, J. L. Wise and J. R. Asay
Sandia National Laboratories,† Albuquerque, New Mexico 87185

ABSTRACT

Reshock and release wave profile measurements in beryllium over the stress region 6-35 GPa have indicated no evidence of an hcp-bcc phase transition as previously suspected. The present measurements show that the yield strength of the material is 1.8 GPa at 34 GPa, a substantial increase from the ambient value of 0.3 GPa.

INTRODUCTION

Based on previous shock wave investigations,[1,2] hydrostatic resistivity measurements,[3] and the low-pressure, high-temperature phase diagram of beryllium,[4] the existence of an hcp-bcc phase transition in the stress range of 6-30 GPa has been suspected.[5] In an attempt to resolve this question, a series of experiments was performed on beryllium[6] in which the specimens were first shocked to stresses ranging from 6-35 GPa and then either reshocked or released from the shocked state. The beryllium samples were provided and prepared by Lawrence Livermore National Laboratory.

The motivation behind performing reshock and release experiments in beryllium is indicated in Fig. 1. The volume change associated with an hcp-bcc phase transition in beryllium could be expected to be less than 2%,[5] and would be represented as a cusp in the stress-volume

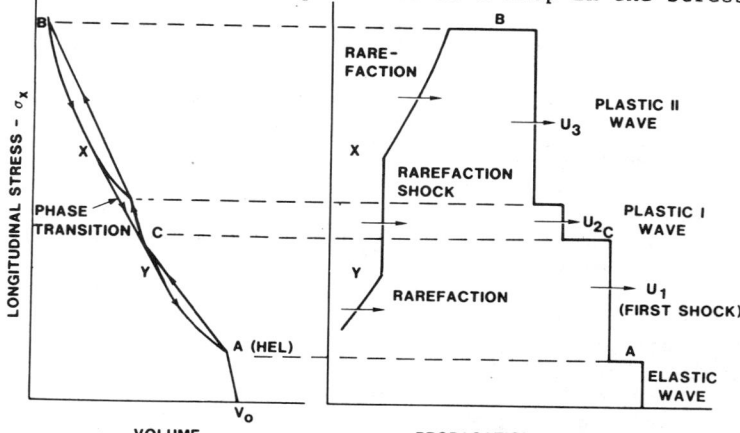

Fig. 1. A cusp in the stress-volume curve due to a phase transition would give rise to a multiple wave structure upon reshock and a rarefaction shock upon release.

---

*This work was supported by the U. S. Department of Energy under Contract DE-AC04-76-DP00789.
†A U. S. Department of Energy facility.

response. For such a small volume change, a single shock, loading the material to state B, would overdrive the phase transition, resulting in no observable effects on the shock wave structure. However, if the material is shocked to state B in two stages, i.e., a single shock up to a state C and then a second shock to state B, a two-wave structure will be observed during reshock from state C to state B. Normally for most materials, the stress-volume curve is concave upward (indicated as BX in Fig. 1), and would result in a rarefaction fan upon release. However, the region XY for a material that undergoes a phase transition is convex upward, and it would produce a rarefaction shock upon release, allowing additional identification of a phase transition. For the purposes of illustration, elastic-plastic effects upon reshock and release have been ignored in Fig. 1. Thus, reshock and release experiments allow an accurate assessment of whether phase transformations occur and are also useful for determining strength effects at high pressures.

RESULTS AND DISCUSSION

The experimental impact conditions and results are given in Ref. 6. In general, single steady wave profiles are observed[7] in all experiments with no evidence for phase transition upon single-shock compression over the stress range studied. This is illustrated in Fig. 2 which shows the particle velocity profiles for release wave experiment BE1 and reshock experiment BE9 from an initial stress of 16.6 GPa. An elastic-plastic structure is observed upon loading, although the elastic wave is not well defined. The final release and reshock states are ~ 1.5 GPa and 25.0 GPa, respectively. Analyses of both release and reshock profiles indicate that the first disturbance is propagating at an elastic wave speed in the shocked material. The plastic portions of the wave profiles provide estimates of release wave velocities and second shock velocities

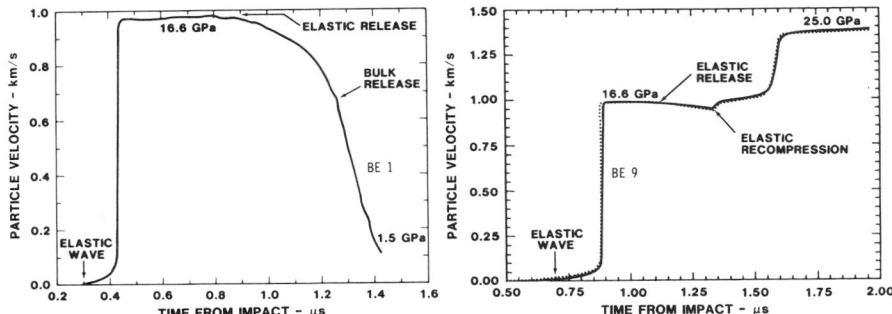

Fig. 2. Wave profiles from release experiment BE1 and reshock experiment BE9. The elastic release in reshock experiment BE9 is due to a 0.013 mm void between the copper and beryllium interface.

which are in good agreement with computed values based on equation of state data.[8] In addition, there is no evidence of a rarefaction shock upon release, nor a multiple-plastic wave structure associated with a phase change upon reshock, indicating an absence of a phase transition over the stress range 1.5 to 25 GPa. The two-wave structure observed in experiment BE9 upon reshock is due to elastic-plastic behavior[9,10] of the material in the shocked state.

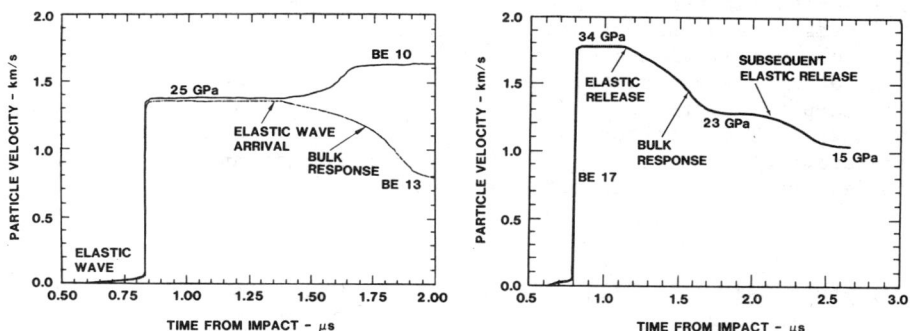

Fig. 3. Wave profiles from reshock experiment BE10 and release experiments BE13 and BE17.

Results of reshock and release experiments BE10 and BE13 are indicated in Fig. 3. The initial shocked state is 25 GPa and the final release state for shot BE13 is 12.5 GPa. The release wave profile for BE13 indicates an elastic release followed by plastic unloading, and the measured release wave velocities agree well with computed values. More importantly, the results of reshock experiment BE9 (Fig. 2) and release experiment BE13 are self-consistent with no indications of a phase change to 25.0 GPa stress.

The particle velocity profile obtained for reshock experiment BE10 at 25 GPa is shown in Fig. 3. The final reshocked state is ~ 32 GPa. Fine structure in the reshocked profile indicates an elastic wave followed by a plastic wave which is very dispersive and the measured second shock velocity is ~ 10% faster than computed values. Likewise, experiment BE17 which is a release experiment from 34 GPa to 23 GPa also indicates that measured release wave speeds are ~ 10% faster than computed values. Although the experimental uncertainty in estimating release wave speeds and second shock speeds is 10%, the results of experiments BE13 and BE17 are mutually consistent indicating fast second shock and release wave velocities in the stress region 25-34 GPa. This indicates a steepening of the stress-volume loading and unloading curve over the stress range 25-34 GPa. The reason for fast release wave speed measurements is probably due to anisotropy, and preferential texturing of the grains could give rise to faster wave speeds than expected values at high stresses.

An absence of a multiple-wave structure associated with a phase change in reshock experiment BE10 and a rarefaction shock release in experiment BE17 does not, however, totally exclude the existence of an hcp-bcc phase boundary which intersects the Hugoniot curve over the stress range 25-34 GPa, since the reloading or release rate at the impactor/beryllium interface is finite and the phase transition in beryllium may possibly be extremely sluggish, and thus not observable on the time scale of the shock experiments.

The measured elastic and bulk release wave speeds over the stress range 6-34 GPa are indicated in Fig. 4. The dashed and solid lines are calculated velocities based on the equation of state of beryllium[10] assuming that $\rho\gamma$ is constant, where ρ and γ are the density and the Gruneisen coefficient, respectively. The elastic release wave velocities are obtained assuming that the Poisson's ratio is constant at 0.05.[6] As indicated in the figure, the measured bulk velocities are higher than computed values at low stresses, which is believed to be due to considerable strain hardening effects in the material. The increase in release wave speed above 25 GPa is also indicated in Fig. 4, although the effect is within the uncertainty (~ 10%) of the calculation and the experiment.

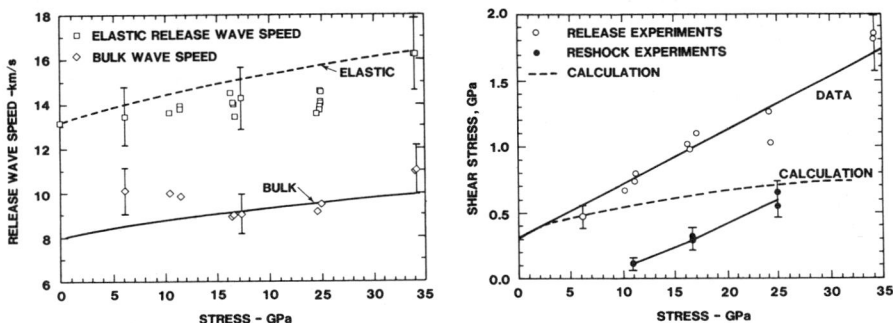

Fig. 4. Release wave speed measurements and shear stress increase in beryllium as functions of shock stress.

Briefly, the shear stress build-up upon release, τ_u, or upon reshock, τ_r, measured from the shocked state can be obtained using,[10]

$$\tau_u = -3/4 \; \rho_o \int_{e_o}^{e_1} (c^2 - c_B^2) de$$

or (1)

$$\tau_r = 3/4 \; \rho_o \int_{e_o}^{e_2} (c^2 - c_B^2) de \quad ,$$

where ρ_o is the density of the material, c is the measured Lagrangian elastic release wave velocity obtained from wave profiles, and c_B is the computed Lagrangian bulk velocity at the corresponding strain e. e_1 and e_2 are the final strain states up to which elastic-plastic effects are observed upon release or reshock, respectively. The shear stress estimated using eqns. (1) is shown plotted in Fig. 4. The experimentally obtained values are much larger than the calculations based on a constitutive model[11] in which the shear modulus and the yield strength are expressed as functions of pressure, temperature and plastic strain using the parameters obtained from low-pressure experiments. Thus, the experiments indicate that the extrapolation of yield strength based on low-pressure data is not accurate for beryllium. The experiments also indicate a considerable increase in strength from the ambient value. In addition, the shear stress increase in beryllium upon reshock is also shown in Fig. 4. This increase in shear stress upon reshock and release is not unique to beryllium, and has been observed in tungsten,[9] aluminum[10] and copper.[12]

To summarize, reshock and release experiments in beryllium over the stress range 6-35 GPa do not indicate an hcp-bcc phase transition, although release wave speeds determined over the stress range 25-35 GPa are faster than expected values. It is conceivable that either no phase transformation exists or the phase transition is extremely sluggish or metastable, and is therefore not observable during the time scale of the experiments. In addition, the shear stress build-up upon release (normally known as the yield strength) increases to 1.8 GPa at 35 GPa, which is a factor of six increase from the ambient value of 0.3 GPa.

REFERENCES

1. L. W. Davison and J. N. Johnson, Sandia Laboratories Report No. SC-TM-70-634 (1970).
2. T. Neal, in High Pressure Science and Technology, edited by K. D. Timmerhaus and M. S. Barber (Plenum, NY, 1979) V1 p. 80.
3. A. R. Marder, Science 142, 664 (1963).
4. M. Fracois and M. Contre, in Proc. Int. Conf. Metallurgy of Beryllium, Grenoble (1965) and the references therein.
5. H. C. Graboske, R. Grover, and K. S. Long, Lawrence Livermore Laboratory Report No. UCRL-50028-78-3, (1978).
6. J. L. Wise, L. C. Chhabildas and J. R. Asay, "Shock Compression of Beryllium," this proceedings.
7. J. R. Asay, L. C. Chhabildas and J. L. Wise, "Strain Rate Effects in Beryllium under Shock Compression," this proceedings.
8. R. G. McQueen, S. P. Marsh, J. W. Taylor, J. N. Fritz and W. J. Carter in High Velocity Impact Phenomena, edited by R. Kinslow (Academic Press, NY, 1970).
9. J. R. Asay, L. C. Chhabildas and D. P. Dandekar, J. Appl. Phys. 51, 4774 (1980).
10. J. R. Asay and L. C. Chhabildas, in Shock Waves and High-Strain-Rate Phenomena in Metals, edited by M. Meyers and L. Murr (Plenum, NY, 1981) p. 417.
11. D. J. Steinberg, S. G. Cochran and M. W. Guinan, J. Appl. Phys. 51, 1498 (1980).
12. L. C. Chhabildas (unpublished).

STRAIN RATE EFFECTS IN BERYLLIUM UNDER SHOCK COMPRESSION*

J. R. Asay, L. C. Chhabildas and J. L. Wise
Sandia National Laboratories,† Albuquerque, New Mexico 87185

ABSTRACT

Shock wave experiments were conducted on beryllium over the range of 6-25 GPa. A steady wave analysis of the shock front was used to estimate effective viscosities achieved during the loading process, which were found to vary approximately as the inverse square root of strain rate for strain rates greater than about 3 μs^{-1}.

INTRODUCTION

Shock wave techniques provide a useful tool for studying material processes which occur at high rates of deformation. A variety of physical processes have been investigated with these methods,[1] including studies of rate-dependent material deformation, such as dynamic yielding and viscoplastic behavior.[2] For example, studies of elastic precursor decay[3,4] have provided information about physical mechanisms, such as dislocation slip or twinning, important to initial yielding, and steady wave analyses of plastic waves following the elastic precursor have been useful for assessing dissipative effects of shock compression.[5,6]

In the present investigation, laser interferometric methods were used to study the structure of elastic-plastic waves in beryllium over the stress range of 6-34 GPa. Plastic waves in beryllium are resolvable with the present instrumentation for stress loads to about 25 GPa, and are observed to be steady for propagation distances greater than 4 mm. These results allow determination of the effective viscosity of beryllium for steady shock compression. Viscosity determined in this way is found to decrease with increasing strain-rate.

EXPERIMENTAL TECHNIQUE AND RESULTS

The experimental technique for producing high amplitude shock waves in beryllium samples provided by Lawrence Livermore National Laboratory has been described.[7,8] Briefly, a propellant-driven projectile is faced with an impactor and impacts polycrystalline beryllium samples to produce longitudinal stresses ranging from about 6-34 GPa. For impact pressures to 12 GPa, Z-cut aluminum oxide single crystals were used as impactors; whereas for higher stress levels, either beryllium, copper or tungsten carbide was used as an impactor.

Interferometer particle velocity histories were measured at the rear surface of beryllium specimens using LiF windows.[8] In most experiments, two interferometers were used to determine particle

---

*This work was supported by the U. S. Department of Energy under Contract DE-AC04-76-DP00789.
†A U. S. Department of Energy facility.

velocity histories at different locations on the specimen. Velocity measurements at these two locations agreed to 0.5 - 1.0%; the overall accuracy judged to 1 - 2%. The interferometer time resolution was approximately 2 ns.

The compression wave structure in all experiments consisted of a dispersed precursor with final amplitude of about 0.3 GPa, followed by a single plastic wave with a characteristic velocity and risetime. The steadiness of the plastic wave for each stress level was evaluated by comparing wave profiles measured at different propagation distances. Fig. 1 illustrates representative results for the plastic portion of the steady wave. Entire wave profiles for various experiments are presented in Refs. 7 & 8. The interferometer time-resolution was sufficient to resolve the plastic wave profiles for stress levels only to about 25 GPa; plastic wave risetimes were not resolvable at a driving stress of 34 GPa. Only one propagation distance (8 mm) was studied at 6 GPa, which precludes a definite conclusion about wave steadiness. However, experiments previously conducted on S-200-E beryllium at a driving stress of 2.4 GPa by Christman and Feistmann[9] indicate that steady waves are formed for propagation distances of 6 mm or greater.

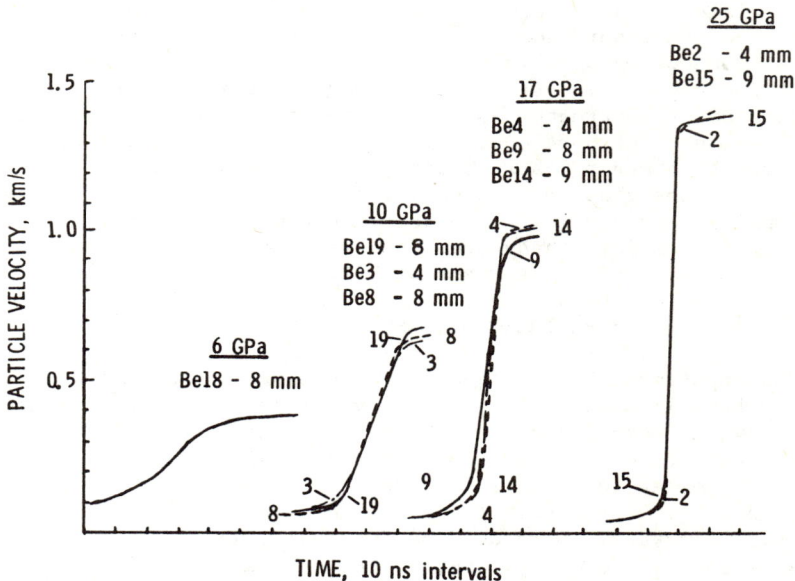

Fig. 1. Typical interface wave profiles used to determine peak strain rates during steady wave compression. A slight impedance correction (refs. 7 & 8) was applied to the data to estimate the true strain rate.

DISCUSSION

Steady waves result from a balance between the competing effects of non-linear material response and dissipative processes occurring in the shock. Measurements of the shock structure can therefore be combined with information on material non-linear response to evaluate the dissipative effects. Several approaches have been developed for characterizing the dissipation produced by shock compression.[5,6,10-12] The approach used here is to describe dissipative effects in terms of an effective viscosity.[10,11] In this approximation, the non-equilibrium stress difference, $\Delta\sigma$, achieved in the steady wave is related to strain rate as [11]

$$\Delta\sigma = \eta\dot{\varepsilon} , \quad (1)$$

where η is an average or effective viscosity operative during the wave risetime. If the Hugoniot is taken as the equilibrium curve from which non-equilibrium departures are measured, the maximum non-equilibrium stress is given to a good approximation as

$$\Delta\sigma \doteq \frac{1}{2} s e_f \sigma_f , \quad (2)$$

where σ_f and e_f are the final values of stress and strain achieved in the wave and s is the coefficient in the shock-particle velocity relation ($U = c_0 + su$). Thus, measurement of the peak strain rate achieved in the steady wave can be combined with the maximum overstress to estimate the effective viscosity.

Fig. 2 shows the measured maximum strain rates achieved in the steady waves at different stress levels. Also shown are data by

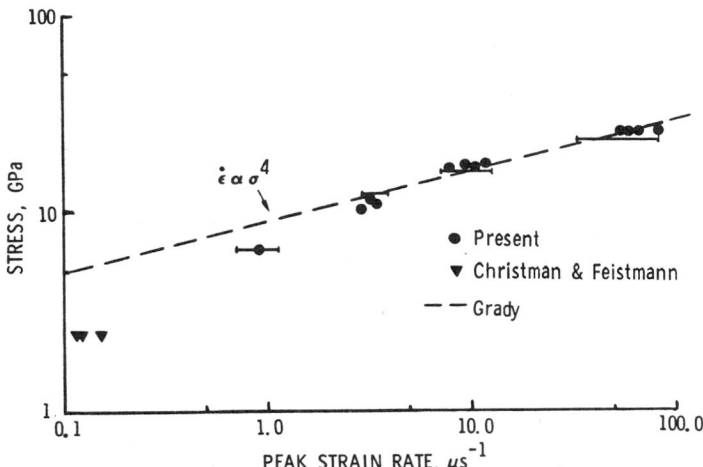

Fig. 2. Hugoniot stress vs peak strain rate achieved during shock compression.

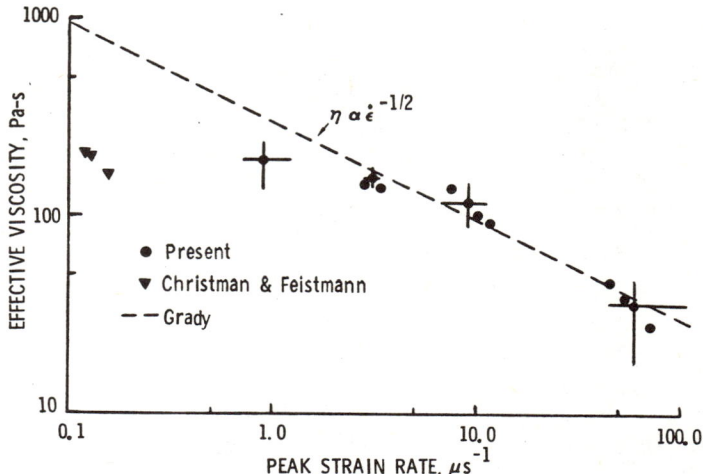

Fig. 3. Effective viscosity vs peak strain rate. Typical error bars are shown.

Christman and Feistmann.[9] These values of strain rate were combined with the maximum overstress to calculate the viscosities, which are shown in Fig. 3. The viscosity is observed to be essentially constant for strain rates to about 3 μs^{-1}, but decreases with strain rate for larger rates.

Godunov et al.[13] have determined the dependence of viscosity on strain rate in several metals over a large strain rate region and observed similar behavior. At low strain rates, the viscosity was observed to be a slowly varying function of strain rate, whereas at higher rates it decreased rapidly with strain rate, with a nearly inverse square root dependence.

Grady[12] has recently shown that the observed variation in effective viscosity can be restated as constancy of a quantity referred to as the "shock invariant," which is the product of dissipative energy and shock wave risetime. This assumed constancy also leads to strain rate depending on the fourth power of Hugoniot stress. These predictions are shown as the dashed lines in Figs. 2 and 3, and to within the error bars shown, are consistent with experimental results at the higher rates. However, a definite departure from these predictions is observed for strain rates less than about 3 μs^{-1}. The underlying physical processes which give rise to the results shown in Figs. 2 and 3 are as yet unknown. However, Hayes and Grady[14] have recently shown that microscopic processes of localized deformation can result in effective viscosities in shock wave experiments which have a nearly square inverse root dependence on strain rate.

Assuming that the physical effects of deformation are important for determining effective viscosity, the transitional behavior shown in Fig. 3 may signify a change in the microstructural mechanism of deformation. In this regard, it is important to note that large departures in the strength of beryllium from low pressure extrapolations have been observed[8] for shock pressures near 6 GPa, which also suggests a change in the mode of deformation. Recovery experiments on beryllium over the stress range of about 4-15 GPa would be extremely helpful in resolving this question. However, in any case, the fact is that similar changes in the strain rate dependence of effective viscosity have been observed in several other materials.[13]

CONCLUSION

Shock wave experiments conducted on beryllium indicate that steady wave profiles are produced for shock pressures ranging from 10-25 GPa and for propagation distances greater than 4 mm. The peak strain rate in the steady wave is observed to vary approximately as the fourth power of the Hugoniot stress for strain rates greater than about 3 μs^{-1}. These results were used to estimate the effective viscosity for steady wave compression in beryllium, which is observed to vary approximately as the inverse square root of strain rate for rates greater than 3 μs^{-1}.

REFERENCES

1. L. W. Davison and R. A. Graham, Phys. Rpt. Vol. 55, 255-379 (1979).
2. W. Herrmann, in *Propagation of Shock Waves in Solids*, ASME, NY (1976).
3. J. R. Asay, G. R. Fowles, G. E. Duvall, M. H. Miles and R. F. Tinder, J. Appl. Phys. 43, 2132 (1972).
4. Y. M. Gupta, G. E. Duvall, and G. R. Fowles, J. Appl. Phys. 46, 532 (1975).
5. J. N. Johnson and L. M. Barker, J. Appl. Phys. 40, 4321 (1969).
6. F. E. Prieto and C. Renero, J. Appl. Phys. 44, 4013 (1973).
7. J. L. Wise, L. C. Chhabildas, and J. R. Asay, "Shock Compression of Beryllium," this proceedings.
8. L. C. Chhabildas, J. L. Wise, and J. R. Asay, "Reshock and Release Behavior of Beryllium," this proceedings.
9. D. R. Christman and F. J. Feistmann, "Dynamic Properties of S-200-E Beryllium," General Motors Corporation Report MSL 71-23 (1972).
10. D. C. Wallace, Phys. Rev. 22, 1487 (1980).
11. L. C. Chhabildas and J. R. Asay, J. Appl. Phys. 50, 2749 (1979).
12. D. E. Grady, Appl. Phys. Lett., in press.
13. S. K. Godunov, A. A. Deribas and V. I. Mali, Combustion, Explosion and Shock Waves 11, 3 (1975).
14. D. B. Hayes and D. E. Grady, "A Thermal-Viscous Model for Heterogeneous Yielding in Aluminum," this proceedings.

SHOCK WAVE PROPAGATION IN BERYLLIUM AT SMALL IMPACT STRESSES AND ELEVATED TEMPERATURES

M. D. Bjorkman and J. E. Shrader
Boeing Aerospace Co., Seattle Wa. 98124

ABSTRACT

Plate impact experiments were conducted with hot isostatically pressed beryllium specimens, shocked transversely to the pressing direction. Impact stress was varied from 0.2 GPa to 1.1 GPa, with initial specimen temperatures of 290 K, 500 K and 1000 K. Elastic precursor shocks were observed in every case, instead of the dispersed precursor wave fronts reported by others. In some cases the arrival of three waves was observed.

METHODOLOGY

Gas gun experiments with commercially pure beryllium were performed. Specimens were heated to 500 K by heating the target holder from the edges. Other specimens were radiatively heated to 1000 K using a carbon sheet resistor placed near the specimen face opposite impact. The specimens were impacted with 1.3 mm thick PMMA flyers, unsupported on the rear face, with the Boeing 2.5 in. gas gun facility.[1] Impact stress was measured using a carbon piezoresistive gauge[2] embedded in the flyer, and the rear free surface response measured using the Boeing combined VISAR/displacement interferometer.[3]

The BG-170-B commercially pure beryllium was produced by the hot-pressed powder metallurgical process. The material was hot pressed, then over aged, to produce a 0.2% offset yield strength between 166 to 270 MPa. Specimens were taken from two billets so that shock propagation was transverse to the pressing direction. Maximum impurity content was specified to be: 1.5 wt.% for BeO, 0.085 wt. % for Al and 0.17 wt % for Fe. The as-pressed density was specified not to vary more than 1% from the theoretical density.

RESULTS

Two specimens were unheated, for comparison with previous work with unheated beryllium specimens. The rest were heated to approximately 500 K and approximately 1000 K. Impact stress was either approximately 0.2 GPa, 0.5 GPa or 1.1 GPa. Specimen thickness was either 5 mm or 10 mm.

The rear free-surface velocity histories of the room temperature shots are shown in Fig. 1. The rear free surface velocity history (hereafter referred to as the history) from the 0.5 GPa impact stress shot, no. 10, shows the arrival of a sharp shock first wave followed by a dispersed second wave. (The term sharp shock is used for waves with risetimes less than a few ns, the resolution of the VISAR used to measure the histories.)

The history from the 1.1 GPa impact stress room temperature shot, no. 11, shows the arrival of a sharp shock first wave, followed approximately 40 ns later by a dispersed second wave, and a portion of a third wave.

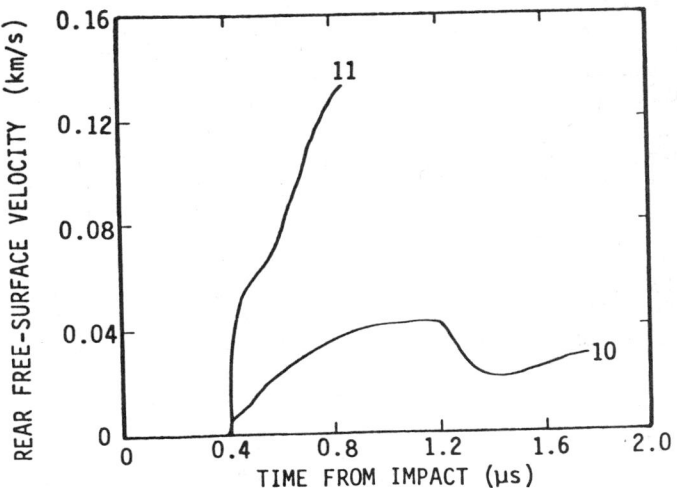

Figure 1 Room temperature wave profiles. Wave profiles are labeled with the shot number. The specimens were 5 mm thick.

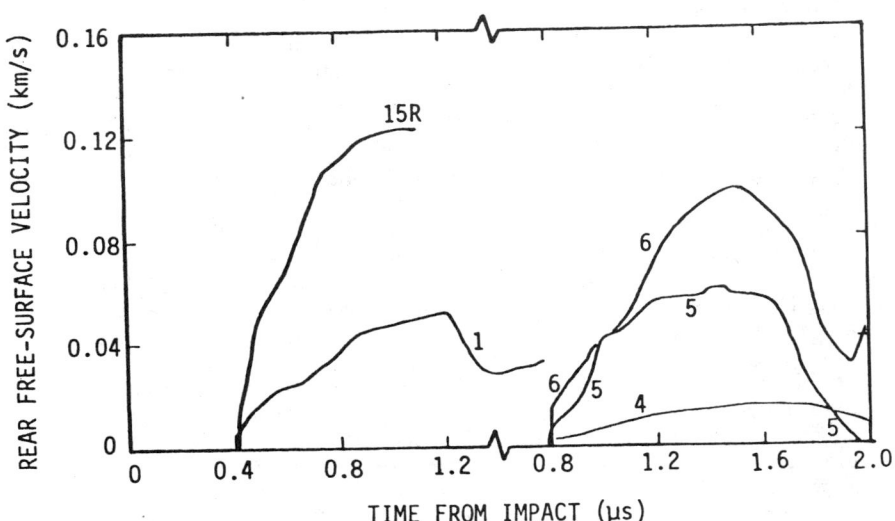

Figure 2 500 K wave profiles. Wave profiles are labeled with the shot number. Note that the time axis has been split to separate the 5 mm thick specimen wave profiles from the 10 mm thick specimens.

The histories of the 500 K shots are shown in Figs. 2 and 3. The histories show the arrival of a sharp shock first wave in every case, and the arrival of 3 waves for the 0.5 and 1.1 GPa impact stress cases.

The repeatability of the histories from specimens from each billet, impacted under similar conditions (5-mm thick specimens, 1.1 GPa impact stress, at 500 K initial temperature) is shown in Fig. 3. The arrivals of three-waves appear on each history. Other than the amplitude of the first wave, the history from shot 2 is not significantly different than shot 15R.

Figure 4 shows the histories from the 1000 K shots. The history from the 0.3 GPa impact stress shot, no. 14, is characteristic of an elastic plastic waveprofile with strain hardening. The 0.5 GPa impact stress shots, nos. 12 and 13, show incipient formation of a particle velocity drop behind the elastic precursor, while the 1.1 GPa impact stress shot, no. 11, shows a fully developed elastic precursor, with a particle velocity drop following.

DISCUSSION

The first wave appears to be an elastic shock because the measured average first wave velocity is consistent with the reported longitudinal elastic acoustic velocity of 12.8 km/s.

Elastic shocks are not usually seen in hot pressed beryllium. Previous work has shown ramped precursors.[4-9] The more well-defined precursors had 50 ns risetimes to peak amplitudes,[5] while some of the more poorly defined precursors were nearly indistinguishable from the plastic wave.[7]

One exception was textured beryllium sheet, which Pope and Stevens[7] found propagated a 0.7 GPa amplitude sharp shock precursor. They interpreted this as evidence for a distribution of yield strengths among the crystal grains of the specimen. The distribution of frozen-in thermal stresses from the hot pressing process would, for large thermal stresses, bring the crystal grain near yield and for small thermal stresses leave yield strength unchanged. A distribution of yield strengths would result in a concave downwards stress-volume R-H curve at stresses near yield thus resulting in dispersive precursor wave fronts. Textured beryllium sheet has crystal grains aligned by plastic deformation during rolling. The plastic deformation relaxes the thermal stresses narrowing the distribution of yield stresses, resulting in sharp shock precursors.

Our data may be consistent with crystal grain yield strength distributed amongst two values; the smaller yield point causing the first wave, and the larger yield point causing the second wave in three wave profiles.

An interesting feature of our data is the first wave amplitude roughly doubling with each doubling of impact stress. This behavior is similar to overdriven steady wave motion in an elastic-viscoplastic solid.

The second wave may result from reflections from the impactor gauge plane, multiple yield points, or interaction of the elastic wave with the free surface. Each will be discussed in turn.

Perturbations on the wave profile will travel at the elastic wave speed. Therefore, gauge reflections will arrive approximately 200 ns after the elastic shock. The difference between the measured arrival times of the first and second waves is 170 $\pm$ 140 ns. The average value is comparable to the expected 200 ns, however the standard deviation is large.

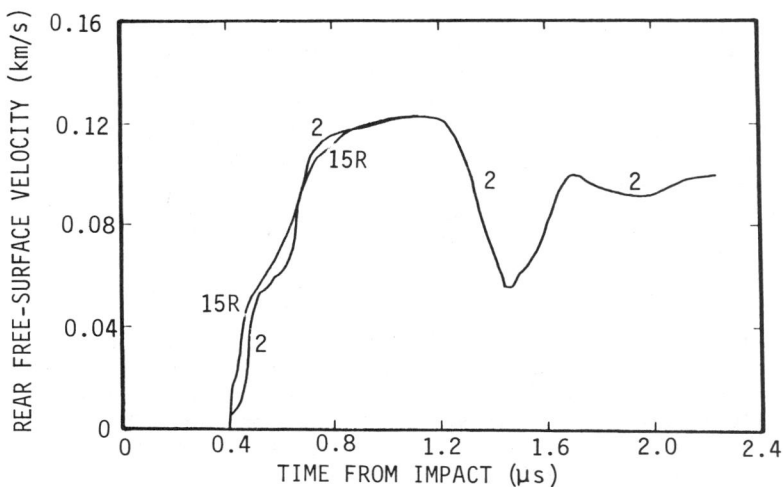

Figure 3 Two 500 K initial temperature shots at the same impact conditions. Impact stress was 1.1 GPa, specimen thickness was 5 mm.

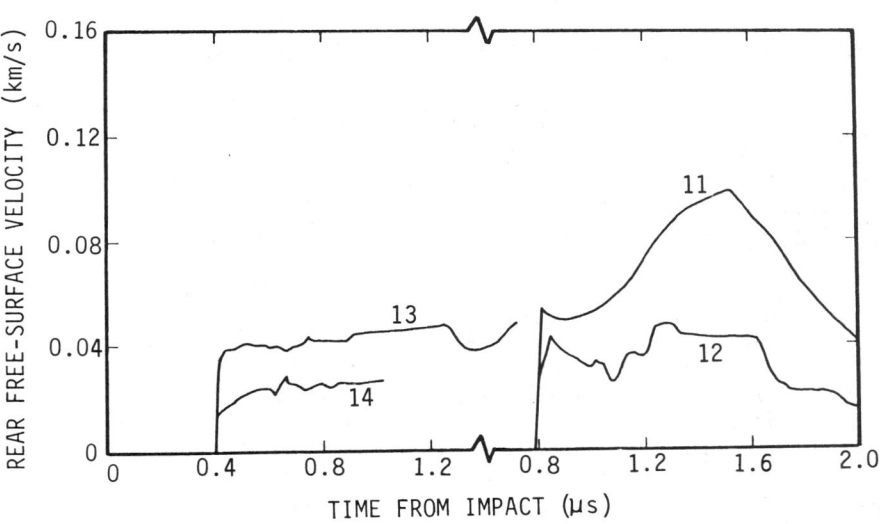

Figure 4 1000 K wave profiles. Wave profiles are labeled with shot number. Note that the time axis has been split to separate the 5 mm thick profiles from the 10 mm thick specimens.

There are several problems, besides the above problematical time of arrival correlations, with interpreting the second wave, in three wave profiles, as a reflection from the gauge plane: the carbon gauge R-H curve lies below the PMMA R-H curve,[10] so the gauge reflection should be a decrease in free-surface velocity and not the increase observed in the experiments, the amplitude of the second wave in three wave profiles is comparable with reported Hugoniot elastic limits, and the three wave structure is not apparent on the 1000 K initial temperature shots.

An alternative explanation is that the first and second waves in three wave profiles result from multiple yield points. Swanson[11] has calculated wave propagation in beryllium assuming two and more yield points; the two yield point calculations are reminiscent of our free-surface velocity histories.

Another possibility is interaction of the elastic precursor with the free surface. The elastic wave will reflect from the free surface as a release wave. The release wave will collide with the plastic wave. At this time a new precursor will form and propagate ahead of the plastic wave. The second elastic wave will interact with the free-surface and double the free-surface velocity from the first elastic precursor. This process has been observed in iron by Barker [12]. Construction of a x-t diagram shows that the second wave arrives 110 ns after the first wave at 5 mm propagation distance, and 250 ns after the first wave at 10 mm propagation distance. These times are comparable with those measured in our experiments.

REFERENCES

1. See R. M. Schmidt, "Boeing Shock Physics Laboratory," this proceedings.
2. J. A. Charest, Shock Pressure Sensors and Connector/Cable Assemblies, DynaSen, Inc. Catalog (1980).
3. L. M. Barker and R. E. Hollenback, J. Appl. Phys 43, 4669 (1972).
4. N. H. Froula, General Motors Materials and Structures Laboratory MSL-68-16 (1968).
5. D. R. Christman and F. J. Feistmann, Defense Nuclear Agency, DNA 2785F (1972).
6. D. R. Christman and N. H. Froula, AIAA J 8, 477 (1970).
7. A. L. Stevens and L. E. Pope, in Metallurgical Effects at High Strain Rates, ed. by R. W. Rhode, B. M. Butcher, J. R. Holland and C. H. Karnes (Plenum Press, New York, 1973) p. 459.
8. D. E. Munson, Sandia Laboratories, SC-RR-67-368 (1967).
9. D. P. Dandekar, J. F. Dignam and A. G. Martin, in High Pressure Science and Technology, ed. by B. Vodar and Ph. Marteau (Pergamon Press, New York, 1980), p. 344.
10. F. H. Ree, W. M. Isbell, and R. R. Horning, Lawrence Livermore Laboratory, UCRL-51682, Pt.4 (1974).
11. R. E. Swanson, Los Alamos Scientific Laboratories, LA-5943-MS (1975).
12. L. M. Barker and R. E. Hollenbach, J. Appl. Phys. 45, 4872 (1974).

LARGE AMPLITUDE COMPRESSION AND SHEAR WAVE PROPAGATION IN AN ELASTOMER.[*]

Y. M. Gupta, W. J. Murri, and D. Henley
SRI International

ABSTRACT

Experimental techniques have been developed to measure the high strain-rate compression and shear response of Solithane 113.[1] Compression and shear wave profiles have been measured in specimens compressed to 20% (compressive stresses $\sim$ 1.2 GPa). The compressive profiles are nearly steady and the compressive stress-strain response is typical of a compliant material. The shear wave profiles are dispersive and show attenuation with propagation. Analyses of these wave profiles will be presented. Shear moduli vary from 0.35 GPa to 0.8 GPa for the compression range examined to date. These values are within a factor of two of the static shear moduli in the glassy state. The data described here have been used to calculate the high strain rate compressive and shear stress-strain curves for Solithane 113.

[*]Work supported by ONR Contract No. 00014-78-C-0549.

[1]Manufactured by Thiokol Corporation, Trenton, NJ.

0094-243X/82/780437-01$3.00 Copyright 1982 American Institute of Physics

SPALLATION BY DUCTILE VOID GROWTH*

J. N. Johnson
Los Alamos National Laboratory, Los Alamos, NM 87545

ABSTRACT

A mathematical model of ductile void growth under the application of a mean tensile stress is applied to the problem of spallation in solids. Calculation of plate-impact spallation in copper (peak compressive stress ∿29 kbar) shows good agreement with the dynamically measured spall signal. A second calculation, using identical material parameters, of explosively produced spallation in copper (peak compressive stress ∿250 kbar) does very well in reproducing experimentally observed multiple spall thicknesses as observed by dynamic x-radiographic techniques. This theoretical model thus appears applicable to a wide range of dynamic uniaxial-strain loading conditions, bridging a gap that has been thought to exist for some time.

INTRODUCTION

For every spallation experiment that is conducted, an ad hoc model can be developed to reproduce damage levels (in the form of residual porosity), spall location, growth rates, and so on. What is presently lacking is a single model of ductile fracture capable of reproducing the experimental results obtained under widely varying conditions. For example, in the work of Breed, Mader, and Venable[1] on explosively produced spallation of copper, a computational model of fracture was developed to correlate spall strength with spall thickness. This model has been useful in reproducing the observed spall layers in explosive events, but is not applicable to low-pressure plate-impact experiments. Likewise, models developed for plate-impact situations seem to be inadequate in the high-pressure regime.[2] These models were developed to represent accurately the onset of fracture in engineering design and no attempt was made to see how they worked under the very extreme conditions of explosive loading at 200-300 kbar.

In the present work, the results of a microscopic model for ductile hole growth are presented which relate the material porosity (an internal state variable) to its initial value, the time history of the tensile pressure (or mean stress), and a single scalar parameter representing the rate-dependent plastic flow properties of the solid material surrounding the voids. This introduces a minimum number of adjustable parameters—also, the ones that are used have the possibility of being determined experimentally.

---
*Work supported by the U.S. Department of Energy.

VOID-GROWTH RELATIONS FOR DUCTILE MATERIALS

Carroll and Holt[3] describe a very useful model of ductile void collapse that lends itself directly to a theory of void growth under tensile loading conditions--the only difference is that the pressure, p, is negative in the void-growth case and the porosity increases. In addition to the obviously trivial replacement of p with -p in Carroll and Holt's model, a rate-dependent plastic flow term is added that was not in the original development.[4]

Ductile void growth is expressed in terms of the distention ratio $\alpha \equiv V/V_s$, where V is the average specific volume of a region containing voids and V_s is the specific volume of the solid material surrounding the voids. The void-growth rate due to an average mean tensile stress p is given by

$$(\rho a_0^2/3)(\alpha_0 - 1)^{-2/3} Q(\ddot{\alpha}, \dot{\alpha}, \alpha) = \alpha p + (2Y/3) \ln \frac{\alpha}{\alpha - 1}$$
$$+ \eta(\alpha_0 - 1)^{-2/3}(\alpha - 1)^{-1/3}\dot{\alpha} \quad , \quad (1)$$

where

$$Q(\ddot{\alpha}, \dot{\alpha}, \alpha) \equiv -\ddot{\alpha}[(\alpha - 1)^{-1/3} - \alpha^{-1/3}]$$
$$+ \frac{1}{6} \dot{\alpha}^2 [(\alpha - 1)^{-4/3} - \alpha^{-4/3}] \quad . \quad (2)$$

In Eq. (1) a_0 is the average initial void radius giving an initial distention ratio α_0, Y is the yield strength of the solid, and η is the "viscosity" of the solid (i.e., the proportionality constant between shear stress and *plastic* strain rate).

PLATE-IMPACT AND EXPLOSIVELY GENERATED SPALL IN COPPER

As an application of the foregoing theory, two quite different spallation experiments on copper are calculated by the finite-difference method. The first is a plate-impact experiment[5] in which a 0.6-mm-thick copper plate strikes a 1.6-mm copper target backed by a relatively thick plate of PMMA (polymethylmethacrylate) in which a manganin pressure gauge is embedded approximately 0.5 mm from the copper (target)/PMMA interface. The impact velocity of 0.016 cm/µs produces a 29-kbar peak stress in the copper. Application of the dynamic hole growth analysis to the problem of time-dependent spallation in copper is shown in Fig. 1. The peak shock amplitudes are in some disagreement, but the spall signals (t > 0.8 µs) show good agreement.

The initial distention α_0 = 1.0003 is taken to be the measured porosity in the recovered sample at locations far from the spall plane: the actual porosity prior to shock loading was not reported.[5] The average initial pore radius is determined to be 1.9 x 10^{-4} cm

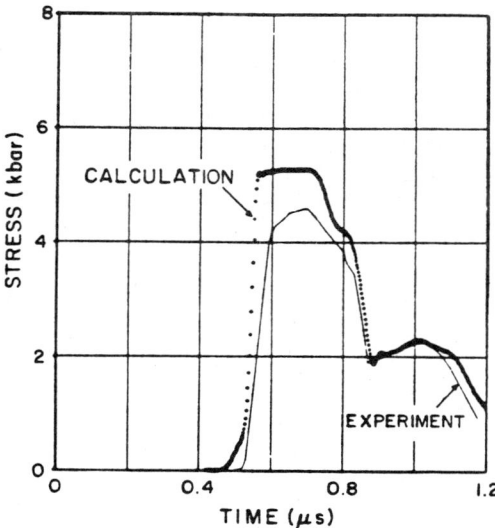

Fig. 1. Calculated spall signal in copper compared with experimental measurement.

from α_0 and the measured void number density (10^7 cm^{-3}).[5] Other void growth parameters used here are Y = 2.6 kbar and η = 10 poise.

To demonstrate the generality of the foregoing model of dynamic ductile fracture of copper, as determined by a single plate-impact experiment, a finite-difference calculation is made of explosively produced spall fracture in copper. A 12.7-mm-thick piece of Composition B in contact with a 25-mm-thick copper plate produces two distinct spall planes--the first (closest to the free surface) plane is approximately 2 mm thick and the second plane is approximately 3 mm thick.[6] A spallation calculation is made with the same pore-growth model used for the plate-impact experiment: the material parameters remain exactly the same, only the loading conditions are changed. The results of this calculation are shown in Fig. 2 (t = 6.0 μs) where a region about 2 mm thick is continually fractured, but the major discontinuities in particle velocity define two spall planes, one 2 mm thick, closest to the free surface, and the second 3 mm thick as observed experimentally.

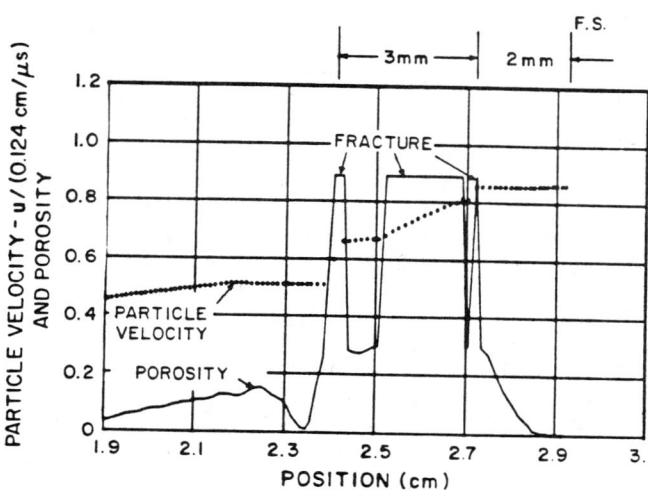

Fig. 2. Final calculated fracture and porosity distribution for Los Alamos PHERMEX shot 500. The region between the two visible spall planes (i.e., those with particle velocity discontinuities) continues to break up after spall plane formation. The free surface (F.S.) is located at the 2.92-cm position.

SUMMARY

A ductile hole-growth model is applied to the problem of spall fracture of copper for plate-impact and explosive loading conditions. Previous models have been found to work well in the incipient stages of ductile fracture, while others seem to be more applicable to complete separation. The description presented here thus tends to bridge the gap between the low-damage and complete separation regimes of spall fracture.

REFERENCES

1. B. R. Breed, C. L. Mader, and D. Venable, J. Appl. Phys. $\underline{38}$, 3271 (1967).

2. C. L. Mader, personal communication, 1979.

3. M. M. Carroll and A. C. Holt, J. Appl. Phys. $\underline{43}$, 1626 (1972).

4. J. N. Johnson, J. Appl. Phys. (to be published).

5. L. Seaman, T. W. Barbee, Jr., and D. R. Curran, Stanford Res. Inst. Tech. Report No. AFWL-TR-71-156, Dec. 1971 (unpublished).

6. C. L. Mader, T. R. Neal, and R. D. Dick, LASL PHERMEX Data, Vol. II (University of California Press, Berkeley, 1980), pp. 178-179.

ANALYSES OF DUCTILE FLOW AND FRACTURE IN TWO DIMENSIONS*

M. E. Kipp and Lee Davison
Sandia National Laboratories,** Albuquerque, NM 87185

ABSTRACT

Previous work on analysis of viscoplastic flow and spall fracture by ductile void growth has been extended to configurations of two-dimensional plane strain and axisymmetric deformation. Current work includes reduction of the general theory to a form applicable to these cases, and incorporation of the resulting equations into a Lagrangian finite-difference code suitable for analyzing nonlinear wave propagation. Results of applying this code to study of symmetric axial impact of aluminum bars are presented.

INTRODUCTION

Davison, Stevens, and Kipp[1] have described a theory of deformation and damage accumulation in ductile metals. Originally specialized to the case of uniaxial strain[1,2], the numerical treatment is now extended to include plane and axisymmetric motions. This paper provides a cursory description of the two-dimensional development, with the details to be made available elsewhere.[3] The motions of particular interest here arise under impact conditions, and are conveniently analyzed through use of an existing finite difference Lagrangian wave propagation code, TOODY.[4] Axial impact of an aluminum rod is included as an example in which internal damage accumulates.

ELASTIC-VISCOPLASTIC-DAMAGE MODEL

The two-dimensional motions considered can be represented by the equations

$$x^1 = x^1(X^1, X^3, t), \quad x^2 = X^2, \quad x^3 = x^3(X^1, X^3, t), \quad (1)$$

where the coordinates x^i designate the position at time t of the material point originally residing at the place X^I. The nonvanishing components of the deformation gradient associated with the motion (1) are

$$F^1{}_1 = \partial x^1/\partial X^1, \quad F^1{}_3 = \partial x^1/\partial X^3, \quad F^2{}_2 = 1, \quad F^3{}_1 = \partial x^3/\partial X^1,$$

$$F^3{}_3 = \partial x^3/\partial X^3. \quad (2)$$

---

* This work was supported by the U.S. Department of Energy Contract DE-AC04-76-DP00789.

** A U.S. Department of Energy Facility

In formulating the theory under consideration, the deformation is attributed to three physical effects: (1) viscoplastic flow, (2) nucleation and growth of voids (damage), and (3) thermoelastic deformation. To facilitate developing the constitutive equations, the velocity gradient tensor $\underline{L} = \dot{\underline{F}}\,\underline{F}^{-1}$ associated with the motion (1) is decomposed into parts $\underline{L}^{(P)}$, $\underline{L}^{(M)}$, and $\underline{L}^{(E)}$ associated, respectively, with the three deformation mechanisms cited:

$$\underline{L} = \underline{L}^{(P)} + \underline{L}^{(M)} + \underline{L}^{(E)}. \qquad (3)$$

The development of voids in the material is characterized by their volume fraction, $\mathscr{D}$, and the rate $\dot{\mathscr{D}}$ at which it increases. This damage accumulation rate is evaluated on the basis of formulae first presented by Barbee, Seaman, and Crewdson[5], and is kinematically related to $\underline{L}^{(M)}$.

The construction of the TOODY code is such that, within a numerical cycle, a prescribed deformation is used to calculate new values for the stress components. A properly invariant stress rate is determined from the part $\underline{L}^{(E)}$ of the velocity gradient by an elastic constitutive relation that takes the damage into account. From the stress rate, we obtain the stress, $\underline{t}$, the mean stress, σ, the shear stress vector $\underline{S}^{(a)}$ on each member of a set of glide systems, the member of which is characterized by its normal $\underline{N}^{(a)}$ and the Burgers' vector $\underline{S}^{(a)}/|\underline{S}^{(a)}|$ in its plane. From these quantities and relations for the dislocation density, D, and dislocation velocity, $V^{(a)}$, the part $\underline{L}^{(P)}$ of the velocity gradient can be determined. This calculation depends on subsidiary determination of the dislocation density and back stress τ from relations for their evolution. When incorporated into the wave propagation code, these relations comprise a coupled system of ordinary differential equations. They are solved within each zone-cycle using an ODE integrator package.

EXAMPLE

The tensile stresses produced when plane decompression waves collide are often of sufficient magnitude to produce damage within the material.[1,2] Tension sufficient to produce internal damage can be generated in two-dimensional geometries under a variety of circumstances. One such circumstance arises when the radial decompression wave that develops in a bar following axial impact converges to, and reflects from, the bar centerline. This problem has been analyzed using the theory discussed, along with material parameters from our previous one-dimensional study.[1] The case considered involves an aluminum rod of 10 mm diameter impacting a similar rod at a relative velocity of 300 m/s.

The formation and convergence of the decompression waves is illustrated by the rod outline and pressure contours in Figure 1(a). Two aspects of the further evolution of this solution are illustrated in Figs. (1b) - (1d). In each case, the top half of the figure shows the pressure contours at the time indicated and the bottom half of the figure shows the corresponding damage contours. An axial section of a bar tested under similar, but undocumented, conditions is shown in Fig. 2. The qualitative similarity of the calculated and observed behavior is apparent.

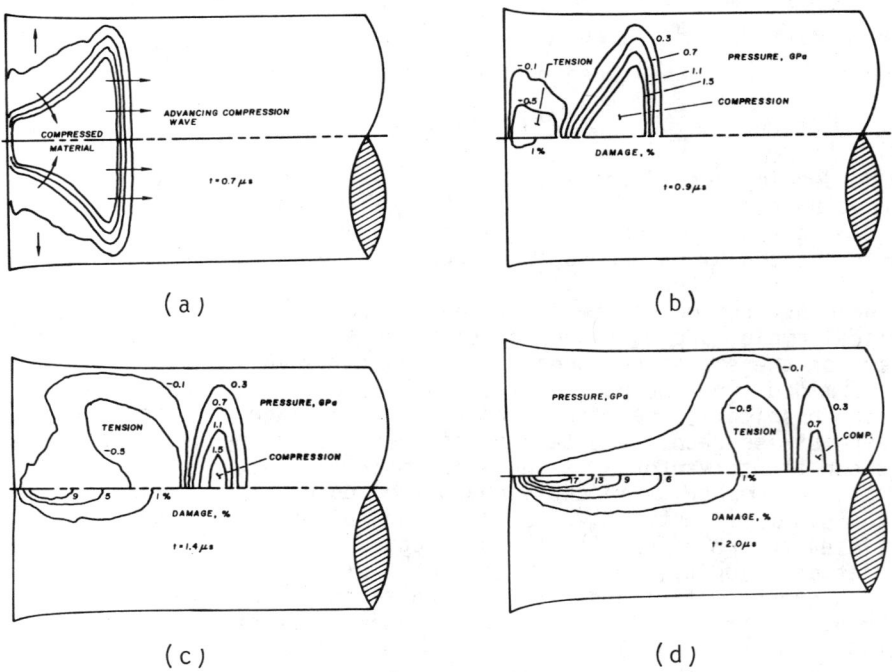

Figure 1. When a bar impacts a rigid boundary (or an identical bar) a compression wave is produced and advances along the length of the bar as indicated by the pressure contours (a). Tension develops along the axis of the bar when radial decompression waves reflect from the centerline (see the top half of part (b) of the figure,) and may produce void growth as delineated by the damage contour in the lower half of this illustration. Further evolution of the solution, including both developing damage and the attendant stress relief, is shown in the same manner in parts (c) and (d) of the figure.

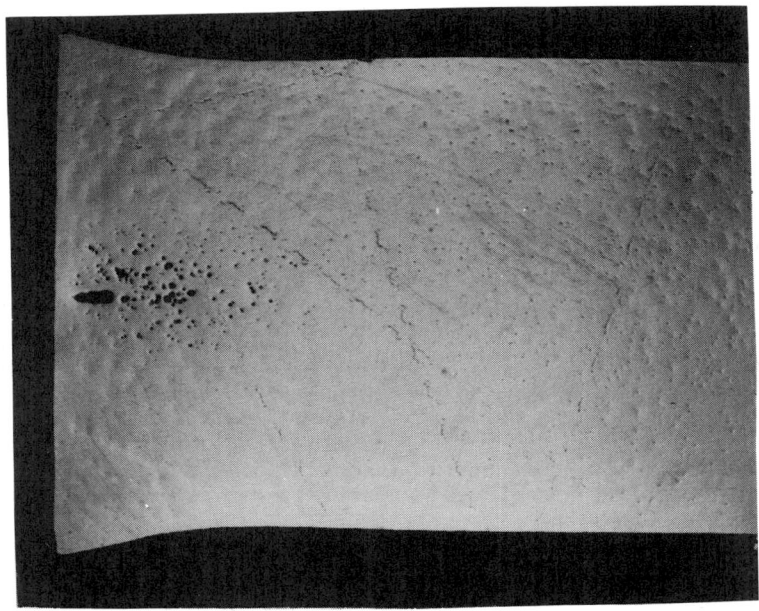

Figure 2. Axial section of the end part of an aluminum alloy bar impacted on a similar bar. This experimental result was obtained through the courtesy of Prof. W. N. Sharpe, Jr.

REFERENCES

1. L. Davison, A. L. Stevens, and M. E. Kipp, J. Mech. Phys. Solids, 25, 11 (1977).

2. L. Davison and M. E. Kipp, High Velocity Deformation of Solids, edited by K. Kawata and J. Shioiri (Springer-Verlag, Berlin, 1978), p. 161.

3. L. Davison and M. E. Kipp, to be published.

4. J. W. Swegle, TOODY IV Report No. SAND78-1552, Sandia Laboratories, Albequerque, NM 1978. (unpublished).

5. T. W. Barbee, Jr., L. Seaman, R. Crewdson, and D. Curran, J. Materials, 7, 393 (1972)

NUCLEATION THRESHOLD STRESSES FOR THE DYNAMIC
FRACTURE OF A LOW-ALLOY Ni-Cr STEEL

G. L. Moss and P. H. Netherwood, Jr.
Ballistic Research Laboratory, Aberdeen Proving Ground, MD 21005

L. Seaman
SRI International, Menlo Park, CA 94025

ABSTRACT

The stresses σ_{no} required for the nucleation of cracks with tensile stress waves were determined as a function of the strength of the steel. These threshold levels were established with the crack densities developed with parallel-plate impacts and the corresponding tensile stresses. The tensile stresses were determined with a procedure that accounts for the effects of elastic-plastic wave interactions and void development on the intensity of the tensile stresses. Two new results were discovered. First, it was established that, within the strength range investigated, σ_{no} decreases as the yield strength σ_Y increases. It is shown that this occurs because the mode of failure changes with σ_Y. Second, it was discovered that the crack nucleation rate is negligible at σ_{no} and increases gradually with stress. An equation that describes this behavior is given.

INTRODUCTION

Two previous determinations of the crack nucleation threshold stress σ_{no} in a low-alloy Ni-Cr steel led to values differing by a factor of 3.4.[1,2] The present study was initiated to clarify this situation by examining the nucleation process in more detail--especially through investigations of cracking at stress levels near the threshold stress and as a function of the strength, or extent of tempering, of the steel.

PROCEDURE

The material investigated was a low-alloy 0.22C-3Ni-1Cr tempered martensitic steel, and σ_{no} determinations were completed for three different rolling and tempering conditions. These corresponded to Brinell hardnesses of 270, 320, and 370 and yield strengths of 0.65, 0.80 and 1.02 GPa respectively.

Partially broken samples were created for investigation with parallel-plate impacts (plate-slap tests) accomplished with a light-gas gun. The degree of damage in the samples was varied by changing the impact velocity. In all tests of a particular material condition, identical impactor and sample thicknesses were used to insure approximately the same load duration.

Crack densities were established with microscopic observations of metallographically prepared sections of the partially broken samples. In the plate-impact test, the load duration depends on the location in the plate. Hence, only voids in the central region--a strip 0.021 to 0.127 cm wide--of each sample were counted. This insured that voids in the regions investigated were initiated over approximately equal time intervals. It also allowed the use of data from the low-pressure tests where no noticeable cracking occurred in the outer regions of the plates.

The nucleation threshold stress was determined iteratively by first estimating an approximate threshold stress σ'_{no} by extrapolating curves of crack density versus the maximum compressive stress to the stress corresponding to no cracking. If this resulted in a stress less than the Hugoniot elastic limit HEL, σ'_{no} was approximated with the HEL. Subsequently, the maximum tensile stress attained in each test was computed with the one-dimensional stress wave-propagation computer code PUFF with the brittle-fracture subroutine BFRACT, and by using σ'_{no} and related material fracture parameters from independent tests. Such a computation automatically accounts for the elastic-plastic wave interactions as well as the effect of void development on the intensity of the tensile stresses computed. Finally, σ_{no} was determined by extrapolating curves of crack density versus the maximum tensile stress to the tensile stress corresponding to no cracking. The stress at no cracking was assumed to be σ_{no}.

Crack morphology was examined at each strength level to aid in interpreting the results of the threshold determinations.

RESULTS

Microscopic observations revealed that failure invariably started at inclusions which either cracked or separated from the matrix. Eventually, cracks extended from these regions into the matrix. Clearly, there are several distinct stages in the failing process, and nucleation can be described in several ways. Here, nucleation was associated with the beginning of the crack extensions into the steel matrix.

Graphs of the crack densities versus stress are shown in Fig. 1 for two heat treatments. It can be seen that the curves based on the tensile and compressive stresses do not extrapolate to the same no-damage levels. This is partly because there is insufficient cracking at stresses just above σ_{no} to get statistically significant crack densities. Since cracking is activated by tensile, rather than compressive, stresses, σ_{no} was related to the tensile stress at which cracking began. A new result shown in Fig. 1 is that σ_{no} decreases as σ_Y increases over the stress range investigated.

The reason for this behavior is revealed by the appearance of the cracks. Examples are shown in Figs. 2 and 3. It is readily seen in Fig. 2 that when σ_Y equals 1.02 GPa, the cracks tend to extend along the edges of inclusions and appear as fine lines in the matrix. They are typical sharp cracks. In contrast, there is approximately spherical void growth around the inclusions in the lower strength steel (σ_Y = 0.65 GPa) as shown in Fig. 3. Eventually, matrix cracks form, but these are clearly nucleated with more plastic deformation than the cracks in the higher strength steel.

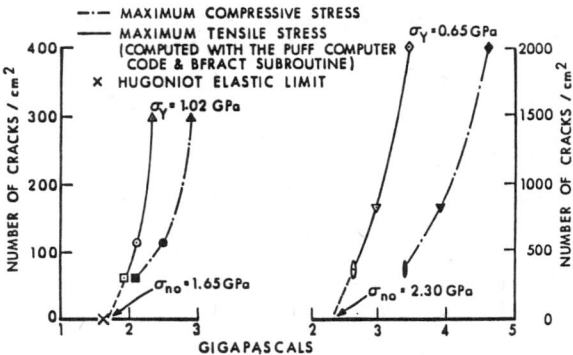

Fig. 1. Crack density dependence on stress. Symbols with the same shape correspond to the same test.

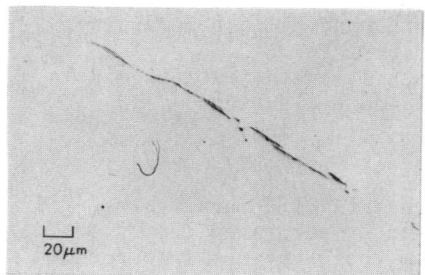

Fig. 2. Sharp cracks at inclusion-matrix interfaces and in the steel matrix (σ_Y = 1.02 GPa).

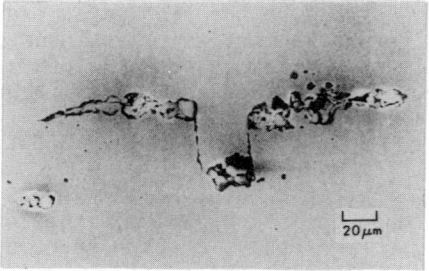

Fig. 3. Approximately spherical void growth at inclusions. Vertical lines are shear cracks (σ_Y = 0.65 GPa).

The nature of the cracking is further emphasized in Fig. 4 where the data for the Ni-Cr steel and several other materials are shown along with curves that approximate bounding conditions for the development of failure.[3-7] The lower limit on threshold stresses for cracking was assumed to be the stress required to develop sharp cracks. This was approximated with the stress just sufficient to initiate plastic deformation. For plane-strain conditions, as encountered in the plate-impact test, this stress is proportional to the yield strength of the standard tensile test and is given by the relation $\sigma = (1-\nu)\sigma_Y/(1-2\nu)$. This curve is shown in Fig. 4 for a Poisson's ratio ν of 0.27, and it is apparent that the threshold stresses for cracking in brittle materials in which sharp cracks form, i.e., Lexan, S-200 Be, Armco Fe and the Ni-Cr steel (σ_Y = 1.02 GPa), almost coincide with this line. Hence, increasing σ_Y of the Ni-Cr steel above about 1 GPa should result in an increase in σ_{no}.

An upper bound on the stress to initiate cracks was assumed to be the stress to develop a perfectly blunted crack, i.e., a spherical

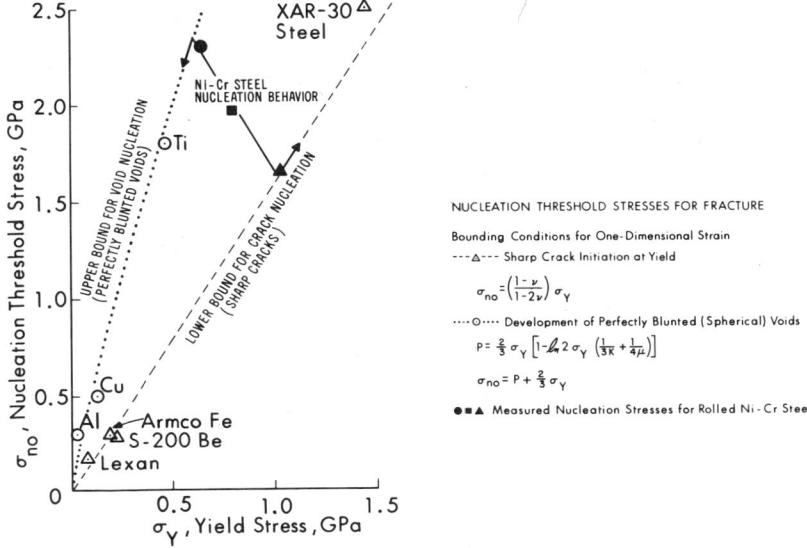

Fig. 4. Nucleation threshold stresses for fracture with stress waves. The curves bounding the possible threshold stresses correspond to the development of perfectly sharp and perfectly blunted cracks.

pore. Hill has shown that the hydrostatic pressure P required to enlarge a spherical void in an infinite elastic-perfectly plastic solid is given by $P = (2\sigma_Y/3)[1-\ln 2\sigma_Y(1/3K-1/4\mu)]$ where K and μ are the bulk and shear moduli respectively.[8] The stress component σ_{11} in the direction the stress wave propagates is $\sigma_{11} = P + (2/3)\sigma_Y$. This is the stress component usually related to fracture with stress waves. When P is taken as the critical stress for void growth, $\sigma_{11} = \sigma_{no}$. This is plotted in Fig. 4 for average values of K and μ for ductile materials and is identified as the upper bound on σ_{no}. Measured threshold stresses for the nucleation of voids in ductile materials (Al, Cu, apparently Ti and the Ni-Cr steel when $\sigma_Y = 0.65$ GPa) are also shown in Fig. 4, and these are in close agreement with the upper limit for σ_{no}, i.e., the curve for perfectly blunted cracks. Since the critical condition for void growth is defined by the expression for P, the agreement between the data and the bounding curve is a quantitative indication that the initial approximately spherical void growth in ductile materials is governed by all the principal stress components rather than by σ_{11} alone.

It is apparent that there is a maximum in the σ_{no}-σ_Y curve for the Ni-Cr steel at about 0.6 GPa because the limiting curve for perfectly blunted cracks is an increasing function of yield stress while in the interval $0.60 \leq \sigma_Y \leq 1.0$ GPa the threshold stress for cracking the Ni-Cr steel is a decreasing function of yield stress. This maximum should be an important feature in the design and selection of tempered martensitic steels that must resist fracture due to stress waves. The implication is that for some loads there may be a tempering condition that will result in optimum fracture resistance.

The data in Fig. 1 are also helpful in establishing appropriate functions for the description of crack nucleation rates $\dot{N}$. Previous results have shown that at stresses appreciably greater than σ_{no}, $\dot{N}$ is approximately given by $\dot{N} = \dot{N}_o \exp(\sigma-\sigma_{no})/\sigma_1$ where $\dot{N}_o$, σ_{no} and σ_1 are material parameters. However, the graph shown in Fig. 1 suggests the behavior of the high-strength steel ($\sigma_Y = 1.02$ GPa) is actually consistent with[9]

$$\dot{N} = \dot{N}_o \left\{ \exp[(\sigma-\sigma_{no})/\sigma_1]^{1.25} - 1 \right\}. \qquad (1)$$

Hence, when σ equals σ_{no}, the nucleation rate is zero and not $\dot{N}_o$. At stresses appreciably above σ_{no}, Eq. 1 and the relation for $\dot{N}$ that has been used in the past are approximately the same.

CONCLUSIONS

In an investigation of the fracture of a 0.22C-3Cr-1Ni tempered martensitic steel with stress waves, the following was established:

1. σ_{no} decreases as σ_Y increases on the interval $0.65 \leq \sigma_Y \leq 1.02$ GPa.

2. There is a tempering condition that should result in optimum resistance to fracture with stress waves.

3. The nucleation rate at stresses near σ_{no} is given by

$$\dot{N} = \dot{N}_o \left\{ \exp[(\sigma-\sigma_{no})/\sigma_1]^{1.25} - 1 \right\}, \quad \sigma_Y = 1.02 \text{ GPa.}$$

REFERENCES

1. L. D. Bertholf, L. D. Buxton, B. J. Thorne, A. L. Stevens and S. L. Thompson, J. Appl. Phys., 46, 3776, 1975.
2. D. A. Shockey, L. Seaman, D. R. Curran, P. S. DeCarli, M. Austin and J. P. Wilhelm, BRL CR 222, BRL, Aberdeen Proving Ground, MD, 1975.
3. L. Seaman, T. W. Barbee, Jr. and D. R. Curran, AFWL-TR-71-156, Kirtland Air Force Base, NM, 1972.
4. D. A. Shockey, K. C. Dao and R. L. Jones, Mechanisms of Deformation and Fracture (Pergamon, Oxford, 1979), p. 77.
5. D. R. Curran and D. A. Shockey, BRL CR 91, Aberdeen Proving Ground, MD, 1973.
6. L. Seaman and D. A. Shockey, AMMRC CTR 75-2, AMMRC, Watertown, MA, 1975.
7. D. A. Shockey, L. Seaman and D. R. Curran, ARWL-TR-73-12 AFWL, Kirtland AFB, NM, 1973.
8. R. Hill, Plasticity (Oxford U. Press, London, 1950), p. 104.
9. G. Moss and L. Seaman, Nucleation Threshold Stress for the Dynamic Fracture of a Low-Alloy Ni-Cr Steel, Submitted to Mech. Matls.

AN INVESTIGATION OF INCIPIENT FRACTURE IN SHOCK-LOADED LAMELLAR COBALT-ALUMINUM EUTECTIC

William E. Thompson and William W. Predebon
Michigan Technological University, Houghton MI 49931

ABSTRACT

Lamellar cobalt-aluminum eutectic provides an in-situ composite in which shock induced fracture may be studied. The lamellae consist of alternating layers of the two constituent phases of the eutectic. Using the Hugoniot equations-of-state for each constituent phase, the eutectic is modeled using a two dimensional finite-difference code for the case of an initially planar pulse traveling parallel to the interphase boundary. The as-grown eutectic is not a perfectly lamellar structure, but rather, it contains terminations and branching of the lamellae. The effects of terminations and branchings on incipient dynamic fracture of the eutectic are considered and compared to the case without these imperfections. Individual layers of the eutectic are coupled through boundary interaction with two extreme cases, perfect bonding and perfect lubrication, considered.

INTRODUCTION

Generally in the past, shock-induced fracture studies in laminates used manufactured composites in which two materials are artificially bonded together. Another type of laminated composite material is the naturally occurring "as formed" metallic eutectics. These eutectics form an in situ laminated composite medium in which shock induced fracture may be studied. The lamellar Co-Al eutectic is the metallic system chosen for this study. The lamellae exist in alternating layers of the constituent phases of the eutectic. In this study, the constituent phases retain their individual character and are not approximated by a homogeneous material.

PROBLEM OUTLINE

The microstructure of the eutectic is shown in Figure 1, and as Figure 1a illustrates, both terminations and branchings of the CoAl phase are common throughout the structure. A portion of the more perfect lamellar structure is shown under higher magnification in Figure 1b. Brawley and Predebon[1] investigated incipient fracture in the lamellar Co-Al eutectic system without branchings and terminations as shown in Figure 1b. The problem considered in this study, and the previous study[1], is one in which an initially planar pulse is traveling parallel to the interphase boundary of the eutectic, and is illustrated in Figure 2.

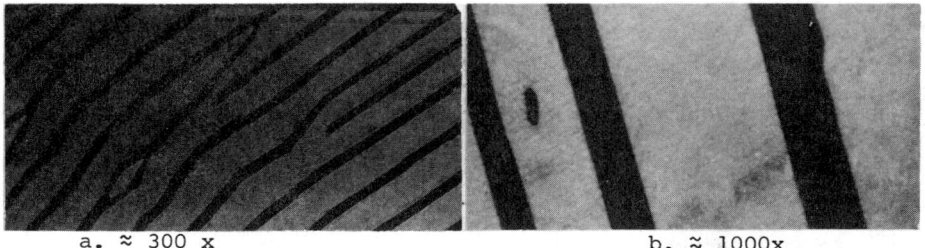

a. ≈ 300 x   b. ≈ 1000x

Figure 1. Lamellar Eutectic of Cobalt-Aluminum System

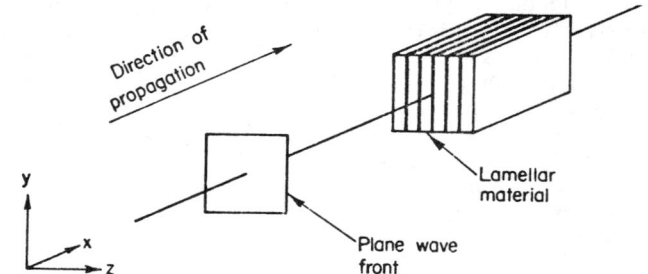

Figure 2. Orientation of Lamellar Eutectic to Wave Front

MATERIAL DESCRIPTION AND FRACTURE MODEL

Some properties of the eutectic and its two constituent phases, designated Co(Al) and CoAl, which were initially determined by Cline[2] and later refined by Stout, et al[3] are summarized in Table 1.

Table I Characteristics of Cobalt-Aluminum Eutectic

Alloy Designation	Co(Al)	Eutectic	CoAl
Composition, wt. % Co	93.1	90.5	81.9
Density kg/m^3	7920		6900
Volume Fraction	0.705		0.295
Crystal Structure	FCC		Ordered-BCC# (B_2)

* See Reference 2

The Hugoniot equation-of-state of the two constituent phases have been determined by Lemar and Duvall[4]. All of the samples used in the Hugoniot experiments and the fracture experiments mentioned later were single crystals or near single crystals grown at Michigan Tech.

The fracture model used is an extension of the damage accumulation model of Tuler and Butcher[5]. This model is extended to two dimensions by using the maximum principal stress, σ_m, in

$$D = \int (\sigma_m - \sigma_o)^\lambda \, dt$$

where D is the damage and σ_o the damage threshold stress. For convenience, λ is assumed equal to 2. This investigation is limited to the prediction of incipient fracture or the onset of microfractures (unspecified voids or cracks) through the damage threshold stress σ_o.

The fracture properties of the individual phases are currently being determined through plate-impact experiments. In the meantime, approximate values are used. Lemar and Duvall[4] observed that the Hugoniot of the Co(Al) phase was similar to that of copper. Based on this, their similar structure (FCC), and similar behavior under conventional loading, it was assumed that σ_o = 0.95GPa, which is the value for copper reported by Smith[6] for a pulse duration of 3/4 μsec. For the CoAl phase, the damage threshold stress σ_o was taken as 2.069 GPa, which is the fracture stress from a static tensile test (inferred from eutectic measurements). Note that this assumed damage threshold stress for CoAl is more than double that assumed for Co(Al).

SIMULATION GEOMETRY AND RESULTS

The computer simulation is conducted with a modified version of HEMP[6], a two dimensional finite-difference code. Two different geometries are studied, the ideal eutectic as shown in Figure 3a, and the eutectic with a termination in the CoAl phase, Figure 3b.

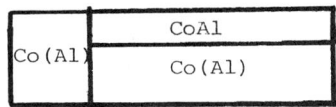 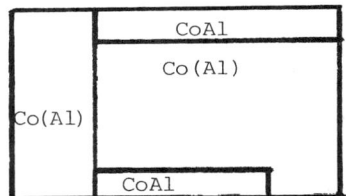

a. Ideal Eutectic   b. Eutectic with CoAl Termination

Figure 3. Simulation Geometry

In both problems, the flyer consists entirely of the Co(Al) phase, as was done in the experimental work, with an impact velocity of 0.122 km/sec. This impact velocity produced compressive stresses of about 2.2 GPa in Co(Al) and about 2.7 GPa in CoAl in the ideal eutectic case, which are greater than their corresponding damage threshold stresses. The CoAl termination was placed in two different positions, 1/2 H_T and 3/4 H_T, where H_T is the thickness of the eutectic target. In all three geometries, two extreme models of the interphase boundary were used, perfect bonding or rigid boundary, and perfect lubrication or slip boundary.

The results of the simulation for the ideal eutectic with a

rigid boundary, which are reproduced in Figure 4b, have been compared to the dynamic fracture experiments in reference 1. The simulation and the experimental results agreed and indicated fractures in the CoAl phase only and along the interphase boundary. The slip boundary model caused the level of damage to increase by two orders of magnitude from that occurring in the rigid boundary model. In addition, a larger number of zones were damaged as illustrated in Figure 4.

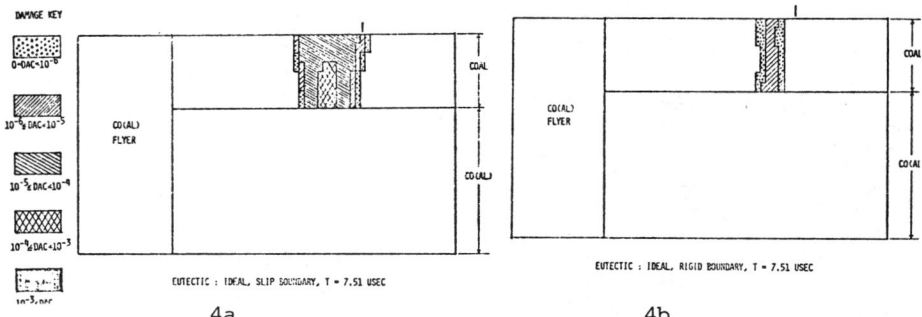

Figure 4. Damage in Ideal Eutectic

The CoAl phase termination positioned at $1/2\ H_T$ did not substantially alter the pattern of damage from the ideal eutectic and is shown in Figure 5. The highest level of damage was again for the slip boundary model and all damage was in the CoAl phase, but none in the termination. When the CoAl termination was positioned at $3/4\ H_T$, a significant change in the damage pattern occurred.

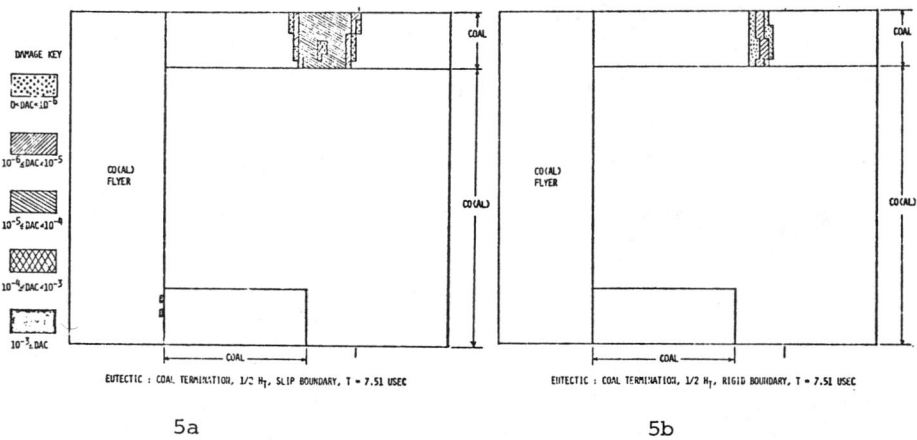

Figure 5. Damage in Eutectic with CoAl Termination at $1/2\ H_T$.

Damage in the CoAl termination is now present for the slip boundary model, as well as in the full CoAl layer as shown in Figure 6. It is also noted that the termination is now in close vicinity of the theoretical spall plane. In all simulations, all initial damage occurred in the CoAl phase and was highest at the interphase boundary. However, at a later time damage is observed in the Co(Al) phase in both interphase boundary model cases. It is noted that in a simulation of a like-like impact of the Co(Al) phase at the same impact conditions, damage was observed, whereas it did not occur in the eutectic configuration. Thus it is found that the CoAl phase reinforces the Co(Al) phase of the eutectic.

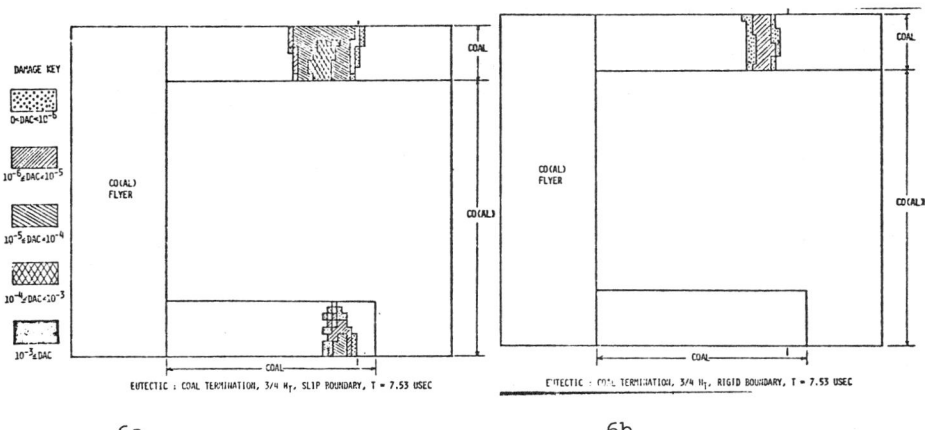

6a   6b

Figure 6. Damage in Eutectic with CoAl Termination at 3/4 H_T.

REFERENCES

1. G. H. Brawley and W. W. Predebon, Eng. Fracture Mech. J., (accepted for publication).
2. H. E. Cline, Trans. TMS AIME 239, 1906 (1967).
3. M. G. Stout, T. H. Courtney and M. A. Przystupa, Met. Trans. A 8A, 1316 (1977).
4. E. R. Lemar and G. E. Duvall, Washington State University Report, MTU-3569, 1979 (unpublished).
5. F. R. Tuler and B. M. Butcher, Intern. J. Fracture Mech. 4, 431 (1968).
6. J. H. Smith, ASTM Special Technical Publication No. 336, p. 264, (1963).
7. M. L. Wilkens, in Methods of Computational Physics, Vol. 3, edited by B. Alder, S. Fernback and M. Rotenburg, (Academic Press, N. Y., 1964).

FRAGMENT SIZE PREDICTION IN DYNAMIC FRAGMENTATION*

D. E. Grady
Sandia National Laboratories†, Albuquerque, NM 87185

ABSTRACT

A general definition of dynamic fragmentation can encompass any impulsive process which partitions a body of material into discrete domains. Included, for example, would be fragmentation due to brittle fracture under impact loading or fragmentation due to shear banding in shock-compression plastic deformation. An energy approach is proposed whereby surface or interface area created in the fragmentation process is governed by an equilibrium balance of the surface or interface energy and a microstructural inertial or kinetic energy term. Relations provided by the analysis compare well with experimental dynamic fracture and shock-wave shear-band results.

INTRODUCTION

On-going efforts at this laboratory and elsewhere to explain and model dynamic fracture and fragmentation have focused on relationships between flaws or fluctuations leading to internal fracture and loading conditions. Evidence is mounting, however, which suggests that, as the fragmentation event becomes increasingly energetic, the sizes and distribution of fragments produced depend less on details of the initiating flaws or fluctuations and more on underlying energy balance principles. This observation leads to a simpler method for calculating fragmentation which relies on fewer and more fundamental material properties. The method is new and must be subjected to more demanding experimental examination; however, comparisons with oil shale and exploding cylinder fragmentation studies have been very encouraging.

In a more general sense the process of shear banding in a shock loaded solid can be considered as a fragmentation event. When a material, initially shearing homogeneously, suddenly transforms to deformation by discrete, regularly spaced shear bands a characteristic fragment size is readily recognized. The same energy balance principles apply here, and we have developed an expression for calculating shear band spacing directly from properties of the loading shock wave.

FRAGMENT ANALYSIS

The analysis is based on a description of the energies governing the fragmentation process. These consist of surface energy, which is created during the formation of new fragment surface area, and a local kinetic energy, which measures the intensity of expansion of the body

---

* This work was supported by the U. S. Department of Energy (DOE), under Contract DE-AC04-76-DP00789.
† A U. S. Department of Energy Facility.

and is responsible for driving the fragmentation process. The crucial point is recognition that the total kinetic energy of the body is not available for fragmentation. To locally conserve momentum only a portion of the kinetic energy is available, while the remainder must continue to reside in the rigid body flight of the fragments. By considering the homogeneous expansion of a spherical fragment within a body of density ρ and density rate $\dot{\rho}$ the following energy expression can be derived,

$$E(A) = \gamma A + \frac{3\dot{\rho}^2}{10\rho A^2} \qquad (1)$$

The two terms on the right are the surface energy and the local kinetic energy, respectively. The variable A is the fragment surface area per unit volume of material and γ is the surface tension. The basic premise assumes that forces brought about during the fragmentation process will minimize the energy in Equation 1. Thus, at equilibrium, $dE/dA = 0$, the calculated fragment surface area depends on surface tension, density, and density rate according to,

$$A = \left(\frac{3\dot{\rho}^2}{5\rho\gamma}\right)^{1/3} \qquad (2)$$

Equation 2 provides a quantitative measure of the surface area created in fragmentation in terms of fundamental material and thermomechanical properties.

APPLICATION

Typically, a nominal fragment size rather than the total fracture area is the result of a fragmentation experiment, and an exact relation between the former and Equation 2 cannot be made without knowledge of the fragment size distribution curve. However, we will assume spherical fragments of equal size, in which case the fragment diameter is related to total surface area through

$$d = 6/A \quad . \qquad (3)$$

The analysis has been applied to calculate experimental fragment sizes in several cases. Although not exhaustive, the results have been encouraging. In both cases the materials underwent brittle fracture. By using $\varepsilon = \dot{\rho}/3\rho$ and $\gamma = K_{Ic}^2/2\rho c^2$, with Equations 2 and 3, an expression for the nominal fragment diameter can be obtained,

$$d = \left(\frac{\sqrt{20}\, K_{Ic}}{\rho c \dot{\varepsilon}}\right)^{2/3} \quad . \qquad (4)$$

Here, $\dot{\varepsilon}$ is the nominal strain rate and K_{Ic} is the material fracture toughness.

Fragment size data for oil shale have been obtained as a function of strain rate[1]. Material properties for the oil shale studied are $K_{Ic} = 0.9 \times 10^6$ N/m$^{3/2}$, $\rho = 2300$ kg/m^3, and $c = 4000$ m/s. The prediction of Equation 4 is compared with the data and agrees with the data mean within about a factor of two over approximately three decades of strain rate.

Further data are available from the explosive fragmentation experiments of Weimer and Rogers[2] on high-strength steel cylinders. In that work four cylinders of FS-01 steel with intrinsic fracture toughness (K_{Ic}) ranging from about 20 to 60 MNm$^{-3/2}$ were loaded with composition B explosive. Fragments were collected and the Mott size and number parameters were determined for the distribution. The tensile hoop strain rate can be calculated with the Gurney Equation and is approximately $\dot{\varepsilon} = 4 \times 10^4$/s. The remaining properties are $\rho = 7840$ kg/m^3 and $c = 5000$ m/s. The Mott size parameter (approximately the mean fragment diameter) agrees with the prediction of Equation 4 within about 30% and the data are consistent with a 2/3 power dependence on K_{Ic}.

It is also worthwhile to emphasize the partitioning of kinetic energy discussed in the opening section by comparing relative magnitudes of the two terms. The total kinetic energy of the cylinder is determined by the radial velocity imparted by the explosive load (or, equivalently, the strain rate) and is $\frac{1}{2}\rho\dot{\varepsilon}^2 r^2$ where r is the radius of the cylinder - about 25 mm in the work of Weimer and Rogers. In contrast the local kinetic energy available for fragmentation is related to the fragment size through $\frac{3}{40}\rho\dot{\varepsilon}^2 d^2$. For d = 2 mm the local kinetic energy available is less than 0.1% of the total radial kinetic energy - a small influence on the overall ejection velocity.

The concepts of energy balance presented here should also be important to the localization of shear deformation thought to occur within the rise time of a large amplitude shock wave in some materials. Partitioning of the kinetic energy to determine the portion which is available to drive the localization process still applies. The relations derived previously also apply if A is interpreted as the interface area of shear bands, γ is the interface energy expended during the localization process and d is a mean spacing of shear bands. Accounting for the different geometry we obtain

$$d = \left(\frac{12\gamma}{\rho\dot{\varepsilon}^2}\right)^{1/3}, \qquad (5)$$

for the mean shear band spacing.

The difficulty in application of Equation 5 is that γ, the energy dissipated in the shear band during localization, is not known. The known quantity in a steady-wave shock-compression process is the total specific energy dissipated, $\mathcal{E}$. If we assume that all of $\mathcal{E}$ is dissipated in shear bands (an upper bound) then the energy dissipated per unit area of shear band is approximately $\gamma = \mathcal{E}d$. Combining with Equation 5 results in,

$$d = \left(\frac{12\mathcal{E}}{\rho\dot{\varepsilon}^2}\right)^{1/2}. \qquad (6)$$

We have applied Equation 6 to shock-compression studies of Asay and Chhabildas[3]. In their work samples of aluminum were shocked to approximately 10 GPa and recovered. Sectioned and polished samples indicated that deformation occurred by shear localization with spacings between about 4 to 20 µm. Wave profile measurements in aluminum indicated that the strain rate is about $\dot{\varepsilon} = 5 \times 10^7$/s and the Hugoniot energy at 10 GPa is $\mathscr{E} = 0.4 \times 10^8$ J/m^3. Using these values in Equation 14 provides a predicted spacing of d = 8.5 µm. The consistency between observation and prediction is encouraging but is regarded as tentative. Much yet remains to be unfolded in terms of cause and effect among local inertia (kinetic energy), adiabatic heating, and shear localization effects in the shock compression of solids.

SUMMARY

The work presented here has attempted to bring out the importance of the local inertial forces which may determine in part the microstructure scale resulting from dynamic fracture or deformation of material. The approach is based on an equilibrium balance of competing energy terms and consequently is not expected to apply if strong viscous or dissipative forces attend the fragmentation process. Application to specific fragmentation and shear localization experiments described here have been encouraging, however. It is suspected that these concepts have not been adequately explored and that further studies will reveal that local inertial effects influence many physical phenomena which occur during shock or intense impulse loading.

REFERENCES

1. D. E. Grady and M. E. Kipp, Int. J. Rock Mech. Min. Sci., 17, 147 (1980).
2. R. J. Weimer and H. C. Rogers, J. Appl. Phys. 50, 8025 (1979).
3. J. R. Asay and L. C. Chhabildas, Proceedings of the International Conference on the Metallurgical Effects of High Strain Rate Deformation and Fabrication, Albuquerque, NM, June 22-26, 417 (1980).

FRACTURE MODEL FOR HIGH ENERGY PROPELLANT

W. J. Murri, D. R. Curran, and L. Seaman
SRI International, Menlo Park, CA 94025

ABSTRACT

A model was developed to predict the fracture and fragmentation of high energy propellants during dynamic loading. The model treats crack nucleation, growth, coalescence, and fragment formation consecutively, describing each as a rate process dependent on the instantaneous value of the tensile stress and the material properties that govern the dynamic fracture response. Impact experiments were performed to determine the fracture parameters and the constitutive properties of the propellant. The model was used in a wave propagation code to simulate several types of dynamic experiments. The agreement between the simulations and experiments was good.

INTRODUCTION

A computational model for dynamic tensile fracture and fragmentation has been developed at SRI for the detailed simulation and study of fracture processes in several materials. The model has been successful in quantitatively predicting the number, sizes, and locations of microfractures or fragments in a wide variety of materials.[1,2] The objective of the work described in this paper was to develop a predictive model for fracture, fragmentation, and resultant formation of new surface area for high energy propellants.

FRACTURE MODEL

The fracture model treats the four stages of the dynamic fragmentation process (crack nucleation, crack growth, crack coalescence, and fragment formation) consecutively, describing each as a rate process dependent on the instantaneous value of the tensile stress and on the material properties that govern the dynamic fracture response. The model also includes stress relaxation associated with the developing damage.

Cracks occur in a range of sizes at nucleation and throughout the calculation. At each material point, these cracks can have only one orientation in the present model. The orientation is fixed by the stress state that exists when the tensile stress first exceeds the initiation criterion at that point.

Cracks are assumed to be penny-shaped, i.e., circular and flat like those observed in polycarbonates.[2] Previous SRI work on fracture of various materials has shown that the concentration of activated cracks can be described by

$$N_g = N_o \exp(-R/R_1) \tag{1}$$

---

*Work supported by Lawrence Livermore Laboratory (Contract P.O. 7250109 under W-7405-Eng 48).

where N_g = the number/cm^3 of cracks greater than R, N_o = the total number/cm^3, R = the crack radius, and $R_{\bar{1}}$ = the shape parameter for the distribution.

In our plate impact experiments[3] fracture began when the stress exceeded a critical level σ_{no}. Hence, a threshold stress criterion is used in the model.

As the dynamic tensile pulse passes through the propellant and exceeds the stress criterion, cracks/voids nucleate in a range of sizes with a distribution given by Eq. 1. The number of voids nucleated, N, is governed by a nucleation rate function.

$$\dot{N} = \dot{N}_o \exp\left[(\sigma - \sigma_{no})/\sigma_1\right] \quad \sigma > \sigma_{no} \tag{2}$$

where $\dot{N}_o$ = the threshold nucleation rate, σ = the normal tensile stress, σ_{no} = the threshold stress for nucleation, and σ_1 = the stress-sensitivity parameter.

In our impact experiments[3] the cracks/voids did not exhibit significant growth until the local stress intensity factors exceeded the time-dependent dynamic fracture toughness of the binder material. The dynamic fracture toughness is expressed in the form

$$\sigma_{go} = \left(\frac{\pi}{R}\right)^{1/2} \frac{K_{1d}}{2} \tag{3}$$

where σ_{go} = the growth threshold stress for cracks of radius R and K_{1d} = the dynamic fracture toughness. If the normal tensile stress σ exceeds σ_{go}, then the cracks are assumed to grow by a viscous growth law

$$\frac{dR}{dt} = \left(\frac{\sigma - \sigma_{go}}{4\eta}\right) R \tag{4}$$

where $1/4\eta$ = a growth coefficient.

The cracks are allowed to open gradually as they are subjected to tensile stresses. The opening time of the crack is controlled by the finite velocity of the crack sides, the time necessary for the tensile wave to engulf the crack, and the time necessary for the tensile wave to travel between cracks.

Coalescence occurs when the cracks become large enough to intersect other cracks. Cracks may intersect in the same plane, and form larger cracks, or they may intersect at right angles and form corners of fragments. Cracks in the same orientation, but different planes, are assumed to coalesce by development of crack extensions out of the plane to join nearby cracks. Thus, a family of cracks in one orientation can coalesce and form a rough, multi-faceted spall plane. Because each crack forms a side of two fragments, three or four cracks are associated with each fragment. The model provides only for a gradual transition from undamaged to fully fragmented material and an accounting of the fragment size distribution at the end. Further details of the model are given elsewhere.[4]

Several gas-gun experiments were conducted on a high energy propellant[3] with impact stresses from 0.08 to 0.37 GPa. No fracture occurred in the lowest velocity impact, but various levels of spall and fragmentation took place in the others. The fracture

parameters were obtained from iterated simulations of these
experiments. Other impact experiments with multiple in-material
particle velocity gages were performed to obtain the constitutive
relations.[3]

The fracture parameters and constitutive relations were used
with the fracture model in a wave propagation code to simulate the
observed damage in several experiments as detailed in the following
paragraphs.

The first experiment, No. 1420, showed no damage in experiment
or simulation.

In the second experiment, No. 1458, a tapered flyer produced
a spalled portion of varying thickness. Good agreement is shown
in Figure 1 between computed and observed spall locations and widths
of the spalled region. The location of the spall plane is readily
predicted because it corresponds to the plane of the first tension
and can be identified even without a fracture model. The thickness
of the fractured region producing the separation, however, depends
on the damage-induced stress-relaxation processes in the model
and is therefore a better discriminator between models.

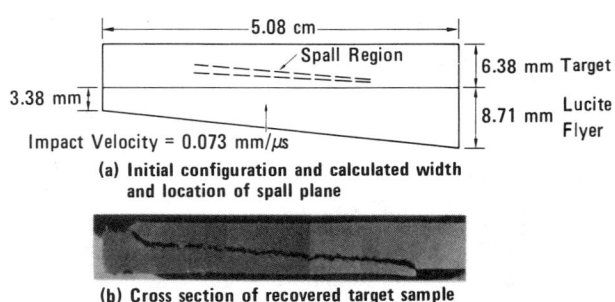

Figure 1 Comparison of computed and observed spall plane
locations in Experiment 1458 in propellant.

In the third experiment, No. 1460, a double spall was seen in
the cross section of the target, as shown in Figure 2. The computations showed that a large region of material was damaged during
the first period of tensile stresses following impact, and the
uppermost spall separation formed. Subsequent reflections led to
a later spall plane closer to the impact plane. The calculation
of the double spall again indicates that the damage-caused stress-
relaxation process in the model is behaving properly. The correspondence between computed and observed regions of damage suggests
that the computed sequence may truly represent the actual sequence.

The fourth experiment, No. 1462, exhibited complete fragmentation. Figure 3 shows the cumulative fragment size distributions;
the radius shown is the radius of a spherical fragment of the same
mass. Both the computed and observed distributions show a change
of slope around 0.1 to 0.2 cm; this effect is probably associated
with the fracture mechanics concept that allows only cracks larger
than a critical size to grow. The computations predict an increased
surface-area-to-volume ratio of about 20 cm^2/cm^3.

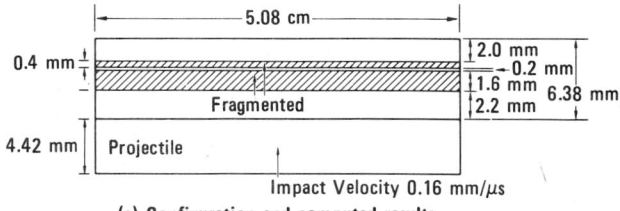

(a) Configuration and computed results showing two fragmented regions

(b) Cross section of specimen recovered from Experiment 1460 showing "double" spall

Figure 2   Comparison of computed and observed fragmented regions in Experiment 1450 on propellant.

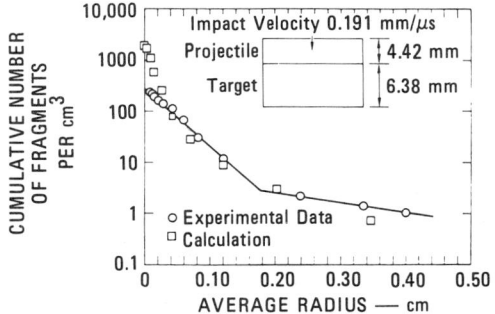

Figure 3   Comparison of computed and experimental fragment size distributions for Experiment 1462.

The fracture model was also used to simulate a two-dimensional impact experiment performed at Lawrence Livermore Laboratory. In this experiment, a 3-in.-long by 3-in.-diameter cylinder of propellant was impacted by a steel projectile whose diameter was greater than the propellant.

This experiment was simulated with a two-dimensional stress wave propagated code called TROTT using the same fracture parameters and constitutive relations as above. Figure 4 shows a stress-time profile for a computational cell in the center of the propellant and 1.35 in. from the impact surface. The initial compression is followed by a tensile pulse and subsequently by recompression. The shape and duration of the tensile pulse are controlled by the stress relaxation that occurs as damage is produced. A simulation of the propellant cylinder 71.2 μs after impact is shown in Figure 5.

The model predicts a surface-area-to-volume ratio of about 540 cm^2/cm^3 in this experiment. With such extensive damage, hot spots could form during the recompression cycle and if burning starts, the conditions are ideal for a deflagration-to-detonation transition.

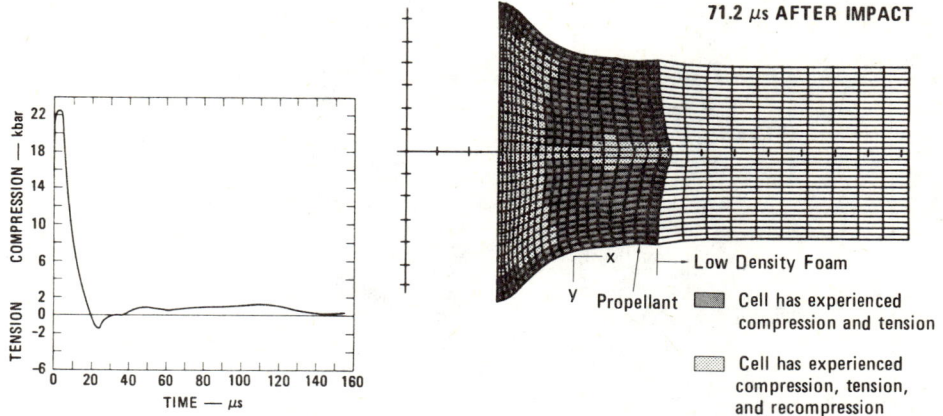

Figure 4 Total stress in direction of impact versus time for a computational cell in the center of the specimen and 1.35 inches from impact surface.

Figure 5 Computer simulation of two dimensional impact test showing computational cells and deformation produced.

In this particular test, detonation occurred about 85 μs after impact.

CONCLUSION

A fracture model, which describes crack nucleation, growth, coalescence, and fragmentation, has been developed for propellant impact. A set of experiments were performed to determine the fracture parameters and constitutive relations. These data were used in wave propagation codes to simulate a series of experiments. Reasonable agreement with experiment was achieved in simulating all of the following features: a no damage experiment; the location and orientation of internal spall planes in a tapered flyer experiment and a "double spall" experiment; fragment size distribution; and the tension and recompression in a two-dimensional impact experiment.

REFERENCES

1. T. W. Barbee et al., Journal of Materials, JMLSA 7, 393 (1972).
2. D. R. Curran, D. A. Shockey, and L. Seaman, J. Appl. Phys. 44, 4025 (1973).
3. W. J. Murri et al., "Fracture and Fragmentation of High Energy Propellant," SRI Final Report to Lawrence Livermore Laboratory, Contract No. 112 under ERDA E(04-3)-115 (April 1979).
4. L. Seaman, "A Computational Model for Dynamic Tensile Microfracture with Damage in One Plane," Poulter Laboratory Technical Report 001-80, SRI International (February 1980)

CALCULATIONS OF CRATERING EXPERIMENTS
WITH THE BEDDED CRACK MODEL

L. G. Margolin
Los Alamos National Laboratory, Los Alamos, NM 87545

The YAQUI/BCM stress wave code is being developed for the oil shale program. Its purpose is to characterize explosive fracture in a brittle, or quasi-brittle, material. YAQUI is a finite difference hydrodynamic code, based on the ALE[1] technique, which we have converted to a solid mechanics code. The BCM (bedded crack model) is the constitutive model. It is a statistical model that describes fracture in terms of the growth of microcracks present in the competent material.

In addition to computer modeling, there is an active program of large-scale field tests in oil shale. In this presentation, we will begin by describing the physical models upon which the BCM is based. Then we will describe calculations of a cratering event in the Colony Oil Shale mine in Colorado. Based on the computer simulations, we will outline the process of crater formation.

Current technology for extracting oil from the rock includes mining the rock, crushing it, and processing the fragments in a surface retort. For environmental as well as economic reasons, we are pursuing an alternate approach in which the rock is broken and retorted *in situ*.

In addition to stress wave propagation, it is crucial to correctly describe fragment size, porosity, and permeability of the fractured rock. Because of these requirements, we have chosen to model the microscopic process of fracture in terms of crack growth rather than to use a macroscopic model based on analogy with plasticity theory.[2] There are many advantages to this type of approach. The input to the model are physical quantities that can be determined by experiment. Also, by focusing on physical processes, the problem of scaling from laboratory tests to field tests does not arise. Finally, a detailed knowledge of crack statistics can provide a basis for calculating fragment size and permeability.

The approach is similar in concept to the BFRACT model developed at SRI by Seaman et al.[3] However, the physical theory on which the BCM is constructed is different. There are three principal elements in this type of approach:
1) a criterion for crack growth,
2) elastic moduli reduction, and
3) a statistical description of the crack distribution as it evolves in time.

Crack growth in the BCM is described by a generalized Griffith criterion.[4] The standard Griffith criterion[5] is based on energy arguments and applies to two-dimensional slits in normal tension. We have extended Griffith's analysis to apply to penny shaped cracks in an arbitrary stress field. When the normal stress is tensile, a crack of radius c may grow if

0094-243X/82/780465-05$3.00 Copyright 1982 American Institute of Physics

$$\sigma_{zz}^2 + \left(\frac{2}{2-\nu}\right)\sigma_{rz}^2 \geq \frac{4\pi TE}{c} \quad . \tag{1}$$

Here, T is the surface energy of the material, E is Young's modulus, and ν is Poisson's ratio.

What is new and different about our generalized criterion is that it predicts that a crack may grow in normal compression if the resolved shear is sufficiently large. In this case, the crack may grow if

$$\left(|\sigma_{rz}| - \tau\right)\left(|\sigma_{rz}| - 3\tau\right) \geq \left(\frac{4\pi TE}{c} \frac{2-\nu}{2}\right) \quad . \tag{2}$$

Here, τ is the frictional stress between the crack faces, which we take as

$$\tau = \mu \, |\sigma_{zz}| \quad ,$$

where μ is the coefficient of friction.

The presence of cracks alters the elastic moduli of the material. The effective moduli[4] are found from static solutions for the displacement field for a body containing a statistical distribution of cracks, and subjected to a spatially uniform, but otherwise arbitrary stress field. The elastic moduli are the elements of a fourth order tensor

$$C_{ijkl} = \frac{\partial \varepsilon_{ij}}{\partial \sigma_{kl}} \quad . \tag{3}$$

For example, for a material containing cracks bedded parallel to the x - y plane, the effective component of compliance

$$C_{zzzz} = \frac{C^o_{zzzz}}{1 + \gamma} \quad . \tag{4}$$

Here C^o_{zzzz} is the modulus of the material matrix and

$$\gamma \equiv \frac{8}{3}(1-\nu^2)\,\overline{Nc^3} \quad . \tag{5}$$

The bar implies integration over the crack density distribution $N(c,t)$.

As the cracks grow, the distribution evolves, and so γ and the effective moduli vary in time. The constitutive relation takes the form

$$\frac{d\varepsilon_{ij}}{dt} = \frac{d}{dt}(C_{ijkl}\sigma_{kl}) \quad , \tag{6}$$

or

$$\frac{d\sigma_{kl}}{dt} = (C^{-1})_{klij} \frac{d\varepsilon_{ij}}{dt} - (C^{-1})_{klij} \overset{\circ}{C}_{ijmn} \sigma_{mn} \ . \qquad (7)$$

Thus, the constitutive relation has the form of a Maxwell solid with a variable relaxation time.

The initial distribution of cracks is assumed to be exponential--the number of cracks with radius greater than c is

$$N_o \exp\left(-\frac{c}{\bar{c}}\right) \ . \qquad (8)$$

N_o and $\bar{c}$ are material constants which are determined by measurement. This is the form assumed in BFRACT. However, the evolution of the distribution function in BCM differs in two essential features from BFRACT.

First, in BCM, only those cracks with radius greater than the critical radius, as determined by the generalized Griffith criterion, are allowed to grow. Thus, the distribution does not remain exponential. This is in contrast to BFRACT in which all cracks grow when the stress exceeds a threshold value.

Second, in the BCM, all cracks grow with the same speed, which is two-thirds of shear wave speed. This is an asymptotic approximation to Mott's result.[6] The existence of a finite velocity for crack propagation makes strain rate effects an intrinsic feature of the model.

The principal restriction of our model lies in the assumption of bedded cracks. Oil shale is a bedded material, and we believe that we are modeling the wave propagation accurately. However, to predict fragment size and permeability, it will be necessary to allow for cracks of different orientations. We are currently generalizing the model in this regard.

We have simulated many actual cratering events with YAQUI/BCM. In a typical experiment, a 10 cm diameter cylindrical borehole is drilled vertically into the mine floor. The hole is partially filled with explosive, and the top of the hole is grouted. The explosive is bottom-detonated, generating a shockwave that fractures the rock, and, depending on scaled depth of burial, usually produces a crater.

Figure 1 shows the initial configuration of a particular experiment, 78-1. The dotted line is the observed crater. The crater is roughly a truncated cone with an apex half-angle of 45°. In Fig. 2, the predicted region of fragmentation is shown with the observed crater superposed. One must distinguish between these profiles, for a fragment will be part of the crater only if it has sufficient vertical velocity for ejection.

The computer simulation predicts the width and general shape of the crater. Also, the YAQUI/BCM is in reasonable agreement with (limited) stress gauge and free surface velocity measurements made in several other cratering experiments. However, the inhomogeneity of the rock--that is, site specific geology,--leads to a large scatter in the data which makes detailed comparisons difficult.

The cratering process can be understood in the following manner. The propagating burn in the explosive produces a conical shock front in the rock. The angle of inclination to the horizontal is the arc tangent of the ratio of detonation velocity to sound speed in the rock. The wave is compressive, and so the fracture is due to shear (Equation 2). In the cratering event 78-1, these speeds are about equal, and so this shock front inclination is about 45°. The maximum shear acting on horizontally bedded cracks is found on a conical surface, also with apex half angle of 45°.

The compressive wave produces a heart-shaped region of fracture around the charge (Fig. 3). The wave continues to the free surface where it reflects as a tensile relief wave. The tension produces a spall region of additional fracture. When the two regions overlap, a crater is produced.

In Fig. 4, we show a computer experiment in which the detonation velocity of the explosive is varied, keeping the total energy release fixed. The case with the larger detonation velocity is predicted to form a wider crater, with less damage above the borehole.

REFERENCES

1. C. W. Hirt, A. A. Amsden, and J. L. Cook, J. Comp. Phys. $\underline{14}$, 227 (1974).
2. J. N. Johnson, Proc. 20th U. S. Symposium on Rock Mechanics, Austin, Texas, June 4-6, 1979.
3. L. Seaman, D. R. Curran, and D. A. Shockey, J. Appl. Phys. $\underline{47}$, 11 (1976).
4. J. K. Dienes and L. G. Margolin, 21st U.S. Symposium on Rock Mechanics, Rolla, Missouri, May 28-30, 1980.
5. J. F. Knott, <u>Fundamentals of Fracture Mechanics</u> (John Wiley and Sons, New York, 1973).
6. E. N. Dunlaney and W. F. Brace, J. Appl. Phys. $\underline{31}$, 2233 (1970).

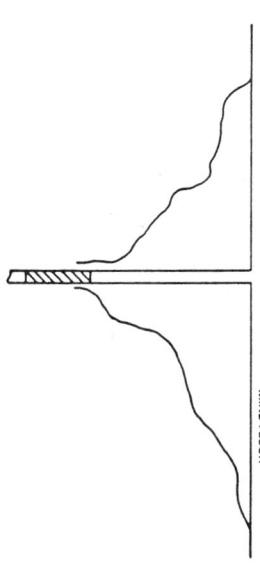

Fig. 1. Initial configuration of cratering experiment 78-1 with final crater profile.

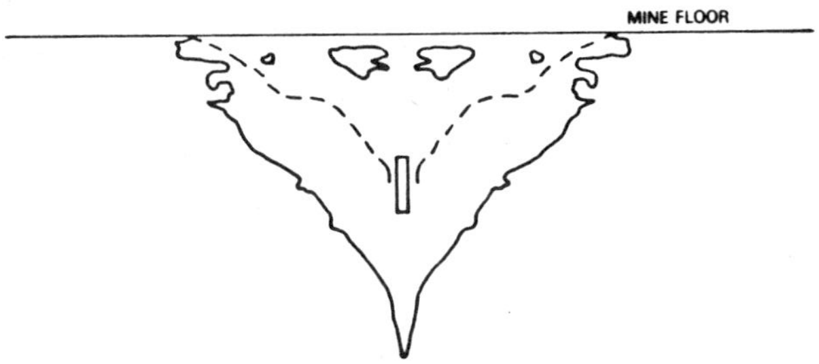

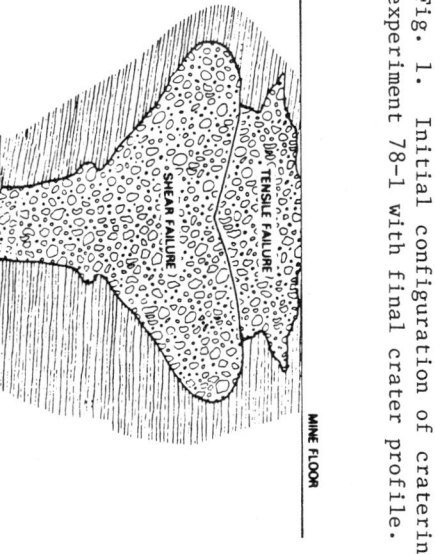

Fig. 2. Calculated fractured region of experiment 78-1. Dotted line shows the actual final crater profile.

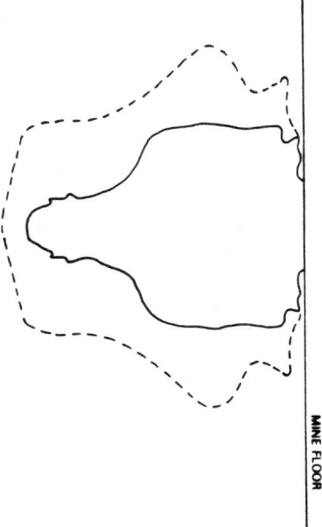

Fig. 3. Calculated fractured region of experiment 79-2. The crater is formed from a shear failed region and a tensile failed region.

Fig. 4. Computer comparison of fractured regions for two experiments in which only the detonation speed of the explosive varied. The wider region corresponds to a higher detonation speed.

FRACTURE INITIATION USING TAILORED-PULSE LOADING*

A. S. Kusubov and R. P. Swift
Lawrence Livermore National Laboratory
Livermore, California 94550

ABSTRACT

Conditions to produce multiple fracturing of boreholes are examined for sandstone samples having a dry density of 2.42 Mg/m^3 and 8.9 percent porosity. Tailored-pulse loading experiments are performed with borehole pressures ranging from 10 to 100 MPa, loading rates from 10^{-4} to 10^2 MPa/ms and pulse durations from 10 to 50 ms. Samples are tested for initial states of dry, quasi-dry, and total saturation with confinement pressure varying from 0.1 to 50 MPa. Data show sensitivity of multiple fracture initiation on loading rate, water content, and confinement pressure.

INTRODUCTION

Over the past few decades, techniques to enhance the recovery of resources from reservoirs have mostly been hydrofracturing and explosive shooting of wells.

For hydrofracturing, a section of a well is pressurized at rates normally lower than 1 MPa/s producing two fractures on opposing sides of the wellbore that tend to propagate in a direction perpendicular to the least principal stress[1]. The induced fractures can usually be extended to large distances[2].

Detonating an explosive in a wellbore generally creates a large number of fractures of short extent. The rate of pressurization is equal or greater than 10^7 MPa/s. The high explosive stress environment can cause compaction or pulverization of a finite zone around the wellbore to such a degree that permeability is significantly decreased as demonstrated by Kusubov et al.[3]

If we consider the limitations of hydrofracture and explosive shooting, it is apparent that a technique for initiating and sustaining multiple fractures in a wellbore should generate a pulse with a rise time fast enough to initiate multiple fractures and of a duration long enough to sustain multiple fractures. As demonstrated by Warpinski et al.[4], tailored-pulse loading at intermediate rates of loading can create multiple fractures without damaging the wellbore region.

TAILORED-PULSE LOADING TECHNIQUE

The generation of tailored-pulse loading to boreholes of samples that may be confined under a specified confining pressure

---

*Work performed under the auspices of the U.S. Department of Energy by the Lawrence Livermore National Laboratory under contract number W-7405-ENG-48.

is achieved with the pressure vessel and flywheel-rotating shaft apparatus described by Swift and Kusubov[5]. The borehole of a cylindrical rock sample confined in a pressure vessel is filled with water, oil, or some other fluid. The reservoir fluid is displaced by a piston to provide prescribed loading conditions of the borehole. Borehole pressure is monitored by a piezoelectric pressure transducer located at the bottom of the borehole, and its output is recorded on an oscilloscope and a high-speed digital recorder. A few of the tailored-pulse loading conditions that can be obtained with the flywheel-rotating shaft apparatus are shown in Figure 1.

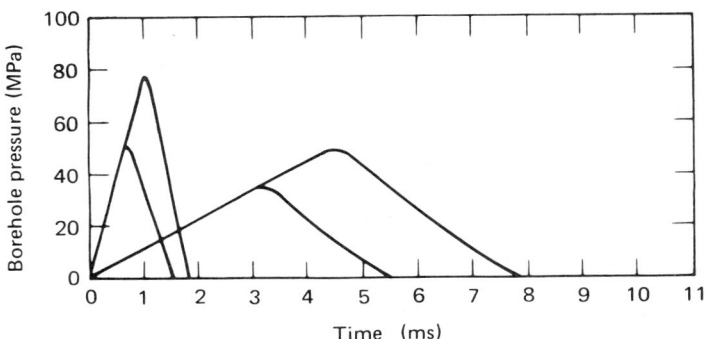

Figure 1. Typical "sawtooth" tailored-pulses using water.

TAILORED-PULSE LOADING OF NUGGET SANDSTONE

Over 50 experiments were performed on samples of Nugget sandstone to determine its susceptibility to multiple fracturing at intermediate loading rates and to define threshold conditions for multiple fracture initiation. The effects of different in situ conditions that could be encountered in borehole wells in the field were examined by varying the initial environmental conditions of the samples to include dry, quasi-dry and fully saturated states along with pressure confinement of 0.1, 15, and 25 MPa. Quasi-dry samples, although originally dry prior to fluid being introduced into the borehole during the test setup, have a thin region around the borehole that is partially-to-fully saturated by the intrusion of fluid from the borehole. For the dry case, the initial fluid intrusion was prevented by either a thin coating of epoxy around the borehole or a 0.05-mm-thick copper liner.

PREPARATION AND POSTSHOT TREATMENT OF SAMPLES

Cylindrical samples, 13 cm in diameter by 15.3 cm long, were cored perpendicular to the bedding planes from blocks of Nugget sandstone quarried near Salt Lake City, Utah. Each sample had a borehole of 1.43 cm in diameter drilled along its axis. The samples were coated on the top and sides with an 0.7-cm-thick layer of epoxy to maintain the water content of fully saturated samples and to prevent intrusion of the confining fluid into the sample during confined experiments. The average dry density of the samples was 2.42 ± 0.01 g/cm^3 with a gas porosity of 8.9 percent. The grain density of sandstone was 2.67 ± 0.01 g/cm^3.

Two techniques were used to examine the fracture pattern created in the samples. The first technique was to cut 1-cm-thick sections perpendicular to the axis of the cylinder, impregnate them with epoxy, and polish. The second technique was to add a dye, flourescein ($C_{20} H_{10} Na_2O_5 \cdot 2H_2O$) to the borehole fluid. Immediately after sectioning, the cut sample was exposed to a black light to provide a detailed picture of the fracture pattern and the depth of fluid intrusion.

DISCUSSION OF RESULTS

Data from the tailored-pulse loading experiments indicate that the initiation phase of multiple fracture is dependent on the initial environment around the borehole, loading rate, and sample confinement. These effects are exhibited in Figures 2a and 2b which plot peak borehole pressure versus loading rate for quasi-dry and wet samples, respectively. Data for initially dry samples are superimposed with the quasi-dry data in Figure 2a. Bands estimating the threshold conditions for fracture and demarking the onset region of multiple fracturing from that of two fractures are also shown. The data shown here are for "sawtooth" pulses only. Examination of these figures illustrates the following important features: (1) threshold breakdown pressure increases with an increase in loading rate, (2) threshold breakdown pressure increases proportionately with sample confinement, (3) the onset loading rate for multiple fracturing tends to decrease for an increase in water content, (4) increase in confinement pressure reduces the onset loading rate for multiple fractures for quasi-dry samples (this may also be true for wet samples but tests have not yet been done to ascertain this), and (5) for dry samples where fluid intrusion is initially prevented, only two fractures occur regardless of the loading rate and confinement conditions.

The influence of the higher water content under confining pressure of 0.1 MPa is more marked at loading rates below about 30 MPa/ms, where multiple fractures still occur for wet samples, but only two fractures (i.e., typical-like hydrofractures) occur for quasi-dry samples. At higher loading rates the character of the multiple fracture initiation is very similar for both quasi-dry and wet samples. Furthermore, the number of fractures tends to increase with loading rate; about 10-12 fractures were initiated at a loading rate of 100 MPa/ms.

The threshold breakdown pressure dependence on loading rate is attributed to transient pore-pressure effects in a thin region around the borehole. For the quasi-dry samples the diffusion of fluid through the permeable borehole wall couples the response of the pore structure with the fluid flow through the pore space causing a pore pressure buildup. For wet samples, the action of the

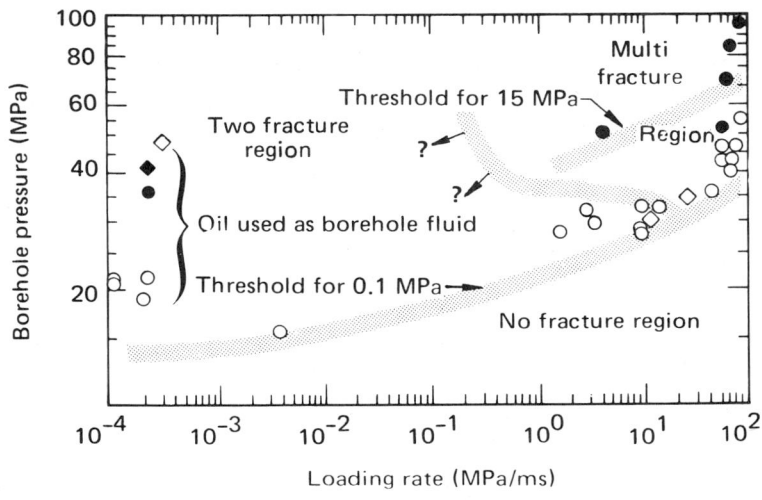

(a) Quasi-dry sample, ○ −0.1 MPa and
● − 15 MPa confining pressure
Dry ◇ −0.1 MPa and ◆ − 15 MPa confining pressure

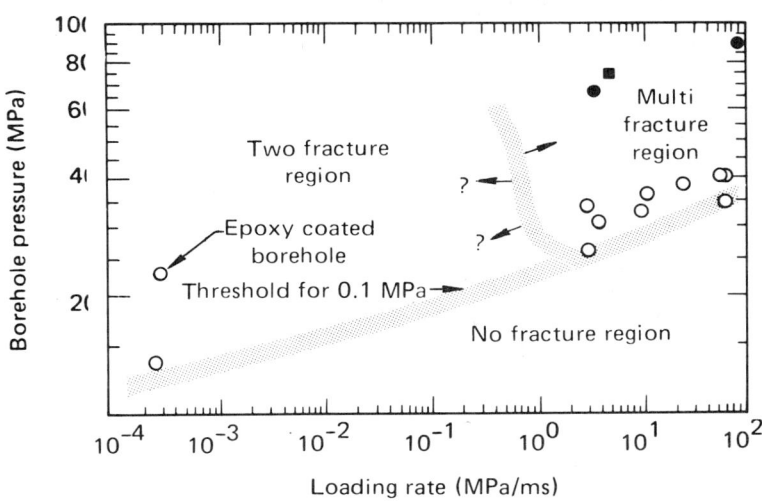

(b) Wet samples, ○ −0.1 MPa, ● − 15 MPa,
and ■ − 25 MPa confining pressure

Figure 2. Fracture threshold characteristics for quasi–dry and wet Nugget sandstone.

borehole fluid displacing the pore fluid leads to a pore pressure buildup. Based on the characteristic diffusivity of the porous material-fluid combination, there exists a thin region around the borehole that may or may not (depending on loading rate and saturation of the sample) be in pressure equilibrium with the borehole pressure. The lower the loading rate, the more likely pressure equilibrium will exist. From this reasoning, we infer that the borehole pressure associated with the onset of fracture increases with loading rate, which is in accord with observed data. A more definitive discussion of pore-pressure influence on multiple fracture initiation is given by Swift and Kusubov[6].

CONCLUSIONS

The observations described here demonstrate that the multiple fracturing process at intermediate loading rates is governed by fluid diffusion through the porous solid coupled to the structural deformation of the solid. The mechanism responsible for multiple fracturing appears to be the effect of transient pore-pressure buildup in the near vicinity of the borehole resulting from the diffusive intrusion of the borehole fluid. In the field, a select use of propellant can provide tailored pulses with loading rates comparable to those reported in this paper thus providing an alternative for present day techniques used to stimulate underground reservoirs.

REFERENCES

1. A. A. Daneshy, SPE Paper 3226, Society of Petroleum Engineers, Dallas, Texas (1971).
2. G. C. Howard, and C. R. Fast, <u>Hydraulic Fracturing</u>, Society of Petroleum Engineers of AIME Monograph Series, Dallas, Texas (1970).
3. A. S. Kusubov, W. B. Durham, R. N. Schock, J. F. Schatz, and T. J. Ahrens, "Explosive Induced Damage in Porous Sandstone," in <u>High Pressure Science and Technology</u>, Vol. 2, edited by B. Vodar and Ph. Marteau (Pergamon Press, Oxford, England, 1980), pp. 917-923.
4. N. R. Warpinski, R. A. Schmidt, P. W. Cooper, H. C. Walling, and D. A. Northrop, Proceedings 20th U.S. Symposium on Rock Mechanics, Austin, Texas 1979, pp. 143-152.
5. R. P. Swift and A. S. Kusubov, Proceedings 21st U.S. Symposium on Rock Mechanics, Rolla, Missouri, 1980, pp. 682-690.
6. R. P. Swift, and A. S. Kusubov, Proceedings 22nd U.S. Symposium on Rock Mechanics, MIT, Cambridge, Massachusetts, June 1981. Also UCRL-85189, March 1981.

CORRELATION OF IMPULSIVE STRAIN EFFECTS IN ROCKS WITH DETONATION PARAMETERS OF THE STRAIN-GENERATING EXPLOSIVE CHARGES

J. Roth, Consultant
308 Canyon
Portola Valley, CA 94025

ABSTRACT

A critical examination of existing explosion-generated strain data in granite-gneiss, salt and limestone indicates that these strain effects are proportional to the stress in the rock at the explosive/rock boundary. Appropriate extrapolation techniques had to be developed for published strain data, most of which were measured at some distance from the borehole and not at the explosive/rock interface. To estimate the stress in the rock, a new semi-empirical method of computing the detonation pressure of explosives was used. The correlation of strain effects and explosive parameters thus obtained is consistent with published Hugoniot data for these materials and does not violate conservation of energy as do some previously published correlations. For example, in salt the total strain energy is about 3-6% of the chemical energy of the explosive used. In the materials examined, the "crushed zone" immediately around the borehole appears to be formed by an intense fracture network rather than by plastic deformation of the rock.

INTRODUCTION

It has been shown[1-9] that explosively-generated strain in mine media correlates with blasting efficiency in such media, and with the environmental effects of blasting such as ground vibrations, etc.[7] This study develops previously lacking physically realistic correlations between strain effects and detonation characteristics of the strain-generating explosives. Experimental strain data are taken from U.S. Bureau of Mines publications.[1-8,10]

CORRELATION MODEL

A typical array for strain measurement is shown in Fig. 1.[3] A list of symbols used in the model is tabulated below.

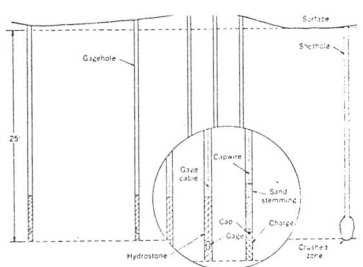

The necessary equations are developed below and the notation used is defined in Appendix A.

Fig. 1. DETAIL OF LINEAR ARRAY TEST

Strain at the explosive/medium boundary is expected to be proportional to the stress in the medium which in turn should vary with the detonation pressure of the explosive or more probably with pressure in the borehole containing the explosive charge. Thus:

$$\varepsilon_m = K'P_m = 1 - \rho_m/\rho'_m \; ; \; \rho'_m = \text{density of the stressed medium} \quad (1)$$

$$P_m \simeq 2P_b(1+\rho_o D/\rho_m C)^{-1} \quad (2)$$

where $P_b \simeq P_j/2$ and $P_j = \rho_o D^2(1-\rho_o/\rho_j)$ \quad (3)

and $\rho_j \simeq 1.303\rho_o + 0.0908$ \quad (4)

Equ. 4 is a semi-empirical correlation developed by the writer and its basis will be presented elsewhere.

DATA

Properties of the media in which measurements were made are given in Table I. As in previous studies,[2,3,6,7] strain was scaled with distance according to:

$$\varepsilon = K''(R/W^{1/3})^{-n} = K'''(R/r)^{-n} \quad (5)$$

where K'' and $K''' = \varepsilon_m$ are intercepts (at $R/W^{1/3} = 1$ ft/lb$^{1/3}$ and $R/r = 1$ respectively)* and n is the slope of "linear" $\log\varepsilon$ vs. log (scaled distance) plots. As shown in Fig. 2, such plots are not linear over the entire range of measurements. Furthermore, there is considerable scatter in the measurements. Consequently extrapolation of strains measured at some distance from the explosive to the explosive/medium boundary is difficult. In several of the previous studies,[6,7,8] values of K'' or K''' were based on the lower steep portions of plots such as the heavy-line shown in Fig. 2. In the present study, we estimated K'' or K''' only from measurements relatively close to the explosive charge (e.g., from data at $R/W^{1/3} < 10$ in Fig. 2). Paucity of data and data scatter in this region make some of these extrapolations rather uncertain. Our best estimates of ε_m and P_m are shown in Table II.

Table I Rock Properties

Rock	Location (State)	ρ_m (kg/m^3)	C (km/sec)
Granite-Gneiss	GA	2.63	5.60
Salt	LA	2.20	4.39
Porous Limestone	OH	2.35	2.90

---

* The relation between K'' and K''' is $K'' = K'''(4/3\pi\rho_o)^{-1/3}$ if r is taken to be the equivalent spherical radius of the explosive charge and ρ_o is in lbs/ft^3.

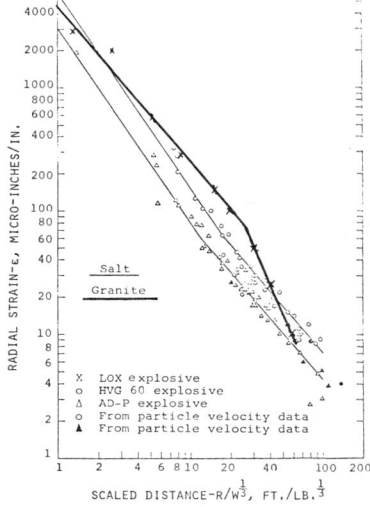

Fig. 2. STRAIN VS. SCALED DISTANCE

Table II. Correlation of Explosives Properties and Peak Strain at the Explosive/Media Boundary

Explosive	ρ_o (kg/m^3)	D^* (km/sec)	P_b (GPa)	P_m(GPa) Granite	Salt	Lime-stone	Strain (n/m) Granite	Salt	Lime-stone
Comp B	1.65	7.69	26.0	14.0	--	--	0.104 a/	--	--
Comp B	0.83	5.16	6.58	5.13	--	--	≈0.047 b/	--	--
50/50 Pentolite	1.64	7.50	24.5	13.5	--	--	0.103 a/	--	--
N$_2$O$_4$/AN/Kerosine**	1.29	6.77	16.3	10.3	--	--	0.078	--	--
LOX	1.25	5.87	11.9	8.00	--	--	0.059	--	--
Comp A	1.16	6.50	13.7	9.12	--	--	0.071	--	--
Comp A	0.81	5.00	6.07	4.77	--	--	0.038	--	--
TNT	1.00	4.59	6.00	--	4.00	--	--	0.085	--
Gelatin Dynamite 60%	1.40	6.18	14.5	9.23	7.78	6.50	0.056	0.138	≈0.40
Gelatin Dynamite 80%	1.28	6.20	13.5	8.88	--	--	--	--	--
Semi Gelatin Dynamite-1	1.30	5.37	10.3	--	--	5.10	--	--	≈0.33
Semi Gelatin Dynamite-2	1.11	4.83	7.30	5.37	4.70	--	0.039	0.112	--
Permissible Dynamite	0.71	2.80	1.71	1.52	1.42	--	≈0.023 b/	0.038	--
ANFO	0.90	≈3.4	≈2.7	≈2.3	--	--	≈0.025 b/	--	--
ANFO	0.95	3.75	3.91	--	--	2.56	--	--	≈0.20
Slurry Explosive-1	1.68	5.12	11.7	--	--	5.17	--	--	≈0.33
Slurry Explosive-2	1.60	5.27	11.9	--	--	5.30	--	--	≈0.33
Slurry Explosive-3	1.51	5.21	11.0	--	--	5.13	--	--	≈0.33

* For the borehole diameter used
** 30.3/60.4/9/3 N$_2$O$_4$/AN/kerosine
a/ Corrected for decoupling
b/ Lack of data close to borehole makes these values uncertain

RESULTS

In Fig. 3 we compare strain-stress data (from Table II) with Hugoniot data obtained in the laboratory.[11,12] Note the similarity in the shape of the curves. That laboratory data show greater strain at a given stress is not unexpected. Laboratory data were obtained for "head-on" explosive or flyer plate impact on the test sample, whereas tangential impact obtained in field measurements. From impedance mismatch considerations, it is known[13] that tangential shocks generate considerably lower stress in media than head-on shocks. For porous limestone ε_m data are questionable (see Table II). Taken at face value these ε_m values, at a given stress, appear to exceed Hugoniot ε_m's.

There is a zone of crushed material (cavity) close to the borehole. The measured ratio[2-8,10] of the volume of this zone to the volume of the explosive charge is a linear function e_m as shown in Fig. 4. It is instructive to speculate about the mechanism by which such cavities are formed. Two extreme cases are considered: plastic deformation or creation of a dense fracture network. In cylindrical geometry in an ideal plastic medium the work of plastic deformation per unit length for a charge of volume V_c and radius R_c is given by:[14]

$$w \simeq \pi R_c^2 Y[(m-1)\ln(\mu m/Y) - m\ln(m)] \tag{6}$$

where $m = V_{cav}/V_c \simeq (R_{cav}/R_c)^2$ since the length of the cavity and charge are approximately equal. For typical granite values of m=14, Y=2kb, μ=260kb; $w/\pi R_c^2 \simeq 1.2 \times 10^4$ J/cm^3 which is more than twice $\rho_o Q \simeq 5 \times 10^3$ J/cm^3 the total chemical energy of the explosive. Thus plastic deformation cannot be the major cavity forming mechanism. On the other hand, the threshold pressure to initiate intense crack formation is:[15,16]

$$P_c = K_{IC}(2.24\sqrt{\pi \ell}) \qquad (7)$$

If we take $K_{IC} \simeq 7000$ psi$\sqrt{\text{in}}$ and $\ell \simeq 0.04$ in, for fine grained granite,[15] $P_c \simeq 8900$ psi $\simeq 60$ J/cm^3 and $P_c/\rho_o Q \simeq 0.012$. Thus creation of an intense fracture network is a likely mechanism from an energy point of view.

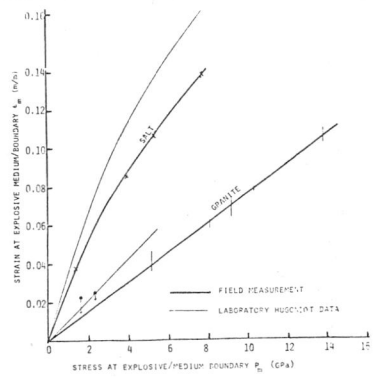

Fig. 3. STRAIN-STRESS CURVES FOR SALT AND GRANITE

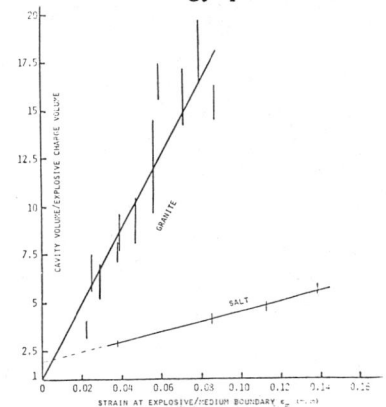

Fig. 4. RELATIVE CAVITY VOLUME AS A FUNCTION OF STRAIN ε_m

Finally, we want to estimate the total strain energy imparted to the medium. According to Atchison and Pugliese,[6] E_a the radial strain per unit area is given by:

$$E_a = (\rho_m C^3/2) \int_0^t \varepsilon^2 dt \qquad (8)$$

and the total radial strain energy per unit volume of explosive is:

$$E_r = 3(R/r)^2 E_a/r = 3E_a/r \text{ at } R/r = 1 \qquad (9)$$

Values of E_r for salt are given in Table III which also shows that the ratio $E_r/\rho_o Q$ varies from about 3 to 6%. Note that E_r/P_m is roughly constant. For granite $E_r/\rho_o Q > 1$ which of course violates conservation of energy. For granite extrapolations of ε and consequently of $\int \varepsilon^2 dt$ to $R/r = 1$ are based on measurements rather distant from the explosive/medium boundary, and correspond to the steep portion of the heavy line in Fig. 2. Undoubtedly these extrapolations give values of E_a which are much too large.

Table III Total Radiated Strain Energy in Salt

Explosive*	$\rho_o Q$ cal/cm^3	E_r cal/cm^3	$E_r/\rho_o Q$	E_r/P_m
Gelatine Dynamite	1040	63.7	0.061	0.034
Semi Gelatine Dynamite	770	43.9	0.057	0.039
TNT	860	27.5	0.032	0.028
Permissible Dynamite	≃450	11.6	≃0.026	≃0.034

*Same explosives as in Table I

CONCLUSIONS

· At the explosive/medium boundary, peak strain is directly proportional to stress in granite, but in salt strain is a non linear function of stress. Laboratory Hugoniot data for these two media provide analogous strain-stress relations as the field measurements.

· Scaled crushed zone volume around a borehole is directly proportional to the strain at the explosive/medium boundary.

· In salt the total strain energy is a small fraction of the total chemical energy of the strain-generating explosive charge.

REFERENCES

1. W.I. Duvall and B. Petkof, BuMine RI 5483 (1959)
2. T.C. Atchison and W.E. Tournay, BuMine RI 5509 (1959)
3. H.R. Nicholls and V.E. Hooker, BuMine RI 6041 (1962)
4. T.C. Atchison, W.I. Duvall and J.M. Pugliese, BuMine RI 6333 (1964)
5. T.C. Atchison and J.M. Pugliese, BuMine RI 6395 (1964)
6. T.C. Atchison and J.M. Pugliese, BuMine RI 6434 (1964)
7. H.R. Nicholls and V.E. Hooker, BuMine RI 6693 (1965)
8. H.R. Nicholls and W.I. Duvall, BuMine RI 6806 (1966)
9. M.A. Cook, The Science of Industrial Explosives (Graphic Service and Supply Inc., 1974), Chapt. 10
10. W.I. Duvall and J.M. Pugliese, BuMine RI 6700 (1965)
11. M. VanThiel, Compendium of Shock Wave Data, UCRL-50108 (1977)
12. D.B. Larson and G.D. Anderson, J. Geophys. Res. 83 B6, 2839 (1978)
13. G.E. Duvall, Proc. Conf. Response of Metals to High Vel. Deformation (1960), p. 179
14. W.J. Carter, Prog. Rept. LASL, Los Alamos, NM, LA-6521-PR,3 (1976)
15. J.W. Dally and W.L. Fourney, Rept. for NSF by M.E. Dept., Univ. of Md. (1976)
16. D.B. Barker and W.L. Fourney, Rept. for NSF by M.E. Dept., Univ. of Md. (1978)

APPENDIX A: LIST OF SYMBOLS

C = Longitudinal sound velocity
D = Detonation velocity
E_r = Radiated strain energy per unit volume of explosive
ε = Strain
ε_m = Strain at explosive/medium boundary
K_{IC} = Fracture toughness of rock
ℓ = Pre-existing crack length
μ = Shear modulus
P_c = Threshold pressure to initiate cracks in medium
P_j = Detonation pressure
P_b = Borehole pressure
P_m = Stress at explosive/medium boundary
ρ_o = Density of explosive
ρ_j = Density of detonation products
ρ_m = Density of medium
Q = Heat of detonation
R = Radial distance
r = Equivalent spherical radius of explosive charge
W = Weight of explosive
w = Work of plastic deformation
Y = Compressive yield strength

DYNAMIC AND STRESS INTENSITY IN ELASTIC STRIPS*

W. G. Hoover and B. Moran

University of California, Lawrence Livermore National Laboratory
and Department of Applied Science, UC Daivis/Livermore
Livermore, California 94550

ABSTRACT

We are studying a series of lowand high-temperature atomistic dynamic crack propagation simulations to reproduce and interpret shadow-pattern measurements of dynamic stress intensity, carried out at the Fraunhofer Institüt by Kalthoff, et al. We analyze crack propagation in a long fixed-width strip with displacement boundary conditions. Despite complications imposed by the free surface and boundary effects, we obtained static values of the material property K with at least three-figure accuracy. In the atomistic calculations it is not necessary to specify an arbitrary damage model or failure rule. Instead, fracture proceeds as a consquence of Newton's microscopic energy-conserving equations of motion. Our results clearly show the importance of boundary effects such as those observed by the Freiburg group. In our dynamic fixed-width strip simulations, crack propagation can occur under loading conditions where static analysis predicts arrest. In a typical case the difference between the dynamic propagation stress and the larger static one is 15 percent.

---

*Supported by the Electric Power Research Institute.
**Work performed under the auspices of the U.S. Department of Energy by Lawrence Livermore National Laboratory under contract #W-7405-Eng-48.

0094-243X/82/780480-01$3.00 Copyright 1982 American Institute of Physics

Chap. 10 Shock Phenomena and Experimental Technique

RAMP-WAVE GENERATORS STUDIES*

M. GERMAIN-LACOUR and M. de GLINIASTY
Centre d'Etudes de Gramat, 46500 Gramat, France

ABSTRACT

Ramp-wave generators are materials which turn a discontinuous shock wave into a continuous acceleration wave. A brief theoretical discussion emphasizes the importance of the Grüneisen parameter with respect to ramp generation capacity. Two commercial glass-ceramics and a titanium silicate glass have been tested : typically a 5.5 mm thick glass-ceramic (Zerodur or Cervit C 101) sample shock loaded up to 13 GPa generates a ramp wave with a rise time of about 600 ns.

INTRODUCTION

Ramp-wave generator materials have the peculiar property to turn a discontinuous shock wave into a continuous acceleration wave. This property is of interest with respect to high pressure loading techniques : the nearly isentropic compression, generated within a direct contact sample, gives access to a new region of the state surface, off Hugoniot's curve ; through time-resolved *in situ* measurement techniques, physical phenomena can be detected and valuable Lagrangian analysis performed ; and at last the interest of continuous loading in order to decrease mass ejection from surfaces, as shown by Asay[1], must be mentioned. The purpose of the present work is to characterize materials which intrinsically generate ramp-waves.

THEORETICAL BACKGROUND

Bethe[2] has derived two assumptions in order to prove the existence and uniqueness of the solutions of Hugoniot's equations :

$$\left(\frac{\partial^2 P}{\partial V^2}\right)_S > 0 \quad (1) \quad \text{and} \quad \Gamma(V) = V \left(\frac{\partial P}{\partial E}\right)_V > -2 \quad (2)$$

where P, E, V and Γ stand for the pressure, specific energy, specific volume and Grüneisen parameter. (1) is a necessary condition for the building of a compressive shock wave, while (2) is only a sufficient requirement to prove that compressive shock waves correspond to an increase in entropy and are therefore thermodynamically stable.

Ramp-wave generators, of course, do not fulfill at least the first condition : for these materials the velocity of elementary compressive waves "a" increases as the specific volume increases :

$$\left(\frac{\partial a}{\partial V}\right)_S = \frac{\partial}{\partial V}\left[-V^2 \left(\frac{\partial P}{\partial V}\right)_S\right]^{1/2} = -\frac{a}{V} - \frac{V^2}{2a}\left(\frac{\partial^2 P}{\partial V^2}\right)_S > 0 \quad (3)$$

which of course implies that $(\partial^2 P/\partial V^2)_S$ be negative. The second assumption of Bethe is met, but that does matter because it is only a sufficient condition. Therefore we shall focus our discussion on Bethe's first condition.

Within the framework of the Born-Oppenheimer model for crystals, the Grüneisen parameter is expressed by :

$$\Gamma(V) = -\frac{d \ln \bar{\omega}}{d \ln V}$$

---

*Work supported by French Ministry of Defense.

where $\bar{\omega}$ stands for an average value of the frequencies of the oscillators constituting the crystal. At moderate temperatures $\bar{\omega}$ may be replaced by ω_D, the Debye frequency. Following the Debye model, we assume that the phase velocity "a" of the acoustic waves – the optical waves are neglected at moderate temperatures – is constant. This yields for ω_D the following expression:

$$\omega_D = (6\pi^2 \, N/V)^{1/3} \, a$$

N is the number of cells, V the volume of the crystal. Then we get:

$$\Gamma(V) = - \frac{d \ln \omega_D}{d \ln V} = \frac{1}{3} - \frac{d \ln a}{d \ln V} \qquad (4)$$

As $d \ln a/d \ln V$ must be positive for a ramp wave generator, we get:

$$\Gamma(V) < \frac{1}{3} \qquad (5)$$

This condition could have been obtained from Slater's expression:

$$\Gamma(V) = - \frac{2}{3} - \frac{V}{2} \frac{d^2 P_c/dV^2}{dP_c/dV} \qquad (6)$$

Assuming that the sound speed on the standard isentrope is close to the sound speed on the cold compression curve Pc (V), we may write:

$$a^2 = - V^2 \left(\frac{\partial P}{\partial V}\right)_s \approx - V^2 \frac{dP_c}{dV}$$

Introducing the last value of a^2 and its first derivative we get:

$$da/dV = a \, (1/3 - \Gamma(V))/V$$

which is the same as (4) and leads to (5).

We deduce that, in a first approach, a ramp wave generator should be characterized by a Γ value lower than 0.33. In fact this maximum value depends on the model (Slater ...) and let's retain from this short discussion that such a material should have a low or even negative Grüneisen parameter.

As usually the values of Γ are not in handbooks, we have converted this condition through the well-known thermodynamic relation:

$$\Gamma = \alpha V/\beta C_V \qquad (7)$$

where α, β and C_V stand respectively for the thermal coefficient of expansion, the isothermal compressibility and the molar specific heat. From (7) we deduce that to a negative Γ can be associated a negative α. In order to search candidate materials we have extended this remark to small values of Γ and α. This is the case for fused silica which is a ramp generator up to 3 GPa[3] : it shows a Γ_0 of 0.036[4] and an α of 45.10^{-8} (°C)$^{-1}$ in standard conditions[5]. It is also the case for Pyroceram 9608, a glass-ceramic first discovered and tested by Benedick and Asay[6] : this Corning product is well-known for its low expansion coefficient. Small or negative thermal expansion coefficient is a condition often requested for optical components and radomes ; among the standard materials used in that field, two commercial glass-ceramics and a titanium silicate glass have been selected for this study.

MATERIALS PRESENTATION

Glass-ceramics are polycrystalline solids prepared by the controlled crystallisation of glasses[7]. Crystallisation is accomplished by subjecting suitable glasses to a carefully regulated heat treatment schedule which results in the nucleation and growth of crystal phases within the glass. In many cases, this process can be taken almost to completion but a small proportion of residual glass-phase is often present. These materials are non-porous, macroscopically homogeneous and isotropic. Two commercial glass-ceramics with a very low (or negative) thermal expansion coefficient have been tested :
- ZERODUR glass-ceramic developed by SCHOTT research labs. The crystal structure is h-quartz (high-temperature quartz). The average size of a crystal is 50 nm. This material has already shown anomalous thermoelastic properties with ultrasonic measurements[8].
- CERVIT C101 glass-ceramic is manufactured by OWENS ILLINOIS. The crystalline phase is a solid solution of β-quartz.

A third commercial material with a low expansion coefficient has been tested : a titanium silicate glass developed by CORNING GLASS COMPANY under the code 7971 and the trade name ULE (Ultra Low Expansion). This is a synthetic amorphous silica glass manufactured by the same basic process as fused silica.

Some thermomechanical properties of these materials are compared in table I:

Table I. Materials characteristics

Commercial designation	Nature	Initial density	Young's modulus (GPa)	Shear modulus (GPa)	Poisson's ratio	Therm. exp. coef. [a]
ZERODUR	Glass-ceramic	2.53	92	37	0.24	- 0.20
CERVIT C101	Glass-ceramic	2.50	92.4	36.5	0.25	0.00
CORNING ULE (7971)	Titanium silicate glass	2.205	67.6	29.0	0.17	0.0 ± 0.6

a) Mean linear thermal expansion coefficient 0-200°C $(10^{-7}(°C)^{-1})$

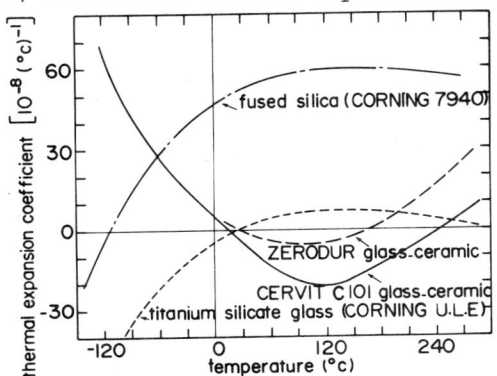

Fig. 1. Thermal expansion coefficient versus temperature variations for different materials. These curves have been experimentally measured with a frequency stabilized laser beam[5].

EXPERIMENTS

The loading technique of the sample is a direct contact explosive charge experiment : a plane wave lens (typical simultaneity of 50 ns) associated with a high explosive pad (baratol 70/30) generates a planar pressure pulse through a 5 mm copper driver plate into the sample. This one is instrumented with eight (4 x 2) 50 Ω manganin piezoresistive gauges (HBM type PMS 6/50 YT1) which have been previously calibrated under shock loading and unloading up to 20 GPa[9]. Signals provided by unbalanced Wheatstone bridges are recorded by a Thomson 693 transient recorder and automatically converted into stress histories with a H.P. 45 calculator. Recorded profiles are presented on figures 2, 3 and 4. Gauges depths are reported near the stress profiles in mm : zero is copper disc-sample interface.

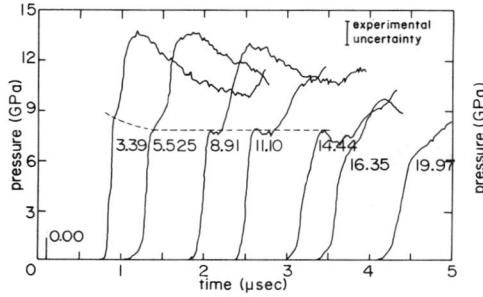

Fig. 2. Stress histories in ZERODUR. Dotted line indicates elastic limit.

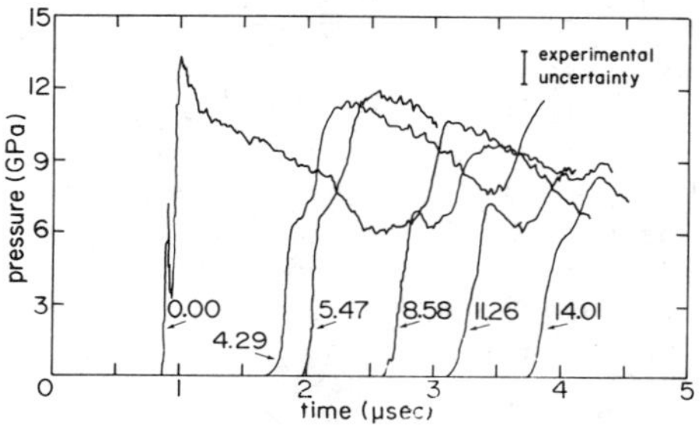

Fig. 3. Stress histories in CERVIT.

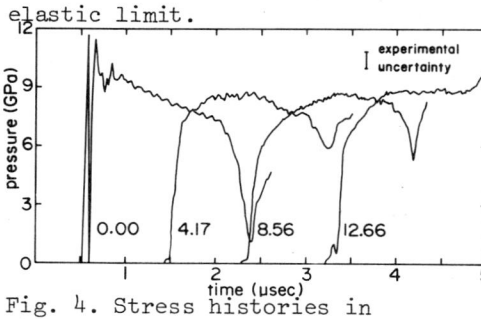

Fig. 4. Stress histories in titanium silicate glass.

In the two glass-ceramics profiles are similar : a non linear elastic area is followed by a plastic compression, the duration of which increases as the loading wave propagates through the specimens. For a 5.525 mm Zerodur sample loaded up to 13 GPa, the total rise time is 700 ns and for 5.47 mm Cervit sample loaded up to 12 GPa it is 600 ns. About titanium silicate glass, profiles show a little step ($\sim$ 0.5 GPa) followed by a rapid compression ; ramp-wave effect is not so noteworthy as for glass-ceramics.

Although gauges are not perfectly adapted with these materials, Lagrangian analysis of Zerodur and titanium silicate glass experiments have been performed. Pressure-volume curves resulting are shown on figures 5 and 6. We intended to derive Grüneisen parameter from this analysis along isochoric curves in the P-E plane, in order to ascertain Bethe's second assumption. Unfortunately we had not enough data about the state surface of these materials to do that.

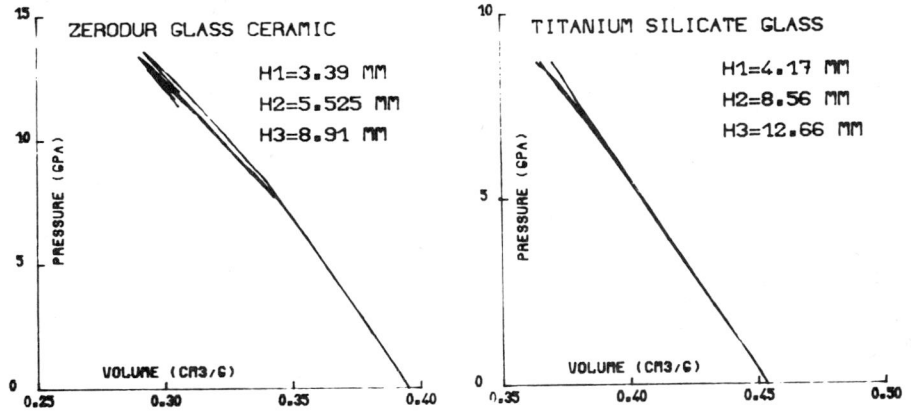

Fig. 5. Loading curve in Zerodur.   Fig. 6. Loading curve in titanium silicate glass.

Sound speed in Zerodur analysis decreases from 5083 m/s for the elastic precursor, to 2800 m/s at 13 GPa : this is in accordance with profiles presented on figure 2. But the elastic velocity calculated with the bulk and shear modulus is 6458 m/s. The large difference between the two elastic sound speeds is worth studying. This material might be viscoelastic : the last value doesn't take in account viscous effects. On the other hand there is an excellent agreement in the case of titanium silicate glass (5733 m/s).

CONCLUSION

This short study has shown that ramp-wave generators may be characterized by a small or even negative Grüneisen parameter, i.e. by a small or negative thermal expansion coefficient. Two commercial glass ceramics and a titanium silicate glass with this property have been tested : glass-ceramics seem promising materials in this field. Probably an adequate heat-treatment on these materials would improve their ramp generation capacity.

REFERENCES

1. J. R. Asay, "Effect of Shock Wave Risetime on Material Ejection from Aluminum Surface", Sand. Lab. Rep. SAND 77-0731 (1977).
2. H. A. Bethe, "On the Theory of Shock Waves for an Arbitrary Equation of State", OSDR n° 545, May 4, (1942).
3. L. M. Barker and R. E. Hollenbach, J. Appl. Phys. 41, 4208 (1970).
4. J. Wackerle, J. Appl. Phys. 33, 922 (1962).
5. J. W. Berthold III and S. F. Jacobs, Appl. Opt. 15, 2344 (1976).
6. W. B. Benedick and J. R. Asay, Bull. Am. Phys. Soc. 21, 1298 (1976).
7. P. W. Mc Millan, in Glass-ceramics, Academic Press, 2nd ed. (1979).
8. D. Gerlich and M. Wolf, J. Non-Cryst. Solids 27, 209 (1978).
9. M. Perez, thèse docteur-ingénieur, n° 714, University Paul Sabatier, Toulouse, France, October 21 (1980).

EXPERIMENTS ON THE ATTENUATION OF SHOCK WAVES IN CONDENSED MATTER

Ch. Klee, M. Kroh, D. Ludwig

Messerschmitt-Bölkow-Blohm GmbH, Werk Schrobenhausen, West-Germany

ABSTRACT

The paper provides results concerning the attenuation of shock waves induced in Cu, Al and PMMA by contact detonation. The results are presented as empirical description of the peak pressure as a function of shock travel distance: $p = p_o \exp(-\alpha x)$. Besides that we investigated the influence of the lateral rarefaction wave using PMMA-samples with different diameters. There was a significant difference in the attenuation rates of the used materials and a marked dependence on the donor charge.

INTRODUCTION

The study of the laws governing propagation and attenuation of shock waves in condensed matter is of great theoretical and practical importance. Particularly such studies are necessary in the fields of detonation physics and terminal ballistics.

Besides that there are a number of applications for shock wave physics in warhead development where functional and safety problems are at hand.

In this report we investigated the behaviour of various bodies with respect to the application for explosive train design such as
- to meet requirements of functional and safety criteria
- to determine shock wave sensitivity of secondary explosives in a wide pressure regime by means of statistical methods like gap-test.

The last case calls for especially precise knowledge of the decrease of shock pressure versus travel distance.

EXPERIMENTAL

The electrically initiated donor charge (60 mm dia., 50 mm long) creates a shock wave within the test material. The transient time is measured by self shorting pins which are fixed to the entrance and outlet of the test block (60 mm dia.). Running time t is recorded vs distance x from 2 to 40 mm by counter and simultaneously by oscilloscope in steps of 1 to 5 mm. Thus the x(t)-dependence will be derived. Fig. 1 shows the test arrangement which was chosen for the samples Cu and Al.

For the measurement of the shock-attenuation in the transparent PMMA we used a rotating mirror camera.

Table 1 quotes the essential physical properties of the inert materials. Table 2 shows the characteristics of the high explosives.

All combinations explosive/inert material used in the tests are listed in table 3. The Chapman-Jouget pressures P_{CJ} and velocities D_{CJ} were calculated from the BKW equation of state[1]. The interface pressure was found by the intersection of the Hugoniot curves of the test material [2] and the reaction products [1].

0094-243X/82/780486-05$3.00 Copyright 1982 American Institute of Physics

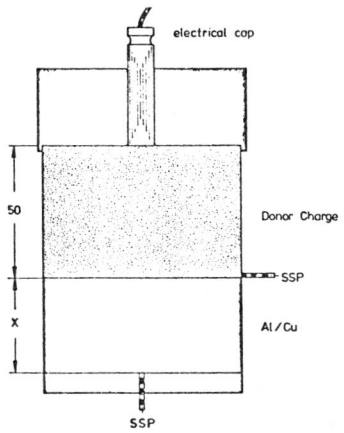

Fig. 1: Test arrangement

Table 1: Properties of the inert materials

Material	ρ [g/cm³]	Hugoniot $U_s = (A + B\,u_p)$ [mm/μs]
Copper	8.87	3.944 + 1.514 u_p
Aluminium	2.78	5.355 + 1.345 u_p
PMMA	1.18	2.561 + 1.595 u_p

Table 2: Properties of the high explosives

H.E.	ρ [g/cm³]	P_{CJ} [kbar]	D_{CJ} [mm/μs]
RDX/Wax 95/5	1.665	285	8.470
HMX/TNT 75/25	1.806	320	8.556
RDX/TNT 65/35	1.720	284	8.084
Tetryl	1.557	206	7.160

Table 3: Tested Donor/Acceptor combinations

Donor	Acceptor	⌀(Acceptor) [mm]
RDX/Wax 95/5	Copper	60
RDX/Wax 95/5	Aluminium	60
RDX/Wax 95/5	PMMA	⌀ 70
HMX/TNT 75/25	Copper	60
HMX/TNT 75/25	Aluminium	60
HMX/TNT 75/25	PMMA	⌀ 70
Tetryl	PMMA	⌀ 70
RDX/TNT 65/35	PMMA	⌀ 70
HMX/TNT 75/25	PMMA	⌀ 50
HMX/TNT 75/25	PMMA	⌀ 40
HMX/TNT 75/25	PMMA	⌀ 20

The shock attenuation in plane waves is essentially due to two mechanisms: The irreversible heating of the material and the pressure reduction by rarefactions.

In the case of spherical or cylindrical waves there is an additional pressure decrease due to the expansion of the wave.

The irreversible heating $\Delta\varepsilon$ of the test material by the shock wave is given by

$$\Delta\varepsilon = \frac{1}{2} P (V_o - V) - \int_{V_o}^{V} (P(V)\,dV)_{s=const.} \quad (1)$$

with V_o the specific volume of the unshocked material. The integral in eq. (1) is taken along the isentropic through V_o which is given in [4], together with [5] and [6].

The portion of the internal energy $\varepsilon/(\varepsilon - \varepsilon_o)$ which causes the irreversible heating of the samples yields approximately 24% or 23% for Cu or Al respectively and 63% for PMMA. The minimal time Δt after which the lateral rarefaction catches up with the shock front can be obtained by a simple geometrical consideration and is given by

$$\Delta t = d/2 \cdot \sqrt{c^2 - (U_s - U_p)^2} \quad (2)$$

c, U_s and U_p being sound velocity, shock velocity and particle velocity, respectively. Sound velocity has been determined approximately by (3)

$$c = [A + U_s (B-1)] \times [2 U_s - A] / B \times U_s \quad (3)$$

A and B being the constants in the linear Hugoniot relation
$U_s = A + B\, U_p$.

The calculated values of U_s, U_p and c are listed in table 4. The minimal distance x where the lateral rarefaction overtakes the shock front is given by $x_{min} \geq 45$ mm for the test samples with a diameter ≥ 60 mm. In our tests therefore only the rear rarefaction influences the pressure-time history.

Table 4: Properties of the shock waves produced by the different Donor/Accepter combinations

H.E.		test material	interface pressure P [kbar]	shock velocity U_s [mm/μs]	Particle velocity U_p [mm/μs]	Local sound velocity C [mm/μs]
RDX/Wax	95/5	Copper	490	5.473	1.010	5.709
RDX/Wax	95/5	Aluminium	360	7.634	1.695	7.713
RDX/Wax	95/5	PMMA	230	7.011	2.790	6.900
HMX/TNT	75/25	Copper	570	5.662	1.135	5.901
HMX/TNT	75/25	Aluminium	380	7.735	1.770	7.801
HMX/TNT	75/25	PMMA	240	7.122	2.860	6.992
RDX/TNT	65/35	PMMA	215	6.787	2.650	6.740
Tetryl		PMMA	150	5.974	2.140	6.024

RESULTS AND DISCUSSION

The evaluation procedure consists of the following steps: Approximation of (x_i, t_i) by suitable differentiable functions (we used polynomials of 2nd and 3rd degree) - Differentiation $dx(t)/dt$ provides the velocity-time history - Application of the Hugoniot relations results in (p_i, x_i) - Approximation of (p_i, x_i) by an exponential $p(x) = p_o \exp(-\alpha x)$.

A summary of the results is given in table 5. A criterion for the adequacy of the chosen approximations is the correlation coefficient r which in all cases was >0.999, implying a good choice of the approximation.

The experimental data of PMMA had to be fit piecewise by two different functions of the above kind.

Fig. 2 shows a plot for p versus the travel distance x. For small distances x the maximal attenuation takes place in PMMA, succeeded by copper. A quantitative picture of this behaviour is given in Fig. 3 which shows $-dp/dx$ versus travel distance x.

Table 5: Experimental results

H.E.		inert material		interface pressure [kbar]	P_0 [kbar]	α [1/mm]	range of validity [mm]
RDX/Wax	95/5	Cu	⌀60	490	393	-0.0362	5 - 50
RDX/Wax	95/5	Al	⌀60	360	308	-0.0172	2 - 45
RDX/Wax	95/5	PMMA	70	230	235/155	-0.07257/-0.03237	2-9/9-40
HMX/TNT	75/25	Cu	⌀60	570	388	-0.0324	2 - 20
HMX/TNT	75/25	Al	⌀60	380	262	-0.0212	2 - 65
HMX/TNT	75/25	PMMA	⌀70	240	131/211	-0.02426/-0.05210	18-40/2-18
Tetryl		PMMA	⌀70	155	159/112	-0.07514/-0.03275	2-8/8-40
RDX/TNT	35/65	PMMA	⌀70	215	205/135	-0.08560/-0.03578	2-8/8-40
HMX/TNT	75/25	PMMA	⌀50	240	238/172	-0.07629/-0.03633	2-6/6-40
HMX/TNT	75/25	PMMA	⌀40	240	238/189	-0.07629/-0.04388	2-6/6-30
HMX/TNT	75/25	PMMA	⌀20	240	244/383	-0.08338/-0.1255	2-10/10-20

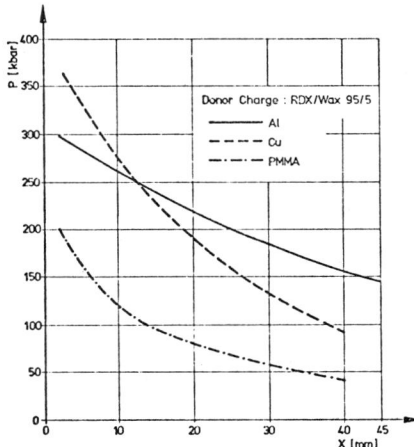

Fig. 2: p(x) for different test materials

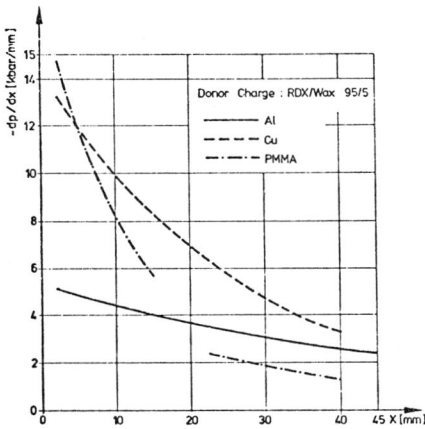

Fig. 3: Attenuation rate versus travel distance for different materials

For x < 6 mm one gets the maximal attenuation in PMMA. The initial steep decrease in the PMMA curve with the following transition into a fairly flat part hints towards different attenuation mechanisms. We suppose that in this case the irreversible heating in the initial regime influences the pressure decrease. The different attenuation rates as a function of p can be compared to each other using the coefficients α of table 5.

For larger x the pressure decrease in PMMA is determined, similar to Cu and Al, almost exclusively by the rear rarefaction, which reflects the pressure profile at the interface.

Comparison of the interface pressures (Table 5) with the peak-pressures after 2 mm (Fig. 2) reveals a considerable pressure decrease within the first 2 mm, a result which should be expected from the 1-D-detonation wave model [3]. Jameson and Hawkins [7] investigated this initial regime and got similar results. Concerning PMMA this effect does not appear because of the rarefaction reflected into the reaction products. A comparison of p(x) for different donor charges is shown in Fig. 4. The relative position of the curves depends upon the interface pressures. The dependence of the attenuation on the different donor charges can be seen comparing the coefficients α of table 5.

The attenuation is not correlated to the magnitude of the interface pressure as would be expected for equally shaped and extended pressure profiles of different height.

The considerable influence of the lateral rarefaction is obvious from Fig. 5. Application of (3) together with the data of Table 4 yields minimal travel distances x_{min} which are required for the lateral rarefaction to catch up with the shock front of x_{min} = 6.4, 12.8 and 16.1 mm for a diameter of 20, 40, 50 mm respectively. This is in good agreement with our experimental results.

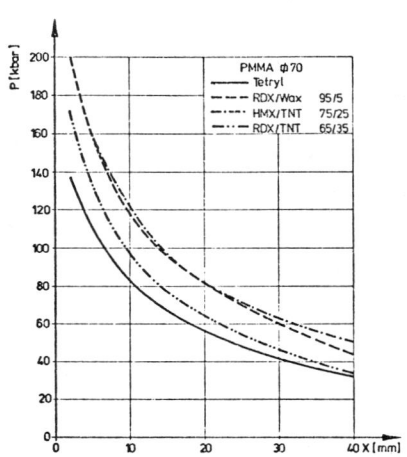

Fig. 4: p(x) in PMMA for different Donors

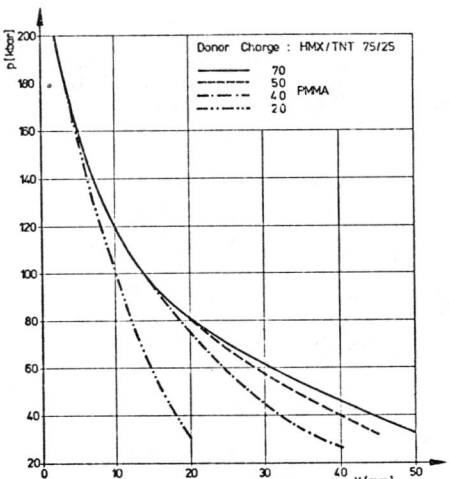

Fig. 5: p(x) in PMMA test samples with different diameters

REFERENCES

1. Charles L. Mader, Detonation Properies of Condensed Explosives using the Becker-Kistiakowsky-Wilson Equation of State, Los Alamos Scientific Laboratory, Los Alamos, New Mexico (1963).

2. M. van Thiel, Compendium of Shock Wave Data, URCL-50/08 (1966).

3. J. von Neumann, O.S.R.D. Report 549 (1942).

4. Ya. B. Zel'dovich, Yu. P. Raizer, Physics of Shock Waves and High Temperature Phenomena, Vol. II, (Academic Press, New York and London, 1967).

5. A. P. Rybakov, Fizika Goreniya i Vzryva, Vol. 14, No. 1, Jan. - Febr. 1978, pp. 109 - 113.

6. L.V. Al'tshuler et al., Soviet Physics JETP page 573, (3) (1960).

7. R.L. Jameson, A. Hawkins, 5th. Symposium on Detonation, page 23, Pasadena, Calif. (1970).

CALIBRATION OF PIEZORESISTIVE GAUGES*

G. L. Nutt and J. O. Hallquist
Lawrence Livermore National Laboratory
University of California
Livermore, California 94550

I. INTRODUCTION

We model the dynamic response of two ytterbium gauges using the two-dimensional Lagrangian finite difference code DYNA2D.[1] The gauges are embedded in a PMMA medium. The stress and strain state of each element of the gauge is then used to calculate the resistance change as a function of time.

The basis for modeling the electrical response of the gauge are piezoresistance equations of the type discussed by Grady and Ginsburg.[2] These equations allow a separation of the gauge signal into a stress related component (reversible) and a strain related component (possibly hysteretic). The calculated signal is then compared with experiments reported by Gupta et. al.[3]

This approach allows us to calculate the separate stress- and strain-related signals. Although the calibration equations, applied to the gauge as a whole, are invariant under rotation about the direction of current flow, experiments show different signals for differenct orientations. We are able to resolve this paradox by calculating the dynamic response of both configurations.

II. CALIBRATION EQUATIONS

Under a uniform applied stress a sample of piezoresistive material will undergo a resistance change given by[2]

$$\frac{\delta R}{R_o} = \pi_t (P_1 + P_2) + \pi_1 P_3 + \frac{\delta l}{l_o} - \frac{\delta A}{A_o} \tag{1}$$

where $P_{1,2,3}$ are the three principal stresses in the gauge and π_t, π_1 are the nonvanishing elements of the piezoresistance tensor; is the gauge length, and A is its cross sectional area. The last two terms, Eq. (1), are strains in the gauge dimension parallel and normal to the current flow respectively. We will call these terms the strain signal. The stress signal comes from the piezoresistance terms proportional to the π's.

Under shock conditions, according to Eq. (1), a gauge will have the resistance change

$$\frac{\delta R}{R_o} = (\pi_1 - \pi_t)\frac{\nu}{1-\nu} + \pi_t \frac{1+\nu}{1+\nu} P_1 - \frac{1}{3B}\frac{1+\nu}{1-\nu} P_1. \tag{2}$$

---

*Work performed under the auspices of the U.S. Department of Energy by the Lawrence Livermore National Laboratory under contract No. W-7405-ENG-48.

0094-243X/82/780491-04$3.00 Copyright 1982 American Institute of Physics

P_1 is the principal stress normal to the shock front and B is the bulk modulers of the gauge material. Equation (2) assumes stress equilibrium between the gauge and the medium, and makes no assumption about orientation except that the current flow is in the plane of the shock.

According to the Von Mieses yield criterion, above the onset of plasticity

$$\nu = \frac{1-|P_1|/Y}{1-2|P_1|/Y} \qquad (3)$$

and below yield ν takes on its low stress value. Equations (2) and (3) together can be used to estimate the response of a gauge under shock conditions above and below the Hugoniot elastic limit.

III. CODE CALCULATION

The length of the gauge is much greater than the lateral dimensions. In DYNA2D we must approximate this shape with an infinite length.

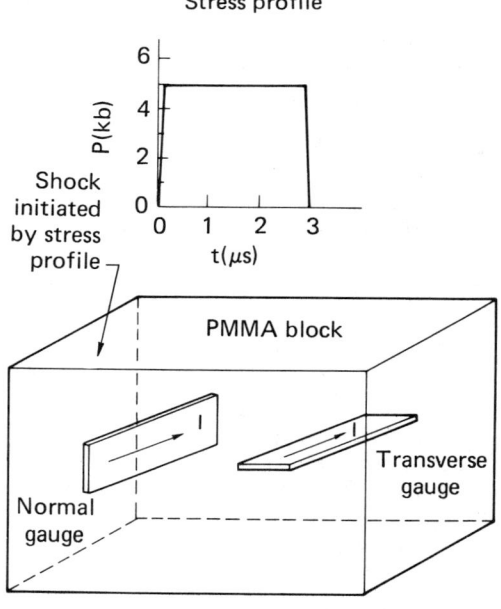

Fig. 1. Diagram of computer model of experiments.

Figure 1 shows two experiments which we modeled on DYNA2D. An elastic-plastic model is used to describe the constitutive properties of PMMA and Yb. For PMMA, the pressure-volume dependence is taken from high strain-rate experiments,[4] and a Hygoniot elastic limit is assigned of 4.2 kb.[5] The ytterbium pressure volume relationship is derived from measurements of Gust & Royce,[6] and a .5 kb yield strength is used.[7]

The electrical response of the gauge is modeled by computing the time dependent stress and strain state in the gauge, computing the resistance change from Eq. (1), and summing the contribution of each zone as if it were a circuit element in parallel with the rest of the gauge elements.

The result is

$$\frac{\delta R}{R_o} = n\left(\sum_{i=1}^{n} \frac{1}{1 + \frac{\delta\rho_i}{\rho_o} - \delta A_i}\right)^{-1} - 1 \qquad (4)$$

where the sum is taken over every element in the gauge.

Using Eqs. (1) and (4) we calculate the gauge signal where we use[2,8]

$$\pi_t = \pi_1 = -.0205 - .00078P \qquad (5)$$

IV. CONCLUSIONS

The results are summarized in Fig. 2. The ratio of the calculated transverse to normal signals is .68; in excellent agreement with the measurements. The magnitude of the calculated signal is at most 6% low for the normal gauge as compared with the experiments.

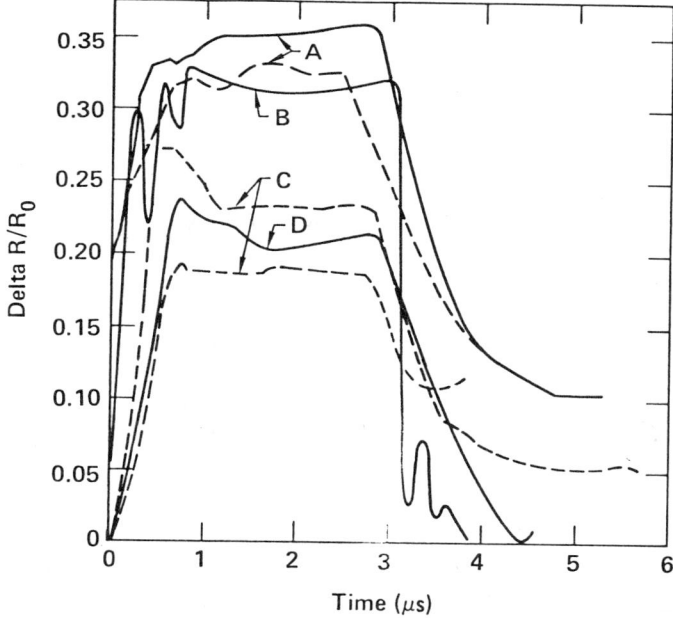

Fig. 2. Calculated signals compared with experiments (A) experimental normal (B) calculated normal (C) eperimental transverse (D) transverse

The strain component of the signal is calculated as about 0.02 for the transverse gauge and about 0.028 for the normal gauge. The plastic part of the deformation is quite large on the surface of the gauge but very small in the inerior zones. Over all, the plastic deformation contributes a negligible component to the signal. Neglect of the change in gauge length due to the two-dimensional approximation probably is small simply because the strain signal itself is so small.

The calculations show a stress state higher than 5 kb on the leading gauge surfaces where the incoming shock stagnates. The stress state of the zones in the gauge are much different from the "free field" stress in the PMMA. The gauge is well above its HEL so the stress state is approximately isotropic whereas the longitudinal stress in the PMMA is 5 kb, the transverse stress is 2.9 kb, and the average stress 3.6 kb.

In the transverse gauge the average stress (which is responsible for the gauge signal under the assumption $\pi_t = \pi_1$) is 4.3 kb at the leading edge and around 1.8 kb at the trailing edge. Thus the transverse gauge is not uniformly stressed. The normal gauge on the other hand has a uniform average stress over its entire volume. Eq. (2), for a normal gauge in a uniaxial strain experiment, gives a signal of 0.31 using 5 kb normal principal stress.

Work is underway modeling two manganin gauges in the normal and transverse orientation at a stress level of 27.5 kb. The material properties are much different in this case as is the experimental results.

REFERENCES

1. John. O. Hallquist "User's Mannual for DYNA2D--An Explicit Two-Dimensional Hydrodynamic Finite Element Code with Interactive Rezoning," UCID-18756 (1980).
2. D. E. Grady, M. J. Ginsburg, J. Appl. Phys., 48(6), 2179 (1977).
3. Y. M. Gupta, D. D. Keough, D. Henley, D. F. Walker, Appl. Phys. Lett., 37(4), 395 (1980).
4. Y. M. Gupta, J. Appl. Phys., 51(10), 5352 (1980).
5. D. Steinberg (private communication).
6. W. H. Gust, E. B. Royce, Phys. Rev., B8(8) 3595 (1973).
7. M. J. Ginsburg et. al. Final Report Contract DNA001-72-C-0146, SRI International, Menlo Park, CA (1973).
8. E. M. Lilley, D. R. Stephens, UCRL-51006 LLNL, Livermore, CA (1971).

ELECTROMAGNETIC GAUGE FOR MEASURING THE RADIAL PARTICLE VELOCITY IN 2-D FLOW

G. Rosenberg, D. Yaziv and M. Mayseless
Ministry of Defence, ADA, P.O. Box 2250, 31021 Haifa, Israel.

ABSTRACT

Applications of existing EMV gauges are limited for uniaxial strain configurations, since the gauge length must remain fixed during the motion. A modification of the electromagnetic technique which provides measurements in 2-D flow is presented. When the problem of a projectile impacting a target is described in cylindrical coordinates (r,z) with z as axis of symmetry, the flow can be defined by the particle velocity components U_z, U_r. A new gauge is made of a thin copper wire having a circular turn shape. The gauge is embedded in the target material in a plane normal to the z axis axisymetrically. Magnetic field is generated by a solenoid wrapped around the target so that the field lines are parallel to the z axis. In a configuration like this, only the radial motion contributes to the EMF, therefore in a uniform field, B, the measured EMF depends on the radial particle velocity: $E = 2\pi B r(t) U_r(t)$, where the circular turn radius, $r(t)$, is obtained by integrating the velocity $U_r(t)$. This new method has been demonstrated by experiment, impacting a rod shaped projectile made of PMMA into a target of the same material. Results are compared with 2-D calculation.

INTRODUCTION

The electromagnetic technique of measuring particle velocity is known for over 15 years[1,2]. Lately this method appeared in a variety of problems: EMV gauges were used by Koller[3] to measure longitudinal and shear waves in compressed Arkansas Novaculite, by Rosenberg & Duvall[4] to monitor stress relaxation in <111> shocked L_iF, by Sheffield and Duvall[5] to determine decomposition of shocked liquid CS_2, and by Gupta et al.[6] to measure longitudinal and transversal motion in PMMA at inclined impact experiments.

Most EMV gauges used were shaped like a short straight active segment with straight leads turned to the back or sides of the target. Applications of these gauges are limited to configurations where the length of the active element remains fixed during the deformation. Such conditions are maintained in planar and inclined-plate impact configurations.

In the present work a modification of the existing EMV gauge is suggested. The new gauge has a circular turn shape, made particularly for measuring the radial particle velocity in 2-D flow. Description of the EMRV (Electro Magnetic Radial Velocity) gauge and the experimental set-up is presented in the next section. Then this method is demonstrated by comparing experimental results with 2-D code calculations.

EXPERIMENTAL PROCEDURE

A problem of a PMMA rod shaped projectile impacting normally on a PMMA flat target was chosen for experimental demonstration of the method. Projectile and target at the moment of impact are illustrated in Fig.1. Velocity is along the z axis, which is also the axis of symmetry. The gauge is made of a thin (0.13 mm) copper wire, shaped like a circular turn of radius r_o. The method of embedding the gauge in the target material is shown in Fig. 2. The wire loop was sited in a premachined groove and the parts were bonded together. Initial coordinates for the gauge location, z_o; r_o, were chosen where a considerable radial flow had been expected from 2-D code calculations.

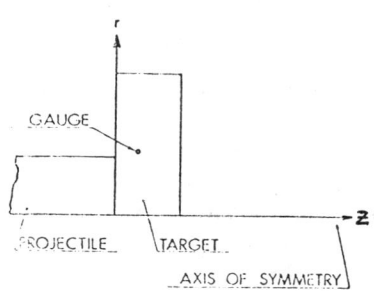

Fig. 1. The impact problem described in cylindrical coordinates.

A magnetic field was generated by a solenoid wrapped around the target. The solenoid was made of 700 turns on a spool of diameter 110 mm and length 200 mm. A field of 630 gauss was generated by 10 Amps driven into the solenoid. The field was uniform over the gauge area, and field lines were parallel to the z axis. Four velocity pins were located prior to the solenoid. The complete set-up is illustrated in Fig.3.

The 2-D flow which follows the impact can be described as a combination of axial motion along the z axis and a radial motion. Any segment along the gauge has a velocity vector which can be split to its radial component, U_r, and to its axial component U_z. Since U_z is parallel to the field lines, only the radial motion, U_r, contributes to the EMF generated at the gauge.

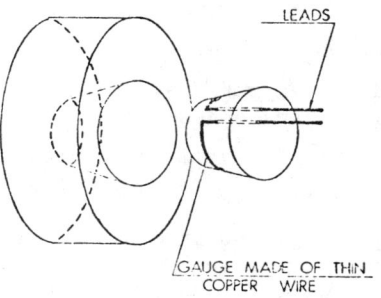

Fig. 2. Target Construction

Thus, in a magnetic field B the measured EMF is expected to be:

$$\varepsilon = 2\pi B r_{(t)} U_r(t) \tag{1}$$

where the radial coordinate $r_{(t)}$ is obtained from experimental data using step by step integration of the radial particle velocity $U_r(t)$.

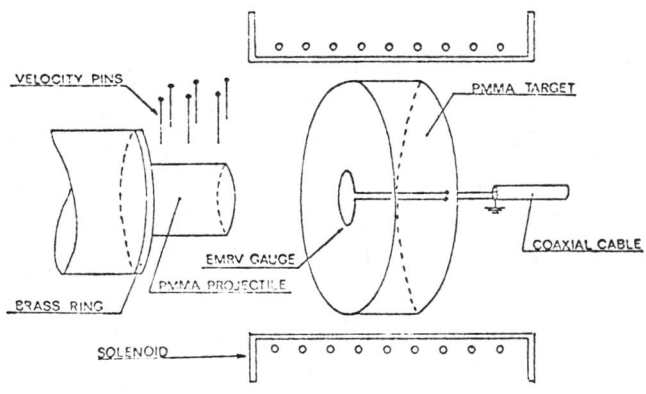

Fig. 3. The experimental set up for EMRV gauge

EXPERIMENTAL RESULTS

Two shots were conducted. Rod and target diameters were 40 mm and 100 mm respectively. In the first shot the EMF signal stopped at about 11 microseconds probably due to a rarefaction which reached the gauge's plane from the target back. For this reason the target thickness was increased from 19 mm in the first shot to 39.5 mm for the second. The signal recorded in the second shot shows that the gauge really lasted for a longer time. Oscillograms of the signal recorded are shown in Figs. 4 and 5 for the first and second shot respectively. Complementary data is given in Table I.

Table I. Complementary data

Shot No.	Projectile Velocity (mm/μs)	Gauge Coordinates (mm)		Magnetic Field (gauss)	Type of Bond
		r_o	z_o		
3-80	0.532 ±0.015	45.9	4.0	625 ±15	Epoxy
1-81	0.456 ±0.016	38	3.6	630 ±15	Liquid plexiglass

DISCUSSION

The EMF records previously displayed show that the EMRV technique is feasible. Application of this method to derive material properties can be done via numerical calculations. The Lagrangian elastic-plastic 2-D code DISCO[7] has been used to obtain a calculated EMF wave profile. This task is not completed, i.e., a good match

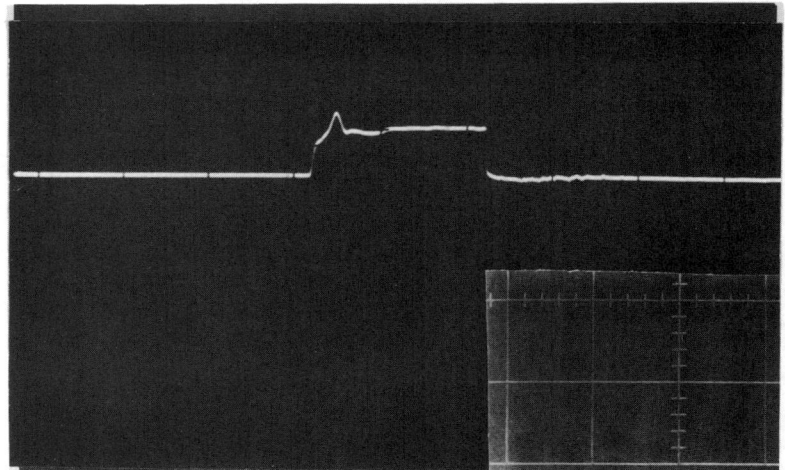

Fig. 4. Signal recorded from shot 3-80.
Vertical 1V/Div. Horizontal 5 µs/Div.

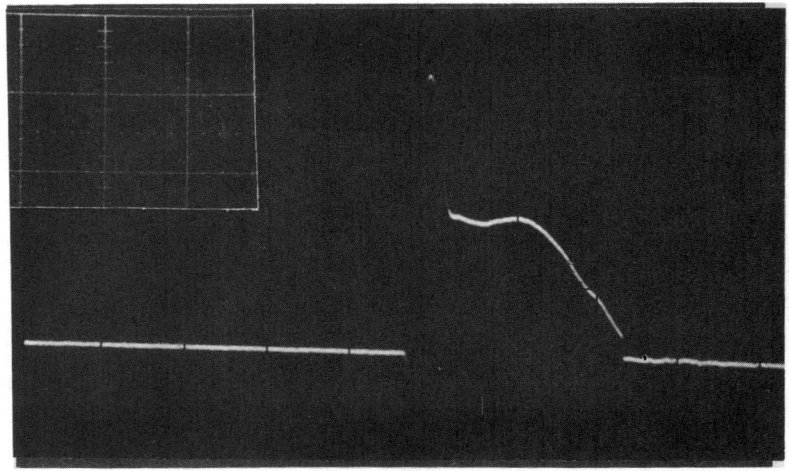

Fig. 5. Signal recorded from shot 1-81.
Vertical 0.2 V/Div. Horizontal 5 µs/Div.

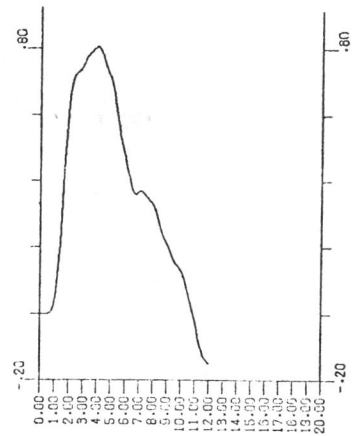

Fig. 6. The calculated EMF for shot number 1-81.

of calculated and experimental wave profiles is not achieved yet. Some information, however, is obtained from several runs already done. We find that the calculated EMF wave profile is very sensitive to the dynamic yield strength, and that the spike seen on both experimental records is related to material flow. An example of a calculated wave profile for shot number 1-81 is shown in Fig. 6.

REFERENCES

1. A. N. Demin & G. A. Adadurox, Sov. Phys. Solid State, 6, 1379 (1964).
2. G. R. Fowles, Dynamic Response of Materials to Intense Impulse Loading, Edited by P. C. Chou & A. K. Hopkins, Air Force Materials Lab., 1972.
3. L. R. Koller, Ph. D. Thesis, Washington State University, 1978. (Unpublished).
4. Gideon Rosenberg & G. E. Duvall, J. Appl. Phys. 51, 319 (1980).
5. S. A. Sheffield and G. E. Duvall, Symposium H.D.P., Paris 1978.
6. Y. M. Gupta et al. Rev. Sci. Instrum. 51, 183 (1980).
7. Y. Kivity & D. Tzur, Proc. of the 4th International Symposium on Ballistics, Monterey, California, 1978.

EFFECTS OF THIN GLUE BONDS ON SHOCK WAVES IN LiF

P. Majewski
Washington State University,* Pullman, Wa.   99164

ABSTRACT

The attenuation of shock waves was measured in a laminate of alternating layers of <100> single crystal LiF and Epon 815 epoxy. Four experiments were performed using samples consisting of five layers of LiF, each layer $\simeq$ 0.5 mm thick, separated by four layers of epoxy, thickness $\simeq$ 7 µm. Electromagnetic velocity gauges occupied the same planes as the epoxy layers. The samples were shocked to 28 kbar using fused quartz impactors. The wave profiles recorded were compared to those obtained in five experiments in which the sample was 2 mm of solid LiF, all other parameters being the same. The attenuation of the elastic precursor from 0.5 to 2 mm behind the impact surface was $\simeq$ 0.02 km/s for the solid LiF and $\simeq$ 0.06 km/s for the laminate. No large scale effect was noted for the plastic wave. The process was modeled using a one dimensional flow code, and it was found that the high frequency components were attenuated much more rapidly than the low frequency components. The effect appears to be due largely to reflections at the LiF-epoxy interfaces.

INTRODUCTION

In the past, not much attention has been given to the effects of glue bonds on the incident shock waves. When the effect was calculated the usual assumption was that the reflected wave did not change the incident wave to any appreciable extent.[1] The result of this assumption was that the glue "rang up" to the incident wave values in pressure and particle velocity in a short time compared to the duration of effects of interest in the incident wave.

Much of the work done in shock dynamics involves the use of rear surface measurements. A major drawback to these methods is that they permit measurements to be taken on only one plane of the sample. A secondary drawback is that the bulk response of the material must be inferred from the data taken at the rear surface of the material. In the case of transducers, this rear surface is not a free surface and impedance matching calculations are needed to reduce the experimentally acquired data.

Embedded gauges overcome these two problems. The underlying assumption being that the material behavior is the same in the presence of a sufficiently small gauge as it would be if no gauge were present. Clearly the validity of this assumption depends upon the physical characteristics of the gauge and the sample.

The main drawback in using these gauges is the necessarily thick glue bond resulting from the gauge separating the layers of the sample. The necessity of attaching leads to the gauge makes it impractical to substantially decrease the gauge thickness. Even at the present gauge thickness, one out of four gauges breaks either during the process of attaching the leads or due to stresses set up in the curing of the epoxy used to pot the signal cables.

Because of the thickness of the gauges themselves, the typical space between layers of LiF in this series of experiments is approximatly 7 microns. Since the gauges take up only a small portion of the area of contact, this space acts like a thick glue bond. In the case of a single gauge, or of the first gauge in a multiple gauge sample, the effect of the glue bond can be ignored, since the gauge itself is in close contact with the LiF. However, in the case of gauges after the first in multiple gauge shots, and in all records of waves reflected from the back surface of the sample, the effect cannot be dismissed.

EXPERIMENTAL TECHNIQUES

The material used in these experiments was ultra-pure LiF doped with magnesium fluoride. The concentration of magnesium fluoride in the melt for the boules used was 160 parts per million. Boules approximately 30 mm in diameter and 50 mm long were purchased from the Crystal Growth Laboratory.[2]

The boules were cleaved along <100> planes perpendicular to the growth axis. Samples of the desired size were normally 0.8 to 2.5 mm thick and about 25 mm square. These samples were then heat treated in accordance with the scheme called Anneal III by Gupta.[3]

Table I Shot Data

Shot No.	Projectile Velocity (Km/sec)	Gauge No.(a)	Gauge Position (mm)	Shot No.	Projectile Velocity (Km/sec)	Gauge No.(a)	Gauge Position (mm)
79-008	0.373	(b)	2.060	79-034	0.418	1	0.620
						2	1.151
79-014	0.414	(b)	1.995			3	1.572
						4	2.100
79-015	0.388	(b)	2.028				
				79-035	0.415	1	0.728
79-025	0.411	(b)	2.012			2	1.228
						3	1.702
79-028	0.416	(b)	2.053			4	2.198
79-033	0.416	2	1.051	79-036	0.416	1	0.822
		3	1.527			2	1.348
		4	2.009				

(a) All multiple gauge shots originally had 4 gauges. Only the gauges that yielded records are listed.
(b) Single gauge shot.

Then the samples were lapped until the desired thickness and degree of parallelism was obtained. The tolerance in parallelism was two microns. Thickness was not held to any definite tolerance. It was found that the <100> axis could be held perpendicular to the sample surface to within one degree.

The samples were then hand polished until all signs of the lapped surface were removed. The polished surfaces were tested against a quartz optical flat, and polishing was continued until three or less interference fringes were obtained.

The gauges used were manufactured by Micro-Measurements, and consisted of a 7 micron copper layer attached to a kapton backing. The attachment of the gauge to the sample and the removal of the backing kapton have been described by Sheffield.[4]

For the experiments requiring a single gauge, after the gauge had been attached, a backing piece of LiF was epoxied over all but the ends of the gauge. The samples for the experiments requiring more than one gauge were made similarly, with the backing for one gauge becoming the substrate for the next. In the multiple gauge samples, the gauge leads were taken out alternate sides.

The preparation of a projectile for use with an EMV gauge was covered in detail by Rosenberg.[5] The only difference in the present work was the use of an 11 mm thick by 25 mm diameter disk of fused quartz as the impacting surface.

Table I gives the pertinent experimental data for the single and multiple gauge shots.

ANALYSIS OF EXPERIMENTAL RESULTS

By considering the records for single gauge shots and for the first gauge of multiple gauge shots, an idea of the behavior of the elastic precursor as a function of distance can be obtained. The line marked A in Figure 1 is a linear fit to these data. The actual precursor decay function (at least for pressure) does not appear to be linear, however the quality of these data does not permit a more accurate fit. Since the linear fit is used only as a rough guide in this work, it is sufficient.

Also in Figure 1 is shown the precursor amplitude as a function of distance for the multiple gauge shots. The numbers in this figure are the last two digits of the shot number. It is clear that all gauges except the first in each shot give a low value for the precursor compared to the linear fit.

From the figure, it appears that the drop in the precursor amplitude is greater at the first and second gauges than it is at the third. A study of the shot records showed that the precursor is generally much sharper at the first two gauges than it is at the last two. It appears that the duration of the precursor has some effect on the amount of attenuation it receives in the glue.

In order to better understand the observed results, a LiF and Epon 815 target was modeled using a one dimensional artificial viscosity flow code.[6] The thickness of the glue bond was kept at 7 microns for all the runs of the code. In the first run, LiF was approximated as an elastic material using the constitutive relation

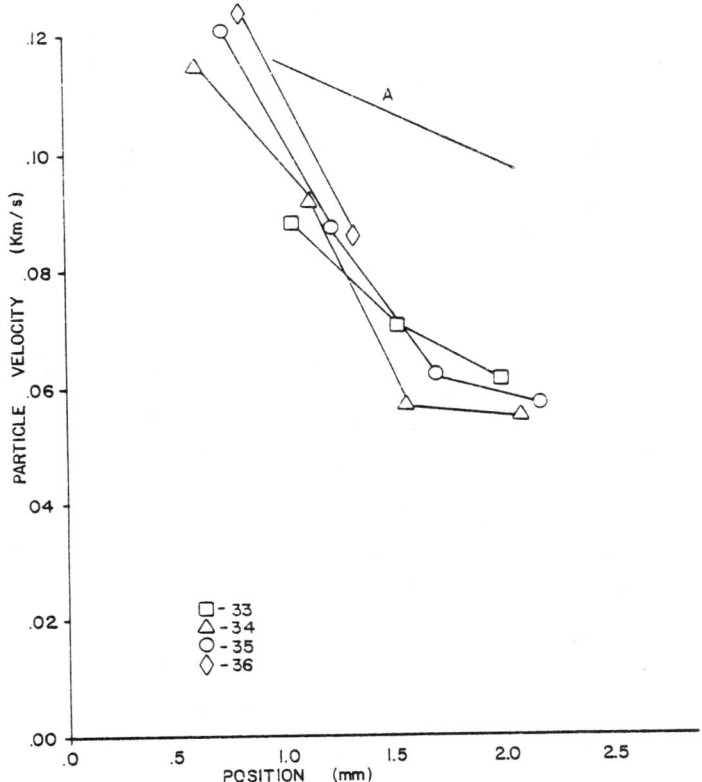

Figure 1. Elastic precursor decay including glue bond effects. The line labled A is a linear fit to the decay without glue bonds. The numbers are the last two digits of the shot number.

given by Asay.[7] The results of this calculation agree with previous results in that the amplitude of the wave remains constant, and the leading edge is only slightly rounded. A strong reflection from the LiF-glue interface was seen at approximately 0.13 microseconds.

In subsequent runs of the code, an elastic-plastic constitutive relation derived by Bjorkman was used for the LiF.[1] Two cases were examined; one with the glue bond located 0.5 mm from the impact surface, the other with the glue bond at 2.0 mm.

In both cases, it appears that the plastic portion of the wave remains relatively unchanged. The elastic precursor, and the stress relaxation region behind it, show a pronounced difference in passing through the glue bond. In both cases, the precursor is attenuated and rounded. The stress relaxation region is shallower. These are the same effects noted in the second through fourth gauges of the multiple gauge shots.

CONCLUSIONS

It appears that the effect of the glue bond is determined to a large extent by the rate of change of the incoming wave. This is reasonable since the values of pressure and particle velocity in the bond in a sense lag the values at the incoming interface. This would cause the values at the outgoing (second) interface to be somewhat time averaged. In the case of a rapid variation in the incoming wave, the outgoing wave would be considerably different. If the incoming wave varied slowly or not at all, the outgoing wave would not show much difference.

The effect of the glue bond is to time average (in some sense) the incoming wave. Another way of looking at this, is to consider the glue bond as a low pass filter. The low frequency components of the incident wave are passed unchanged, the higher frequencies are attenuated. Both means of considering the situation lead to the same conclusions, namely that the rapidly changing aspects of the incoming wave are changed or lost beyond the glue-LiF interface.

No attempt was made to correlate the amount of attenuation with the thickness of the bond. Since the rapidly changing portion of the incident wave was the elastic precursor, and since the precursor decays with distance in LiF, the amount of attenuation ascribable to the glue bond was not clearly identifiable. Also, the range of thicknesses of the glue bonds involved was small. However, if the attenuation is a function primarily of the thickness of the bond, then in the case of very thin bonds the attenuation should be negligible. This conclusion assumes that reflections from the glue-LiF interface also decrease as the bond thickness decreases. The validity of this assumption could be tested in two ways. One way would be to make a sample consisting of a LiF-glue laminate, using a bonding method that gave very thin bonds. By measuring the particle velocity at some plane past the bonds and comparing this to the particle velocity obtained at the same position in a solid LiF sample, the effect of thin bonds could be determined directly.

Since the waves reflected from the LiF-glue interfaces are very easily seen in the multiple gauge shots, a second method to test the assumption of small reflections would be to place a gauge in a plane between the impact surface and a thin glue bond. In this manner the reflections from the glue bond would be directly observable.

*This work was supported by N.S.F. Grant No. DMR79-06122.

REFERENCES

1. M. D. Bjorkman, Ph. D. Thesis, W.S.U., Pullman, Wa. (1980).
2. Crystal Growth Laboratory, Department of Physics, University of Utah, Salt Lake City.
3. Y. M. Gupta, Ph. D. Thesis, W.S.U., Pullman, Wa. (1973).
4. S. A. Sheffield, Ph. D. Thesis, W.S.U., Pullman, Wa. (1978).
5. G. Rosenberg, Ph. D. Thesis, W.S.U., Pullman, Wa. (1978).
6. J. Von Neumann and R. D. Richtmyer, J. Appl. Phys. $\underline{21}$, 232 (1950).
7. J. R. Asay, Ph. D. Thesis, W.S.U., Pullman, Wa. (1971).

EJECTION OF MATERIAL FROM SHOCKED SURFACES OF TIN, TANTALUM AND LEAD-ALLOYS

P. Andriot, P. Chapron, F. Olive[*]

C.E.A. - B.P.7 - 93270 SEVRAN - France

ABSTRACT

We present an experimental study of some mechanisms of material ejection from free surfaces during the reflection of a strong shock wave. Basic phenomenon is momentum transfer from the ejecta to a thin foil whose velocity is recorded by means of a Doppler Laser Interferometry technique. Geometrical defects of the surfaces appear to be an important source of jetting. A marked influence of surface melting has also been evidenced. Moreover, in some materials, density inhomogeneities and localized fusion can also significantly contribute to ejection.

INTRODUCTION

It has been known for a long time that a strong shock reflecting at a free surface can cause ejection of material from the surface. These ejected particles often induce inconvenient perturbations in various experimental techniques such as high speed photographic observations and free surface velocity measurements. Thus, in the last few years, a number of experimental[1,2] and theoretical[3] studies were performed in order to understand the physical mechanisms responsible for material ejection.

In this paper, we present our own experimental results obtained with different materials. The main investigated parameters are surface geometrical defects, density inhomogeneities and localized fusion.

EXPERIMENTAL TECHNIQUE

Our experimental technique for measuring flux of ejected mass versus time is based upon the principle of momentum transfer between the ejecta and a thin foil whose distance to the free surface is initially specified. The main necessary assumptions are : instantaneous ejection during shock reflection at the surface ; uniform distribution of ejected mass perpendicularly to shock propagation direction ; completely inelastic collision between ejecta and the thin foil.

We present in figure 1 a sheme of our experimental device, while figure 2 sums up the momentum transfer interpretation and the corresponding equations.

---

[*]Communication presented by F. Olive.

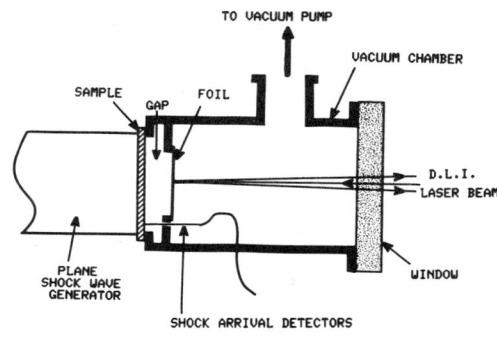

Figure 1 : Experimental device.

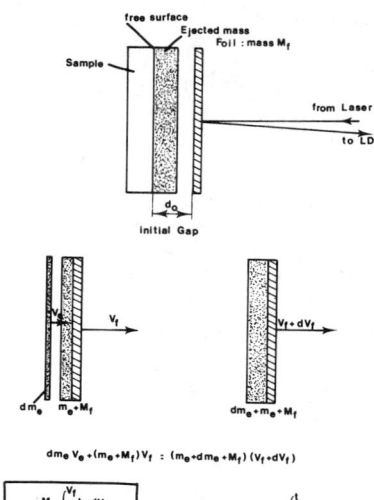

Figure 2 : Principle of momentum transfer interpretation.

Initial time is determined by shock arrival detectors placed on the free surface, coupled with a digital chronometer. Foil velocity is recorded by means of a Doppler Laser Interferometry system[4,5] associating an adjustable Fabry-Perot interferometer and a streak electronic camera. This mass measurement method has been tested on a predetermined ejection : a well defined thin layer of epoxy resin deposed on a polished copper plate. Shock wave reflection in this low shock impedance resin induces its complete spallation associated with a depolymerization. Thus we obtain, quasi-instantaneously, a cloud of particles flying ahead of the copper plate.

In figure 3 we show the results of a mass measurement performed on such a device. The experimental values for total ejected mass are quite in good agreement with the predetermined resin mass.

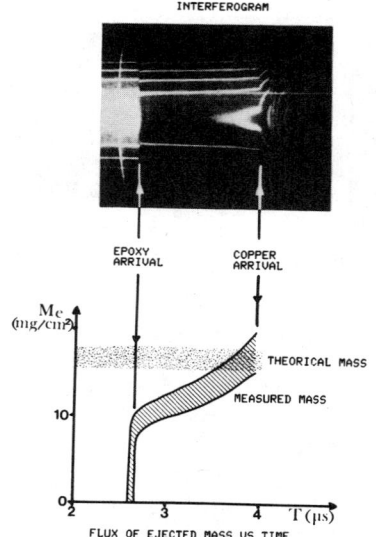

Figure 3: Test of mass measurement method on a precalibrated epoxy ejection.

INFLUENCE OF SURFACE GEOMETRICAL DEFECTS

Tantalum samples: We compared mass ejections occurring under same loading (shock pressure P=60GPa) for tantalum samples with two different free surface preparations : polishing (no scratch or pit larger than 0.1μm) and machining with parallel grooves (arithmetic average deviation from centerline $x_{AA} = 1.7\mu m$).

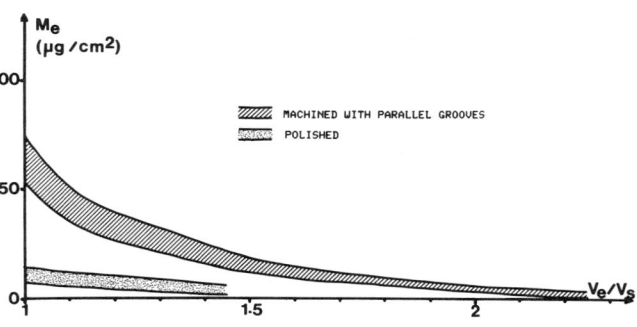

Figure 4 : Ejected masses from two tantalum surfaces.

The corresponding results are displayed in figure 4: the amount of specific mass m_e travelling faster than velocity v_e is plotted versus normalized velocity v_e/v_s.

Tin samples : Four different surface preparations of tin samples were tested using the same shock loading(P=50GPa) rough cast ($x_{AA} = 7\mu m$) ; roughly machined with parallel grooves ($x_{AA} = 5\mu m$) ; finely machined with parallel grooves ($x_{AA} = 0.5\mu m$) and polished ($x_{AA} = 0.02\mu m$).
Another roughly machined sample, identical with the preceding one, was submitted to a weaker shock wave (shock pressure P=20 GPa). In Figure 5, the flux of ejected mass appears related to the surface rugosity : the total amount of mass, under a 50 GPa shock wave, varies from 1 mg/cm² for a polished surface up to 25 mg/cm² for a roughly machined one. In those experiments free surface melting occurs during the shock reflection. For a weaker shock wave (20 GPa), where the material remains solid, the mass flux -with the same surface preparation- is reduced by a factor of 3.

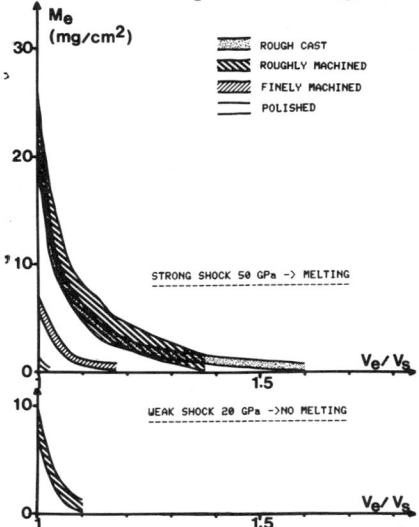

Figure 5 : Ejected masses from different tin surfaces at two shock levels.

Discussion : Our experimental results for tin and tantalum clearly point out the importance of microjetting from surface geometrical defects already evidenced by J.R. Asay[1,2]. We also find a marked influence of free surface melting on the jetting process. We have tried to apply Asay's jetting model[3] to our results : In table 1 we compare experimental and theorical total ejected masses for our two materials when remaining solid after shock reflection. The jetting factor R is assumed to be 1 and the total cavity volume per surface unit for the saw tooth grooves is estimated from the arithmetic average deviation from centerline x_{AA}.

Table 1 : Comparison between experimental results and Asay's model[3].

Material	Initial density	Shock pressure	x_{AA}	Theorical mass	Experimental mass
Tin	7.2 g/cm^3	20 GPa	5µm	7 mg/cm^2	10 mg/cm^2
Tantalum	16.6 g/cm^3	60 GPa	1.7µm	6 mg/cm^2	0.05 mg/cm^2

In the case of tin, Asay's model provides a pretty good estimation of the total ejected mass, but for tantalum the predicted mass is approximately 100 times greater than the measured one. We presently have no explanation for this notable disagreement.

INFLUENCE OF DENSITY INHOMOGENEITIES

Samples of two different SnPb alloys (One SnPb 14 weight % and one SnPb 38 weight %) were elaborated in order to study the influence of density inhomogeneities. Those two alloys exhibit a two-phase structure composed by pure tin grains (density $\rho = 7.2$ g/cm^3) about 100µm in diameter included in an entectic SnPb matrix of higher density. Free surface were carefully polished (no scratch or pit larger than 0.1µm). In figure 6 we compare the ejected mass flux for these two alloys and for polished pure tin under the same 50 GPa shock wave. The total amount of ejected mass appears to increase significantly with increasing density of inhomogeneities. However the first detected particles remain close to the free surface ($v_e/v_s \leqslant 1.1$). A mechanism of spalling based upon shock impedance difference between pure tin and SnPb entectic could reasonably explain those mass values and velocity distributions.

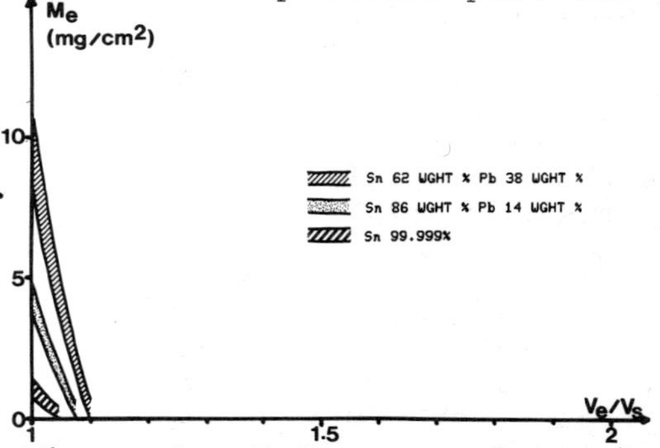

Figure 6 : Influence of density inhomogeneities on mass ejection.

INFLUENCE OF LOCALIZED FUSION

Two CuPb alloys (one Cu Pb 15 weight % and one CuPb 36 weight %) were developed for this particular study. Their structure also presents two phases : copper grains around 100μm in diameter in a Cu Pb entectic matrix.These samples and a pure copper one were carefully polished.

In figure 7 the corresponding ejected mass flux measurement for a 50 GPa shock wave are plotted. Important mass ejection for the two Cu Pb alloys should be noted, specially when compared with those for pure Copper. Moreover the first particle velocities almost reach twice the free surface velocity. When shock wave reflects at the free surface, the copper grains remain solid while the Cu Pb entectic matrix probably melts. This mechanism could cause instabilities to develop during surface acceleration and then provide an important source of ejection.

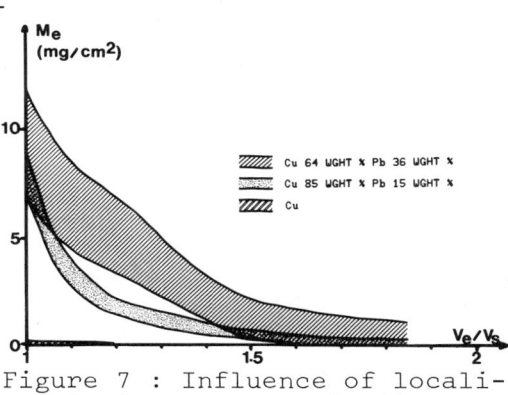

Figure 7 : Influence of localized fusion on mass ejection.

CONCLUSION

Three different sources of material ejection during the reflection of a strong shock wave at a free surface have been pointed out in our experiments. The first one is microjetting from geometrical surface defects, already evidenced and modeled by J.R. Asay[2,3]. A second one is spallation of lower shock impedance inclusions. The last one is localized fusion whose contribution to fast ejected mass should not be neglected in some alloys.

ACKNOWLEDGEMENTS : The authors wish to thank S.Bedere and R.Cottin for their participation in the metallurgical parts of this work .

REFERENCES

1. J.R.Asay, L.P.Mix and M.C.Perry - Appl. Phys.Lett. Vol.29, n°5, sept.1976.
2. J.R.Asay, SAND 77-0731, 1977.
3. J.R.Asay, L.D.Bertholf, SAND 78-1256, 1978.
4. M.Durand and al. Rev.Sci.Instrum.,vol.48,n°3,March 1976
5. P.Andriot, P.Chapron, B.Laurent, F.Olive. Communication 5th intern.congress Laser 81 Opto-Elektronik (MUNCHEN) June 1981.

SHOCK ATTENUATION IN AN
INERTIAL CONFINEMENT FUSION REACTOR

L. A. Glenn
University of California, Lawrence Livermore National Laboratory
Livermore, California 94550

ABSTRACT

It has been proposed that the first wall of an inertial confinement fusion reactor be protected by a thick blanket of liquid lithium surrounding the exploding DT fuel pellet. Several schemes have been analyzed in detail and the results of this work are reviewed.

INTRODUCTION

Depending on the specific design, an ICF reactor will generate from 200-4000 MJ of thermonuclear energy per pulse, at repetition rates of from 1-20 Hz. The impulse from such an explosion is, to first approximation, proportional to the square root of the pellet debris mass and is therefore much smaller than that from a TNT explosion of the same energy; an early estimate gave the fusion explosive mass as $<10^{-6}$ that of the TNT. Nevertheless, a 4000 MJ TNT explosion involves almost a metric ton of that substance, so that our ICF reactor must still withstand up to the equivalent of 1 kg of high explosive detonated each second for, say, 30 years-- the estimated lifetime of the reaction chamber.

Most of the energy (from 65-75%) derived from the pellet is in the form of high energy (up to 14 MeV) neutrons. To moderate these neutrons, and to reduce the fluence on the first wall, it has been proposed to employ a thick, flowing annulus of liquid lithium between the pellet and wall.[1] The lithium would serve as well to breed tritium for pellet re-supply, to act as an energy sink and heat exchange medium with an external power loop, and as a shock absorber to protect the wall from excessive blast effects.

LONG-RANGE ENERGY DEPOSITION

The energy would not be uniformly distributed within the lithium. The mean free path for 14 MeV neutrons in liquid lithium[2] is ~ 0.3 m and calculations have shown that, for 4000 MJ, acceptable wall fluence levels obtain with liquid thicknesses of ~ 1 m.[3] The energy release is virtually instantaneous in the sense that when the DT pellet is imploded to the thermonuclear 'burn' condition, the neutron penetration throughout the liquid occurs in a time period small in comparison with the time for release waves to move into the bulk of the fluid from any free surfaces. The liquid is therefore heated at constant volume, producing pressures as high as 1 GPa at the inner annular radius; at the outer annular radius, the initial pressure may be lower by two orders of magnitude. Rarefactions centered at these two locations then move into the interior fluid, reflect at roughly the geometric mid-point of the annulus, and produce two slugs of fluid, one moving inward, and the other outward towards the first wall. As the initial pressure gradient would indicate, calculations have shown that the velocity distribution within these slugs is decidedly non-uniform and that liquid fragments can reach the wall with velocities high enough to cause severe erosion, if not direct hoop failure due to impact loading.[4] The situation can be very much improved if the continuous annulus is replaced with a series of concentric annuli, separated by gaps. The reason is that the gaps allow pressure relief from surfaces which are interior to the main annular boundaries. This promotes momentum exchange between fluid elements moving in opposite directions, and effectively smears or averages out the velocity distribution in the resulting outward-moving fluid. Moreover, the concentric ring design should reduce the impulse delivered to the wall, as well. The inability of liquids to support significant tensile stresses will result in the formation of incipient cavities in the outward-moving segments of each ring. When the segments contact the inward-moving segments, shock compression of the "porous" liquid will dissipate energy, and the collision pressure can accordingly be much less than in the corresponding slab impact.

The physical picture may be clarified somewhat by the P-V diagram sketched in Fig. 1, in which we assume uniform energy density everywhere. Point A corresponds to the initial state attained upon energy deposition. The fluid is then isentropically expanded along the path ABC. Point B corresponds to the contact point for a nondivergent (slab) impact. For the concentric ring

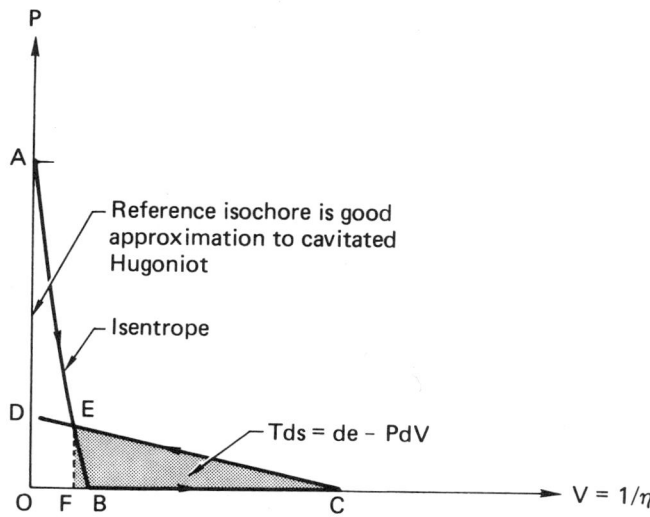

Fig. 1 Disassembly and collision of two concentric liquid rings, sketched in the pressure-volume plane.

design, collision occurs at point C. The path CD is a Rayleigh line, and no intermediate states are possible. Conservation of energy requires that the areas AEBOD and CDO be equal; it is therefore clear that the collision pressure, P_D, decreases rapidly as V_C increases. The rebound velocity varies directly with this pressure, so that an inelastic collision results and the outward-directed momentum (which constitutes the impulse imparted to the first wall) is reduced. Calculations indicate that reduction of at least a factor of 5 can be attained with practical designs.[5]

SHORT-RANGE ENERGY DEPOSITION

Thus far we have considered only the energy transported from the pellet by neutrons. From 25-35% of the total output is however normally in the form of x-rays and ionic debris which, in contrast to the neutrons, would penetrate only a very thin layer of the liquid lithium at the innermost annular radius. As a result, an intense shock, with a peak pressure on the order of 100 GPa initially forms in this layer. The shock then moves into the body of the inner ring, but is quickly attenuated by a rarefaction centered at the free inner surface. This rarefaction typically traverses the initial penetration layer in a few nanoseconds. In so doing, only a small fraction of the short range energy deposition couples to the main body of liquid; most of the energy is blown back into the inner cavity in the form of lithium gas kinetic energy. The blow-off is primarily directed inward and the flow collapses on the cylinder axis with leading edge velocities that may exceed 100 km/s. The implosion creates a hot, high-pressure lithium plasma, which then expands out against, and imparts momentum to, the collection of liquid rings. The process is Taylor-unstable, so that the plasma will eventually break through the liquid, but not before the latter is accelerated to unacceptably high radial velocity.[4]

This problem may be eliminated by substituting for the concentric rings a close-packed annular array of discrete jets;[6] an end view of a typical design is shown in Fig. 2. With this arrangement, the plasma can flow around, or in between, the jets and vent the cavity. In order for the discrete jets to function properly, the drag exerted on them by the venting plasma must not cause too great an impulse to be delivered since this impulse will eventually be transmitted to the wall. Moreover, the plasma must transfer energy to the liquid at a rate sufficient to preclude excessive gas pressure build-up on the cavity wall. Finally, the plasma must eventually cool and condense on the liquid surfaces, or be removed from the cavity, before the next fuel pellet is injected, typically in a second or less.

We devised a quasi one-dimensional method for calculating the drag impulse exerted on the jet array by the venting plasma.[7] The method was based on the topological correspondence between an impulsive divergent flow, constricted by the presence of jets, and an equivalent flow between intersecting symmetric surfaces. The

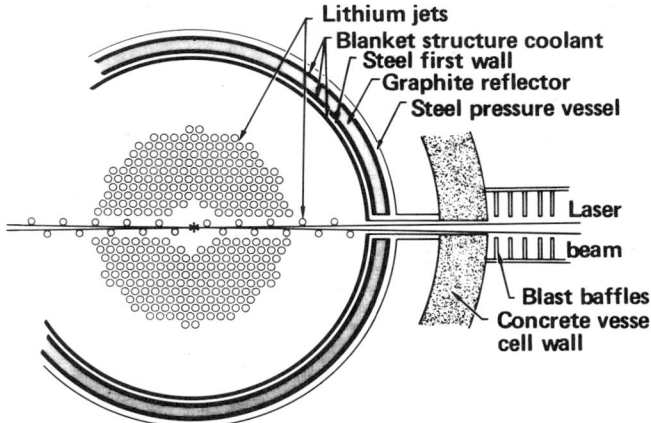

Fig. 2 End view of the jet array in a typical ICF reactor design.

impulse delivered to the jets is due partly to skin friction and partly to profile drag, and both of these effects were accounted for, as was radiation flow in the plasma. The method was subsequently extended to include heat and mass transfer between plasma and liquid[8] so that the direct impact of the plasma on the wall could be studied. The calculations showed that the crosswise momentum imparted by the plasma to the jets represents only a small fraction of the mainly-neutron-induced total momentum that arrives at the wall (the resulting impulse is readily withstood by conventional wall design).

Heat transfer from the plasma to the jets (mainly by radiation) was found to have a significant role in reducing the pressure build-up by direct impact of the plasma. Depending on the jet break-up*, up to 3/4 of the energy initially contained in the plasma diffused into the liquid jets in the wall loading period. The peak hoop stress under these conditions was calculated as 160 MPa (for a 10 cm thick solid steel wall with a radius of 5 m). This is quite adequate for developmental purposes, but probably still too high for full cycle duty ($\sim 10^9$ cycles over a 30 year period) unless the wall is regeneratively cooled. It should be noted, however, that our quasi one-dimensional analysis of necessity ignores axial pressure relief. For initial conditions, we conservatively employed for the radial energy density profile that which obtains at the axial plane of the fusion pellet, which is approximately 10 times the overall energy density in the cavity. Our results then provide only an upper bound on the wall loading; a more accurate estimate can presumably be made with higher dimensional calculations, and these are in progress.

*In the time period of interest, at least the inner rows of jets will entirely disassemble as a result of isochoric neutron heating.

Cooling, and eventual condensation of the plasma on the lithium jet fragments, was shown to be governed mainly by forced convection. We found that, as long as the average jet fragment size was less than ~ 25 mm, the plasma would cool and condense to the limiting vapor pressure required for (laser driver) beam propagation in the time it took for the jet fragments to arrive at the cavity wall. Moreover, this time period is at least one order of magnitude less than that required for the envisioned pulse repetition rate of 1 Hz.

CONCLUDING REMARKS

We have analyzed the problem of protecting the first wall of an inertial confinement fusion reactor. Our analysis indicates that a close-packed annular array of liquid lithium jets surrounding the DT fuel pellet affords adequate protection from excessive neutronic and hydrodynamic loading. The equivalent of a 1 meter thickness of lithium should suffice for 4000 MJ micro-explosions repeated at 1 second intervals (to yield at least 1000 MW net electric power, representative of an economically viable power station).

The sensitivity of our results to important, but difficult to estimate, parameters (such as the plasma emissivity and disassembled jet particle size distribution) was determined insofar as this was possible. Moreover, we were careful, in performing this study, to make very conservative assumptions, where required. Nevertheless, it must be strongly emphasized that there have been no experiments to confirm or refute our analysis. The relevant experiments (e.g., measurements of the particle size and spatial distribution derived from isochorically heated liquid lithium, or of the opacity of lithium, especially near the condensed phase) are difficult and costly and, thus far, there has been insufficient support for their undertaking.

REFERENCES

Work performed under the auspices of the U.S. Department of Energy by Lawrence Livermore National Laboratory under contract #W-7405-Eng-48.

1. J. Maniscalco et al, Lawrence Livermore National Laboratory Report UCRL-52349 (1977).
2. E. F. Plechaty, D. E. Cullen, R. J. Howerton and J. R. Kimlinger, Lawrence Livermore National Laboratory Report UCRL-50400, Vol. 16, Rev. 2 (1978).
3. W. R. Meier, Nuclear Technology 52, 22 (1981).
4. L. A. Glenn and D. A. Young, Nucl. Engrg. Des. 54, 1 (1979).
5. L. A. Glenn, Nucl. Engrg. Des. 60, 327 (1980).
6. M. Monsler et al., Trans. ANS 30, 21 (1978).
7. L. A. Glenn, Nucl. Engrg. Des. 56, 429 (1980).
8. L. A. Glenn, Lawrence Livermore National Laboratory Preprint UCRL-85061 (1980); accepted for publication in Nucl. Engrg. Des.

CAVITATION IN WATER INDUCED BY
THE REFLECTION OF SHOCK WAVES

P. L. Marston and G. L. Pullen*
Physics Department, Washington State University
Pullman, Washington 99164

ABSTRACT

Negative pressures were produced in distilled water by reflecting 5 MPa - 7 MPa shock pulses off a quasi-free water-Mylar-air interface. An interferometer measured the displacement history of the interface. After the pulse's trailing edge reaches the Mylar, the velocity does not tend toward zero as it would for an elastic solid. The velocity reduction or "pullback" was approximately 4 m/s. An approximation is derived which relates the pullback to the tensile strength. It appears that the Mylar separated from the water during pullback and the data places only a lower limit (2.8 MPa or 28 bars) on the tensile strength.

INTRODUCTION

The tensile strength of distilled, deionized, or degassed water at room temperature has been measured by several investigators.[1-4] The largest strength reported, 43 MPa, was for shock induced cavitation.[1] The strength was indirectly inferred from the motion of a PMMA target which was struck by a spalled water layer. Other shock experiments[2] give strengths $\simeq 2$ MPa. An acoustic cavitation experiment (with highly purified water) gave a strength of 21 MPa while others give 2 MPa. A strength of 28 MPa was measured using a spinning capillary[4]; however, there was an anomalous reduction in strength at lower temperatures.

A theory for the homogeneous nucleation of bubbles in metastable liquids is known to be useful for certain clean, superheated,[5] and/or tensilely stressed[6] liquids. The theory, which attributes nucleation to thermally stimulated crossing of an energy barrier, predicts a static strength $\simeq 150$ MPa for room temperature water.[1,6] Because the theory does not describe observations for water, nucleation is thought to occur at impurities. Models of heterogeneous nucleation[7] successfully describe acoustic cavitation at low frequencies; however their quantitative application to shock induced cavitation[1,2] is complicated due to the dynamics of cavity growth. Consequently, it is desirable to measure the early-time response of water to well defined pulses of negative pressure.

This paper describes the first attempts to use conventional impact techniques to study cavitation in water.

---

*Present address: Naval Undersea Warfare Engineering Station, Keyport, Washington 98345

THEORY

Modeling of the experiment shown in Fig. 1 is facilitated by considering the waves induced by projectile impact. The impactor and buffer were either both PMMA or both aluminum. The pressure (p) and particle velocity (u) states may be approximated[8] by taking the material impedances to be the acoustic impedance $z_n = \rho_n c_n$ of the impactor ($n = 1$), buffer ($n = 2$), and water ($n = 3$). States generated are shown in Fig. 2 for a projectile velocity $u(\hat{A})$ of 10 m/s with a PMMA impactor; p_0 is the initial pressure (0.1 MPa for the water). The longitudinal coordinate (x) and time (t) description of states is shown in Fig. 3(a) for the case of an aluminum impactor. Shocks (upward p transitions) are denoted by solid lines, rarefactions by dashed lines. The sequence parallels that in Fig. 2 and corresponding states are denoted by capital letters; primes denote states induced by cavitation. Fig. 3(b) shows the predicted (u,t) history of the air-water interface. The material thicknesses Δx_n for Fig. 3 were 3.3 mm, 7.5 mm, and 3.8 mm for $n = 1$, 2, and 3.

Behind the water (Fig. 1) was a 25 µm thick aluminized Mylar film. The time for a wave to travel across the Mylar $\simeq 0.01$ µs; consequently, ring-up and ring-down of the Mylar was not included in Fig. 2 and 3 which assume the pressure at the water surface is p_0.

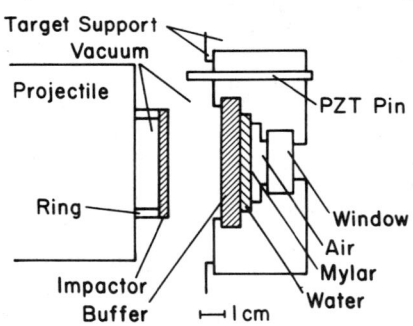

Fig. 1 Impact experiment

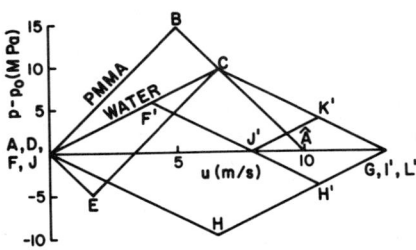

Fig. 2 Wave transitions in the (p,u) plane. The slopes are the acoustic impedances

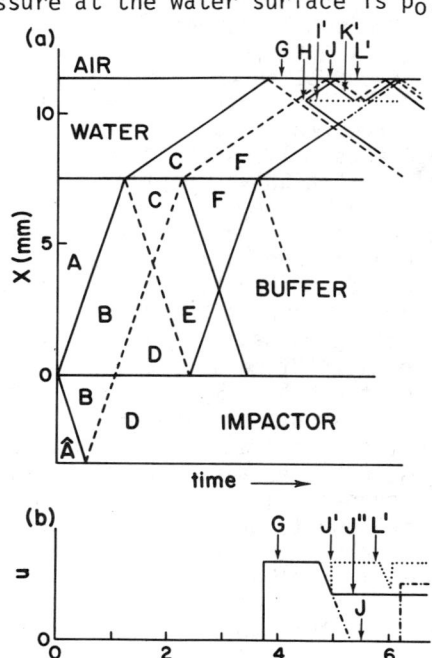

Fig. 3(a) (x,t) diagram and (b) free-surface (u,t) diagram. The time axes in (a) and (b) correlate.

That Fig. 2 and 3 give the essential features of states (in the absence of cavitation and Mylar effects) has been verified with a flow code. The rarefaction which terminated G in Fig. 3(b) has been idealized; E is terminated after a rarefaction pulls the buffer away from the impactor.

Intersection of the C→G rarefaction with the relatively gradual C→F rarefaction produces a transition to a state of tension H if there is no cavitation. Intersection first occurs at a distance $\frac{1}{2}\Delta t c_3 \simeq 0.8$ mm from the back surface where the pulse length $\Delta t = 2\Delta x_1/c_1 = 1.1$ μs. Cavitation would be expected to occur in this plane first when the intermediate state H' reaches the tensile strength. If we neglect the time for cavity growth, p is reduced to near zero (state I') via a radiated upward transition (assumed to be a plane wave). The surface transition toward $u(J) = 0$ is terminated at $u(J')$ when that wave reaches the surface and $u \rightarrow u(L') \simeq u(G)$. Inspection of Fig. 2 gives for the tensile strength:

$$-p(H') \simeq \frac{1}{2} \rho_3 c_3 [u(G) - u(J')] - p_o \qquad (1)$$

Equation (1) has been applied to spall data for solids.[9]

EXPERIMENTS

The Mylar displacement was measured with a laser interferometer operating with a wavelength $\lambda = 632.8$ nm and monitored with a phototube. The interferometer (Fig. 4) was similar to a Fizeau interferometer in that the reference beam was the reflection from the flat window surface closest to the Mylar. The output voltage V cycled for each displacement of $\lambda/2$. Oscilloscopes were triggered either with a filtered version of V (low frequencies rejected) or with a PZT Pin. The projectile and target were optically aligned and tilt measurements made with a substitute target gave impact tilts $\theta \simeq 0.5$ mrad. Due to the small value of $u(\hat{A})$, the projectile was driven by atmospheric pressure. Electrical pins gave measurements of $u(\hat{A})$ consistent with predicted[10] projectile dynamics.

The water sample was distilled and degassed. Dissolved gases were driven from the water by holding it in a low-pressure flask while the flask was immersed in an ultrasonic cleaner. The water was filtered through a 0.2 μm Millipore filter. The water temperature was typically 22°C prior to impact.

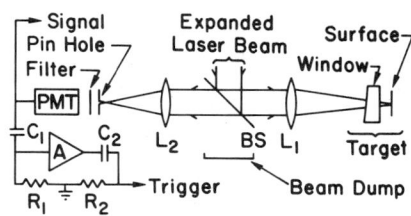

Fig. 4 Displacement Interferometer

RESULTS AND DISCUSSION

Table I summarizes the shot data. The value of $p(C) - p_o$ on the left was computed with a (p,u) diagram from measured $u(\hat{A})$ [except for shot 4 where $u(\hat{A})$ was inferred from launch data]; the value on the right was computed from the measured u(G). The discrepancy may be attributed to tilt. The impactor and buffer were aluminum except in shot 4; Δt is computed from the impactor properties.

Fig. 5 is a velocity record which is representative of shots 4, 6, and 8. The dots and triangles were obtained by differencing the displacement times at opposite extremes of V; t = 0 corresponds to the start of the sweep which was near the midterm of state G. The end of G is at $t \simeq 1$ μs. The salient feature of shots 4, 6, and 8 is that the velocity is closer to the solid curve in Fig. 3(b) than either of the modeled options. The absence of the predicted large rebound in u suggests that the Mylar has separated from the water and moves away in a nearly force free environment. Because the Mylar's $z \simeq 2.4\ z_3$, a small amount of tension may be created at the water-Mylar interface which leads to separation. A similar separation of hi-z gauges from the back of low-z elastomer samples has been observed.[11] Due to the separation, -p(H') [Eq. 1 with u(J") replacing u(J')] must be interpreted as a lower limit on the tensile strength. If the water was not at least this strong, the rebound signal would have arrived prior to the observed Mylar separation. If water's response to tensions > 3 MPa is to be investigated, the Mylar must be eliminated.

Shot 7 was anomalous in that the Mylar moved away with u(J") > u(G). Evidently the tilt was large since u(G) was only 53% of the value expected for parallel plane impact with the measured u(Â). The tilt may be due to a technical problem unique to this shot.[12]

In summary, we have presented a model for surface motion associated with rapid cavitation in the plane at which tension first appears. The experiments (shots 4 and 6) place a lower limit $\simeq 2.8$ MPa on the dynamic tensile strength of moderately clean water. If bubbles are to nucleate and grow significantly on a submicrosecond time scale, the tension induced by reflection of shock pulses should exceed this dynamic strength. The technique described here may be applicable to hydrocarbons and other

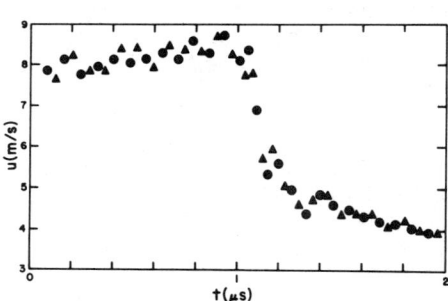

Fig. 5 Mylar's u from the displacement record of shot 4.

Table I Shot Data

Shot No.	u(Â) (m/s)	Δt (μs)	p(C) - p₀ (MPa)		u(G) (m/s)	u(J") (m/s)	-p(H') (MPa)
4	(6.2)	2.2	(6.2)	6.2	8.5	4.5	2.9
6	5.9	1.1	8.1	7.0	9.5	5.7	2.7
7	4.8	1.5	6.6	3.5	4.7	6.0	-
8	5.4	1.5	7.3	5.3	7.1	4.7	1.7

liquids for which a low strength is anticipated. We recommend, however, that if the liquid is to be initially confined by a thin film, the acoustic impedance of the film should be less than that of the liquid.

Acknowledgement is made to the Donors of the Petroleum Research Fund (administered by the American Chemical Society), the Washington State University Research and Arts Committee, and the Office of Naval Research, for partial support of this research. P. L. Marston is an Alfred P. Sloan Research Fellow.

REFERENCES

1. A. N. Dremin, G. I. Kanel, and S. A. Koldunov, "Investigation of Splitting off in water, ethyl alcohol, and plexiglass," in Third All-Union Symposium on Combustion and Explosion, Moscow, 1972, (NTIS Translation, No. AD/A-001 861).
2. Y. Urabe, T. Otani, and K. Komatsu, J. Phys. (Paris) $\underline{40}$, C8-315 (1979); B. E. Richards, D. H. Trevena, and D. H. Edwards, J. Phys. D. $\underline{13}$, 1315 (1980).
3. M. Greenspan and C. Tschiegg, J. Res. Nat. Bur. Std. $\underline{C71}$, 299 (1967).
4. L. J. Briggs, J. Appl. Phys. $\underline{21}$, 721 (1950); $\underline{26}$, 1001 (1955).
5. M. Blander, Adv. Coll. Interface Sci. $\underline{10}$, 1 (1979).
6. R. E. Apfel, J. Acoust. Soc. Am. $\underline{49}$, 145 (1971); Nature (Phys. Sci.) $\underline{233}$, 119 (1971).
7. L. A. Crum, Nature $\underline{278}$, 148 (1979).
8. G. E. Duvall, in Physics of High Energy Density, edited by P. Caldirola and H. Knoepfel (Academic, New York, 1971), p. 7.
9. S. Cochran and D. Banner, J. Appl. Phys. $\underline{48}$, 2729 (1977).
10. J. Vorthman, Ph.D. Thesis, Washington State University (1979).
11. Y. M. Gupta and W. J. Murri, SRI Report No. PYU-7802, 1981 (unpublished).
12. In the acoustic approximation, the tilt of the $A \rightarrow C$ shock in the water is $\arcsin[c_3 \sin(\theta)/u(\hat{A})]$. Since $u(\hat{A}) \ll c_3$, this tilt is much larger than the impact tilt θ.

SHOCK WAVE STABILITY

G. R. Fowles
Washington State University, Pullman, Wa 99164

ABSTRACT

This paper summarizes conditions for the stability of shock waves based on theories due to D'yakov, Kontorovich and Whitham. Various equivalent conditions on the equation of state are shown.

D'YAKOV METHOD

In an important paper published in 1956, S. P. D'yakov treated the problem of the instability of shock waves by investigating the behavior of a linear perturbation of the shock of the form,

$$v(x,y,t) = v_o e^{i(\omega t - kx - \ell y)}$$

where v represents a perturbation of any of the dependent flow variables: density, entropy, particle velocity, or pressure, and the unperturbed shock travels in the negative x direction in laboratory coordinates.[1] The perturbation thus represents a plane wave traveling away from the shock at an angle determined by the real parts of k and ℓ.

D'yakov found that the imaginary parts of ω and k are negative and the shock is therefore unstable whenever,

$$j^2 (dv/dp) < -1 \tag{1}$$

or,

$$j^2 (dv/dp) > 1 + 2M \tag{2}$$

In these expressions j^2 is the negative slope of the Rayleigh line joining the equilibrium initial and final states, the derivative is the slope of the Hugoniot curve, and M is the Mach number of the shock. Thus,

$$M = \left| \frac{D - u_1}{c_1} \right| \tag{3}$$

where D is the shock velocity, u_1 the particle velocity, and c_1 the sound speed of the shocked material. It is assumed throughout that $M < 1$ so that the shock is always subsonic with respect to the material behind.

These results were checked by Swan and Fowles,[2] and also by Erpenbeck[3] who used a somewhat different mathematical technique. We shall refer to instabilities resulting from violation of (1) and (2) as "corrugation instabilities".

In addition to the absolutely unstable regions specified by Ineqs. 1 and 2, D'yakov found a region for which the imaginary parts of ω and k are zero, thus admitting a steady, non-decaying, perturbation. An error in the expression for this limit was corrected by Kontorovich.[4] Such perturbations can occur unless

$$j^2 (dv/dp) < \frac{1 - M^2 - M^2(V_o/V)}{1 - M^2 + M^2(V_o/V)} \qquad (4)$$

Since $0 \leq M \leq 1$, and $(V_o/V) > 1$ this inequality is a smaller bound than is Ineq. 2 and is therefore controlling.

The physical basis for Ineq. 4 can be understood by examining the reflection coefficients of small amplitude acoustic waves obliquely incident on the shock from behind.[5] It is found that amplification occurs whenever Ineq. 4 is violated and that, moreover, whenever there is an admissible angle of incidence for which the amplification is greater than one, there is also an admissible angle for which the reflection coefficient is infinite. Consequently, spontaneous emission, or splitting, can occur and the resulting flow behind the shock becomes turbulent in a way that isn't yet fully understood. Some experimental verification of this limit in gases has been obtained by Griffiths, Sandeman, and Hornung,[6] and by Glass and Lieu.[7] Buchanan has obtained schlieren flash photographs of the wave structure in the unstable range that appear to match the theory at least qualitatively.[8] We refer to instabilities resulting from violation of Ineq. 4 as "splitting" instabilities.

WHITHAM'S METHOD

The stability problem can be approached differently using a method due to Whitham.[9] In this method the behavior of convergent or divergent shocks is treated by adding to the continuity equation a term that admits changes in area of a "ray tube". For small changes from an initially straight tube, Whitham derives the relation

$$\frac{1}{A} \left(\frac{dA}{dM_o} \right) = -g(M_o)$$

where A is the cross-sectional area and M_o is the unperturbed shock velocity normalized by the sound speed.

The function $g(M_o)$ is an expression of the equation of state of the medium. For an arbitrary equation of state it is easily shown that,

$$g(M_0) = \left[\frac{u_1 + c_1}{M_0 \quad 1}\right]\left[\frac{1 + 2M - j^2(dv/dp)}{1 + j^2(dv/dp)}\right] \tag{5}$$

This function is negative whenever the D'yakov absolute limits, Ineq. 1 and 2, are exceeded. A shock then speeds up in a diverging channel and slows down in a converging one so that corrugations tend to grow. The results of the two theories are thus consistent, except that as might be expected, Whitham's method does not predict splitting instabilities.

SUBSONIC-SUPERSONIC CONDITION

In each of the theories described it is assumed throughout that the shock velocity is subsonic with respect to the material behind, i.e., $M \leq 1$. Violation of this condition is therefore clearly not necessary for instabilities to occur. In non-reactive materials, however, such violation is tantamount to violation of the Second Law and is therefore a sufficient condition for instability.[10] When this occurs, splitting of a single shock into two shocks traveling in the same direction is observed.

No such restriction to subsonic flow applies to shocks in reactive materials. Consequently, weak detonations are theoretically possible. The stability of these waves must be approached from an entirely different point of view than for non-reactive waves, and at present no satisfactory solution appears to exist.

CONDITIONS ON THE EQUATION OF STATE

The stability limits can be expressed in terms of other thermodynamic derivatives that are sometimes more convenient. To this end we can show,

$$j^2(dv/dp) = \frac{M^2(a - 1)}{1 - M^2 a}$$

where a is defined as

$$a = \frac{\Gamma}{2V}(V_0 - V)$$

and Γ is the Gruneisen parameter, $\Gamma = V(\partial P/\partial E)_V$. The parameter, "a", is normally less than one; values greater than one are possible however, and correspond to $dv/dp > 0$ in the stable range.

The subsonic condition is,

$$(\partial P/\partial V)_S < -j^2 = (P - P_0)/(V_0 - V) \tag{6}$$

Ineq. 4 can be expressed, for $M^2 a < 1$, as,

$$(\partial P/\partial V)_S < -\frac{(P - P_0)(1 + \Gamma)}{V} \tag{7}$$

or, equivalently,

$$\left(\frac{\partial P}{\partial V}\right)_K < -\left(\frac{P - P_0}{V}\right)$$

where $K \equiv E + P_0 V$, is the adiabatic free energy. Similarly, Ineq. 1 is, for $M^2 < 1$,

$$\left(\frac{\partial P}{\partial V}\right)_S < -\frac{(P - P_0)\Gamma}{2V} \qquad (8)$$

Examination of these relations shows that it is not possible to satisfy both 6 and 7 while violating 8. Consequently, for $M^2 a < 1$, corrugation instability cannot occur. Conversely, when $M^2 a > 1$, splitting instability cannot occur.

For ideal gases the equation of state in the form,

$$P(V,S) = A(S) V^{-\gamma}$$

gives,

$$(\partial P/\partial V)_S = -\gamma(P/V)$$

where $\gamma = \Gamma + 1$ is the ratio of specific heats.

For this equation of state the condition for splitting instability resulting from violation of Ineq. 4 requires,

$$-(1 + \Gamma) P/V > -(1 + \Gamma)(P - P_0)/V$$

or,

$$\gamma P_0 < 0 \ .$$

Neither γ nor P_0 can be negative, however, so that instability of this type cannot occur in ideal gases. Corrugation instability also cannot occur.

For condensed matter a widely used empirical equation derived from shock experiments is,

$$D = C + Su$$

where D is shock velocity, u is particle velocity, and C and S are constants. The corresponding Hugoniot pressure - volume relation is,

$$P/(\rho_0 c^2) = \eta/(1 - S\eta)^2$$

where $\eta = 1 - V/V_0$. Differentiating this relation gives

$$j^2(dv/dp) = -(1 - S\eta)/(1 + S\eta)$$

and when this is combined with Ineq. 7, we get,

$$(\Gamma - S + 1)/(\Gamma - S + 2) < V/V_0 \qquad (9)$$

as the condition for stability.

As an example we choose typical values for Γ and S: say $\Gamma = S = 1.5$. Splitting will then occur for $V/V_0 < 0.5$, corresponding to a pressure, $P/(\rho_0 C^2) = 8$. Pressures of this magnitude are readily accessible for many materials. Experiments are clearly in order to establish the splitting instability limit as well as to study the subsequent behavior.

ACKNOWLEDGEMENT

I wish to thank Dr. V. E. Fortov for pointing out the consistency of Whitham's method with that of D'yakov.

REFERENCES

1. S. P. D'yakov, Zh. Eksp. Teor. Fiz., 27, 288, (1956).
2. G. W. Swan and G. R. Fowles, Phys. Fluids 18, 28, (1975).
3. J. J. Erpenbeck, Phys. Fluids, 5, 1181, (1962).
4. V. M. Kontorovich, Zh. Eksp. Teor. Fiz., 33, 1525, (1957) [Sov. Phys. - JETP 6, 1179, (1957)].
5. G. R. Fowles, Phys. Fluids 24 (2), 220, (1981).
6. R. W. Griffiths, R. J. Sandeman, and H. G. Hornung, J. Phys. D, 9, 1681, (1976).
7. I. I. Glass and W. S. Lieu, J. Fluid Mech., 84, 55, (1978).
8. W. L. Buchanan, UTIAS Technical Note No. 222, Institute for Aerospace Studies, University of Toronto, (1980).
9. G. B. Whitham, Linear and Non-Linear Waves (Wiley and Sons, New York, 1974), p. 270ff.
10. G. R. Fowles, Phys. Fluids 18, 776 (1975).

PIEZOELECTRIC SHEAR STRESS GAGE FOR DYNAMIC LOADING.[*]

Y. M. Gupta and W. J. Murri
SRI International

The objective of this work was to develop a piezoelectric gage that would be sensitive only to shear stresses and could be used in dynamic loading situations involving large amplitude compression and shear stresses. A linearized analysis of the mechanical and piezoelectric response was carried out to provide criteria for a suitable gage. Application of this analysis to alpha quartz and lithium niobate ($LiNbO_3$) crystals showed that only $LiNbO_3$ crystals could be cut to the desired orientation needed for the shear gage. Impact experiments using a recently developed experimental facility[1] were conducted to examine the gage response under one-dimensional compression and shear loading. The result of these experiments showed that the gage, as desired, had negligible sensitivity to compressive stresses and a very large sensitivity to shear stress. Further development to obtain detailed calibration data and check cross axis shear effects is currently under way.

[*] Work supported by the Defense Nuclear Agency.

[1] Y. M. Gupta et al., Rev. Sci. Instr. 51, 183 (1980).

0094-243X/82/780525-01$3.00 Copyright 1982 American Institute of Physics

ANALYSIS AND MODELING OF PIEZORESISTANCE RESPONSE

Y. M. Gupta
SRI International

A phenomenological model of piezoresistance was formulated to calculate the resistance change of a gage element subjected to mechanical deformation. This model incorporates the tensor nature of piezoresistivity, elastic-plastic gage response to include mechanical and electrical hysteresis, and dimensional contributions. The phenomonological model was used to analyze past data on Manganin and ytterbium. The stresses and strains in the gage element, needed for this analysis, were approximated from geometrical considerations. The results showed that the piezoresistance model could explain some of the experimental results. Improved analysis requires a rigorous determination of the stresses and strains in the gage element. To rigorously model the gage response, the gage element was represented as an inclusion and the corresponding boundary value problem was solved. The Eshelby method for elastic inclusions was modified to include an elastic-plastic inclusion and to obtain solutions for loading and unloading. This procedure gave good agreement with experimental data. Thus, the elastic-plastic inclusion formulation in conjunction with the phenomenological model can explain piezoresistance response.

A THEORY FOR THE SHOCK-LOADING RESPONSE OF AN ALUMINA-FILLED EPOXY MIXTURE*

D. S. Drumheller
Sandia National Laboratories,[†] Albuquerque, NM 87185

ABSTRACT

Alumina-filled epoxy is an engineering material which is used as a potting compound in impulsively-loaded ferroelectric power supplies. It is a composite material composed of tightly packed alumina particles (8 μm in diameter) in an epoxy matrix. No voids are present in the composite.

Extensive experimental data characterizing the dynamical response of this material is available. These data include ultrasonic shear and longitudinal wave speeds,[1] large-amplitude shock and release wave profiles,[2] and a large-amplitude shear wave profile.[3] From these observations it is clear that contact between neighboring alumina particles significantly influences the behavior of the composite. At low pressures, measured shock velocities are 50 percent greater than predictions which ignore interparticle contact. At high pressures, particle contact results in both large release wave velocities and an enhancement of the effective shear modulus of the composite.

The mixture theory of Drumheller and Bedford[4] is used to develop a three-dimensional model for alumina-filled epoxy.[5] Interparticle contact is treated by partitioning the bulk strain of the particles between two effects: that due to the loads imposed by the surrounding epoxy; and that due to the loads imposed by contact with neighboring particles.

The portion of the confining pressure which arises from interparticle contact results in an enhancement of the shear modulus of the composite. To account for this effect, the shear modulus is assumed to be a linear function of this portion of the confining pressure. Through careful consideration of thermodynamical principles, it is shown that this assumption results in a kinematical constraint on the material response. The material must dilate during pure shearing motion. Schuler[6] has observed this dilatancy phenomenon in static triaxial tests on this material.

The available experimental data is sufficient to allow both a unique evaluation of the model constants and an independent evaluation of the predictive capability of the model. Good comparisons are achieved for both the loading and unloading behavior of longitudinal and shear waves.

---

*This work was supported by the U. S. Department of Energy under Contract DE-AC04-76-DP00789.
[†]A U. S. Department of Energy facility.

REFERENCES

1. H. J. Sutherland, J. Comp. Matls. $\underline{13}$, 35 (1979).
2. D. E. Munson, R. R. Boade, and K. W. Schuler, J. Appl. Phys. $\underline{49}$, 4797 (1978).
3. L. C. Chhabildas and J. W. Swegle, J. Appl. Phys. (submitted).
4. D. S. Drumheller and A. Bedford, Arch. Rat. Mech. Anal. $\underline{73}$, 257 (1980).
5. D. S. Drumheller, J. Appl. Phys. (submitted).
6. K. W. Schuler, Sandia National Laboratories, Albuquerque, NM, private communication.

NUMERICAL MODELING OF OBLIQUE HYPERVELOCITY IMPACT USING TWO-DIMENSIONAL PLANE STRAIN MODELS *

William T. Brown
Sandia National Laboratories,† Albuquerque, New Mexico 87185

ABSTRACT

A numerical procedure has been developed to obtain information regarding the hypervelocity impact of a sphere onto a plate by considering the corresponding plane strain problem. The method consists of treating a rod as a superposition of an infinite chain of spheres and then numerically unfolding information about a single sphere. This technique has been used to calculate the peak pressure distribution when the impact is normal to the plate surface and excellent results were obtained. When extended to oblique impact problems, physically reasonable results are obtained but at this time no independent checks are available.

INTRODUCTION

The computer analysis of the impact of a spherical projectile onto a target surface at an oblique angle requires the use of three-dimensional techniques. However, three-dimensional computer codes for such problems are very expensive to use and the analysis of results can be very cumbersome and time consuming. Consequently, two-dimensional plane strain models are often employed to treat oblique impact problems.

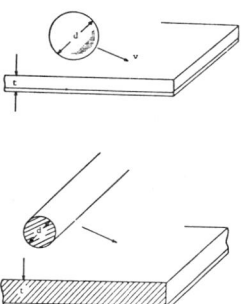

Fig. 1. Comparison of actual impact plane of sphere with plane strain approximation.

Two-dimensional plane strain codes are much easier to use and are less expensive than three-dimensional codes. However, the two-dimensional analog of an impacting sphere is an infinitely long rod of circular cross section as shown in Figure 1. These are fundamentally different problems and some important physical information can be lost[1] when plane strain models are substituted for the actual three-dimensional problems. Even for the case of normal impact, the plane strain model ignores any out-of-plane deformation and quantitative details are incorrect.

---

*This work was supported by the U. S. Department of Energy under Contract DE-AC04-76-DP00789.
†A U. S. Department of Energy facility.

One approach which has been used to deal with this difficulty is to scale the initial conditions of the problem in order to achieve an equivalence of either energy[1] or damage[2] between the plane strain and three-dimensional models. In the calculations discussed in the current work, plane strain modeling was performed to determine peak pressures at detectors embedded in a target plate. For the purposes of this analysis, the only variable of interest is the spatial distribution of peak pressure at some depth in the plate.

A numerical scaling has been developed to obtain the peak pressure distribution of the actual three-dimensional case from an analysis of results of the plane strain problem.

DESCRIPTION OF THE METHOD

The scaling procedure used in these studies involves a very simple application of linear superposition. The impact of an infinitely long rod (Figure 1) is approximated as the impact of an infinite chain of spheres having the same radius as that of the rod (Figure 2).

With this approximation, the peak pressure at a given depth in a plate which is impacted by a long rod can be written as a discrete sum and the sum can then be approximated as an integral,

$$P^c(x) \cong 2W \int_0^\infty P^s(x^2 + y^2)dy \quad (1)$$

where W is a constant.

A simple change of variables, $\eta = (x-x_o)^2 + y^2$, allows Eq. 1 to be written as

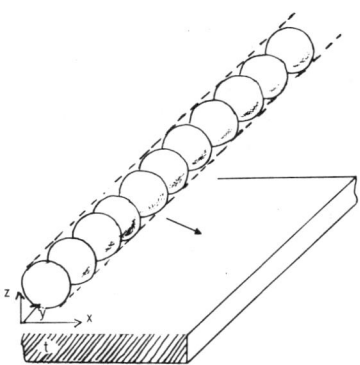

Figure 2. Approximation of cylinder as an infinite chain of spheres.

$$P^c(x) = W \int_{(x-x_o)^2}^\infty \frac{P^s(\eta)d\eta}{[\eta-(x-x_o)^2]^{1/2}} \quad (2)$$

where the parameter x_o has been introduced to account for the oblique impact angle situation in which the pressure distribution is not symmetric about the line $x = 0$. Equation 2 is Abel's integral equation and has a solution[3]

$$P^s(x-x_o) = \frac{-2}{\pi W} \frac{d}{dx} \int_{x-x_o}^\infty \frac{x'P^c(x')dx'}{[x'^2 - (x-x_o)^2]^{1/2}} \quad (3)$$

By requiring that the average mass per unit length for the cylinder and the infinite chain of spheres be the same, the parameter W is determined to be $W = 3/4a_0$ where a_0 is the radius of the sphere.

The procedure outlined here is a very simple one. The function P^c is determined by solving a two-dimensional plane strain problem. This funtion must be fit to a form which can be integrated to infinity. P^c is then substituted into Equation 3 and the integral is evaluated followed by a numerical differentiation.

APPLICATION TO NORMAL IMPACT

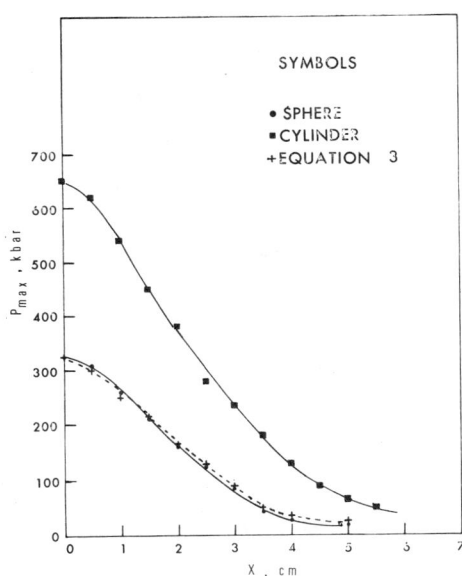

Figure 3. Comparison of peak pressure distribution for spherical and plane strain calculations.

The two-dimensional finite-difference code CSQII[4] was used to solve the problem of a steel sphere of radius 9.4 mm impacting a phenolic-resin target at approximately 7 km/s. The normal impact problem has been solved in both plane strain and axisymmetric geometries.

Pressure histories at a depth of about two projectile radii below the impact surface are plotted as a function of time and the maximum value at each location is obtained from each calculation to provide curves of of the type shown in Fig. 3.

The curve in Figure 3 representing the result of the plane strain calculation is substituted into Eq. 3 and the solution is checked against the result obtained from axisymmetric calculation. Results from Eq. 3 are given by the dashed line (--+--+); excellent agreement is obtained between the two solutions. This provides some confidence in attempting to extend the method to non-normal impact situations.

EXTENSION TO OBLIQUE IMPACTS

The same problem discussed above was repeated with the impact taking place at incidence angles of 20°, 38° and 53°. Of course, these calculations were done in plane strain geometry.

Peak pressures are obtained in the same manner as before and plotted for the three impact angles in Figure 4. An important observation from this figure is that the curves are still symmetric but the point of symmetry along the x-axis varies with impact angle. This symmetric point is the parameter x_0 used in Eqs. 2 and 3; it is critical in applying the technique to oblique impact.

In order to extend results to oblique impact situations, two important assumptions must be made. The first assumption is that the variation of x_0 with impact angle is the same for both the actual impact of a sphere and for the impact of the infinite cylinder. The second assumption is that the peak pressure distribution due to a sphere still retains full rotational symmetry about the point $(x_0,0)$. The first assumption can be justified by a simple heuristic argument based on the corrrespondence between the three-dimensional and plane strain problems at early times. The second assumption is a more drastic one and has not been investigated in any rigorous fashion. Nevertheless, Eq. 3 has been applied to the three cases depicted in Figure 4. The solutions for the spherical impact as obtained from the plane strain cylinder calculations are displayed in Figure 4 for each angle (dashed lines).

No independent check has been obtained for the results of Figure 4; this would require either experiments or three-dimensional calculations. However, the basic form of the solutions obtained from Eq. 3 is encouraging. In particular, the solution at the origin ($x=x_0$) is always such that the pressure due to a sphere is exactly half of that due to the cylinder. This is in agreement with the normal impact results.

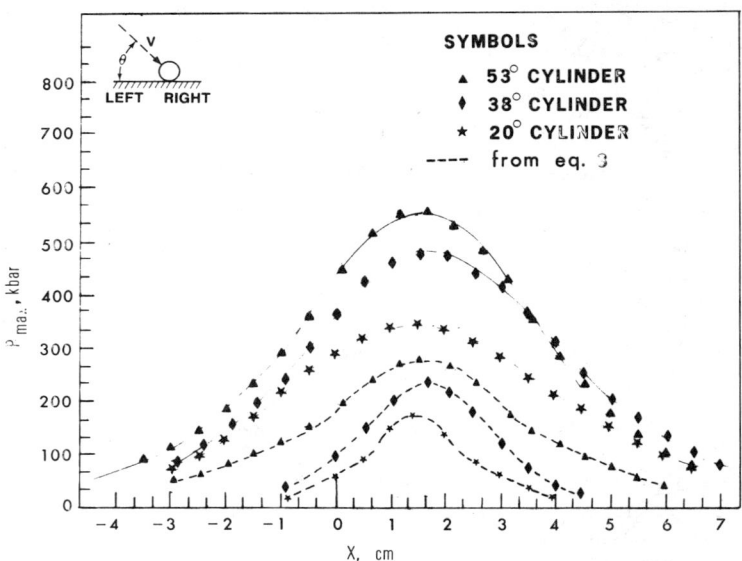

Figure 4. Plane strain solutions for impacts at 20°, 38° and 53°.

SUMMARY AND CONCLUSIONS

A numerical scaling technique has been developed for estimating peak stress distributions in flat-plate targets impacted by spherical projectiles. In the absence of methods for modeling three-dimensional impacts, the scaling technique provides a means for making estimates from two-dimensional plane strain computer simulations. The technique has been verified from two-dimensional calculations of normally impacting projectiles and has been extended to oblique impacts. The oblique impact calculations are based on some assumptions which have not been justified in a rigorous sense; however, the basic form of the solution is encouraging.

ACKOWLEDGMENTS

The finite-difference solutions used in these calculations were provided by J. M. McGlaun. Discussions with D. L. Hicks and F. R. Norwood on many aspects of this problem were extremely useful. Some valuable information relevant to integral equation techniques was provided by P. B. Bailey and F. Biggs.

REFERENCES

1. G. H. Jonas, J. A. Zukas and J. J. Misey, Report No. ARBRL-MR-02969 Ballistic Research Laboratories, Aberdeen Proving Grounds, MD. Oct, 1979.
2. L. D. Bertholf, M. E. Kipp and W. T. Brown, Report No. SAND 75-0541, Sandia Laboratories, Albuquerque, NM, January 1976.
3. R. P. Kanwal, <u>Linear Integral Equations</u>, Academic Press (1971) pp. 167-173.
4. S. L. Thompson, Report No. SAND 77-1339, Sandia Laboratories, Albuquerque, NM, January 1979.

MEASUREMENT PROBLEMS IN HIGH VELOCITY IMPACT EXPERIMENTS

William Lawrence

US Army Ballistic Research Laboratory, USA ARRADCOM, Aberdeen Proving Ground, MD 21005

I. INTRODUCTION

Reverse-ballistic experiments have been performed at the Ballistic Research Laboratory using a light-gas gun to launch targets against stationary long-rod projectiles instrumented with foil type resistance strain gages. Signals from the strain gages are recorded, measured and analyzed to obtain information about the dynamic behavior of the projectile as it penetrates the target. Signals from the strain gages are commonly divided so that part of the signal may be displayed at higher amplification for better resolution of detail at low strains. Error in the data reduction, especially error in time, is apparent from the poor agreement of the combined data. Different sources of error have been identified but the primary source is optical distortion in the oscilloscope cameras and in the copying camera. This report briefly reviews the instrumented rod experiment and the analytical procedure but primarily considers sources of error. A modified calibration procedure is introduced to minimize the error in the data reduction.

II. BACKGROUND

The general nature of the experimental approach will be reviewed before problem areas are introduced. This section briefly describes the experiments, the reduction of data, and the analytical procedures. Greater experimental detail has been reported by Hauver[1], and the theoretical basis for the analysis has been reported by Wright[2].

EXPERIMENTATION AND ANALYTICAL PROCEDURE

The experimental arrangement of a reverse-ballistic experiment at 710 m/s is shown in Figure 1. A long steel rod is shown instrumented with pairs of strain gages which are located 20, 40, 60, and 80 mm from the tip of the rod. This rod is impacted at normal incidence by a plate of rolled homogeneous armor launched from the gun. A compression wave travels back the rod and is detected by changes in gage resistance. These changes in gage resistance are measured by bridge circuits, and the output signal from each bridge is displayed by a cathode-ray oscilloscope and recorded photographically on polaroid type 410 film. Signals from the strain gages are commonly divided so that part of

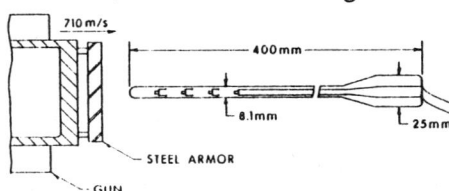

Fig. 1. Experimental Arrangement of Normal Impact

the signal may be displayed at higher amplification for better resolution of detail at low strains. Divided signals from a strain gage in an early experiment are shown in Figure 2. With diametrically opposed gages at each rod location, and each signal divided, four sets of data must be reduced and combined to provide a single strain-time history at each location on the rod. These combined strain-time plots are shown in Figure 3. The strain-time histories from different locations on the rod are used to perform a simple-wave analysis. From the strain-time histories, curves for constant strain may be plotted in distance-time space, along these curves the slope is defined as C_p, the plastic wave velocity. Based on the strain-rate independent theory for plastic wave propagation, particle velocity and stress are calculated. Stress-strain and particle velocity-strain plots are shown in Figure 4.

Fig. 2. Divided Signals at 5% and 20% Strain

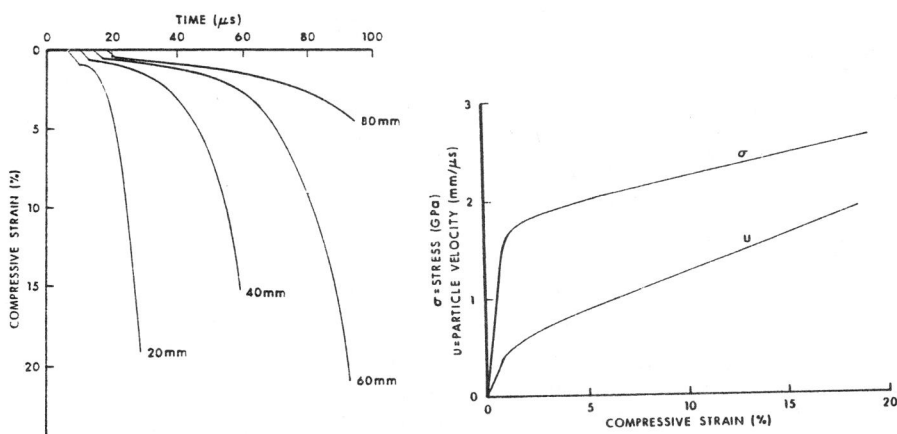

Fig. 3. Combined Strain-Time Plots at Difference Locations

Fig. 4. Stress-Strain and Particle Velocity-Strain Plots

III. PROBLEMS AND DISCUSSION

Figure 3 shows strain-time histories measured at four gage locations on a long-rod penetrator. Problems associated with the measurement and analysis of oscilloscope records become evident when the

TABLE I
WAVE ARRIVAL TIME AT DIFFERENT LOCATIONS

GAGE NO.	%	LOCATION	ARRIVAL TIME (µs)	TIME DIFF (µs)
4	5	20mm	21.665	0.614
	20		21.051	
5	5		—	
	20		21.593	
3	5	40mm	24.511	0.467
	20		24.217	
6	5		24.249	
	20		24.044	
2	5	60mm	28.368	1.028
	20		27.340	
7	5		28.008	
	20		27.680	
1	5	80mm	31.426	0.108
8	5		31.318	

details of Figure 3 are considered. These measured signals are commonly displaced from each other as shown in Table 1 and cannot be combined without arbitrary adjustments; these displacements provide evidence of measurement error.

In an effort to use the data from tests already performed, a procedure for combining test data was developed. At each rod location the four sets of strain-time data were forced into agreement at the mean arrival time for the midpoint of the elastic fronts. The data in the distance-time diagrams exhibited significant scatter which suggested that the procedure developed to combine data did not overcome the measurement errors.

Before examining sources of error, it should be noted that the errors in strain or time to be considered result primarily from errors in measurements of distance on photographic records. It will also be useful to consider the magnitude of error which is a concern. Errors in the measurement of time influence the accuracy of wave velocities and consequently the accuracy of analytical results.

Measurement errors can arise from different sources. The major sources are bleieved to be optical distortion, film distortion, and camera instability. Errors also relate to image quality and the ability of a human operator to read and interpret images correctly[3]. Additional error may also be introduced by the comparator used for record measurements. The major sources of error have been considered and will be discussed in the following sections.

A. OPTICAL DISTORTION

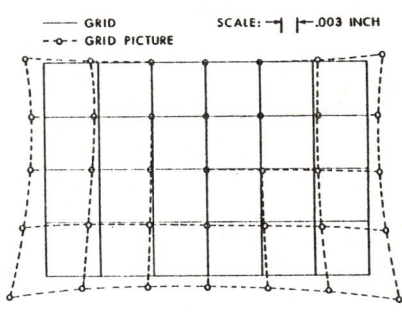

Fig. 5. Asymmetry Because of Camera, and Distortion Plot

Optical distortion introduced by the oscilloscope and copying cameras was determined by measuring photographs of a precision glass grid. Distortion plots such as the one shown in Figure 5 were prepared from these measurements. The distortion in these plots was exaggerated by measuring the displacement between the intersection points on the photograph and corresponding intersection points on the precision grid, multiplying this displacement by a factor of 1000, and plotting this magnified displacement on the scale of the precision grid. Figure 5 shows distortion produced by an oscilloscope camera.

B. FILM DISTORTION

Film distortion is a factor which limits the accuracy of a measurement, and it is difficult to eliminate. The primary causes of distortion are dimensional changes of the film base which undergoes various temporary and permanent changes between exposure to the image and measurement of the processed record[4]. Experimental study of oscilloscope records has revealed measurable distortion which results from lack of film flatness. When the record is removed from the camera it curls during the drying process. Then, when the record is flattened for a measurement, the emulsion stretches introducing elongation of an image already distorted. Although the distortion produced by the Type 410 film is more symmetrical, it is also somewhat greater than the distortion of Type 146-L film. This suggests that the paper base of the Type 410 film may be dimensionally less stable than the plastic base of Type 146-L film. Although it is difficult to separate film distortion from optical distortion, the ten microns value is judged to be reasonably consistent with the film distortion.

C. CAMERA STABILITY

Mechanical characteristics may also contribute to measurement error. These cameras may not be rigidly constructed and exact alignment may be difficult to maintain. There is a tendency for cameras to droop, and there may be a considerable movement in the assembly that attaches the camera to the oscilloscope. Poor camera rigidity can conceivably introduce asymmetry into the distortion as shown in Figure 5 where the distortion is obviously greater at the bottom of the record than at the top.

IV. MODIFIED CALIBRATION PROCEDURE

The format of the oscilloscope records in Figure 2 does not permit accurate measurements when records suffer from the distortion shown in Figure 5. However, the record format can be modified to reduce the influence of distortion. The reference sweep, necessary for record alignment, can be located in the central region where distortion produces minimum curvature. The distortion plot also suggests that the timing error, to a close approximation, increases linearly with distance from the horizontal axis. To evaluate this error, the calibration must establish time as a function of distance along the horizontal axis, and corresponding time as a function of distance at extreme locations above and below the horizontal axis.

Figure 6 shows the calibration format used to determine time from distorted test records. In this format, time marks are superimposed on square waves which are recorded above and below the horizontal axis. Each square wave extends from near the horizontal axis to the extreme off-axis distance where data are recorded. In practice, only maxima and minima appear on a photographic record; the writing rate in rising and falling portions of a square wave is too high to produce an exposure. The superimposed time marks occur at five-microsecond

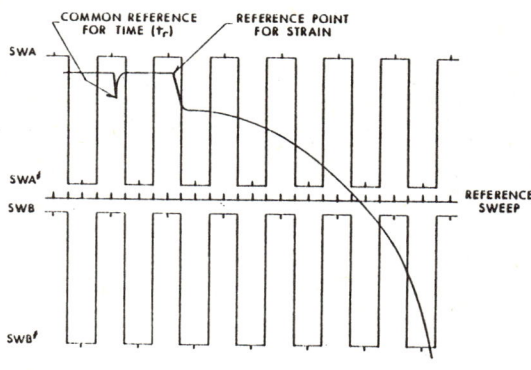

Fig. 6. Modified Calibration Procedure

intervals. The period of the square wave is long enough that each extremum always has at least one superimposed time mark. A record is aligned for measurement using the reference sweep located along the horizontal axis. The time marks on extrema of the square waves provide the data by which time is determined over the entire record. Time marks near the horizontal axis show an almost constant spacing in distance. However, it is found that the projection from an axial time mark usually does not pass through the mid-point between off-axis time marks. This timing error depends on the horizontal location, X, and on the vertical distance from the horizontal axis. We assume the timing error is zero between SWA´ and SWB (in Fig. 6), a maximum at both SWA and SWB´, and varies linearly with vertical distance. The position, X, of time marks at SWA, SWA´, SWB, and SWB´ is related to time, t, by the cubic

$$t = a + bX + cX^2 + dX^3 .$$

With these time corrections an adjusted signal history is obtained.

V. CONCLUSION

Any significant improvement in the accuracy of data from photographic records has to come through gradual elimination of these various sources of inaccuracies of the image coordinates. There has been an improvement from the use of a modified calibration procedure, but 0.1 μs accuracy is difficult to achieve. Image coordinates are still afflicted by these systematic errors even after the correction. The best solution to these problems may be to replace the analog oscilloscopes by digital recorders which do not require optical devices for recording or reducing test data.

ACKNOWLEDGEMENT

I would like to express my sincere appreciation to G. E. Hauver for his guidance, valuable advice, and for directing the project.

REFERENCES

1. G. E. Hauver, Int. J. Eng. Sci., 16, 871-877 (1978).
2. T. W. Wright, 12th Annual Army Science Conf., West Point, NY, 1980.
3. D. D. Preonas and R. F. Prater, Journal of the SMPTE, V. 79 (1970).
4. J. Vlcek, Photogrammetric Engineering, V. 35, 585-593 (1969).

EXPERIMENTAL AND NUMERICAL INVESTIGATIONS CONCERNING THE DYNAMICS OF PENETRATION PROCESSES

H. Senf, U. Hornemann, H. Rothenhäusler
Fraunhofer-Institut für Kurzzeitdynamik, Ernst-Mach-Institut, Abteilung für Ballistik, Weil am Rhein, FRG

F. Scharpf, A. Poth, W. Pfrang
Industrieanlagen-Betriebsgesellschaft mbH, Ottobrunn, FRG

ABSTRACT

The perforation of aluminium plates by means of fast steel penetrators shows a definite change in cratering as the impact velocity is increased beyond a critical value. It is supposed that this change is due to the additional occurrence of strong shock waves. For substantiation shock wave propagation during perforating was determined by means of a manganin probe technique. Experiments were supported by computational simulations using the Lagrange-code DYSMAS/L. First results are presented.

INTRODUCTION

Up to now the phenomena during penetration of metallic and non-metallic projectiles striking metallic or non-metallic targets with high impact velocities can be physically explained or numerically described in some special cases only. A complete penetration model should be able to explain the influence on projectile-target interaction of the striking velocity and of other material and geometric parameters. But, the material response under impact velocity is insufficiently known and further, each model is installed under simplified assumptions only. Therefore it is reasonable to investigate special perforation phenomena which admit a comparison between experiment and calculation. Such tests of the numerical simulations are suitable for improvement of model assumptions.

EXPERIMENTAL RESULTS

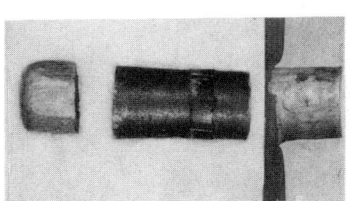

Fig. 1: Plate failure by plugging

Blunt cylindrical steel projectiles (C40) were accelerated against normally inclined aluminium targets (AlCuMg1-F40). In the range of ballistic limit velocity the observed plate failures in thin targets could be described by the plugging mechanism (Fig.1). However, if the striking velocity was more than 50 m/s higher than the ballistic limit velocity, the crater no longer showed smooth walls attributed to high shear loading but scaly bursted wall structures were observed (Fig. 2). The cracks extend in

Fig.2: Crater with scaly bursted walls

the target material parallel to the surface reaching lengths of some centimeters. Simular results were obtained if steel cylinders of 20 mm-diameter and of length to diameter ratio of 3 were fired against 50 mm thick aluminium plates. Figure 3 shows penetration appearance as a function of striking velocity. At striking velocities $v_p > 600$ m/s a mushroom-like head is seen at the projectile's tip and nevertheless plugs were formed at ballistic limit velocity. Enhancing the striking velocity scale formation becomes more pronounced and the cracks run further into the target material.

Fig.3: Impact phenomena as a function of striking velocity

If a blunt steel cylinder hits an aluminium target with high velocity a shock wave is generated at the interface. For a striking velocity of e.g. 1050 m/s the pressure amplitude in the target can be calculated to be 13 GPa using Hugoniot data. Initially it was expected that a shock wave of this strength would initiate spalling effects after reflection at the rear free surface of the target and that the change of the feature in cratering from smooth to scaly bursted walls could be interpreted as a superposition of spallation and plugging. Therefore the shock wave attenuation in the target was determined by means of piezoresistive manganin pressure gages. Figure 4 shows the profile of shock waves observed in distances of 10 mm and of 40 mm from the impact point in flight direction. In all distances an elastic precursor with a constant amplitude of 0.6GPa was observed. Figure 5 shows the attenuation of the shock wave amplitude versus the distances travelled in the aluminium target. The amplitude decreases from the calculated impact pressure of 13.0 GPa to 1.02 GPa at the rear surface. The spread in data can be explained

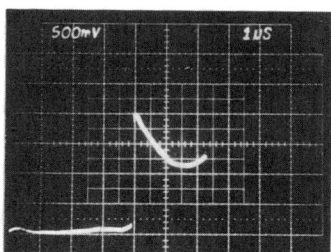 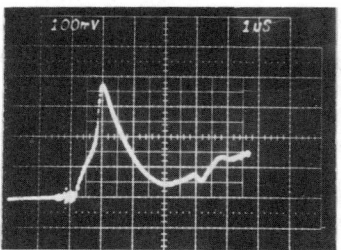

Fig. 4: Shock wave profiles at 10 and 40 mm distances from impact surface

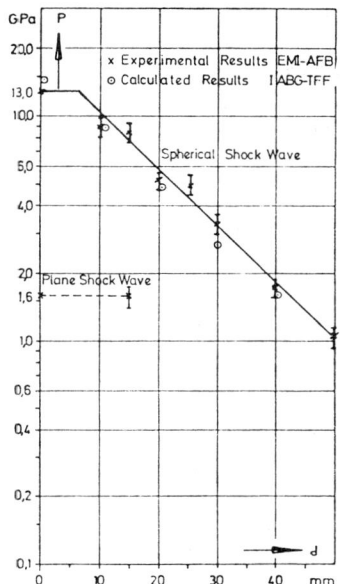

Fig.5: Shock wave attenuation in aluminium

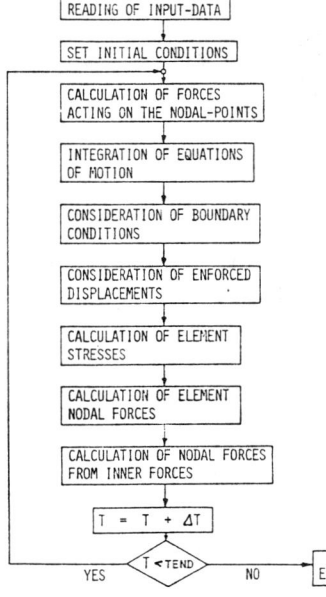

Fig.6: Flow-chart

by the individual changes in pressure gage response of 5 to 10 per cent. Furthermore each test point corresponds to a single round for which the impact velocity may vary up to ± 40 m/s from 1050 m/s.

To find out the dynamic behaviour of aluminium under one dimensional shock wave loading with very low plastic deformation 10 mm steel flyer plates were accelerated against aluminium plates of 15 mm thickness. First cracks were observed in the target material at a striking velocity of 145 m/s. For the same impact conditions the measurement of the pressure time history at the rear free surface of the target yielded a shock wave amplitude of 1.60 GPa. This value represents the minimum loading at the rear which led to spalling.

NUMERICAL RESULTS

A computer code called DYSMAS/L is available at IABG that was developed to describe the dynamics of structures with multiple material behaviour and arbitrary shapes. It is a Lagrangian Code. The structure is discretized by finite elements. The explicit time integration is performed by means of difference formulas. So DYSMAS /L is called a mixed Finite Element/Finite Difference Code. The flow chart in Figure 6 gives a view of the algorithm of DYSMAS /L. First the input data are read and the initial input values are calculated, that means input of coordinates of the nodal points, element-type specification, calculation of the nodal point masses, material properties, initial velocities, boundary conditions, etc. This time loop starts with the calculation of the forces acting on the nodal points. The acceleration of the nodal points results from the sum of all nodal forces divided by the nodal mass. The integration of motion is performed using a central difference scheme, where the velocity is calculated at "half time" while the displacements are calculated at the end of the time step. This is a simple, but stable integration procedure.

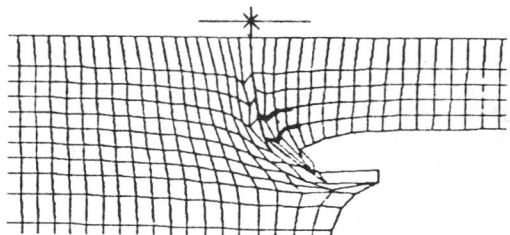

Fig.7: Penetration of an aluminium plate at the moment of impact (real time 0μs). Element numbers are shown.

Figure 7 shows one half of the cross section of the two dimensionally discretized projectile and target, just at time of first contact. The projectile penetration 10.3 μs after impact is shown in Figure 8. The interface of the two bodies is marked. At this plane of impact the lateral motion of the projectile is delayed by friction. The elements above this interface show an enhanced extension to the side compared to the extension of the element in the next upper row. Figure 9 shows in comparison the micrograph of an etched projectile which was fired with 1050 m/s against the target. The lines of texture show similar features. That means the numerical model describes the material behaviour in a very realistic way. Further, in Figure 8 is to be seen the beginning of the development of a conical region with maximal shear loading in the target material. The interior of this cone suffers only low deformation and remains

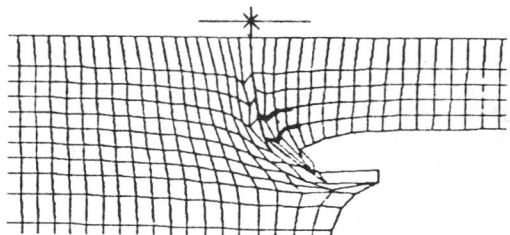

Fig.8: Penetration. Real time 10.3.μs

Fig.9: Micrograph of a steel projectile

Fig.10: Recovered cone and projectile

stable in shape during the whole perforation process, as it can be seen by the recovered cone (Fig. 10).

The calculated shock wave attenuation using the DYSMAS/L-code is in good agreement with experimental results as shown in Fig. 5. The numerical results are within the range of the experimental data at the corresponding depth.

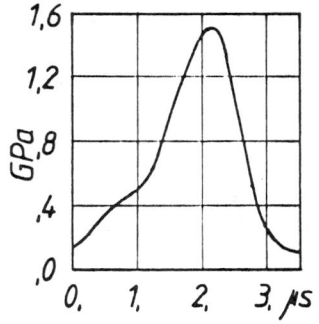

Fig. 11: Calculated shock wave profile at 40 mm distance from impact surface

The agreement between experiment and calcuation is evident in the slope of the shock wave maxima, furthermore even the shock wave profiles are well reproduced by the numerical model. Fig. 11 shows the calculated shock wave profile at 40 mm distance from impact surface. The computed profile clearly shows even the measured elastic precursor in shape and quantity (Fig. 4).

CONCLUSIONS

First results were represented of a comparison between a penetration in aluminium targets by steel cylinders and computational simulation using DYSMAS/L-code. Calculation was extended to a real time of 12 µs to ensure that the reflected shock wave would not be able to cause spallation. Good agreement was found between experimental and numerically determined shock wave attenuation and plastic projectile deformation.

TOODY-WONDY CALCULATIONS OF PENETRATION EVENTS*

D. L. Hicks, F. R. Norwood and T. G. Trucano
Sandia National Laboratories,† Albuquerque, New Mexico 87185

ABSTRACT

One and two-dimensional Lagrangian wavecode calculations simulating penetration of a conical-nosed steel penetrator into Antelope Lake Tuff are summarized. These calculations are used to investigate the validity of the cylindrical cavity expansion theory, which has been used to simplify the analysis of earth-penetration events.

INTRODUCTION

For the purpose of providing a simple and fast tool for analyzing penetration problems, Norwood[1] has introduced the cylindrical cavity expansion theory. This theory assumes that the target may be separated into infinitesimal layers, that the motion in each layer is primarily radial, and that the force produced by the radial stress component is the radial component of the total force acting on the penetrator.

The cylindrical cavity expansion assumption has been widely incorporated in one-dimensional models of penetration events,[2-6] with various target material models. While agreement with experiment has generally been encouraging,[2-5] it is desirable to examine the basic assumption underlying the cylindrical cavity expansion models. A sequence of computer studies of normal impact penetration events has been performed, using the one-dimensional Lagrangian wavecode WONDY-NORML,[8] and the two-dimensional Lagrangian wavecode TOODY.[9] NORML specifically uses the assumptions of the cylindrical cavity expansion theory. The target material, Antelope Lake Tuff, was modeled in two different ways. First, it was treated as a hydrodynamic material. Second, it was given shear strength via the linear τ-P model.[5-7]

In the following section, we briefly describe the actual computational situation, present some results from the two different material studies, and attempt to draw certain conclusions about the validity of the cylindrical cavity expansion.

THE COMPUTATIONAL CONFIGURATION

Fig. 1 shows the relationship between the steel penetrator and the soil one cycle into the TOODY calculation. The penetrator was started with its nose half-buried for these calculations. The zoning is such that there are eight zones per penetrator radius, and that the vertical spacing is preserved as we move from the penetrator to the soil. The impact velocity of the penetrator was 457.2 m/sec.

*This work was supported by the U. S. Department of Energy under Contract DE-AC04-76-DP00789.
†A U. S. Department of Energy facility.

The problem time for each calculation was 400 μsec. The nose of the penetrator was completely buried at 300 μsec. Stress histories taken from the penetrator nose show a sharp rise followed by a decay to the values at 400 μs which were plotted in Figs. 2 and 3.

HYDRODYNAMIC SOIL COMPUTATIONS

To model hydrodynamic Antelope Lake Tuff, a constitutive relation of the form

$$P = K\eta \qquad (1)$$

was assumed. Here, η is the volumetric strain, defined by

$$\eta = 1 - \rho_0/\rho \qquad (2)$$

K is the bulk modulus, found to be 2×10^9 Pascals from hydrostatic data, and ρ_0 is the initial tuff density, given as 1750 kg/m^3.

The results of the TOODY and WONDY calculations of normal stress on the penetrator nose are found in Fig. 2. TOODY uniformly predicts smaller values of normal stress than does the cylindrical cavity expansion as implemented in WONDY.

LINEAR τ-P MODEL COMPUTATIONS

This model results from coupling equations (1) and (2) with a yield law of the form

$$Y = \mu P \qquad (3)$$

Y is the yield strength and μ is a material constant found from triaxial test data.

In Fig. 3 comparison between the normal stress calculations of WONDY and TOODY for this material model have again been made. Again, TOODY predicts smaller values of normal stress than WONDY, except at the nosetip. Comparison with Fig. 2 also shows that the predicted normal stresses for the linear τ-P target are higher than for the hydrodynamic target as expected.

In Fig. 4. typical data for the target velocity field in the vicinity of the penetrator nose are shown, at t = 300 μsec, just when the nose of the penetrator has been completely buried. We observe that the target particle velocity decays radially outward, but has an approximately fixed direction, at a given depth in the neighborhood of the penetrator. In general, the target velocity is largely radial, but the radial component becomes less dominant as the nosetip is approached.

CONCLUSIONS

The major difference between the fully two-dimensional axisymmetric TOODY calculation and the cylindrical cavity expansion calculation (via WONDY-NORML) was the lower normal stress on the penetrator, for both hydrodynamic and non-hydrodynamic soil. We believe this is due to the fact that TOODY allows for side-relief in the axial direction, while WONDY does not. WONDY closely models the cylindrical cavity expansion model and therefore does not allow for free surfaces. Also, in the TOODY calculation the penetrator is modeled as a deformable body. The acoustic impedance of a rigid penetrator would be infinite, while that of a deformable penetrator is not. Using a deformable penetrator would lower the impact stress by a factor of approximately $a_s/(a_p + a_s)$, where a_p and a_s are, respectively, the acoustic impedances of the penetrator and the soil. For steel and Antelope Lake Tuff, we find that the impact stress is thus lowered by approximately 5%.

Fig. 4 illustrates that the soil velocity field in the vicinity of the penetrator nose is not strictly radial. Within the limits of comparison of the two-dimensional transient solution, for a deformable penetrator, with the essentially steady-state, one-dimensional cavity expansion methods this suggests caution in applying the one-dimensional models. Additional complexities in penetration phenomena, such as target and penetrator material behavior and frictional effects, must be addressed. These may be more significant than the question of whether or not a radial target flow field occurs. The current lack of experimental data precludes identifying the relative importance of the effects of material behavior, friction or radial flow fields.

ACKNOWLEDGMENTS

The authors would like to acknowledge the many valuable contributions, suggestions and criticisms which were made by L. D. Bertholf, R. K. Byers, A. J. Chabai, J. W. Swegle, P. Yarrington and M. E Kipp.

REFERENCES

1. F. R. Norwood, Report No. SLA-74-0201, Sandia Lab's., Albuquerque, NM, 1974.
2. P. Yarrington, Report No. SAND77-1126, Sandia Lab's., Albuquerque, NM, 1977.
3. D. Z. Yankelevsky and J. Gluck, Int. J. Mech. Sci. 22, 297 (1980).
4. D. Z. Yankelevsky and M. A. Adin, Int. J. Num. Anal. Mech. Geomechanics 4, 233 (1980).
5. M. J. Forrestal, D. B. Longcope and F. R. Norwood, J. Appl. Mech. 48, 25 (1981).
6. M. J. Forrestal, F. R. Norwood and D. B. Longcope, "Penetration into Targets Described by Locked Hydrostats and Shear Strength," to appear in Int. J. Solids Structures.
7. F. R. Norwood, "Constitutive Models in WONDY for Use with Geological Materials," to appear (1981).
8. F. R. Norwood, Report No. SAND80-0249 Sandia Lab's., Alburquerque, NM, 1980.
9. J. W. Swegle, Report No. SAND78-0552 Sandia Lab's, Albuquerque, NM. 1980.

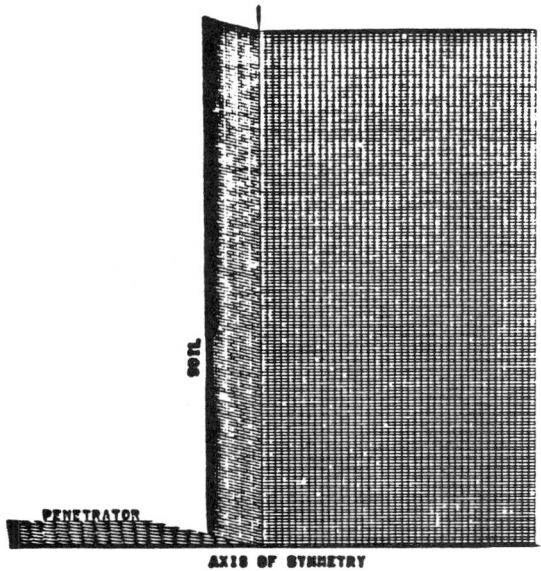

Fig. 1. Initial penetrator - soil configuration.

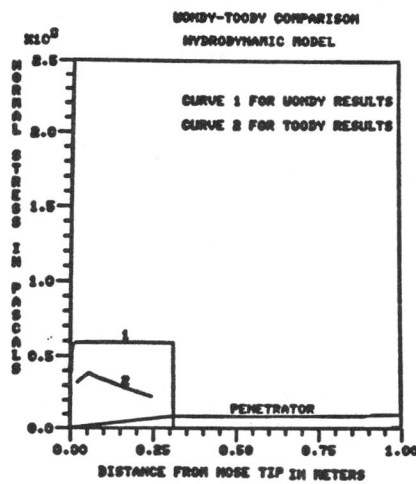

Fig. 2. Normal stress on the penetrator for the hydrodynamic model.

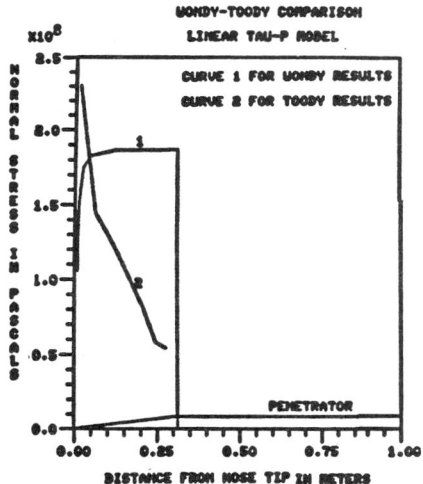

Fig. 3. Normal stress on the penetrator for the τ-P model.

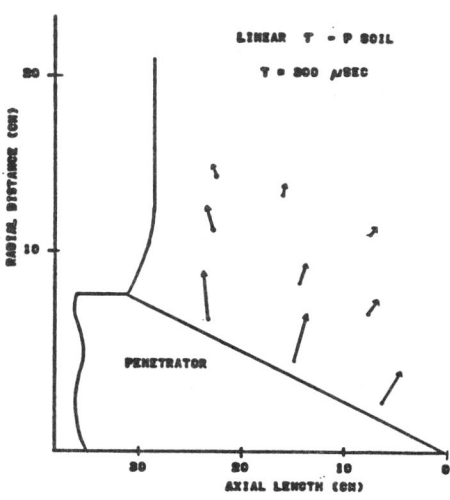

Fig. 4. Soil velocity field at t = 300 μs.

HYPERVELOCITY IMPACT RESPONSE OF TI AND BE

S. J. Bless
University of Dayton Research Institute

ABSTRACT

Thick Grade 2 titanium has a volume resistance to hypervelocity penetration of 3.87 J/mm^3 at 20°C. Crater volume increases with temperature and purity. Crater depth is well described by the Charters-Summers equation. Cratering in beryllium is brittle. Projectile scaling was also investigated.

INTRODUCTION

Impact of small particles with velocity in excess of 15 km/s is an environmental hazard in space. Laboratory testing is generally limited to velocities below 7.5 km/s. Therefore, a physical understanding of hypervelocity impact phenomena is necessary in order to use experimental results to predict consequences of meteoroid impact.

The simplest impact configuration is cratering in a thick target. In our work, impacts have been quite successfully modeled with the Charters-Summers equation, which can be written as:

$$p = \left(\frac{81}{4\pi} \frac{\rho_p}{\rho_t} \frac{E}{S_t} \right)^{1/3} \qquad (1)$$

where E = the kinetic energy of the projectile, p = crater depth in a thick target, ρ_p = particle density, ρ_t = target density, S_t = cratering resistance parameter (a material property). In this formulation, it can be readily seen that p is inversely proportional to $S_t^{1/3}$. The crater volume, V, can be written in terms of a volumetric cratering resistance parameter, ε, as:

$$V = E/\varepsilon \qquad (2)$$

Again, crater dimensions are inversely proportional to $\varepsilon^{1/3}$. There are several useful formulas for describing impacts on structural elements. For plates, the threshold penetration thickness, T_p, is that thickness which just provides protection against an incoming particle. The plate may spall, but not be perforated. A large body of empirical experience in ballistics and hypervelocity impact supports the approximation $T_p = 1.5\ p$.

The important phenomenology of impacts into multiplate structures concerns generation of projectile debris that may breech critical inner elements. There are published procedures for analysis of multiplate (bumper) configurations[1,2,3]. However, inspection of the data and analysis will reveal that the conclusions are almost exclusively based on double aluminum plates. Extrapolation to multiplate structures and other materials is not justified. It has been shown that for fixed projectiles and variable bumper density, impact effects on double targets scale as the areal density ratio of the projectile and first plate; equal damage occurs when:

$$\rho_L d_L / T_L = \rho_m d_m / T_m \qquad (3)$$

where d is the particle diameter and the subscripts L and m denote a laboratory test and an actual meteoroid impact, respectively[4] (In equation (3) both bumpers have the same volume density.). An assumption in this similarity analysis is that in both instances the projectile material is in the same state after impact. This requirement can usually be met in laboratory tests by using cadmium projectiles. Limited data for projectiles of various densities[3] indicate that equation 3 is conservative; for aluminum and foam spheres (ρ_p = 0.70 g/cm^3) against dual sheet aluminum structures, it was found that the most effective (optimum) bumper design was $\rho_L d_L / T_L$ = 10.8 g/cm^3 for aluminum and $\rho_L d_L / T_L$ = 28 g/cm^3 for foam.

EXPERIMENTS

Particle launch was carried out using a two-stage light-gas gun. Projectile velocity was measured redundantly using a Hall camera and a high speed framing camera. Values were between 6 and 7 km/s. The titanium target materials used in this program were "Grade 2" and hardness was RB 82-90. Two alloys were tested; Alloy A contained 5% Al and 2.5 Sn, Alloy B contained 6% Al, 2% Sn, 4% Zr, and 2% Mo. They had a hardness of RB 82. The glass projectiles were spheres with ρ = 3.99 ± 0.05 g/cm^3. The metal projectiles were spherical, or cylinders with 68° conical ends which have the same mass as equal diameter spheres.

RESULTS

Crater data were obtained for impacts on thick targets of Grade 2 titanium at 20° and 500°C, and Alloy A and Alloy B titanium at 20°C. The crater volume data span two orders of magnitude in projectile energy. Projectile materials were glass, aluminum cadmium, and copper.

A curious feature of the craters is that they were smooth for $\rho_p < \rho_t$ and rough for $\rho_p > \rho_t$. The best values of $S_t^{1/3}$ and $\varepsilon^{1/3}$

for various impacts were calculated from the data by using inverse variance weighting of experimental uncertainties. The variations from shot to shot were interpreted as representative of the intrinsic variability of the cratering process, so the sample standard deviation was retained for the means, shown in Table 1.

Table I. Cratering Parameters for Grade 2 Ti $(J/mm^3)^{1/3}$

T (°C)	20	500
$S_t^{1/3}$	3.82 ± 0.31	3.20 ± 0.18
$\varepsilon_t^{1/3}$	1.57 ± 0.10	1.46 ± 0.12

It is surprising how consistent those values of $S_t^{1/3}$ and $\varepsilon_t^{1/3}$ are, considering the large span of projectile energy and density encompassed by the data. The total mean relative uncertainty is only 7 percent. This is remarkably small. The consistency of the data for room temperature titanium is graphically portrayed in Figure 1 for $\varepsilon^{1/3}$ and in Figure 2 for $S_t^{1/3}$.

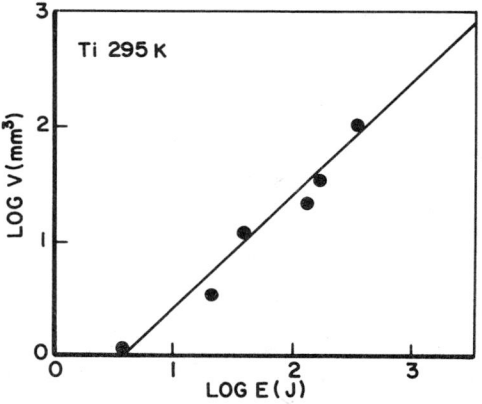

Fig. 1. (Above) Crater volume vs. impact energy for 20°C titanium. Line is for $\varepsilon = 3.87$ J/mm^3.

The two titanium alloys were significantly more resistant to impact cratering. For alloy A, $S_t^{1/3} = 4.17$ (J/mm^3), which is 30 percent higher than the pure

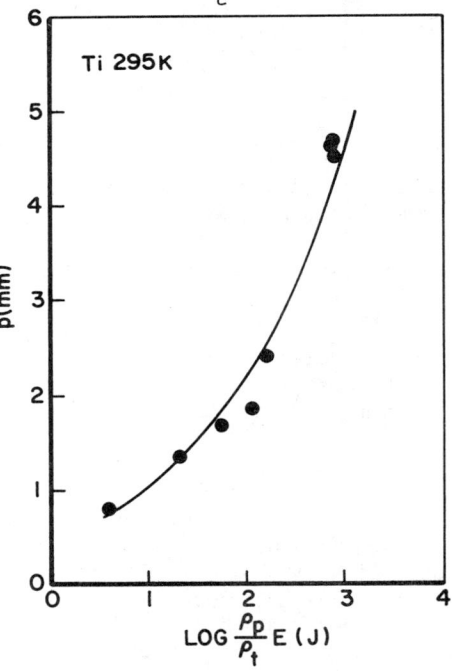

Fig. 2. (Above) Crater depth vs. energy parameter for 20°C titanium. Line is equation 1 with $S_t = 55.7$ J/mm^3.

metal. For Alloy B, the improvement was 25 percent. The improvement in the volume parameter $\varepsilon^{1/3}$ was only about half the percentage as for $S_t^{1/3}$. The improvement seems to be associated with less plastic flow. The crater rim featrues common in the pure metal were wholly absent in the alloys.

In contrast to titanium, beryllium target blocks behaved in a very brittle fashion. Figure 3 shows a post-impact view. There was massive surface spall. The target blocks split along radial cracks. Subcrater crack networks were evident on polished cross sections. This observation is consistent with previous observations of impacts into beryllium tubes[5]. Table II presents the data. The values of $\varepsilon^{1/3}$ were computed from the values for $S_t^{1/3}$ by assuming hemispherical craters. Data for 700°C were computed from Reference 5.

Fig. 3. Post impact view of beryllium target struck by 1.6 mm diameter copper projectile. Note surface spall.

Shots were conducted against single thin titanium target plates. Figure 4 shows three frames from the framing camera record for which T = 1.5 p for 20°C titanium. A spall plate was launched at 213 m/s; however, target perforation did not occur. Other shots substantiated this equation for heated titanium struck by low-density projectiles. The data indicated that the equation underestimated T_p for high density projectiles, but those results were not quantitative.

Table II. Cratering Parameters for Be $(J/mm^3)^{1/3}$

T (°C)	20	500	700
$S_t^{1/3}$	4.59 ± 0.09	3.64 ± 0.08	3.35 ± 0.20
$\varepsilon_t^{1/3}$	1.14	0.90	---

Many shots were executed against structural models. One series of these is especially useful because it shed light on proper projectile scaling for complex targets. In these experiments a 0.8 mm pressurized titanium hull was protected by a mylar blanket

Fig. 4. Four frames from a shot with 1.0 mm glass bead at 2.78 mm thick titanium plate. $T_p = 1.5$ p. Target spalled but was not perforated.

with an areal density of 0.24 g/cm^2. Projectiles were cadmium and 0.95 g/cm^3 plastic. For cadmium, the critical mass, M_L, causing plate rupture was 20 mg. For plastic, 45 mg $< M_L <$ 71 mg. According to equation 3, $M_L = 16$ g. This result may be summarized by $M_L \rho^{.31} =$ constant for impacts of similar lethality.

REFERENCES

1. J. W. Gehring, "Engineering Considerations in Hypervelocity Impact," <u>High Velocity Impact Phenomena</u>, ed. R. Kinslow, Academic Press, 1970.
2. B. G. Cour-Palais, "Meteoroid Protection by Multiwall Structures," AIAA Hypervelocity Impact Conference, May 1969.
3. R. J. Arenz, "Influence of Hypervelocity Projectile Size and Density of the Ballistic Limit of Dual-Sheet Structures," AIAA Hypervelocity Impact Conference, May 1969.
4. H. F. Swift and A. K. Hopkins, <u>J. of Spacecraft & Rockets</u>, Vol. 7, 73-77, 1970.
5. J. H. Diedrick, I. J. Loeffler, and H. R. McMillan, NASA-TN-D-3018, 1965. (Available NTIS)

ACKNOWLEDGEMENT

This work was supported by US DOE under Contract 4-X49-1156 H-1 and the Jet Propulsion Laboratory under PO No. 955828.

FREE-SURFACE VELOCITY MEASUREMENTS OF PLATES DRIVEN BY REACTING AND DETONATING RX-03-BB AND PBX-9404

L. M. Erickson, H. G. Palmer, N. L. Parker and H. C. Vantine
Lawrence Livermore National Laboratory
P. O. Box 808, L-368
Livermore, CA 94550

ABSTRACT

Copper plates 80 mm in diameter, of thickness 0.25 mm and 0.5 mm, were accelerated by adjacent 17 mm thick, 90 mm diameter cylinders of RX-03-BB or PBX-9404-03. The explosive was initiated by impact of a thick flyer from the LLNL 102 mm gun, providing either a reactive or fully detonating wave, by appropriate choice of flyer velocities up to 1.30 mm/µs. The free-surface velocity of the plates was measured with a Fabry-Perot velocimeter. We have obtained excellent experimental free-surface velocity histories. Our calculations of this history employing beta-burn and nucleation and growth high explosives models are in good agreement with fully detonating experiments. For reacting RX-03-BB, adjustments in the parameters are needed. We have an experimental technique that gives records whose agreement with calculation is sensitive to the model and is therefore a good way of testing new high explosive models. Also, this method allows us to infer information about the reaction zone length.

INTRODUCTION

Several workers have reported the measurement of the jump-off velocity of metal plates accelerated by high explosives.[1,2,3] For thick plates, thess measurements allow the inference of the Chapman-Jouget pressure. With thin plates, that is plates of the order of the reaction zone thickness, inferences about the reaction zone structures can be made. The interpretation of these records has been the subject of some controversy.[3,4,5] At issue is the interpretation of breaks in a plot of jump-off velocity versus plate thickness. These issues are resolved only when one considers the structure of the evolving wave and the interaction of the HE/plate interface with the reaction zone. We sought to measure the entire velocity history of the plate and not just the jump-off velocity. This would indicate the wave structure seen in the plate.

"This work was performed under the auspices of the U.S. Department of Energy by the Lawrence Livermore National Laboratory under contract number W-7405-ENG-48."

0094-243X/82/780553-05$3.00 Copyright 1982 American Institute of Physics

We measured the velocity history of copper plates pushed by PBX-9404 and RX-03-BB explosives. The plate velocity was measured by reflecting an Argon ion laser off the plate and measuring the wave-length shift with a Fabry-Perot interferometer. This method allows us to infer both information about the reaction zone structure and information about explosive performance.

EXPERIMENT

Figure 1 is a sketch of the experimental setup. A 0.6 cm thick copper flyer traveling at speeds up to 1.30 mm/μs was launched into the target chamber by the 102 mm gun. The targets consisted of 90 mm diameter by 17 mm thick disks of high explosive, PBX-9404 of density 1.840 g/cm^3 or RX-03-BB of density 1.907 g/cm^3. On the back of each high explosive disk was bonded a 80 mm diameter flat metal disk of oxygen free high conductivity copper of nominal thickness 0.5 mm, or 0.25 mm. The adhesive used was Shell 815 epoxy with T-1 catalyst.[6] For the experiments reported herein, Table 1 gives the experimental configurations. In Experiment No. 3769, a PBX-9404 disk was bonded onto the front of the RX-03-BB in order to achieve full detonation.

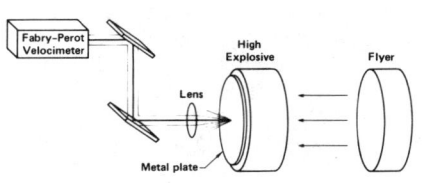

Figure 1. Experimental setup of free-surface velocity measurements using a Fabry-Perot velocimeter.

Table I Experimental configuration

Experiment Number	High Explosive Type	Thickness (mm)	Plate Thickness (mm)	Flyer Velocity (mm/μs)
3815	PBX-9404	17.007±.003	.501±.001	1.27
3816	PBX-9404	16.851±.001	.250±.002	1.25
3769	RX-03-BB	14.015±.002	.433±.005	1.14
	PBX-9404	3.025±.002		
3777*	RX-03-BB	17.00 ±.02	.4 ±.1	1.30

*Reacting

The flyer velocity was measured by a flash x-ray photographic technique, and independently with crystal pins near the target impact plane. The velocity measurements were accurate to 1 percent.[7] Flyer tilt was measured by crystal pins placed on a 95 mm radius circle with respect to the target axis at the target impact plane, and was typically on the order of 1 mrad. Plate jump-off time was recorded by anodized aluminum shorting pins, in contact with the metal plate, on a 17 mm radius circle. Time measurements are typically good to ±0.01 μs.

The Fabry-Perot velocimeter[8] is an instrument for measuring and recording Doppler-shifting light from moving targets. The beam spot size at the target is ~0.3 mm. The lensing system converts the usual Fabry-Perot output ring pattern into a series of bright dot pairs, which is streaked to produce a set of lines whose separation is proportional to the target velocity. At velocities of a few mm/μs, resolution of 2 percent in velocity and 7 ns in time are achieved. Figure 2 is the raw data Fabry-Perot velocity record obtained in Experiment No. 3815.

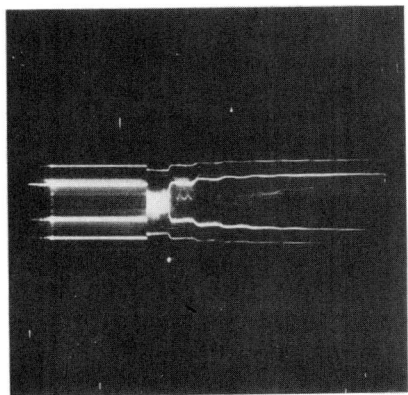

Figure 2. Raw-data Fabry-Perot velocity record of experiment No. 3815.

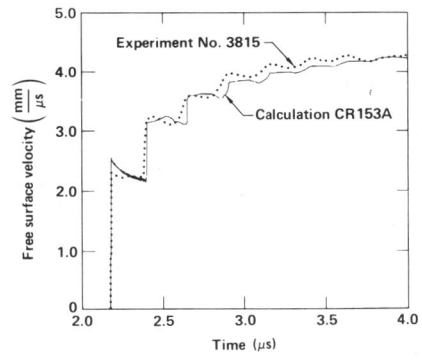

Figure 3. Free-surface velocity record of experiment No. 3815, and nucleation and growth calculation.

CALCULATIONS

All calculations of the front surface velocity experiments were done with the computer code KO[9]. The metal was modeled by a Gruneisen equation-of-state with a sophisticated elastic-plastic model. The high explosives were modeled either by the "beta-burn" model,[10] or by a "nucleation and growth" model.[11] The copper flyer was given the experimentally observed velocities.

Figure 3 shows free surface velocity V_{fs} versus time for Experiment No. 3815, and calculation CR153A. We see excellent agreement in the free-surface jump-off time (2.20±0.03 µs, experimentally; 2.174 µs, calculationally), and the shock reverberation time in the copper plate. Also late time velocities agree to within a few percent. Thus, we believe that the copper equation-of-state is reasonable, and the explosive equation-of-state gives the correct energy delivery to the plate to within a few percent. The spike seen in the first velocity jump in the calculation is due to the spike that is present in the nucleation and growth model.[12] The adhesive layer, typically 0.05 mm thick, lowers the calculated velocity by ~2 percent.

We obtained agreement to within approximately 10 percent with Experiment No. 3777 (reacting RX-03-BB) at calculational flyer velocity of 1.37 mm/µs. Calculations also indicate that we may have observed an experimental "reverse detonation," i.e., detonation of the high explosive by the shock reflected from the plate.

CONCLUSIONS

We have obtained excellent experimental free-surface velocity records using the Fabry-Perot velocimeter. Our calculations of this velocity employing beta-burn and nucleation and growth high explosive models are in good agreement with fully detonating PBX-9404 and RX-03-BB experiments. Improved agreement will require adjustments in the high explosives equation-of-state, and in the nucleation and growth parameters. We have an experimental technique that gives records whose agreement with calculations are sensitive to the model and is therefore a good way of testing high explosive models.

Measurement of the reaction zone length is possible using a technique reported by Duff and Houston.[1] By use of their equation, we have made a rough estimate of the reaction zone length in PBX-9404 to be less than 0.25 mm. Experiments with thinner plates of various materials are planned.

ACKNOWLEDGEMENTS

The authors wish to thank Dr. Richard C. Weingart for suggesting these experiments and for valuable discussions. We also thank Leona Meegan, who assembled the experiments, and the 102 mm gun crew: Bill Mumper, Bill Duguid and Mel Bainter. We extend our appreciation to Dr. Lawrence L. Marino, S Division Leader for his encouragement and support of this work.

REFERENCES

1. R. E. Duff and E. Houston, J. Chem. Phys. $\underline{23}$, 1268 (1955).

2. W. E. Deal, J. Chem. Phys. $\underline{27}$, 796 (1957).

3. W. C. Davis and D. Venable, "Pressure Measurements for Composition B-3," Fifth Symposium (International) on Detonation, Passadena, CA, 13 (1970).

4. F. J. Petrone, Phys. of Fluids $\underline{11}$, 1473 (1968).

5. V. A. Veretennikov, Khim. Fiz. Protsessov Goreniya i Yzryva: Detonatsiza (Chem. Phys. of Processes of Combustion and Explosion: Detonation), Chernogolovka, 3 (1980). (English Translation: LLNL Ref. 02925 (1980)).

6. "Reference to a company or product name does not imply approval or recommendation of the product by the University of California or the U.S. Department of Energy to the exclusion of others that may be suitable."

7. L. M. Erickson, "The Detonation Physics Program," Lawrence Livermore National Laboratory Internal Document No. UCID-182655, 1979 (unpublished).

8. H. H. Chau, D. R. Goosman, J. W. Lyle, and M. A. Summers, "A Simple Velocity Interferometer System," Conference on Laser and Electro-Optical Systems, OSA/IEEE, Abstract TUHH4, 20 (1978).

9. M. L. Wilkins, Lawrence Livermore National Laboratory Report No. UCRL-7322, REV. 1, 1969 (unpublished).

10. J. P. Woodruff, "KOVEC User's Manual," Lawrence Livermore National Laboratory Internal Document No. UCID-17306, 1976 (unpublished).

11. S. G. Cochran and J. Chan, "Shock Initiation and Detonation Models in One and Two Dimensions," Lawrence Livermore National Laboratory Internal Document No. UCID-18024, 1979 (unpublished).

12. J. Chan, Private Communication.

THIN PULSE INITIATION OF PBX-9404

H. C. Vantine, J. Chan and L. M. Erickson
Lawrence Livermore National Laboratory
Livermore, California   94550

ABSTRACT

We studied the evolution of a critical energy thin pulse in PBX 9404 using both experimental and calculational techniques. We showed that our temperature-dependent model is more successful for calculating the thin pulse evolution than our pressure-dependent model. Detailed structure of the wave shape was not calculated by either model and requires further development.

INTRODUCTION

A pressure pulse of amplitude P and duration τ will cause initiation of PBX-9404 explosive if it satisfies the relation $P^2\tau > 5.5$ $GPa^2\mu s$. We have investigated the evolution of such a pulse with $P^2\tau = 5.6$ $GPa^2\mu s$ as it builds to a detonation front using embedded velocity gauges[1].

We also calculated the evolution of the shock pulse with two high explosive (HE) ignition models[2]. One of the models is pressure-dependent; the other model is temperature-dependent. We expected the temperature-dependent model to agree more closely with experiment. This is because after the passage of the thin pulse the explosive is left near zero pressure but the hot spots are left at a finite temperature. With the pressure-dependent model the reaction rate drops to nearly zero while the temperature-dependent model continues to have a finite reaction rate behind the shock front.

We introduced a second wave with a value of $P^2\tau$ about one-tenth of the first pulse. The second wave grows dramatically in amplitude and overtakes the first pulse at a speed of 25 km/s -- three times the detonation velocity. Neither model can predict the details of the evolution of the second wave.

EXPERIMENTAL

We placed velocity gauges in the flow at depths of 0, 2, 4, 8, 14 and 22 mm (see Figure 1).

We wanted to introduce a thin pulse into the 9404 explosive that quenched quickly. To do this we used a flyer material, dynasil, that has a "rarefaction shock;" the front of the wave is a ramp; the back of the wave is a step. We preferred this to a high impedance flyer which would produce a staircase release in the explosive. We believe that a sharp pulse is a more difficult test of the HE

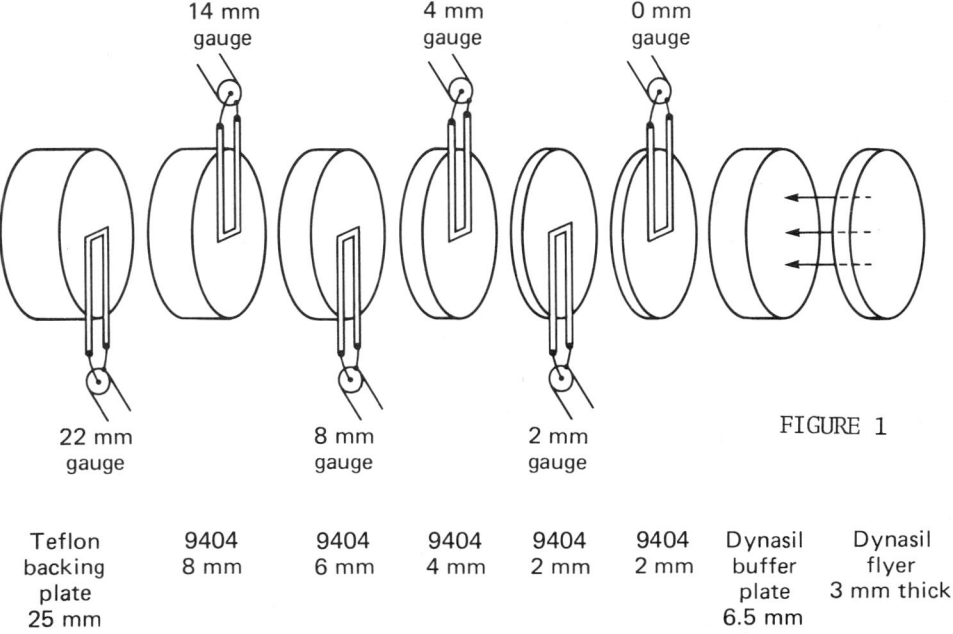

FIGURE 1

initiation models. Analysis of wave propagation in the dynasil flyer and buffer plate, shown in Figure 2, shows that two pulses are introduced into the HE. Figure 3 shows the early time evolution of both pulses. Note that the first pulse travels with a near constant shock velocity of 3.8 km/s. The amplitude of the first pulse is nearly constant at 0.4 km/s but is decaying slowly. The second pulse also travels initially at a velocity of about 4 km/s. A comparison of Figures 4a and b, however, shows that the second pulse overtakes the first pulse with an apparent velocity of 25 km/s.

CALCULATIONS

We calculated the flow with two HE initiation models and a 1D hydrodynamic code. The pressure-dependent model had a rate law that depended only on pressure. This rate law is shown in equation 1 (P is in GPa). F is the fraction reacted and varies between 0 and 1. The temperature-dependent model has a rate law that depends only on the hot spot temperature. This rate law is shown in equation 2.

$$R = A(1-F)P^n + BF(1-F)P^m \qquad (1)$$
$$A = 0.756\text{E-}3 \; \mu s^{-1} \quad n = 3.2$$
$$B = 0.24 \; \mu s^{-1} \quad m = 1.0$$

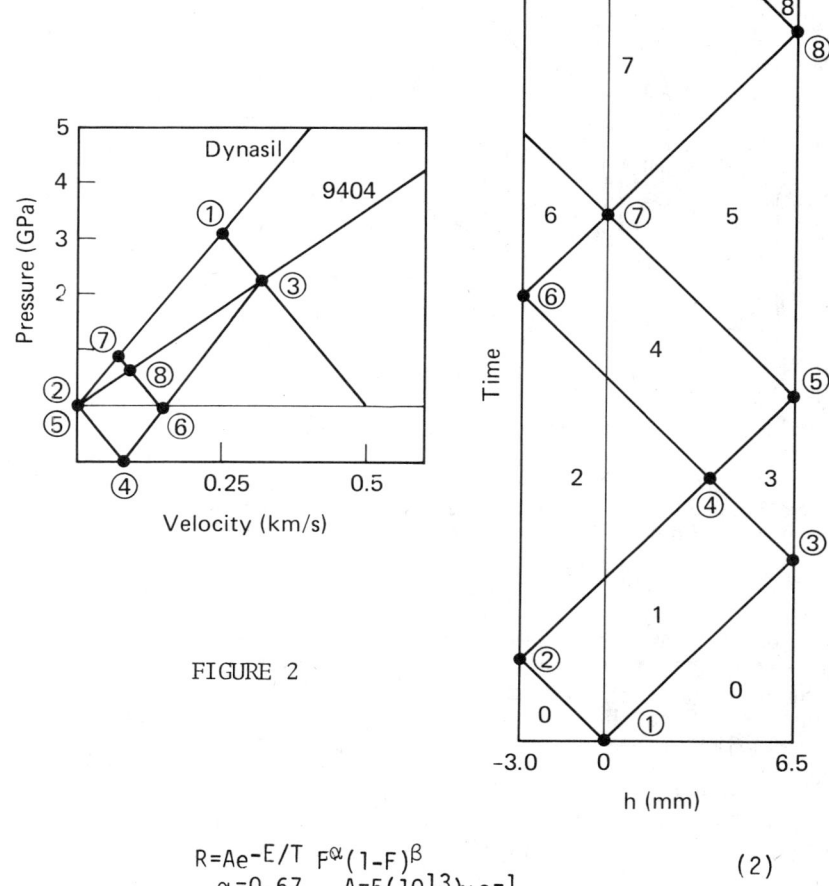

FIGURE 2

$$R = Ae^{-E/T} F^{\alpha}(1-F)^{\beta} \qquad (2)$$
$$\alpha = 0.67 \quad A = 5(10^{13}) \mu s^{-1}$$
$$\beta = 0.2 \quad E = 3(10^4) K$$

The particle velocity records predicted by the two rate laws are shown in Figures 3 and 4. Each model is fairly successful in calculating the first gauge position. However, at 2 mm there is already a difference in the model predictions. Behind the first pulse the pressure has fallen to zero. The hot spot temperature is estimated to be about 700 K in the pulse and behind the pulse. The pressure-dependent model predicts that growth should occur near the front. The predicted rate is too large and the first pulse grows to a detonation. The temperature-dependent rate predicts growth behind the front and this is more in tune with the observed record. This model predicts about the right time to detonation.

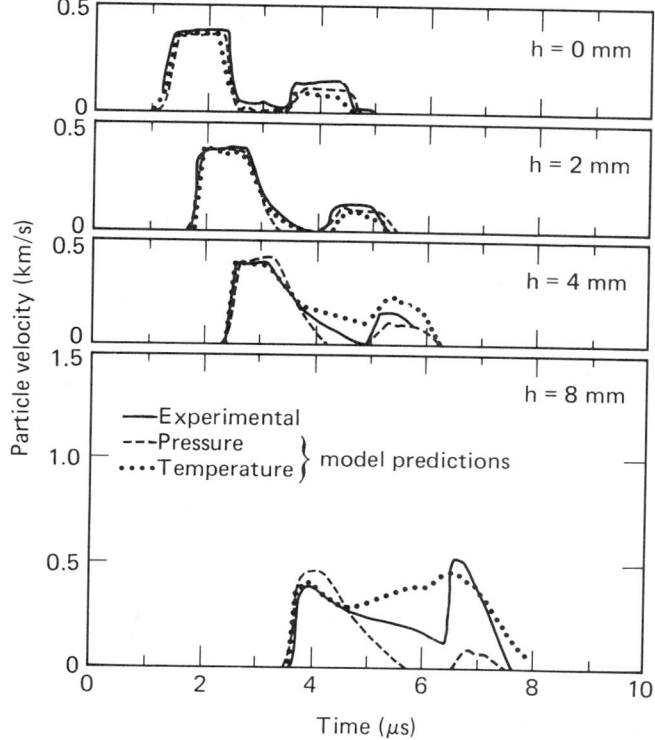

FIGURE 3

DISCUSSION

It appears that the temperature-dependent model predicts too much reaction behind the front in the early stages and the pressure-dependent model predicts too little reaction. This suggests that both pressure and temperature need to be included in one model. Physically this is reasonable since the hot spot temperature controls the chemical kinetics and the pressure controls the hydrodynamic growth of the hot spots.

We may speculate on the added physics that is needed to calculate the behavior observed in Figure 4. It seems unlikely that the second shock wave travels at 25 km/s. This being the case, it seems likely that the observed velocity of 25 km/s is more in the nature of a phase velocity. If this is true then the behavior shown in going from Figure 4a to 4b is some sort of "thermal explosion" phenomena.

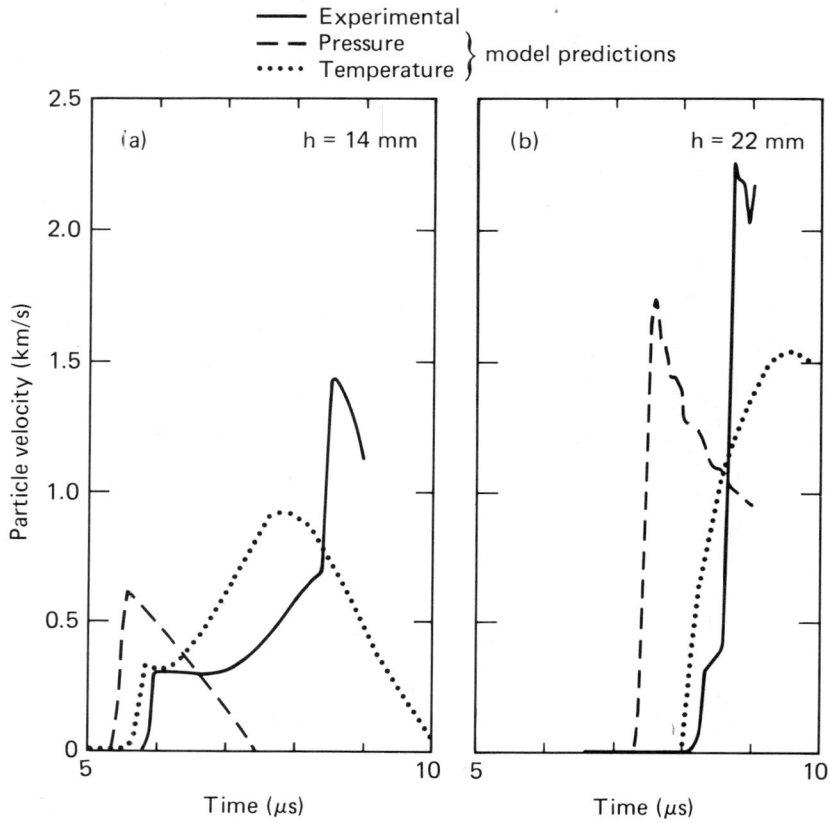

FIGURE 4

(1) H. C. Vantine, W. D. Curtis, L. M. Erickson and R. S. Lee, "A Comparison of Stress and Velocity Measurements in PBX-9404 Explosive," Proceedings of the Eighteenth Symposium on Combustion (in press), The Combustion Institute (1980).

(2) S. G. Cochran and J. Chan, "Shock Initiation and Detonation Models in One and Two Dimensions," Lawrence Livermore National Laboratory Report UCID-18024 (1979).

HIGH EXPLOSIVE DETONATIONS IN VARYING OXYGEN ATMOSPHERES

F. R. Kovar, K. R. Trigger, L. G. Guymon,
and J. R. Harvey
Lawrence Livermore National Laboratory*
P. O. Box 808, Livermore, CA 94550

ABSTRACT

Small spheres of PBX9404 and TNT were detonated in atmospheres of varying oxygen content to study the magnitude of the effects of oxygen on afterburn of high explosive (HE) detonation products (DPs). Each of the fireballs (FBs) formed by the HEDPs was observed by high-speed cameras and radiometers in the visible and near infrared. It was found that afterburn dominates the FB radiation, while having much less influence on FB size than do instabilities and turbulence. One-dimensional nonreactive fluid flow calculations of the FB radius-time (R-T) history and shock characteristics within the FB were performed. The R-T results are in good agreement with the experiment at early times, while the internal shock agreement extends over a much longer period.

INTRODUCTION

The DPs of HE such as TNT or HMX contain large quantities of H_2, CO, and solid carbon (C_s).[1] The FB formed by these DPs in air undergoes secondary combustion, or afterburn, converting some of the H_2 to H_2O and some of the C_s and CO to CO_2. This process is expected to affect the size and radiation of the FB. To investigate the magnitude of this effect, we performed a series of experiments in which we varied the oxygen content in both the atmospheric gas and the HE.

An accurate analysis of these experiments requires detailed calculations of complex phenomena, particularly instabilities, turbulence, and hydrocarbon chemical kinetics. This was beyond the scope of our effort. However, calculations of the FB R-T history and shock characteristics within the FB for one-dimensional nonreactive fluid flow were made.

After a brief description of the experiments and the diagnostic apparatus, the results are presented and compared with the calculations. This is followed by a discussion of the measurements of the radiation from the FB and the deduced time dependence of the intensity, which is closely related to afterburn. Identification of the specific sources of the radiation from the FB was not considered, since it was beyond the scope of this work.

---

*Work supported by the U.S. Department of Energy.

EXPERIMENTS

The experiments are listed in Table I. Oxygen variation in the HE was accomplished through the use of two different explosives, TNT ($C_7H_5N_3O_6$) and PBX9404 ($C_{1.4}H_{2.75}N_{2.57}O_{2.69}Cl_{0.03}P_{0.01}$) which is 94% HMX. The oxygen balance[1] of these explosives is -74% for TNT and -24% for PBX9404.

Table I. Type of HE and atmosphere in each experiment. (All spheres had an initial radius of 4.0 cm.)

HE type	Mass (g)	Atmosphere, % O_2 content	HE density (g/c^3)	Detonation point
PBX9404	493	Air (21% O_2)	1.84	Center
*PBX9404	493	N_2 (3.5% O_2)	1.84	Center
*PBX9404	493	O_2 (96% O_2)	1.84	Center
*PBX9404	493	Air (21% O_2)	1.84	Center
TNT	440	Air (21% O_2)	1.64	Surface
*TNT	440	N_2 (1% O_2)	1.64	Surface

*These shots were fired inside an ~10 foot cubical sealed "tent" of 4 mil mylar.

The HE FBs were observed with the following cameras and radiometers.
  Cameras: Locam at 400 frames/s, Hycam at ~7200 frames/s (2 with different apertures), Model 6 at 1 µs/frame, Model 6 at 5 µs/frame (2 with different apertures).
  Photodiode detector radiometers (band-pass filters are specified with half-width at half-maximum): 3 channels, recorded at discrete time intervals--Si, 0.626 ± 0.047 µm; Si, 0.840 ± 0.048 µm; PbS, 2.2 ± 0.166 µm.
  3 channels, continuously recorded on 4 oscilloscopes with sweep lengths of 100 µs, 500 µs, 5 ms, and 50 ms--Si, 0.565 ± 0.0145 µm; Si, 0.860 ± 0.0235 µm; Si, no filter.

The photographic data were used to measure the FB growth and to estimate the onset of significant instabilities and the subsequent turbulence at the HEDP-gas interface. The signals detected by the radiometers provided a relative measure of the FB radiation dependence on the variations of the oxygen content in the atmosphere and the HE.

The PBX9404-air experiments were conducted with and without an enclosure to determine its effect on the radiometer signals. After accounting for the mylar attenuation, the differences between the signals were ~10%, our limits of reproducibility.

DISCUSSIONS

Calculations of the HEDP expansion in air were made using KOVEC, an LLNL Lagrangian, one-dimensional hydrodynamic, finite-difference code. The FB radii from the PBX9404 experiments and KOVEC have excellent agreement at times $\lesssim 50$ µs/cm as shown in Fig. 1, where the geometric similarity law for scaling has been used, i.e., both the experimental times and radii have been divided by the initial radius of the HE sphere.

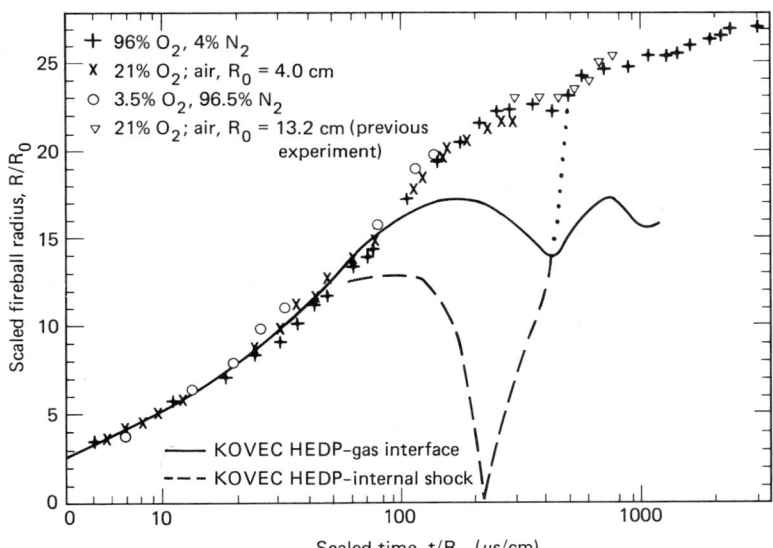

Fig. 1. FB radius vs time plots in units of R_0, the initial radius of the HE sphere, for the expansion of PBX9404 DPs in various atmospheres of oxygen content. For $t/R_0 \lesssim 50$ µs/cm, the uncertainty in R/R_0 is $\lesssim 5\%$, whereas for greater times it is $\lesssim 10\%$. The FB spherical asymmetry and turbulent surface are the main sources of error.

The Model 6 photographs showed that the FB surface had developed a large number of "lumps" by a scaled time of ~50 µs/cm. These "lumps" are an indication of the turbulent mixing[2] of the HEDP and gas. Since KOVEC ignores this phenomenon, the later divergence between calculation and experiment is not surprising.

Figure 1 also shows a difference between the interface trajectories for the different PBX9404-gas cases at times <100 µs/cm which we attribute primarily to the difference in gas densities.

The KOVEC results show that the compressed atmosphere first reverses the FB expansion at a maximum radius of 17.5 IR (initial HE sphere radius) at ~175 µs/cm. The experimental results in Fig. 1 show that the FB reaches its first maximum of ~23 IR at ~250 µs/cm and then stagnates. The calculations also show that an internal FB shock is generated during the initial fluid motion at the HEDP-gas interface. This shock travels to the center, is reflected, and travels out to the surface, causing the FB to start to expand at 420 µs/cm, whereas the experimental data show the start of the expansion at ~500 µs/cm. This difference of times of shock arrival at the FB surface is consistent with the difference of the FB radii in KOVEC and the experiments.

KOVEC shows a repeated process of FB expansion and contraction, followed by another internal shock interaction at its surface at ~1 ms/cm. Similarly, there is some experimental indication that the FB starts another expansion at about the same time.

Since the differences in FB radii between the PBX9404 experiments during the time 200 to 700 µs/cm are within experimental uncertainties, we conclude that, before ~700 µs/cm, afterburn has a much smaller effect on the FB size than instabilities and turbulence do. While we are tempted to draw this conclusion for the entire FB history, insufficient experimental data at late times prevents this.

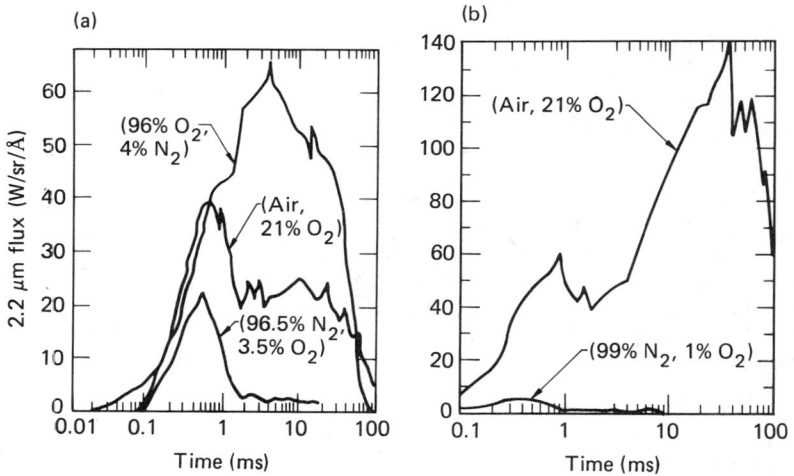

Fig. 2. Radiometer signals at 2.2 µm from the FBs of HE detonation products shown in various atmospheres of oxygen content with (a) centrally-detonated spheres of PBX9404 and (b) surface-detonated spheres of TNT.

The effect of afterburn can be more easily seen by the marked increase in radiation intensity as the oxygen concentration in the gas increases. This is observed in all of the radiometer data. Those obtained at 2.2 µm for PBX9404 and TNT are shown in Figs. 2(a) and 2(b) respectively. These clearly indicate that the radiation is dominated by afterburn. In addition, comparison of these figures shows considerably more radiation from TNT in air than from PBX9404 in air. This is consistent with the afterburn phenomenon, since the more negative oxygen balance of TNT allows for more combustion of its DP. We can obtain a better description of the radiation intensity, I, by eliminating the effect of the FB area with radius r on the detected flux, $F \propto Ir^2$. Using the continuously recorded Si detector data at 0.86 μm from 10 μs/cm $\lesssim t \lesssim$ 100 μs/cm, we found that $I \propto \exp(-t/\tau)$, where τ increases as the oxygen content in the gas increases. This time dependence of the intensity is about the same as that for the oxidation of fuel at the HEDP-gas interface using a global reaction rate[3] with the appropriate, time-dependent densities and energies from the KOVEC calculations. After the first peak of the detected flux decays, the signal levels off, as does the radius until a smaller radiometer peak occurs. It appears that the initial rise of the signal is caused by the appearance of the internal shock within an optical depth of the FB surface. This shock raises the temperature, enhancing the combustion and the ensuing radiation. The shock then causes the FB to expand with further mixing of the oxidizing gas and the DP because of the turbulent FB surface. The intensity then decays as the shock disappears from "view" and the additional surface mixing slows.

The later increase in the FB radius, commencing at ~1 ms/cm, accounts for much of the corresponding flux increase. This occurs with the rise of another radiation peak, which decays with the end of the afterburn.

ACKNOWLEDGMENTS

We wish to thank Terry Herther of Sandia National Laboratories, Albuquerque, for providing the use of his photodiode radiometers and their data analysis.

REFERENCES

1. D. L. Ornellas, The Heat and Products of Detonation of Cyclotetramethylenetranitramine, 2, 4, 6-Trinitrotoluene, Nitromethane, and Bis[2, 2-dinitro-2-fluoroethyl] formal, J. Chem. Phys. 72, 2390 (1968).
2. J. O. Hinze, Turbulence (McGraw-Hill Book Company, Inc., New York, 1959), p. 277.
3. C. K. Westbrook and F. L. Dryer, Simplified Reaction Mechanisms for the Oxidation of Hydrocarbon Fuels in Flames, Lawrence Livermore National Laboratory, Livermore, Calif., UCRL-84943 (1980).

CALCULATED SHOCK PRESSURES IN THE AQUARIUM TEST*

J. N. Johnson
Los Alamos National Laboratory, Los Alamos, NM 87545

ABSTRACT

A new method of analysis has been developed for determination of shock pressures in aquarium tests on commercial explosives. This test consists of photographing the expanding cylindrical tube wall (which contains the detonation products) and the shock wave in water surrounding the explosive charge. By making a least-squares fit to the shock-front data, it is possible to determine the peak shock-front pressure as a function of distance from the cylinder wall. This has been done for 10-cm and 20-cm-diam ANFO (ammonium nitrate/fuel oil) and aluminized ANFO (7.5 wt% Al) aquarium test data.

INTRODUCTION

The aquarium test of explosive performance consists of optical measurement of detonation velocity, shock-wave position, and expansion rate of the pipe containing the test product. These experiments have been described in a report by Craig, et al.[1] The expansion of the tube wall gives information on the equation of state of the detonation products, and this has been used to good advantage in determining the performance properties of ANFO.[1] The shock wave in water contains information on peak shock pressure delivered to the surrounding medium, and much less has been done with this information. The angle between the shock front and the cylinder axis is maximum at the pipe wall and monotonically decreases with increasing distance from the pipe wall. This is a consequence of higher shock pressures near the cavity wall and the dependence of shock velocity on compression. This information is used here to calculate pressure as a function of radius at the expanding shock front in a number of ANFO and aluminized ANFO aquarium tests.

Explosive initiation is assumed to take place at the top of a cylindrical column. In a downward-moving coordinate system traveling with steady detonation velocity D, the detonation wave appears to be stationary, with material below flowing upward at velocity D, as shown in Fig. 1. The shock velocity at point A is given by U. The unit normal and unit tangent vectors at point A are $\hat{n}$ and $\hat{t}$, respectively. If the shock wave in the water is steady (that is, unchanging in shape), it is also stationary in this coordinate system and the relationship between U and D becomes

$$U = D \cdot \sin \theta. \tag{1}$$

In this stationary coordinate system, material ahead of the shock is moving at velocity $-D \sin \theta$ and $D \cos \theta$ relative to the $\hat{n}$ and $\hat{t}$ axes.

*Work supported by the U.S. Department of Energy.

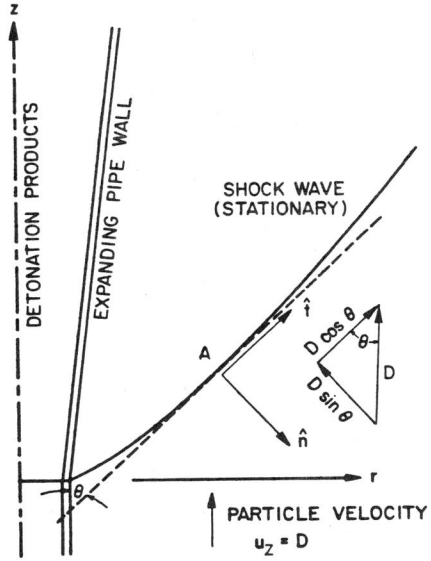

Fig. 1. Stationary shock front and expanding pipe wall in coordinate system moving downward at velocity D.

The Rankine-Hugoniot jump conditions for the cylindrically diverging, curved shock front shown in Fig. 1 can then be written as[2]

$$\rho u = -\rho_0 D \sin \theta \quad , \qquad (2)$$

$$v = D \cos \theta \quad , \qquad (3)$$

$$p + \rho u^2 = \rho_0 D^2 \sin^2 \theta \quad , \qquad (4)$$

$$(u^2 + v^2)/2 + e + p/\rho = D^2/2 \quad , \qquad (5)$$

where u and v are velocity components in the $(\hat{n},\hat{t})$ coordinate system, ρ is the material density, p is the pressure, and e is the internal energy per unit mass, all evaluated behind the shock, and ρ_0 is the undisturbed density of the material ahead of the shock (at zero pressure and zero internal energy). From Eqs. (2)-(5), the internal energy change across the oblique shock of Fig. 1 is given by

$$e = (p/2)(1/\rho_0 - 1/\rho) \quad , \qquad (6)$$

which is exactly the same as that for a normal shock. Therefore, the pressure-volume states for the oblique shock lie on the Hugoniot curve determined by one-dimensional plane shock-wave experiments;

$$p = \frac{\rho_0 c_w^2 \varepsilon}{(1 - s\varepsilon)^2} \quad , \qquad (7)$$

where $\varepsilon \equiv 1 - \rho_0/\rho$, $\rho_0 = 1.0$ g/cm^3, $c_w = 0.148$ cm/µs is the acoustic (low amplitude sound) wave velocity in water, and $s = 2.0$ is the slope of a straight-line fit to the shock velocity-particle velocity data for water.[3] From Eqs. (2) and (4), it is found that

$$p = \rho_0 D^2 \sin^2 \theta \varepsilon \quad , \qquad (8)$$

and hence, from Eqs. (7) and (8),

$$p = \rho_0 D^2 \sin^2 \theta (1/s)[1 - c_w/(D \sin \theta)] \quad , \qquad (9)$$

which gives the pressure behind the shock wave when the steady detonation speed D and the angle θ between the shock and the cylinder axis are known.

The angle θ is obtained from a fit of the shock front data to an expression of the form

$$r = r_0 + \tan\theta_{min}\left[z + \frac{1}{a}\left(\frac{\tan\theta_{max}}{\tan\theta_{min}} - 1\right)\left(1 - e^{-az}\right)\right] , \quad (10)$$

where

$$\tan\theta_{min} = \frac{c_w/D}{\sqrt{1 - (c_w/D)^2}} , \quad (11)$$

and a and $\tan\theta_{max}$ are the two parameters determined by the method of least squares.[4] The angles θ_{max} and θ_{min} have the physical interpretation of being the values of θ at (z = 0, r = r_0) and as r (and z) → ∞, respectively.

The expanding-pipe-wall data is fit to a linear expression of the form

$$r = r_0' + (\bar{V}/D)z , \quad (12)$$

where D is again the detonation speed, and r_0' and $\bar{V}$ are constants determined by the method of least squares. For a steady propagating detonation (without change in shape of pipe wall or shock-front positions) in the negative z-direction,

$$r(z,t) = r_0' + (\bar{V}/D)(z + Dt) , \quad (13)$$

and the outward pipe velocity is given by

$$dr/dt = \bar{V} . \quad (14)$$

Shock front and pipe-expansion data for an aquarium test on ANFO (Gulf N-C-N 100, Gulf Oil Chemicals Co., Miriam, Kansas) contained in a 10-cm-i.d. clay pipe are shown in Fig. 2a along with the least-square fits; the ordinate in this figure is Z/D, the time behind the assumed steady propagating detonation front. The calculated shock pressure as a function of radial position is shown in Fig. 2b. A summary of initial density, detonation velocity, $\bar{V}$, and P_{wall}, the calculated shock pressure at the pipe wall, is given in Table I.

DISCUSSION

The calculated shock pressures at the pipe wall depend on initial density and charge diameter, as expected. The effect of aluminization on explosive performance is difficult to access from the data presented here, but remains an important one to try to quantify.

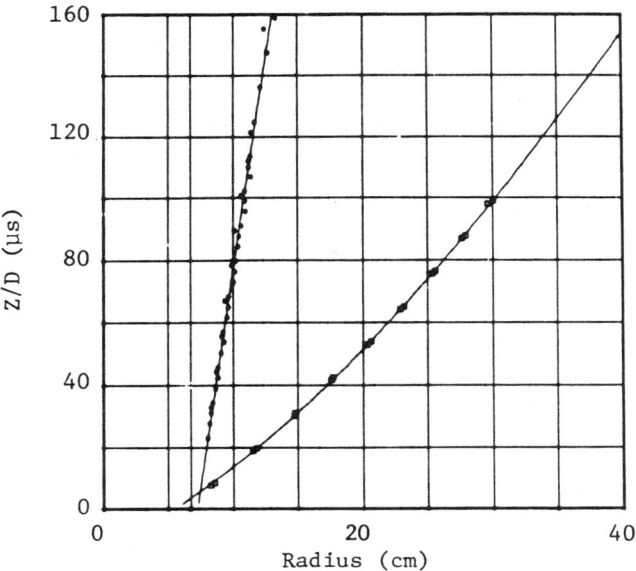

Fig. 2a. Shock front and pipe expansion data for aquarium test 4652 ($\rho_0 = 0.90$ g/cm^3) with least-square fits (solid lines).

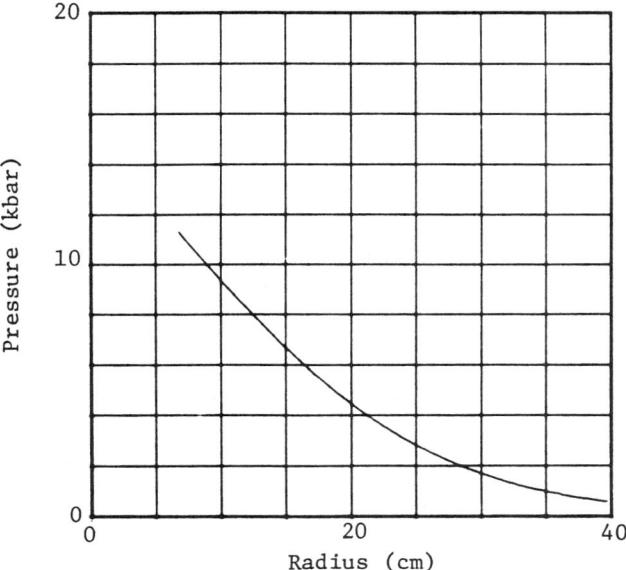

Fig. 2b. Calculated shock pressure as function of radial position.

Table I. Summary of ANFO aquarium tests: clay pipe confinement

Experiment	Diameter (cm)	ρ_0 (g/cm^3)	D (km/s)	$\bar{V}$ (km/s)	P_{wall} (kbar)
4678	10	0.79	3.27	0.34	8.9
4724*	10	0.87	3.63	0.42	9.3
4652	10	0.90	3.47	0.37	11.3
4768	10	0.93	3.60	0.39	11.7
4688	20	0.79	3.78	0.39	12.3
4707*	20	0.88	3.98	0.50	16.0
4664	20	0.90	4.12	0.43	15.0
4700	20	0.90	4.15	0.47	15.8
4752*	20	1.11	4.24	0.51	19.5

*7.5 wt% aluminum (Gulf N-C-N 750).

Experimental work is continuing on the actual time-resolved measurement of shock pressures for well characterized aluminized ANFO explosives. Combination of these measurements with the analysis presented here will provide a check on data consistency as well as unambiguous information on the role of aluminum in improving explosive performance.

REFERENCES

1. B. G. Craig, J. N. Johnson, C. L. Mader, and G. F. Lederman, "Characterization of Two Commercial Explosives," Los Alamos Scientific Laboratory report LA-7140 (May 1978).

2. L. D. Landau and E. M. Lifshitz, Fluid Mechanics (Pergamon Press, London, 1959), pp. 317-319.

3. S. P. Marsh, LASL Shock Hugoniot Data (University of California Press, Berkeley, 1980), pp. 573-574.

4. H. Margenau and G. M. Murphy, The Mathematics of Physics and Chemistry (D. Van Nostrand Co., Inc., Princeton, 1956), pp. 517-518.

EFFECT OF CHARGE DIAMETER ON DETONATION PRESSURE MEASURED BY AQUARIUM TECHNIQUE

Kang Xu, De-yang Yu, Yun-xiang Xu, Xiung-fei Zeng
Lanzhou Modern Chemistry Research Institute, China

ABSTRACT

In this work, the aquarium technique is used for explosive detonation pressure measurement and the effect of charge diameter (from 10 to 100 mm) on the measured results is studied. The obtained results show that, when different charge diameters are used, we get nearly the same initial shock velocities in water, and thus nearly the same detonation pressures. Based on this fact, the aquarium technique can be used for detonation pressure measurements with small size samples of explosive. The process of shock wave attenuation in water is discussed briefly.

INTRODUCTION

Detonation pressure is one of the main characteristics of explosives. Ordinarily, detonation pressure cannot be determined directly. In most methods for determining detonation pressure, one measures the shock wave parameters in an inert medium adjacent to the detonating explosive charge and then, using the condition of continuity at the interface, calculates the detonation pressure. One of these methods is the aquarium technique[1-3] in which the shock velocity in water is measured. By this technique, the attenuation process of shock waves in water can be determined in one shot; this result can scarcely be obtained by other methods.

The purpose of this work is to study the effect of explosive charge diameter on the detonation pressure measured by the aquarium technique, and to examine the possibility of determining detonation pressure of high explosives with small samples.

EXPERIMENTAL

The experimental setup used is shown in Fig. 1, which is similar to that in Reference 2. A wooden aquarium with two windows is used. A fine quality plate glass of about 1.5 mm thick is mounted on the front window and a lens on the back one. There are two differences between our setup and that of Rigdon.[2] In our work an ordinary explosive light source is used for photographing. Between the light source and the lens is a wooden plate with a 3-4 mm diameter hole. By setting the hole at the focus of the lens, one obtains parallel light, which greatly improves the quality of the photographic image. With this method a clearcut sweep picture can be obtained. Second, in experiments with charges of diameters less than 60 mm, we use a new type simple plane wave generator,[4] its diagram is shown in Fig. 2. A plexiglas plate with central hole is placed between the explosive charges. By adjusting plate thickness and hole diameter, we can find their optimum values, with which we get a quite good plane detonation waves at the end of the explosive charge. Arrival times of the detonation wave over the central 60% of the end surface of the charge are within 0.03 µs.

In our experiments a Model GSJ high speed streak camera made in China is used. The sweep rate is 3 mm/µs and the photographic ratio is about 1:2.2. A typical sweep picture is shown in Fig. 3.

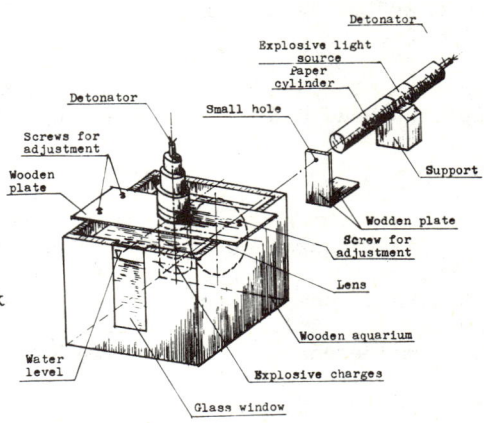

Fig. 1 Experimental setup in detonation pressure measurement by the aquarium technique.

The pictures are measured by an instrumented microscope. From the picture we can see that the initial part of the sweep line is nearly straight. For charges of different diameters the length of this nearly straight part is not the same; the larger the diameter, the longer the length of the straight part. For charges of 10 mm diameter, this length is only about 0.9 mm (length on photograph). We use two measuring methods: In the first method, after reading out X,Y coordinates of points on the sweep line with 0.1 mm intervals along Y-axis, we use a linear regression method to calculate the slope of the straight part of the sweep line. In the second method, we determine the angle between this straight line and the horizontal axis, and then calculate its slope. From the slope we get the shock velocity in water. For sweep pictures of good quality these two methods give results in agreement with each other within experimental error. The method of angle measurement is simpler and more convenient. For examining the attenuation rate of shock wave in water we make angle measurements along a sweep line with 0.2 to 0.5 mm intervals to get the shock velocities at different shock wave transit distances.

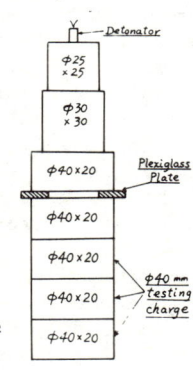

Fig. 2 Simple plane wave generator and charges.

Using the Hugoniot data of water given by Rice and Walsh,[5] we calculated the particle velocity and shock pressure in water; and using the method given by Deal in 1958,[6] we calculate the detonation pressure.

Figure 3. Sweep picture (Charge diam. 40 mm)

EXPERIMENTAL RESULTS AND DISCUSSION

The effect of charge diameter was studied mainly with RDX/Paraffin 95/5 explosive. Charges of diameter 100, 80, 60, 40 and 25 mm were used. The obtained results are shown in Table 1.

Table 1. Effect of RDX/Paraffin 95/% charge diameter on detonation pressure
(Density of charge: 1.655 g/cc. Detonation velocity: 8.358 mm/μs)

Charge diam. (mm)	Charge length (mm)	Number of shots	Initial shock velocity in water (mm/μs)	Detonation pressure (GPa)	Note
100	120	5	6.09 ± 0.04	26.2 ± 0.3	with ordinary PWG
80	120	4	6.11 ± 0.06	26.4 ± 0.5	
60	90	4	6.06 ± 0.04	26.1 ± 0.3	
40	80	9	6.05 ± 0.06	26.0 ± 0.4	with simple PWG
25	70	8	6.11 ± 0.08	26.4 ± 0.4	
Average of 30 shots			6.08 ± 0.06	26.4 ± 0.4	Error: 1.5%

Data are converted to $\rho = 1.655$ g/cc by following methods:
For detonation pressures: $P_{\rho_1} = P_{\rho_2}(\rho_1/\rho_2)^2$
For detonation velocities: $\Delta D = 0.0045$ mm/μs/g/cc.

In addition, we measured detonation pressures with RDX/Binder 99.5/0.5 charges of diameters 40, 25, 10 mm and charges of 10 mm diameter with confinement. Results are shown in Table 2.

Tables 1 and 2 show that for high explosives like RDX, charges of diameters from 10 to 100 mm give initial shock velocities in water and thus the detonation pressures are in agreement with each other within experimental error, which is about 2% for detonation pressures. The only difference is, when small diameter charges are used, shock velocities attenuate faster, i.e., sweep line curved notably at the earlier part of sweep line as compared with the larger diameter charges. Fig. 4 shows the sweep line obtained with 10 mm diameter charges. (Compare with Fig. 3).

If we fit the equation $U_s = a'e^{bY}$ or $\ln U_s = a + bY$ to the sweep line, (where U_s is the shock velocity in water; Y is the shock wave transit distance ;

Fig. 4 Sweep picture (Charge diam. 10 mm)

Table 2. Effect of RDX/Binder 99.5/0.5 charge diameter
on detonation pressure
(Density of charge: 1.700 g/cc. Detonation velocity: 8.415 mm/μs)

Charge diam. (mm)	Charge length (mm)	Number of shots	Initial shock velocity in water (mm/μs)	Detonation pressure (GPa)	Note
40	80	7	6.39 ± 0.07	29.4 ± 0.7	
25	75	5	6.31 ± 0.08	28.7 ± 0.7	
10	55	4	6.31 ± 0.08	28.7 ± 0.7	
10	55	5	6.33 ± 0.06	29.0 ± 0.5	with steel confinement
Average of 21 shots:			6.34 ± 0.07	29.0 ± 0.7	Error: 1.5%

Data are converted to ρ = 1.700 by following methods:
For detonation pressures: by the same method as in Table 1.
For detonation velocities: ΔD = 0.0035 mm/μs/g/cc.

a, a' and b are constants), then in $\ln U_s$ - Y plot, straight lines can be obtained with a slope b. Fig. 5 is the plot of experimental results obtained with charges of diameters 25 and 10 mm. Fig. 5 shows that for charges of 25 mm diameter, there is a deflection point at about Y = 15 mm, and for charges of 10 mm diameter (with or without confinement), there are deflection points at about Y = 5 mm. We think that these deflection points indicate the moment at which the side rarefaction wave in water arrives at the axis of the charge and starts to decelerate the shock wave. For charges of 40 mm diameter, within the measured range, the $\ln U_s$- Y plot is a straight line because the rarefaction wave has not arrived at the axis of charge in this period. On the $\ln U_s$- Y plot of 10 mm diameter charges, charges, we can see some additional deflection points before the arrival of a side rarefaction wave. We think that at the very beginning, the shock wave in water is attenuated only by dissipation, so charges of different diameters, with or without confinement, should give the same value of b.   (see Fig. 5). This is followed by a region where the shock wave attenuates more quickly, which, perhaps, is due to the rear rarefaction wave. For correct determination of the detonation pressure, only the velocity of shock wave uninfluenced by rear or side rarefaction waves should be measured. It is seen from Fig. 5 that for 10 mm diameter charges with or without confinement the length of this region is about 2 mm, which is only 0.9 mm on our photographs. This is the nearly straight part of the sweep line. If a longer region were taken, a lower slope and thus a lower detonation pressure would be obtained.

The detonation pressure of RDX measured in this work is lower than the value in the literature (for example see Reference 7). The reason is not clear yet.

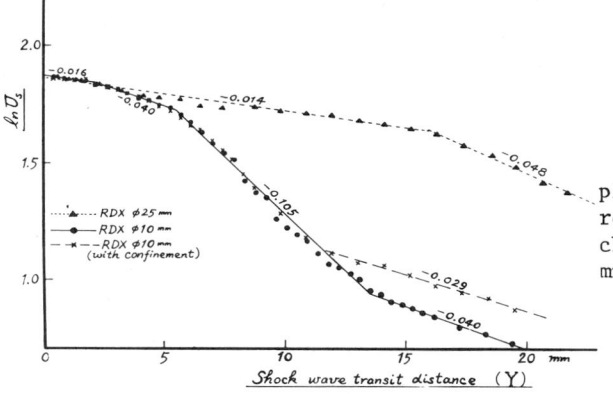

Fig. 5 ln U_s - Y plot of experimental results obtained with charges of 25 and 10 mm diameters.

CONCLUSION

Experimental results obtained in this work show that, when we measure the detonation pressure of high explosives by the aquarium technique, explosive charges of different diameters larger than 10 mm give nearly the same initial shock velocities in water, so the detonation pressures calculated from them are also the same within experimental error. But their attenuation rates are quite different. Since with small charges the attenuation rate of the shock wave is larger than with larger charges, when small charges are used the initial shock velocity must be measured within a very small region in order to get the same value obtained with larger charges. Based on these results, we propose that the aquarium technique can be used for detonation pressure measurement with small samples of explosives.

By the aquarium technique we get some preliminary information about the attenuation process of shock waves in water. These results have some significance in evaluation of explosive performances. We are going to do further work on this problem.

REFERENCES

1. M. A. Cook et al., Proc. 3rd ONR Symposium on Detonation, (1960) p. 357-385.
2. J. K. Rigdon, I. B. Akst, Proc. 5th Symposium on Detonation, (1970) p. 59-65.
3. P. E. Kramer, R. W. Ashcraft, MHSMP-73-6 (1972).
4. D. Y. Yu et al., "A simple small plane wave generator", Explosion and Shock Waves (China), 1981 (in press).
5. M. H. Rice, J. M. Walsh, J. Chem. Phys., 26, 824-827 (1957).
6. W. E. Deal, Phys. Fluids, 1, 523-527 (1958).
7. UCRL-51319 Rev. 1 (1974) p. 8-6.

SHOCK WAVES IN FRESH WATER GENERATED BY DETONATION OF PENTOLITE SPHERES

T. P. Liddiard and J. W. Forbes
Naval Surface Weapons Center
White Oak, Silver Spring, Maryland 20910

ABSTRACT

Shock velocities as a function of distance were determined by differentiating distance-time paths of the shock waves. The distance-time information was obtained from high-speed photographs of the shocks obtained by a shadowgraphing technique. In addition, peak pressure and pulse time histories were recorded using lithium niobate gages. Pentolite spheres of three different initial radius R_0 were used in the gage experiments. The shock velocities were obtained for relative distances of R/R_0 ranging from 1.2 to 12.3 while the pressure data were obtained for R/R_0 ranging from 4.4 to 12.4. Scalability of peak pressure and pulse widths (at one-half peak pressure) as a function of R/R_0 was successfully demonstrated.

BACKGROUND

Spherical shock-generating systems in water have been designed for use in inducing reaction in explosive samples[1-4]. Initiation of burning and detonation threshold information in terms of peak pressure and pulse half-widths can be obtained with these underwater systems for a pressure range of 1-30 Kb and pulse half-widths from 10-50 μs. In addition, the measured shock pulse parameters provide useful information on the underwater performance of pentolite.

The shock pressures and time histories of shocks in water from pentolite spheres have been calculated by Walker and Sternberg[5]. These calculations are compared to the measured parameters of this work.

MEASUREMENTS

Velocity of Spherical Shocks in Water

Most of the velocity information was obtained from underwater explosive sensitivity experiments where the donor explosive is a 82.3 mm diameter cast pentolite sphere, (50% TNT, 50% PETN), nominally weighing 472 g. Shocks in water are observed by a high speed framing camera[6]. Diffuse reflected backlighting is obtained by an argon flash bomb which illuminates a white cardboard in back of the tank. The shock wave velocity in the fresh water can be obtained over a range of R/R_0 distances. Velocities from 2.5 cm diameter pressed pentolite spheres were obtained at R/R_0 of 8.9 and 14.3 using a shadowgraph technique with a streak camera. The shock velocities are given by the following equation

$$U - C_o = (4.57 \text{ mm}/\mu s) \left(\frac{A+1}{A+R_1^B} \right) \qquad (1)$$

where $R_1 = R/R_0$, $A = -0.3854$, $B = 1.3184$, and $C_0 = 1.483$ mm/μs. This equation gives the correct limits of 6.053 mm/μs[7] at $R = 1$ and C_0 at $R = \infty$. Combining equation (1) with the conservation of momentum relation and the shock velocity-particle velocity relation for water[8] at a density of 0.998 g/cm^3 results in

$$P = 76.829 \left[1 + \frac{1.9772}{R_1^{1.3184} - 0.3854} \right] \left[\exp\left(\frac{0.2556}{R_1^{1.3184} - 0.3854} \right) - 1 \right] \qquad (2)$$

where P is in Kb and R_1 ranges from 1.24 to 14.3.

Pressure-Time Measurements of Spherical Shocks in Water

Lithium niobate gages were used in some tests to measure peak pressures and time histories as a function of distance from the pentolite sphere. The lithium niobate gage has the advantage of being made from commercially grown synthetic crystals and its calibration is independent of manufacturer.[9]

The lithium niobate gage is capacitively loaded. The electrical charge generated by hydrostatic pressure charges the capacitor which can be recorded by an oscilloscope. The oscilloscope input impedance must be large so that the discharge time of the capacitor thru the scopes impedance is small compared to the event time. Another consideration is to load the gage down with capacitance near 100 times that of the gage cable capacitance so that any changes in cable capacitance due to shock loading can be ignored.

The peak pressures and pulse half-widths were measured by lithium niobate gages for a number of cast and pressed pentolite spheres with radii ranging from 1.27 to 6.84 cm. Those gage results are given in Table 1. The measured pressures span a range of 0.8 to 5.9 Kb for R/R_0 ranging from 12.3 to 3.5 respectively, as shown in Fig. 1. The peak pressures are the same within experimental error for the same R/R_0. The pulse widths at one-half peak pressure ranged from 10 to 66 μs. The pulse half-widths divided by R_0 give the same results for different size spheres when the measurements are made at the same R/R_0. These pulse half-widths are given within experimental error for R/R_0 from 4 to 13 by

$$\tau = 2.053 \, R_0 \, (R/R_0)^{0.731} \qquad (3)$$

where τ is in μs. The pulse half-width of 10.3 μs for a 4.1 cm cast pentolite sphere at R/R_0 of 3.49 does not fit this equation. The record has a different general shape than the other records. Possible gage breakdown is suspected. In fact an attempt to measure the shock pulse at R/R_0 of 1.89 failed to give a reasonable record. This implies that the gages are failing for peak pressures near 6 Kb.

DISCUSSION AND CONCLUSIONS

Shock wave velocities, peak pressures and pulse widths at one-half peak pressure have been measured as a function of distance in fresh water from pentolite spheres. An analytic function that reproduces the pressure data within experimental error over the relative distance (R/R_0) interval from 1.5 to 12 has been found. This expression is empirical and resulted from the observation that the shock velocity minus the sound velocity was proportional to $1/(R^B+A)$. The pressure equation goes to the correct limits of 167 Kb for $R/R_0 = 1$ and 0 Kb for $R/R_0 = \infty$. The measured peak pressures agreed within experimental error with equation (2).

The calculated shock pressures and time histories of shocks in water from pentolite spheres by Walker and Sternberg[5] are in fair agreement with the present experimental results as shown in Fig. 1. Extrapolation of the present data by equation (1) to R/R_0 greater than 13 results in pressures lower than calculated[5].

The pressure pulse in water from the pentolite sphere can be scaled as a function of the sphere radius. This indicates that most or all of the pentolite detonates even for very small spheres of 1.2 cm radius. This is consistent with the known short run distance to detonation for pentolite.

ACKNOWLEDGEMENTS

The authors wish to acknowledge the assistance of J. Marshall and A. Brown in conducting the lithium niobate gage experiments. D. Gillmore and R. Baker for conducting the underwater sensitivity experiments from which the shock velocities in water were obtained. H. Jones obtained the gage information on the 6.84 cm radius pentolite sphere. R. A. Graham of Sandia National Laboratories suggested the use of lithium niobate gages.

REFERENCES

1. C. H. Winning, Proc. Roy Soc. 246, 288 (1958).
2. T. P. Liddiard, Fourth Symp. on Det., 487, (1965).
3. L. W. Hantel and W. C. Davis, Fifth Symp. on Det., 599, (1970).
4. F. E. Walker and R. J. Wasley, Comb. and Flame, 22, 53 (1974).
5. W. A. Walker and H. M. Sternberg, Fourth Symp. on Det., 27, (1965).
6. S. J. Jacobs, J. D. McLanahan, Jr. and E. C. Whitman, Jour. SMPTE 72, 923 (1963).
7. N. L. Coleburn, NOLTR 64-58, (1964).
8. M. H. Rice and J. M. Walsh, J. of Chem. Phys., 26, 824 (1957).
9. R. A. Graham, Ferroelectrics 10, 65 (1976).

TABLE 1 -- SHOCK WAVE PROPERTIES IN WATER

Charge Radius (cm)	Type	Pentolite Density (g/cm³)	R/R_0	Peak Pressure (kbar)	Pulse Half-Width τ (μs)	τ/R_0 (μs/cm)	Water[a] Temp. (°C)
1.27	Pressed	1.61	12.27	0.803±0.030	15.23±0.61	12.01±0.48	21.1
1.27	Pressed	1.62	8.04	1.447±0.043	12.46±0.93	9.83±0.58	22.2
4.11	Cast	1.65	4.39	3.823±0.126	24.56±0.85	5.98±0.21	17.8
4.11	Cast	1.64	3.47	5.862±0.251	10.32±1.93	2.51±0.47	17.8
4.11	Cast	1.64	4.97	3.048±0.080	24.66±1.16	6.00±0.28	17.8
4.11	Cast	1.63	8.01	1.543±0.076	48.19±2.96	11.73±0.72	20.6
4.11	Cast	1.62	12.28	0.817±0.033	50.08±2.42	12.18±0.59	20.6
6.84	Cast	1.62	6.50	2.038±0.075	57.15±2.87	8.36±0.42	26.0
6.84	Cast	1.62	8.01	1.449±0.061	65.88±1.48	9.63±0.22	16.9

[a]Distilled water was used for the first two experiments and tap water was used in the rest of the experiments.

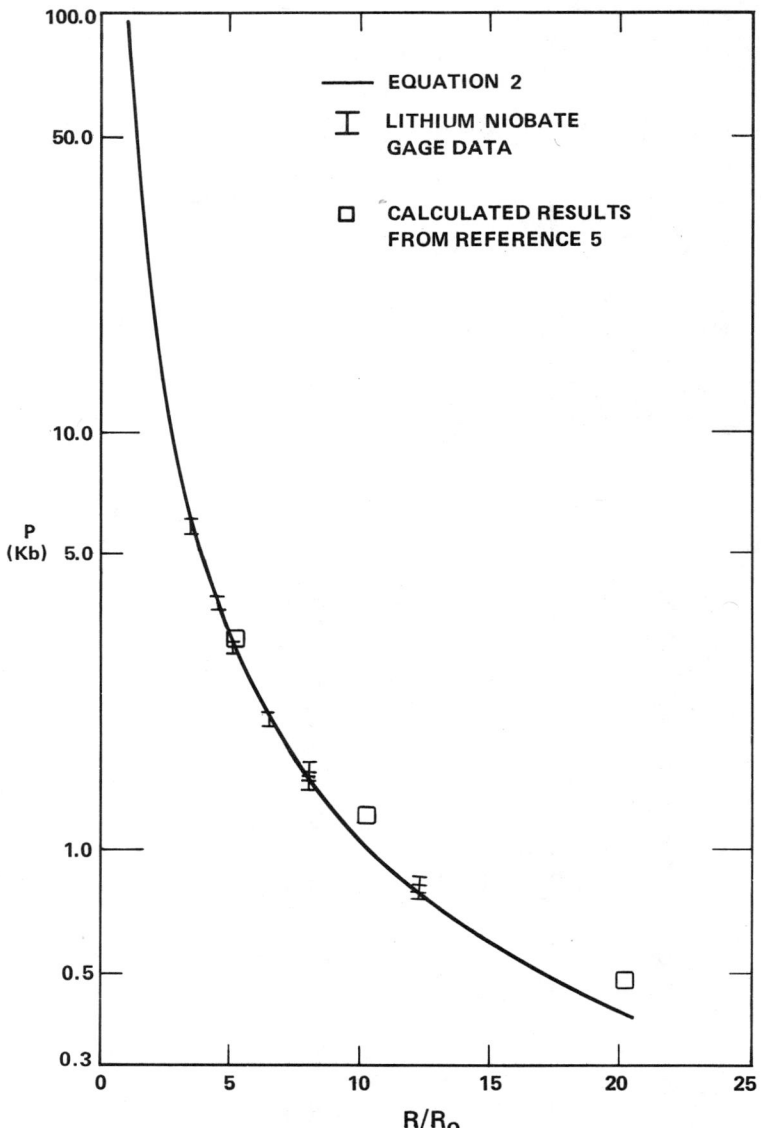

FIGURE 1 PEAK PRESSURE VERSUS RELATIVE DISTANCE FROM PENTOLITE SPHERES

COMPARISON OF AXIAL LONGITUDINAL VELOCITY MEASUREMENTS DETERMINED ULTRASONICALLY AND BY A WEAK SHOCK VELOCITY TECHNIQUE ON AN ALUMINIZED MELT CAST EXPLOSIVE

J. W. Forbes and W. L. Elban
Naval Surface Weapons Center
White Oak, Silver Spring, Maryland 20910

ABSTRACT

The axial longitudinal sound velocity of two cylindrical samples of an aluminized melt cast explosive was measured ultrasonically and by a weak shock velocity technique. The melt cast explosive contained RDX, TNT, and aluminum and it was brittle. The purpose of this work is to ascertain the difference in longitudinal sound velocity obtained for the same sample as measured by the two different methods. A difference in velocity may occur because of the differences in wave amplitude and frequency that the sample experiences with the two techniques. A further difference in velocity would be expected if the amplitude of the weak shock exceeded the dynamic yield strength of the explosive. However, the magnitude of the weak shock did not exceed the dynamic yield strength of this material.

Within experimental error, a good correspondence was found between the results obtained by the two techniques. This agreement suggests that either no measurable frequency dependence for longitudinal sound velocity occurs for this material or the frequency spectrum in the weak shock is near the frequency (1.00 MHz) used in the ultrasonic measurements.

INTRODUCTION

The amount of work appearing in the open literature on sound velocity of explosives, propellants, and simulants is fairly limited[1-6]. The most thorough study[3] measured the temperature and frequency dependence of sound velocity of X-0219 and PBX-9404. For PBX-9404 good agreement was found between the velocities measured ultrasonically and the weak-shock value[4]. However, these measurements were made on different samples. No direct comparison was possible for X-0219 since a weak-shock value was not measured.

Good correlation between longitudinal velocity measured ultrasonically and by the weak shock method would not necessarily be expected because of differences in frequency spectrum and the amplitude of the wave. It is the intent of this work to compare directly axial longitudinal velocity measurements determined by the two methods for the same sample. An assessment is given of the frequency content and the wave amplitude that occurs for both measurement techniques. A further difference in velocity would be expected if the strength of the weak shock exceeded the dynamic yield strength of the explosive.

PROCEDURE

Ultrasonic Measurements

Ultrasonic longitudinal velocity measurements were obtained using a radio frequency tone burst system[7] operating in through-transmission with a delay rod. A block diagram of the apparatus appears in Fig. 1. This set-up uses a high frequency sweep generator (Wavetek Model 144) operating in the gated mode. A sine wave tone burst is produced, consisting of a few cycles, say five to nine, having a single frequency (at least in the center portion of the wave) that is continuously variable. The sweep generator is employed with two closely matched broadband contact longitudinal transducers (taken from the Panametrics Videoscan series). For this work, these transducers had an approximate center frequency of 1.00 MHz and an element diameter of either 12.7 or 25.4 mm. Since the acoustic zero was not known precisely, a cylinder of low attenuating material, serving as a time delay, was inserted between the transmitting transducer and the sample. When transducers having a 12.7 mm element diameter were used, an 18.0 mm diameter by 25.4 mm high cylinder of polystyrene was incorporated, while a 50.8 mm diameter by 12.7 mm high cylinder of polymethylmethacrylate was used with the 25.4 mm diameter transducers. All of the interfaces were coupled with a viscous liquid (Fisher Scientific Nonaq stopcock grease).

A pre-amplifier (Panametrics Model 5050AE-160B) was added providing additional gain to the transmitted signal of either 40 or 60 dB. A toggle switch attenuator (Alan Industries Model 50HT82.5) was included to allow transit time determinations for transmitted signals (both delay and sample plus delay) that had been adjusted to the same height. This unit provides for signal attenuation ranging from 0 to 82.5 dB in 0.5 dB increments.

Signal amplitude versus time output is displayed on an oscilloscope (Hewlett Packard Model 1743A), and transit time measurements are made directly from the cathode ray tube screen using a digital readout of the time interval. Transit time measurements on the sample of interest were obtained by difference: the transit time for the delay rod was subtracted from the transit time for the sample plus delay rod. Each determination was repeated either five or seven times allowing the statistical nature of the time of flight velocity measurements to be assessed. An error analysis[8] was performed and it is estimated that the accuracy of this velocity measurement is ±0.7%. All of the ultrasonic velocity measurements reported in this work were made at a frequency of 1.00 MHz and were obtained at room temperature (22 ± 3°C). The wave amplitude is estimated not to exceed 0.01 bar.

Weak Shock Velocity Measurements

Fig. 2 gives the experimental set-up for the weak shock velocity technique[5] using optical shadowgraphing to observe the transit time of the shock wave through the explosive sample. The sample was placed underwater at a distance of 25 cm from a submerged small detonator which provided a shock wave of 50 bar in water at the sample. The water temperature was 20 ± 3°C. The shock in water refracts the light going to the high speed streak camera (Cordin Model 132), yielding a record of the shock incident to the sample, a reflected shock in water, and a transmitted shock in water on the opposite or back side of the sample. Reference 5 gives just such a streak camera record.

The transit time through the sample is obtained from the photographic records. A correction is made to the transit time if the thickness of the shadow of the samples edge is greater than the known thickness. The samples shadow thickness can be larger if the sample's flat surfaces are misaligned even slightly from the camera's axis. An error analysis[8] was performed and it is estimated that the accuracy of these weak shock velocities is about 2.5%.

RESULTS

The results for the melt cast explosive from the sequentially and interleaved measurements obtained by the two techniques appear in Table 1. The average ultrasonic longitudinal velocity for the first sample was 2485 ± 17 m/s, while the weak shock technique gave a value of 2550 ± 61 m/s. The average ultrasonic velocity for the second sample was 2553 ± 18 m/s, while the average weak shock value was 2550 ± 60 m/s. No systematic difference in the velocities was observed for the two samples. Within experimental error, a good correspondence is found between the results obtained by the two techniques.

A lithium niobate gage was used to measure the pressure pulse in water at 25 cm distance from the WOX-76A detonator. The peak pressure in the shock in water was 50 bar. An upper limit for rise time of the shock front can be obtained from the gage records. The time for the recorded pressure pulse to reach a maximum was 450 ± 150 ns. This time is comparable to the time required for the shock wave to travel a distance equal to the gage thickness of 0.63 mm. This suggests that a large part of the measured response time is the equilibration time of the pressure in water surrounding the gage. This information infers that the frequency content of the weak shock wave in water for this measurement technique is greater than or equal to 2 MHz.

DISCUSSION AND CONCLUSIONS

The purpose of this work was to compare the longitudinal velocities measured on the same explosive sample using both ultrasonic and weak shock techniques. A brittle melt cast explosive was chosen with the anticipation that the yield strength would not be exceeded by the weak shock wave. The ultrasonic velocity did not change, within experimental error, after either sample had been subjected to the weak shock technique. This indicates that the weak shock had not altered the elastic properties of the samples and is consistent with the expectation that the dynamic yield strength had not been exceeded. Therefore, the velocity measured by both techniques is the longitudinal elastic wave velocity. The frequency was 1.00 MHz for the ultrasonic technique and greater than or equal to 2 MHz for the weak shock technique. Differences in velocity can be expected due to differences in wave frequency and amplitude between the two techniques. The measured velocities are all in good agreement. This agreement implies that the wave amplitude and frequency component difference inherent between the two measuring techniques does not affect longitudinal velocity for this material.

ACKNOWLEDGEMENTS

It is a pleasure to acknowledge the efforts of R. Baker and H. Gillum in conducting the weak shock velocity measurements. A number of helpful discussions on the ultrasonic measurement technique and the resulting velocity measurements were held with G. V. Blessing and A. L. Bertram. R. R. Bernecker, N. L. Coleburn and S. J. Jacobs provided many insights into the error analysis, the weak shock technique, and the comparison of measurement techniques.

REFERENCES

1. J. A. van der Schroeff and R. S. de Boer, Prop. and Expl, $\underline{1}$, 2 (1976).
2. G. C. Knollman, R. H. Martinson, and J. L. Bellin, J. Appl. Phys., $\underline{50}$, 111 (1979).
3. H. J. Sutherland and J. E. Kennedy, J. Appl. Phys., $\underline{46}$, 2439 (1975).
4. N. L. Coleburn and T. P. Liddiard, Jr., J. Chem Phys., $\underline{44}$, 1929 (1966).
5. N. L. Coleburn, J. Acoust. Soc. Am., $\underline{47}$, No. 1 (Part 2), 269 (1970).
6. S. P. Marsh, LASL Shock Hugoniot Data, (U. Cal., Berkeley, 1980).
7. E. P. Papadakis, Physical Acoustics, Vol. XII, (Mason and Thurston, Academic Press, N.Y., 1976) pp 308-312.
8. Y. Beers, Introduction to the Theory of Error, (Addison-Wesley, Mass., 1962).

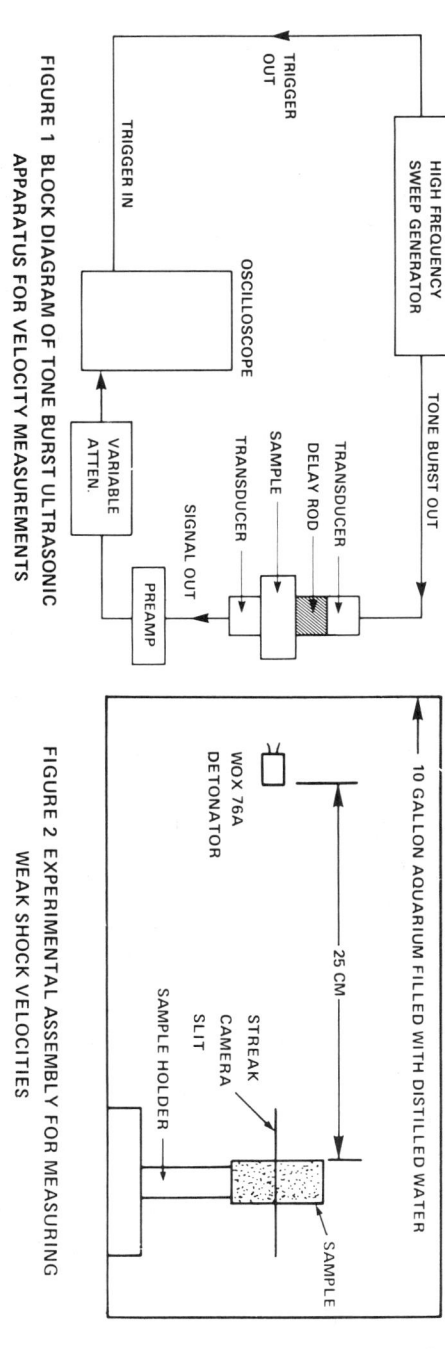

FIGURE 1 BLOCK DIAGRAM OF TONE BURST ULTRASONIC APPARATUS FOR VELOCITY MEASUREMENTS

FIGURE 2 EXPERIMENTAL ASSEMBLY FOR MEASURING WEAK SHOCK VELOCITIES

TABLE 1 COMPARISON OF AXIAL VELOCITY MEASUREMENTS DETERMINED ULTRASONICALLY AND BY A WEAK SHOCK TECHNIQUE ON AN ALUMINIZED MELT CAST EXPLOSIVE

SAMPLE CONFIGURATION	TIME OF MEASUREMENT (RELATIVE TO WEAK SHOCK MEASUREMENT)	ULTRASONIC VELOCITY (m/s)	WEAK SHOCK VELOCITY (m/s)
CYLINDER 1 (88.96 mm DIA; 12.74 mm HIGH; ρ =1.68 g/cm^3)	BEFORE	2490 ± 17[a]	— — —
	AFTER	2480 ± 17[a]	2550 ± 61
CYLINDER 2 (50.88 mm DIA; 12.87 mm HIGH; ρ =1.68 g/cm^3)	BEFORE	2550 ± 18[b]	— — —
	AFTER I	2560 ± 18[b]	2460 ± 61
	AFTER II	2550 ± 18[b]	2540 ± 59

[a] AVERAGE OF FIVE MEASUREMENTS, [b] AVERAGE OF SEVEN MEASUREMENTS

GRÜNEISEN PARAMETER MEASUREMENTS FOR HIGH EXPLOSIVES*

George H. Bloom
Lawrence Livermore National Laboratory,
Livermore, California 94550

ABSTRACT

The Grüneisen parameter of the explosives PBX-9404 and RX-03-BB have been measured at full and reduced density. The samples were heated in 40ns with a 2 MeV electron beam and the pressure developed was determined with a laser interferometer.

INTRODUCTION

The Mie-Grüneisen equation of state

$$P - P_o = \Gamma (E - E_o) \rho \qquad (1)$$

is often used to describe the behavior of materials under pressure. P is the pressure, E is the specific internal energy, and ρ is the material density. More specifically, the Grüneisen parameter

$$\Gamma = \frac{1}{\rho} \left(\frac{\partial P}{\partial E} \right)_V \qquad (2)$$

is useful in determining the response of materials to rapid heating at constant volume. The Grüneisen parameter is dimensionless.

The Grüneisen parameter can be calculated from lattice dynamics, from thermoelastic constants, or from shock-wave data. There is good agreement on the values for metals, but less work has been done on other materials. In some cases the value of Γ may be uncertain by an order of magnitude. This has prompted us to measure the Grüneisen coefficient for two high explosives of interest to us: PBX-9404 (94% HMX) and RX-03-BB (92.5% TATB), at both full and reduced density. This work is covered in detail in an LLNL report.(1) Aluminum was used to check our system.

The Grüneisen parameter for many materials does not change

---

* This work was performed under the auspices of the U. S. Department of Energy by Lawrence Livermore Laboratory under contract No. W-7405-Eng-48.

significantly with energy. Our measurements show this is true for explosives at full density and for aluminum. The Grüneisen coefficient for the porous explosives decreases with increasing energy because the material yields.

EXPERIMENT

To measure the Grüneisen parameter of a material we heated the sample rapidly with an electron beam while simultaneously measuring the pressure and energy. Our Febetron 705 has a nominal maximum particle energy of 2 MeV and a pulse length of about 40 ns. The maximum charging voltage is 35 kV. Our irradiations were performed at 20 and 30 kV charging voltage, with correspondingly reduced particle energies. This resulted in energy densities in the target on the order of 8 and 80 kJ/kg respectively. No energy release from the high explosives was observed. The beam propagated best in air at one torr.

To achieve uniform heating, we used samples 1 mm thick and 5 mm diameter. For the higher energy irradiations a carbon plate 0.17 mm thick was placed in front of the target to move the energy density maximum near the center of the target.

To measure the energy deposited in the sample we developed a carbon calorimeter with nine elements arranged in two rings of four around a central element. The central element could be replaced by the explosive target. We periodically checked the calorimeter for each charging voltage by putting in the central element. The calorimeter was calibrated by comparing the temperature rise of the central element with the temperature rise of the other eight elements. Then when the central element was replaced by the target, the temperature rise in the target was calculated from the temperature rises of the other elements. The reading was corrected for the differences in energy absorption for different target materials. The outside diameter of the outer ring of calorimeter elements was 20 mm, and the target was 5 mm diameter.

The pressure developed in the sample was calculated from the surface velocity of the sample measured with a HeNe laser interferometer. Our interferometer was set up in a Michelson configuration which gives an interference fringe for each half-wave moved by the target face.

The peaks of the oscilloscope records were digitized and the time values were plotted (y axis) against the distance moved (x axis), which is proportional to the fringe number. A least-squares fit was then made to a selected portion of this curve. The slope of this line, with suitable corrections for units and scale factors, gives the velocity of the surface.

The pressure developed in the sample can then be calculated from the surface velocity with the formula:

$$\Gamma = \frac{P}{E} = \frac{vC_l}{E} \qquad (3)$$

Here v is the velocity of the rear surface. E is the specific energy at time t in a sample, and C_l is the dilatational wave speed.

A small correction should also be applied for the finite time of energy deposition. We have not done this, but the change would be less than our experimental errors, since we ignore early times in our velocity determination.

We measured the longitudinal sound speeds of our samples using ultrasonic techniques. The values used in the present work were:

Table I. Target Densities and Sound Velocities.

Material	Density Mg/m^3	Longit. Vel. km/s	Shear Vel. km/s
Aluminum	2.70	6.46	3.04
Pyrex glass	2.32	5.64	3.28
PBX-9404	1.84	2.86	1.33
PBX-9404	1.70	2.30	1.30
RX-03-BB	1.89	2.81	1.37
RX-03-BB	1.70	1.54	0.90

DISCUSSION

The strength of the material is important for aluminum at these modest pressures. It is unfortunate that more is not known about the material properties of explosives so that we could estimate the effects and the errors in our measurements for the materials of most interest to us.

The Grüneisen coefficient can be calculated from thermoelastic data with the formula:

$$\Gamma = \frac{B_T(3\alpha)}{\rho C_v} \qquad (4)$$

where B_T is the isothermal bulk modulus, 3α is the volume coefficient of thermal expansion, ρ is the mass density, and C_v is the specific heat.

Using these formulae and the best thermoelastic data we could find in various handbooks, we have calculated the Grüneisen constant for several materials of interest. For many materials it is difficult to find the necessary data.

The Grüneisen constant can also be calculated from shock-wave data as described by Rice, McQueen and Walsh. They show that if there is a linear relation between the particle velocity U_p and the shock velocity U_s, that is that if:

$$U_s = C + SU_p \qquad (5)$$

then the Grüneisen constant can be found from the relation:

$$\Gamma = 2S - 1. \qquad (6)$$

The linear U_s-U_p relation holds reasonably well for some materials but for some materials it does not. For most of the materials of interest to us in this study, it does not.

The agreement between thermoelastic calculations and measured values of the Grüneisen parameter is good, indicating that calculation of the Grüneisen Γ using thermoelastic data is more reliable than other methods of calculation. This is one important question that prompted our experimental work.

We have made a study of the effects of the mirror on the Grüneisen results. This was greatly hampered by a lack material properties data for the explosive materials. From this work, the best we can conclude is that our measurements are probably between 3 and 20% too low.

RESULTS

Table II summarizes the results of our Grüneisen parameter measurements for the two high explosives, PBX-9404 and RX-03-BB, at full and reduced density.

Table II. Measured Grüneisen Parameter Γ

Material	Density Mg/m^3	Charge kV	Ave. En. Dep. kJ/kg	Ave. Γ
6061 T-6 Al	2.70	20	11.0	1.90 ± 0.07*
		30	73.6	2.31 ± 0.12*
		20 & 30	–	2.04 ± 0.08*
PBX-9404	1.84	20	9.32	0.881 ± 0.039*
		30	86.3	0.854 ± 0.019*
		20 & 30	–	0.871 ± 0.026*
PBX-9404	1.70	20	11.3	0.434 ± 0.053*
		30	91.7	0.156 ± 0.002*
RX-03-BB	1.89	20	8.66	1.014 ± 0.045*
		30	80.8	0.830 ± 0.045*
		20 & 30	–	0.962 ± 0.041*
RX-03-BB	1.70	20	11.3	0.210 ± 0.011*
		30	84.7	0.173 ± 0.010*

*The errors of the mean quoted consider only the variations in the final results and are much smaller than the actual uncertainties due to systematic errors. We estimate the actual errors in the results are closer to twenty percent.

REFERENCE

1. G. H. Bloom and J. H. Claar, Grüneisen Parameter Measurements for High Explosives, Lawrence Livermore National Laboratory, Livermore. Rept. D-81-2 (31 March 1981)

PRESSURE DEPENDENT VIBRONIC RELAXATION IN SHOCKED EXPLOSIVES

M. J. Frankel[†]
Naval Surface Weapons Center
White Oak, Silver Spring, Maryland 20910

ABSTRACT

To a first approximation we consider the energy of an initiating shock wave to be distributed entirely amongst the intermolecular lattice modes with the relatively strong intramolecular vibrational modes initially undisturbed. The liberation of chemical energy follows a process of energy infiltration into the internal molecular vibrations. At the high shock levels characteristic of detonation waves, low order perturbation techniques which were previously developed for accoustic absorption are inadequate. We have generalized a technique due to Diestler on isolated molecule relaxation, wherein the instantaneous interaction potenial evaluated at the equilibrium molecular positions and the pressure and temperature dependent relaxation rate is then calculated.

INTRODUCTION

A need to understand the mechanisms of shock dissipation and chemical activation in a condensed explosive medium has provided a stimulus for the study of thermal relaxation behind a propagating detonation front. The initiating shock energy is deposited in an explosive in the form of a strong compressional wave. To a first approximation we may consider this input energy to be distributed entirely amongst the intermolecular lattice modes with the relatively strong intramolecular vibrational modes initially undisturbed. The liberation of chemical energy follows a process of energy infiltration into the internal molecular vibrations and the subsequent excitation of some critical mode for bond breaking.

We consider the relaxation of energy into the internal modes to result from a vibration-vibration (V-V) interaction process. V-V interaction has been treated previously in the theory of ultrasonic absorption in benzene by Herzfield[1] and Lieberman[2] and in CCl_4 by Rasmussen.[3] While successful with these compounds, the approach employed suffers from the need to resort to ever higher orders of perturbation theory and/or higher order potential function expansions to explain multiphonon affects when the gap between the lattice spectrum and the internal molecular vibrations exceeds three Debye frequencies. At high shock compressions large phonon densities may

[†]Present address: Defense Nuclear Agency, Washington, D. C. 20305

be expected and the first order perturbation treatment used by these investigators is no longer adequate. A number of approaches which obviate the need for high order perturbation theory have been developed[4,5] and applied to the problem of isolated molecule impurity relaxation in a host lattice. We have generalized an approach due to Diestler[6,7] wherein the external lattice modes see an interaction potential averaged over the state of the molecular system to the case of two interesting subsystems, and obtain expressions for the pressure dependent transition rate from the ground to the first excited state. It is assumed that the value will constitute a rate determining step.

THE MODEL

The total Hamiltonian for the system may be written

$$H^{TOT} = H^M(q) + H^L(Q) + H^{INT}(q,Q), \qquad (1)$$

where H^M, H^L and H^{INT} refer to the molecular, lattice and interaction energies, respectively. Q and q refer to lattice mode and internal molecular mode coordinates, respectively. The molecule-molecule coupling thus appears indirectly through the lattice-molecule interaction term $H^{INT}(q,Q)$.

Let Φ represent an eigenstate of H^M. Then Φ will be given by

$$\Phi = \prod_i \phi_{\alpha_i}(q_i), \qquad (2)$$

where ϕ_{α_i} satisfy

$$\left\{ \frac{p^2}{2m_i} + V^M(q_i) \right\} \phi_{\alpha_i}(q_i) = \epsilon_{\alpha_i} \phi_i(q_i). \qquad (3)$$

The index i runs over individual molecular sites and α_i represents the αth eigenstate of the ith molecule. V is an unspecified internal molecular potential binding the molecule and m_i is a molecular mass. Let $\chi(Q)$ represent a wave function governing the external lattice modes where $\vec{\alpha} = (\alpha_i, \ldots, \alpha_n)$ and let $\chi^\alpha(Q)$ be expanded as

$$\chi^\alpha(Q) = \prod_{k=i}^{3N} \chi_{n_k}^\alpha(Q_k), \qquad (4)$$

where n_k is the occupancy of the kth normal mode. Let $\psi(q,Q)$ represent an eigenstate of the total Hamiltonian

H^{TOT} such that

$$H^{TOT} \psi(q,Q) = E\psi(q,Q), \quad (5)$$

and expand $\psi(q,Q)$ as a linear combination of direct product states

$$\psi(q,Q) = \sum_\alpha C_\alpha \phi^\alpha(q) \chi^\alpha(Q). \quad (6)$$

Substituting (6), (4) and (3) into (5) one obtains the equation

$$H^L \chi^{\alpha'}(Q) + \sum_\alpha (C_\alpha/C_{\alpha'}) H^{INT}_{\alpha'\alpha} \chi^\alpha(Q) = (E - \epsilon^n_{\alpha'}) \chi^{\alpha'}(Q). \quad (7)$$

Neglecting diagonal elements in (7) we obtain

$$H^{EFF}_L \chi^\alpha(Q) = \epsilon^L_{n_\alpha} \chi^\alpha(Q), \quad (8)$$

where

$$H^{EFF}_L = H^L + H^{INT}_{\alpha\alpha}, \quad (9)$$

and

$$\epsilon^L_{n_\alpha} = E - \epsilon^n_\alpha. \quad (10)$$

Physically, elimination of the off diagonal elements corresponds to the case where the slow, external lattice modes respond only to an average of the instantaneous internal potentials of the molecular subsystem which is vibrating so rapidly as to appear blurred to the external modes. The development followed thus far is formally similar to that in Reference 4 with the understanding that ϕ now refers to states of the many molecule subsystem with the assumption that the lattice modes may be adequately described by a harmonic oscillator approximation and assuming that lattice equilibrium positions and frequencies undergo a parallel shift when the internal molecular system is in an excited state.

TRANSITION RATE

The transition rate W for the non-radiative transition may be estimated by using the first order Fermi "Golden Rule"

$$W_{\alpha,\beta} = \frac{2\pi}{\hbar} \sum_{n_\alpha} \sum_{n_\beta} \frac{e^{-\beta\hbar\omega_{n,\alpha}}}{\sum_{n_\alpha} e^{-\beta\hbar\omega_{n,\alpha}}} |\langle \chi^\alpha \phi_\alpha | H^{INT} | \chi^\beta \phi_\beta \rangle|^2 \delta(E^\alpha - E^\beta);$$

$\beta = 1/k_B T$ and T is understood to be a lattice temperature. (11)

Using a generating function technique, Engleman and Jortner[8] have converted the double sum in (11) to the form

$$W_{\alpha,\beta} = \frac{2\pi}{\hbar} |H^{INT}_{\alpha,\beta}(Q)|^2 \exp\left[-G(0) \int_{-\infty}^{\infty} dt\, e^{-i\omega_{\alpha,\beta} t + G_+(t) + G_-(t)}\right], \quad (12)$$

where
$$G(0) = \sum_k \Delta_k^2 (2n_k+1) , \tag{13}$$

$$G_+(t) = \sum_k \Delta_k^2 (n_k+1) e^{i\omega_k t} , \tag{14}$$

and
$$G_-(t) = \sum_k \Delta_k^2 n_k e^{-i\omega_k t} . \tag{15}$$

Δ_k is a reduced displacement defined by

$$\Delta_k = \left(\frac{m\omega_k}{2\hbar}\right)^{1/2} \Delta Q_k , \tag{16}$$

and n_k is the usual Bose occupation number for the kth lattice mode. In order to carry out an explicit evaluation of (13) the sums over the different ω_k are replaced by a single mean frequency ω_m such that

$$\sum_k f(\omega_k) \to N f(\omega_m) . \tag{17}$$

With this approximation the integral (13) can be evaluated by a saddle point technique. A formal expression for the transition rate may then be written:

$$W_{\alpha,\beta} = \left|\frac{H_{\alpha\beta}^{INT}}{\hbar^2}\right|^2 \frac{\{2 \sinh(\frac{\beta\hbar\omega_m}{2})\}^{1/2}}{\{N\Delta^2\omega_m^2 (1+\delta^2)^{1/2}\}^{1/2}} \exp\left[-N\Delta^2(2\eta_m+1)\right]$$

$$\times \exp\left[-\frac{\beta\hbar\omega}{2}\alpha\beta + \frac{N\Delta^2(1+\delta^2)^{1/2}}{\sinh(\frac{\beta\hbar}{2}\omega_m)} - \frac{\omega_{\alpha\beta}}{\omega_m}\log\left[\delta+(1+^2)^{\frac{1}{2}}\right]\right] \tag{18}$$

where
$$\delta \equiv \frac{\omega_{\alpha\beta}}{\omega_m} \frac{\sinh(\frac{\beta\hbar\omega_m}{2})}{n\Delta^2} . \tag{19}$$

PRESSURE DEPENDENCE

To first order the pressure dependence of the transition rate can be accounted for solely through the mean frequency ω_m. To higher order one would have to make some estimate of the pressure dependence of the parameters H^{INT} and Δ. To estimate the change in ω_m we recall the definition of the Gruneisen parameter γ:

$$\gamma = -\frac{d(\log \omega)}{d(\log v)} . \tag{20}$$

Taking a mean Gruneisen parameter $\bar{\gamma}$ assume we have

$$\bar{\gamma} \int d(\log v) = -\int d(\log \omega), \tag{21}$$

or
$$\omega_m = \omega_{mo} \left(\frac{V_o}{V}\right)^{\bar{\gamma}} . \tag{22}$$

Taking a typical explosive $\bar{\gamma} \approx 4$ we have $\dfrac{\omega_m}{\omega_{mo}} = 1.5$ at 10% compression. We shall estimate the change in W due solely to pressure (at T = 300°K). We take a published value for CO^9 of $\omega_{mo} \approx 7\times10^{12}$ sec^{-1} and $N\Delta_{mo}^{\,2} \approx 19$, $\omega_{\alpha\beta} \approx 4\times10^{14}$ sec^{-1}. Substituting into (18) we obtain

$$W/W_o \approx 10 , \qquad (23)$$

an increase of just one order of magnitude in the reaction rate due to pressure.

CONCLUSION

We have outlined a model calculation of the rate determining step in explosive energy relaxation from the external lattice modes where the initial shock energy is concentrated to internal molecular vibrations. The calculation seems to suggest that the reaction acceleration due to pressure would not be as significant as the temperature effect. It should be emphasized that the exponential form of the transition rate is extremely sensitive to values of the material constants, small shifts in the constants giving rise to possible orders of magnitude change in the transition rate.

Recently Toton[10] has calculated transition rates for shocked explosives using the adiabatic method and has demonstrated explicitly the pressure dependence for excitation of the normal modes of nitromethane. Considerable experimental work remains to be done, however, before overall reaction rate dependence on pressure can reliably be predicted.

REFERENCES

1. Herzfeld, K. F., J. Chem. Phys 20, 288 (1952).
2. Liebermann, L. N., Phys. Rev. 113, 1052 (1959).
3. Rasmussen, R. S., J. Chem. Phys. 46, 211 (1967).
4. Nitzan, A., and Silbey, R. S., J. Chem. Phys. 60, 4070 (1974).
5. Nitzan, A., and Jortner, J., Molec. Phys. 25, 713 (1973).
6. Diestler, D. J., J. Chem. Phys. 60, 2692 (1974).
7. Diestler, D. J., J. Chem. Phys. 7, 394 (1975).
8. Englman, R., and Jortner, J., Molec. Phys. 18, 145 (1970).
9. Diestler, D. J., Radiationless Processess, p. 169 (Ed. K. Fong, Springer-Verlag, 1976).
10. Toton, E. T., "Thermal Relaxative Rates in Condensed Explosives," NSWC TR 80-101.

SHOCK-INDUCED INSTABILITY DUE TO CRACK-LIKE DEFECTS IN A SOLID PROPELLANT

B. M. Belgaumkar
Technical Consultant, IAEC (Bom.) Ltd., "Vainatheya"
S.V. Road, Irla, Vile-Parle (West), Bombay 400 056 India

ABSTRACT

The presence of crack-like defects in a propellant, may lead to unstable combustion, and cause improper functioning of a rocket engine or a gun. Methods to predict allowable levels of crack-like cavities in propellants without undue risk of abnormal functioning are developed.

INTRODUCTION

A propellant or propellant system as used in rockets and guns is generally considered and treated as any other explosive material. In its normal behavior, it should deflagrate in a controlled, repeatable, and predictable manner. Nonetheless, in addition to what is expected during normal operation and safe storage, it is also prone to uncontrolled deflagration or detonation, depending upon the intensity of an unexpected impulsive stimulus. Thus, safety considerations become as important as with any other explosive.

A review of accidental explosions is given in reference 1 and studies of the detonation process in reference 2. It is the object of this paper to propose methods to predict allowable crack-like defects in propellants.[3-5]

BRITTLE FRACTURE AND COMBUSTION INSTABILITY

There is a wide variety of materials, environments, and technology in which fracture of one kind or another may affect the integrity and functioning of a system. Brittle fractures are not so common as fatigue, yielding or buckling failures, but when they do occur they may be more costly in terms of human life and/or property damage. Many brittle failures have occurred and still do occur, where subsequent combustion of the products is not involved. Where the failure is followed by combustion, the brittle fracture analysis is still the same but the effects of combustion are studied separately.

A well-agreed basic principle of fracture mechanics is that unstable fracture occurs when the stress-intensity factor at the crack tip reaches a critical value, which depends on many factors like stress-strain relationship, size and shape of specimen, crack geometry, nominal stress, flaw size, temperature, rate of loading, environment, etc. Recent development of fracture mechanics[6,7] has established that there are three primary factors, material toughness, crack size, and nominal stress level, that control the susceptibility of any equipment to brittle fracture. Knowing the value of

the critical-stress at failure for a given material of particular
thickness and at a specific temperature, loading rate and environment, one can determine the flaw sizes that can be tolerated in the
material.

When triggered by a suitable sufficiently energetic impulse an
explosive or marginally explosive substance reacts chemically with
liberation of energy. The input of energy needed to start reaction
is very different for different substances, and also different for
different reaction rates that follow.

As the combustion front approaches the edge of a crack-like
cavity, the combustion rapidly spreads all over the cavity surface
since the pressure in the chamber is expected to be much higher than
the initial pressure in the cavity. This impedes the gas off-take
resulting in the local pressure and temperature to increase abruptly
with a probability of shock-wave initiation. Because of the specific
structure of solid propellants in the tip of the cavity, volume combustion may occur and jointly with fracture mechanism in this region
it may cause a burn-through or even explosion.

LINEAR ELASTIC FRACTURE MECHANICS

The work required to propagate a crack which is done as a
result of increasing external forces or due to reduction in the
elastic energy of a body with increasing crack size is expressed in
terms of a coefficient called the intensity factor which plays a
fundamental role in the whole theory. Confining our attention to
brittle fracture, we need to consider only the theory of linear
elastic fracture mechanics.[7]

Fracture stress S_f for a brittle material is

$$S_f = S\sqrt{2Er/\pi a} \qquad (1)$$

where r is the surface energy, S the nominal stress, a, half length
of crack or flaw, and E the elastic modulus.

In the absence of an adequate knowledge of bond strengths,
Griffith, using an energy argument, derived the condition for propagation of a brittle crack. Orowan and Irwin modified this condition
to formulate a force criterion and also to show the equivalence of
the two (energy and force) criteria and put it in the form:

$$S_{22}^2 + S_{12} \geq K^2 \qquad (2)$$

where S_{22} and S_{12} are the normal and tangential stresses, respectively, at the crack tip and $K = [4rG/\pi(1-p)]^{1/2}$, G and p are respectively shear modulus and Poisson's ratio.

The LHS of Eq. (2) has a maximum when shear stress is zero,
that is, when the normal stress is one of the principal stresses. If
this stress happens to be compressive, such a crack will not propagate. Thus, Eq. (2) does not lead to a satisfactory criterion for
the propagation of cracks under shear.

Sharfuddin,[8] taking Griffith's crack as a model, has shown that the necessary and sufficient criterion is given by the parabolic relationship

$$S_{12}^2 + 4 K' S_{12} = 4K^2 \qquad (3)$$

where K', the measured tensile strength, equals the minimum principal stress.

Equation (3) is the envelope of the stress circle in the Mohr's diagram. Equation (2) is represented by a circle of radius K with center at origin and Eq. (3) is represented by a parabola. As long as the distance to any point on the parabola is greater than or equal to K, Eq. (2), is satisfied for that point. This is true for all points on the parabola if $K' \geq K$.

Equation (3) has been derived from the necessary and sufficient criterion that there must be a local stress sufficient to rupture the atomic bonds at the edge of the crack, that is, if the measured tensile strength is greater or equal to that given by Orowan's equation, Eq. (2). The fact there is more than enough energy available for the crack propagation process is perhaps the explanation of the phenomenon of explosive shattering of some of these materials when they are fractured.

SHOCK INITIATION AND THE CRITICAL ENERGY CONCEPT

In understanding the process by which shock waves initiate reaction and build up to detonation in explosives, the concept of critical energy has an important role. There are however still many loose ends to tie up. The energy per unit area just necessary to cause detonation is simple referred to as critical energy, which can be formulated as the initiation criterion:

$$E_c = Put = \frac{P^2 t}{R_o U} = \text{constant} \qquad (4)$$

where t is the time of application of the shock pressure P, u the particle velocity, U the shock velocity and R_o is the initial density. An alternative criterion is:

$$P^2 t = \text{constant} \qquad (5)$$

There is a difference of opinion[9] whether Eq. (4) may be relied upon universally. Similarly there is disagreement over the applicability of Eq. 5.

The author proposes a criterion that a minimum amount of energy per molecule, called the activation energy or energy barrier, must be supplied before any chemical reaction can proceed, and that a minimum shock strength is necessary for initiation and overcoming the shock impedance $R_o U$. The first condition is satisfied by Put = constant, (4a), which is an expression for energy, and the second condition is satisfied by Eq. (5) whose origin is associated with a shock impedance term in the denominator. In general, these two conditions should be satisfied simultaneously as necessary and sufficient conditions, as in the case of brittle fracture. That the minimum steady detonation pressure to overcome the shock impedance is so high compared with the pressure required to supply the activation energy, explains the phenomenon of detonation.

FRACTURE MECHANICS APPROACH TO LIMIT DETONATION

The Griffith-Irwin-Orowan concept is applicable to cracks under thermal loading, as shown by Sih,[10] It follows that the concept of a stress intensity factor is preserved for problems concerned with crack development under a thermal stress field.

The general fracture mechanics design practice for terminal fracture is approached by ascertaining the critical stress intensity factor, K_c, corresponding to the expected mode of failure (there are three modes) for the particular material at a given temperature and loading rate. This factor is related to nominal stress, and flaw size, thus:[7]

$$K_c = CS\sqrt{\pi a} \qquad (6)$$

where S is the nominal (gross section) tensile stress perpendicular to the plane of the crack at the location of the crack; a is a characteristic crack dimension; and C is a nondimensional constant, the value of which depends on the crack geometry, ratio of crack size to the size of the structural member, and type of loading.

The limiting size of crack or crack-like defect in a propellant is given by substituting the shock pressure, P, (from Eq. (4a) or Eq. (5), the one which is maximum) for S in Eq. (6) and rearranging, thus:

$$a = \frac{1}{\pi}\left[\frac{K_c}{CP}\right]^2 \qquad (7)$$

This criterion has been proposed to limit the size of crack-like flaws in the finished grain and to fix norms for nondestructive tests for detection of flaw sizes.

REFERENCES

1. W. E. Baker and R. A. Strehlow, Prog. Energy Combust. Sci., 2, pp. 27-60 (1976).
2. C. H. Johansson and P. A. Persson, Detonics of High Explosives (Academic Press, London and New York, 1970).
3. B. M. Belgaumkar, "Fracture Evaluation of Accidental Explosion", Proc. Symp. Fracture Mechanics, ISTAM, Rourkela, Feb. 8-10, 1980.
4. B. M. Belgaumkar, "Shock Retardation and Presence of Cracks in Materials," 6th Int. Symp. on Detonation, San Diego, USA (1976).
5. B. M. Belgaumkar, "Shock Propagation of Cracks in Materials," Int. Conf. on Fracture Mechanics and Tech. Fracture Mechanics, Hong Kong (1977).
6. V. Z. Parton and E. M. Morozov, "Elastic-Plastic Fracture Mechanics and Tech. Fracture Mechanics", (MIR pub, Moscow, 1978).

7. S. R. Rolfe and J. M. Bartom, Fracture and Fatigue Control in Structures, Application of Fracture Mechanics, (Prentice-Hall, Inc., N.J., USA, 1977).
8. S. M. Sharfuddin, Jour. Enrg. Maths. $\underline{9}$ (3), July 1975.
9. F. E. Walker, "Discussion on Shock Initiation and P^2t," Session I, 6th Int. Symp. on Detonation, San Diego, USA (1976).
10. G. C. Sih, Trans. ASME, Ser. E, $\underline{29(3)}$, 587-598 (1962).

THE DIVERGENT QUASISTATIONARY DETONATION WAVE

G. Damamme
Commissariat a l'Energie Atomique
B.P. N° 7 93270, Sevran, France

ABSTRACT

Taking account of the width and structure of a detonation wave, we establish that, for small curvature of the wave front, the wave is stationary and equivalent to a one dimensional wave within terms of the order of 1/R. We give the wave celerity and the corresponding end of the detonation state.

INTRODUCTION

In a perfect fluid a diverging sonic detonation wave is stable because the characteristic lines of the downstream flow go away from the discontinuity[1] (so the downstream flow is an expansion wave). As it is observed experimentally that the celerity of such a detonation wave is less than the C-J celerity (the minimum celerity if detonation is schematized by a discontinuity), the spreading and the structure of the reaction zone must necessarily be taken into account to explain this phenomenon.

DEVELOPMENT IN FUNCTION OF THE CURVATURE

Consider a diverging detonation wave, the radii of curvature of which are great compared with the width of the reaction zone. Suppose that it goes through a uniform medium and that the downstream flow is an expansion wave. Choose a reference frame attached to the shock initiating the chemical reactions (we have adopted the Z-N-D scheme[2] in which the shock precedes chemical reactions). So at the origin of our reference frame the flow is normal to the shock.

In a first approximation the width of the reaction zone can be considered as negligible compared with the radius of curvature of the detonation wave so the flow is plane and stationary, i.e., it is the C-J detonation.

In a second approximation that width can no longer be neglected and one must also take into account the structure of the reaction zone. Locally, within terms of the order of 1/R, the flow continues to be stationary (the time derivatives are of second order in 1/R) and it can be considered as one dimensional (in the first order term only the averaged values enter). On the other hand it can no longer be considered plane because the divergence of the flow lines is related to 1/R. More precisely the evolution of the area of a normal section of flow tube is given by:

$$dA/A = (1/R) (D/u(z) - 1) dz$$

where z is the distance to the shock, D its celerity and u(z) the corresponding particle velocity which can be evaluated with the associated zero structure (the C-J structure).

GENERALIZED JUMP RELATIONS

As the flow is stationary and one dimensional, the conservation of mass, momentum and energy enables one to establish generalized jump relations which take into account flow divergence. Give or take terms of the order of 1/R, for a perfect fluid without conduction, we have:

$$\rho\, u\, A = m = \text{constant}$$

$$E + u^2/2 + p/\rho = \text{constant}$$

$$\{mu + pA\}_o^z = \int_o^z p\, dA$$

ρ = density, p = pressure, E = internal energy per unit mass.

If the fluid is dissipative, the last two equations are modified and complementary terms in 1/R appear. In particular, in the last equation, the pressure p is replaced by the component $-\sigma_{zz}$ of the stress tensor.

By themselves, these equations do not suffice to determine the detonation structure that we are looking for. To do that it is necessary to consider the complementary hypothesis (the downstream expansion wave). It requires that relative to the wave front, the flow is sonic at the end of the reaction zone.

SONIC STRUCTURE OF THE QUASISTATIONARY DETONATION

Among the sonic structures which can be thought of, only one is possible and its celerity is minimal. This is due to the fact that if a structure is sonic at the distance z from the shock, then its celerity is the slowest among the ones of the flows algebraically possible at such a distance (flows which are obtained by use of the jump relations).

The true slowest detonation celerity is the greater of these different ones; so there exists only one sonic structure, the one which corresponds to the greater celerity. It generalizes the C-J structure of plane and stationary detonation. However, the sonic zone no longer corresponds exactly with the end of reaction but with an anterior state determined by the slowest end of reaction kinetics. We shall call ε the extent of the chemical reactions and z_* the shock distance. To determine these two parameters it is necessary to introduce the function M(Z) which gives the minimal celerity of a plane and stationary detonation in regards to the algebraically possible flows at the distance z from the shock. Then z_* is given by:

$$\frac{1}{M^2} \frac{dM^2}{dZ}(z_*) = \frac{\gamma^2}{1 - G/2\gamma} \frac{1}{R}$$

γ = polytropic coefficient, G = Gruneisen coefficient.

To precisely determine the functions $M^2(Z)$ and z_*, it is necessary to define the end of reaction kinetics. When only one parameter ε is sufficient to do that, it can be written as

$$d\varepsilon = h(\varepsilon) \, F(p, V; \text{shock}) \, dt \quad \text{if } \varepsilon \sim 1$$

Because
$$\frac{d \, \text{Log} \, M^2}{d\varepsilon} = \left(\frac{\delta \, \text{Log} \, p}{\delta \varepsilon}\right)_{V,E} \bigg/ \left(1 - \frac{G}{2\gamma}\right)$$

for a kinetic relation which ends asymptotically, ε_* and z_* are given by:

$$\left(\frac{\delta \, \text{Log} \, p}{\delta \varepsilon}\right)_{V,E} \left(\frac{F}{u}\right)_{C-J} h(\varepsilon_*) = \frac{1}{R}$$

and
$$z_* = \int \varepsilon^* \frac{d\varepsilon}{\frac{F}{u}_{C-J} h(\varepsilon_*)}$$

Downwards in the sonic zone, the reactions which take place no longer contribute to the propagation of the wave.

QUASISTATIONARY DETONATION CHARACTERISTICS

As with the C-J detonation, its principal characteristics are its celerity and the thermodynamic state of the sonic zone. They differ from the corresponding C-J quantities by terms in z_*/R, $1 - \varepsilon_*$ and $1/R$. The coefficients of the first two terms only depend on the C-J state. That of $1/R$ depends on the whole C-J structure; in particular for a dissipative fluid, it depends on viscosity and thermal conduction. This limited development is established within terms of the order of $1/R$, which take into account the initial conditions and the boundary ones (geometric details, detonation initiation, confinement).

The detonation celerity is given by:

$$D_* = D_{C-J} \left\{ 1 - \frac{1}{1-G/2\gamma} \left[\gamma^2 \frac{z_*}{R} \right.\right.$$
$$\left.\left. + \left(\frac{\delta \, \text{Log} \, p}{\delta \varepsilon}\right)_{V,E} (1 - \varepsilon_*) + \frac{J}{R}\right] + 0\left(\frac{1}{R}\right)\right\}$$

where the coefficient J depends on the whole C-J detonation structure. If we adopt for the explosive decomposition kinetics hot spots followed by rapid rearrangement reactions, $h(\varepsilon)$ is proportional to $(1-\varepsilon) |\text{Log}(1-\varepsilon)|^{2/3}$, z_* in $|\text{Log } R|^{1/3}$ and $1 - \varepsilon_*$ in $0(1/R)$.

EXAMPLES AND APPLICATIONS

These relations are interesting because they apply to the spherical or cylindrical diverging detonations as well as to the stationary or planar detonations. In that case, although detonation celerity depends on form, dimensions (diameter) and confinement of the cylinder, on the axis (the place where detonation wave is normal to the cylinder direction) the relation between celerity and curvature does not depend on these parameters within second order terms. A celerity and sonic state - curvature relation is very interesting because it permits determination of the detonation wave movement and downstream conditions (the sonic zone) independent of the subsequent flow of detonation products. An analogous relation was established for a converging detonation,[1] where surface contraction replaced curvature.

The properties are the basis of a simple numerical method to simulate detonation. The wave propagation is determined by use of the celerity-curvature relation and behind it, the thermodynamical and mechanical (u) state is given by the celerity of the wave. Such a relation requires incorporation of the model boundary conditions of empirical nature with the curvature or the tangent of the shock front.

Further, to have a more realistic model, there is interest in the use of (in the celerity-curvature relation) an average curvature taken on a convenient surface (to introduce the lateral sonic propagation inside detonation wave). Such a procedure does not contradict our initial relation which has been established within second order terms and permits us to take into account lateral gradients.

Since about the time corresponding to the end of reaction, sound celerity and particle velocity are quasi-identical, if the reactions are assumed to end asymptotically for a reaction width of z_* the radius of the influence disk is also z_*. Here for a step-time Δt, the radius of the influence disk must be $C_* \Delta t$ (C = sound celerity).

REFERENCES

1. G. Damamme, C. R. Acad. Sc. Paris, 292, 4, 381 (1981).

2. R. Courant and K. O. Friedrichs, Supersonic Flow and Shock Waves (Interscience Publishers, New York, 1948), p. 430.

RADIOGRAPHIC STUDY OF IMPACT IN POLYMER-BONDED EXPLOSIVES*

by

Erik Fugelso, J. D. Jacobson, Robert R. Karpp
Los Alamos National Laboratory, Los Alamos, NM 87545

Russ Jensen
Hercules Corporation, Magna, Utah

ABSTRACT

Computer-tomography generated material-density maps from flash x-ray radiographs of the impact of cylinders of mockup polymer-bonded explosive (PBX) striking a steel plate. Comparison of the density fields with computer simulation allowed discrimination of rather complex deformation and flow models for insensitive explosives to be used in further studies of chemical reactions initiated by shock waves.

INTRODUCTION

An initial step to determine the mechanism of detonation initiation in the ballistic impact of insensitive high explosives is the measurement and computational modeling of the mechanical deformation. To this end, impact parameters of mockup polymer bonded explosives (PBX) cylinders striking a steel plate were measured through the axisymmetric adaptation of computer assisted tomography (CAT) and compared with several numerical simulations involving different models of the PBX. CAT generates a density map throughout the cylinder at selected time intervals from low energy ray radiographs.[1,2] Features such as shock fronts, cracks and recompression zones are identified and quantified. Variations in the mathematical modelling to match early and late time behavior are somewhat complex and lead to improved characterization of the PBX mechanical models, to which reactive descriptions can later be applied.

EXPERIMENT

Two flash radiographs from a set of radiographs of the impact of an inert PBX mockup on a steel plate were selected for study. The projectile was a right circular cylinder 8.85 mm radius, L/D ~1.2, the impact velocity was 677 m/s. The composition of the cylinder is listed in the table. The x-ray source was 150 kev; radiographic times after initial contact with the steel plate were 4.8 and 11.2 μs.

The transmitted intensity, I, of an x-ray beam through the projectile is related to the incident intensity, I_0, of the x-ray from the source,

---

*Work performed under contract W-7405-ENG-36

$$I = I_0 \left\{ \int_0^\infty s(e) \exp\left[-\int_0^L \rho(\xi)\mu(e,\xi)d\xi\right] de \right\} / \left\{ \int_0^\infty s(e)de \right\}, \quad (1)$$

where $\rho(\xi)$ is the material density, $\mu(e, \xi)$ is the x-ray absorption coefficient, which is a function of x-ray energy, e, s(e) is the distribution function of the energy of the x-ray source and ξ is the path of the x-ray.

Equation (1) reduces to a simple exponential if either μ is independent of the x-ray energy or the x-ray source is monochromatic. If the absorption coefficient is not independent of the x-ray energy and the source is not monochromatic radiation, it is possible to calculate an effective constant absorption coefficient through a nonlinear stretching transformation, utilizing the property that the transmitted intensity decreases monotonically with increasing thickness.

In either case the intensity can be written in a simpler form,

$$\ln I = \left(-\int \rho \mu d\xi\right) + \ln I_0. \quad (2)$$

The negative logarithm of the x-ray intensity, after corrections for nonuniform incident intensity and geometric beam spreading, is proportional to the integral of the product of the material density and the x-ray absorption coefficient. The profile has axisymmetry and, if we select values of this integral on a plane slice perpendicular to the axis of symmetry, elementary application of the techniques of computer assisted tomography will allow reconstruction of the material density along the radius. Denote by g(x) the value of the negative logarithm of the intensity at fixed axial distance, at a distance x from the axis. Denote by f(r) the value of the product $\rho\mu$ at distance r from the axis. Then

$$g(x) = 2 \int_x^a \frac{rf(r)dr}{\sqrt{r^2 - x^2}}, \quad (3)$$

where a is some radius beyond which f(r) vanishes. We have measured g(x) and can calculate numerically the value of f(r).

The two radiographs were digitized on a PDS scanning microdensitometer at 50 micron spacing with a 50 micron by 50 micron aperture; 512 lines with 1024 pixels per line were obtained. Figure 1 shows the digitized radiographs for 4.8 μs.

These radiographs were then displayed on the Comtal 8000 digital image display and an interactive program was utilized to determine the centerline of the projectile. The maximum tilt of the projectile from ideally normal impact was measured and was less than 2^0. The needed projections were then extracted from the digitized radiograph.

To reduce film grain noise, a median or Tukey filter was used on each line, its negative logarithm taken and then corrected for the polychromatic source. The x-ray absorption coefficient for this material in the range of the incident x-ray spectrum are essentially independent of x-ray energy;[3] therefore, this effect of beam hardening is small. The intensity outside the projectile image is nearly constant throughout this line and throughout the entire radiograph and was subtracted from the entire digitized projection. The resulting projection was then inverted numerically line by line to give the material density along that line (Fig. 2).

Since the expected variation of density in a shock wave or rarefaction wave is of the order of 0.5%-10%, the images were enhanced by a linear stretch from 87.5% peak density to 100% peak density in eight grey level increments (Figs. 3 and 4). Each grey level represents a 1.56% relative density step in $\Delta\rho/\rho$. This enhancement brings out significant detail. In the 4.8 μs picture, a slightly curved shockwave is visible near the top of the projectile and a rarefaction wave is seen to extend from the intersection of the shockwave and the outer edge to the axis and then continues on from the axis (in the picture this looks as if the wave reflects from the axis). There is also a annular high compression zone whose center is at the intersection of the edge of the projectile and the plate. A fine resolution enhancement showed a very narrow ridge of slightly higher density extending from the center recompression zone. Higher density regions appear behind the top shock and the reflected rarefaction. Densities in the spray are very low. The enhanced picture for the 11.2 μs radiograph shows no definable wavefronts, but shows a pronounced density structure. The recompression zone has shifted towards the axis. A different enhancement, concentrating on lower densities, shows a narrow crack extending from the center line and bounding the lower density bubble.

COMPARISON WITH THEORY

The density fields obtained from radiographs can be compared to those obtained from computer simulations and the adequacy of the assumed material response functions judged from the extent of agreement. A Lagrangian program (SALE), using artificial viscous pressure in the shock, gave the solutions shown in Fig. 5 and 6 for a simple fluid model in symmetric impact (677 m/s). In this calculation the pressure is given by the Mie-Gruneisen form referred to the shock Hugoniot, but modified to give negligible rupture strength.

The prominent features of the radiographs appear also in the simulation: the inital shock (21 kb) rapidly attenuated by lateral expansion, followed by jetting and a recompression to near the stagnation pressure (5 kb), beginning near the periphery and converging to the center, finally producing at the impact face a disc of warm material at normal density and, at the back face, a dome of spalled material at low density. Some details of the simulations are less plausible and would be regarded as numerical artifacts were they not present also in the resolved radiographs: the slightly higher density at the center of the impact face and the shell of normal density material rising from the compacted disc near the edge. The recompression is of particular interest in this system, because it has been proposed as a trigger for the delayed detonation often observed in propellants under these conditions.

TABLE I
COMPOSITION BY WEIGHT OF THE INERT POLYMER-BONDED
EXPLOSIVE (PBX) MOCKUP CYLINDERS

Constituent	Wt.%
Cross linked polymers	19.00
Plasticizer	10.00
Powdered Aluminum (5 μm)	19.50
Talc	14.12
Salt (100 μm)	17.38
Salt (50 μm)	20.0

REFERENCES:

1. R. P. Kruger, G. W. Wecksung, and R. A. Morris, Opt. Engr., 19, 273, (1980).

2. E. Fugelso, "Material Density Measurements from Dynamic Test X-ray Radiographs Using Axisymmetric Tomography," Los Alamos Scientific Laboratory, LA-8785-MS (1981).

3. Alpha, Beta, and Gamma Ray Spectroscopy, Vol. I, edited by K. Seigbahn (North Holland, Amsterdam, London, 1965).

Fig. 1. Digitized radiograph of the mockup PBX cylinder striking a steel plate at 4.8 μs after impact. Striking velocity is 677 m/s.

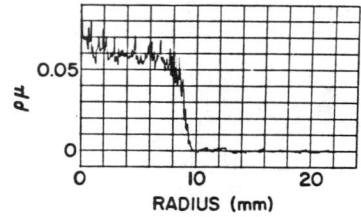

Fig. 2. Typical tomographic reconstruction of the density profile along a line at 1 mm below the top of the projectile.

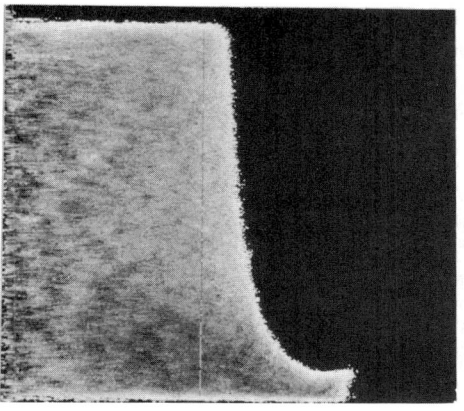

Fig. 3. Enhanced image (linear stretch) for the projectile at 4.8 μs after impact. A shock wave and two rarefaction waves are now visible.

Fig. 4. Enhanced image (liner stretch) for the projectile at 11.2 μs after impact. No specific wavefront structure is visible, but the density gradients are quite pronounced.

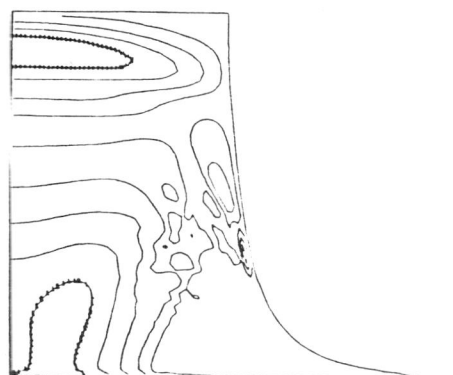

Fig. 5. Calculated density contours for impact of a mockup PBX cylinder striking a steel plate at 677 m/s at 6.0 μs after impact. Maximum and minimum density ratios are 1.140 ad 0.815. Density contours are 0.033.

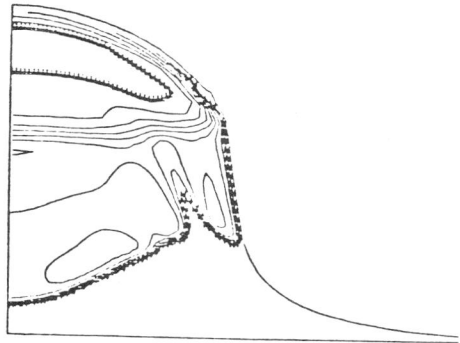

Fig. 6. Calculated density contours for impact of a mockup of PBX cylinder striking a steel plate at 677 m/s at 12.0 μs after impact. Maximum and minimum density ratios are 1.027 and 0.304. Density contours are 0.073.

PLANE SHOCK INITIATION OF GAMMA-IRRADIATED PETN SINGLE CRYSTALS

J. J. Dick
Los Alamos National Laboratory, Los Alamos, NM 87545

ABSTRACT

Pentaerythritol tetranitrate (PETN) single crystals of (110) orientation about 1 cm thick were subjected to room temperature irradiation from a 450 Ci ^{60}Co source. The time and distance to detonation were measured in plane shock initiation wedge experiments after doses of up to 1 MR. Shock strength was 8.6 GPa. The crystals were sensitized by the irradiation. The initiation is apparently still of the homogeneous type.

Chap. 13. Experimental Facilities for Shock Compression Research

THE LAWRENCE LIVERMORE NATIONAL LABORATORY TWO-STAGE LIGHT-GAS GUN*

A. C. Mitchell, W. J. Nellis, and R. J. Trainor

University of California, Lawrence Livermore National Laboratory
Livermore, California 94550

ABSTRACT

The diagnostics and experimental program of our facility are described.

DESCRIPTION

This facility is used to study condensed matter at high pressures. The gun launches a planar impactor plate at a velocity up to 8 km/s. The operation of the gun is described and illustrated elsewhere in these proceedings.[1] When the impactor strikes a target specimen, shock pressures up to several Mbars and temperatures up to a few tens of thousands of Kelvins are generated. Maximum compression factors of 1.5-6 over initial specimen density are achieved, depending on the specimen material.

The duration of the experiments is a few hundred nanoseconds. The facility has a fast diagnostic system[2] with nanosecond or less time resolution to measure material properties. Shock velocity and impactor velocity are measured to obtain equation-of-state data in solids[3] and liquids.[4] A fast optical pyrometer was developed to measure temperature in initially transparent materials.[5,6] Electrical conductivity has been measured.[7] Flash x-ray diffraction has been performed in crystals at shock pressures up to about 1 Mbar.[8,9]

In the past few years many materials have been studied including Al, Cu, Ta[3] and liquid H_2, D_2,[10] N_2, O_2, Ar,[4,11] CO, CH_4,[12] Xe,[13] NH_3, H_2O,[6,7] CH_2 (polybutene) C_6H_6 (benzene),[14] I_2,[15] graphite,[16,17] BN,[18] steel,[19] and UO_2.[20] We are now planning experiments on liquid He and the alkali metals.

The facility is summarized in Table I.

*Work performed under the auspices of the U.S. Department of Energy by Lawrence Livermore National Laboratory under contract #W-7405-Eng-48.

Table I. The LLNL Two-Stage Light-Gas Gun Facility

Facility:	Launch velocities of 1-8 km/s; launch tube inner diameter of 22, 28, 38, 51, or 64 mm.
Diagnostics:	Pulsed x-ray shadowgraphy for impactor velocity; shorting, self-shorting, and piezoelectric shock-wave detectors with 300 psec time resolution; 6-channel optical pyrometer with 5 ns time resolution; electrical conductivity with 2 ns time resolution; flash x-ray diffraction; rotating mirror framing and streak camera.
	Tektronix 519 and 7903 oscilloscopes and 7912 transient digitizers in development.
Experiments:	Equation of state Pyrometric temperature Electrical conductivity Flash x-ray diffraction High explosive response to high-velocity impact
Scientific Staff:	A. C. Mitchell, R. J. Trainor, W. J. Nellis
Support Staff:	D. Bakker, M. Brooks, W. Thomas, R. Neatherland, W. Graham, G. Devine, J. Chmielewski, J. Rego.

REFERENCES

1. A. C. Mitchell, W. J. Nellis, and B. Monahan, "Enhanced Performance of a Two-Stage Light-Gas Gun", these proceedings.

2. A. C. Mitchell and W. J. Nellis, Rev. Sci. Instrum. **52**, 347 (1981).

3. A. C. Mitchell and W. J. Nellis, J. Appl. Phys. **52**, 3363 (1981).

4. W. J. Nellis and A. C. Mitchell, J. Chem. Phys. **73**, 6137 (1981).

5. G. A. Lyzenga and T. J. Ahrens, Rev. Sci. Instrum. **50**, 1421 (1979).

6. G. A. Lyzenga, T. J. Ahrens, W. J. Nellis, and A. C. Mitchell, "The Temperature of Shock-Compressed Water", Lawrence Livermore National Laboratory, UCRL-86791 (1981), submitted for publication.

7. A. C. Mitchell and W. J. Nellis, "Equation of State and Electrical Conductivity of Water and Ammonia Shocked to the 100 GPa (1 Mbar) Pressure Range", Lawrence Livermore National Laboratory, UCRL-86478 (1981), submitted for publication.

8. Quintin Johnson, A. C. Mitchell, and Ian D. Smith, Rev. Sci. Instrum. 51, 741 (1980).

9. Q. Johnson and A. C. Mitchell, in High Pressure Science and Technology, edited by B. Vodar and Ph. Marteau (Pergamon, Oxford, 1980), p. 977.

10. W. J. Nellis, M. Ross, M. van Thiel, A. C. Mitchell, G. J. Devine, and N. Brown, "The Shock Compression of Liquid H_2 to 10 GPa (100 kbar)", these proceedings.

11. M. Ross, W. Nellis, and A. Mitchell, Chem. Phys. Letters 68, 532 (1979).

12. W. J. Nellis, F. H. Ree, M. van Thiel, and A. C. Mitchell, J. Chem. Phys. 75, 3055 (1981).

13. M. van Thiel, A. C. Mitchell, and W. J. Nellis, Bull. Am. Phys. Soc. 25, 513 (1980).

14. R. J. Trainor, A. C. Mitchell, M. B. Boslough, W. J. Nellis, to be published.

15. A. K. McMahan, B. L. Hord, and M. Ross, Phys. Rev. B 15, 726 (1977).

16. W. H. Gust and D. A. Young, in High-Pressure Science and Technology, Vol. I, edited by K. D. Timmerhaus and M. S. Barber (Plenum, New York, 1979), p. 944.

17. W. H. Gust, Phys. Rev. B 22, 4744 (1980).

18. W. H. Gust and D. A. Young, Phys. Rev. B 15, 5012 (1977).

19. W. H. Gust, D. J. Steinberg, and D. A. Young, High Temp.-High Pressures 11, 271 (1979).

20. W. H. Gust, in High-Pressure Science and Technology, Vol. II, edited by B. Vodar and Ph. Marteau (Pergamon, Oxford, 1980), p. 1025.

SHOCK WAVE PHYSICS GROUP (M-6)

Charles E. Morris
Los Alamos National Laboratory, Los Alamos, NM 87545

The Shock Wave Physics Group located at Ancho Canyon was formed in the early 1950's to provide equation-of-state (EOS) data for the high pressure physics community. One of the early advances which greatly facilitated obtaining accurate EOS data was the impedance matching technique of J. M. Walsh[1]. With this technique the Hugoniots of over 500 materials[2,3,4] have been measured. Currently the EOS effort at M-6 is largely confined to the two-stage gun facility where many of the previously measured Hugoniots are being extended to higher pressures. Most of the other shock wave studies are directed toward a better understanding of the shock process and physical properties of the shocked state. The experimental facilities and personnel being utilized for these investigations are listed in Table I along with some of the most recent studies.

Phase changes under shock loading have been observed for many materials. A manifestation that a phase change has occurred is a discontinuity both in magnitude and slope of the U_s-U_p relation. For some phase changes with small volume differences and similar elastic constants there are no observable discontinuities either in value or slope of the U_s-U_p curve. To detect these changes of state a more sensitive probe is needed. The optical analyzer developed by McQueen[5] is a detector with sufficient sensitivity to see small wave velocity discontinuities in the shock state. Consequently some of the more subtle phase changes can be detected. Recently the solid-solid ($\varepsilon \rightarrow \gamma$) phase transition at 200 GPa in iron has been observed along with the γ iron to melt transition at 250 GPa[6]. Both these transitions are undetectable from the U_s-U_p Hugoniot curve. The Lagrangian wave interaction diagram for the optical analyzer is shown in Fig. 1. Upon symmetric impact of a

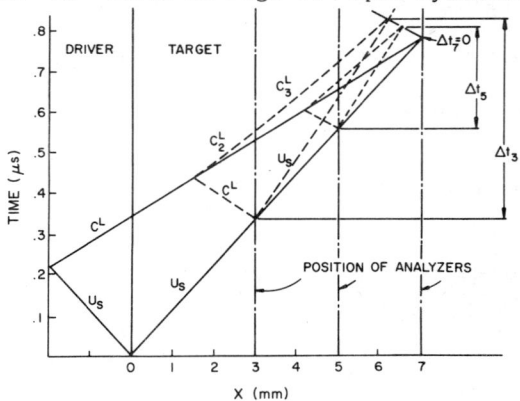

Fig. 1. Lagrangian Wave Interaction Diagram

TABLE I Compressed Gas Gun, Two-stage Gas Gun and
High Explosive Facilities

Facilities:	51-mm-diam. single-stage, compressed-gas gun (0.02 → 1.00 km/s), 29-mm-diam. two-stage, light-gas gun (3 → 8 km/s) and HE firing site using explosive lenses from 102 mm to 305 mm diam.
Diagnostics:	Tektronix 7903 (0.7 ns resolution) and Tektronix 519 (0.3 ns resolution) oscilloscopes, radiation detectors (< 1 ns response), ASM probe (1 ns/cm coil diam.), Manganin gauges (> 15 ns response), free surface capacitor (15 ns response), flash radiography, rotating mirror framing and streak cameras, framing and streak image converter camera.
Recent Studies:	• Melting of iron under core conditions. • High pressure equation-of-state of SiO_2. • Thermodynamic properties for shocked porous iron. • High pressure strength of metals at pressures exceeding 100 GPa. • Radiation technique for very accurate shock velocity measurements. • Melting of Al and Cu on the Hugoniot. • Optical technique for determining rarefaction wave velocities at multimegabar pressures. • Synthesis of superconducting A15 Nb_3Si by shock compression. • Mechanisms for rapid phase transformation of Nb_3Si under shock conditions.
Principal Investigators:	Joseph N. Fritz, Stanley P. Marsh, Robert G. McQueen, Jerry A. Morgan, Charles E. Morris, Bart W. Olinger, Steve C. Schmidt, John W. Shaner.
Technical Support:	Austin D. Bonner, Walter W. Quintana, Dennis L. Shampine, Robert S. Medina, Charles W. Caldwell, Felix A. DePaula, Connie N. Gomez, Maria T. Roybal.

thin flyer with a step wedge target shock waves are propagated both in the target and flyer plate. When the shock wave reaches the backside of the flyer plate a forward rarefaction is generated which ultimately overtakes the forward going shock front in the transparent analyzer causing a decrease in the radiation. The time interval Δt between initial emergence of the shock at the target-analyzer interface until overtake by the rarefaction wave is a linear function of target thickness and when extrapolated to zero determines the target thickness where the rarefaction would have overtaken the shock wave at the target-analyzer interface. Through this extrapolation procedure the hydrodynamic perturbation due to the presence of the analyzer is eliminated and therefore a true in situ overtake velocity is obtained. The rarefaction wave velocity C at pressure is given by [5]

$$C = U_s \, (\rho_o/\rho) \, (R+1)/(R-1) \qquad (1)$$

where R is the ratio of target to flyer plate thickness for Δt equals zero. Experimentally determined R values of a few tenths of a percent precision are typical.

Another technique useful for studying the physical properties of the shock state is the axially symmetric magnetic (ASM) probe. A paper describing measurements of the quasi-elastics structure in 2024 Al to 90 GPa and in OFHC Cu to 140 GPa is presented in these conference proceedings. The ASM probe has also been used to study insulators, detonating explosives[8] and HE driven metal plates. Illustrated in Fig. 2 is the construction of the ASM probe assembly designed to measure the quasi elastic structure in metals. A permanent magnet provides a nonuniform steady magnetic field through which the target surface is accelerated. Because of the magnetic field, eddy currents are generated in the moving target surface which produce a time-varying magnetic field. These time-varying fields induce a voltage signal in a pickup coil from which the velocity of the metal plate may be obtained.[7] A wave

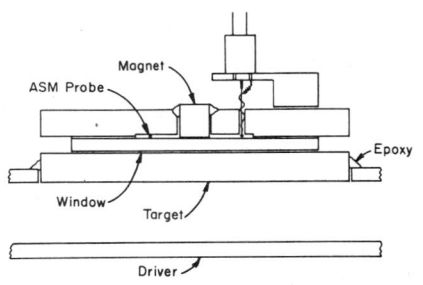

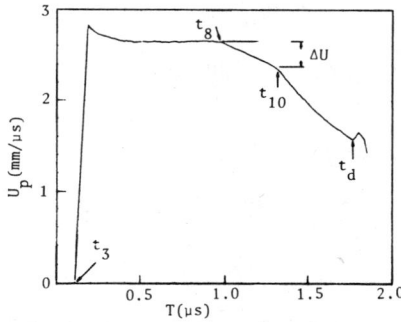

Fig. 2. ASM Probe Assembly.   Fig. 3. Wave Profile in 2024 Al.

profile in 2024 Al at 55 GPa is shown in Fig. 3. From this wave profile the longitudinal modulus, bulk modulus and shear strength of the material at pressure may be calculated.[9]

Another major interest at M-6 is the use of pulsed megagauss fields as a tool for solid state research for high-energy pulsed power sources and as pressure sources. Max Fowler, who is the developer of the cylindrical flux compression generator,[10] directs this research effort. Some physical characteristics of the explosive flux compression generators developed at M-6 are listed in Table II and illustrated in Fig. 4. All the generators have relatively large working volumes. Some of the recent studies using

TABLE II Pulsed Megagauss Field Facility

Facilities:	Single-stage flux compressor: fields to 150 T, > 1 μs 90% to peak, experimental volume 76 mm long to 19 mm diam. Two-stage flux compressor: fields to 250 T, 5 μs 90% to peak, experimental volume 76 mm long to 19 mm diam. Cylindrical flux compressors: fields > 1000 T, < 1 μs 90% to peak, experimental volume 10-20 mm long to 7 mm diam. Capability of cryogenic experimentation.
Diagnostics:	Time resolved rotating mirror spectrographs, coverage 200-680 nm, various light sources including quasi-continuous, and variety of laser sources, wide variety of electrical measuring techniques, flash radiography, image intensifiers, rotating mirror framing and streak cameras, framing and streak image converter camera.
Recent Studies:	• Magneto-resistance of Bismuth. • Isentropic compression of deuterium. • Faraday rotation of many materials. • Excitonic absorption spectrum of GaSe. • Preliminary mapping of B-T spin-flip to paramagnetic phase line for MnF_2. • Zeeman spectra at 500 T fields.
Principal Investigators:	Robert S. Caird, Dennis J. Erickson, C. Max Fowler, Bruce L. Freeman, Jim H. Goforth.
Technical Support:	Jim C. King, A. Richard Martinez, Ralph R. Roy, Steve E. Salazar, David T. Torres.

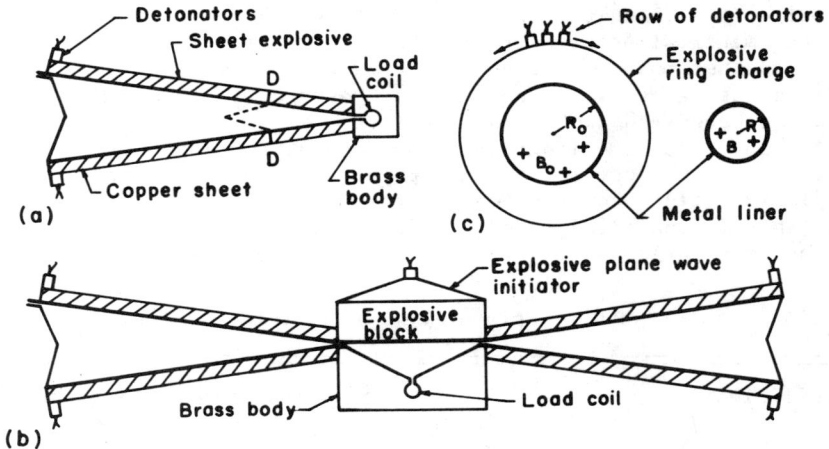

Fig. 4. Flux Compression Generators: a) Single Stage; b) Two-Stage; c) Cylindrical.

these systems are listed in Table II along with the principal investigators and supporting personnel. Presently much of the effort in the magnetic program is directed toward the development of pulsed power sources for physics research.

REFERENCES

1. J. M. Walsh and R. H. Christian, Phys. Rev. $\underline{97}$, 1544 (1955).
2. M. H. Rice, R. G. McQueen, and J. M. Walsh, Solid State Phys. $\underline{6}$, 1 (1958).
3. R. G. McQueen, S. P. Marsh, J. W. Taylor, J. N. Fritz and W. J. Carter, in High Velocity Impact Phenomena, edited by R. Kinslow (Academic Press, New York, 1970).
4. Los Alamos Shock Hugoniot Data, S. P. Marsh, Compiler (University of California Press, Berkeley, 1980).
5. R. G. McQueen, J. W. Hopson and J. N. Fritz, "An Optical Technique for Determining Rarefaction Wave Velocities at Very High Pressures," Submitted Rev. Sci. Instrum. (1981).
6. J. M. Brown and R. G. McQueen, Geophys. Res. Let. $\underline{7}$, 533 (1980).
7. J. N. Fritz and J. A. Morgan, Rev. Sci. Instrum. $\underline{44}$, 215, (1973).
8. W. C. Davis, in Proceedings of the Sixth Symposium on Detonation, edited by D. J. Edwards (Office of Naval Research, Arlington, Va., 1976), p. 637.
9. J. R. Asay and L. C. Chhabildas in Shock Wave and High Strain Rate Phenomena in Metals, edited by M. A. Meyers and L. E. Murr (Plenum Press, New York, 1981).
10. C. M. Fowler, W. B. Garn and R. S. Caird, J. Appl. Phys. $\underline{31}$, 588, (1960).

THE SANDIA SHOCK THERMODYNAMICS APPLIED RESEARCH FACILITY*

Lalit C. Chhabildas
Sandia National Laboratories,† Albuquerque, New Mexico 87185

ABSTRACT

A brief description of the Sandia Shock Thermodynamics Applied Research Facility is given. Three different smoothbore guns are used to launch projectiles for controlled planar impact studies. Combined pressure-shear loading conditions can also be obtained through use of an intervening anisotropic crystal. Time-resolved interferometric and holographic instrumentation is used for diagnostics. This facility has the unique capability of providing time-resolved measurements in shock loaded specimens over the stress range 0.1 to 700 GPa.

INTRODUCTION

The present article provides a brief description of the Shock Thermodynamics Applied Research (STAR) Facility located at Sandia National Laboratories, Albuquerque, NM. This research facility has evolved from many contributions by various Sandia personnel over the last two decades. The facility is presently equipped with three different smoothbore guns which produce controlled plane impact conditions at velocities ranging from 0.02 to 8 km/s. A variety of both time- and space-resolved diagnostic techniques, including velocity, displacement, holographic and microwave interferometry is available. Thus, the facility has the unique capability of providing time-resolved measurements in shock-loaded specimens over the stress range 0.1 to 700 GPa.

The principal investigators who use the facility for research and applications are listed in Table I. The group consists of both experimentalists and theorists, thereby allowing a close interaction in developing material models for different applications. The facility is also used by other Sandia personnel in support of various other programs. The members of the technical staff whose support is crucial for the smooth operation of the facility are also listed in Table I. The loading and diagnostic capabilities with some applications are described in the following sections.

SHOCK-LOADING CAPABILITIES

The STAR facility has three guns that are used to launch projectiles. The single-stage compressed gas gun[1] is commonly used when impact velocities in the range 0.02 km/s to 1 km/s are required. The 100-mm bore diameter allows long duration experiments (8 µs) to be performed while preserving one dimensionality, free from edge effects.

*This work was supported by the U. S. Department of Energy under Contract DE-AC04-76-DP00789. †A U. S. Department of Energy facility.

TABLE I. Personnel and Applications of the Sandia Shock
Thermodyamics Applied Research Facility

Principal Investigators:
  James R. Asay (Division Supervisor), Lynn M. Barker,
  Timothy J. Burns, Lalit C. Chhabildas, Douglas S. Drumheller,
  Dennis E. Grady, Jack L. Wise

Technical Support:
  Carl Konrad (Facility Supervisor), Dave Cox, Robert Hardy,
  Ronald Moody, John Nevers, Charles Whitney (Ktech),
  Brian Warner (Manpower, Inc.)

Recent Studies:
  • Localized Shear Deformation under Dynamic Loading
  • Shock-Induced Vaporization
  • Pressure-Shear Loading of Metals and Viscoelastic Materials
  • High Pressure Strength of Shocked Metals
  • Dynamic Fracture and Fragmentation
  • Experimental Studies and Modeling of Geological Materials
  • Microwave Properties of Shocked Dielectrics
  • High Pressure Interferometer Window Studies
  • Mass Ejection from Shocked Surfaces
  • Shock-Induced Melting and Polymorphic Phase Changes
  • Wave Propagation in Composite and Viscoelastic Materials
  • Shock Compaction of Porous Materials
  • Rain Drop Erosion Studies
  • Impact Cratering

Exceptionally good alignment between the two impacting surfaces is attained with typical values of tilt ranging from 0.2 to 0.5 mrad.

The 89-mm bore diameter propellant gun[2] is a useful impulse loading device for experiments where impact velocities ranging from 0.6 km/s to 2.3 km/s are desired. A maximum 9 kg propellant charge is used to launch a 1.5 kg projectile at 2.3 km/s. Typical values of tilt acquired on this gun range from 1 - 2 mrad. Since the bore is smaller and tilts typically larger than the compressed gas gun, the propellant gun is not used in the lower velocity range.

The two-stage light gas gun[3] with its small bore diameter (29 mm) and light projectiles (~ 25 gms) provide impact velocities over a range of 2.5 km/s to 8 km/s. Impact tilts range from 6 - 9 mrad close to the muzzle. The small bore size, however, limits working areas over which shock loading is planar and one dimensionality is preserved.

These guns produce a high-stress, uniaxial-strain pressure pulse in specimens under study during dynamic loading. Combined pressure-shear loading techniques have also been developed[4] for use on the single-stage compressed gas gun. In this technique, the coupled longitudinal and transverse motion generated by the normal impact of Y-cut quartz is transmitted into a sample bonded to the rear surface of the quartz. Techniques have been developed so that 0.6 GPa shear stress can be transmitted at 1.5 GPa longitudinal stress.

DIAGNOSTIC CAPABILITIES

A wide variety of both discrete and time-resolved instrumentation[5] is available for diagnostic purposes. Impact tilt is determined by measuring the simultaneity of the projectile surface arrival at an array of co-planar charged pins, co-axial charged pins or anodized pins. Typically, the impact velocity is measured to 0.2% accuracy on the single-stage compressed gas and propellant guns and 1% accuracy on the two stage light gas gun. In each case, velocity is inferred by time intervals over which the projectile impacts precharged pins arranged at premeasured spacing. The signal generated from these pins provides a convenient fiducial, which is used to correlate the impact time to wave arrival times to determine shock velocities.

A velocity interferometer,[7] VISAR, is the most commonly used diagnostic tool at the STAR facility. This interferometer superposes the doppler-shifted light from the moving surface, at two slightly different times, producing a fringe number proportional to the surface velocity. By varying the delay time, the fringe frequency can be controlled. The time resolution of the interferometer is 3 ns and the velocity can be measured to a precision of 0.2%. There are two VISARs available at the facility. They are often both used, with widely varied sensitivity on a single experiment to yield unique and unambiguous velocity measurements,[8] thereby reducing the number of experiments to be performed.

An air delay leg VISAR is also available at the facility which is five times more sensitive than the VISARs which employ etalons and is particularly suitable for low-velocity measurements.[9] More recently, the VISAR has also been used to determine the dynamic shear-wave particle velocity in pressure-shear loaded materials.[10] In this technique, two VISARs are used to monitor different directions of gathered light beams from a surface which undergoes both longitudinal and shear motion. Fringes produced in the interferometer are proportional to a linear combination of both the longitudinal and shear components of the surface velocity.

Since the VISAR is widely used at the facility, it has been interfaced with waveform digitizers and a PDP-11 computer for data acquisition and immediate data reduction. The flow diagram for automatic interferometer data reduction is given in Fig. 1.

The displacement (Michelson) interferometer[11] is well suited for particle velocity measurements less than 0.1 km/s. Fringes produced in the interferometer by superposing the reflected beam from the moving surface with the reference laser beam are proportional to the displacement of the surface. This system has a resolution of ~ 20 ns and typical accuracies in velocity measurements are ~ 1%. The maximum surface velocity that can be measured is limited to 0.2 km/s where the fringe frequency begins to exceed the frequency response of the recording systems.

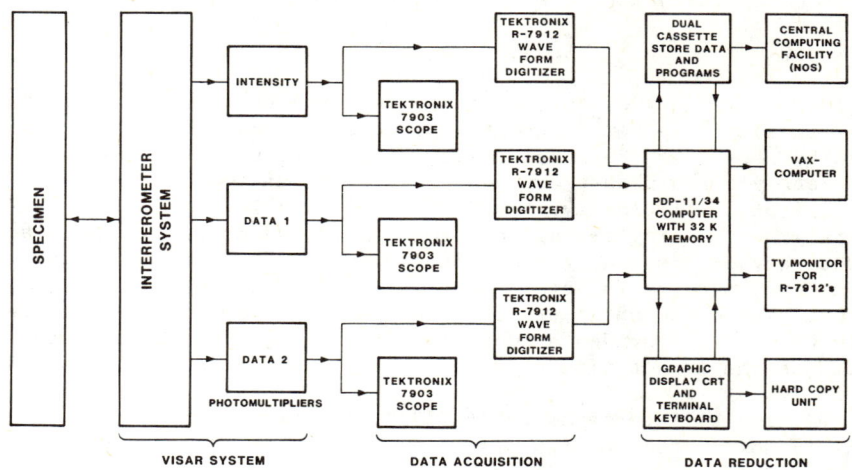

Fig. 1. Interferometer data acquisition and reduction scheme.

A holographic system[12] is used for time- and space-resolved measurements of the surface motion. This system utilizes a 3 ns optical signal from a pulsed ruby laser which is split into four individual pulses with 26 ns delay between them. After reflection from the target, each pulse is combined with the reference optical pulse at corresponding times to produce holograms on the photographic plate. Images of the target can then be reconstructed from these holograms. Holographic interferometry is accomplished by combining holographic images taken before and during the dynamic event. The time and space resolutions of this system are 3 ns and 20 µm, respectively.

In addition, the facility is also equipped with a flash x-ray unit having a 30 ns exposure pulse. A microwave interferometer[13] is used to determine surface motion through insulating materials.

APPLICATIONS

Experimental studies involving the use of both impact and diagnostic capabilities at the STAR facility are numerous;[14] some are listed in Table I. The time-resolved techniques at the facility have allowed detailed studies of phenomena associated with shock processes. Experimental studies on different types of materials have led to the development of material models for porous, composite, geological and anisotropic materials. A few highlights involving applications will be briefly summarized.

Fine structure in the shock loading wave profiles for different forms of calcite have indicated the importance of microstructure on the calcite I-II transition. Both loading and release wave profiles in armco-iron indicate the rate effects associated with the bcc-hcp polymorphic phase transition in iron and the hysteresis effect associated with the reverse phase transition. Rise-time measurements in aluminum, copper and beryllium have led to estimates of shock thicknesses and viscosity, and have also allowed development of models relating viscosity to strain rate. Pressure-shear loading of aluminum and alumina-filled epoxy have allowed a critical evaluation of existing material models. The development of lithium fluoride as a window material up to 120 GPa for interferometric measurements will permit time-resolved material property studies (e.g. strength, melting) up to 700 GPa in some high-impedance materials. Time- and space-resolved holography have been used to determine the rate effects associated with shock induced vaporization in lead up to 300 GPa. Finite shear-stresses measured in beryllium and aluminum upon reshock are being attributed to recovery from heterogeneous yielding processes and demonstrate the limitation of elastic-plastic models.

FUTURE DIRECTIONS

There has been an ongoing capability development program at the facility. Items planned for the facility include the installation of an oblique impact capability to provide more flexibility for combined pressure-shear loading; development of a dual air delay leg velocity interferometer to provide better precision for pressure-shear studies; development of picosecond interferometry to allow estimates of risetimes and shock thicknesses at high stresses and acquisition and testing of a multi-channel analyzer to allow measurements of temperatures at hot spots believed to be due to localized inhomogeneities.

REFERENCES

1. C. D. Lundergan, Sandia Corporation Report No. SC-RR-4421 (1960, unpublished).
2. J. R. Asay, "The Sandia Shock Thermodynamics Applied Research Facility," Sandia Laboratories Report (1981, unpublished).
3. D. E. Munson & R. P. May, J. AIAA 14, 235 (1976).
4. L. C. Chhabildas & J. W. Swegle, J. Appl. Phys. 51, 4799 (1980).
5. R A. Graham & J. R. Asay, "High Temp.-High Press. 10, 355 (1978).
6. L. M. Barker & R. E. Hollenbach, Rev. Sci. Instrum. 35, 742 (1964).
7. L. M. Barker & R. E. Hollenbach, J. Appl. Phys. 43, 4669 (1972).
8. R. A. Lederer, S. A. Sheffield, A. C. Schwarz, & D. B. Hayes, in Sixth Symposium on Detonation, Office of Naval Research Report No. ACR 221 (Washington, D.C.) p. 668.
9. B. T. Amery, Ref. 8, p. 681.
10. L. C. Chhabildas, H. J. Sutherland, & J. R. Asay, J. Appl. Phys. 50, 5196 (1979).
11. L. M. Barker & R. E. Hollenbach, Rev. Sci. Instrum. 36, 1617 (1965).
12. J. G. Kelly & L. P. Mix, J. Appl. Phys. 46, 1084 (1975).
13. J. L. Wise, Sandia National Labs. (private communication).
14. References pertaining to applications are in Ref. 2.

SHOCK-WAVE EXPERIMENTS USING EXPLOSIVES AND LIGHT-GAS GUN FACILITIES

R. Cheret, P. Andriot, P. Chapron, C. Le Drean,
J. M. Lezaud, R. Loichot, J. Martineau, F. Olive

C. E. A. Vaujours, B.P.7, 93270 Sevran, France

ABSTRACT

We describe shock wave experiments using explosive generators and a two-stage light-gas gun. Experimental results are presented to illustrate the facilities, the diagnostics, and the measurement techniques.

INTRODUCTION

A part of our experimental program deals with the behavior of shocked materials: mainly equation of state, shock-induced phase and state changes, mass-ejection from shocked surfaces, dynamic fracture and fragmentation. The purpose of this paper is to describe the different experimental capabilities available in this field in our laboratory. Thus, we present the shock-wave facilities, the experimental arrangement for the different diagnostics, and then some recent representative results.

EXPERIMENTAL FACILITIES

In the low pressure range (below 15 GPa) we use a powder gun, 2 meters in length, 90 mm in diameter. The projectile is guided along a 1.5 meter long launch tube.

Pressures from 15 GPa to greater than 100 GPa are provided either by explosive generators or by a two-stage light-gas gun. This last tool[1] is one of the most powerful drivers presently in operation in the world. It accelerates a 0.85-1.0-kg and 80-mm-diameter projectile from 1800 m/s up to 4500 m/s with helium driving gas.

DIAGNOSTICS

Most of the fundamental parameters for understanding the hydrodynamic behavior of shocked materials can be derived from shock-wave and free-surface velocities. Moreover, in the case of dynamic fracture, global observation and specific rupture diagnostics are needed. In order to fulfill all these conditions, we have developed different experimental methods involving optics, electronics, radiography and magnetic techniques.

OPTICAL METHODS

We use high speed cameras: a Barr and Stroud CP5 streak or framing camera and a Cordin operating as streak and framing camera simultaneously. We also have TSN 503 electronic streak cameras with 10 ps time resolution. For light-gas gun experiments, projectile velocity is currently obtained by means of electronic probes. Another technique measures the projectile time-of-flight between two He-Ne laser beams a measured distance apart. In our laboratory, the most-used diagnostic for the measurement of free surface velocities is the Doppler Laser Interferometry technique (D.L.I.). The reflection of a monochromatic laser beam on a moving surface induces a change in wavelength, which enables one to deduce the velocity time history of the surface. Our experimental set-up consists of a 4 watt argon ion laser of wavelength 5145Å associated with a Fabry Perot interferometer. The maximum adjustable optical path length L is 50 cm and the reflectivity R is 0.94. The image of the interferometric pattern, composed of concentric rings, is formed on the electronic streak camera slit. The variations of the ring diameters are signature of the velocity vs time evolution. This system gives velocity measurements over the wide range of $1-10^4$ m/s with an uncertainty of less than 1%.[2,3]

SHOCK WAVE VELOCITY ELECTRONIC MEASUREMENTS

Shock-wave velocity is commonly obtained by measuring the shock transit time between two parallel planes a measured distance apart. The interplanar distance is typically of the order of a few mm and is known with a precision of 1 μm. The transit times are measured with an accuracy better than 1%, provided by subnanosecond probes and 100-ps-time-resolution electronic recording system. According to the pressure range and to the studied material, different kinds of detectors are available:

Mylar probe (pressure > 25 GPa). The isolation of the inner conductor is insured by a 5 μm thick mylar foil. This probe has a jitter of ±0.8 ns.

Silicon probe (pressure > 25 GPa). The specimen material and the probe are separated by few microns of vacuum-evaporated silicon. The jitter of this probe is ±0.3 ns.

Piezoelectric probe (for pressure < 20 GPa). Small ceramic disks are depolarized when strongly shocked and an electric signal of a few tens of volts is generated. The jitter is ±1.1 ns.

MAGNETIC FRACTURE DIAGNOSTICS

An original magnetic method of measurement[4] has been developed in order to follow the motion of explosively deformed metallic shells and to detect their fractures. The moving shell is placed in a stationary field. The variations of magnetic field is related either to the section changes of the conducting shell or to the appearance of fractures. Variations in the magnetic field are recorded vs time by an open coil surrounding the shell and connected to an oscilloscope.

DIAGNOSTIC DEVELOPMENT

Presently we are trying to develop new tools. For the projectile velocity in the light-gas gun experiment, in addition to optical and electronic probing measurements, we plan to install DC or pulsed x-ray generators. In addition, shock-wave velocity measurements will be performed in the near future by using a new generation of detectors consisting of an ionization micro-chamber coupled to photo-detectors through optical fibers. In order to illustrate the different facilities described in this paper, we have chosen to present three typical examples of our experimental studies.

ISENTROPIC COMPRESSION

A projectile impacting onto a target with a velocity V_p induces a shock through a copper plate in the sample and in the AuNi thin foil which reaches pressure A in the (u_p, P) diagram (Figs. 1 and 2). The projectile is accelerated by the two-stage gun.

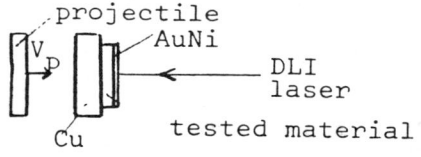 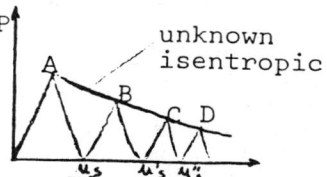

Fig. 1 Experimental configuration.   Fig. 2 $P(u_p)$ diagram.

Then the tested material expands along its own release isentrope while the AuNi foil is successively compressed and released. The D.L.I. diagnostic provides the different AuNi free surface velocities u_s, u'_s, u''_s, etc. If the shock adiabats and isentropes of the thin foil are known, we can determine the different states A, B, C, D which belong to the tested-material isentrope. This method remains valid as long as the shock impedance of the tested material is low enough compared to the accelerated thin-foil shock impedance.

EJECTION OF MATERIAL FROM SHOCKED SURFACES

Some mechanisms for material ejection from shocked surfaces of metals have been studied recently in our laboratory.[5] The experimental device uses a thin foil whose velocity is recorded by the D.L.I. A simple momentum transfer interpretation allows us to derive the mass flux of ejected material vs time. The main sources of ejection are surface geometrical defects (with a strong influence of a shock-induced surface melting) and melting temperature inhomogeneities in some alloys.

RUPTURE BEHAVIOR OF METALS AT HIGH RATES OF STRAIN

Rupture behavior of metals during expansion at high rates of strain has been investigated.[4] Explosive devices have been developed in order to perform ideal types of deformation. Several experimental methods were used simultaneously: mainly high speed photography, magnetic rupture diagnostics and micrographic observation of recovered fragments.

CONCLUSION

In this paper we have presented some of our experimental capabilities (drivers and diagnostics). Having explosive generators, powder guns and a more sophisticated tool as the light-gas gun is certainly an advantage which allows to explore a large range of pressure for various experimental applications in the shock-wave field. A summary is included in Table I.

ACKNOWLEDGEMENTS

The authors are indebted to J. Marilleau for this original experimental contribution to the development of the laboratory in last past years.

REFERENCES

1. H. Bernier, M. Valadon, M. Zembri, and J. M. Lezaud, "Experimental Hugoniot of Copper in the Shock Pressure Range 40 to 150 GPa", submitted for publication (1981).
2. M. Durand, P. Laharrague, P. Lalle, A. Le Bihan, J. Morvan, and H. Pujols, Rev. Sci. Instrum. $\underline{48}$, 275 (1977).
3. P. Andriot, P. Chapron, B. Laurent, and F. Olive, Communication to the Fifth International Laser Congress, Munich, 1981.
4. F. Olive, A. Nicaud, J. Marilleau, and R. Loichot, in Institute of Physics Conference Series No. 47 (Institute of Physics, Bristol, 1980), p. 242.
5. P. Andriot, P. Chapron, and F. Olive, these proceedings.

Table I. Shock wave experiments using explosives and two-stage light-gas gun facilities.

===

Laboratory Facilities:

   Powder gun 2 m long, 90 mm in diameter,
      pressure range 5-15 GPa.
   Two-stage light-gas gun, 80 mm in diameter,
      projectile velocity: 1800-4500 m/s (with He),
                          up to 6500 m/s (with H_2),
      pressure to 250 GPa.
   Explosive generators
      pressure range 10-100 GPa.

Diagnostics:

   High speed cameras, electronic streak cameras
      (10 ps time resolution).
   Fast recording Tektronix oscilloscopes.
   Various electronic probes associated with a high precision
      digital chronometry (accuracy < 1%).
   D.L.I. system using an adjustable Fabry Perot interferometer
      (velocity range $1-10^4$ m/s, accuracy better than 1%).
   Projectile velocity, time-of-flight measurement technique
      using He-Ne laser.

Recent Studies:

   Equation of state.
   Shock compaction of porous materials.
   Mass-ejection from shocked free surfaces.
   Shock induced melting.
   Dynamic fracture and fragmentation.

Principal Investigators:

   R. Cheret
   P. Andriot, P. Chapron, C. Le Drean, J. M. Lezaud,
   R. Loichot, J. Marilleau, J. Martineau, F. Olive.

Technical Support:

   P. Boutillier, M. Fievet, J. Foucart, D. Hombourger,
   M. Joly, B. Laurent, R. Lefevre, J. Leyris,
   G. Loison, J. Mathias, S. Robert, S. Roux,
   R. Tcherniaeff, J. Trouchenko.

===

SHOCK WAVE APPARATUS FOR STUDYING MINERALS AT HIGH PRESSURE AND IMPACT PHENOMENA ON PLANETARY SURFACES

Thomas J. Ahrens, Mark B. Boslough, Warren G. Ginn,
Mario S. Vassiliou, Manfred A. Lange, J. Peter Watt, Ken-ichi Kondo,
Robert F. Svendsen, Sally M. Rigden and Edward M. Stolper
Division of Geological and Planetary Sciences,
California Institute of Technology
Pasadena, California 91125

ABSTRACT

Shock wave and experimental impact phenomena research on geological and planetary materials is being carried out using two propellant (18 and 40 mm) guns (up to 2.5 km/sec) and a two-stage light gas gun (up to 7 km/sec). Equation of state measurements on samples initially at room temperature and at low and high temperatures are being conducted using the 40 mm propellant apparatus in conjunction with Helmholtz coils, and radiative detectors and, in the case of the light gas gun, with streak cameras. The 18 mm propellant gun is used for recovery experiments on minerals, impact on cryogenic targets, and radiative post-shock temperature measurements.

FACILITIES

Two propellant (18 and 40 mm) diameter and one two stage light gas (160 mm 1st stage; 25mm, 2nd stage) guns are being operated to obtain equation of state, impact and shock recovery data of geophysical and planetary science interest.

(a) Propellant gun (18 mm)

This apparatus employs up to 20 g of propellant and accelerates ∼ 20 g projectiles to speeds of ∼ 2.5 km/sec in approximately 2 m. The barrel and a portion of the impact assembly is evacuated. Impact velocity is determined to a precision of ∼ 1% by means of the projectile obscuring two laser beams which are detected with photodiodes. Although some photographic recording has been carried out, this apparatus has been principally used for shock recovery experiments[1-8], impact experiments at low temperatures[9] post-shock radiative temperature measurements[10,11], and gas recovery experiments from volatile-bearing minerals[12].

(b) Propellant gun (40 mm)

This 6m long apparatus employs up to 0.5 kg propellant to launch 100g projectiles to 2.5 km/sec. Three laser beam observations, over 1.5 m baseline, determine projectile velocity to better than ± 0.3%. Double flash x-ray radiography using two 150 KV Field

Emission, 30 nsec duration sources are used to carry out impact studies or provide additional impactor speed data. Recovery experiments on larger samples[13] and spall studies at low speeds[14], have been carried out with this apparatus. The latter employ a compressed gas breech which permits operating apparatus at speeds of as low as $\sim$100 m/sec. For optical recording the apparatus utilizes a Beckman and Whitley 339 continuous writing streak camera which permits in conjunction with a 75 µF 5KV xenon light source, shock velocities and free surface velocities to be determined using optical recording methods[15-18]. Measurements on low temperature materials have been undertaken[19,20] and preheated molten oxide measurements are in preparation[21]. Absorption and thermal emission spectroscopy is carried out with a 0.5 m Ebert, Jarrell-Ash Monochrometer apparatus[22-24].

(c) Light Gas Gun Apparatus

This apparatus utilizes a continuous x-ray source and photomultiplier to detect[25] the emergence of the $\sim$ 20 gram projectile from the muzzle of the 6m long, H_2 driven, 25 mm diameter second stage of this gun and provide a light source and x-ray trigger. Two Field Emission 15 nsec 150 KV flash x-ray generators provide images of the projectile in flight and yield projectile velocities to an accuracy of better than ± 0.2%. Upon impact with a mineral target, shock and free surface velocity are recorded with a Model ID TRW Image Converter Streak Camera writing at 35 mm/µsec. A 30 MHZ oscillator driving a Pockel's cell provide time marks on the streak record at 16.67 nsec intervals. Hugoniot and release adiabat data for a series of oxides, silicates and sulfides have been obtained to pressures of $\sim$ 200 GPa[26-32].

REFERENCES

1. Hörz, F. and Ahrens, T. J., Am. J. Science, **267**, 1213 (1969).
2. Ahrens, T. J., Fleischer, R. L., Price, P. B. and Woods, R. T., Earth Planet. Sci. Lettr., **8**, 420 (1970).
3. Gibbons, R. V. and Ahrens, T. J., J. Geophys. Res., **76**, 5489 (1972).
4. Ahrens, T. J. and Graham, E. K., Earth Planet. Sci. Lettr., **14**, 87 (1972).
5. Kleeman, J. D. and Ahrens, T. J., J. Geophys. Res., **78**, 5954 (1973).
6. Gibbons, R. V., Ahrens, T. J. and Rossman, G. R., Am. Mineralogist, **59**, 177 (1974).
7. Gibbons, R. V. and Ahrens, T. J., Phys. and Chem. of Minerals, **1**, 95 (1977).
8. Jeanloz, R., Ahrens, T. J., Lally, J. S., Nord, G. L. Jr., Christie, J. M., and Heuer, A. H., Science, **197**, 457 (1977).
9. Lange, M. A. and Ahrens, T. J., Proc. 12th Lunar and Planetary Sci. Conf. (in press).
10. Raikes, S. A. and Ahrens, T. J., High Pressure Science and Technology, Vol. II, ed. by K. D. Timmerhaus and M. S. Barber, Plenum Press, 889 (1979).

11. Raikes, S. A. and Ahrens, T. J., Geophys. J. Royal Astron. Soc., 58, 717 (1979).
12. Boslough, M. B., Vizgirda, J. and Ahrens, T. J. (unpublished data).
13. Vizgirda, J., Ahrens, T. J., and Tsay, F-D., Geochim. et Cosmochim. Acta., 22, 1059 (1980).
14. Cohn, S. N. and Ahrens, T. J., J. Geophys. Res., 86, 1794 (1981).
15. Ahrens, T. J., Lower, J. L. and Lagus, P. L., J. Geophys. Res., 76, 518 (1971).
16. Graham, E. K. and Ahrens, T. J., J. Geophys. Res., 78, 375 (1973).
17. King, D. L. and Ahrens, T. J., J. Geophys. Res., 81, 931 (1976).
18. Vizgirda, J. and Ahrens, T. J. (unpublished data for calcite and aragonite).
19. Lagus, P. L. and Ahrens, T. J., J. Chem. Phys., 59, 3517 (1973).
20. Gaffney, E. S. and Ahrens, T. J., Geophys. Res. Lettr., 7, 407 (1980).
21. Rigden, S. M., Stolper, E. M. and Ahrens, T. J. (in progress).
22. Goto, T., Ahrens, T. J., Rossman, G. R. and Syono, Y., Phys. Earth Planet. Interiors, 22, 277 (1980).
23. Goto, T., Ahrens, T. J. and Rossman, G. R., Phys. and Chem. of Minerals, 4, 253 (1979).
24. Kondo, K. and Ahrens, T. J., this volume.
25. Long, J. R. and Mitchell, A. C., Rev. Scient. Instr., 43, 914 (1972).
26. Ahrens, T. J., J. Geophys. Res., 84, 985 (1979).
27. Jackson, I. and Ahrens, T. J., Phys. Earth Planet. Inter., 20, 60 (1979).
28. Jeanloz, R., Ahrens, T. J., Mao, H. K. and Bell, P. M., Science, 206, 829 (1979).
29. Jeanloz, R. and Ahrens, T. J., Geophys. J. Royal Astron. Soc., 62, 529 (1980).
30. Ibid, 505 (1980).
31. Sommerville, M. R. and Ahrens, T. J., J. Geophys. Res., 85, 7016 (1980).
32. Vassiliou, M. S. and Ahrens, T. J., Geophys. Res. Letters, 8, 729 (1981).

BOEING SHOCK PHYSICS LABORATORY

R. M. Schmidt
Boeing Aerospace Co., Orgn. 2-3646, Seattle, WA 98124

ABSTRACT

The experimental capability of the Boeing Shock Physics Laboratory is described. Various laboratory facilities include a 64-mm bore light-gas gun, a 1.5-mm and a 3-mm bore two-stage hypervelocity projectile gun, an exploding-foil facility, an FX-75 flash X-ray and electron beam accelerator, and a 600-G geotechnical centrifuge with a 100-kg payload capacity. A short bibliography is included which summarizes recent research activities.

INTRODUCTION

The Boeing Shock Physics Laboratory has been performing experimental and theoretical work on the dynamic response of materials and structures due to mechanical, radiation and thermal shock loading for more than twenty years. The basic philosophy of the Laboratory has been to use analytic modeling and scaling (verified by experiment) for evaluating response to conditions not achievable or difficult to achieve in tests. Such procedures eliminate the necessity for complete full-scale simulation experiments which are generally difficult and expensive to perform. A variety of experimental and analytical techniques have been developed to support these goals.

The Laboratory has access to a unique combination of equipment which includes a 64-mm light gas gun, two small bore two-stage hypervelocity projectile guns, an exploding foil apparatus, high explosive test facilities, pulsed electron beam facility, and conventional test machines. Extensive instrumentation supports each of these facilities. Dual VISAR and displacement interferometers, strain interferometers, rotating mirror and image converter streak and framing camera, piezoelectric and piezoresistive stress gages and optical pyrometry are regularly used. A feature of recent studies has been the determination of the material properties of metals at high temperatures (3000 K) extending into the melt regime of refractory metals.

Small-scale structural response experiments under explosive loading are conducted at remote test sites or in the laboratory. Recent tests have included the dynamic response to explosives of typical aircraft panels, the functioning of a blast plug, and the development of a crushable cable for measuring ground shock velocity.

Analytic capability includes finite difference, finite element and method of characteristics computer programs which describe wave propagation, and structural and thermal responses. The wave propagation programs allow a variety of constitutive relations including elastic-plastic and strain rate dependent descriptions, and can include the effects of temperature.

0094-243X/82/780634-05$3.00 Copyright 1982 American Institute of Physics

FACILITIES

The **Boeing Light Gas Gun** is a smooth bore with the following specifications:

Bore	64 mm	Maximum breech pressure	42 MPa
Overall length	30 m	Maximum impact velocity	1.4 km/sec

It is a self-contained pressure vessel consisting of a breech, barrel, test chamber and an impact fragment catcher tank. Firing is accomplished by rupture of a diaphragm separating the pressurized breech from the evacuated barrel. The design allows photographic observation of the target as well as electronic instrumentation of the impacts.

The **Boeing Hypervelocity Projectile Guns** are powder-driven and have a hydrogen second stage to attain velocities up to 8 km/sec for projectiles of 1.5 mm diameter. Specifications and conditions achievable are listed below. These guns are versatile in application, having been fired vertically at soil targets and on a centrifuge at 500 G.

	Small gun	Large gun
Bore	12 mm	16 mm
Propellant mass	50 grains	250 grains
Typical projectiles	1.5 mm dia Lexan at 8 km/sec 1.5 mm dia Al at 6 km/sec	3 mm dia Al at 5 km/sec 6 mm dia Al at 3 km/sec

The **Boeing Exploding Foil Facility** utilizes the electrical explosion of thin metal foils to generate high-pressure waves in solids and to accelerate thin flyer plates to high velocity. The foils are exploded by high currents obtained from the discharge of several 15-microfarad, 20-kilovolt, NRG (Corness-Dubilier) capacitors. The generated pressure pulses are substantially plane and accelerate the flyer plates with very little tilt. A velocity range from about 0.1-km/sec to 3.0 km/sec is achievable with a simultaneity of impact of 100 nanoseconds over a 2.5-cm square by varying such parameters as capacitance, voltage, foil dimensions, flyer dimensions and foil confinement. Flyer plates up to a 5-cm square are commonly used.

The **Boeing FX-75 Flash X-Ray/Electron Beam Accelerator** provides a source of high intensity relativistic electrons for material and structure response to thermomechanical shock, exoatmospheric nuclear weapon radiation effects testing (SGEMP), and system level prompt gamma simulation. The Boeing FX-75 was designed by Ion Physics Corp. and installed in 1967. It has been continuously upgraded, and now provides a magnetically controlled beam for improved intensity and uniformity. Some of its characteristics are:

5 kJ delivered to 10 cm^2 target

Fluence range 10 - 400 cal/cm^2

Mean energy range 1.5 - 3.5 MeV

Front and rear optical access

30 shots/day

The energy storage is a gas insulated coaxial d.c. line charged by a Van deGraaf generator. The stored energy is 16 kJ and energy delivery time is 55 nsec. The single capacitor and single switch design offers the advantage of reliability and reproducibility making it an excellent scientific tool for the investigation of new beam diagnostic techniques. The coaxial energy storage will couple to a wide range of diode load impedances without producing a prepulse or postpulse.

The **Boeing 600-G geotechnical centrifuge** has a maximum dynamic load rating of 60,000 G-kg (66 G-tons) at 620 RPM. The aerodynamic housing and the main shaft assembly are from a Gyrex Model 2133 centrifuge. The rotor, designed and fabricated by the Boeing Company, incorporates symmetric swing baskets to allow testing of noncohesive soil materials. The arm radius to the fully extended base plate is 139.7 cm with a maximum payload mass of 250 kg on each rotor end. The centrifuge is powered by a 30-horsepower Eaton Dynamic Model ACM-326-910B drive unit incorporating an adjustable speed, constant torque eddy-current clutch. The unit also has electrical dynamic braking allowing shutdown from maximum RPM in less than 30 seconds. The constant speed motor and variable drive unit are shock mounted and coupled to the main shaft with a belt to minimize vibration.

The rotor shaft is equipped with 24 slip rings for instrumentation channels, three 200 V.a.c. power slip rings and a hydraulic slip ring which can accommodate either gas or liquid. A pair of motor driven Nikon F2 35-mm still cameras are hub mounted in a stereo configuration. These cameras provide stereo photo coverage of the number one rotor end with a maximum framing rate of six per second. High-speed framing cameras are also rotor mounted.

Associated Test Equipment includes controlled environment systems for each of these facilities. Vacuum or inert atmospheres and a wide range of temperatures are achievable. Radiant and resistive heating systems have been developed with heating times ranging from milliseconds to tens of seconds.

Instrumentation developed and maintained by the Shock Physics Laboratory for the examination of stress-wave and detonation phenomena includes ultra-high-speed photographic and flash x-ray equipment; displacement, VISAR and strain interferometer systems; piezoelectric and piezoresistive stress transducer systems; ballistic pendulums and pyrometers.

The photographic equipment includes a Barr & Stroud framing/streak camera which will give 28 frames at 1.6×10^6 frames/second, or, in the streak mode, a writing speed of 32 km/sec and a minimum total recording time of 20 microseconds; and a TRW model ID image-converter camera which has both framing and streak capability. In addition, a Field Emission Flash X-ray Unit can take three 30-nanosecond exposures, each with variable time delay between exposures. The synchronous use of two or more of these photographic systems is a standard feature of experiments performed in the laboratory.

PERSONNEL

Principal Investigators

Michael D. Bjorkman	**B. J. Henderson**	**John E. Shrader**
David W. Cruikshank	**Keith A. Holsapple**	**Robert M. Schmidt**
Ken D. Friddell	**Brian M. Lempriere**	**Alan B. Zimmerschied**

Technical Support

Harris E. Watson	**Charles R. Wauchope**	**Richard A. Zilbert**

SELECTED BIBLIOGRAPHY

Bjorkman M.D., J.E. Shrader and B.M. Lempriere **"Structural Validation and Fragility Technology - Volume 2."** Defense Nuclear Agency Report DNA 5399F-2, Washington D.C., 1980.

Bjorkman M.D., **"Effect of Risetime and Surface Hardness on Precursor Decay in Shocked Lithium Fluoride,"** Ph.D. Dissertation, Washington State University, Pullman WA, 1980.

Davies F.W., **"Hardness Evaluation of Refractory Alloys Operating at Elevated Temperatures - Volume 1."** Boeing Document T2-4163-1, Seattle WA, 1977.

Davies F.W., A.B. Zimmerschied, F.G. Borgardt and L. Avrami, **"The Hugoniot and Shock Initiation Threshold of Lead Azide."** Proc. of the Sixth Int. Symp. on Detonation, Coronado CA, Aug 1976.

Davies F.W., J.E. Shrader, A.B. Zimmerschied and J.F. Riley, **"The Equation of State and Shock Initiation Characteristics of HNS II."** Proc. of the Sixth Int. Symp. on Detonation, Coronado CA, Aug 1976.

Friddell K.D., R.H. Holze and J.C. Pyle, **"Evaluation of Industrial Plant Hardness Techniques."** Proc. of the Sixth International Symposium of Blast Simulation, Cahors, France, 1979.

Gerstle, J.H. and R.M. Schmidt **"Radiation Effects on Fiberglass Mechanical Properties, Phase II."** Boeing Document D2-19914-3, Seattle WA, 1973.

Henderson B. J. (with H.E. Bates), **"Maximum Theoretical Efficiency for Passive Pulse - Shaping Systems."** Journal of the Optical Society of America, **68**, p. 919, July 1978.

Holsapple K.A., **"Thermodynamic Properties of Materials in the Melt Regime."** Journal of Applied Physics, **48**, 1509-1515 (1977).

Holsapple K.A., **"On the Melt Line Slope of Alloys."** International Journal of Physics and Chemistry of Solids, **38**, 55-58, 1977.

Holsapple K.A., "The Equivalent Depth of Burst for Impact Cratering," Proc. Lunar and Plant. Sci. Conf. 11th, Pergamon Press, p. 2379-2401, 1980.

Holsapple K.A. and R.M. Schmidt, "Theory and Experiments on Rapid Melting of Metals Including Alloy Effects." Journal of Applied Physics, 49, No. 11, p. 999, Nov. 1978.

Holsapple K.A. and R.M. Schmidt, "A Material Strength Model for Apparent Crater Volume," Proc. Lunar and Planet Sci. Conf. 10th, Pergamon Press, p. 2757-2777, 1979.

Holsapple K.A. and R.M. Schmidt, "On the Scaling of Crater Dimensions 1: Explosive Processes," Journal of Geophysical Research, 85, No. B12, p. 7247-7356, Dec. 10, 1980.

Lempriere B.M. and F.W. Davies, "Extendible Nozzle Exit Cone (ENEC) Hardening Technology." Defense Nuclear Agency Report DNA 5103F-1, Washington D.C., 1979.

Lempriere B.M. and F.W. Davies, "Structural Validation and Fragility Technology - Volume 1." Defense Nuclear Agency Report DNA 5399F-1, Washington D.C., 1979.

Schmidt R.M., "A Centrifuge Cratering Experiment: Development of a Gravity - Scaled Yield Parameter." Impact and Explosion Cratering (D.J. Roddy, R.O. Pepin, and R.B. Merrill, eds.), p. 1261-1278, Pergamon Press, New York, 1977.

Schmidt R.M., "Centrifuge Simulation of the JOHNIE BOY 500 Ton Cratering Event," R. M. Schmidt, Proc. Lunar and Planet Sci. Conf. 9th, Pergamon Press, p. 3877-3889, 1978.

Schmidt R.M., "A Viscoelastic Wave Propagation Model for Tungsten at Near-Melt Conditions (3000 K)," Journal of Applied Physics, 50, No. 4 p. 2600-2606, April 1979.

Schmidt R.M. "Meteor Crater—Implications of Centrifuge Scaling," Proc. Lunar and Planet. Sci. Conf. 11th, Pergamon Press, p. 2099-2128, 1980.

Schmidt R.M., F.W. Davies, B.M. Lempriere and K.A. Holsapple, "Temperature Dependent Spall Threshold of Four Metal Alloys," Int. Journal of Physics and Chemistry of Solids, 39, No. 4, pp. 375-385, 1978.

Schmidt R.M. and K.A. Holsapple, "Theory and Experiments on Centrifuge Cratering," Journal of Geophysical Resarch, 85, No. 1, p. 999, Jan. 10, 1980.

INVESTIGATIONS OF HYDRODYNAMIC STABILITY USING ELECTRON AND ION BEAMS

F. C. Perry
Sandia National Laboratories, Albuquerque, NM 87185

ABSTRACT

Intense electron and ion beams are being used to study hydrodynamic stability at the terrawatt level. These investigations support the target design and beam focusing efforts for the particle beam fusion accelerators PBFA I and II. Both shock accelerated and ablatively accelerated perturbation growth and convergence effects have been observed and compared with calculations.

INTRODUCTION

In experiments at Sandia National Laboratories the physics of implosion stability is being studied in order to support target design activity for the particle beam fusion accelerator PBFA I and its upgrade, PBFA II. Electron beam experiments have been performed using the Hydra accelerator, and light ion experiments are currently being carried out on the Proto I accelerator. We are particularly concerned with the effects of fabrication irregularities and beam non-uniformities on target performance and the scaling of stability effects to higher beam powers.

DISCUSSION

Rayleigh-Taylor type instabilities, which occur when a heavy fluid is accelerated by a light fluid, are of utmost concern to inertial confinement fusion (ICF) target designers. These instabilities are also important in studying the mixing of fuel and coolant in a hypothetical fission reactor accident, and the generation of supernovae explosions by instabilities in the collapsing stellar core accompanied by violent release of neutrinos.

An example of an electron beam experiment[1] is shown in Fig. 1. This experiment involved a particular class of Rayleigh-Taylor instability, that due to the impulsive acceleration by a shock wave (1-2 Mbar amplitude) traveling from a light fluid into a heavier fluid.[2] We have used 4-pulse holographic interferometry to record the implosion of a double shell cylinder containing initial well-defined perturbation at the buffer (CH_2) - pusher (Au) interface. Electrons deposit energy around the copper waist of the cylinder, thus generating a strong shock wave which propagates toward the axis of the cylinder. The motion of the inner surface of the gold pusher is then viewed by means of four co-linear laser pulses (each 3 ns duration) directed along the axis of the cylinder. One notes that the implosion has a long wave length asymmetry due to non-uniform electron beam deposition. The effects of the smaller wave length (initial $\lambda = 268$ μm) shell perturbation are also apparent in each frame. A comparison of the data with two-dimensional, planar geometry numerical calculations (Fig. 2) which include materials effects indicates shock accelerated unstable growth of the irregularities at the perturbed shell interface. Both the experiment and calculation show a decrease (late time) in amplitude of the free surface perturbations; however, in the experiment this amplitude decrease begins earlier and the magnitude of the decrease is larger because of convergence in the cylindrical geometry.

In Fig. 3 is shown a recent example of a cylindrical implosion driven by an intense proton beam.[3] In this experiment the inner surface of the cylinder was ablatively accelerated during the 30 ns beam pulse to a maximum velocity of 2.0 cm/µs implying that the absorbed power density exceeded 1.0 TW/cm^2. One notes that the implosion boundary is irregular in shape--a result of non-uniform ion beam absorption. These experiments have also been shown to be useful in studying the effects of target perturbations such as fill holes (that would be used for filling a fusion target with fuel), seams and microscopic defects.

A diagnostic technique which will prove especially important for studying implosion stability of spherical targets is high resolution flash x-radiography.[4] This technique is presently under development for application to intense ion beams.

CONCLUSION

Optical and x-ray imaging techniques which are resolved in time and space have been shown to be important tools in studying the hydrodynamic behavior of particle beam targets. These data are being used to evaluate both beam and target quality for particle beam fusion research.

REFERENCES

1. M. A. Sweeney and F. C. Perry, accepted for publication in J. Appl. Phys.
2. R. D. Richtmyer, Comments on Pure and Appl. Math 13, 297 (1960).
3. D. J. Johnson, G. W. Kuswa, A. V. Farnsworth, Jr., J. P. Quintenz, R. J. Leeper, E. J. T. Burns, and S. Humphries, Jr. Phys. Rev. Lett. 42, 610 (1979).
4. J. Chang, D. L. Fehl, M. M. Widner, K. W. Bieg, and M. A. Palmer, to be published in July 1981 issue of Appl. Phys. Lett.

641

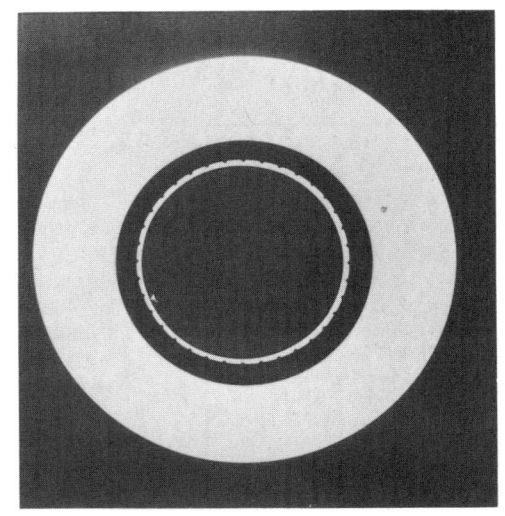

SHOT 8145
Cu = CH_2 - Au TARGET
Au Δr = 100 μm
AMPLITUDE η_0 = 25 μm
WAVELENGTH λ = 268 μm
ASPECT RATIO $\frac{r}{\Delta r}$ = 15

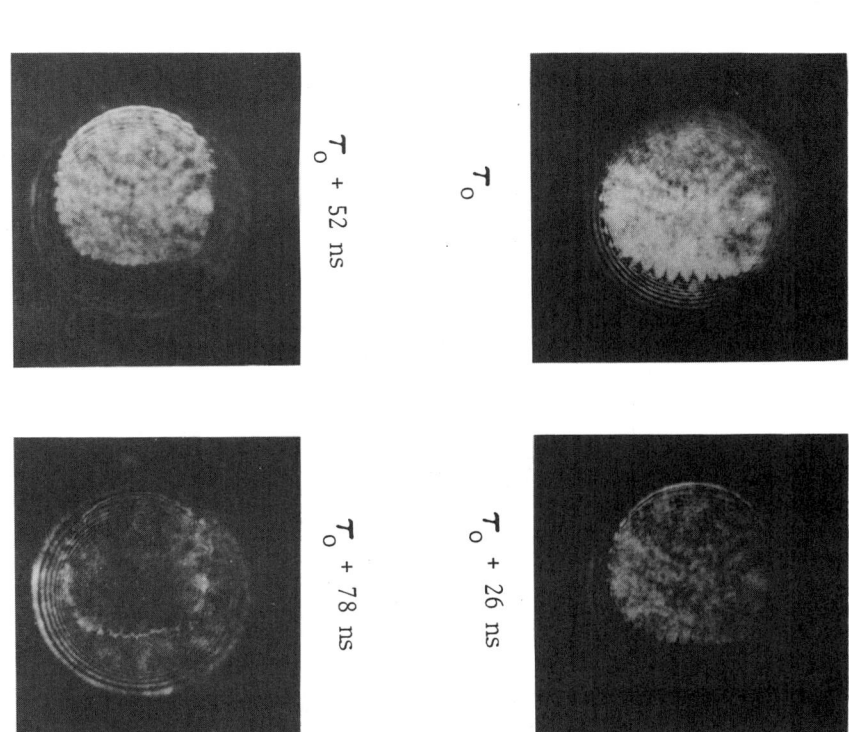

τ_0

τ_0 + 26 ns

τ_0 + 52 ns

τ_0 + 78 ns

Fig. 1. Initial state and holograms of the perturbed Cu/CH_2/Au cylinder for imploded states at 26 ns intervals. The first hologram is taken 100 ns after the shock breaks through the free surface.

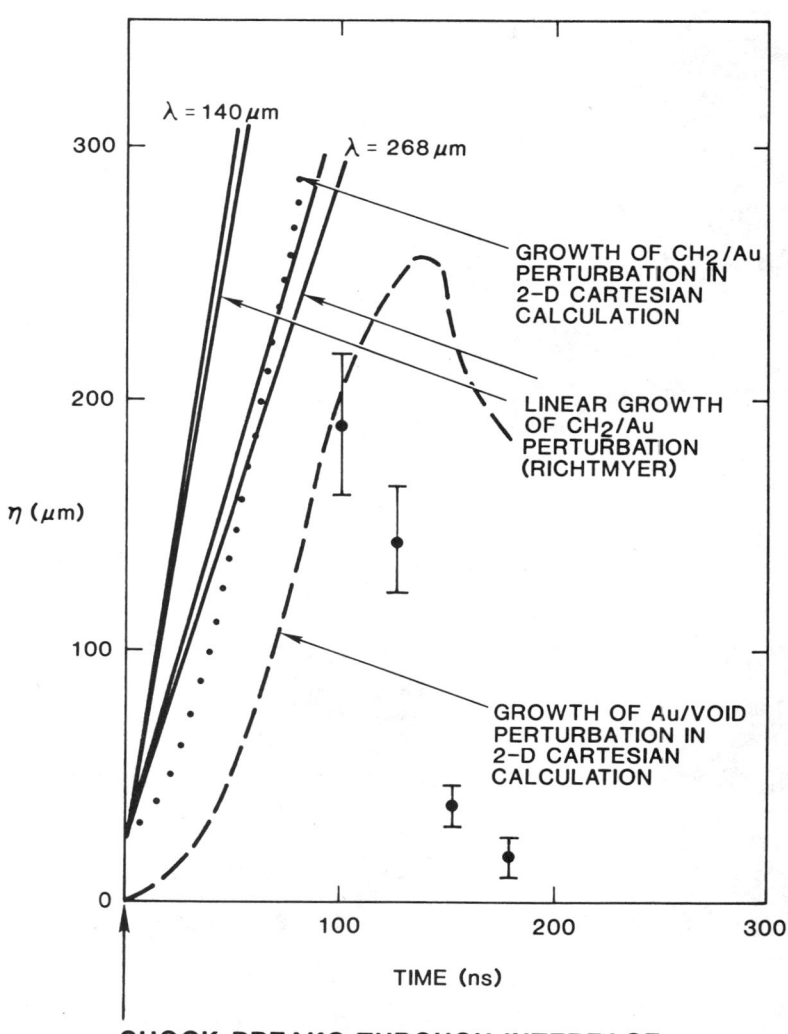

Fig. 2. Experimental data (with error bars) from the holograms of Fig. 1. Also shown are both linear growth calculations and the results of a two-dimensional simulation of the experiment.

643

SHOT NO.

2740

TARGET

Aluminum Cylinder

3.0 mm diam.

50 μm thick

ENERGY

14 kJ

T_0

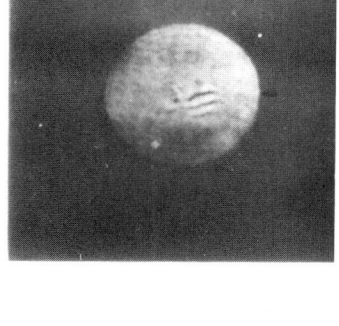

$T_0 + 26$ NS

$T_0 + 52$ NS

$T_0 + 78$ NS

Fig. 3. Holograms of a cylindrical implosion driven by an intense proton beam. In this experiment reference fringes were recorded to improve definition of the implosion boundary.

NUCLEAR-EXPLOSIVE-DRIVEN EXPERIMENTS*

C. E. Ragan
Los Alamos National Laboratory, MS 442, Los Alamos, NM 87545

ABSTRACT

Ultrahigh pressures are generated in the vicinity of a nuclear explosion. We have developed diagnostic techniques to obtain precise, high-pressure equation-of-state data in this unique environment.

INTRODUCTION

Dynamic shock wave experiments using conventional laboratory techniques are limited to pressures considerably below 1 TPa for most materials. Underground nuclear explosives provide a means for reaching considerably higher pressures, but in a much more hostile environment. We have developed experimental techniques that can be used in this environment to obtain precise equation-of-state (EOS) data at pressures from 1 to 100 TPa.

EXPERIMENTAL TECHNIQUES

Shock compression of a material produces a Hugoniot state with properties that can be determined from both absolute and relative measurements. As a first step in our high-pressure EOS program, we obtained[1] a Hugoniot point for molybdenum at 2.0 TPa in an absolute measurement. We then used molybdenum as a standard material to obtain[2-4] Hugoniot data for many samples using the impedance-matching technique.

Table I summarizes our measurement techniques and gives a list of the personnel involved in our high-pressure EOS program. In the measurement at 2.0 TPa we simultaneously determined both the shock velocity and the particle velocity. The particle velocity was deduced from Doppler shifts of neutron resonances in the shocked material, and Fig. 1 illustrates this technique schematically. In the impedance-matching experiments, we used electrical contact pins to determine the shape of the shock front and to obtain shock velocities with 1% to 2% uncertainties. Figure 2 is a photograph of the sample stacks placed adjacent to the molybdenum standard. The holes in the sample are for the electrical contact pins; these are shown in place in Fig. 3.

In addition to these measurements, we recently performed a feasibility experiment to study the problems associated with accelerating a flyer plate to high velocities. This technique is being developed to obtain absolute Hugoniot data at pressures up to 10 TPa from symmetric-impact experiments and is illustrated schematically in Fig. 4. The 5-mm-thick by 180-mm-diam molybdenum

---

*Work supported by the U. S. Department of Energy.

0094-243X/82/780644-04$3.00 Copyright 1982 American Institute of Physics

TABLE I Summary of Experimental Techniques and Personnel

I. Absolute Measurements
   A. Mo at 20 Mbar
   B. Measure Shock Velocity (D) and Particle Velocity (u)
   C. Use Doppler Shift of Neutron Resonances for u ($\pm$2% to 5%)
   D. Use Evacuated Light Pipes for D ($\pm$3% to 5%)
   E. Theoretical Predictions in Agreement with Experimentally Determined Point for Mo at 20 Mbar

II. Relative Measurements
   A. Impedance Matching using Mo Standard
   B. Planar Shock (<1.3 mrad tilt)
   C. Measure Shock Velocities ($\pm$1%)
   D. Use Electrical Contact Pins ($\pm$1 ns)
   E. Data for U, Pb, Fe, Al, Quartz, and low-density Mo
   F. Experimental Hugoniot point for U at 67 Mbar

III. Data Recording
   A. Detector Signals are Transmitted ∼800 m over Coaxial Cables to High-Speed Oscilloscopes
   B. Photographs of the Traces are Recorded along with a Time Base

IV. Personnel
   A. Principle Investigators--Charles E. Ragan and Ben C. Diven
   B. Support
     1. Experimental--William A. Teasdale, Ed R. Robinson, and J. Manuel Anaya
     2. Calculational--Marv Rich and Rod B. Schultz

flyer plate was accelerated to a velocity of ∼40 km/s over a distance of 150 mm, and electrical contact pins were used to monitor its motion and the resulting shock velocity in the molybdenum impact target. The plate was intact upon impact, and jitter in the timing data indicated that perturbations in the surface smoothness were <2 mm. However, the plate was tilted by about 15° and slightly bowed, probably by an additional shock produced in a nearby experiment.

SUMMARY

Experiments using underground nuclear explosives have extended the accessible pressure range up to ∼10 TPa. These measurements have stimulated improved theoretical treatments and provide bench marks for checking sophisticated EOS theories. We are planning an impedance-matching experiment to obtain additional data for a number of sample materials and thus provide consistency checks for the various EOS theories. Future experiments at even higher pressures should provide tests of statistical models that are assumed to be valid at extremely high pressures.

DOPPLER SHIFT TECHNIQUE

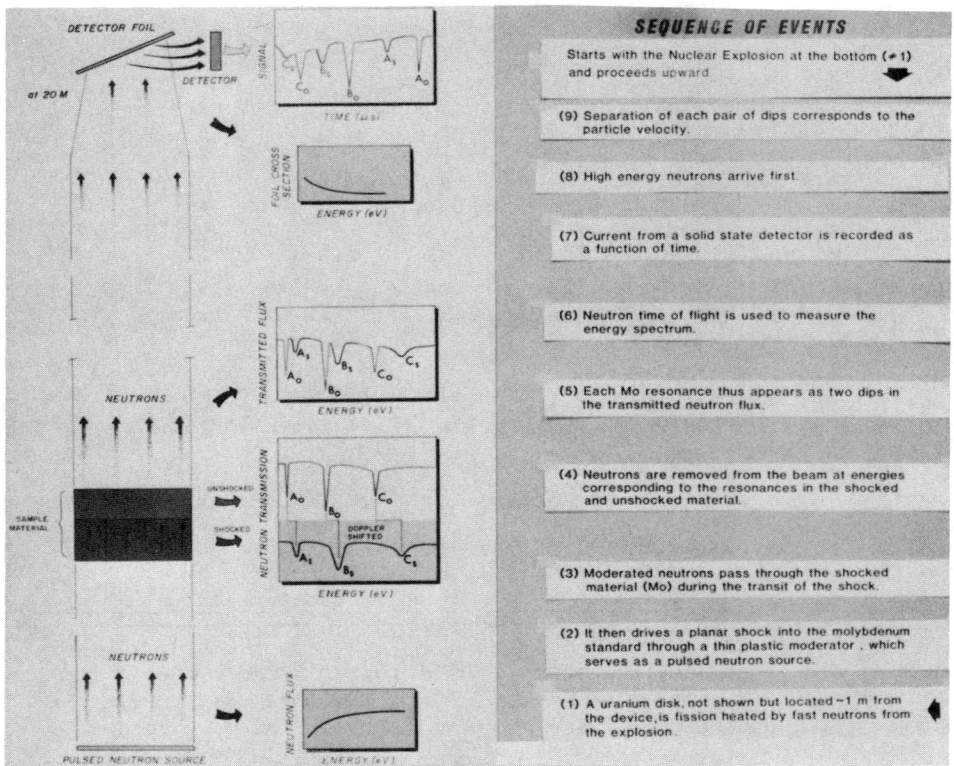

Fig. 1 Schematic illustration of the various phases of the Doppler-shift technique for measuring the particle velocity behind a planar, stable shock. The sequence of events is explained in the right-hand column starting at the bottom with graphic illustrations shown in the left-hand column.

REFERENCES

1. C. E. Ragan, M. G. Silbert, and B. C. Diven, J. Appl. Phys. 48, 2860 (1977).
2. Charles E. Ragan, III, Phys. Rev. A 21, 458 (1980).
3. C. E. Ragan, III, "High Pressure Science and Technology," edited by B. Vodar and Ph. Marteau (Pergamon, New York, 1980), p. 993.
4. C. E. Ragan, B. C. Diven, M. Rich, E. E. Robinson, and W. A. Teasdale, Proceedings 1981 Topical Conference on Shock Waves in Condensed Matter, in press.

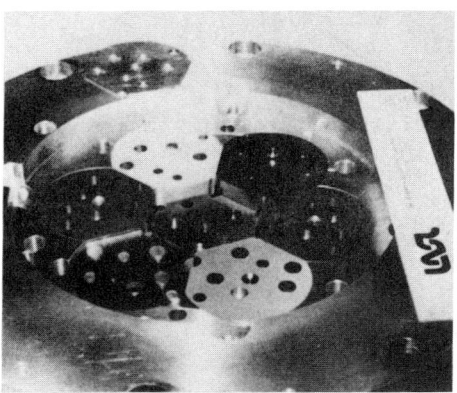

Fig. 2 Photograph of samples arranged in seven stacks with the six outer stacks containing two disks 10-mm thick and the central position containing only one 10-mm thick disk.

Fig. 3 Photograph of entire sample assembly with 75 electrical contact pins embedded in the 13 samples and the molybdenum standard.

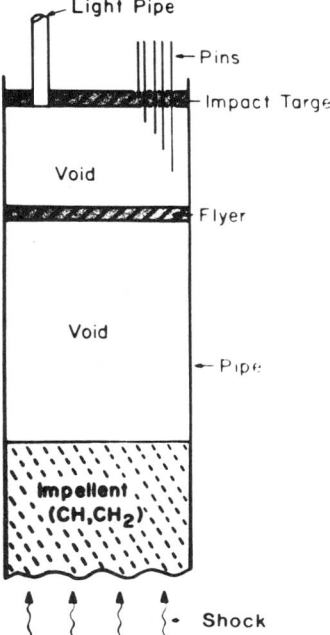

Fig. 4 Schematic illustration of the technique for shockless acceleration of the molybdenum flyer plate. A strong stock vaporizes the impellent, which then fills the void and pushes on the flyer.

HIGH ENERGY LASER FACILITIES AT LAWRENCE LIVERMORE NATIONAL LABORATORY*

N. C. Holmes
University of California, Lawrence Livermore National Laboratory
Livermore, California 94550

ABSTRACT

High energy laser facilities at Lawrence Livermore National Laboratory are described, with special emphasis on their use for equation of state investigations using laser-generated shockwaves. Shock wave diagnostics now in use are described. Future Laboratory facilities are also discussed.

INTRODUCTION

In support of the nation's inertial confinement fusion (ICF) effort, Lawrence Livermore National Laboratory (LLNL) has built a series of high-energy lasers. Recently, each has also been used for ultra-high pressure shock-wave research, as part of the Laboratory's program of equation-of-state (EOS) investigations. I will briefly describe the three largest lasers, with special emphasis on their use as shock-wave drivers; the special diagnostics for shock wave studies using lasers will also be described. Finally, I will discuss the use of new lasers soon to be in operation at LLNL.

LASER FACILITIES AT LLNL

Each of the three high-energy lasers at LLNL were designed as multiple beam systems using Nd: glass as the active amplifying medium at a wavelength of 1.06 µm. The three lasers - Janus, Argus, Shiva - differ mainly in size, energy output, and the sophistication of their operating systems. Table I provides a comparison of the performance specifications for these laser systems.

Table I High Energy Lasers Used at LLNL

Facility	Janus	Argus	Shiva
Energy	100 J	2 kJ	10 kJ
Pulselength	30-300 ps	0.1-1 ns	0.1-30 ns
Wavelength	1.06 µm	1.06 µm	1.06 µm
Beams	1	2	20
Output Aperture	8 cm	28 cm	15 cm

*Work performed under the auspices of the U.S. Department of Energy by Lawrence Livermore National Laboratory under contract #W-7405-Eng-48.

0094-243X/82/780648-04$3.00 Copyright 1982 American Institute of Physics

Janus, while originally a two-beam system for ICF research, is now operated as a single beam laser dedicated solely to EOS research and related experiments. Janus experiments provided the first conclusive evidence that lasers could produce shock pressures over 1 TPa without significant preheat.[1] In other experiments Janus was used to measure preheat,[2] which provided a basis for normalization of target design simulation codes, and tests of analytical preheat models.[3-5] Janus is now being used for a systemmatic study of impedance-match techniques using lasers.[6]

The ultrafast recording technique we use at Janus to measure shock velocities is depicted in Fig. 1. Simply put, a streak camera is used to record an image of the luminosity emitted from the stepped rear surface of a metallic target as a function of time. We can measure the time difference of shock arrival across the two levels of a small (∼5 μm) step to an accuracy of roughly 20 ps. A typical shock transit time of a 10 Mbar shock in aluminum across a 5 μm high step is 200 ps. More advanced streak cameras now becoming available should improve our accuracy to roughly 2-3 ps.

A fiducial signal, derived from the main laser pulse, is also recorded by the camera and provides absolute timing of shock arrival relative to the target irradiation pulse. This allows us to measure the transit time of the shock after passing through the bulk of the target. Diagnostics of similar design are now installed at both Argus and Shiva as well.

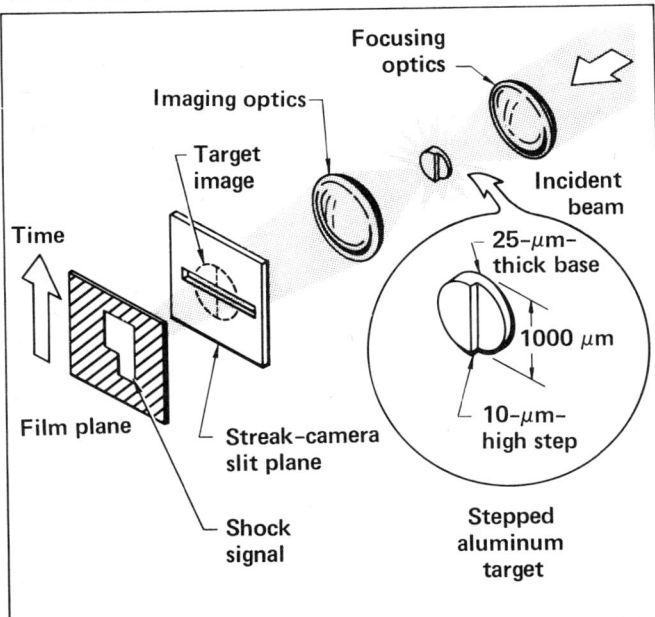

Fig. 1   An ultrafast streak camera is used to measure the velocity of a laser-generated shock wave in a stepped metal target.

The Argus laser is now principally used for laser-matter interaction studies. Starting in mid-1980, Argus was converted to operate at the second harmonic wavelength (532 nm) and after a series of target interaction experiments was converted to the third harmonic (355 nm) in late 1980. Shock wave experiments using the third harmonic capability showed that much lower preheat levels and somewhat higher shock pressures can be achieved using short wavelength irradiation.

Shiva, the world's most powerful laser, is used mainly for fusion target experiments. Experiments using Shiva have produced both the highest recorded neutron yield from a D-T filled target, and also the highest compression, to roughly 100 times liquid density. As a shock wave driver, Shiva can produce very uniform irradiation to drive planar shocks, and its high energy allows relativey large samples to be used. In an experiment in 1980, a pressure over 3 TPa was inferred from the shock velocity measured in a gold target with negligible preheat.

FUTURE LASER FACILITIES

As laser energy is increased, longer pulse irradiation over large samples can be maintained at very high laser intensities, over 10^{15} W/cm^2. This allows increased accuracy in the experiments by providing longer shock transit times across larger steps. Also, the use of short wavelength irradiation can reduce preheat levels. With thinner targets required to provide preheat shielding, shock stability should also be enhanced. Two new lasers are soon to be operating at LLNL; they will provide high energy, long pulses, and irradiation at harmonics of the basic 1.06 μm wavelength.

Shiva will be upgraded in two stages to become the Nova laser. Its 20 beams will provide an output energy of about 250 kJ in 1-3 ns pulses at 1.05 nm, and will also be able to operate with about 150 kJ output at 527 nm. Argus is to be rebuilt using two Nova beams, and will be renamed Novette. It is expected to have the capability of doing target experiments at three wavelengths: 1.05 μm, 527 nm, 351 nm. The expected performance of these two new lasers is specified in Table II.

Table II Performance Specificaions for Nova and Novette

Facility	Novette	Nova I	Nova II
Energy	10 kJ	100-150 kJ	200-300 kJ
Pulselength	1-3 ns	1-3 ns	1-3 ns
Wavelengths	1.053 μm 527 nm 351 nm	1.053 μm 527 nm	1.053 μm 527 nm
Beams	2	10	20
Output Aperture	46 cm	46 cm	46 cm

Using these new facilities for shock wave research should allow reduction of experimental errors to roughly the 1% level at pressures of 5 TPa and above, which will significantly extend our knowledge of material properties under extreme conditions of pressure and temperature.

REFERENCES

1. R. J. Trainor, J. W. Shaner, J. M. Auerbach, N. C. Holmes, Phys. Rev. Lett. $\underline{42}$, 1154 (1979).
2. R. J. Tainor and N. C. Holmes, Lawrence Livermore National Laboratory, UCRL-50028-79-3 (1979)(unpublished).
3. Y. T. Lee and R. J. Trainor, Lawrence Livermore National Laboratory, UCID-18574-79-4 (1979)(unpublished).
4. M. D. Rosen, Lawrence Livermore National Laboratory, UCRL-83022 (1979)(unpublished).
5. G. J. Caparaso and S. S. Wilson, Lawrence Livermore National Laboratory, UCRL-83308 (1979)(unpublished).
6. N. C. Holmes, R. J. Trainor, R. A. Anderson, L. R. Veeser and G. A. Reeves, these proceedings.

SHOCK WAVE FACILITIES AT POULTER LABORATORY OF SRI INTERNATIONAL

W. J. Murri
SRI International, Menlo Park, CA 94025

ABSTRACT

Shock wave research in the Poulter Laboratory covers two broad areas: dynamic material response and dynamic structural response. Workers in both areas use common facilities. The Laboratory has several guns and the facilities to perform various types of high explosive loading experiments. The use of these facilities and experimental techniques is illustrated with examples from research projects.

INTRODUCTION

Poulter Laboratory has been engaged in shock wave research for over 25 years. The staff has performed research in structural dynamics, fracture and fragmentation, stress wave propagation, explosives and explosions, penetration mechanics, and fluid mechanics. Dynamic loading is produced by projectile impact and by explosive loading. Selected examples from the above research areas are used to illustrate the experimental facilities and techniques being used at the laboratory.

EXPERIMENTAL TECHNIQUES

Under proper impact conditions, dynamic tension can be produced in materials and, if the tension is sufficient, damage occurs. Such damage has been studied in a variety of materials.[1-4] Figure 1 shows the experimental arrangement used to produce dynamic tension and recover the sample. After impact, the samples are recovered and sectioned to expose the damage. Micromechanical models have been developed to describe the damage evolution under dynamic tension.

High explosive loading had been used to generate shear bands in metals. The experimental arrangement is shown in Figure 2 and described in Ref. 5.

SRI's 63.5-mm-diameter light-gas gun has been modified so that impact experiments can be done in which combined compression and shear stresses are produced in the target.[6] The combined compression and shear stresses are produced by impacting parallel inclined plates as shown in Figure 3. Shear wave measurements under impact loading have been made in several materials,[7,8] and a piezoelectric gage has been developed to detect only shear stress.[9]

The light-gas guns are also used to determine the equation of state of materials. Multiple in-material gages (either stress or particle velocity) record the shock wave profile at several positions in the sample. The experimental data are then used in a technique known as Lagrangian analysis[10,11] to integrate the conservation of momentum and the continuity equations. In-material gages have also been used in experiments where the specimen was loaded by high explosive to study geologic materials.[12,13]

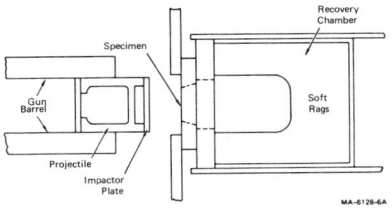

Figure 1 Plate impact for studies of dynamic fracture.

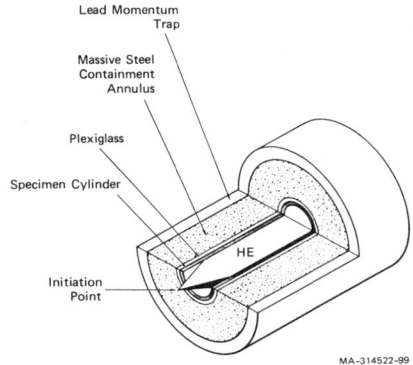

Figure 2. Exploding cylinder experiments for studying shear band kinetics.

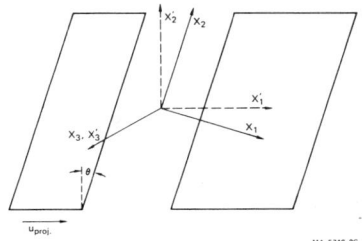

Figure 3 Schematic view of experimental technique to produce compression and shear waves.

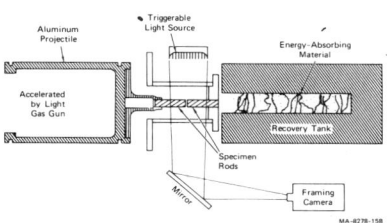

Figure 4 Arrangement for measuring high rate constitutive equations.

To study dynamic plasticity, we devised a symmetric Taylor test using a gas gun. The experimental arrangement is shown in Figure 4. A high speed framing camera was used to record the results.[14]

A compressed air launching system was used in scale model tests of missile impact on reinforced concrete.[15] The experimental arrangement is shown in Figure 5.

A small caliber gun was used to accelerate single spheres, 2.4-mm in diameter, for impacting ceramics.[16] The experimental arrangement is shown in Figure 6.

The Lagrangian gage technique has been used to study the detonation process.[17] A typical experimental arrangement is shown in Figure 7.

A facility to study the residual stress surrounding an explosively formed cavity in a specimen is shown in Figure 8. With this apparatus, specimens are pressurized (to simulate a geologic overburden), and a small spherical explosive charge is detonated in the specimen.[18] The external pressure is maintained and the specimen is hydrofractured to expose the cavity formed by the detonation.

Small cylindrical charges have been used in laboratory cylinder experiments with geologic materials (Figure 9). In-material stress and particle velocity gages were used to measure the stress wave profile produced by the detonation. The cylinders were then sliced to expose the fracture patterns produced. Results on oil and gas bearing shale have been obtained.[19]

SRI has participated in numerous field tests. Generally our work has involved placing stress and/or particle velocity gages in the ground in a specified pattern to measure the stress wave profile produced by the detonating high explosive. Figure 10 shows one type of stress gage and a mutual inductance particle velocity gage commonly used in field experiments.[20]

As an example of work done to study structural response, Figure 11 shows the shock tube facility used to measure loads on the MX Horizontal Shelter at 1/80-scale. A particular pattern of sheet explosive was initiated by a series of detonators so as to produce a blast wave that impinged on the MX shelter components.[21] Kistler gages were used to characterize the blast wave. High speed framing cameras were used to monitor the shelter response.

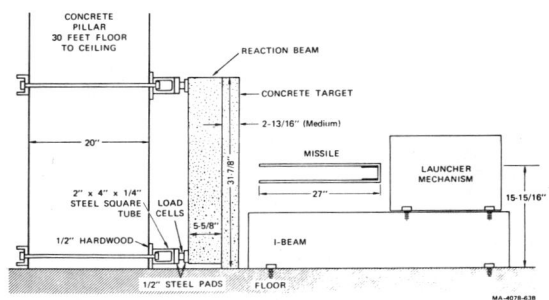

Figure 5 Overall view of scale model test setup.

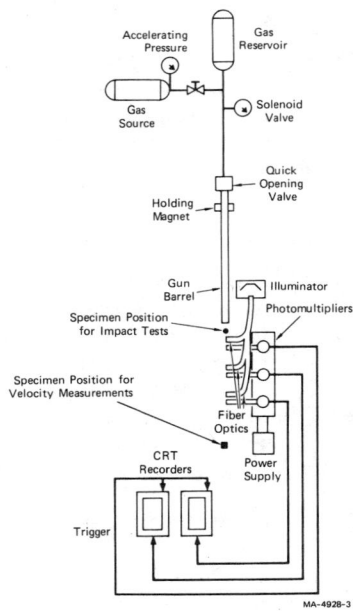

Figure 6 Particle impact test stand with arrangement for particle velocity measurement.

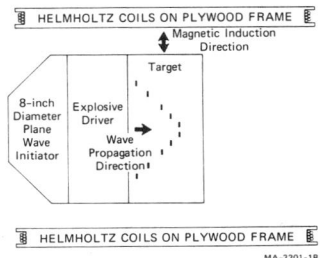

Figure 7 Basic configuration of large-scale Lagrange particle velocity gage experiments.

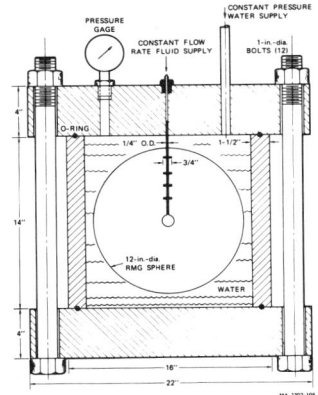

Figure 8 Containment experiment apparatus.

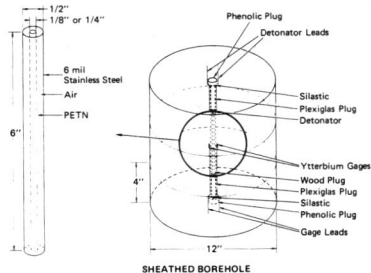

Figure 9 Configuration of laboratory experiments in tuff.

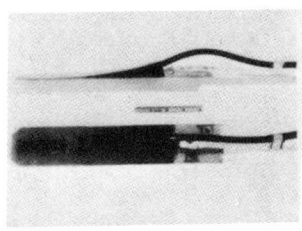

(a) Stress gages — 6Ω Ω

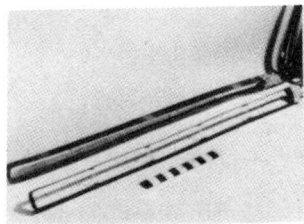

(b) Mutual-inductance particle velocity gages

Figure 10 Field gages.

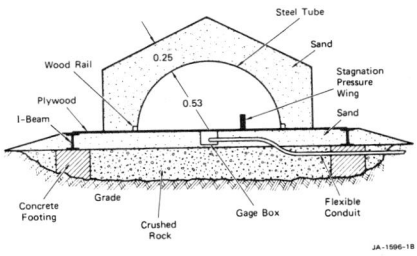

Figure 11 Section view of shock tube facility. (Dimensions in meters)

REFERENCES

1. T. W. Barbee et al., J. Mat. JMLSA 7, 393 (1972).
2. D. R. Curran, D. A. Shockey, and L. Seaman, J. Appl. Phys. 44, 4025 (1973).
3. D. R. Curran, L. Seaman, and D. A. Shockey, Phys. Today (Jan. 1977).

4. D. A. Shockey, C. F. Petersen, D. R. Curran, and J. T. Rosenberg, in New Horizons in Rock Mechanics, Proc. 4th Symp. on Rock Mech. H. R. Hardy, Jr., and R. Stefanko, Eds. (Am. Soc. of Civil Eng., New York, 1973), p. 709.
5. D. A. Shockey and D. C. Erlich, in Shock Waves and High-Strain-Rate Phenomena in Metals, M. A. Meyers and L. E. Murr, Eds. (Plenum Publ. Corp., New York, 1981).
6. Y. M. Gupta, D. D. Keough, D. F. Walter, K. C. Dao, D. Henley, and A. Urweider, Rev. Sci. Instr. $\underline{51}$, 183 (1980).
7. Y. M. Gupta, Appl. Phys. Lett. $\underline{29}$, 694 (1976).
8. Y. M. Gupta, J. Appl. Phys. $\underline{51}$, 5352 (1980).
9. Y. M. Gupta and W. J. Murri, Bull. Am. Phys. Soc. $\underline{26}$, 660 (1981).
10. M. Cowperthwaite and R. F. Williams, J. Appl. Phys. $\underline{42}$, 456 (1971).
11. L. Seaman, J. Appl. Phys. $\underline{45}$, 4303 (1974).
12. D. E. Grady, W. J. Murri, and G. R. Fowles, J. Geophys. Res. $\underline{79}$, 332 (1974).
13. D. E. Grady, W. J. Murri, and P. S. De Carli, J. Geophys. Res. $\underline{80}$, 4857 (1975).
14. D. C. Erlich, D. A. Shockey, and L. Seaman, "Symmetric Rod Impact Technique for Dynamic Yield Determination," presented at Am. Phys. Soc. Mtg., Shock Waves in Condensed Matter, June 23-25, 1981 (proceedings to be published).
15. S. McHugh, Y. Gupta, and L. Seaman, "Scale Model Experiments of Reinforced Concrete Under Pipe Missile Impact," Final Report for Electric Power Research Institute, SRI International (Dec. 1980).
16. D. A. Shockey etal., "Particle Impact Damage in Ceramics," Annual Report Parts 1 through 5, Office of Naval Research Contract No. N00014-76-C-0657, SRI International (January 1979 to April 1981).
17. M. Cowperthwaite and J. T. Rosenberg, in Seventh Symposium (International) on Detonation, p. 835, June 1981 (proceedings to be published).
18. J. C. Cizek and A. L. Florence, "Laboratory Investigation of Stemming and Containment in Underground Nuclear Tests," Draft Final Report, Contract DNA001-80-C-0040, SRI International (January 1980).
19. S. McHugh, "Small-Scale Experiments with An Analysis to Evaluate the Effect of Tailored Pulse Loading on Fracture and Permeability," Final Report Phase 1, Department of Energy, Contract No. DE-AC21-79MC11577, SRI International (June 1980).
20. J. T. Rosenberg etal., "In Situ Constitutive Relations of Rocks and Soils, Spherical Field Tests and LASS Results for Pre-Dice Throw II Materials, LASS Error Propagation Analyses," Final Report, Contract DNA001-76-C-0113, SRI International (April 1977).
21. J. R. Bruce and M. Sanai, "Load Measurements on the MX Horizontal Shelter at 1/80-Scale," Final Report, Contract No. DNA001-80-C-0272, SRI International (April 1981).

SANDIA 25-METER COMPRESSED HELIUM/AIR GUN*

R. E. Setchell
Sandia National Laboratories,† Albuquerque, NM 87185

ABSTRACT

For nearly twenty years the Sandia 25-meter compressed gas gun has been an important tool for studying condensed materials subjected to transient shock compression. Major system modifications are now in progress to provide new control, instrumentation, and data acquisition capabilities. These features will ensure that the facility can continue as an effective means of investigating a variety of physical and chemical processes in shock-compressed solids.

ORIGINAL DESIGN

The 25-meter gas gun became operational in 1962. The basic requirement met by the design was the achievement of impact velocities over a continuous range from 0.03 to 1.6 km/s using 63 mm diameter projectiles having masses from 0.25 to 5 kg.[1-3] A 63 mm diameter by 6 m breech chamber pressurized with air or helium up to 35 MPa was coupled to a 25 m barrel. The gun assembly was installed below ground level adjacent to a laboratory building containing space for a control and instrumentation room, and experiment preparation areas. The breech and impact areas were housed in individual safety bunkers that can be sealed prior to breech pressurization. Firing has been accomplished by either a quick-opening gate valve or a double-diaphragm rupture assembly. Aluminum projectiles sealed with viton O-rings and Teflon backup rings are initially held in place by vacuum behind the projectile, then fired through a barrel vacuum of 1 Pa. Projectile impact velocities are measured by using precisely positioned coaxial switch pins located at 10 mm intervals in the final 10 cm of the barrel. These pins are used with nsec discharge circuits and up to three 10 nsec counters to establish impact velocities with 0.1% precision. Operation of a gas compressor, vacuum pumps, and all control valves is accomplished through relays actuated by switches in the control room. High-pressure and vacuum levels are displayed nearby on digital read-outs or panel meters. The status of all gun systems is displayed on an adjacent light display actuated by the control switches.

Target assemblies are prepared within plexiglass or aluminum cups, and are held in place on a hand-lapped mounting surface by the barrel vacuum. Perpendicularity of this mounting surface to the barrel bore, flatness and parallelism of the target assembly, and perpendicularity and flatness of the projectile face determine the deviation from perfect planarity at impact. Typical care in

*This work sponsored by U.S. DOE under Contract DE-AC04-76-DP00789.
†A U.S. Department of Energy facility.

preparing experiments results in impact tilt angles less than 0.5 mrad, with angles less than 0.2 mrad achievable in high-precision experiments. Following impact, the target and projectile material is largely confined within an open-ended steel catcher assembly containing wood blocks or other energy-absorbing materials. Electrical data signals are transmitted from target assemblies to the control room through 14 m of low-loss coaxial cables running through a connecting tunnel. Optical data signals for velocity interferometry (VISAR) measurements also pass through this tunnel by means of mirrors and windows. In the control room, data recording is provided by a number of 500 MHz oscilloscopes equipped with polaroid cameras, together with counters having resolutions down to 20 psec.

SYSTEM MODIFICATIONS

A number of modifications are in progress this year to improve gun performance and to provide new research capabilities. One change that has been completed is the removal of the slide-valve firing mechanism. Low pressure, double-diaphragm rupture assemblies now are used for operations in the valve firing regime (impact velocities below 0.7 km/s). Velocity reproducibility is now within 1% over most of the gun operating range, with 2-3% reproducibility at the lowest velocities. A second modification, now in engineering design and fabrication, is the addition of a vacuum catcher system. With this improvement, target assemblies need not hold off the barrel vacuum, but rather can be mounted on an adjustable platform within an evacuated chamber. A large catcher tank sealed to the vacuum chamber through a thin diaphragm will contain the overpressure and debris following impact. The existing catcher system will continue as an internal component of this large tank. Optical access to the target will be provided by selectable window ports in the vacuum chamber. This access will permit the use of a double-delay VISAR system, now operational, and also will permit applications of a recently developed intensified image-converter camera having both streak and framing capabilities. The design of the vacuum chamber will also permit the use of electromagnetic gauge techniques. A third modification is the incorporation of a desktop computer and solid-state relays to control all system operations. A serial multiplexer connects the computer with the relays, which in turn operate the compressor, vacuum pumps, and control valves. The computer also controls a light display identifying the status of all gun systems. On hand and ready for installation, the computer modification will allow all preparation and firing sequences to be software determined. A final modification is to introduce programmable transient digitizers as an alternative to oscillograms for data recording. One digitizer interfaced to a PDP 11/10 minicomputer is on hand, with additional digitizers and a more powerful minicomputer to be added subsequently.

NEW DIRECTIONS

Along with new capabilities provided by the modifications, new research interests are developing that will affect the experiments to be conducted on the gas gun in the coming years. Examples of recent studies that can be enhanced by the new capabilities include basic Hugoniot data, electrical properties under shock compression, and shock initiation of granular explosives. In the area of initiation of explosives, for example, a strong need exists for experimental information on hot-spot formation and growth. One application of the new image-converter camera, protected by the enclosed catcher system, will be to determine if these processes can be visualized at an interface with a window material. A related activity currently in progress is to apply thin-foil thermocouple techniques to investigate chemical energy release in shock-compressed explosives. An area of rapidly growing interest is the study of shock-induced, non-thermal chemical processes occurring on very short time scales. Experiments on organic compounds have shown cusps in Hugoniot data, the onset of shock-induced polarization, and sharp increases in thermocouple-measured temperatures all occurring at similar shock-pressure thresholds. A related question is the influence of shock-induced defects on enhanced reactivity in inorganic compounds. Investigations of this topic have required the development of recovery techniques so that shocked samples can be preserved for subsequent analysis. In addition, efforts are underway to develop real-time absorption spectroscopy techniques with nanosecond resolution for observing changes in chemical structure during shock wave passage.

With new capabilities for expanding on-going activities and pursuing exciting new interests, the Sandia 25-meter compressed gas gun should continue as an important research facility for many more years.

REFERENCES

1. S. Thunborg, Jr., G. E. Ingram and R. A. Graham, Rev. Sci. Instr. $\underline{35}$, 11 (1964).
2. G. E. Ingram, Rev. Sci. Instr. $\underline{36}$, 458 (1965)
3. G. E. Ingram and R. A. Graham. in Fifth Symnposium (International) on Detonation, Office of Naval Research Report ACR-184, August, 1980, edited by S. Jacobs and R. Roberts, p 369.
4. R. A. Graham, in High Pressure Science and Technology, edited by B. Vodar and Ph. Marteau, Pergamon Press, NY (1980) p 1032.
5. R. A. Graham, J. Phys. Chem. $\underline{83}$, 3048 (1979).
6. S. A. Sheffield and D. D. Bloomquist, this Proceedings.
7. D. D. Bloomquist and S. A. Sheffield, J. Appl. Phys. $\underline{51}$, 5260 (1980).
8. D. D. Bloomquist and S. A. Sheffield, in Seventh Symposium (International) on Detonation, Office of Naval Research (in press).
9. R. E. Setchell, Comb. Flame (in press).
10. R. E. Setchell, in Seventh Symposium (International) on Detonation, Office of Naval Research (in press).
11. See papers by Morosin, Graham and coworkers, this Proceedings.

APPENDIX: SANDIA 25-METER GAS GUN SUMMARY

Dimensions:
- bore diameter 63 mm
- barrel length 25 m
- breech length 6 m

Performance:
- impact velocities from 0.03 to 1.6 km/s, with breech pressurized up to 35 MPa using air or helium

Firing mechanism:
- double-diaphragm rupture assemblies

Instrumentation:
- Tektronix 7912 AD digitizer / PDP-11/10 mini-computer
- Tektronix 7903 and 7844 oscilloscopes
- HP pulse generators and counters (20 psec resolution)
- Sandia-designed double-delay velocity interferometer (VISAR)

Recent studies:
- Impact-induced electrical switching of polymer films[4]
- Shock-induced polarization in polymers[5]
- Hugoniot properties of polymers[6]
- Thin-foil thermocouple measurements in polymers[7] and granular explosives[8]
- Ramp wave and short-pulse shock initiation of granular explosives[9,10]
- Recovery experiments for examining consequences of shock-induced defects in inorganic compounds[11]

Facility personnel:
Douglas Bloomquist
Merri Brown
Jamey Browning
Brian Dodson
Douglas Dugan
Robert Graham
Michael Russell (MRL)
Robert Setchell

PLATE IMPACT FACILITY AT BROWN UNIVERSITY

R. J. Clifton
Brown University, Providence, RI 02912

ABSTRACT

The facility includes two gas guns: 63.5 mm diameter x 2.5 m long and 101.6 mm diameter x 3.4 m long. Maximum projectile velocities for usual projectiles and normal operating pressures are 0.3 mm/μs or 0.5 mm/μs, respectively. the barrel of the 63.5 mm diameter gun is slotted to prevent projectile rotation. Wave profiles are measured with three interferometer systems: a normal displacement interferometer (NDI), a normal velocity interferometer (NVI), and a transverse displacement interferometer (TDI). The light source for all three systems is a one watt argon ion laser. Principal applications of the facility have been to investigations of pressure-shear waves, stress-strain curves at strain rates of 10^5 s^{-1}, and dislocation mobility in single crystals.

DESCRIPTION OF THE FACILITY

The main features of the facility are summarized in Table I.

REFERENCES

1. P. Kumar and R. J. Clifton, J. Appl. Phys. 50, 4747 (1979).
2. K. S. Kim, Ph.D. Thesis, Brown University (1979).
3. K. S. Kim, R. J. Clifton, and P. Kumar, J. Appl. Phys. 48, 4132 (1977).
4. K. S. Kim and R. J. Clifton, J. Appl. Mech. 47, 11 (1980).
5. A. Gilat (manuscript in preparation).
6. P. Kumar and R. J. Clifton, J. Appl. Phys. 48, 1366 (1977).
7. P. Kumar and R. J. Clifton, J. Appl. Phys. ,
8. C. H. Li and R. J. Clifton (in these Proceedings).
9. G. Meir (work in progress).

TABLE I Brown University plate impact facility

Driver Capabilities:	(i) 63.5 mm diameter compressed gas gun, 2.5 m barrel (0.02-0.30 km/s); slotted barrel to prevent projectiles rotation; anvil to stop projectile in recovery experiments (ii) 101.6 mm diameter compressed gas gun, 3.4 m barrel (0.02-0.50 km/s); anvil for stopping projectile
Diagnostics	system of shorted wires and 10 nsec counters for measuring projectile velocity; optical alignment system with sensitivity of 2×10^{-5} radians; tilt monitoring system, interferometer systems (NDI, NVI, TDI) with one watt argon ion laser, photodiode and photomultiplier detectors (3 ns); oscilloscopes: Tektronix 556, 519 (2), 7904 (2); computer programs for plane waves in elastic/viscoplastic materials, including the effects of finite deformations and combined pressure-shear loading
Recent Studies	• Recovery experiments on LiF and MgO crystals to measure the stress dependence of dislocation velocity and the dislocation densities produced by known pulses[1,2] • Development of a transverse displacement interferometer[3] • Measurement of pressure-shear waves in 6061-T6 aluminum[4,5] and α-titanium[5] • Development of an optical alignment procedure[6] • Development of an eight-pointed star-shaped flyer to minimize the effects of unloading waves from the sides in recovery experiments[7] • Development of a pressure-shear technique for obtaining stress-strain curves at strain rates of 10^5 s^{-1}, including initial results for 1100-0 aluminum and 6061-T6 aluminum[8] • Development of a precursor decay experiment in which surface damage is minimized by including fluid layers at both faces of the specimen[9]
Personnel	R. J. Clifton (Principal Investigator) A. Gilat, C. H. Li, G. Meir (Graduate Students) L. Hermann (Senior Research Engineer) R. Reed (Technician) R. Beck, R. Theriault (Undergraduate Research Assistants)

NSWC/WO LIGHT GAS GUN AND EXPLOSIVE FACILITY

E. R. Lemar, J. W. Forbes, J. O. Erkman, and J. W. Watt
Naval Surface Weapons Center
White Oak, Silver Spring, Maryland 20910

ABSTRACT

The facility contains a single stage light gas gun and various methods of using high explosives to drive shock wave experiments.

FACILITY

Table I gives the highlights of the facility, recent work, and main contributors. The light gas gun has a bore of 89 mm and is capable of driving projectiles at velocities from 0.1 to 0.8 km/sec. It was installed with its muzzle in a test chamber (bombproof) so that targets may include very reactive materials. Up to 2.3 Kg of explosive can be used in the test chamber for a single experiment. The gun is used to drive dynamic high pressure experiments in which the targets may be made of reactive materials (explosives or propellants) or inert materials. The facility is used to obtain the equations of state of inert materials and of explosives in the unreacted state, and to measure shock wave evolution within these materials.

Various explosive systems are used to produce shocks in targets in several ways. Table II lists the basic characteristics of a number of driver systems. Starting a detonation in a charge of explosive at a point generally results in a spherical expanding detonation wave. Configurations called plane wave generators (PWG) are used to convert the spherical wave so that its front lies in a plane. The PWG is usually in contact with a slab of explosive which has been selected for its output characteristics. In some shock driver systems, the explosive drives a flyer plate which impacts on the target. The waves induced by the impact of a flyer plate have a short but useful region of steady or uniform flow behind the shock front which is frequently described as being a flat-topped wave. This uniform flow makes it easier to interpret the records obtained from an experiment because the relief wave in the explosive gases (i.e., the Taylor wave) does not have to be dealt with. Explosively driven flyer plates are subject to relatively high pressure shocks that alter the material in the plates prior to the impact of the plates on the test sample, which may be a disadvantage for some applications. Various combinations of inert materials and explosives allow the plane wave systems to induce pressures in material from 10 to 800 Kb. The specific pressures obtained are discrete values within this range. Continuous variation between these discrete values is not possible with these explosive systems.

The underwater and modified gap shock systems in Table II use shock attenuation as a method of obtaining specific incident peak pressures in test samples. These decaying shock waves are accurately reproducible from experiment to experiment, which makes them generally useful without requiring instrumentation to measure shock profiles for each experiment. These particular systems span the range

Table I NSWC/WO Light Gas Gun and Explosive Facility

Laboratory Facilities	89 mm diameter Single Stage Compressed Gas Gun (0.1-0.8 Km/s), various Explosive Shock Generators (1-800 Kb), Low Velocity Ramp Generators
Diagnostics	VISAR Interferometry (3 ns resolution, 0.2% precision), Three Channel X-ray (30 ns pulse), Optical Framing and Streak Cameras, Electronic Framing and Streak Cameras, Manganin and Quartz Gages
Recent Studies	•Elastic-plastic properties of Comp B-3 •Hugoniots of explosives and inerts •$\alpha \rightarrow \epsilon$ phase transitions in Armco iron •Shock sensitivity of explosives and propellants •Deflagration-to-detonation transition in porous explosives and propellants •Electrical properties of single crystal silicon •Microwave and electrical properties of explosives •Metal jets •Projectile impact studies •Electromagnetic energy coupling to explosives
Principal Investigators	Richard Bernecker, A. Robert Clairmont, Nathaniel Coleburn, David Demske, John Erkman, Jerry Forbes, Gordon Hammond, E. Ray Lemar, Thomas Liddiard, Harold Sandusky, and J. William Watt
Technical Support	Robert Baker, Alfred Brown, Dempsey Gillmore, Herman Gillum, Carl Groves, Keith Harrison, Bruce Holland, Bernard Snowden, Nathaniel Snowden, and Charles Sorrels

Table II Summary of Shock Generator Systems

Shock System	Pulse Characteristics	Range of Pressure (Kb)	Range of Pulse Half-Widths (μs)
Gas Gun	one-dimensional flow with flat top shock	1-50	0.2-10
Explosive Plane Wave Generators/ Flyer Plate Systems	one-dimensional flow with flat top shock	10-800	0.5-10
Underwater System (470 g Pentolite sphere)	one-dimensional flow with decaying spherical shock	5-25	20-50
Modified Gap Test (MGT)	two-dimensional flow with decaying triangular-shaped shock	20-80	1-3
Low Velocity Ramp Generator	(under development)		

of shock stimuli to which explosives would normally be subjected.

A low velocity ramp generator is under development which is expected to push a piston from near zero velocity to 0.3 mm/μs with a ramp rise time of nearly 100 μs.

DIAGNOSTICS

A VISAR interferometer is used for recording the motion of surfaces of targets that have been shocked by a gas gun driven flyer. The equipment gives 3 ns resolution with a precision of 0.2%. Pressure gages, such as quartz, manganin, and others, are also used. Flash radiography is used to record the state of a target during the time it is being dynamically deformed. Conventional and electronic high-speed framing and streak cameras are used to record the progress of a detonation in an explosive charge in order to measure the detonation velocity of the explosive. In such experiments, the cameras can either be synchronized with the event by sending a signal from the camera to the firing circuits when the camera rotor is in a predetermined position, or be used in the continuous access mode.

In the experiments where the time between the firing of the detonator and the event being studied is unknown, electronic cameras can also be used. A sensor, such as an ionization pin, is placed in the experiment. When the sensor is activated, the electronic camera is turned on to record the event.

RECENT STUDIES

The gas gun and explosive facility are used to obtain Hugoniot and phase transformation data for both reactive[1] and nonreactive materials.[2,3] These studies measure the shock velocity-particle velocity relationships of these materials. Measurements of the evolution of the shock fronts reveal information on the kinetics of the phase transformation process.[3] For the most part, explosives have been treated as fluids in modeling computations. In recent gas gun experiments on pressed Comp B-3, it has been observed that the material shows elastoplastic behavior with a Hugoniot elastic limit near 1.7 Kb. This elastoplastic behavior requires that the equation of state relationships used in computer codes be changed so that propagation elastic waves are in the solutions. Present concepts of initiation of explosives, especially in cased charges, may have to be revised for those explosives with non-trivial elastic limits.

A common application of explosive drivers is in the study of sensitivity of a test explosive by placing the donor and acceptor explosive charges in water. This underwater method[4] has been calibrated[5] so that specimens can be shocked to different pressures by changing the distance between the two charges. The interaction of the shock in the water with the explosive target is recorded by a camera. The motion of the faces of the explosive test sample are recorded. An enhancement of the expansion rate of the surfaces of the test sample with respect to expansions measured with weaker shocks is a good test of the low level reaction in the test sample.

Another method commonly used to study the sensitivity of explosives is called the gap test. In these tests, the donor charge is separated from the acceptor by a plastic attenuator. A plastic length is experimentally found which transmits a shock into the test sample of just sufficient amplitude to cause detonation. For greater lengths of plastic, the transmitted shock is too weak to cause detonation, but may cause burning. Many explosives have been tested in this manner so that they can be ordered according to threshold values for burning and detonation.[6,7]

There are current studies being conducted to understand the deflagration-to-detonation transition (DDT) process in porous explosives.[8,9] The compaction of porous inert columns is also being studied to understand the mechanical compaction process.

There are a variety of other studies of interest to the shock wave community which are only being mentioned briefly. Among these are the electrical breakdown and microwave permittivity properties of a number of explosives[10,11] and electrical properties of shocked single crystals of silicon.[2] In addition, studies are being conducted on metal jet formation, the initiation of explosives by projectile impact, and electromagnetic energy coupling to reacting explosives.[12]

REFERENCES

1. N. L. Coleburn and T. P. Liddiard, J. Chem. Phys. $\underline{44}$, 1929 (1966).
2. N. L. Coleburn, J. W. Forbes, and H. D. Jones, J. Appl. Phys. $\underline{43}$, 5007 (1972).
3. J. W. Forbes and G. E. Duvall, Proc. 4th AIRAPT Conf., Kyoto, Japan (1974).
4. M. J. Frankel, T. P. Liddiard, and J. W. Forbes, accepted, Comb. and Flame.
5. T. P. Liddiard and J. W. Forbes, Proc. of this Conf.
6. T. P. Liddiard and J. W. Forbes, Proc. 7th Det. Symp., Annapolis, Md. (1981).
7. D. Price, A. R. Clairmont, and J. O. Erkman, NOLTR 74-40 (1974).
8. D. Price and R. R. Bernecker, Prop. and Expl. $\underline{6}$, 5 (1981).
9. R. R. Bernecker, H. W. Sandusky, and A. R. Clairmont, Proc. 7th Det. Symp., Annapolis, Md. (1981).
10. J. W. Forbes, Bull. A.P.S. $\underline{25}$, 496 (1980).
11. G. L. Hammond, Bull. A.P.S. $\underline{25}$, 496 (1980).
12. D. L. Demske, E. T. Toton, and E. Zimet, Bull. A.P.S. $\underline{24}$, 712 (1979).

IMPACT PHYSICS FACILITIES AT THE UNIVERSITY OF DAYTON RESEARCH INSTITUTE

S. J. Bless
University of Dayton Research Institute
Dayton, Ohio 45469

ABSTRACT

The University of Dayton Research Institute maintains a versatile impact facility for study of fundamental and engineering aspects of impact mechanics. Extensive instrumentation is available, including high repetition rate pulsed lasers.

INTRODUCTION

Impact physics research was initiated at the University of Dayton in 1965. In that year, the University began operation of the Impact Physics Laboratory of the Air Force Wright Aeronautical Laboratories/Materials Laboratory (AFWAL/ML) at Wright-Patterson Air Force Base. The present Impact Physics Group is a direct descendent of that effort. Today the Impact Physics Group has grown to a staff of thirteen full-time employees. It operates one of the most versatile indoor ballistics facilities in the world, and it draws support from over a dozen different government agencies concerned with impact phenomena. The subjects investigated span a remarkably wide range of basic and applied research.

RECENT STUDIES

Some of the technical areas encompassed by recent efforts are shown in Table I.

TABLE I. RECENT STUDIES

Spall formation	Splash mechanics
Hydrodynamic ram	Armor evaluation
High performance gun design	Rod penetration mechanics
Electric gun technology	Transparency design
High speed camera design	Aircraft vulnerability analysis
Fan blade design	Rifling design
Water blasting	Explosive initiation

FACILITIES

The launch facilities include both compressed gas and powder guns. Eight ranges are available, plus facilities for explosive detonation. The gun ranges are designed for interchangeability of launch tubes and blast tanks. A list of the available launch tubes is provided in Table II.

TABLE II. A PARTIAL LIST OF AVAILABLE LAUNCH TUBES

Tube Diameter	Propellant	Barrel Length (m)	Typical Package Weight (projectile plus sabot) (g)	Maximum Velocity (m/sec)
7.62	powder	2.4	6	2000
7.62	powder	1.5	standard AP	1000
12.7	powder	1.5	6.6	1980
12.7	powder	1.5	standard AP	1500
20	powder	2.4	19.0	1800
20	powder	1.5	standard AP	1800
40	powder	3.6	150	2200
50	powder	6.0	100	2860
			200	2400
			400	1800
51	compressed air	7.3	300	365
51	compressed He	7.3	300	510
89	compressed air	6.0	680	260
178	compressed air	9.1	4500	270
7.6	two stage light gas gun, 40 mm pump tube	1.9	0.6	7000
20	two stage light gas gun, 40 mm pump tube	1.9	9	7500

The Impact Physics Group specializes in the highly precise measurement of impact and ballistic phenomena. Modern and sophisticated data-acquisition equipment is routinely employed. The well-controlled indoor environment makes possible the use of delicate equipment which could not survive exposure to the ambient conditions on an outdoor ballistic range. The available instrumentation is summarized in Table III.

The framing cameras currently available include a Beckman and Whitley 300 (4.5×10^6 ips), a WF1 and a WF2 Fastax (6000 ips), a Hycam 40-004 (800 ips), a Dynafax Model 326-3 (26000 ips), and a Photec-4A (4×10^4 ips). These are employed to provide both qualitative and quantitative three-dimensional (2 in space, 1 in time) information concerning material deformation and movement during the course of many types of impact experiments. At the highest framing rates a high intensity 10 kJ spark source is employed for illumination of the target. At the lower framing rates, either continuous tungsten flood lamps or photoflash illuminators are used.

Streak cameras are employed for analysis of extremely rapid phenomena when information in one spatial direction will suffice. Such devices provide a continuous two-dimensional (1 space and 1 time) display of an event. Writing speeds as high as 15 mm/μsec can be achieved which provides a good temporal resolution of an impact event.

A VISAR system (velocity interferometer for any reflecting surface) is employed to provide an extremely accurate and convenient method of performing interior ballistics measurements or other measurements of the velocity of a test object. A one-watt single-longitudinal mode Argon ion laser is used as an illuminator.

High intensity and short duration (30 nsec) x-ray exposures are employed to determine projectile and target integrity, velocity, and orientation during the course of various impact experiments. Twelve channels are on hand, ranging from 105 keV and 300 keV; they are usually employed in a configuration of 2 channels per range.

During the course of various experimental programs involving impacts with large panels of metal or plastic, it is often necessary to measure the out-of-plane displacement of the surface of the panel. This is accomplished by the use of a special apparatus which operates on the principle of the moiré fringe. A combination of a light source, camera, and Ronchi ruling is employed to project a series of parallel dark lines onto the surface to be examined. This surface is viewed by a fast framing camera in conjunction with a second Ronchi ruling to produce a moiré fringe display that is dependent upon the characteristics of the optical system and the panel shape. During the course of the impact, the panel displacement can be determined from an analysis of the moiré fringe pattern.

TABLE III. INSTRUMENTATION

Framing cameras	VISAR
6000 to 20,000 ips	20 mm/μs/fringe
2 35 mm	1W power
3 16 mm	300 mHz
4.5×10^6 ips	Flash Radiography
Beckman & Whitley 300	2 105 kV
Streak cameras	8 150 kV
16 mm Hall Camera	4 300 kV
70 mm 10 mm/μs Cordin	pulsed lasers
Dynamic displacement	4J Nd-glass
moiré fringe	1J Ruby
6000 ips	25 kHz Mn-vapor
Transducers	signal recordings
quartz, Li-niobate	10 channels Zonics DMS
strain gauges	Nicolet Explorer III
	7844 Tektronix

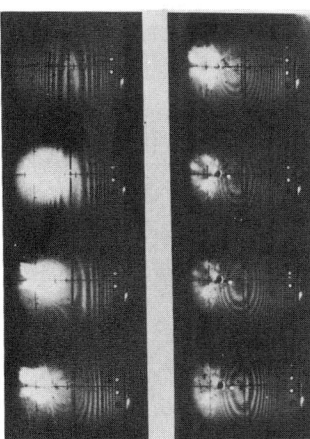

Figure 1. Sample moiré fringe record of impact. Frames are 165 μs apart.

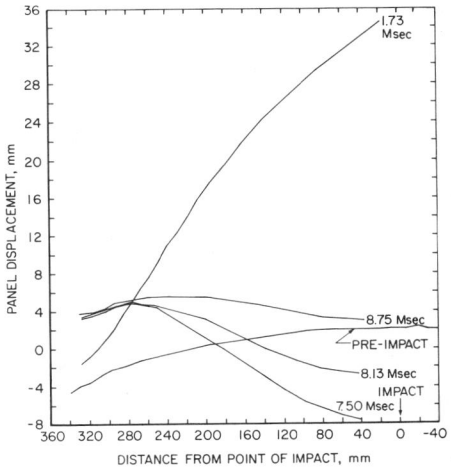

Figure 2. Displacement at selected times of panel shown in Figure 1.

Figure 1 illustrates the appearance of fringes on a simulated aircraft fuel tank penel during impact and Figure 2 illustrates the panel displacement data obtained from the photographic record.

Very short exposure-time still photographs can be employed for high resolution recording of impact processes. Q-switched ruby and Q-switched, frequency-doubled neodymium laser systems are employed with conventional cameras. An image converter camera is also available. The 10 nsec exposure time afforded by these illuminators is capable of reducing the motion blur due to the displacement of a particle during the photographic exposure to a very small level even in a hypervelocity impact. Figure 3 illustrates one of the very high resolution photographs which have been obtained via laser illumination. A 25 kHz pulsed Mn-vapor laser has also been developed.

Several types of transducers for the measurements of pressure and displacement are routinely employed. These include quartz and tourmaline high-speed pressure gauges and strain gauges. University personnel are skilled in the application of pressure and strain gauges. Data are normally recorded in a digital mode, so as to facilitate analysis. Special transducers have been designed for underwater application, and special apparatus has been constructed for dynamic calibration of pressure gauges.

A variety of signal recording systems are available. Signals below 100 kHz bandwidth are normally recorded on a ten channel Zonic Technical Laboratories DMS. Each channel is 2048 ten-bit words. Higher bandwidth signals can be recorded with a two-channel Nicolet digital oscilloscope (20 mHz bandwidth) or on oscilloscopes. Oscilloscopes to 300 mHz bandwidth are available. Digital data can be directly transferred to the University's VAX 11-780 computer.

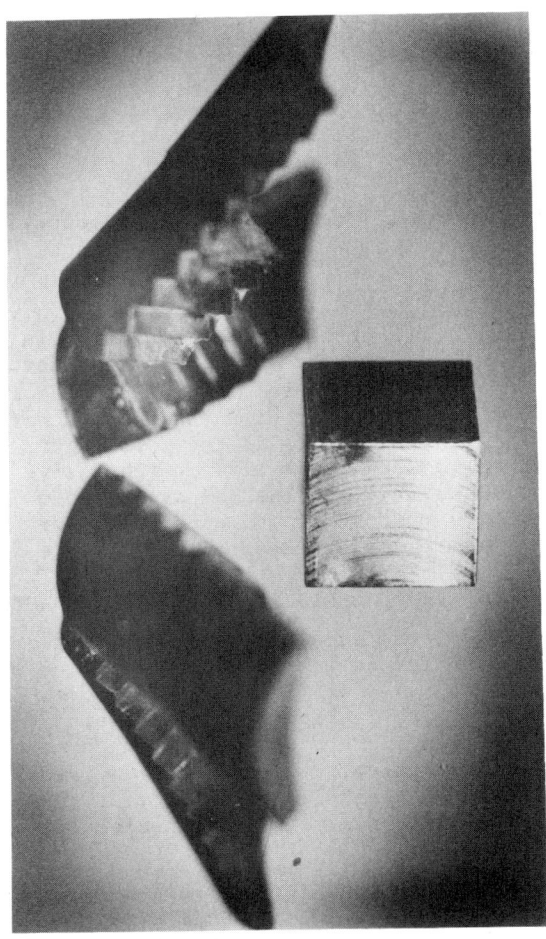

Figure 3. Pulsed-laser photograph of sabot releasing an 11-mm steel cube at 1500 m/s.

ISL SHOCK WAVE FACILITIES

presented by F. BAUER

Institut Franco-Allemand de Recherches (ISL), 68301 St-LOUIS, France

ABSTRACT

Shock wave facilities and experimental techniques for diagnostics are presented.

INTRODUCTION

The "Institut Franco-Allemand de Recherches" (ISL) is active in cooperative research in such areas as aerodynamics, ballistics, detonics, and metrological recording of transient events. Mean-size facilities are especially used for investigating the effects of shock waves on condensed matter. These shock facilities such as the powder gun, light-gas gun, and other conventional facilities like the explosive driven plane shock wave generators as well as flash X-ray and holographic devices used for investigation and diagnostic purpose (for instance measurement of shock pressures) are presented here.(Table I).

I. CALIBRATED SHOCK WAVE GENERATOR: SMALL CALIBER POWDER GUN. F.BAUER

The powder gun shown in figure 1, was especially built for study of the behavior of PVF_2 polymers and ferroelectric materials under shock loading [1,2]. The caliber of this gun is equal to 15 mm. The projectile velocities range from 50 m/s to 700 m/s. In Figure 1 are shown, [1] tube, [2], [3] and [4] breech, [5] connection piece, [6] combustion chamber, [7] [8] electric contacts, [9] electric plug, [10] [11] [12] insulators, [13] springs. A gas seal is mounted on the cylinder projectile. The specimens to be studied are in a holder mounted at the muzzle. Tests are performed in vacuum (10^{-2} Torr).

For determination of Hugoniot curve, quartz gauges are used in the projectile and behind the sample. Measurement of the projectile velocity is made through two pins inserted in the tube.

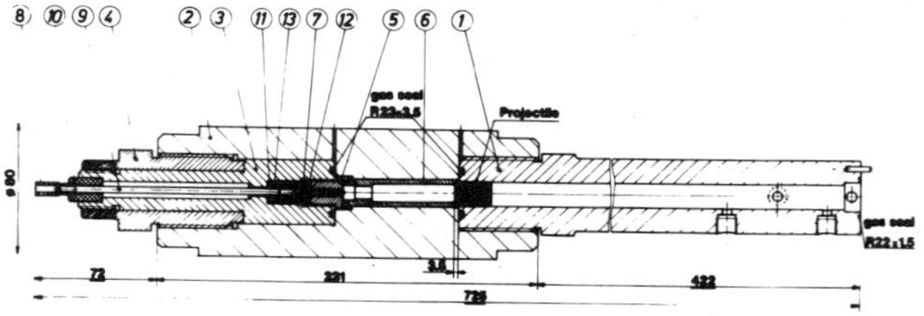

Diagnostic equipements consists with Tektronix 7844 and 7623 oscilloscopes Transient recorder 7912, ISL Flash radiography (see paragraph III.3.) and Schlumberger Time counters.

Recent studies can be as follows:

Fig. 1: Powder Gun for Shock Studies

- Behavior of ferroelectric PZT ceramics under shock wave action.
- Hugoniot equation of state of PZT 96.5/3.5 undergoing ferroelectric → antiferroelectric phase change.
- Properties of piezoelectric polymer PVF_2 under shock loading: equation of state.

II. ISL TWO-STAGE HYDROGEN GAS GUN. P.Y. CHANTEREET, J.P. HANCY

The light gas [3,4] gun is principally used for terminal ballistics applications. In ISL, we study meteoroid impacts up to 9 km/s and long rod penetrations at velocities between 2 at 5 km/s [5].

Recently a new method has been experimented to measure the depth of penetration caused by multiple fragments in order to simulate broken up shape charge jet penetration.

As it is almost impossible to launch several long projectiles with controlled geometrical and timal impact parameters, we have extended to high velocities, a method which was developped in ISL with a powder gun [6]. It consists to produce the interaction with a moving target instead of a moving projectile.

Fig. 2

Flash radiograph of a flying aluminium target (∅ 15 mm v_o = 3100 m/s) after interaction with a steel rod:
$\emptyset_p$ = 0,8 mm
L_p = 10 $\emptyset_p$

The depth of the crater is measured either by soft recovering the launched target or more easily by taking a flash radiograph of the flying target after interaction with the pre-placed fragments.

III. DIAGNOSTIC TECHNIQUES

III.1. Double exposure interferometric holography and range of practical use. P.SMIGIELSKI, H.FAGOT, F.ALBE [7]

Investigation techniques:

The holographic camera uses two pulsed ruby lasers. Characteristics :
- Pulse time : 2 ns to 20 ns
- Pulse energy: 100 mJ

- Time interval separating 2 pulses: 10 ns to ∞
- Maximum area visualized 1 m^2.

Investigation conducted:

- Study of the local displacement of the free surface of ceramics subjected to point shape shock loading.

III.2. Use of holography for kinetic energy measurements of target splinters [8]. H. ROYER, M. GIRAUD

The estimation of the residual energy of high kinetic energy projectiles perforating armored targets is of topical interest. The method is based on the double holographic recording of the splinters and on their smooth recovery in order to identify them afterwards. Thus coupling the velocity v and the mass m becomes possible for each splinter recognized in the two holographic images [8].

The light source is a double ruby laser which delivers two pulses of 20 mJ in 20 ns each with an adjustable delay (λ = 6943 Å).

Holographic images in flight of recollected splinters can be achieved [8].

Experimental facilities: 90 mm powder gun
Velocity of projectile : 1300 m/s
Projectile mass : 1,9 kg
Target thickness : 120·10^{-3} m
Target strength : 1000 N/mm^2.

III.3. Flash X-ray measurement of a shock pressure.
F. JAMET, F. BAUER

III.3.1. Absorption method[10,11]

The absorption of a homogeneous X-ray beam is given by the following well-known expression

$$I = I_0 \exp\left(-\frac{\mu}{\rho} \rho d\right)$$

where I_0 and I denote the incident and emergent intensities, respectively, of the emitted radiation; μ/ρ is the mass absorption coefficient and d is the thickness of the absorber.

As the coefficient μ/ρ is independent of the physical state of the absorber, the above formula shows thickness variation to be equivalent to density variation.

The principle of the method is a simple one. Beside the specimen subjected to a variation in density, a calibration step is placed which is made of the same material as that used for the sample at rest. The images of both the calibration step and the sample are recorded together on the same film, which is then analyzed with the aid of geometric and photometric measurements. Consider now an example illustrating the range of possible uses of this method.

We analyze the detonation wave of a foam explosive stick 1,2 inch diameter, the density of which, at rest, equals $\rho_0 \approx 0.60$ g/cm^{-3} (fig. 3). It should be noted here that because of the edge effects, the wave is not strictly a plane one. The curvature of the wave can be taken into account by introducing a corresponding correction into the analysis.

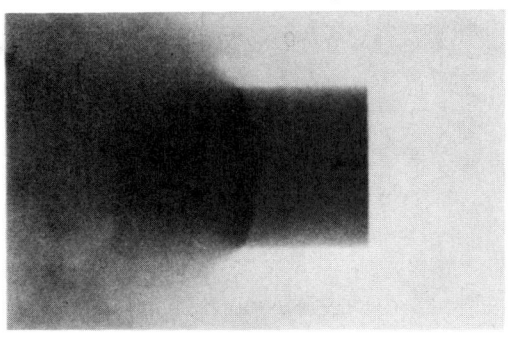 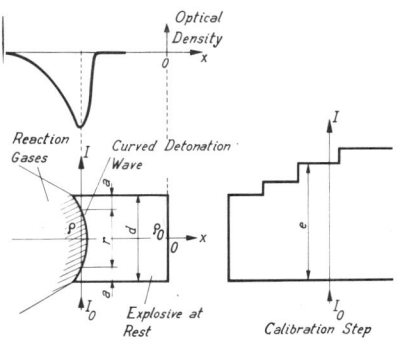

Fig. 3: Detonation Wave in a Foam Explosive Stick

Fig. 4: Evaluation of Compression Rate in a Curved Detonation Wave

Let e be the thickness of the high-explosive calibration step (figure 4) whose absorption equals that occurring as the radiation passes through the curved zone of the detonation wave. The equality of the expression for emerging intensities I in both casis is

$$I = I_0 \exp\left[-\frac{\mu}{\rho_0}\rho_0 e\right] = I_0 \exp\left[-\frac{\mu}{\rho_0}(2a\rho_0 + \rho r)\right]$$

It allows one to determine the density with respect to the curvation of the wave:

$$\rho = \rho_0 \frac{e - 2a}{r}$$

where r and a can be measured directly on the radiographs, whereas e is obtained by the photometric analysis. At the minimum of the optical density (figure 4), the density in the detonation wave is found to be $\rho = 1.06$ g/cm^{-3}. The detonation velocity measured with the aid of two successive flash radiographs is $D \approx 2.2 \cdot 10^3$ ms^{-1}. The detonation pressure is then given by the following classical expression:

$$p = \rho_0 D\left(1 - \frac{\rho_0}{\rho}\right) \approx 12.3 \text{ kilobars}.$$

III.3.2. Flash X-ray diffraction method [9,11]

The detection of distances variations between lattice planes of a material under shock compression can be studied by means of diffraction patterns.

Fig. 5 shows the direct application of Bragg's law to the determination of distance variation between the (200) lattice planes of a shock loaded sodium chloride single crystal.

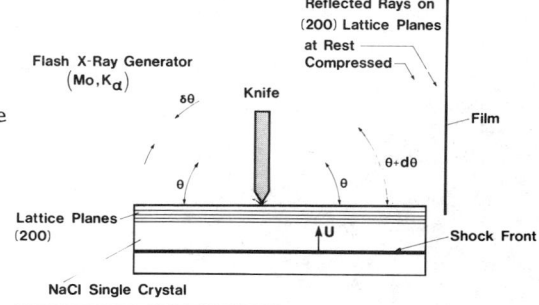

Fig. 5: Measurement of the Distance Variation between the (200) Lattice Planes

REFERENCES

1. BAUER F., Top.Conf. on Shock Wave in Condensed Matter Menlo Park, june 1981
2. BAUER F., EYRAUD L., FETIVEAU Y., VOLLRATH K., J. of Am. Ceram. Soc., 5-6, 63, 1980
3. HANCY J.P., LECOMTE C., ISL report T 50/68 unpublished
4. LECOMTE C., ISL report T 51/65 unpublished
5. PEREZ E., HAEUSEL R., 2nd Int. Symp. on Ballistics, Daytona Beach 1976
6. PEREZ E., GIRAUD M., ISL report CO 217/78 unpublished
7. ALBE F., SMIGIELSKI P., FAGOT H., Jpn, J. Appl. Phys., 14, 14-1, 229 - 264, 1975
8. ROYER H., GIRAUD M., ISL report CO 219/78 unpublished
9. JAMET F., BAUER F., Conference on Shock Waves in Condensed Matter Pullman, Washington, 1979
10. JAMET F., THOMER G., Flash radiography, Elsevier Sci. Publishing Company, Amsterdam 1976
11. JAMET F., Proc. of Flash Radiography Symp. Ed. by L.E. Bryant, American Society for nondestruction Testing 1976

TABLE I Institut Franco-Allemand de Recherches (ISL) powder gun single stage gas gun and two stage light gas gun facilities

Laboratory Facilities	- 15 mm caliber powder gun (50 - 700 m/s) - 90 mm single stage compressed gas gun (50 - 1000 m/s) - Two-stage light gas gun (2 - 9 km/s)
Dignostics Techniques	Tektronix 7844, 7912, 7633... Flash X-ray radiography Shadowgraph Framing and streak cameras Rotating mirror camera Holography...
Recent Studies	- Behavior of ferroelectric and piezoelectric materials - Target penetration - Impact cratering - Shock compression of porous materials - Equation of state of material - Phase change of shock loaded ferroelectric - Sensitivity of explosives - Radiocrystallographic analysis of crystal lattice deformation - Mass ejection from shocked targets
Principal Investigators	F. BAUER, P.Y. CHANTERET, F. JAMET, M.GIRAUD J.P. HANCY, H. MOULARD, H. ROYER, P. SMIGIELSKI

FACILITIES FOR THE STUDY OF SHOCK INDUCED DECOMPOSITION OF HIGH EXPLOSIVES

J. E. Vorthman
Los Alamos National Laboratory, Los Alamos, New Mexico 87545

ABSTRACT

This paper briefly describes facilities used by the Los Alamos Explosives Technology group to study the shock-induced decomposition of high explosives.

INTRODUCTION

Shock wave experiments have been done at the DF-Site of Los Alamos National Laboratory (Table I) for almost 30 years. The primary mission of this section of the Laboratory is to gain a fundamental understanding of shocked induced decomposition of high explosives. Work done in the past has included experimental technique development, equation of state measurements, and reaction rate formulations. This paper briefly describes our facilities and research program. All references are to work that has been done here even though the methods may have been developed elsewhere.

COMPRESSED GAS GUN

Our most successful modeling of shock induced decomposition of explosives[1,2] has been based on embedded Manganin pressure gauge data gathered using a compressed gas gun. The gun is 7.6-m long, its bore is 70 mm, and it has been used to fire projectiles at velocities in excess of 1.6 km/s (see Table II). Recent modifications to the double-diaphragm breech are expected to increase the maximum projectile velocity by several hundred meters per second. Projectile velocity is measured using shorting pins (Fig. 1) of various heights to start and stop electronic counters (10-ns resolution).

The principal use of the gun is to produce a planar shock of known strength in a high explosive (HE). High speed oscilloscopes (Tektronix 7844) are used to measure the output signal from a Manganin pressure gauge embedded in the HE (Fig. 2). A series of these experiments is done, varying only the location of the gauge (Fig. 3). Finally, a Langrangian analysis[1,2] of the data is performed, an equation of state assumed, and the degree of reaction obtained (Fig. 4).

Other instruments used at the gun facility include quartz gauges, electromagnetic stress and particle velocity gauges, a rotating mirror streak camera, and image intensifier cameras. A technique is being developed to combine electromagnetic stress and particle velocity gauges (Fig. 5).

TABLE I

SHOCK WAVE FACILITY

Laboratory Facilities

70-mm-bore single stage compressed gas gun (0.05 to > 1.6 km/s)
2 sites for firing high explosives
Electrically-driven flyer facility (56 µF, 20 kV)

Instruments

Tektronix 7844 Oscilloscopes
   (1) Manganin pressure gauges
   (2) Combined electromagnetic stress and particle velocity gauging
   (3) Quartz gauges
Rotating Mirror Cameras
     Writing speed ≤ 16 mm/µs
Gear Electronic Streak Camera
     Writing speed ≤ 5 mm/ns
Image Intensifier Cameras
     10-ns resolution, intensification gain $\leq 10,000$

Principal Investigators

Jerry Wackerle, Section Leader (HE, Gas Gun, Theory)
Allan Anderson (Theory)
Mike Ginsberg (Gas Gun, HE)
Ron Rabie (Propellants, Theory)
Wendell Seitz (HE, Electric Flyers)
Garry Schott (HE, Gas Gun, Theory)
Robert Spaulding (HE)
John Vorthman (Gas Gun, Theory)

Recent Studies

A Shock Initiation Study of PBX 9404
Shock Initiation of Porous TATB
Pulsed Laser Stereo Photography of Electrically Exploded Bridges
Precursors in Detonations in Porous Explosives
The Polymorphic Detonation
Three-Dimensional Shock-Change Relations for Reactive Fluids
An Empirical Model to Compute the Velocity Histories of Flyers
    Driven by Electrically Exploding Foils

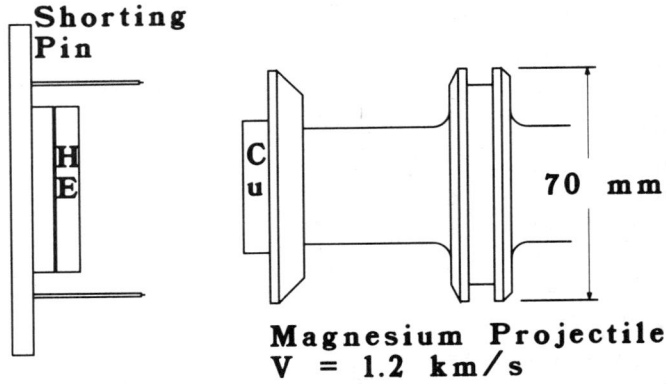

Fig. 1. Typical gun experiment.

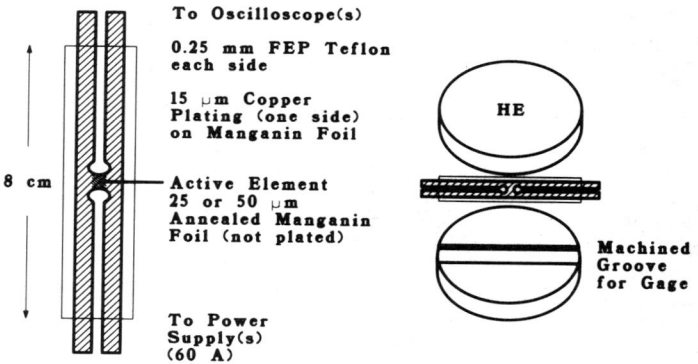

Fig. 2. Manganin gauge.

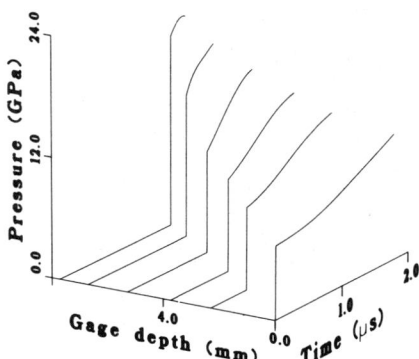

Fig. 3. Manganin-gauge data.

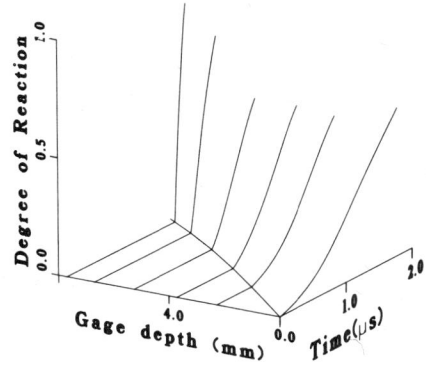

Fig. 4. Analyzed data.

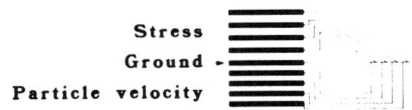

Fig. 5. Combined E-M gauges.

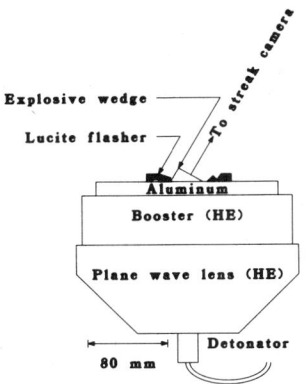

Fig. 6. Typical HE experiment.

TABLE II

Compressed Gas Gun

Recoiling barrel
Bore   70 mm
Length 7.6 meters

Wraparound breech
  Volume 23 liters
  Maximum gas pressure 43 MPa (6250 psi)

Double-diaphragm breech
  Volume 28 liters
  Maximum gas pressure 103 MPa (15,000 psi)

EXPLOSIVELY PRODUCED SHOCKS

One datum that is used to characterize each HE is the distance a shock must travel before the onset of detonation for a given input shock strength. This datum is usually collected using another HE to generate the input shock (Fig. 6). The HE of interest is wedge-shaped and a rotating mirror camera is used to record the position of the shock wave as it breaks out of the sloping wedge face. Detonation is characterized by a (usually) sharp change in the shock velocity. Reference 3 is another example of research being done that uses one HE to study another.

ELECTRICALLY-DRIVEN FLYER FACILITY

Our electrically-driven flyer facility is capable of accelerating a one-half-inch-square Mylar flyer to 3.8 km/s. Energy is stored in capacitors (56 µF at 20 kV total). Instruments used in conjunction with this facility include a pulsed ruby laser for stereo photography[4] and an electronic streak camera. A Fabry-Perot interferometer is being assembled for use with all the facilities described above.

REFERENCES

1. Jerry Wackerle, R. L. Rabie, M. J. Ginsberg, and A. B. Anderson, Proceedings of the Symposium on High Dynamic Pressures, Paris, France, p. 127, 1978.
2. Allan B. Anderson, M. J. Ginsberg, W. L. Seitz, and Jerry Wackerle, Seventh Symposium on Detonation, Office of Naval Research (to be published) (1981).
3. R. L. Spaulding, Jr., Seventh Symposium on Detonation, Office of Naval Research (to be published) (1981).
4. W. L. Seitz and S. D. Gardner, SPIE $\underline{94}$, 100 (1976).

LAWRENCE LIVERMORE NATIONAL LABORATORY SINGLE-STAGE 101 MM GUN*

LeRoy Erickson

University of California, Lawrence Livermore National Laboratory
Livermore, California 94550

TABLE I.

Laboratory Facilities	101 mm diameter, single-stage gas gun, 0.1-2 km/s.
Diagnostics	Electromagnetic particle velocity measurement system, equipment for 6 manganin pressure gauge measurements, Fabrey-Perot velocimeter, continuous access framing and streaking cameras, timing pins.
Experiment Tank	• Capability to use 200 g of explosive in the experiment tank, soon to be upgraded to 400 g. • Electromagnet in experiment tank capable of 0.1 Tesla with field uniform to 0.1 percent in the region of interest. • Double flash X-ray for Sabot velocity and integrity check.
Recent Studies	• Detonation Physics • Inductive Energy Storage • Infrared Radiometry • High Explosive Sensitivity Studies • Velocity Studies on Explosive-Driven Plate • Equation-of-State of Metal Alloys • Calibration of Stress and Particle Velocity Gauges • Studies of Propellants Under Shock Stimulus
Principal Investigators	Kerry Bahl Harold Palmer Robert Barlett Norval Parker LeRoy Erickson William Von Holle

RAIL GUN DEVELOPMENT FOR EOS RESEARCH*

C. M. Fowler and D. R. Peterson
Los Alamos National Laboratory, Los Alamos, NM 87545

R. S. Hawke and A. L. Brooks
Lawrence Livermore National Laboratory, Livermore, CA 94550

ABSTRACT

We give the status of a railgun program for equation-of-state (EOS) research in progress at Los Alamos and Lawrence Livermore National Laboratories.

INTRODUCTION

We will discuss the operating principle of rail guns, the power supplies used to drive them, diagnostic techniques used to monitor their performance and initial efforts to develop projectiles suitable for EOS research. We conclude with a table that contains available and projected facilities and diagnostics, recent studies and personnel presently associated with the program.

OPERATING PRINCIPLES

Figure 1 shows the basic components of a square-bore rail gun. The projectile is placed in the gun breech, just forward of a metallic fuse. When power is supplied, the fuse vaporizes and a current arc forms behind the projectile. The resulting force on the projectile is approximately $F = LI^2/2$, where I is the current and L is the rail inductance per unit length. For a square bore L is typically a few tenths µH/m.

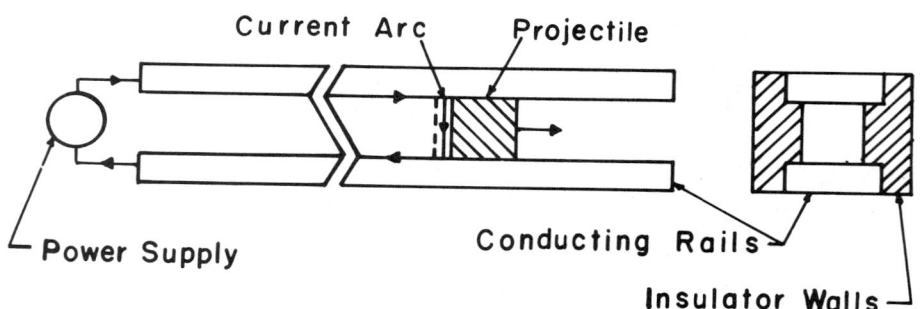

Fig. 1. Two views of a square-bore rail gun.

---

* Work done under the auspices of the US Department of Energy.

0094-243X/82/780686-05$3.00 Copyright 1982 American Institute of Physics

POWER SUPPLIES

Railgun power supplies should be able to deliver currents up to several MA for times up to several ms. Among those used are homopolar generators, capacitor banks and flux-compresssion generators. The latter two systems are used in the present program.

1. Capacitor Bank

    A 400-kJ, 30-mF, 5-kV capacitor bank is used as a direct power source. Large-capacitance, low-voltage banks are better suited for railgun applications because the current pulses are longer and voltage breakdown between rails is less likely than for higher voltage banks of equal energy.

2. Flux Compression Generators

    Figure 2 shows the flux compressor presently used. The explosive strip overlaid on the upper copper conductor is detonated after magnetic flux is introduced into the generator by a capacitor bank (940 kJ, 25 kV max). As detonation proceeds, the upper copper conductor is driven into the lower conductor pushing the flux into the railgun load.

RAILGUN PERFORMANCE DIAGNOSTICS

Diagnostics include magnetic probes and light pipes to monitor projectile motion in the gun, flash x-radiography to photograph the projectiles in free flight, Rogowski loops to obtain the current time history, and muzzle voltage measurements.

Figure 3 shows a 1.2-m-long rail gun. The rails are potted in a strong pipe for strength with cables from magnetic probes (top) and light pipes (below). Figure 4 shows an x-radiograph of a cubical lexan projectile in free flight. In Fig. 5 are plotted current-time records obtained for 3-g lexan projectiles accelerated to about 5.5 km/s (lower curve) and one estimated to be about 10 km/s (upper curve).

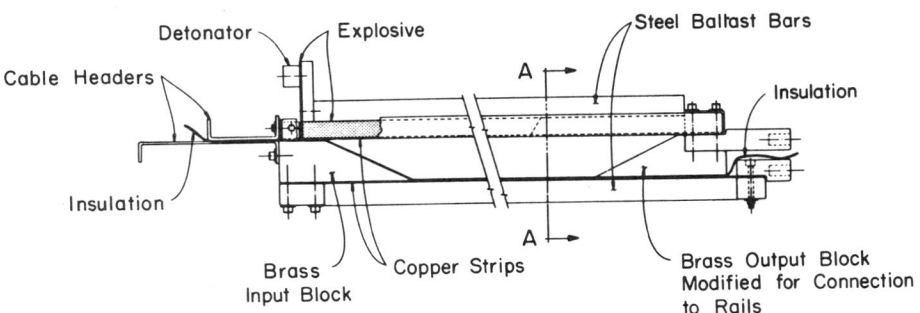

Fig. 2. Schematic of flux-compression strip generator.

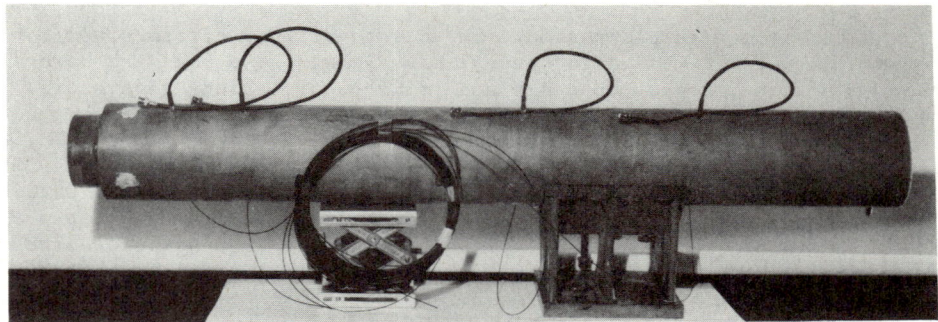

Fig. 3. 1.2-m-long rail gun showing various probes.

Fig. 4. Flash x-radiograph of lexan cube in free flight.

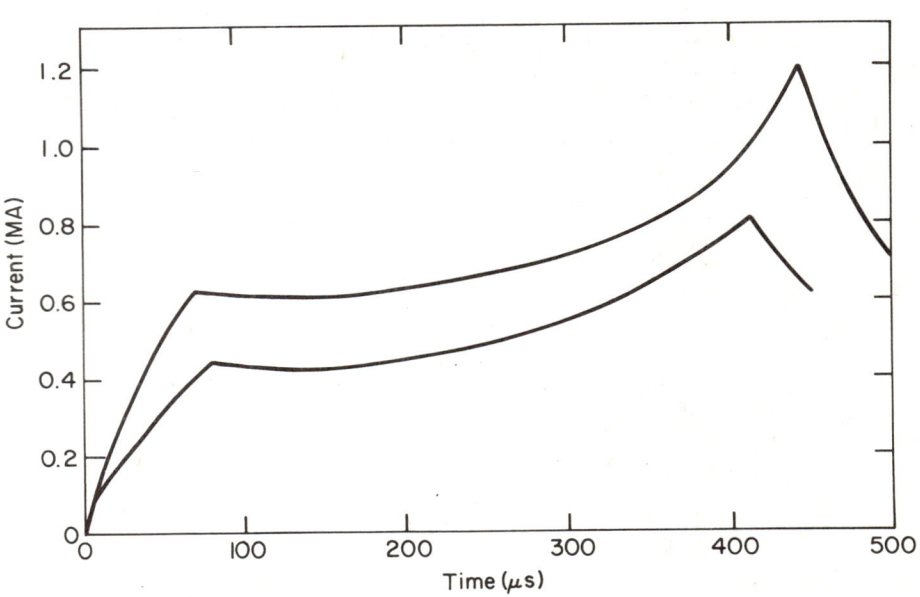

Fig. 5. Current-time curves.

Figure 6 shows the internal consistency of some of the diagnostic data. The solid curves were calculated by integrating the force equation using measured current-time records. Other diagnostics fit these curves well including the flash x-ray diagnostics, muzzle voltage discontinuities, and signals from magnetic probes and light-diodes (not shown). Similar agreement was obtained for the estimated 10 km/s shots. However, intact projectiles were not seen on the x-radiographs and it was concluded that the projectiles broke up.

Figure 7 shows a lexan sabot with an inset Ta impactor, subsequently accelerated to 2.4 km/s. However, a flash x-radiograph showed that the impactor separated from the sabot.

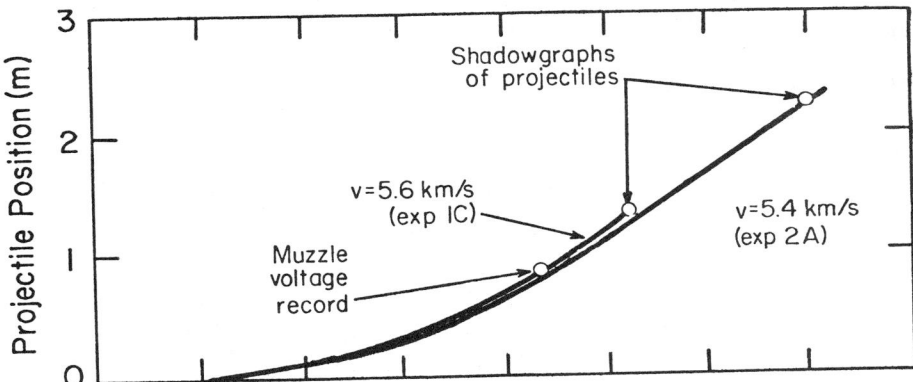

Fig. 6. Distance vs time for 3-g lexan cubes. The horizontal time ordinates are 100 µs per division.

Fig. 7. Photograph of lexan sabot (0.5 in x 0.5 in x 0.75 in) with Ta insert.

TABLE I Joint Los Alamos Lawrence Livermore National Laboratories Rail Gun Effort

Laboratory Facilities:	8-30 mm bore electromagnetic rail guns (expected velocity range 5-25 km/s); powered by 940 kJ, explosive-driven, magnetic-flux-compression generators.
	8-12 mm bore electromagnetic rail guns (expected velocity range, 5-20 km/s); powered by 400 kJ, 30 mF and 5 kV capacitor bank.
Diagnostics:	• Pins for shock arrival. • Fiberoptic-coupled photodiodes for plasma arc passage. • Railgun current and voltages for gun performance. • Laser beam for projectile launch data. • X-ray and optical diagnostics.
Recent Studies:	• Railgun performance studies. • EOS diagnostics and railgun compatability studies. • Impactor launch demonstration. • Rail and dielectric lifetime studies. • 5.5 km/s intact projectile launch.
	Los Alamos / Lawrence Livermore
Principal Investigators:	C. M. Fowler / R. S. Hawke D. R. Peterson / A. L. Brooks J. W. Shaner / A. C. Mitchell
Technical Support:	Jim King / Neil Gibson Richard Martinez / Peggy Lorton Barbara Fox / Martin Roeben Dennis Shampine / Craig Wozynski Jerry Kerrisk / Gin Yee Charles Cummings

PERFORMANCE OF A 100 KV, 78 KJ ELECTRIC GUN SYSTEM

H. Chau, G. Dittbenner, K. Mikkelsen, and R. Weingart
Lawrence Livermore National Laboratory

K. Froeschner
Martin, Froeschner and Associates

R. Lee
Kansas State University

June 23, 1981

ABSTRACT

We have constructed a new electric gun system for use in high-pressure EOS studies. The system is powered by a 100 kV, 15.6 µF capacitor bank. At 100 kV charging voltage the system inductance is 23 nH. This system has driven 0.3 mm-thick Kapton projectiles to >20 km/s and 0.3 mm Kapton 30 µm Ta projectiles to ~10 km/s. Projectile velocity is modeled phenomenlogically by an electrical Gurney model.

INTRODUCTION

At LLNL we have made extensive use of a device, which we call an electric gun, to generate planar shock waves for high explosive initiation experiments. An electric gun system operates by discharging a capacitor bank through a thin metallic foil. Ohmic heating of the foil deposits a significant portion of the stored capacitor bank energy into the foil, causing the foil to explode. The explosion of the foil drives a thin plate of material, placed on top of the foil, down a barrel to impact a target. We have discussed the operation of electric gun systems in some detail in a recent publication.[1]

As a result of our experience with electric gun systems in high-explosives experiments, we have come to believe that electric gun systems offer potential for ultra high-pressure research (pressures greater than 0.5 TPa). We have developed a new electric gun system for use as a shock wave generator for equation-of-state (EOS) experiments in the 0.5-1 TPa pressure range. With further development we feel that substantially higher pressures can be achieved.

EXPERIMENTAL ARRANGEMENTS

The heart of our new system is a 100 kV, 15.6 µF capacitor bank produced by Maxwell Laboratories, Inc. The bank is connected

---

"This work was performed under the auspices of the U.S. Department of Energy by the Lawrence Livermore National Laboratory under contract number W-7405-ENG-49."

to the exploding-foil load through a rail-gap switch and a large, flat-plate transmission line. The exploding-foil load is aluminum, typically 51-152 μm thick and 9.5 mm wide. The transmission line and load are sketched in Fig. 1. Total system inductance at current start varies from 23 nH at 100 kV to 32 nH at 35 kV bank charging voltage. Variation in system inductance with charging voltage is due to the operation of the rail-gap switch, which breaks down more uniformly at high voltages. Total system resistance up to the time the foil bursts is about 10 mΩ. At 100 kV the bank stores 78 kJ of energy.

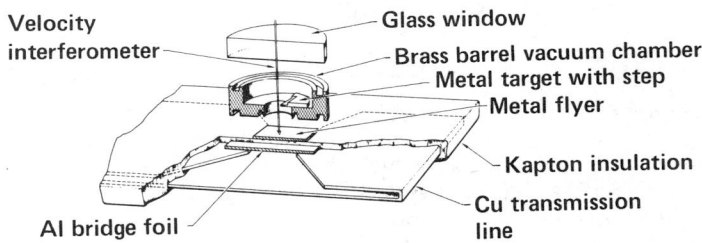

FIGURE 1: Sketch of Transmission Line and Load

Also shown in Fig. 1 are the evacuated barrel assembly and target configuration used for our high-pressure EOS experiments. The transmission line from the bank tapers down to a width of 15.2 cm and is terminated by a disposable 15.2 cm-width, flat-plate transmission line, shorted at one end. The exploding-foil element (bridge foil) bridges a gap in a disposable section of the line. The evacuated barrel assembly is usually made of brass. The barrel itself may be an insert of some other material, e.g. alumina, boron carbide (B_4C). We have found that barrels with high shock impedance give the highest-quality flyer plate impact. For most of our experiments we use 3.2 mm-long, 8.3 mm-diameter barrels. The projectiles accelerate all the way down the barrels, so we can obtain substantial increases in velocity by using longer barrels (5-7.5 mm) but there is some degradation of projectile flatness with longer barrels. The exploding foils are aluminum, 51-152 μm thick and with the 8.3 mm-diameter barrels we use foils 9.5 mm in width. Depending on the desired projectile diameter and velocity, foil widths can vary from 3.2-38 mm.

Electrical diagnostics are fairly standard. Bank current is obtained by electronic integration of the signal from a calibrated Rogowski probe. We also routinely measure bank charging voltage and voltage across the transmission line.

Shock transit times are recorded with an Imacon image converter camera. Projectile velocities are measured with a Fabry-Perot laser velocimeter or by recording impact times on a glass step.

Tantalum projectiles are of rolled foil 30-76 μm thick, bonded to 0.3 mm Kapton with epoxy adhesive. Tantalum targets are also of rolled foil, 76-152 μm thick with a step cut in the outer edge as shown in Fig. 2.

RESULTS AND DISCUSSIONS

The specifications for a capacitor bank used to power an electric gun depend on its end use. The intended purpose for the bank described here is to generate ultra-high-pressure shock waves suitable for equation-of-state measurements. For this purpose, projectile velocity and simultaneity of projectile impact (flatness) are of paramount importance. Our experience with the smaller bank used for high-explosives work has been very useful, both in deciding on initial specifications and in understanding the behavior of projectiles powered by the new bank.

In our work with the smaller, 40 kV banks we found that bank characteristics and projectile velocity were related through a phenomenological theory developed by Gurney[2] and extended to projectiles accelerated by exploding foils by Tucker and Stanton[3]. The electrical Gurney theory relates projectile velocity, V_F, to projectile mass/area, M, mass/area of the exploding foil, C, and current density in the foil at burst time, J_B, through the equation

$$V_F = f(J_B)(M/C + 1/3)^{-1/2}, \quad (1)$$

where $f(J_B)$ is experimentally determined from measured projectile velocities and burst current densities. Eq. (1) describes electric gun performance over a remarkable range of foil/projectile masses and foil geometries[1,4]. To complete the model one must relate J_B to the bank characteristics and charging voltage. This is done through the expression for an underdamped capacitor discharge,

$$I = I_0 \sin\omega t \; e^{-t/\tau} \quad (2)$$

and the empirical result of Anderson and Nielson[5],

$$g = \int_0^{t_B} J^2(t)dt = \text{constant}. \quad (3)$$

In Eq. (2), $I_0 = V_0/L\omega$, $\tau = 2L/R$ and $\omega = (1/LC + R^2/4L^2)^{1/2}$, where V_0, R, L, and C are bank charging voltage, system resistance, inductance and capacitance, respectively, and t_B is the time the foil explodes.

To achieve high projectile velocities, Eq. (1) indicates the desirability of large values of J_B. Eq.s (2) and (3) show that to achieve large values of J_B one needs a fast-rising current, i.e., small L and large V_0. The value of J_B can also be optimized by reducing the conductor cross section. For our EOS experiments we have chosen Al foils with 9.5 mm width and 51–153 μm thickness.

Fig. 2 shows $f(J_B)$ for the 100 kV bank, determined for a range of M/C values using Eq. 1 and measured values of V_F. For the 40 kV banks used in earlier work, we found that $f(J_B)$ was best described by a power law, but the data for the 100 kV bank are better described by a linear function,

$$f(J_B) = 11.01 \, J_B + 2.64, \quad (4),$$

where J_B is in TA/m². The fit to the data worsens at large values of J_B, so Eq. (4) should not be used beyond the range of the data.

Fig. 3 shows projectile velocity, V_F, as a function of charging voltage, V_0 for 0.3 mm-thick Kapton projectiles and composite, 30 μm Ta/0.3 mm Kapton projectiles. Solid lines are calculation using Eq. (1) and values of $f(J_B)$ determined from experimental data, e.g. Fig. 3. Data are shown for both 100 kV and 40 kV systems. To compare the performance of the banks, we compare data for the same type of projectiles and the same barrel length. From the two bottom curves we see that at 35–40 kV the performance of the two systems is virtually identical.

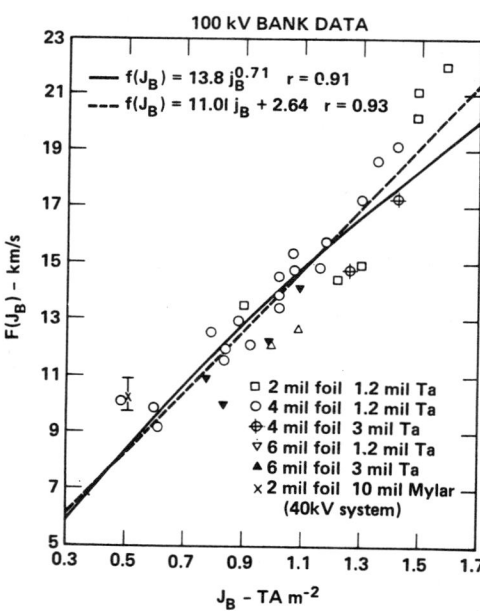

FIGURE 2: Graph of F(J_B) for the 100 kV Bank

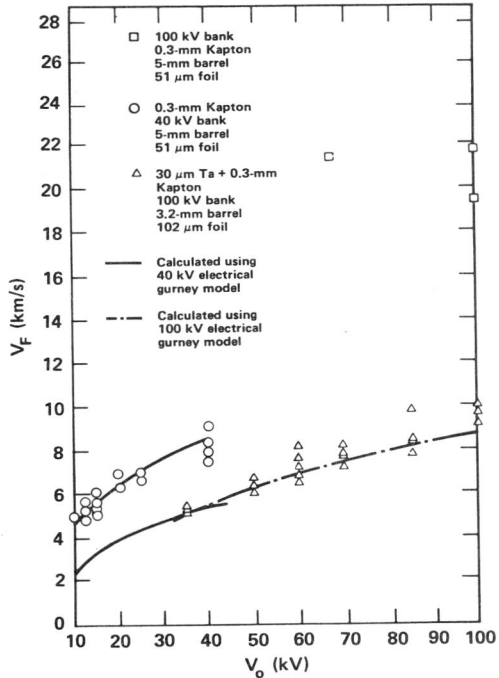

This is because of high switch inductance at low charging voltages. As charging voltage is increased to 100 kV, there is a substantial difference in performance of the two systems. Data for 0.3 mm-thick Kapton projectiles shows an even greater difference in performance. Using 5 mm barrels, the projectile velocity from the 100 kV system is more than double that of the 40 kV system.

FIGURE 3: Flyer Velocity vs. Charge Voltage

REFERENCES

1. H. Chau, G. Dittbenner, W. Hofer, C. Honodel, D. Steinberg, J. Stroud, R. Weingart, Rev. Sci. Instrum. $\underline{51}$,12 (1976).
2. R. W. Gurney, Battelle Memorial Institute, Columbus, Ohio, Report 405, (1943).
3. T. J. Tucker and P. L. Stanton, Sandia National Laboratories, Albuquerque, NM, SAND 74-0244 (1975).
4. "Acceleration of Thin Flyers by Exploding Metal Foils: Application to Initiation Studies," R. Weingart, R. Lee, R. Jackson, and N. Parker, in Proceedings of the Sixth Symposium (International) on Detonation (Office of Naval Research, Washington, DC, ACR-221, 1976), p. 653.
5. "Exploding Wires," G. Anderson and F. Neilson, edited by W. Chase and H. Moore (Plenum, New York, 1959), Vol. 1, p. 97.

SHOCK-WAVE FACILITY AT TOKYO INSTITUTE OF TECHNOLOGY

A. Sawaoka and K. Kondo
Research Laboratory of Engineering Materials
Tokyo Institute of Technology, Midori, Yokohama 227, Japan

ABSTRACT

The shock-wave facility at the Tokyo Institute of Technology is described. Two double-stage light-gas guns are used for studying material science and technology. Recently construction has begun for a new type of rail gun combined with a double-stage light-gas gun.

SHORT HISTORY AND OUTLINE OF FACILITY

Very high pressure research using the dynamic technique was started at the Research Laboratory of Engineering Materials, Tokyo Institute of Technology, in 1968 by Prof. S. Saito. A projectile accelerator with a 20-mm-diameter bore using an electric discharge in liquid was constructed and used for the study of mechanochemical effects of ceramic powder.

A powder gun with a 15-mm-diameter bore was constructed in 1972 and used also for the study of mechanochemical effects in alumina, magnesia and boron nitride. The maximum velocity of this gun was 2 km/s.

The double-stage light-gas gun (HS-2A) with a 15-mm-diameter bore, 6 m-long was constructed and a velocity of 4.2 km/s was obtained in 1973.[1] The recovery technique for shocked materials was developed. Shock synthesis of diamond and the high-density form of boron nitride were studied. The bore diameter of this gun was changed to 20 mm. Flash X-ray diffraction during shock compression of LiF was performed up to 50 GPa.[2,3] The magnetoflyer technique for precise measurement of projectile velocity was developed and combined with this gun (HS-2A).[4]

The second double-stage light-gas gun (HS-3A) with a 20-mm-diameter bore, 10 m long was constructed and a velocity of 6 km/s was obtained in 1976. A new type of CW X-ray system for measuring projectile velocity has been combined with this gun.[5] Particle velocity[5] and optical measurements[6] were conducted by using the gun HS-3B,[7] which was an improved type of HS-3A. A description of our facility and a summary of recent results are shown in Table I.

FLASH X-RAY DIFFRACTION

Flash X-ray diffraction study was performed during shock-compression by using the Blumlein flash X-ray source and an image intensifier.[2,3] The X-ray source is similar to that of Q. Johnson et al. The dimension of this Blumlein body is approximately 6 m in length and 0.2 m in diameter. The voltage pulse is 80 ns wide but the X-ray pulse is estimated to have a width of a few tens of ns.

TABLE I. TIT double-stage light-gas gun facility

Laboratory Facilities	20-mm-diameter double-stage light-gas gun, HS-2B (1-5 km/s), HS-3B (1.5-6 km/s)
Diagnostics	Electromagnetic particle velocity measurement system, flash X-ray diffraction system, projectile velocity meter using CW X-ray, optical measurement system using OMA and PIN diode
Recent Studies	• Flash X-ray diffraction during shock compression[2,3] • Conductivity of shocked dielectrics[8] • Polarization effects in shocked dielectrics • EMF study of Cu-constantan junction[9] • Dynamic response of fused silica[10] • Shock-reverberations in a layer structure[11] • Spectral photometry for radiation from shock-compressed materials[6] • Development of new type of rail gun
Principal Investigators	Akira Sawaoka Ken-ichi Kondo, Hiroshi Sugiura, Shu USUBA

Fig. 1 Over-view of two double stage light gas guns at Tokyo Institute of Technology. Left small gun is HS-2B and rear big one is HS-3B. Now, HS-2B was modified and combined with rail gun as HS-2C.

The schematic layout of the experimental arrangement is shown in Fig. 2. The flash X-ray generator was combined with the double stage light gas gun (HS-2A) at an angle of 80° to the projectile flight direction. The distance from the center of the back surface of a specimen to the fluorescent screen of the detector is 80 mm.

Microphotometer tracings of typical diffraction records on a LiF single crystal taken before and during shock conditions are shown in Fig. 3. These were taken using unfiltered molybdenum radiation. The projectile with a copper plate impacted the crystal at 1.8 km/s parallel to the (200) plane. The shock pressure is estimated to be 24.0±0.3 GPa.

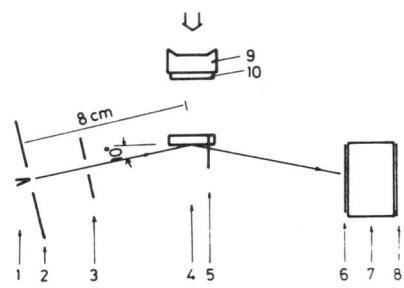

Fig. 2 Schematic diagram of the experimental arrangement.
1. Anode, 2. Cathode, 3. Collimeter, 4. Specimen, 5. Trigger pin, 6. Fluorescent screen, 7. Image intensifier, 8. Emulsion film, 9. Projectile, 10. Metal impactor.

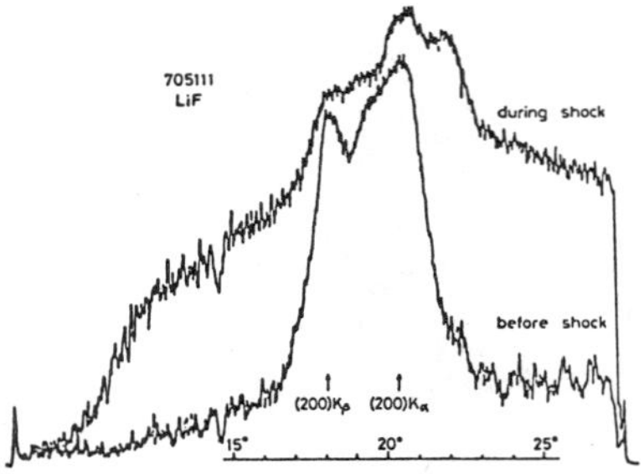

Fig. 3 Density pattern of the FXD record of LiF single crystal obtained by microphotometer; the density scale is about 2.5 from image intensifier back-ground to peak and additional peak corresponds to the shock-compressed state at 24.0±0.3 GPa.

SPECTRAL PHOTOMETRY AND CW X-RAY VELOCIMETER

The new optical system consists of a wide range spectrometer, with function of memory whose minimum time of exposure is 200 ns, and a two-color pyrometer having high time resolution.[6,12] A schematic diagram of this system is shown in Fig. 4. The light flux reflected on a mirror surface and aligned by an achromatic lens is divided for each detector by prism beam splitters. As the image of the specimen on the slit of the spectroscope is 4.2 mm in diameter, and the slit size was 0.05 x 2 mm^2, the detected light is that emitted from the central region of the object only. A holographic grating with 152.6 lines/mm is used to cover the whole visible spectrum. The spectrum is detected and recorded using an optical multi-channel analyzer (OMA) with a silicon intensified target (SIT) detector (PAR 1205A and 1205D). 500 SIT's were lined up in a length of 12.5 mm. The SIT detector is used in the gating mode, with a high-voltage pulse generator (PAR 1211).

The other part of the light flux is further split into two parts. 620- and 810-nm bands are selected by interference filters, with a half-height bandwidth, peak transmittance, and actual diameter of 10 nm, 50%, and 22 mm, respectively. The image of the object on the PIN photodiode (NEC LSD 39A) is 1.8 mm in diameter, and the effective area of the diode is only 0.04 mm². The signals (load resistance 500Ω) are amplified 22 times through IC amplifiers (SN 72733N), with a rise time and internal delay of about 10 ns.

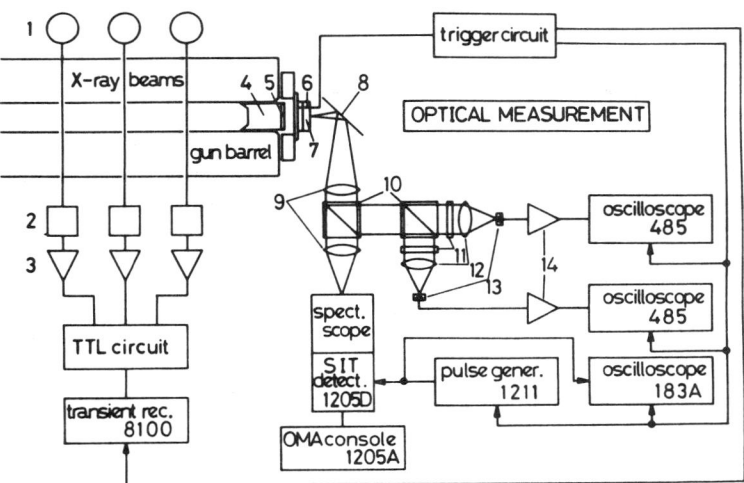

Fig. 4 Schematic diagram of the systems for the projectile velocity measurement and the optical measurement. 1. X-ray sources with Mo target, 2. X-ray detectors with NaI scintillator and photomultiplier, 3. AC linear amplifier, 4. Projectile, 5. Flyer plate, 6. Pin contactor for trigger pulse, 7. Sample, 8. Aluminum coated mirror, 9. Achromatic lenses, 10. Non-absorbing prism beam splitters, 11. Interference filters, 12. Lenses, 13. PIN photodiodes (LSD 39A), 14. Wide range video amplifiers.

The impact velocity is measured by a dc X-ray technique. Three pairs of X-ray sources and detectors, consisting of a NaI scintillator and photomultiplier, are aligned at X-ray windows on the final part of the launch tube, at intervals of 200 mm. Three signals generated by a lead foil in the projectile interrupting the X-ray beams and a trigger signal from the shock wave shorting pin-contactors in the target assembly are recorded on a transient recorder (Biomation 8100). The three intervals between the trigger signal and the X-ray signals are used for velocity determination. The acceleration is assumed to be constant. The accuracy of this system is better than 1%.

ELECTROMAGNETIC PARTICLE VELOCIMETER

The principle of this technique is a simple application of Faraday's law of electromagnetic induction. The electromagnetic gauge consists of a rectangular copper foil backed with a polyamide film.

The length, thickness and resistance of the active element of the gauge are 4.15 mm, 12.5 μm and 0.2 Ω, respectively, and the applied magnetic field is 658 G. The outputs from the gauges were terminated with 50 Ω resistors close to the gauges and the signals were observed on oscilloscopes (Tektronix 485). Changes of particle velocity in fused silica were observed clearly in the shock pressure to 22 GPa.

A RAIL GUN COMBINED WITH DOUBLE STAGE LIGHT GAS GUN

We are constructing a new type of accelerator which consists of a two-stage light-gas gun (HS-2C) and an electromagnetic rail gun to accelerate a projectile to the velocity of 15 km/s. The rail gun is 2 m long and has a round bore 8 mm in diameter. The projectile with an aluminum fuse is accelerated by double-stage light-gas gun and enters into the rail gun at the velocity of ∿5 km/s. Then the capacitor bank (500 kJ, 10 kV) supplies an electric current through a brass rail and the aluminum fuse, which explodes and forms a plasma. The projectile is accelerated by electromagnetic force. According to numerical calculations, we expect that the value of maximum current is ∿1 MA and the final velocity of a 1-gram projectile is higher than 15 km/s.

REFERENCES

1. A. Sawaoka, T. Soma, and S. Saito, in Proceedings of the Fourth International Conference on High Pressure (Physico-Chemical Society of Japan, Kyoto, 1975), p. 739.
2. K. Kondo, A. Sawaoka and S. Saito, ibid., p. 845.
3. K. Kondo, T. Mashimo, A. Sawaoka, and S. Saito, in High-Pressure Science and Technology, Vol. II, edited by K. D. Timmerhaus and M. S. Barber (Plenum, N.Y., 1979) p. 883.
4. K. Kondo, A. Sawaoka and S. Saito, Rev. Sci. Instrum. $\underline{48}$, 1581 (1977).
5. T. Mashimo and A. Sawaoka, Jpn. J. Appl. Phys. $\underline{20}$, 963 (1981).
6. H. Sugiura, K. Kondo and A. Sawaoka, Rev. Sci. Instrum. $\underline{51}$, 750 (1980).
7. H. Sugiura, T. Mashimo, K. Kondo and A. Sawaoka, Report of RLEMTIT, No. 6, 93 (1981).
8. K. Kondo, T. Mashimo and A. Sawaoka, J. Geophys. Res. B2 $\underline{85}$, 977 (1980).
9. I. Imaoka, K. Kondo and A. Sawaoka, Jpn. J. Appl. Phys. $\underline{19}$, 1011 (1980).
10. H. Sugiura, K. Kondo and A. Sawaoka, J. Appl. Phys., in press.
11. K. Kondo, Y. Yasumoto, H. Sugiura and A. Sawaoka, J. Appl. Phys. $\underline{52}$, 772 (1981).
12. H. Sugiura, K. Kondo and A. Sawaoka, Report of RLEMTIT, No. 5, 129 (1980).

SHOCK WAVE FACILITIES FOR HIGH-PRESSURE EXPERIMENTS AT TOHOKU UNIVERSITY

Yasuhiko Syono and Tsuneaki Goto
The Research Institute for Iron, Steel and Other Metals,
Tohoku University, Sendai 980, Japan.

RIISOM shock facilities including a two-stage light gas gun and a single-stage propellant gun are described. Recent research activities are summarized.

Shock wave research using gun method at Tohoku University began in 1978, although some shock experiments, concurrently with flux compression experiments for producing ultrahigh magnetic field, were done by means of explosive method in early 70s[1-7]. Our major facilities include a two-stage light gas gun (2SG-TH1) and a single-stage propellant gun (1SG-TH2). The two-stage light gas gun (Fig. 1) consists of a 20 mm-bore and 3.4 m-length launcher and a 60 mm-bore and 3.6 m-length pump tube[12]. The apparatus is capable of accelerating a 10 g-projectile to velocity of 4.5 km/s (Fig. 2). The single-stage propellant gun (Fig. 3) has a 25 mm-bore and 2 m-length launcher and is now being used for the velocity range of 1.0-2.4 km/s (Fig. 2). These guns have been used for both equation-of-state studies and shock recovery experiments. Hugoniot measurements can be done by optical method using a rotating mirror type streak camera with a writing speed of 10 mm/μs[13] (Fig. 3). Block diagram of the measurement system is shown in Fig. 4. Time-resolved spectroscopy (Fig. 5) is now being developed in order to observe the aborption spectrum under shock compression.

The central issue of our research activities lies in the shock-induced phase transition in solids. We have determined shock compression curves of various materials in the pressure range of 15-200 GPa. Precise evaluation of the high pressure phase, including determination of the phase transition pressure and the volume change accompanied by the phase transition, permits identifications of semiconductor-metal transition in GaP, high-spin low-spin transition in α-Fe_2O_3 and disproportionation in Mg_2SiO_4. Shock residual effects have been observed by means of various modern diagnostics such as single-crystal x-ray diffraction, Mössbauer spectroscopy and transmission electron microscopy so as to elucidate the mechanism of the shock-induced phase transition. First observation of disproportionation in Mg_2SiO_4 under intense shock loading provides support for the relevance of shock data to geophysical problems. Application of shock wave technique for the synthesis of new materials has also been carried out for the Nb-Si system.

The authors express their sincere thanks to Professor Yasuaki Nakagawa for continuous support and warm encouragement. They benefited from the machine shop of RIISOM for many expendable parts necessary for experiments. The work has been supported by Grant-in-Aid for Special Project Research (# 321503, 420902 and 510104) and for Scientific Research (# 546016 and 554064) given by the Ministry of Education, Science and Culture, Japan.

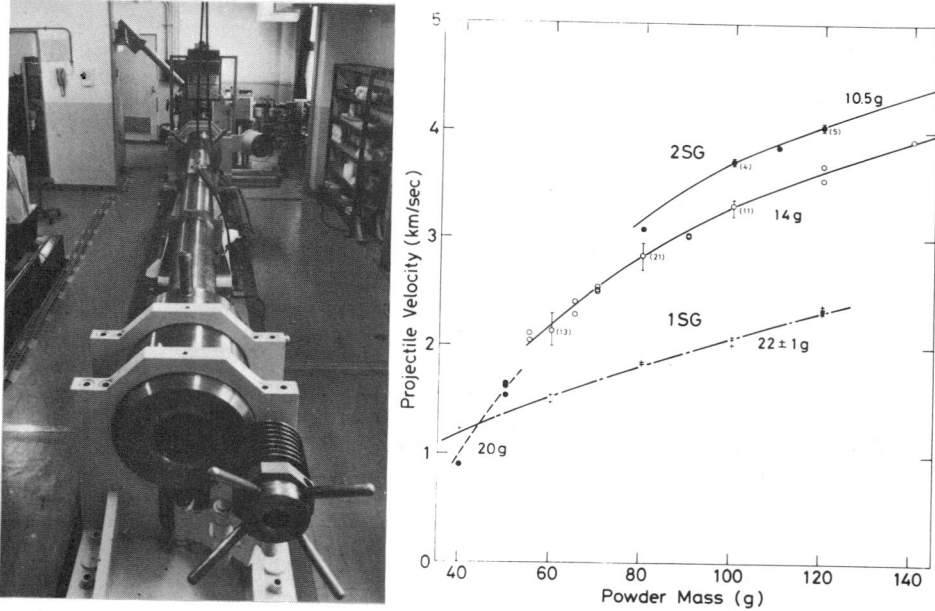

Fig. 1 Two-stage light gas gun

Fig. 2 Performance characteristics of the two-stage light gas gun and single-stage propellant

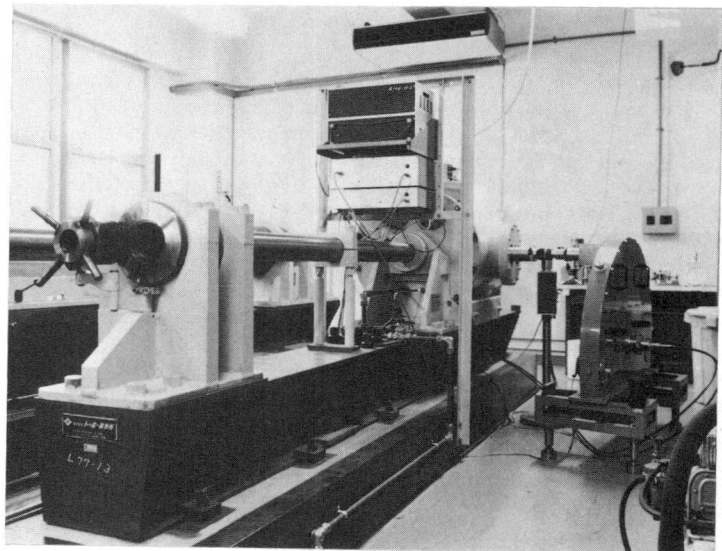

Fig. 3 Single-stage propellant gun with the equipment of Hugoniot measurement

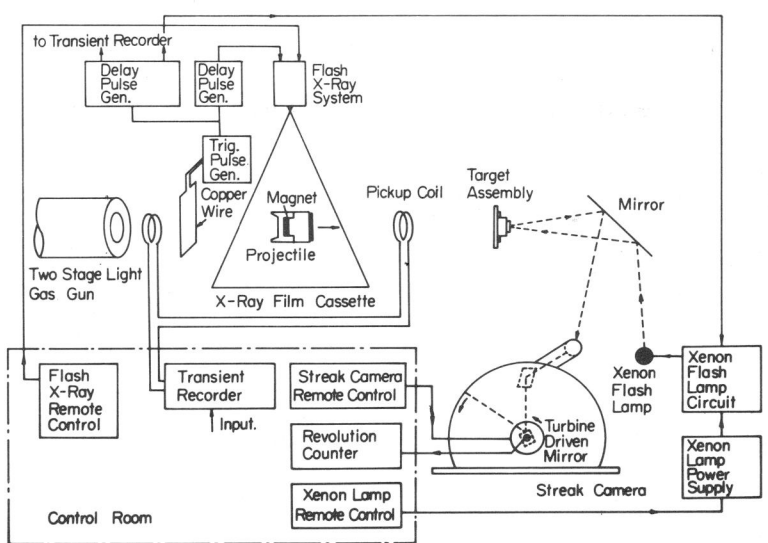

Fig. 4 Block diagram of the optical measurement system

Fig. 5 Time-resolved spectrograph installed at the two-stage light gas gun

TABLE I RIISOM* shock wave facility

Laboratory Facilities	25 mm-Diameter Single-Stage Propellant Gun (1.0 - 2.4 km/s), 20 mm-Diameter Two-Stage Light Gas gun (1.5 - 4.5 km/s).
Diagnostics	Two Rotating-Mirror Streak Cameras (f. 34, 10 mm/µs), Transient Recorder (Biomation 8100), Two Synchroscopes (Tektronix 475, 200 MHz), Flash X-Ray Radiography (Hewlett Packard, 150 kV, 30 ns), 2 kJ Xenon Flash Lamp System, Spectrometer (Nikon G 250 Modified), Microphotometer (Shimazu, 802).

Recent Studies	• Equation of State Studies at Very High Pressures (TiOx (1), Fe_3O_4 (2, 6), α-Fe_2O_3 (2, 6, 16, 19), GaAs (4, 6), GaP (6, 20), Mg_2SiO_4 (15, 18)).
	• Shock-Induced Phase Transitions
	  • Semiconductor-Metal Transition (GaAs (4, 6), GaP (6, 21)).
	  • High-Spin Low-Spin Transition (α-Fe_2O_3 (2, 6, 16, 19)).
	  • Disproportionation (Mg_2SiO_4, (15, 18)).
	• Measurements of Physical Properties under Shock Compression.
	  • Electrical Conductivity (MnO & CoO (2), Fe_2SiO_4 (11), GaP (21)).
	  • Absorption Spectrum (Al_2O_3:Cr^{3+} (8, 9), MgO:Fe^{2+} (10)).
	• Observation of Shock Residual Effects.
	  • Anisotropic Heterogeneous Yielding ($FeTiO_3$ (14)).
	  • Disproportionation or Incongruent Melting (Mg_2SiO_4 (18, 20)).
	  • Glass Formation ($CaAl_2Si_2O_8$ (5, 7), Mg_2SiO_4 (18, 20)).
	• Synthesis of Superconducting Materials (Nb-Si System (17, 22)).

Principal Investigators	Yasuhiko Syono (Acting Supervisor) Tsuneaki Goto Toshiyuki Sato (Graduate Student)

* The Research Institute for Iron, Steel and Other Metals, Tohoku University, Katahira, Sendai 980, Japan.

Contribution from RIISOM shock wave laboratory

1. Y. Syono, T. Goto, J. Nakai, Y. Nakagawa and H. Iwasaki, J. Phys. Soc. Japan 37, 442 (1974).
2. Y. Syono, T. Goto, J. Nakai and Y. Nakagawa, Proc. 4th Intern'l Conf. on High Pressure (1975), p. 466.
3. T. Goto, Y. Syono, J. Nakai and Y. Nakagawa, Sci. Rep. RITU A25, 186 (1975).
4. T. Goto, Y. Syono, J. Nakai and Y. Nakagawa, Solid St. Commun. 18, 1607 (1976).
5. M. Kitamura, T. Goto and Y. Syono, Contrib. Miner. Petrol. 61, 299 (1977).
6. Y. Syono, T. Goto and Y. Nakagawa, in High Pressure Res: Appl. in Geophys. (M. H. Manghnani and S. Akimoto, eds., Academic, New York, 1977), p. 463.
7. Y. Syono, T. Goto, Y. Nakagawa and M. Kitamura, ibid. (1977) p. 477.
8. T. Goto, G. R. Rossman and T. J. Ahrens, High-Pressure Science & Technology: 6th AIRAPT Conf. (K. D. Timmerhaus and M. S. Barber, eds., Vol. 2, Plenum, New York, 1979), p. 895.
9. T. Goto, T. J. Ahrens and G. R. Rossman, Phys. Chem. Minerals, 4, 253 (1977).
10. T. Goto, T. J. Ahrens, G. R. Rossman and Y. Syono, Phys. Earth Planet. Interiors 22, 277 (1980).
11. T. Mashimo, K. Kondo, A. Sawaoka, Y. Syono, H. Takei and T. J. Ahrens, J. Geophys. Res. 85, 1876 (1980).
12. Y. Syono and T. Goto, Sci. Rep. RITU A29, 17 (1980).
13. T. Goto and Y. Syono, ibid, 32 (1980).
14. Y. Syono, H. Takei, T. Goto and A. Ito, Phys. Chem. Minerals 7, 82 (1981).
15. Y. Syono, T. Goto, J. Sato and H. Takei, J. Geophys. Res. 86, (1981), in press.
16. T. Goto, Y. Syono, J. Sato and Y. Nakagawa, Proc. 3rd Intern'l Conf. on Ferrite, (1981), in press.
17. S. Ohshima, N. Sone, T. Wakiyama, T. Goto and Y. Syono, Solid St. Commun., (1981), in press.
18. Y. Syono and T. Goto, in High Pressure Res. in Geophys., (S. Akimoto and M. H. Manghnani, eds., Center for Academic Publ. Tokyo, 1981), in press.
19. T. Goto, J. Sato and Y. Syono, ibid. (1981), in press.
20. Y. Syono, T. Goto, H. Takei, K. Nobugai and M. Tokonami, Science, (1981), in press.
21. T. Goto and Y. Syono, This volume.
22. Y. Syono, T. Goto, W. K. Wang, H. Iwasaki, S. Ohshima and T. Wakiyama, This volume.

RAFAEL TERMINAL BALLISTICS LABORATORY - FACILITY AND CAPABILITY

Gideon Rosenberg
Ministry of Defence, A.D.A., P. O. Box 2250, 31021 Haifa, Israel.

ABSTRACT

Brief descriptions are given of three guns and diagnostic methods used in this laboratory. The guns include a 64 mm diameter powder gun and 64 mm and 101.6 mm compressed gas guns. A small caliber terminal ballistics range is also included in the Facility.

INTRODUCTION

Our program is aimed at understanding penetration mechanisms of armor. A basic approach of studying materials response to impact loading is used. We have one powder gun and two gas guns, the first is used primarily to drive penetrators and the others for planar impact experiments.

DESCRIPTION OF GUNS

- The powder gun is of 64.00 mm bore diameter and 10 m barrel length. This gun is designed to fire a 3500 grams projectile up to velocity of 1500 m/sec. A target chamber mounted at the muzzle, with vacuum arrangements, target holder and velocity pins, allows optional use for planar impact studies. The major task of this gun is, however, shooting long rod penetrators. A tube assembly is set up from the target chamber through the building's wall to provide a capability of shooting against heavy targets located outside. The gun is shown in fig.1 and the target set up in fig. 2 .
- A short-barrel gas gun is used to conduct most of our planar impact studies. The gun is 4.4 meters long and its bore diameter is 64.00 mm. A 400 gram projectile is driven to a velocity of 550 m/sec at a Nitrogen pressure of 25 MPa (250 bar). Further details about this gun are given elsewhere[1]. This gun is shown in fig. 3 .
- A four-inch gas gun is now installed in our lab. This gun was purchased from Physics International Company and has been fully described in their report[2]. This system, not yet in operation, is shown in fig. 4.
- Our small caliber range is an open range with the following guns: 0.5" rifled, o.5" smooth bore, 14.5 mm rifled, 20 mm smooth.

DIAGNOSTIC METHODS

Three major methods are used in our experiments:
- The projectile velocity is measured by velocity pins located at the muzzle, or by velocity screens ahead of the target. The pins are used in planar impact shots and the screens for all other shots. Time intervals between the electrical pulses are measured by high frequency digital counters.
- Shock parameters are measured using various gauges. Manganin gauge and shorted quartz gauge are presently in use. A set up for measuring particle velocity with an electromagnetic gauge is in preparation. The signals generated by the gauges are recorded by Tektronix oscilloscopes with Polaroid cameras.
- Penetration phenomena and behind target effects is monitored by a radiography method.

A three channel 300 kV flash X-ray system is used for shots conducted by the 64 mm powder gun. A two channel 150 kV system is set for the small caliber shots. Both systems are HP (Field Emission) products.

The instrumentation list given below demonstrates the diagnostic capability of our laboratory:

Time interval counters

- A 3 channel, Tektronix 503A, 100 MHz
- A 3 channel, ADA product, 100 MHz.
- A 3 channel, ADA product, 50 MHz.
- A 4 channel, ADA product, 10 MHz.

Oscilloscopes & recording instrumentation

- High frequency oscilloscopes, Tektronix 7904, 7844 and 7704 of 500 MHz, 400 MHz and 250 MHz types respectively.
- Moderate frequency oscilloscopes, Tektronix two dual beam 556, two storage 549 and one 547.
- A 5 channel digital waveform recorder, 0.5 µs sampling intervals, 2 k-bits memory capacity.

Additional general instrumentation

- A 3 channel regulated pulse power supply for Manganin gauges. ADA product.
- A 4 channel digital delay pulse generator. ADA product.
- Power supplies, pulse generators, Time marker, etc.

SUPPORTING FACILITIES

The following aids are associated with our program:
- Preparation room with lapping machine, diamond disc saw, jeweler's lathe, etc.
- Fully equipped machine shop.
- Ultrasonic system, PEO method.
- Dark room.
- Computing terminal, hooked to the ADA Computer Center.

Research topics and associated staff

The publications listed below were chosen to present some of the studies conducted in our laboratory and the active investigators as well.
- Nonlinear response of a penetrating warhead under impact, Y. Kivity & D. Peretz, 5th Symposium on Ballistics, Toulouse, France, 1980.
- Release wave calibration of manganin gauges, D. Yaziv, Z. Rosenberg and Y. Partom, J. Appl. Phys. $\underline{51}$, 6055 (1980).
- Extension of measurement times of quartz transducers by the use of a fused quartz housing, Z. Rosenberg, J. Phys. E, $\underline{11}$, 401, (1978).
- Shock wave impedance match at low pressures, M. Avinor, Z. Rosenberg and Y. Oved, J. Phys. E, $\underline{11}$, 300 (1978).
- Resistivity measurements in silicon compressed by shock waves, Gideon Rosenberg, J. Phys. Chem. Solids, $\underline{41}$, 561, (1980).
- Variation of the elastic constants of 2024T351 Aluminum under dynamic pressures, D. Yaziv, Z. Rosenberg and Y. Partom, To be published in J. Appl. Phys.
- Shear waves in yawed rod impacts, M. Mayseless, Y. Kivity, G. Rosenberg, A. Betser, submitted to the sixth International Symposium on Ballistics.

REFERENCES

1. Y. Porat and M. Gvishi, J. Phys. E, $\underline{13}$, 504 (1980).
2. P. Holton, Physics International Report: PIIK-20-69, March 1970.

Fig. 1: 2½" powder gun.

Fig. 2: External target set-up.

Fig. 3: Short barrel 2½" gas gun.

Fig. 4: 4" gas gun.

IMPACT FACILITIES AT THE ERNST-MACH-INSTITUTE

A. J. Stilp, V. Hohler, E. Schneider
R. Tham, M. Hülsewig, G. Kuscher, W. Junckermann

Ernst-Mach-Institute, Freiburg i.Br.
Germany

ABSTRACT

The Ernst-Mach-Institute has nine guns at five different indoor ranges. Typical studies and instrumentation are described.

INTRODUCTION

Impact facilities with a wide range of capabilities are employed for a variety of impact experiments. For convenience, several guns are located together with common instrumentation in five different indoor ranges. Areas of research include: penetration mechanics, modeling of terminal ballistic effects, penetration in concrete, sand and earth materials, hypervelocity impact, wave propagation in solids and fluids, the behavior of materials under impact loading and metallography studies.

RECENT STUDIES

The most recent studies are concerned with: rod penetration, temperature measurements, penetration in laminated targets, glass and ceramics, metallographic studies on hydrodynamically deformed tungsten sinter alloys, hydraulic ram studies, penetration of fragments into sand, rock and concrete, planar impact effects on fiber-reinforced concrete, impact of highly deformable projectiles, Taylor tests at various initial temperatures, perforation resistance of fibre-reinforced plastic materials, hypervelocity impact in geologic materials, hypervelocity impact in low density materials, impact flash studies, meteorite bumper design and testing, wave propagation in transparent materials, and measurement of rod deceleration.

GUN FACILITIES

Range I

Range I is a closed indoor range with three guns whose bore diameters range from 10 to 25 mm. A compressed gas gun has an impact velocity range from 0.1 to 0.5 km/s. A powder gun has an impact velocity range from 0.5 to 2 km/s, while a two-stage light gun has a velocity range from 2 to 6 km/s. These guns fire into a blast tank with a volume of 2.5 m^3 and which contains two shadowgraph stations. Velocity measurement is with two laser beam barriers.

There are three tank constructions for different target sizes. Diagnostics include four 180 KV stereo X-ray channels, one 300 KV stereo X-ray channel, one 600 KV X-ray channel with super radiant light equipment and a rotating prism high speed camera.

Range II

Range II is a closed indoor range with a 100 mm diameter compressed gas gun with a velocity capability from 0.1 to 0.5 km/s and a 50 mm diameter powder gun capable of producing impact velocities from 0.5 to 1 km/s.

The blast tank has a 2 m^3 volume and provides a capability for planar impact studies. Velocity measurements are with laser beam barriers and the tank contains two shadowgraph stations.

The impact tank allows targets of 1 meter by 1 meter size and contains a dust exhauster. A Dynafax camera is available.

Range III

Range III has a compressed gas gun with velocity capability from 0.1 to 0.5 km/s and a powder gun with a velocity capability from 0.5 to 1.5 km/s. Both guns have a bore diameter of 50 mm.

The impact tank has a volume of 2 m^3 and has provision for planar and reverse impact studies.

Instrumentation includes a VISAR with a 5 Watt argon-ion laser and a Tektronix 7912AD programmable digitizer with a 4052 desktop computer. Pulsar and K-Line Manganin and carbon gauge power supplies are available.

Range IV

Range IV has a powder gun with velocity capability from 0.5 to 2 km/s with a bore diameter of either 20 or 30 mm. The blast tank volume is 1 m^3 and the impact tank has dust and gas filters for use with toxic materials. The impact tank also has provisions for special environmental conditions on the target such as elevated temperature, reduced temperature and prestress.

Velocity measurement is with two laser beam barriers or with two image converter cameras with xenon flash lamps for back lighting. The range has four 150 kV X-ray channels.

Range V

This range is a closed high vacuum range with a pressure capability of 10^{-6} mm of mercury. The two-stage light gas gun has a 4.5 mm diameter and a velocity capability from 2 to 8 km/s.

Velocity measurement is with optical radiation detectors. The stainless steel impact tank has provision for use of a TRW image converter camera with both streak and framing capabilities. A high aperture transmission spectrograph is also available.

Author Index

Ahlstrom, Harlow G. 155
Ahrens, T. J. 236, 299
Ahrens, Thomas J. 231, 631
Anderson, R. A. 145, 160
Ando, M. 325
Andriot, P. 505, 626
Asay, J. R. 188, 417, 422, 427

Babare, L. V. 27
Banner, D. L. 164
Banner, David L. 155
Batsanov, S. S. 1, 14
Bauer, F. 251, 674
Belgaumkar, B. M. 598
Bellamy, P. M. 282
Bjorkman, M. D. 310, 432
Bless, S. J. 548, 668
Bloom, George H. 588
Bloomquist, D. D. 57, 304
Boslough, M. B. 236
Boslough, Mark B. 631
Brooks, A. L. 179, 686
Brown, N. 223
Brown, William T. 529
Bukowinski, M. S. T. 218
Burns, T. J. 277, 372

Cagnoux, J. 392
Chan, J. 558
Chartagnac, P. F. 397
Chapron, P. 505, 626
Chau, H. 174, 691
Cheret, R. 626
Chhabildas, L. C. 417, 422, 427, 621
Clark, G. E. 277
Clifton, R. J. 360, 407, 661
Costin, L. S. 372
Cottet, F. 130
Cunningham, W. G. 92
Curran, D. R. 135, 460

Damamme, G. 603
Davison, Lee 67, 442
Devine, G. J. 223
Dick, J. J. 612
Dittbenner, G. 174, 691
Diven, B. C. 168
Dodson, B. W. 42, 62, 335
Dremin, A. N. 27
Drumheller, D. S. 527
Duvall, G. E. 282, 292, 296

Elban, W. L. 583
Erickson, L. M. 553, 558
Erickson, LeRoy 685
Erkman, J. O. 663
Erlich, D. C. 342

Forbes, J. W. 578, 583, 663
Fowler, C. M. 179, 686
Fowles, G. R. 520
Frankel, M. J. 593
Fritz, J. N. 193, 382
Froeschner, K. 691
Froeschner, K. E. 174
Fugelso, Erik 607

Germain-Lacour, M. 481
Ginn, Warren G. 631
Glenn, L. A. 510
Gliniasty, M. de 481
Golden, John 72
Goto, T. 87
Goto, Tsuneaki 320, 701
Grady, D. E. 372, 412, 456
Graham, R. A. 4, 42, 52, 67, 72, 77, 82, 277, 330
Gupta, Y. M. 437, 525, 526
Guymon, L. G. 563

Hallquist, J. O. 491
Hankey, D. L. 82
Hardy, J. R. 92
Harrach, R. J. 164
Harris, P. 309
Hartman, J. K. 277
Harvey, J. R. 563
Hawke, R. S. 179, 686
Hawken, D. 345
Hayes, D. B. 412
Henley, D. 437
Herrmann, W. 346
Hicks, D. L. 544
Hixson, R. S. 282
Hohler, V. 711
Holian, Brad Lee 241, 382
Holmes, N. C. 145, 160, 164, 648
Holmes, Neil C. 155
Hoover, W. G. 480
Horie, Y. 315
Hornemann, U. 539
Hülsewig, M. 711

Author Index Continued

Iwasaki, H. 87

Jacobson, J. D. 607
Jensen, Russ 607
Johnson, J. N. 438, 568
Johnson, R. O. 277
Jung, I. 140
Jünckermann, W. 711

Kamegai, Minao 381
Karo, A. M. 92
Karpp, Robert R. 607
Keck, J. D. 82
Kerley, G. I. 208
Kim, K. 376
Kipp, M. E. 442
Klee, Ch. 486
Kleiman, Y. 345
Kobierecki, Marian 155
Kondo, K. 299, 325, 696
Kondo, Ken-ichi 631
Kuscher, G. 711
Kovar, F. R. 563
Kroh, M. 486
Kusubov, A. S. 470

Lange, Manfred A. 631
Lawrence, William 534
Lee, R. 691
Lee, R. S. 174
Lee, Y. T. 164
Le Drean, C. 626
Lemar, E. R. 663
Lezaud, J. M. 626
Li, C. H. 360
Liddiard, T. P. 578
Loichot, R. 626
Long, K. S. 213
Ludwig, D. 486
Lutze, A. B. 246
Lyzenga, G. A. 268
Lyzenga, Gregory A. 231

Majewski, P. 500
Mar, H. 345
Margolin, L. G. 465
Marston, P. L. 515
Martineau, J. 626
Mayseless, M. 495
McMahan, A. K. 340
McQueen, R. G. 193

Mikkelson, K. 174, 691
Mikkola, D. E. 98
Mitchell, A. C. 97, 179, 184, 223, 613
Monahan, B. 184
Moran, B. 480
Morosin, B. 4, 72, 77, 82, 330
Morris, C. E. 382
Morris, Charles E. 616
Moss, G. L. 446
Murri, W. J. 437, 460, 525, 652

Nellis, W. J. 184, 223, 613
Nellis, William J. 97, 226
Netherwood, Jr., P. H. 446
Norwood, F. R. 544
Nutt, G. L. 491

Ogilvie, K. 292, 296
Oh, S. I. 376
Ohshima, S. 87
Olive, F. 505, 626
Olness, R. J. 164

Palmer, H. G. 553
Parker, N. L. 553
Partom, Y. 387
Perry, F. C. 188, 639
Peterson, D. R. 179, 686
Pfrang, W. 539
Poth, A. 539
Predebon, William W. 451
Presles, H. N. 309
Price, Robert H. 155
Pullen, G. L. 515

Ragan, C. E. 169, 644
Ree, F. H. 97, 213
Reeves, G. A. 160
Rich, M. 169
Rigden, Sally M. 631
Robinson, E. E. 169
Romain, J. P. 130
Rosen, M. D. 164
Rosen, Mordecai D. 155
Rosenberg, G. 495
Rosenberg, Gideon 706
Rosenberg, Z. 387
Ross, M. 223
Ross, Marvin 226
Roth, J. 475
Rothenhäusler, H. 539

Author Index Continued

Salansky, N. 345
Sato, K. 325
Sawaoka, A. 299, 325, 696
Scharpf, F. 539
Schmidt, R. M. 634
Schneider, E. 711
Schneider, H. 140
Seaman, L. 118, 402, 446, 460
Senf, H. 539
Setchell, R. E. 657
Shaner, J. W. 179
Sharp, Jr., Richard W. 367
Sheffield, S. A. 57, 304
Shockey, D. A. 402
Shrader, J. E. 310, 432
Steinberg, D. 174
Steinberg, Daniel J. 367
Stilp, A. J. 711
Stolper, Edward M. 631
Straub, Galen K. 241
Svendsen, Robert F. 631
Swanson, Richard E. 241
Sweeney, M. A. 188
Swift, R. P. 470
Syono, Y. 87
Syono, Yasuhiko 320, 701

Teasdale, W. A. 169
Tham, R. 711
Thompson, William E. 451
Tokheim, R. E. 246
Trainor, R. J. 145, 160, 164, 613

Trigger, K. R. 563
Trucano, T. G. 544

van Thiel, M. 97, 223
Vantine, H. C. 553, 558
Vassiliou, Mario S. 631
Veeser, L. R. 160
Venturini, E. L. 72, 77, 335
Von Holle, William G. 287
Vorthman, J. E. 680

Wakiyama, T. 87
Walker, F. E. 92
Wang, W. K. 87
Watt, J. Peter 631
Watt, J. W. 663
Webb, D. M. 67
Weingart, R. C. 174, 691
Widner, M. M. 188
Williams, Frank 72
Wilson, C. R. 282, 296
Wise, J. L. 277, 417, 422, 427
Wright, R. N. 98

Xu, Kang 573
Xu, Yun-xiang 573

Yaziv, D. 387, 495
Young, D. 213
Yu, De-yang 573

Zeng, Xiung-fei 573
Zickuhr, James R. 155

AIP Conference Proceedings

		L.C. Number	ISBN
No.1	Feedback and Dynamic Control of Plasmas	70-141596	0-88318-100-2
No.2	Particles and Fields - 1971 (Rochester)	71-184662	0-88318-101-0
No.3	Thermal Expansion - 1971 (Corning)	72-76970	0-88318-102-9
No.4	Superconductivity in d-and f-Band Metals (Rochester, 1971)	74-18879	0-88318-103-7
No.5	Magnetism and Magnetic Materials - 1971 (2 parts) (Chicago)	59-2468	0-88318-104-5
No.6	Particle Physics (Irvine, 1971)	72-81239	0-88318-105-3
No.7	Exploring the History of Nuclear Physics	72-81883	0-88318-106-1
No.8	Experimental Meson Spectroscopy - 1972	72-88226	0-88318-107-X
No.9	Cyclotrons - 1972 (Vancouver)	72-92798	0-88318-108-8
No.10	Magnetism and Magnetic Materials - 1972	72-623469	0-88318-109-6
No.11	Transport Phenomena - 1973 (Brown University Conference)	73-80682	0-88318-110-X
No.12	Experiments on High Energy Particle Collisions - 1973 (Vanderbilt Conference)	73-81705	0-88318-111-8
No.13	π-π Scattering - 1973 (Tallahassee Conference)	73-81704	0-88318-112-6
No.14	Particles and Fields - 1973 (APS/DPF Berkeley)	73-91923	0-88318-113-4
No.15	High Energy Collisions - 1973 (Stony Brook)	73-92324	0-88318-114-2
No.16	Causality and Physical Theories (Wayne State University, 1973)	73-93420	0-88318-115-0
No.17	Thermal Expansion - 1973 (lake of the Ozarks)	73-94415	0-88318-116-9
No.18	Magnetism and Magnetic Materials - 1973 (2 parts) (Boston)	59-2468	0-88318-117-7
No.19	Physics and the Energy Problem - 1974 (APS Chicago)	73-94416	0-88318-118-5
No.20	Tetrahedrally Bonded Amorphous Semiconductors (Yorktown Heights, 1974)	74-80145	0-88318-119-3
No.21	Experimental Meson Spectroscopy - 1974 (Boston)	74-82628	0-88318-120-7
No.22	Neutrinos - 1974 (Philadelphia)	74-82413	0-88318-121-5
No.23	Particles and Fields - 1974 (APS/DPF Williamsburg)	74-27575	0-88318-122-3
No.24	Magnetism and Magnetic Materials - 1974 (20th Annual Conference, San Francisco)	75-2647	0-88318-123-1
No.25	Efficient Use of Energy (The APS Studies on the Technical Aspects of the More Efficient Use of Energy)	75-18227	0-88318-124-X

No.	Title	LCCN	ISBN
No. 26	High-Energy Physics and Nuclear Structure - 1975 (Santa Fe and Los Alamos)	75-26411	0-88318-125-8
No. 27	Topics in Statistical Mechanics and Biophysics: A Memorial to Julius L. Jackson (Wayne State University, 1975)	75-36309	0-88318-126-6
No. 28	Physics and Our World: A Symposium in Honor of Victor F. Weisskopf (M.I.T., 1974)	76-7207	0-88318-127-4
No. 29	Magnetism and Magnetic Materials - 1975 (21st Annual Conference, Philadelphia)	76-10931	0-88318-128-2
No. 30	Particle Searches and Discoveries - 1976 (Vanderbilt Conference)	76-19949	0-88318-129-0
No. 31	Structure and Excitations of Amorphous Solids (Williamsburg, VA., 1976)	76-22279	0-88318-130-4
No. 32	Materials Technology - 1976 (APS New York Meeting)	76-27967	0-88318-131-2
No. 33	Meson-Nuclear Physics - 1976 (Carnegie-Mellon Conference)	76-26811	0-88318-132-0
No. 34	Magnetism and Magnetic Materials - 1976 (Joint MMM-Intermag Conference, Pittsburgh)	76-47106	0-88318-133-9
No. 35	High Energy Physics with Polarized Beams and Targets (Argonne, 1976)	76-50181	0-88318-134-7
No. 36	Momentum Wave Functions - 1976 (Indiana University)	77-82145	0-88318-135-5
No. 37	Weak Interaction Physics - 1977 (Indiana University)	77-83344	0-88318-136-3
No. 38	Workshop on New Directions in Mossbauer Spectroscopy (Argonne, 1977)	77-90635	0-88318-137-1
No. 39	Physics Careers, Employment and Education (Penn State, 1977)	77-94053	0-88318-138-X
No. 40	Electrical Transport and Optical Properties of Inhomogeneous Media (Ohio State University, 1977)	78-54319	0-88318-139-8
No. 41	Nucleon-Nucleon Interactions - 1977 (Vancouver)	78-54249	0-88318-140-1
No. 42	Higher Energy Polarized Proton Beams (Ann Arbor, 1977)	78-55682	0-88318-141-X
No. 43	Particles and Fields - 1977 (APS/DPF, Argonne)	78-55683	0-88318-142-8
No. 44	Future Trends in Superconductive Electronics (Charlottesville, 1978)	77-9240	0-88318-143-6
No. 45	New Results in High Energy Physics - 1978 (Vanderbilt Conference)	78-67196	0-88318-144-4
No. 46	Topics in Nonlinear Dynamics (La Jolla Institute)	78-057870	0-88318-145-2
No. 47	Clustering Aspects of Nuclear Structure and Nuclear Reactions (Winnepeg, 1978)	78-64942	0-88318-146-0
No. 48	Current Trends in the Theory of Fields (Tallahassee, 1978)	78-72948	0-88318-147-9
No. 49	Cosmic Rays and Particle Physics - 1978 (Bartol Conference)	79-50489	0-88318-148-7

No. 50	Laser-Solid Interactions and Laser Processing - 1978 (Boston)	79-51564	0-88318-149-5
No. 51	High Energy Physics with Polarized Beams and Polarized Targets (Argonne, 1978)	79-64565	0-88318-150-9
No. 52	Long-Distance Neutrino Detection - 1978 (C.L. Cowan Memorial Symposium)	79-52078	0-88318-151-7
No. 53	Modulated Structures - 1979 (Kailua Kona, Hawaii)	79-53846	0-88318-152-5
No. 54	Meson-Nuclear Physics - 1979 (Houston)	79-53978	0-88318-153-3
No. 55	Quantum Chromodynamics (La Jolla, 1978)	79-54969	0-88318-154-1
No. 56	Particle Acceleration Mechanisms in Astrophysics (La Jolla, 1979)	79-55844	0-88318-155-X
No. 57	Nonlinear Dynamics and the Beam-Beam Interaction (Brookhaven, 1979)	79-57341	0-88318-156-8
No. 58	Inhomogeneous Superconductors - 1979 (Berkeley Springs, W.V.)	79-57620	0-88318-157-6
No. 59	Particles and Fields - 1979 (APS/DPF Montreal)	80-66631	0-88318-158-4
No. 60	History of the ZGS (Argonne, 1979)	80-67694	0-88318-159-2
No. 61	Aspects of the Kinetics and Dynamics of Surface Reactions (La Jolla Institute, 1979)	80-68004	0-88318-160-6
No. 62	High Energy e^+e^- Interactions (Vanderbilt, 1980)	80-53377	0-88318-161-4
No. 63	Supernovae Spectra (La Jolla, 1980)	80-70019	0-88318-162-2
No. 64	Laboratory EXAFS Facilities - 1980 (Univ. of Washington)	80-70579	0-88318-163-0
No. 65	Optics in Four Dimensions - 1980 (ICO, Ensenada)	80-70771	0-88318-164-9
No. 66	Physics in the Automotive Industry - 1980 (APS/AAPT Topical Conference)	80-70987	0-88318-165-7
No. 67	Experimental Meson Spectroscopy - 1980 (Sixth International Conference, Brookhaven)	80-71123	0-88318-166-5
No. 68	High Energy Physics - 1980 (XX International Conference, Madison)	81-65032	0-88318-167-3
No. 69	Polarization Phenomena in Nuclear Physics - 1980 (Fifth International Symposium, Santa Fe)	81-65107	0-88318-168-1
No. 70	Chemistry and Physics of Coal Utilization - 1980 (APS, Morgantown)	81-65106	0-88318-169-X
No. 71	Group Theory and its Applications in Physics - 1980 (Latin American School of Physics, Mexico City)	81-66132	0-88318-170-3
No. 72	Weak Interactions as a Probe of Unification (Virginia Polytechnic Institute - 1980)	81-67184	0-88318-171-1
No. 73	Tetrahedrally Bonded Amorphous Semiconductors (Carefree, Arizona, 1981)	81-67419	0-88318-172-X
No. 74	Perturbative Quantum Chromodynamics (Tallahassee, 1981)	81-70372	0-88318-173-8
No. 75	Low Energy X-ray Diagnostics-1981 (Monterey)	81-69841	0-88318-174-6
No. 76	Nonlinear Properties of Internal Waves (La Jolla Institute, 1981)	81-71062	0-88318-175-4
No. 77	Gamma Ray Transients and Related Astrophysical Phenomena (La Jolla Institute, 1981)	81-71543	0-88318-176-2
No. 78	Shock Waves in Condensed Matter - 1981 (Menlo Park)	82-70014	0-88318-177-0